Vegetable Crops

— *Volume 1* —

THE EDITORS

Professor T.K. Bose obtained his M.Sc. (Ag.) and Ph.D. degree from Calcutta University, West Bengal, India. Trained at the Royal Botanic Garden, Kew, UK. Former Secretary of the Agri-Horticultural Society of India; Project Coordinator, All India Coordinated Floriculture Improvement Project (ICAR); Professor and Head, Department of Horticulture; Dean, Faculty of Agriculture; Director of Research, Bidhan Chandra Krishi Viswavidyalaya (BCKV), West Bengal; Member of Scientific Advisory Committees of ICAR, CSIR (Govt. of India) and Government of West Bengal. Awarded (D.Sc.) Honoris Causa from OUAT, Bhubaneshwar and UBKV, Coochbehar, India. Guided around 45 Ph.D. students, published more than 250 research papers and authored 30 books on horticultural crops.

Prof. M.G. Som served as former Vice-chancellor, Dean, Faculty of Agriculture and Professor of Vegetable Science, BCKV, West Bengal. Obtained his M.Sc. and Ph.D. degrees from ICAR-IARI, New Delhi. Received fellowship in 1997 and Plaque in 1998 from the Indian Society of Vegetable Science (ISVS). Served as FAO expert in Bangladesh for increasing the vegetable production. Acted as a member of governing body of the ICAR, New Delhi; member of Advisory Research Council, ICAR-IIVR, Varanasi, and ICAR-CIARI, Port Blair, Andaman. Published more than 100 research papers, written 8 books, 2 monographs and few bulletins and several popular articles. Guided 14 Ph.D. students and more than 50 M.Sc (Ag.) students. In 2019, ISVS felicitated Prof. Som for his outstanding contribution in the field of vegetable research and education.

Prof. Arup Chattopadhyay obtained his M.Sc. and Ph.D. degrees from BCKV, West Bengal. He has 23 years of experience in Research, Teaching and Extension. Acting as the Officer-in-Charge, AICRP on Vegetable Crops, and served as Professor for more than 8 years. He has guided 15 M.Sc. and 9 Ph.D. students in Vegetable Science. Published more than 150 research papers, authored 05 books, 07 book chapters, and many popular articles, folders, leaflets. Developed nine trait specific varieties/hybrids of different vegetable crops. First recipient of 'Dwarikanath Young Scientist Gold Medal Award' for the best Ph.D. thesis in Vegetable Science; 'Fellows' of ISVS, Varanasi, and CWSS, West Bengal. BCKV has been felicitated 'Lt. Amit Singh Memorial Award' by the ICAR for the Best coordinating centre on Vegetable Crops in India for outstanding contribution in vegetable research under his able leadership.

Prof. Tapan Kumar Maity has acclaimed professional career of more than 35 years in the Department of Vegetable Science, BCKV, West Bengal. Acted as Nodal Officer of Quality Control Laboratory for Horticultural Produce, and PI of AINRP on Onion and Garlic (ICAR). Served as the Head of the Department of Vegetable Science and the Dean, Faculty of Horticulture. Published more than 100 original research papers, contributed 13 book chapters, edited 3 books. Supervised 12 Ph.D. and 32 M.Sc. students. Acted as Member of the Working Group of the Agriculture Commission, Government of West Bengal, Member of the Monitoring Team of AINRP on Onion and Garlic for different centers; acting as consultant of National Horticulture Board, Government of India.

Prof. Jahangir Kabir obtained his M.Sc.(Ag) and Ph.D.(Horticulture) degrees from (BCKV) West Bengal. He has 33 years of varied experience in Teaching and Research in Vegetable Crops and Post Harvest Physiology and Technology of Horticultural Crops. Acted as former Head of the Department of Post Harvest Technology, and served as Professor for more than 15 years. He has guided more than 30 M.Sc. and 10 Ph.D. students, authored more than 120 publications including books (edited), research papers, book chapters, review/technical and popular articles. Acted as reviewer of research papers, paper setter, and member of selection committee of Public Service Commission and several State and Central Universities.

Vegetable Crops

— Volume 1 —

— Senior Editor —

M.G. Som

Department of Vegetable Science
Faculty of Horticulture
Bidhan Chandra
Krishi Viswavidyalaya
West Bengal 741 252, India

— Editor Emeritus —

T.K. Bose

Faculty of Horticulture
Bidhan Chandra
Krishi Viswavidyalaya
West Bengal 741 252, India

— Edited by —

A. Chattopadhyay • T.K. Maity • J. Kabir

Faculty of Horticulture
Bidhan Chandra Krishi Viswavidyalaya
West Bengal 741 252, India

2022

Daya Publishing House®
A Division of

Astral International Pvt. Ltd.
New Delhi – 110002

Cataloging in Publication Data--DK
Courtesy: D.K. Agencies (P) Ltd. <docinfo@dkagencies.com>

Vegetable crops / senior editor, M.G. Som ; editor emeritus, T.K. Bose ; edited by A. Chattopadhyay, T.K. Maity, J.
Kabir. -- Fourth revised & illustrated edition.
        volumes ; cm
      Includes index.
      ISBN 9789354616822 (HB)

      1. Vegetables--India.   I. Som, M. G., editor.

LCC SB320.8.V44 2021 | DDC 635.0954   23

Published by   :   **Daya Publishing House®**
                                A Division of
                                Astral International Pvt. Ltd.
                                – ISO 9001:2015 Certified Company –
                                4736/23, Ansari Road, Darya Ganj
                                New Delhi-110 002
                                Ph. 011-43549197, 23278134
                                E-mail: info@astralint.com
                                Website: www.astralint.com

From the Desk of Editor Emeritus

The undersigned, who had initiated the publications entitled 'Fruits: Tropical and Sub-tropical', 'Vegetable Crops' and 'Commercial Flowers' in one volume each in the 1980's in collaboration with co-editors, humbly expresses from the core of his heart that the updated revised editions of the above mentioned publications with colour illustrations in thirteen which was planned in June, 2020, are being printed from October this year, a rare event in Horticultural Science has been possible by the blessing of Almighty God, with the cooperation of dedicated horticulturist as editors and contributors.

Kolkata

T.K. Bose
Editor Emeritus

From the Desk of Senior Editor

In the year 1986, the book entitled "Vegetable Crops in India" was first published. Afterwards, due to tremendous progress in different vegetable crops world-wide, this book was revised and published in the year 1993 and 2003 in three volumes by the name "Vegetable Crops". The previous three editions proved to be very successful and valuable to the students, teachers and researchers both at national and international level. Due to the rapid increase in production and productivity of vegetable crops as a result of genetic improvement through both conventional and non conventional breeding methods including modern biotechnological tools, apart from the proper management of the crops, we decided to revise and publish the fourth edition of this book in four volumes. This edition includes coloured illustrations wherever necessary to benefit the bibliophiles.

The first volume of this edition includes chapters on leguminous crops (9 crops) and root crops (4 crops). I am extremely happy to acknowledge the contributors for their dedicated efforts to make this volume very effective and meaningful. I would like to thank all the co-editors of this volume who tirelessly and sincerely participated in editing and proof reading of the manuscript. It is expected that this revised edition will act as a platform to enrich and strengthen the students and the researchers over their knowledge on vegetable crops.

I would like to specially appreciate Prof. T.K. Bose, who took active initiative and guided us to complete the fourth illustrated edition of this book. It would have been impossible for us to put this book into table without his encouragement and guidance.

Despite of our best efforts, the editors are conscious about the lapses that might have crept in the manuscript, however, we would like to emphasis on the fact that these lapses or errors are unintentional but circumstantial.

Kolkata   **M.G. Som**
Senior Editor

Preface to the Fourth Edition from Editors

The first edition of the book entitled 'Vegetable Crops in India' published in 1986 was greatly appreciated and proved valuable to the students, teachers, researchers and extension specialists in horticulture in general and vegetable science in particular, in different parts of the world. Then the book was revised twice in 1993 and 2003 with the title 'Vegetable Crops' in three volumes due to its increasing demand. All the previous editions, which contained updated scientific information of all commercially grown vegetable crops, were considered as valuable contribution on the subject. After about 17 years of the last revised publication in 2003, voluminous information pertaining to vegetable crop improvement, production, protection and post harvest management has been generated necessitating further revision in four volumes.

This particular edition has been enlarged by incorporation of up-to-date information on various aspects of vegetable crop production, protection and improvement particularly through biotechnological interventions with meaningful illustrations, and published in four volumes. Efforts have been made to cover as much information as possible. First volume in this edition includes chapter on leguminous and root crops only covering nearly 700 pages. Most of the contributors have taken care to review up-to-date research works and prepared the manuscript. The revision has been made by the editors and in few cases new contributors have been associated with some chapters.

We express our sincere thanks to all contributors the Editor Emeritus Prof. T.K. Bose and the Senior Editor Prof. M.G. Som who sincerely participated in revision, editing and proof-reading of the manuscript and we appreciate their commendable efforts to bring out this publication. We are thankful to one of the contributors Prof. P. Hazra for his critical comments on first draft and offering valuable suggestions in strengthening the manuscript.

Apart from own collection of photographs, editors have also taken some photographs which are in public domain from different websites as far as possible. Therefore we acknowledge with enormous thanks to all State Agricultural Universities and Research Institutes under Indian Council of Agricultural Research, New Delhi, India; Public and Private International Agricultural Institutes across the world for taking photographs from their websites to make this book a valuable knowledge for the readers across the world. We think images can be powerful teaching tools, as illustrations to in-class lectures, or for studying concepts outside of the classroom. We also think identification of vegetable crops, their package of practices, cultivars, utilization, diseases and pests with the help of suitable illustrations in a consolidated manner will make a dent for the students and teachers of vegetable science.

Our thanks are due to Shri Prateek Mittal, Director of Astral International (P) Ltd., New Delhi for their keen interest in bringing out this publication. Our special thanks go to Mr. Debashis Roy of Kolkata for processing the manuscript.

Kolkata   **A. Chattopadhyay • T.K. Maity • J. Kabir**

                                                        Editors

Contents

List of Contributors

Bhattacharjee, T.

Department of Horticulture, College of Agriculture, Tripura (West), Lembucherra 799210, Tripura, India

Chakraborty, A.K.

Division of Vegetable Science, Indian Institute of Agricultural Research, Pusa 110012, New Delhi, India

Chattopadhyay, A.

Department of Vegetable Science, Faculty of Horticulture, Bidhan Chandra Krishi Viswavidyalaya (State Agricultural University), Mohanpur 741252, West Bengal, India

Hazra, P.

Department of Vegetable Science, Faculty of Horticulture, Bidhan Chandra Krishi Viswavidyalaya (State Agricultural University), Mohanpur 741252, West Bengal, India

Maurya, P.K.

Department of Vegetable Science, Faculty of Horticulture, Bidhan Chandra Krishi Viswavidyalaya (State Agricultural University), Mohanpur 741252, West Bengal, India

Medhi, B.P.

Division of Horticulture, ICAR Research Complex for NEH Region, Barapani 793103, Meghalaya, India

Mukhopadhyay, S.K.

Department of Agronomy, Faculty of Agriculture, Bidhan Chandra Krishi Viswavidyalaya (State Agricultural University), Mohanpur 741252, West Bengal, India

Nath, R.

Department of Agronomy, Faculty of Agriculture, Bidhan Chandra Krishi Viswavidyalaya (State Agricultural University), Mohanpur 741252, West Bengal, India

Pandita, M.L.

Department of Vegetable Crops, CCS Haryana Agricultural University, Hisar 125004, Haryana, India

Parthasarathy, V.A.

ICAR – Indian Institute of Spices Research, 32/482 C Narmada Nilayam, Bharathan Bazar, Chelavoor, Calicut 673571, Kerala, India

Pratap, P.S.

Department of Vegetable Crops, CCS Haryana Agricultural University, Hisar 125004, Haryana, India

Sadhu, M.K.

College of Agricultural Sciences, Calcutta University, Calcutta 700019, West Bengal, India

Sarkar, K.

College of Agricultural Sciences, Calcutta University, Calcutta 700019, West Bengal, India

Sen, H.

Department of Agronomy, Faculty of Agriculture, Bidhan Chandra Krishi Viswavidyalaya (State Agricultural University), Mohanpur 741252, West Bengal, India

Sureja, A.K.

Division of Vegetable Science, Indian Institute of Agricultural Research, Pusa 110012, New Delhi, India

Leguminous Crops

PEA

A. Chattopadhyay, M.L. Pandita, V.A. Parthasarathy,
P.S. Pratap and P.K. Maurya

1.0 Introduction

Pea is a very common nutritious vegetable grown in the cool season throughout the world. Pea (*Pisum sativum*) is grown as a vegetable crop for both fresh and dried seed. World production of green peas in 2016 was 19.88 mt, and the major producers were China (12.21 mt), India (4.81 mt), and USA (0.31 mt), which accounted for >85 per cent of the total production (FAO, 2016). India ranks second after China, in terms of area and production, however it occupies third position in the world in productivity, after UK and Egypt. Garden pea is grown in an area of 4.2 lakh ha with annual productivity of 9.5 t/ha in India. At the state level, Uttar Pradesh contributes about 47 per cent of the total garden pea production having highest productivity among Indian states (Singh *et al.*, 2018). Other major garden pea growing states are Madhya Pradesh, Jharkhand, Himachal Pradesh and Punjab.

It ranks third or fourth in the worldwide production amongst the grain legumes (Farrington, 1974). In India, it is grown as winter vegetable in the plains of North India and as summer vegetable in the hills. Pea is highly nutritive containing high percentage of digestible protein, along with carbohydrates and vitamins. It is also very rich in minerals (Choudhury, 1967). Being a rich source of protein, it occupies an important place in the vegetarian diet. Due to its low water requirement, it is an important cash crop in water deficit areas. Peas are placed in group III on food production efficiency (MacGillivray, 1961). Being a leguminous crop, it enriches the soil by fixing the atmospheric nitrogen into the soil; thereby its value as a green manure crop has long been recognized. The interest in pea as a soil building crop will increase day by day as the chemical fertilizers are becoming less available and more expensive (Sutcliffe and Pate, 1977). It occupies a pride place in the history of genetics being genetic material used by Gregor Johann Mendel to establish the laws of inheritance.

2.0 Composition and Uses

2.1.0 Composition

It is very rich in protein, carbohydrate, vitamin A and C, calcium and phosphorus, and the nutritive value of fresh green peas has been presented in Table 1.The protein concentration of peas range from 15.5-39.7 per cent (Davies *et al.*, 1985; Bressani and Elias, 1988). Pea is very low in saturated fat, cholesterol and sodium.

Table 1: Composition of Pea (100 g of edible portion)*

Calories	44	Iron	1.2 mg
Water	75.6 per cent	Sodium	6 mg
Protein	6.2 g	Potassium	350 mg
Fat	0.4 g	Carotene	405 ug equivalent
Carbohydrate	16.9 g	Thiamine	0.28 mg
Crude fibre	2.5 g	Riboflavin	0.11 mg

Ash	0.9 g	Niacin	2.8 mg
Calcium	32 mg	Ascorbic acid	27 mg
Phosphorus	102 mg		

* Duke (1981), Hulse (1994).

Dried peas contain: 10.9 per cent water, 22.9 per cent protein, 1.4 per cent fat, 60.7 per cent carbohydrate,1.4 per cent crude fiber and 2.7 per cent ash (Duke, 1981; Hulse, 1994). Flour contains: 343 calories, 10.9 per cent moisture, 22.8 g protein, 1.2 g fat, 62.3 g total carbohydrate, 4.2 g fiber, 2.8 g ash, 72 mg Ca, 338 mg P, 11.3 mg Fe, 0.86 mg thiamine, 0.18 mg riboflavin, and 2.8 mg niacin (Duke, 1981). An average amino acid composition, reported in terms of grams per 100 grams of protein: 6.9-8.2 lysine, 1.4-2.7 methionine + cystine, 3.9 threonine, 0.9 tryptophan, 0.8-1.7 cystine (Huisman and van der Poel, 1994; Bressani and Elias, 1988). Methionine and cystine are the main limiting amino acids. The largest chemical component in peas as in other legumes is carbohydrate which constitute about 56.6 per cent of seed weight (Bressani and Elias, 1988). The most abundant pea carbohydrate is starch, 36.9-48.6 per cent, while amylose is about 34 per cent of seed weight in peas (Bressani and Elias, 1988). Nutrient composition of milled and polished peas as measured per 100 g of edible portion of dried matured whole seeds are 1.4 g oil, 6 g crude fiber, 16.7 g dietary fiber, 54.1 per cent starch, 8.1 per cent sugars, 4.4 mg iron, 0.77 mg thiamine, 0.18 mg riboflavin, 3.1 mg niacin and 330 kcal energy (Newman *et al.*, 1988). Fertilizing peas with sulphur has increased their methionine content from 1.3 to 2.2 g per 100 g protein. Pea hay (at 88.6 per cent DM) contains (zero moisture basis): 10.7-21.6 per cent crude protein, 1.5-3.7 per cent fat, 16.8-36.1 per cent crude fiber, 6.0-9.3 per cent ash, and 41.9-50.6 per cent N-free extract (Duke, 1981).

2.2.0 Uses

Peas are cultivated for the fresh green seeds, tender green pods, dried seeds and foliage (Duke, 1981). Green peas are eaten cooked as a vegetable, and are marketed fresh, canned, or frozen while ripe dried peas are used whole, split, or made into flour (Davies *et al.*, 1985). In some parts of the world, dried peas are consumed split as dahl, roasted, parched or boiled. Green peas are the number one processed vegetable specifically in UK and USA. Sugar peas are flat, tender, and stringless and the whole pod is consumed. They are usually stir-fried in Chinese dishes, steamed or cooked like snap beans. If the seeds are allowed to develop, they are shelled and cooked as garden pea. However, shelling is more tedious compared with garden pea. Snap peas are also edible podded. The pods are round with well-developed grains and are eaten raw in salads, snapped and cooked like snap beans or shelled as garden pea. Pea-straw is a nutritious fodder. Pea is very rich in protein and, therefore, very valuable for the vegetarians.

Muehlbauer and Tullu (1997) presented a comprehensive review on the various aspects of pea including uses. Pea shoots *i.e.* the young, tender vine tips of green pea are also consumed as a vegetable in China, Japan and South East Asia. Consumption of pea shoots is also gaining popularity in the United States. Pea

Uses of Pea

Peas in frozen condition

Canning of peas

Fried snacks prepared from peas   Soup prepared from peas

shoots are generally 5-15 cm long and include 2-4 pairs of leaves, immature tendrils and may also include flower buds and miniature pods. The shoots are eaten fresh, lightly steamed or sautéed and are most commonly served in salads or stir-fried. Green foliage of garden pea is also used as vegetable in parts of Asia and Africa. Leaves are used as a pot herb in Myanmar and parts of Africa (Kay, 1979). Oil from ripened seed has antisex hormonic effects; produces sterility and antagonizes effect of male hormone (Duke, 1981). Pea is being used in a growing snack market. One snack item is prepared by soaking the peas overnight and frying them in palm oil or coating them with other food items such as rice flour before frying for the purpose of imparting different flavours. Another product is prepared by finely grinding the peas and extruding them under pressure to create different shapes. The different shapes are then fried, seasoned and packaged (Jambunathan *et al.*, 1994).

2.2.1 Traditional Medicinal Uses

Seeds are thought to cause dysentery when eaten raw (Muehlbauer and Tullu, 1997). In Spain, flour is considered emollient and resolvent, applied as a cataplasm. It has been reported that seeds contain trypsin and chymotrypsin which could be used for contraceptive, ecbolic, fungistatic and spermicide (Duke, 1981). Smart (1990) reported that there are no significant amounts of toxicity or anti-metabolites in peas. The dried and powdered seed has been used as a poultice on the skin where it has an appreciable effect on many types of skin complaint including acne. The oil from the seed, given once a month to women, has shown promise of preventing pregnancy by interfering with the working of progesterone. The extracts of *P. sativum* have been investigated and found to be pharmacologically active inducing anticancer activity (Clemente *et al.*, 2005, 2012; El-Aassar *et al.*, 2014; Patel, 2014; Stanisavljevic *et al.*, 2016).

3.0 Origin and Taxonomy

3.1.0 Origin

Archaeological studies have indicated that pea has been cultivated since ancient times. Janick *et al.* (1969) described association of genus *Pisum* with man at least from Stone Age. The cultivation of field pea was a very old practice. Renfrew (1973) reported that in Austria its discovery was from Bronze Age, whereas in Germany they were grown in Iron Age. Later, it was introduced into North West Europe. De Candolle (1896) has written in 'Origin of Cultivated Plants' that the species existed in western Asia before it was cultivated. The Aryans introduced it into Europe but existed in North India before the arrival of Aryans. Vavilov (1926) listed different centres of origin of pea. The Central Asia was regarded as the birth place of all legumes including pea whereas Asia Minor is the secondary centre of origin. A further support to this thought was provided by Govorov (1928) and the study of peas collected by Vavilov (1951) from Afghanistan.

Pea probably originated in south western Asia, possibly north western India, Pakistan or adjacent areas of former USSR and Afghanistan and thereafter spread to the temperate zones of Europe (Kay, 1979; Makasheva, 1983). Based on genetic diversity, four centres of origins, namely, Central Asia, the near East, Abyssinia and the Mediterranean have been recognized (Gritton, 1980). Non-pigmented peas to be used as a vegetable were grown in United Kingdom in the middle Ages (Davies *et al.*, 1985). Pea was introduced into the Americas soon after Columbus and a winter types pea was introduced from Austria in 1922. Pea was taken to China in the first century (Makasheva, 1983). Peas were reported to be originally cultivated as a winter annual crop in the Mediterranean region (Smart, 1990).

The cultivated garden pea has not been found in wild state, whereas De Candolle (1896) reported that small seeded field pea referred to *P. arvense* L. occurred in parts of Italy. It is, therefore, suggested that cultivated garden pea arose from the wild field pea or other related wild species. On the basis of the extant taxonomic literature and the historic records of the cultivated peas, Lamprecht (1956) concluded that *P. sativum* L. arose in medieval times through a mutation to white flower and large seeds in cultivated forms of *P. arvense*. Purseglove (1974) regarded *P. elatitus*, a wild species in Russia as an ancestor of *P. sativum* and suggested that high genetic variability must have given a broad genetic base for selection of garden pea.

3.2.0 Taxonomy

The genus *Pisum* includes 6-7 species, mostly found in the Mediterranean area and West Asia out of which only *P. sativum* is cultivated. However, the taxonomic status of these species has often been questioned (Smartt, 1976). Whyte *et al.* (1953) also reported that the genus has many morphologically divergent wild types found in Old World. Most of them are closely related and all the wild and cultivated types have same chromosome number (2n=14). Govorov (1928) suggested that cultivated form of pea was only one species *P. sativum* and subdivided it into two subspecies *sativum* and *arvense*. Menjkova (1954) divided cultivated pea into two species, the field pea (*P. arvense*) and garden pea (*P. sativum*). Duke (1981) reported that garden peas are treated as *P. sativum* ssp. *hortense* Asch. and Graebn., field peas as *P. sativum* ssp. *arvense* (L.) Poir., and edible podded peas as *P. sativum* ssp. *macrocarpon*; early dwarf pea as *P. sativum* var *humile*.

From a practical standpoint, according to Lamprecht (1956) *P. sativum* can be divided into following groups:

    I. *P. saccharatum*– No or very thin lining membrane in the pod walls.

        A. Types with thin pod walls

           1. Type with smooth seeds

           2. Type with wrinkled seeds

        B. Types with thick pod walls

           1. Type with smooth seeds

           2. Type with wrinkled seeds

    II. *P. pachylobum*– Well developed lining membrane in the pod walls.

        A. Types with thin pod walls

           1. Type with smooth seeds

           2. Type with wrinkled seeds

B. Types with thick pod walls

1. Type with smooth seeds

2. Type with wrinkled seeds

The whole group of *saccharatum* and A 2 group of *pachylobum* are known as garden pea, whereas others are called field pea (Whyte *et al.*, 1953).

Later, Smart (1990), based on studies undertaken by Ben-Zeiev and Zohary (1973), and Polhill and van der Maesen *et al.* (1985) reported that pea comprises only two species, *viz.*, *P. sativum* and *P. fulvum* Sibeth. and Mith.

4.0 Botany

The plant is a short lived, herbaceous annual, glaucous, which climbs by leaflet tendrils. The cultivars may be dwarf, semi-dwarf or tall. The root system is not strongly developed except taproot. The stem is slender, circular and weak. The leaves are pinnate with up to 3 pairs of leaflets and terminal branched tendril; stipules large and leaf-like, ovate; stipels absent; leaflets ovate, or elliptic, up to 6 in number, entire with undulate margin. The flowers are solitary axillary or up to 3 flowers per raceme; bracts very small; calyx oblique, lobes unequal; corolla white, pink or purple; keel short, incurved, obtuse, stamens diadelphous, filaments broad, anthers uniform; style falcate, flattened, bearded on inner surface, stigma minute and terminal. Pods are swollen or compressed, straight or curved on short stalks up to 12-15 cm with as many as 10 seeds per pod. Seeds are angular or globose, 4-10 in number, smooth or wrinkled, no endosperm, green, grey or brown, sometimes mottled; seed weight up to 250 mg (Smartt, 1976).

5.0 Cultivars

Large number of pea cultivars has been developed in several countries. Trebuchet *et al.* (1953) described 113 cultivars in France on the basis of seed characters, such as round and wrinkled seed coat; yellow or green cotyledons and plant height. Sneddon and Squibbs (1958) classified 284 cultivars and collected the information on their origin, history and synonyms. Wade (1943) suggested classification of cultivars on characters like node of first bloom, height of vines, type of seed, size of pods, end of pod, foliage and seed colour, resistance or susceptibility to various diseases. Study of the shape, outline, relative size and height of the central and lateral leaflets of the two lowest leaves have been used to differentiate between cultivars (Hillman, 1954). Shape of seeds, internode length, foliage colour, stipules, number of leaflets, peduncle length, flower type, pod shape and pod colour have been used for distinguishing the cultivars (Singh and Joshi, 1970). The following vegetative and reproductive characters of pea cultivars are considered of horticultural importance as described by Singh and Joshi (1970).

5.1.0 Distinguishing Characters

5.1.1 Height of Stem

A great variation in the height of the plant in different cultivars has been observed. Generally, four categories are recognized, such as, dwarf, dwarf-medium,

medium and tall. The height is related to number of internodes, earliness in flowering and productivity.

5.1.2 Branching

It is also a varietal character. Cultivars like Lincoln and Wando are more branched, whereas Alaska has low branching capacity. Branching character is related to yield potential of a cultivar.

5.1.3 Nodes to First Inflorescence

It is closely correlated to day to flower and maturity period. The earlier cultivars flower at lower node.

5.1.4 Flower per Peduncle

The number of flowers per node varies from one to two, whereas in some cultivars they bear three or four or more flowers. Various categories found are (a) singles, (b) mainly singles, (c) singles and pairs, (d) mainly pairs, and (e) pairs.

5.1.5 Pod Size

There is large variability for pod size. Large pods are not always the best since their shelling percentage is generally lower than in smaller pods. Well-filled medium sized pods have better shelling percentage. Sneddon and Squibbs (1958) recognized following classes in respect of length, breadth and width of pods of garden pea.

Length	Below	8.1 cm	Short
		8.1-9.5 cm	Medium-long
		9.6-11.0 cm	Long
	Above	11.0 cm	Very long
Breadth	Below	1.7 cm	Narrow
		1.7-1.9 cm	Medium-broad
	Above	1.9 cm	Very broad
Width	Below	1.4 cm	Thin
		1.5-1.7 cm	Medium-wide
	Above	1.7 cm	Wide

Pod size is almost constant from year to year and first formed pods are generally best developed and most typical of a cultivar.

5.1.6 Seeds per Pod

Number of ovules per pod is a better measure of pod capacity than number of seeds per pod (Sneddon and Squibbs, 1958). The average number of ovules per pod is 7-11 and number of seeds 6-0. The number of seeds per pod is related to pod and seed size.

5.1.7 Shelling Percentage

It is an important consideration from consumer point of view and is determined by filling of pod, boldness of seed and thickness of shell. The range of shelling percentage lies between 30.7 and 56.4 in different cultivars.

5.1.8 Seeds

On the basis of shape of seed coat, peas are broadly divided into wrinkled and smooth-seeded cultivars. Smooth and wrinkled seeds differ in their starch characters: the starch grains from smooth seeds are round and starch sugar ratio is higher at maturity and wrinkled peas whose starch grains are corrugated and lobed. The smooth seeded cultivars have hard texture and less flavour. Kellenbarger *et al.* (1951) reported that wrinkled seeds contain less starch (about 33.7 per cent) than smooth seeds (about 46.3 per cent) and are sweeter.

5.1.9 Testa and Cotyledon Colour

The testa colour in pea cultivars is either bluish green or white. Green colour is affected by excessive sunshine, which bleaches it. The cotyledons are either yellow or green.

5.2.0 Groups

The cultivars in pea are grouped on the basis of various characters (Chauhan, 1968; MacGillivray, 1961).

   a) *According to seed*
       1. Round or smooth-seeded cultivars
       2. Wrinkled-seeded cultivars
   b) *According to height of plant*
       1. Bush or dwarf types
       2. Medium-tall types
       3. Tall types
   c) *According to maturity period*
       1. Early – 65-80 days
       2. Medium – 90-100 days
       3. Main season – 110-120 days
   d) *According to use of pods*
       1. Fresh market types
       2. Freezing types
       3. Canning types
       4. Dehydration types

5.3.0 Some Important Cultivars

A large number of cultivars are under cultivation in different parts of the world. Some important pea cultivars are described below.

5.3.1 Early Smooth-Seeded

Asauji

It is a selection made from the material collected from Amritsar. It is a dwarf, green and smooth-seeded cultivar suitable for early sowing. It flowers in 30-35 days and first blossom appears at 6-7th node. The pods are produced singly, about 8 cm long, curved, dark green, narrow and appear round when fully developed and 7-seeded. It gives a shelling percentage of about 45 (Pal *et al.*, 1956).

Lucknow Boniya

A dwarf white-seeded cultivar flowers in 40 days and first blossom appears at 8-9th node. The pods are borne singly, small, narrow, green, 4-5 seeded. It is the earliest cultivar in the plains (Singh and Joshi, 1970).

Alaska

This is an early canning cultivar with bluish green seeds. Plant is about 45 cm tall and first blossom appears at 8-9th node after 38 days. Pods are borne singly, light green in colour, 7×1.25 cm, contain 5-6 small green seeds; shelling percentage 42 (MacGillivray, 1961).

Early Superb

It is an English dwarf cultivar with yellowish green foliage. It flowers in about 45 days and first blossom appears at 8-10th node. The pods are borne singly, dark green and curved with 6-7 seeds. The shelling percentage is 40. The plant branches from base (Singh and Joshi, 1970).

Meteor

Plants are 35-40 cm tall, dark green, flowers borne generally singly; pods dark green, 8.7 cm long, well filled with 7 seeds, having shelling percentage of 45. Pods mature in 58-60 days, suitable for early October sowing (Choudhury and Ramphal, 1975).

5.3.2 Early Wrinkled-Seeded

Arkel

The plant is dwarf but the growth is vigorous, plant 35-45 cm tall, flowers borne in double on few lower nodes and single afterwards; pods attractive, deep green, about 8.8 cm long, incurved towards suture at distal end, well filled, 7-8 seeds, shelling percentage 40; suitable for fresh market and dehydration; pods harvested in 50-55 days (Choudhury and Ramphal, 1975).

Early Badger

It is a dwarf cultivar evolved at Wisconsin. The plant flowers in 40-45 days and first blossom appears at 10-11th node. The yellowish green pods are borne singly, 7.5 cm long, well filled, 5-6 seeded. It is a good canning cultivar having a shelling percentage of 36, reported to be resistant to *Fusarium* wilt and tolerates heat and drought (Choudhury, 1967).

Little Marvel

This cultivar was bred in England from the cross Chelsea Gem × Suttons Alaska. Plants are dwarf with dark green foliage. It flowers in 40 days and first blossom appears at 9-10th node. Pods are about 8 cm long, borne singly, thick skinned, dark green, straight and broad containing 5-6 seeds. It gives shelling percentage of 40. The peas are very sweet and of fine quality (Singh and Joshi, 1970).

Kelvedon Wonder

Dwarf plant, flowers in about 40 days, and first blossom appears at 8-9th node. Pods are curved, borne singly, green, about 9 cm long and 6-seeded. The shelling percen-tage is about 40.

Early December

Wakankar and Mahadik (1961) produced this cultivar in Madhya Pradesh. It is a selection from the cross T.19 × Early Badger. It is dwarf, producing light green pod, 7 cm long. It has higher number of pods per plant than Early Badger but is somewhat late in flowering.

5.3.3 Wrinkled-Seeded Main Season and Late Types

Bonneville

A medium-tall, double podded cultivar, flowers in 50-60 days and first blossom appears at 13-15th node. Pods are light green, straight; about 9 cm long and 6-7 seeded with shelling percentage of 45.

T.19

Agriculture Department, U.P. developed this cultivar. It is a medium-tall and double-podded cultivar. The plant flowers in 55-60 days and first blossom appears at 12-14th node. Pods are yellowish green, slightly curved, 8.5 cm long and 6-7 seeded having shelling percentage of 45.

Lincoln

It is a dwarf to medium-tall, single podded cultivar, flowering in 55-60 days and first blossom appears at 11-12th node. Pods are dark green, 9.5-10 cm long, 6-7 seeded with shelling percentage of 45 and it is suitable for late sowing. Pods retain good colour after harvest and good for canning.

Delwiche Commando

The University of Wisconsin, from a cross between Admiral and Pride, developed this cultivar. It is reported to be resistant, to common wilt and near-wilt diseases. The plant is medium-tall with dark green foliage, flowers in 55-60 days and first blossom appears at 10-12th node. Pods are 8-8.5 cm long and 5-6 seeded having a shelling percentage of 46.

Khapar Kheda

A tall growing double-podded cultivar, flowers in 65-70 days and first blossom appears at 15-16th node. Pods are 5.5 to 6 cm long and 4-5 seeded with shelling percentage of 50. It is very popular in Madhya Pradesh.

NP 29

This cultivar has been developed through selection at the Indian Agricultural Research Institute (IARI), New Delhi; it is medium-tall, double-podded cultivar with dark green foliage. It flowers in 75-80 days and first blossom appears at 14-16th node. Pods are green, straight; about 7.5 cm long and 6-7 seeded. It has shelling percentage of 50 and is highly suitable for dehydration (Jain and Choudhury, 1963).

Perfection New Line

It is a heavy yielding mid-season cultivar. The plant is medium-tall and pod is about 8 cm long, dark green, sweet and well filled with wrinkled seeds. It is ready for first picking in 80 to 85 days (Nath, 1976).

Thomas Laxton

It is a general-purpose cultivar with deep green foliage. Pods are of excellent quality, broad, blunt, 8.25 cm long having cream and green coloured seed and suitable for freezing (MacGillivray, 1961).

Alderman

This is an excellent cultivar for home garden, shipment and freezing. Plants are dark green, 150 cm high and pods are borne singly, 11.25 cm long with 8 to 10 seeds of excellent quality. This cultivar is suitable for hills (Yawalkar, 1969).

GC-14

This cultivar was developed from a cross T.19 × Greater Progress. Plants are bushy and erect; pods are pale green, 8.7 cm long, 7-seeded, medium maturity and first picking starts after 90 days (Wakankar, 1978).

GC-195

This was developed from a cross between 7.19 and Little Marvel. Plants are 45 cm tall having 7.1 cm long pod. Average number of seeds per pod is 7.9. First picking starts after 75 days (Wakankar, 1978).

5.3.4 Smooth-Seeded Main Season Type

Kanawari

This is a tall-growing double-podded cultivar, flowering in 65-70 days and first blossom appears at 15-17th node. Pods are about 8.5 cm long, yellowish green, and 5.6 seed with a shelling percentage of 40.

5.3.5 Edible-Podded Cultivar

Sylvia

It is a tall growing cultivar flowering in 60 days and first blossom appears at 14-16th node. Pods are borne singly, yellowish, 12 cm long and sickle-shaped. Pods have general appearance of a medium-sized French bean pod. Staking is desirable and it is suitable for late sowing.

During the past few years many new cultivars have been developed in different countries. Some of these promising cultivars are described below:

Matter Ageta 6

It is a dwarf, dark green and early maturing Indian cultivar and is ready for first picking after 40-50 days from sowing, producing 6-7 seeds per pod. It produces 50 per cent of its total yield in the first picking (Dhillon *et al.*, 1989).

Aparna

It is a compact, dwarf, high yielding cultivar, resistant to wilt and pod borer and tolerant to powdery mildew and leaf miners (Kumar *et al.*, 1989).

Oregone '523'

A commercial freezing cultivar of USA with field resistance to pea enation mosaic virus, red clover vein mosaic virus and race 1 of *Fusarium oxysporum* f. sp. *pisi* (Baggett and Kean, 1987).

Pershotsuit

An Ukrainian cultivar, with growth period of 65-70 days, resistant to drought, shed-ding and to *Fusarium* and *Aphanomyces* root rots. Its N fixing capacity is high (Shevchenko, 1987).

KharKovskiiUsatyi

This is a high yielding cultivar bred at Ukraine with tendril type leaf, good lodging resistance and for seed production (Chekrygin, 1987).

Vica

An early, small seeded, high yielding Hungarian cultivar, suitable for processing and resistant to *F. oxysporum* f. sp. *pisi* race 1 (Schmelcz, 1985).

Alaska 81

Alaska 81 out-yielded Alaska by an average of 33 per cent, about 7 cm taller than Alaska and produces heavier seeds. It is immune to pea seed borne mosaic virus and resistant to *F. oxysporum* f. sp. *pisi* race 1, and suitable for canning (Muehlbauer, 1987).

Taichung 12

In Taiwan, Taichung 12 was derived from a cross of 25-2-1 (Taichung 11) with *Erysiphe polygoni* resistant cv. Manoa Sugar. Plants are tall (194 cm) vigorous and produce purple flowers from the 17-18th node. Pods are bright green and of very high quality for the fresh market and freezing. Taichung 11 is tolerant of wet soil and is resistant of *E. polygoni* (Kuo, 1988).

Taichung 13

Another cultivar from Taiwan, Taichung 13 was developed from Sugar Snap × Knight, through pedigree selection. The cultivar is tall with smaller leaves, less branches and a lower pod setting position (57 cm). The pods attain edible maturity in 63-71 days, large and uniform. The cultivar is tolerant to *F. oxysporum* (Anon., 1989).

Pervenets

This Russian cultivar has determinate habit and a high number of productive nodes. The stem is short and lodging resistant; pods are 5-7 cm in length and contain 3-7 dark green seeds of the wrinkled type with good flavour (Tsyganok, 1990).

In Brazil, for dry-seed, cultivars Triofin XPC-88, Caprice and XPC-143 and for green-pea, cultivars Trolly, Klos, Bolero and Bonnaire were found outstanding (Nascimento *et al.*, 1987).

VL Agetimatar 7 (Pant Uphar × Arkel)

This cultivar was developed at Vivekananda Parvatiya Krishi Anusandhan Shala. An early maturing cultivar with a green pod yield potential of 10-12.5 t/ha. It matures 5 days earlier than Arkel, which helps to control damage caused by powdery mildew (*E. polygoni*) and frost. It was released for commercial cultivation in the Uttar Pradesh hills (Mani and Shridhar, 1996).

Trounce

Trounce was selected from a number of powdery mildew (*Erysiphe pisi*) resistant plants found in a processing crop of cv. Small Sieve Freezer in 1987. It is mid-season maturing and, as well as a high level of resistance to powdery mildew and resistant to pea common wilt (*F. oxysporum* f. sp. *pisi* race 1), pea top yellows virus (bean leaf roll luteovirus) and bean yellow mosaic bymovirus. Trounce performed well in quality assessment of its suitability for freezing. The high proportion of small (7.1-8.7 mm) and medium (8.7-10.3 mm) sized peas makes it suitable for use as premium grade baby peas (Goulden and Scott, 1993).

Apex

Apex is derived from a cross (Fasciated × S1) × (Multipod 5) × (William Massey × Victory Freezer). It is a mid-season processing pea cultivar resistant to pea common wilt (*Fusarium oxysporum* f. sp. *pisi* race 1), pea top yellows virus (bean leaf roll luteovirus), and bean yellow mosaic. Apex has a highly determinate fruiting habit, which assists in maximizing vining yield. It is also found to be suitable for dehydration (Scott and Goulden, 1993).

5.3.5 Other Cultivars

VL-Ageti Matar-7 (VL-7)

Developed through advanced generation selection from the cross Pant Uphar × Arkel. Plants are dwarf with dark green foliage and white flowers. Pods are light green, attractive, slightly in curved towards suture at the distal end and medium in size (about 8 cm) containing 6-7 seeds. The seeds are light green, dimpled bold and very sweet with high TSS (16.8 per cent). Average yield is 100 q/ha with 42 per cent shelling.

Jawahar Matar 3 (JM 3, Early December)

This cultivar was developed at JNKVV, Jabalpur, India through hybridization of T19 × Early Badger followed by selections. Plant height 70-75 cm with bushy growth habit; flower colour white, pods light green, roundish-oval in shape with 4-5 wrinkled seeds. It gives high shelling percentage (45 per cent). This cultivar suffers severely from powdery mildew. First picking starts at 50-55 days after sowing. Average yield is 40 q/ha.

Jawahar Matar 4 (JM 4)

This cultivar was developed at JNKVV, Jabalpur, India through advanced generation selections from the cross T19 × Little Marvel. Plant height 65 cm, foliage and stem green. First picking can be taken after 70 days. Pods are green, medium in size (7 cm) with 6-7 green, wrinkled and sweet seeds. It is highly susceptible to powdery mildew. Average pod yield 70 q/ha with 40 per cent shelling.

Harbhajan (EC 33866)

Developed at JNKVV, Jabalpur, India by selection from the exotic genetic stock. It is very early and first picking can be taken in 45 days of sowing. Plant type resembles that offield peas; pods are small with yellow, round and small seeds. It is also highly susceptible to powdery mildew disease. Average pod yield 30 q/ha.

Pant Matar 2 (PM-2)

It was developed at GBPUAT, Pantnagar, India through pedigree selection from the cross Early Badger × IP3 (Pant Uphar). Plant height 50-55 cm; fruit setting starts from 6th node. Pods are green, relatively small in size with 6 sweet and wrinkled seeds. First picking starts 55 days after sowing. It is also highly susceptible to powdery mildew. Average pod yield 70-80 q/ha.

Jawahar Peas 54 (JP 54)

This powdery mildew resistant cultivar was developed at JNKVV, Jabalpur, India through advanced generation selection from a double cross (Arkel × JM5) × ('4bc' × JP 501). Plants are dwarf (45-50 cm) and vigorous, pods are big, incurved towards sutures (sickle shaped) and enclosing 8-9 big, wrinkled, greenish-yellow seeds. Average pod yield 70 q/ha.

Hisar Harit (PHI)

This mid-early cultivar was developed at CCSHAU, Hisar, India through bulk-pedigree method of selection from the cross Bonneville × P23. Plant semi dwarf, first picking after 60 days, foliage green; single to double podded; pod well filled and sickle shaped, large and green; seed green dimpled after drying. Average pod yield 90 q/ha.

Jawahar Peas-4

This powdery mildew resistant and wilt tolerant cultivar for hillocks was developed at JNKVV, Jabalpur, India through advanced generation selections from a triple cross Local Yellow Batri × (6588 × 46C). Plants attain height of around 75 to 80 cm on hillocks and about 1 m in plains; medium size pods with 5-6 big, green seeds. First picking after 60 days in hillocks and 70 days in plains. Average pod yield 30-40 q/ha in hillocks and 90 q/ha in plains.

Jawahar Matar 1 (JMl or GC 141)

This cultivar was developed at JNKVV, Jabalpur, India through advanced generation selections from the cross of T19 × Greater Progress. Plant height 65-70 cm, bushy, foliage green, flower white with two flowers per axil. Pods are straight and big (8.8 cm) with 8-9 big, sweet and wrinkled seeds. It is susceptible to powdery mildew disease. Average pod yield 120 q/ha with 52 per cent shelling.

Jawahar Matar 2 (JM 2)

The cultivar was developed at JNKVV, Jabalpur, India through advanced generation selection from the cross of two exotic lines Greater Progress × Russian-2. Pods are dark green, big, curved with 8-10 sweet seeds. Seeds are wrinkled, green and bigger in size. It is susceptible to powdery mildew.

VL-Matar-3

This cultivar was developed through pedigree selection from the cross Old Sugar × Early Wrinkled Dwarf 2-2-1. Plant height 67 cm, determinate in habit with light green foliage, white flower and bear two pods in a bunch; pods are light green, 6.8 cm long, straight with 5 wrinkled seeds. Tolerant to powdery mildew and wilt; first picking starts 100 days after sowing. Average pod yield is 100 q/ha with 46 per cent shelling.

Pant Uphar (IP-3)

This cultivar was developed at GBPUAT, Pantnagar, India through selection. Plant height 70-75 cm with relatively thin leaflets of light green in colour; flower white and two buds are bome per axil; pods are round and 7-8 cm in length with yellowish and wrinkled seeds. First picking starts 75 to 80 days after sowing. Susceptible to powdery mildew disease, but tolerant against pea stem fly. Average pod yield is 100 q/ha with 52 per cent shelling.

Punjab 88 (P-88)

This cultivar was developed at Ludhiana, India through selections from the hybrid progeny of the cross Pusa-2 × Morrasis-55. Plants are dwarf, vigorous, erect with dark green foliage; one or two flowers per axil. Flowering after 75 days and first picking after 100 days of sowing. Pods are dark green, long (8-10 cm) and slightly curved at centre with 7-8 green, wrinkled and less sweet seeds. Highly susceptible to powdery mildew disease. Average yield is 150 q/ha with 47 per cent of shelling.

Azad P-2

A powdery mildew resistant cultivar developed at Kalyanpur, India through advanced generation selection from the cross Bonneville × 6587. Plants are tall (130-150 cm), erect with light green foliage and white flowers. Pods are medium in size, light green, straight, smooth, firm, borne in cluster of two with 6-7 wrinkled and brownish seeds. Crop duration 90-95 days. Average yield is 120 q/ha.

Ooty-1

Developed at Udhagamandalam, India through pure line selection from the accession PS 33. It is a dwarf type having yield potential of 119 q/ha in 90 days crop duration. It is resistant to white fly.

Jawahar Pea 83 (JP 83)

This powdery mildew resistant cultivar was developed at JNKVV, Jabalpur, India through advanced generation selection the double cross (JMI × JP 829) × (46 c × JP 501). Plants are dwarf (50 cm), pods are big and curved with 8 big, green and sweet seeds. Average pod yield is 120-130 q/ha.

Jawahar Peas 15 (JP 15)

This duel resistant (powdery mildew and fusarium wilt) cultivar was developed at JNKVV, Jabalpur, India through advanced generation selections from the triple cross (JMI × R 98 B) × JP 501 N2. Plants are dwarf (50 cm), having compact intermodes and bigger pods containing 8 seeds. Average pod yield is 130 q/ha.

UN 53 (6)

Developed at IIHR, Bangalore, India which gives 80-90 q pod yield per hectare in a crop duration of 90 days.

NDVP-8

This medium maturity group cultivar has been developed at NDUA and T, Faizabad, India and notified by CVRC Notification no. 843(E) dated 21.09.1998. Plants are 70-75 cm long, pods 9-10 cm long with 6-8 green and sweet grains. Pods are ready to harvest in 70-80 days after sowing. This cultivar has been recommended for cultivation in Punjab, Uttarakhand, U.P. and Bihar and has yield potential of 130-160 q/ha.

VL-8

This mid maturity group cultivar has been developed at VPKAS, Almora, India and notified by CVRC Notification no. 1135(E) dated 15.11.2001. Pods are green, filled with bold seeds and attractive in colour. This cultivar has been recommended for cultivation in J8tK, H.P. and Uttarakhand and has yield potential of 100-120 q/ha.

Vivek Matar-9 (VL -9)

This medium maturity group cultivar has been developed at VPKAS, Almora, India and notified by CVRC Notification no. 664 (E) dated 10.05.2005. Plants are medium with dark green foliage. Pods are straight, 6-7 cm long and green. This cultivar has been recommended for the cultivation in J&K, H.P. and Uttarakhand and has yield potential of 100-105 q/ha.

Kashi Nandini (VRP-5)

This early cultivar has been developed through hybridization (P-1542 × VT-2-1) followed pedigree selection at IIVR, Varanasi, India and notified by CVRC Notification no. 597(E) dated 25.04.2006. Plant height is 47-51 cm, erect and 50 per cent plants bear flowers 34 days after seed sowing foliage is dark green with 7-8 pods per plant. Pods are 8-9 cm long, attractive well-filled with 8-9 seeds, shelling percentage 47-48 per cent. Plants are tolerant to leaf miner and pod borer. This cultivar has been recommended for cultivation in J&K, H.P., Uttarakhand, Punjab, U.P., Bihar, Jharkhand, Karnataka, Tamil Nadu and Kerala and has yield potential of 110-120 q/ha.

Narendra Sabzi Matar-6 (NDVP.12)

This cultivar has been developed through hybridization (KS-123 × Arkel) followed pedigree selection at NDUA and T, Faizabad, India and notified by CVRC Notification no. 597 (E) dated 25.04.2006. Plants are green, 45-55 cm tall, flowering starts in 30-35 days, early maturity, first green pod picking in 60-70 days after seed sowing. Pods are ~8 cm long filled with 7-8 green sweet seeds. It is recommended for cultivation in Punjab, U.P., Bihar, and Jharkhand and has yield potential of 85-95 q/ha.

Narendra Sabji Matar-4 (NDVP.9)

This cultivar has been developed at NDUA and T, Faizabad, India and notified by CVRC Notification no. 597(E) dated 25.04.2006. Plants height 70-75 cm, pods 8-9 cm long with 7-8 green, sweet grains. Plants become ready for harvest in 75-80 days after sowing. It has been recommended for cultivation throughout India and has yield potential of 100-110 q/ha.

Narendra Sabji Matar-5 (NDVP-250)

This cultivar has been developed at NDUA and T, Faizabad, India and notified by CVRC Notification no. 597(E) dated 25.04.2006. Plants are 70-75 long with less dark green leaves. Normally, two pods/axil appear, pods are 7-8 cm long with 7-8 green grains. Mature seeds are creamy and wrinkled. Pods are ready to harvest in 80-85 days. It is resistant to powdery mildew and has yield potential of 80-110 q/ha.

Kashi Udai (VRP-6)

This cultivar has been developed through hybridization (Arkel and FC-1) followed pedigree selection at IIVR, Varanasi, India and notified by CVRC Notification no. 597(E) dated 25.04.2006. Plant height 58-62 cm, short internodes with 8-10 pods per plant with dark green foliage and -50 per cent plants bear flower after 35-37 days of seed sowing. Pods are 9-10 cm long, attractive filled with 8-9 bold seed, shelling percentage 48. This cultivar has been recommended for the cultivation in Uttar Pradesh and has yield potential of 100-110 q/ha.

Kashi Mukti (VRP-22)

This early maturity group cultivar has been developed through hybridization (No.7 × PM-5) followed pedigree selection at IIVR, Varanasi, India and notified by CVRC Notification no. 597(E) dated 25.04.2006. Plant height is 50-53 cm and 50 per cent plants bear flowers at 35-36 days after seed sowing (early group). Pods are 8.5-9 cm long, attractive filled with 8-9 bold, soft textured seeds and shelling percentage 48-49. Plants are resistant to powdery mildew. It has been recommended for cultivation in Uttar Pradesh and has yield potential of 120-140 q/ha.

Kashi Shakti (VRP-7)

This medium maturity group cultivar has been developed through hybridization (Hara Bona × NDVP-8) followed pedigree selection at IIVR, Varanasi, India and notified by CVRC Notification no. 597(E) dated 25.04.2006. Plant height is 90-98 cm and 50 per cent plants bear flowers at 54-56 days after sowing. Plants have dark green foliage with 11-12 pods per plant. Pods are 10-10.5 cm long, attractive filled with 7.5-8.5 bold seed and shelling percentage 48-49. It has been recommended for cultivation in Uttar Pradesh and has yield potential of 140-160 q/ha.

Azad P-5 (KS-225)

This late maturity group cultivar has been developed at CSAUA&T, Kanpur, India and notified by CVRC Notification no. 2035(E) dated 28.11.2006. Plant growth is medium with straight pods full of grains and having extended bearing upto March. It is resistant to powdery mildew and has yield potential of 95-105 q/ha.

Swarna Mukti (CHP-2)

This cultivar has been developed through hybridization followed by pedigree selection at HARP, Ranchi, India and notified by CVRC Notification no. 2035(E) dated 28.11.2006. Pods are mildly concave, obtuse apexed, long, light green. Shelled seeds are light green and sweet with very good cooking quality. Plants are tolerant to powdery mildew and suitable for September-October sowing. It has been recommended for cultivation in Jharkhand, Bihar and Rajasthan and has yield potential of 120-140 q/ha.

Swarna Amar

Developed through pure line selection. Pods are dark green, concave with acute apex having more than 50 per cent recovery of shelled dark green peas. Mature dried seeds are wrinkled green. Resistant to powdery mildew disease. First picking 80-90 days after sowing. Average fresh pod yield is 200-250 q/ha

Arka Ajit (FC-I)

This mid maturity group cultivar has been developed through back cross and pedigree selection involving the Parents Bonneville, IIHR-209, Freezer-656 (FC-1) at IIHR, Bangalore, India and notified by CVRC Notification no. 2035(E) dated 28.11.2006. Pods 8-9 cm long, seeds bold, green and sweet, shelling percent 55. Plants are resistant to powdery mildew and rust diseases. It has been recommended for the cultivation in U.P., Rajasthan and Karnataka and has yield potential of 95-100 q/ha in 90 days.

NDVP-10

This mid maturity group cultivar has been developed at NDUA and T, Faizabad, India and identified for release through AICRP in 1998. Plants are 70-75 cm long, pods 8-9 cm long with 7-8 green, sweet grains. Pods are ready to harvest in 75-80 days after sowing. This cultivar has been recommended for cultivation in Punjab, Uttarakhand, U.P. and Bihar and has yield potential of 115-125 q/ha.

VRP-2

This is an early maturity group cultivar developed at IIVR, Varanasi, India and identified for release through AICRP in 2001. Plants grown upto 50-55 cm height wit vigorous dark green foliage. Pods are straight, medium-size (7-8 cm), light green filled with bold ovules. Flowering starts at 38 days after sowing and green pods may be harvested 55-58 days after sowing. It has been recommended for cultivation in Punjab, U.P. and Bihar and has yield (green pod) potential of 95-105 q/ha.

VRP-3

This mid maturity group cultivar has been developed at IIVR, Varanasi, India and identified for release through AICRP-VC in 2001. This is the first cultivar, which floweres between early and mid season, *i.e.*, 45-50 days after sowing. This cultivar has been recommended for cultivation in J&K, H.P. and Uttarakhand and has yield potential of 80-90 q/ha.

Priya (DPP-68)

This cultivar has been developed at HPKV, Palampur, India and identified for release through AICRP-VC in 2001. Plants are medium tall, having green wrinkled seeds and light green straight pods containing 7-8 seeds per pod. It takes 100-110 days for first picking, Plants are resistant to powdery mildew disease. It has been recommended for cultivation in Himanchal Pradesh and has yield potential of 150-175 q/ha.

DPP-9411

This mid maturity group cultivar has been developed through recombination breeding at HPKV, Palampur, India and identified for release Ii through AICRP-VC in 2002. Plants have deep green dense foliage and are medium tall with an average plant height of 67.6 cm. Pods are deep green, straight, well-filled, medium-sized with pod length of 6-7 cm and containing 7-8 bold grains per pod, sweet with TSS 16.5°B. It has marketable maturity in 130 days in the hills and it is resistant to powdery mildew. It is recommended for cultivation in J&K, H.P. and Uttarakhand and has yield potential of 100-105 q/ha.

Pusa Prabal (GP 473)

Developed at IARI, New Delhi, India. It is a medium maturing cultivar of garden pea.

Pusa Shree

Fusarium wilt resistant early cultivar with dark green color pods having 6–7 seeds/pod. It is also tolerant to high temperature environment prevalent in north Indian plains and lower hills at the time of early sowing (October). Harvesting time 50–55 days after sowing in early October. Pod yield 45–50 q/ha during early sowing and 90–100 q/ha in normal sowing (November) condition.

Palam Priya

Developed through Pedigree Selection at HPKV, Palampur, India. It is medium tall, flowers borne in double almost throughout the plant. Profuse bearer, pods are attractive, light green, straight, 8-9cm long, 7-9 seeds/pod. Shelling percentage is 45-50 per cent, wrinkle seeded, sweet, ready in 90-100 days. Average yield is 120-130q/ha, slow mildewing.

Pusa Pragati

Introduction from Germany (1983) with its pods long (10 cm), green with 9 seeds per pod; first picking 60-65 days; resistant to powdery mildew. Yield potential is 70 q/ha.

Arka Sampoorna

Developed through back cross and pedigree selection. It is mid-season and whole pod edible pea cultivar. Pods are medium long, light green coloured. Seeds are medium bold, light green and sweet. Resistant to rust and powdery mildew. Developed by Pedigree method of selection from F_7 generation involving (Bonneville × IIHR 209) × Freezer 656 × Oregon Sugar). Pod yield is 80 q/ha in 90 days.

Arka Apoorva

It is mid-season and whole pod edible pea cultivar. Pods are medium long, dark green coloured. Seeds are medium bold, dark green and very sweet, crisp in texture. Resistant to rust and powdery mildew. Developed by pedigree method of selection from F_7 generation involving (Arka Pramodh × Oregon Sugar). Arka Apoorva is an improvement over Arka Sampoorna. Pod yield is 100-110 q/ha in 90 days.

Arka Karthik

It is mid-season cultivar. Pods are medium long, green coloured. Seeds are medium bold, green and sweet. Resistant to rust and powdery mildew. Developed by pedigree method of selection from F_7 generation involving (Arka Ajit × IIHR 554). Pod yield is 110 t/ha in 90 days.

Arka Priya

Mid-season cultivar. Both pods and seeds are round and dark green coloured. Seeds are very sweet and bold. Resistant to powdery mildew and rust. Developed by pedigree method of selection from F_7 generation involving (Arka Ajit × IIHR 562). Pod yield is 120 q/ha in 90 days.

Arka Mayur

Early cultivar, suitable for both kharif and rabi. Pods are short, straight and oval. Seeds are dark green, bold and sweet. Developed by pedigree method of selection from F6 generation involving (IIHR 105 × Arka Pramodh) × Oregon Sugar). Pod yield is 80 q/ha in 60 days.

Arka Chaitra

Tall, mid-season cultivar, tolerant to high temperature (upto 350 C). Pods are long and light green coloured. Seeds are light green, round in shape and sweet. Developed by pedigree method of selection from F6 generation involving (Arka

Cultivars of Pea

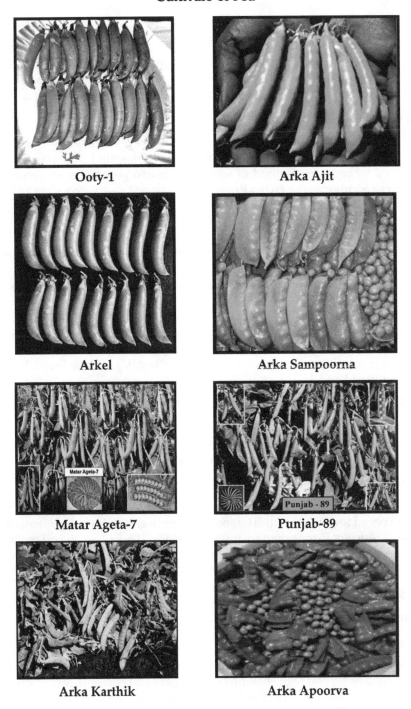

Ooty-1

Arka Ajit

Arkel

Arka Sampoorna

Matar Ageta-7

Punjab-89

Arka Karthik

Arka Apoorva

Ajit × Arka Sampoorna) × (Arka Pramodh × Oregon Sugar). Pod yield is 70 q/ha in 90 days.

Arka Tapas

Medium tall, mid-season cultivar, tolerant to high temperature (upto 35°C). Pods are short and dark green coloured. Seeds are dark green, round in shape and sweet. Developed by pedigree method of selection from F6 generation involving (Arka Pramodh × Oregon Sugar) × Arka Priya). Pod yield is 60 q/ha in 90 days.

Arka Uttam

Tall, mid-season cultivar, tolerant to high temperature (upto 35°C). Pods are long and dark green coloured. Seeds are dark green, round in shape and sweet. Developed by pedigree method of selection from F6 generation involving (Arka Ajit × Arka Sampoorna) × (Arka Pramodh × Oregon Sugar). Pod yield is 70 q/ha in 90 days.

Vivek Matar 8

Developed at VPKAS Almora, India. Light green pods with plant height is 65-70 cm; Maturity -135-140 days (seeding to first picking) during November sowing in mid hills; Tolerant to powdery mildew and white rot, pod yield is 110-120 q/ha.

Kashi Samridhi

Resistant to powdery mildew. Plant type semi-determinate with dark green pods, Medium maturing, Number of seeds per pod is 7-8, average yield is 125 q/ha. Recommended for release and cultivation in the states of Uttar Pradesh, Punjab, Bihar and Jharkhand.

Pant Sabji Matar-3

It is an early maturing cultivar developed through pedigree selection from a cross of Arkel and GC 141. Plants are dwarf with dark green foliage. The pod are long well filled with 8-10 seeds. Its green pod yield is 90-100 q/ha.

Pant Sabji Matar -5

An early-maturing cultivar whose plant is dwarf. Pods are long, well-filled and slightly curved towards the tip. The seeds are green and wrinkled at maturity. This cultivar is resistant to the powdery mildew disease. The first green pod picking can be done within 60 to 65 days and seed maturity is recorded in 100 to 110 days after sowing. Its green pod yield potential is 90-100 quintals per hectare.

PM 85 (Pant Sabji Matar 6)

An early season cultivar, resistant to powdery mildew.

Punjab 89

Medium maturity, bright green, very long pods (9-10 cm) with more number of seeds/pod (9-10) and high shelling percentage (45-50 per cent). More pods/plant (20-25) borne in doubles almost in every node, medium growth habit and sweet in taste (17.2° Brix TSS). Average yield is 135 q/ha

Mithi Phali

An introduced cultivar from the USA. It is a non-shelling cultivar and produces a 12-15 per cent higher green pod yield than shelled varieties Punjab 87 and Punjab 88. On average, it produces a green pod yield of 115-120 q/ha and a seed yield of 10-12 q/ha.

Phule Priya

Pods straight, green, tender, sweet, 8-10 grains per pod, Tolerant to powdery mildew disease, Suitable for Rabi season, Crop duration is100-110 days, Yield is about 100-105 q/ha

Solan Nirog

Developed at Dr. YSPUHF, Solan, Himachal Pradesh, India. Pods are 8-10 cm long, dark green with 8-9 seeds/pod; It matures in 90-95 days and resistant to powdery mildew disease.

Palam Triloki

Early maturing, about 10 days earlier in maturity than recommended cultivar 'Arkel' besides having higher yield potential with an average of 70-75 q/ha. It has long, bright green, round, well filled pods containing 8-10 seeds with 48-50 per cent shelling.

Palam Sumool

Medium in maturity having very long (12-15cm), dark green and flattish round pods containing 8-10 bold seeds. High yield potential (100-120 q/ha), 45-48 per cent shelling, sweet in taste (TSS 18°brix), and resistant to powdery mildew disease.

6.0 Soil and Climate

6.1.0 Soil

According to Purewall (1957), early crop should be sown on light soils, whereas heavier yields are obtained from well-drained, loose, friable and heavy soils, like silt loam or clay loam in which roots can penetrate deep. Soils able to retain sufficient moisture to carry the crop to the stage of maturity with minimum number of irrigation are ideal because frequent irrigation tends to increase vegetative growth at the expense of pod formation (Singh and Joshi, 1970). Fields with uniform soil are good for even growth and maturity of the crop.

Pea plant does not thrive on highly acidic soils. If the soil has pH less than 5.5, liming is essential. It grows under alkaline conditions. The favourable range of pH is between 6 and 7.5 (Singh and Joshi, 1970).

For very early crops, a sandy loam is preferred; for large yields where earliness is not a factor, a well-drained clay loam or silt loam is preferred (Duke, 1981).

6.2.0 Climate

Pea seeds can germinate at a minimum temperature of 5°C but the process is slow. The time required for emergence decreases rapidly as the temperature increases. The optimum temperature for germination is about 22°C (Singh and Kumar, 1979). At higher temperature the germination is rapid but plant stand is affected due to decay (Thompson and Kelly, 1957).

Peas thrive best in cool weather but do not grow well in hot weather. The optimum mean monthly temperature for good growth is 10-18.3°C. It is tolerant to frost at early stages of growth. The flowers and pods are affected, whereas leaves and stem are not damaged by frost. As the temperature rises the maturity is hastened but the yield drops rapidly. The wrinkled-seeded cultivars are more sensitive to high temperature and a temperature of 30°C and above even for a day affects the quality of canning peas. The Alaska cultivar is tolerant to heat (MacGillivray, 1961). Peas require a cool, relatively humid climate and are grown at higher altitudes in tropics with temperatures from 7 to 30°C (Duke, 1981; Davies *et al.*, 1985) and production is concentrated between the Tropics of Cancer and 50°N (Davies *et al.*, 1985). As a winter annual, pea tolerates frost to –2°C in the seedling stage, although top growth may be affected at –6°C. Winter hardy peas can withstand –10°C, and with snow cover protection, tolerance can be increased to –40°C (Slinkard *et al.*, 1994). The optimum temperature levels for the vegetative and reproductive periods of peas were reported to be 21 and 16°C, and 16 and 10°C (day and night), respectively (Slinkard *et al.*, 1994). Temperatures above 27°C shorten the growing period and adversely affect pollination. A hot spell is more damaging to peas than a light frost. Peas can be grown successfully during mid-summer and early fall in those areas having relatively low temperatures and a good rainfall, or where irrigation is practised.

As a general rule, pea cultivation is favoured in dry land areas where the weather is cool and moisture is abundant during early growth and rainfall is minimum or absent during later stages of crop development (Sutcliffe and Pate, 1977).

Peas have specific requirement in respect of changes in temperature during their growth cycle. The species is strongly thermoperiodic because the growth at constant temperature is much inferior in comparison to environment in which day temperature exceeds night temperature by 6-10°C (Highkin, 1960). This deleterious effect is transmitted from one generation to another after growing for 5/6 generations at constant temperature. Germination at high temperature resulted in tall plants, whereas low temperature at early growth stages promoted branching and dwarf growth habit (Highkin and Lang, 1966). Pod yield and leaf area/plant were higher at 10°C than at 17°C. Glaman *et al.* (1989) presented heat sums for the various maturity groups: 870-980°C for the early (60-70 days), 980-1060°C for the mid-early (70-75 days), 1060-1170°C for the mid-late (75-80 days) and >1170° for the late (>80 days).

Temperature has been reported to influence the root length and the root length is inversely correlated with temperature. Root growth decreased with changes of temperature from 25 to 32°C while changes of temperature from 32 to 25°C increased the primary root growth rate. Lateral root distribution was also growth rate dependent among seedlings grown at 25° or less. Periods of exposure (1 or

3 day) to 32° inhibited the initiation of lateral roots, after a lag period, in plants otherwise maintained at 25°C (Gladish and Rost, 1993).

7.0 Cultivation

7.1.0 Soil Preparation

Thorough preparation of soil is essential for pea because it is an exhaustive and short duration crop. It helps in rapid and free spread of roots to collect the nutrients from soil. The germination of seed is also better and even in well-prepared soil free of clods having smooth surface (Thompson and Kelly, 1957). This is achieved by ploughing the field with soil turning plough followed by one or two harrowing. Levelling of the field is also essential (Chauhan, 1968).

There should be optimum moisture at the time of sowing to facilitate seed germination. Application of irrigation after sowing affects the germination and subsequent growth. In case of deficiency of moisture, pre-sowing irrigation should be given followed by field preparation of sowing (Singh and Joshi, 1970). The soil compaction after sowing reduces the yield as a result of poor plant emergence and growth of individual plant (Hebblethwaite and McGowan, 1980).

7.2.0 Sowing

In the plains of North India, the sowing of garden peas starts from first fortnight of October and continues up to the end of November. The early sown crop is prone to wilt and late sown crop suffers from powdery mildew. For early market, the round seeded cultivars are even sown in second fortnight of September (Singh and Joshi, 1970). In peninsular India, pea crop is sown in June-July. In hills of north India, the summer crop is sown in March and autumn crop in May. In Darjeeling region of West Bengal, sowing is done from May to August. Hussain *et al.* (1976) obtained highest yield in mid-October sown pea crop in Punjab. Chaubey (1977) reported adverse effect on yield when crop was sown after 4th December.

7.2.1 Germination and Establishment

Seeds of pea are subject to rapid loss of viability in storage. Under normal storage conditions pea seeds remain viable for three years. High seed moisture and higher temperature have adverse effect (Roberts and Roberts, 1972). Germination of 70-80 per cent in laboratory test is minimum germination requirement for pea seeds. However, laboratory test cannot give any information as to how the seed will germinate and get established under field conditions. The extent of leaching of solutes and electrolytes has been reported to be a better form of assessment of cultivars in the field but the method needs to be universally standardized (Bradnock and Mathews, 1970).

Powell and Mathews (1980) found relationship between imbibition damage and field emergence of seed. Imbibition damage caused by rapid uptake of water was responsible for reduced emergence of pea seedling. They reported a significant negative correlation ($r = -0.72$) between electrical conductivity of seed-soak water with seedling emergence (Powell and Mathews, 1979). Krarup and Ross (1979) found relationship between sugar content of seeds and days to emergence. Pathogens and microorganisms also attack the cotyledons before seedling emergence and affect establishment of plants (Sutcliffe and Pate, 1977).

In temperate region sowing of pea is done from October to March, since at early stage of growth the crop is not affected by frost. Under English conditions, Bland (1971) reported 18th March to 29th April as optimum sowing period. For processing, cultivars adopted for early, mid and late sowing are grown in succession in order to supply produce at proper stage of maturity.

The depth of sowing is recommended from 5 to 7.5 cm depending on soil moisture and texture. The seed sowing is done by hand dibbling and drilling.

In the first method, seed is dibbled with hand. It can be done on a small-scale. The main commercial practice is drilling. The seeds are dropped in furrows opened by a country plough and furrows are covered by planking. Some farmers drill seed by *pora* attached to plough (Singh and Joshi, 1970). In soils where pea is being sown for the first time, seeds should be treated with bacterial culture (*Rhizobium leguminosarum*) to ensure proper nodule formation and early growth. Rhizobia are naturally occurring soil bacteria that fix atmospheric nitrogen into the soil. They live in symbiosis with legume plants and are found in nodules on pea plant root system. The culture material is emulsified in 10 per cent sugar or jaggery solution sufficient to moist the seed. It is mixed thoroughly with seed and dried in shade before sowing.

7.2.2 Spacing and Seed Rate

Bundy (1971) has reported results of experiments conducted on the effect of plant population in obtaining optimum yield. The larger haulm size and longer time to maturity demand a somewhat wider spacing and lower sowing density. Row width of 20 cm with 2.5 ´ 106 seeds/ha was regarded normal by some workers, whereas Meadley and Milbourn (1970) did not find any difference in the plant densities spanning the range of 0.43-1.72 ´ 106 seeds/ha. Weed problem may become acute at low plant population; whereas damage by disease and high cost of seed are important at high plant density (Sutcliffe and Pate, 1977). Singh and Joshi (1970) recommended 22.5-30 cm between rows and 3.75-5 cm between plants in early sown dwarf cultivars and 30-37.5 cm under non-irrigated condition and 45-60 cm under irrigated condition in main and late season sown tall cultivars. In hilly region, peas are generally staked, especially the tall cultivars. However, staking cannot be practised on commercial scale because it is very expensive (Choudhury, 1967). Abd-Alla *et al.* (1972) reported one plant per hill give the highest yield per plant and 3 plants per hill had higher total yield.

Cebula *et al.* (1987) from Poland recommended a spacing of 10 cm between the rows and 3 cm within the row for dwarf cv. Bordi and 20 cm between the row and 3 cm within the row for vigorous cv. Nike.

The untidy growth habit of pea makes it difficult to study the effect of canopy structure on yield. Hence various workers have reported the variable leaf area index (LAI) values in relation to maximum yield. However, Eastin and Gritton (1969) reported that unless a LAI value greater than 3.5 is attained there is little benefit from changing canopy structure.

Seed Germination in Pea

Seed germination and plant establishment of pea

Choudhury (1967) recommended the seed rate of 100-120 kg per hectare in early dwarf cultivars and 80-90 kg per hectare in mid-season and late cultivars.

7.3.0 Nitrogen Fixation, Manuring and Fertilization

The pea being a leguminous crop, it can fix atmospheric nitrogen in the symbiosis with *Rhizobium* bacteria and thus has low nitrogen requirement. The cycling of N from plant residues may reduce the need for N-fertilization in succeeding crops. About 25 per cent of the crop nitrogen is in plant residues (Jensen, 1989).

Mikanova *et al.* (1995) added seven strains of *Rhizobium leguminosarum* (T70, D600, D562, D561, T97, D28 and D559) to a solution containing $Ca_3(PO_4)_2$ in laboratory conditions. All the strains except for D28 were able to solubilize P from $Ca_3(PO_4)_2$. Peas cv. Tyrkys were germinated in inert material and root inoculated with suspensions of different *R. leguminosarum* strains with phosphate added. Strain T97 gave the highest above ground DM yield. Inoculated pea seeds were also sown in soil. Seed DM was affected by inoculant and by fertilizer levels. With the addition of superphosphate fertilizer DM yield fell where P-solubilizing inoculants were used. Yield increased with the use of P-solubilizing inoculants in the absence of fertilizer to a level similar to that obtained with 45 kg P/ha alone.

Prasad and Prasad (1998) found that P (30, 60 or 90 kg P_2O_5/ha) and GA_3 (0, 25 and 50 ppm) with or without seed inoculation with *Rhizobium* increased the contents of protein, carbohydrate and P. The parameters responded significantly to increased levels of P and GA_3 with inoculation during both the years. Interactions between *Rhizobium* and P and P and GA_3 also gave significant response for protein and P content. They (Prasad and Prasad, 1999) also reported that there was a linear increase in the nodulation of plants due to the interaction of *Rhizobium* sp. and P and P and GA_3. The greatest nodulation (72.62 mg/plant, compared to 15.32 mg/plant in the control) was achieved by the associative effect of *Rhizobium* sp. and P (90 kg/ha). The interaction of *Rhizobium* sp. 'P' GA_3 also increased nodulation with inoculation + 90 kg/ha P + 50 ppm GA_3 producing 74.5 mg nodules/plant, compared to 11.6 mg/plant in the control. Srivastava *et al.* (1998) reported that for the cv. Arkel, application of 25.8 kg P, 0.5 kg Mo and seed inoculation with *R. Leguminosarum* and/or phosphate solubilizing bacteria (PSB) resulted in significant increases in nodulation, nitrogenase activity, growth and grain yield, with dual inoculation significantly better than single inoculation. Kanaujia *et al.* (1998) applied 0-90 kg/ha each of P_2O_5 and K_2O to peas which were seed inoculated with *Rhizobium* or not inoculated. Seed inoculation, plus the application of 60 kg each of P_2O_5 and K_2O gave the highest pod yield of 13.17 t/ha. Ozdemir *et al.* (1999) inoculated peas cv. Marmara with *Rhizobium* or given 10 kg N and/or 5 kg P_2O_5/ha. Inoculation significantly increased nodule number and nodule dry weight. N application and inoculation significantly increased total above ground dry matter and seed yields compared with untreated controls. The highest above ground dry matter and seed yields were obtained with N + P, while N and inoculation produced similar seed yields.

Popov *et al.* (1996), using a clay soil, grew peas from untreated seeds or seeds treated with 0.345 g ammonium molybdate/200 g seeds and were given PK, PK + 20 mg N/ha soil as urea or NPK + foliar sprays of the synthetic cytokinin-like growth regulators N-allyl-N-2-pyridyl-thiocarbamide (APTC) and/or N, N-diphenyl carbamide (DPC). Nitrogenase, glutamine synthetase (glutamate-ammonia ligase) and asparagine synthetase (aspartate-ammonia ligase) activity of intact nodulated roots generally decreased between the bud stage and flowering and were increased by Mo seed treatment when no N was applied. The effects of the growth regulators on enzyme activity depended on growth stage and Mo treatment; application of either APTC or DPC increased the activities of the 3 enzymes at the bud stage in the absence of Mo but combined application reduced nitrogenase activity. Glutamate alanine-aminotransferase activity increased as growth progressed and was not much affected by growth regulator treatment. Plant fresh weight at bud formation was highest with NPK + APTC or DPC; at full flowering it was highest with PK + Mo or NPK + Mo + both growth regulators. Effect of biofertilizers and chemical fertilizers on growth and yield of garden pea was studied by Patel *et al.* (1998). In a field trial at Indore, Madhya Pradesh, peas cv. Arkel were given 20 kg N + 80 kg P_2O_5 + 40 kg K_2O/ha, *Rhizobium* inoculation, 3 kg phosphate solubilizing microorganisms (PSM)/ha, *Rhizobium* + PSM, 25 or 50 per cent of the NP rates + *Rhizobium* and/or PSM. Application of 50 per cent NP + *Rhizobium* + PSM increased plant height, number of branches, leaves per plant, number of pods per plant, grains per pod and pod yield significantly compared with recommended level of nutrients (20 kg N + 80 kg P_2O_5 + 40 kg K_2O/ha) applied through chemical fertilizers. Saravanan and Nambisan (1994) observed highest pod yield with 100 kg N + 150 kg P_2O_5 + soil inoculation with phosphobacterium.

A crop giving a yield of 4500 kg of green pea approximately removes 55 kg of nitrogen, 20 kg of P and 40 kg of K. A basal dose of 50 kg N per hectare at planting is sufficient for stimulating early growth.

Higher dose of nitrogen has adverse effect on nodulation and nitrogen fixation (Reynolds, 1960; Choudhury, 1967; El-Behidi *et al.*, 1985). The crop gives good response to phosphorus application. It favours nitrogen fixation by increasing nodule formation (Gukova and Arbuzova, 1969; Jakobsen, 1985). Phosphate increases yield and quality of pea. Potassic fertilizers also have effect in increasing the yield and nitrogen fixation ability of the plants.

Elneklawy *et al.* (1985) reported that application of 60 kg N/0.42 ha or seed inoculation with a local inoculum Okadin or an introduced inoculum TAL increased fresh weight and dry weight of pea, N uptake and green pod yield. Evans *et al.* (1980) reported that efficient nodulation of N-fixation did not occur at pH 4.8 or below (Table 2).

N-fixation rate varied with the bacterial strain. Jarak *et al.* (1989) tested 49 strains of *R. leguminosarum* on peas cv. Mali Provansalac grown on N-free sand culture and observed variation from 12.6-44.6 nodules/plant. Inoculation increased root N content at flowering from 1.88 to 2.69 per cent and shoot N content from 2.37 to 4.29 per cent. Nelson and Edic (1991) suggested that selection of rhizobial strains with enhanced efficiency of C utilization may ensure a symbiosis with greater N2 fixation capacity in the presence of combined N. Patseva and Kosenko (1991) isolated exopolysaccha-rides (EPS) from five strains of *R. leguminosarum* differing in their activity and studied their influence on nodule formation in pea. The introduction of EPS of less active strains to the inoculants of more active bacteria suppressed nodule formation in the latter. EPS of moderately active strains stimulated nodule formation in less active bacteria and inhibited it in more active forms.

Table 2: Dry Mass and Total Nitrogen Content in Roots and Shoots* of *P. sativum*[t]

pH	Dry Mass Plant^{-1} (g)	Roots Nitrogen Dry Mass^{-1} (Per cent)	Nitrogen Root^{-1} (mg)	Dry Mass Plant^{-1} (g)	Shoots Nitrogen Dry Mass^{-1} (per cent)	Nitrogen Root^{-1} (mg)
6.6	0.30 ± 0.02**	3.8 ± 0.11	11.4 ± 1.0	1.46 ± 0.22	2.3 ± 0.1	34 ± 6
5.6	0.18 ± 0.02	2.9 ± 0.1	5.2 ± 1.0	1.13 ± 0.17	2.5 ± 0.1	28 ± 5
4.8	0.17 ± 0.04	2.7 ± 0.1	4.6 ± 1.0	0.74 ± 0.22	2.8 ± 0.2	20 ± 5
4.4	0.20 ± 0.03	2.5 ± 0.1	5.0 ± 1.0	0.88 ± 0.21	2.3 ± 0.1	20 ± 5
3.8	0.09 ± 0.01	3.4 ± 0.0	3.0 ± 0.0	0.51 ± 0.07	3.6 ± 0.07	18 ± 2

* Plants were harvested 50 days after beginning of test; ** Mean and standard error of 4 samples.

t Evans *et al.* (1980).

With a seed yield of 2.2-2.5 t/ha, peas accumulated 35-45 kg N, 7-11 kg P_2O_5 and 27-33 kg K_2O/ha in the roots and stubble (Ageev and Demkin, 1987). Naik *et al.* (1991) observed that application of P from 0 to 120 kg/ha increased total dry matter yield, although the increase was only significant at maturity. Among the different sources of phosphatic fertilizers compared, basic slag at 60 kg/ha produced highest pea yield in mid hill soils of Himachal Pradesh (Bioshnoi *et al.*, 1985).

In India, Dhesi and Nandpuri (1965) recommended 8-10 cart loads of FYM in addition to 50 lb N, 50 lb P and 25 lb K per acre in light soils of Punjab. Yawalker (1969) recommended 20 lb N, 40 lb P and 40 lb K/acre for Nagpur region. Sen and Kavitkar (1958) recommended 36 kg P per hectare for Bihar. Studies at IARI have shown that 45 kg P/ha improved yield in peas, whereas nitrogen application had deleterious effect (Singh and Joshi, 1970). In trial with cv. Bonneville, Naik (1989) obtained highest pod yield with closest spacing of 30 ´ 5 cm and 75 kg N and 100 kg P_2O_5/ha but observed no appreciable response to K at 25 or 50 kg/ha.

Chamberland (1982) recommended 15-30 kg N/ha; 20, 40 and 60 kg P/ha soils with high, medium and low available P, respectively and 20-40 kg K/ha for good yield. The optimum fertilizer rates for pea grown in a crop rotation were 40 kg N, 80 kg P and 30 kg K/ha (Ageev and Demkin, 1987).

Blomfield (1958) reported that besides NPK, application of 140 g of sodium molybdate per hectare either as pre- or post-emergence spray doubled the yield and the treated plants showed more resistance to collar rots. Similar response to molybdenum application was reported in New Zealand (Singh and Joshi, 1970). Sharga and Jauhari (1970) reported that foliar application of 0.1 per cent ammonium molybdate significantly increased the number of root nodules, yield, TSS, and number of grains per pod. Meagher *et al.* (1952), however, observed that pea seeds should contain sufficient molybdenum (0.5-5.0 mg per seed) for normal growth of crop. Singh and Dahiya (1976) reported that dry matter yield increased by Fe treatment along with different levels of $CaCO_3$ and the effect of Fe was more pronounced at 10 ppm when applied 45 days after germination and with 5 ppm at 75 days.

Singh and Raj (1988) found that application of 30-75 kg S/ha increased dry matter accumulation by plants, nodulation and the leghaemoglobin contents in nodules. Leghaemoglobin contents increased with plant age up to 75 days and decreased thereafter. Sulphur increased seed yields per plant and seed protein contents. The optimum S rate was 45 kg/ha. The efficiency of fertilizer application is greatly influenced by method of application. The fertilizer-use efficiency is poor under broadcast method. Best results are obtained by fertilizer placement 6.25 cm away and 2.5 cm deeper than the seed. The germination of seed is adversely affected if fertilizers, especially nitrogen and potash, come in contact with seed (Singh and Joshi, 1970). According to Gupta (1983), the best method of application was placement of P fertilizer 3-5 cm below the seed rather than in contact with the seed or by broadcasting.

Negm *et al.* (1997) conducted a greenhouse experiment on a calcareous soil with pea seeds coated with Fe, Zn, Mn, Mo or the combined application of these elements, or received foliar application of these elements. Seed yields were similar with seed coating and foliar application. The application of Zn, Mn or Mo significantly increased seed yield but not Fe. The combined application of the elements produced the highest mean seed yield of 21.95 g/pot. This was not significantly different from 20.53 g/pot obtained with Mo alone. Seed coating with Mn or Mo produced the highest seed crude protein contents of 22.81 and 22.19 per cent, respectively. Seed sugar, amino acid and Fe, Zn and Mn contents were significantly increased by all the trace elements. In sandy loam soil, Negi (1992) found that application of 20 kg N/ha increased pea cv. Lincoln seed yield (1.66 t/ha in 1989 and 3.32 t in 1990 compared with the control yields of 1.18 and 2.42 g) but a further increment of 20 or 40 kg N had no affect on yield in 1989 and decreased it in 1990. Application of 60 kg P_2O_5 increased seed yield (1.73 and 3.28 t vs. 1.04 and 2.08 t in the controls in 1989 and 1990, respectively) while a further increment of 60 kg P_2O_5 did not increase yields significantly. Petkovas (1997) grew peas on leached smonitza poorly supplied with nitrogen, well supplied with assimilable phosphorus and highly supplied with potassium. It was found that under these conditions, a sowing density of 189 germinating seeds/m^2 ensured the highest yield per unit area.

Yadav *et al.* (1996) reported that for the pea cvs. Lincoln, Bonneville, GC 141 and Kinnauri, N application increased the number of days to flowering and marketable maturity but did not consistently affect the pod yield/plot. Naik (1995) reported that pod yield was not significantly affected by N or K rate, and was highest with 100 kg P_2O_5 (1.30 t/ha). Pod yields from spacing of 30 ´ 5, 10 or 15 cm were 0.81, 1.15 and 1.38 t/ha, respectively.

Gangwar *et al.* (1998) found the requirement of N, P and K for the production of 0.1 t of vegetable pea seed to be 8.25, 1.03 and 5.65 kg, respectively under Pantnagar conditions. The percentage utilization of available N (per cent organic carbon), P_2O_5 and K_2O (ammonium acetate-K) were 36.59, 13.83 and 11.81, respectively.

The contribution from fertilizer as a percentage of its nutrient content was 188.82, 20.79 and 46.57 for N, P and K, respectively. Verma *et al.* (1997) found on a sandy loam soil (inceptisol) the green pod and seed yields increased significantly with N and P applications up to 15 kg N/ha and 60 kg P_2O_5/ha. A decline in yield occurred with higher fertilizer rates. Straw yield increased up to 45 kg N/ha and 60 kg P_2O_5/ha. The N content in seeds and straw increased with increasing levels of N and P (up to 60 kg P_2O_5/ha). The P content in seeds and straw increased with increasing levels of P and N (up to 15 kg N/ha). K, Ca, Mg and S contents increased up to 15 kg N/ha and 60 kg P_2O_5/ha; at higher fertilizer levels a slight decline was observed. It is concluded that 15 kg N and 60 kg P_2O_5/ha were the optimum rates for maximum yield and high nutrient concentrations in peas. Verma and Bhandari (1998) also reported in cv. Lincoln that N and P contents in pea plants generally increased with up to 15 kg N and 60 kg P_2O_5/ha, respectively. K, Ca, Mg and S contents followed a similar pattern, and tended to decrease at higher fertilizer rates. Concentrations of N and P were highest at 90 days and those of K, Ca, Mg and S at 120 days, after which concentrations decreased. Green pod, seed and straw yields increased with up to 15 kg N/ha and 60 kg P_2O_5. Utilization of phosphorus can be enhanced with supplementation of organic or inorganic materials. Singh *et al.* (1997) found that the dry matter yield/pot, total P uptake and percentage P derived from fertilizer increased with rate of P application and were highest with SSP + sodium pyrophosphate + biogas slurry with the percentage P utilization also being highest. Pulung and Harjadi (1994) reported from Indonesia that the productive efficiency (pod yield/kg P applied) and yield increased with increasing P rate. Kohli *et al.* (1992) reported that P significantly increased the number of seeds/pod and total green pod yield from 4.49 with no P to 5.12 t/ha at the highest P rate. Kumar and Rao (1992) grew ten pea cultivar in pots in clay loam soil (3.27 ppm available P) amended with 0, 15 or 0.30 ppm P. Mean seed and DM yields, and plant P uptake were increased by applied P. Without applied P, seed yield was highest in cv. DMR-17 and lowest in cv. Rachna while the yield was highest when P was applied. There was considerable variation between cultivars in response to P and tolerance of P stress.

Physiological and/or physical damage to cell membranes is associated with loss of seed vigour, because of their influence on the deteriorative metabolic changes, which occur as seeds age. The nutrients available to the mother plant, particularly nitrogen (N) and phosphorus (P), because they are important constituents of proteins and phospholipids, may affect cell membrane integrity and hence seed vigour. This hypothesis was examined by Padrit *et al.* (1996) in a field trial with garden pea cv. Pania at Massey University, New Zealand. Plants received either 0, 100 or 200 kg N/ha in combination with 0 or 22.5 kg P/ha, and seed quality was assessed using the standard germination, conductivity, hollow heat and accelerated aging (AA) tests. Germination for all treatments was 94-98 per cent and did not differ significantly. N application significantly improved seed vigour as conductivity readings and hollow heart percentage were reduced, and post-AA germination was increased. P application had no effect on conductivity, but improved seed vigour through reducing hollow heart and increasing post-AA germination. There was a significant N ´ P interaction where-by post-AA germination was increased by 100 kg N plus 22.5

kg P/ha. The effects of N and P differed with pod position. In the absence of N or P, seeds from top pods had significantly higher hollow heart levels than those from bottom pods, but for 200 kg N/ha and 22.5 kg P/ha, hollow heart did not differ with pod position. Saini and Thakur (1996) studied the effect of nitrogen and phosphorus on vegetable pea in cold desert area. In a field experiment on cv. Lincoln, 0-60 kg N and 0-66 kg P/ha were applied. Mean green pod yield increased with up to 30 kg N (17.4 t/ha) and was highest with up to 52.8 kg P (20.9 t). In Indonesia, Hanolo and Pulung (1994) applied 0 or 200 kg P/ha as triple superphosphate and 0, 50, 100, 150 or 200 kg KCl/ha and reported that primary branch number, DM, pods/plant and pod and seed yields were greater with P and increased with increasing K rate. There was no interaction between P and K fertilizers. Shukla *et al.* (1993) reported that yield improvements could be achieved by increasing plant height, shelling percentage, number of pods/plant and seeds/pod.

Jana and Paria (1996) sprayed peas cv. Arkel with partially chelated tracel (0.5 per cent, containing a mixture of Zn, Mn, B, Mg, K_2O, Mo and S in chelated form), tracel for vegetables (0.5 per cent), aminos (0.1 per cent, 0.2 per cent or 0.4 per cent, amino acid based bio-stimulant), PPL-2 (0.1 or 0.2 per cent, not specified) or akomin (0.4 per cent, Na salt of phosphoric acid) at the onset of flowering and pod-development stages. All the treatments produced higher pod yields than the untreated control (4.88 t/ha). The best treatments were 0.5 per cent Tracel for vegetables (9.89 t) and 0.1 per cent PPL-2 (7.58 t).

Podlesna and Wojcieska (1996) inoculated in a pot trial, pea cvs. Ramir and Koral grown in a soil/sand mixture with *Rhizobium* and given 20 or 800 mg N/pot as ammonium nitrate. Plants were harvested at 10-day intervals and uptake and accumulation of Zn, Cu, Mn and Fe were determined. Micronutrient concentrations were influenced by growth stage, plant part and element, but cultivar and N nutrition had less effect. Accumulation was pronounced between the bud stage and formation of the first pods. Maximum amounts of these nutrients were found during seed ripening. Trace element contents decreased at the end of the growing period, especially in the roots.

Some recent studies have found that legume-rhizobial symbiosis may even be cultivar-strain-specific. Different varieties within the same plant species may prefer different optimal rhizobial strains for maximum N fixation. In several legumes including pea (*Pisum sativum* L.) significant plant host, strain, and host × strain interaction effects have been observed on N fixation (Skot, 1983). *Rhizobium*-legume symbioses are also species-specific, and each host plant may be nodulated by one or a few microsymbiont species (Unkovich *et al.*, 2008). Although nodules of the pea (*Pisum*) as well as vetch or bean (*Vicia*) are mostly colonized by *Rhizobium leguminosarum* sv. *viciae* (*Rlv*), other *Rlv*-related rhizobial species were recently reported to nodulate the pea and bean, such as *Rhizobium fabae* and *Rhizobium pisi* (Tian *et al.*, 2008; Ramirez-Bahena *et al.*, 2008). The biodiversity of pea microsymbiont populations is known to be site-specific (Riah *et al.*, 2014). The presence of host plants in the environment (as crops or wild legumes) is favorable for sustaining the genetic diversity of *R. leguminosarum* sv. *viciae* populations, whereas heavy metal pollution of soil or monocropping has been suggested to reduce biodiversity (Martyniuk *et al.*, 2005; Ventorino *et al.*, 2007). Pea microsymbionts are common inhabitants of Polish soils (Martyniuk *et al.*, 2005) and their local population may be diverse; thus, individual pea plants may enter into symbiotic interactions with many rhizobial strains (Wielbo *et al.*, 2011).

The growth and yield of peas cultivated on eight different soils, as well as the diversity of pea microsymbionts derived from these soils were investigated by Wielbo *et al.* (2015). All soils examined contained pea microsymbionts, which were suggested to belong to *Rhizobium leguminosarum* sv. *viciae* based on the nucleotide sequence of the partial 16S rRNA gene. PCR-RFLP analyses of the 16S-23S rRNA gene ITS region and *nodD* alleles revealed the presence of numerous and diversified groups of pea microsymbionts and some similarities between the tested populations, which may have been the result of the spread or displacement of strains. They observed most of the tested populations comprised low-effective strains for the promotion of pea growth. No relationships were found between the characteristics of soil and symbiotic effectiveness of rhizobial populations; however, better seed yield was obtained for soil with medium biological productivity inhabited by high-effective rhizobial populations than for soil with high agricultural quality containing medium-quality pea microsymbionts, and these results showed the importance of symbiosis for plant hosts.

It was found that nodules where terminal bacteroid differentiation takes place are more efficient in terms of energy use. Oono and Denison (2010) reported that legume species with terminal bacteroid differentiation (such as peas (*Pisum sativum* L.) and peanuts (*Arachis hypogaea* L.) invest less in nodule construction but have greater fixation efficiency when compared to species with reversible bacteroid differentiation [such as French bean (*Phaseolus vulgaris* L.) and cowpeas (*Vigna unguiculata* (L.) Walp.]. This effect is probably due to genomic endoreduplication of the bacteroids and full contact of single undivided bacteroid with peribacteroid membrane (some reproductive bacteroids can lose contact with PBM after they divide). Still, this is not known if these useful features of terminal bacteroids differentiation in some legumes could be transferred into other legume species.

7.4.0 Irrigation

The water requirement of pea crop is very low and can be grown even without irrigation. Muthuswami *et al.* (1980) reported that pea cultivars Bonneville and Superfection could be raised successfully under rainfed conditions yielding 5.31 and 4.95 t/ha, respectively. In flat sown crop, the field must have adequate moisture for seed germination. Pre-sowing irrigation is essential if the soil does not have enough moisture at planting. Irrigation immediately after sowing results in poor germination due to formation of hard crust and impediment in the emergence of plants. Second irrigation is recommended at the time of flowering and third, if needed, at pod filling stage (Singh and Joshi, 1970). However, in light and sandy soil more frequent irrigation is necessary for better growth and yield of crop. In case of furrow and bed system of planting watering is required more frequently. In this method, the crop is irrigated immediately after sowing if the soil moisture is not sufficient (Chauhan, 1968). El-Beheidi *et al.* (1978) studied the effect of water regime and reported that weight of green pods per plant; early and total yield and 100 seed weight were highest with irrigation after 70 per cent water depletion throughout the growth. On a clayey dark red latosol soil, pod and seed production of peas were not significantly affected by differing levels of available water (30-90 per cent) up to flowering stage. After flowering, levels greater than 70 per cent adversely affected production (Matos *et al.*, 1984).

Canell *et al.* (1979) studied the effect of waterlogging up to 5 days on the growth of pea at different stages of development. The oxygen content declined to less than 2 per cent within 2-3 days. The leaf senescence was hastened; stem growth slowed and yields decreased. Waterlogging for 24 hours just before flowering restricted growth and yield. The effect of waterlogging at later stage was less marked. Belford *et al.* (1980) also reported similar adverse effect of waterlogging on pea yield. Bisseling *et al.* (1980) also found adverse effect of waterlogging on nitrification by *Rhizobium leguminosarum* in root nodules of pea. Irrigation scheduling can now be done with personal computers and by using a more dynamic description of the factors affecting crop water uptake. Malik and Bhandari (1994) found that for the pea cv. Lincoln on loamy sand (inceptisol), irrigation as 4 cm of water applied at IW: CPE ratio of 0.8 significantly increased pod yield/ha with the highest water-use efficiency. Baswana and Legha (1995) recommended 4 irrigations (presowing, flowering, fruiting and 20 days after fruiting) during the winter with pea cv. PH 1. Malik and Kumar (1997) suggested 0.75 or 1.0 CPE with N application by fertigation for high yield in pea cv. Lincoln in Himachal Pradesh, India.

Fougereux *et al.* (1997) studied the effects of different periods of water stress during reproductive stages on seed yield and seed physiological quality of peas in France. Irrigation during seed filling (IDSF) and irrigation during the period from the start of flowering to the start of seed filling were compared with a non-irrigated (NI) and a season-long, irrigated-as-needed (WI) treatment. The WI treatment showed the best yields. Water stress during the flowering period did not reduce seed quality more than WI, and reduced seed yield only slightly. Water stress during seed filling decreased seed yield but the effect of seed quality was not significant. Individual seed weight was higher and less variable in IDSF treatments than in WI treatments. Changing irrigation strategies for pea seed production by irrigating during seed filling may improve the physiological quality of the seedlots without decreasing the seed yield.

Annandale *et al.* (1999) successfully adapted the soil-water balance (SWB) model, which quantifies water uptake as a water supply or evaporative-demand-limited process, to estimate the water-use of pea cv. Puget under well-watered and water-stressed conditions. Simulations of soil water deficit and canopy growth compared well with independent data sets in a water-stress field trial. The model, developed in a user-friendly format, can be used as a generic crop irrigation-scheduling tool, for full or deficit irrigation conditions, provided that specific crop growth parameteres are known.

Malik and Kumar (1996) found that a drip irrigation level of 75 per cent PAN-E coupled with an N fertilizer (25 kg N/ha) applied though drip irrigation was the optimum combination for maximizing water-use efficiency and yields of peas grown on a sandy loam soil. Raina *et al.* (1998) reported that plastic mulch plus drip irrigation further raised the yield to 10.54 t/ha. Water-use efficiency under drip irrigation alone, drip irrigation plus plastic mulch and surface irrigation was 0.188, 0.221 and 0.106 t pod/ha cm, respectively. Drip irrigation besides giving a saving of 32 per cent water resulted in 49.5 per cent higher yield as compared to surface irrigation. The benefit: cost ratio of pea cultivation under drip alone, drip plus plastic mulch and surface irrigation was 2.06, 2.11 and 1.93, respectively. Szukala *et al.* (1995) found from the field trials with the cultivars Koral (tall), Elektron (semidwarf) and Ramir (leafless) at 3 crop densities (80, 120 and 160 plants/m² in theory, but 79, 114 and 143 plants/m² in practice just before harvest), with and without sprinkler irrigation, there was a significant (61 per cent) increase in yield from irrigation only in the dry season. In years of rain deficiency, Ramir and Elektron significantly out-yielded Koral. Plant density did not affect the seed yields of the tested cultivars.

In field trials in Spain, Rodriguez Maribona *et al.* (1993) subjected thirteen pea cultivars and lines to low, moderate or high degrees of water stress regulated by a line sprinkler system of irrigation. Water stress had the greatest effect on number of pods/m² and least effect on 100-seed weight and number of seeds/pod. There was a positive and linear correlation between 100-seed weight and seed yield under water stress. Harvest index decreased with reduction in water supply and was closely correlated with seed yield under each water regime.

Study conducted by Saha (2011) revealed that water requirement of garden pea estimated by Hargreaves method was in close agreement (4.4 per cent deviation) with the actual water requirement (489.1 mm), followed by Blaney-Criddle method and FAO Pan method with deviations of −90.0 (−22.6 per cent) and −175.0 mm (−55.7 per cent), respectively. The other empirical methods predicted high water requirement than the actual value. Hence, Hargreaves method could be considered the most suitable among evaluated methods for predicting water requirement of pea in hill agro-ecosystem of Meghalaya, India.

The objective of irrigation management is to establish proper timing and amount of irrigation for achieving higher water use efficiency. A study conducted by Ramana Rao *et al.* (2017) revealed that the performance of pea was found to be better under micro sprinkler irrigation, considering the crop growth parameters, crop yield and water productivity in comparison with drip and conventional irrigated pea.

7.5.0 Growth Substances

Application of different forms of exogenous gibberellin and gibberellin containing substances on stem tip at 75 mg/l, twice at 2 days interval increased the growth and yield of the dwarf mutant A202 (Sidorova *et al.*, 1986). Mishriky *et al.* (1990) reported that application of gibberellic acid (GA$_3$) at 50 ppm twice at 30 days after sowing and again 15 days later was most effective in increasing early and total yields. Chlormequat (CCC) also at 50 ppm twice significantly increased total yields. GA$_3$ tended to increase the protein content of the green pods while CCC the dry matter content of green seeds.

Drought resistance of pea was improved and yields were increased by the use of chlormequat either as a foliar spray at 3-6 kg/ha or as a soil application at 6 kg/ha or as seed treatment with 1 per cent solution. For foliar spray the most suitable time was the beginning of flower bud development (Dolgopolova and Lakhanov, 1979).

The responses of garden pea genotypes with contrasting foliage types and stem heights to gibberellic acid (GA_3) and chlorocholine chloride [CCC (chlormequat)]. Irrespective of foliage type, GA_3 induced stem elongation was most pronounced in dwarf pea forms due to an increase in internode length, but not in their number. Treatment with CCC slightly inhibited the stem elongation of short-stemmed pea forms but did not affect tall pea forms. As the result of the conversion of leaflets to tendrils, the area shaded semi-leafless forms decreased several-folds. Seed soaking with 100 ppm gibberellic acid for 6 h, or were treated with 2.5 g captan/kg seeds improved seed emergence as well as green pod yield (Kohli *et al.*, 1992). Seed inoculation with *Rhizobium* along with application of phosphorus at 90 kg/ha and 50 ppm GA_3 has been reported to increase nodulation in pea (Prasad and Prasad, 1999). NAA (50 or 100 ppm), cycocel (chlormequat) (500 or 1000 ppm) or ethrel (ethephon) (200 or 400 ppm) applied to garden pea cv. GC 322 as a 24-h seed soak, a foliar spray at the three- and six-leaf stages or a foliar spray at full bloom increased growth and yield compared with foliar spray treatments. Greatest plant growth and highest green pod yield (158.20 q/ha) were obtained with 50 ppm Planofix applied as a 24-h seed soak (Bisen *et al.*, 1991). Early flowering by the application of 100 ppm GA_3 compared to NAA and control was recorded in variety Bonneville (Thompson *et al.*, 2015) and Pusa Pragati (Singh *et al.*, 2018).

Plant growth regulators such as Rastim 30 DKV [benzolinon; 3-(benzyloxycarbonyl-methyl)-benzothiazolin-2-on] at 17, 33 and 167 ml/ha and Antonik [sodium 5-nitroguaiacolate + sodium 1-nitrophenolate + sodium 4-nitrophenolate] at 2 l/ha were tested by Sarikova (1995) for their effect on crop yield and quality, applied before flowering and at the pod formation stage. Different effects were noted in different years and in soils of different types (3 types were studied). The greatest mean increase in yield (averaged over 3 years) was seen with the application of Rastim300 DKV at 167 ml/ha on old yield (averaged over 3 years) was seen with the application of Rastim 300 DKV at 167 ml/ha on old field soil and of Antonik on old field gley soil (11.23 per cent and 10.08 per cent increase, respectively). Skrobakova (1995) carried out studies with growth regulators *viz.*, Supresivit (a microbial preparation based on *Trichoderma harzianum*), Rastim 30 DKV (benzolinon), Bio-S (Biostim, a microbial preparation), Kadostim (a growth stimulator based on synthetic amino acids, potassium and trace elements) and the herbicide Sencor 70 WP (metribuzin) at 3 stages (before flowering, at the end of flowering and at peak growth). The yield increase from growth regulators was in the range 3.7-5.2 per cent, the greatest increase being achieved with Rastim 30 DKV (5.2 per cent, 190 kg/ha). Biostim increased seed yield by up to 3.7 per cent. Kadostim and Sencor did not have uniformly favourable effects. Supresivit applied to seeds before sowing improved emergence compared with the untreated control.

Effect of plant growth regulators (PGRs) on growth and seed yield in many legumes have been studied (Basuchaudhuri, 2016) but information on garden pea is scarce (Singh *et al.*, 2018). PGRs are believed to be involved in nodulation in garden pea (Chan and Gresshoff, 2009) and may also have potential to improve the growth and seed yield. Foliar application twice at 25 days after sowing (DAS) and another at onset of flowering (35 DAS) with freshly prepared solutions of GA_3 at 75 ppm increased growth, seed yield and seed quality parameters in cv. Pusa Pragati at New Delhi, India (Singh *et al.*, 2018).

The roles of gibberellin and auxin in phytochrome-mediated growth responses were investigated by Behringer (1991) using peas. He found that low frequency red light inhibited dark-grown seedlings by 90 per cent within 3 h. End-of-day far red light promoted growth rates of light-grown seedlings by 50 per cent within 4 h. Growth responses were photoreversible. IAA applications delayed red light inhibition of stem growth, also supporting the conclusions that auxin is involved in mediating the red light response.

A study on the potential protective effects of GA_3, IAA, Ca and citric acid against intracellular Cu accumulation and its induced damage on pea seedling growth, cell viability and nicotinamide coenzymes redox state as well as their related recycling enzyme activities namely, NAD(P)H-oxidases, glucose-6-phosphate dehydrogenase (G6PDH), 6-phosphogluconate dehydrogenase (6PGDH) and malate dehydrogenase (MDH) was conducted by Massoud *et al.* (2019). They concluded that IAA, GA_3, Ca or citric acid application is an effective approach for enhancing pea tolerance to copper toxicity by reducing metal accumulation in tissues through Cu entry restriction, and by maintaining nicotinamides redox homeostasis via modulations of recycling activities.

7.6.0 Weed Control

Weeds compete with crop plants, affect its growth and yield considerably. Several types of weeds are associated with peas but the predominant weeds are *Anagallis arvensis, Avena ludoviciana, Chenopodium album, Convolvulus arvensis, Cyperus rotundus, Fumaria parviflora, Galium aparine, Lepidium sativum, Medicago denticulate, Melilotus alba, Phalaris minor, Poa annua, Polygonum convolvulus, Rumex dentatus, Spergula arvensis, Stellaria media,* and *Trigonella polycerata.*

Weeds cause 37.3 to 64.4 per cent reduction in pea yield (Tewari *et al.*, 1997; Banga *et al.*, 1998; Harker, 2001). Peas are poor competitors, particularly at the seedling stage, avoiding early season weed interference is critical. The critical period for crop weed competition in pea is up to 60 days after sowing (Kumar *et al.*, 2009). Weeds can hamper pea production in many ways. First, weeds can reduce yield through competition for light, moisture, nutrients, and space. Second, weeds may harbour insect pests and pathogens that can affect crop production. Finally, late season weeds can be a nuisance that reduces harvest efficiency (Das, 2016).

Frank and Grigs (1957) reported that *Chenopodium album* competed well with pea and utilized more nitrogen and other nutrients than pea crop. Nelson and Nylund (1962) also reported competition of wild mustard *Brassica hilta* with pea crop. A two-year study showed that weeds reduced pea yield by 30 per cent, whereas the crop reduced the weed yield by 80 per cent. If pea crop was kept weed free for 30-40 days, later emerging weeds were suppressed and yield reduction did not occur (McCue and Minotli, 1979).

Cultivation of Pea

Furrow irrigation in peas

Field view of pea cultivation

Singh and Choudhury (1970) obtained excellent control of weeds in pea by pre-sowing spray of 1 kg/ha treflan. Marocchi (1977) found trifluralin 1 kg/ha very effective in pea. Pre-emergence application of 1.5 kg/ha prometryn was found effective by Hall (1962), whereas Lawson and Rubens (1970) found 1.25 kg/ha prometryn after two days of sowing most effective. Saimbhi*et al.* (1970) found that C-6313 at 2 kg/ha suppressed the weed population but reduced the yield in pea crop. Singh *et al.* (1973) also reported reduction in yield by using C-6313 but prometryn 1 kg/ha was the most effective herbicide in controlling weeds without affecting productivity. Singh and Sharma (1968) obtained best weed control and highest yield with application of dalapon at 4 kg/ha. Pandita *et al.* (1977) reported very good weed control by 2 l/ha nitrofen followed by one weeding which was similar in yield to weed-free plot. Randhawa and Saimbhi (1976) obtained a similar result. In Poland, Radziszewski and Rola (1991) reported that application of pivot (imazethapyr) at 1.1 or 1.2 l/ha or command (clomazone) at 0.4 l/ha as pre-emergence or of pivot at 0.8 or 1.0 l/ha at emergence resulted in 96-98 per cent crop cover and reduced weed cover to 18 per cent or less.

Nitrofen and diuron increased crude protein in pea at 1.25 and 0.4 kg/ha, respectively, compared with weed-free treatment. Nitrofen also increased reducing sugars and starch content of seeds (Paul *et al.*, 1976).

Effective weed control and higher yields of peas with the application of pendimethalin at various doses have been observed by various researchers (Sharma and Vats, 1988; Chauhan *et al.*, 1992; Sekhon *et al.*, 1993). Sharma and Vats (1986) and Kolar and Sandhu (1989) have found methabenzthiazuron to be an effective herbicide for controlling weeds in peas and to get higher yields.

Gogoi *et al.* (1991) evaluated various herbicides for weed control in peas on sandy loam soil in India and found that the application of oxadiazon at 0.5-1.0 kg/ha and pendimethalin at 1.5 kg/ha resulted in higher weed control efficiency (62.7-65.0 and 64.5 per cent, respectively) and grain yield (1170 -1177 and 1196 kg/ha).

Isoproturon at 0.98 kg/ha, methabenzthiazuron at 1.35 kg/ha and their combinations at 0.6 kg/ha with alachlor 1.25 kg/ha gave good weed control and green pod yield of peas (Saimbi *et al.*, 1990).

Kudsk (1992) got marginal to substantial control of nine weed species with split application of bentazone + pendimethalin at 0.24 + 0.25 kg/ha, respectively applied twice when peas were 1-2 cm tall and 8-14 days later than single application.

Few post-emergence (POST) herbicides are available for broad-spectrum weed control in pea. Pea exhibited excellent tolerance to imazamox up to the highest rate applied, 40 g/ha. Pea yield comparable to that of the hand-weeded control was attained with 20 to 30 g/ha of imazamox (Blackshaw, 1998). Imazamox offers growers a POST option for broad-spectrum weed control in pea. Das (2016) recommended that application of Imazethapyr @ 25ml/ha as post-emergence (15-20 DAS) may be a suitable, effective and eco-friendly chemical weed management practice for pea.

Experimental results revealed that nodulation in pea was not affected significantly due to the application of chemical herbicides which could be due to their rapid inactivation in soil or its rapid translocation along with photosynthate, to distant metabolic sink (Fernandez *et al.*, 1992; Gonzalez *et al.*, 1996; Raman and Krishnamoorthy, 2005; Ali *et al.*, 2014; Das, 2016).

Allelopathy is an alternative safe strategy for chemical herbicides in controlling weeds. The study conducted by El-Rokiek *et al.* (2019) suggested that the pea seed powder has allelopathic and phytotoxic effects that controlled investigated weeds in wheat. Analysis of the allelopathic pea seed powder at the applied rates revealed the presence of phenolic compounds and flavonoids. Both allelopathic compounds showed high levels with increasing the pea seed powder rate. The mixing of pea seed powder with the soil surface at 80 g/pot 1 week before sowing was the most efficient treatment in controlling weeds under investigation that consequently achieved the highest wheat yield.

8.0 Harvesting, Yield and Storage

8.1.0 Harvesting

The high quality in pea is associated with tenderness and high sugar content. During maturity sugar contents decrease rapidly and there is an increase in starch and other polysaccharides and insoluble nitrogenous compounds such as protein. Calcium migrates to seed coat and the toughness of skin increases during ripening. The toughness of seed coat and firmness of pulp are the measure of maturity and is determined by mechanical means such as tenderometer. The price of pea varies with tenderometer reading and highest price is offeree for pea with low tenderometer reading (Thompson and Kelly, 1957). Technical maturity can be determined (a) from the alcohol insoluble content, and (b) by a rapid method based on firmness measurements (Hegedus and Trnka, 1989).

Temperature has great effect on the rate of ripening of pea and at higher temperature there should be no delay in harvesting. Hence many workers use a system of accumulated degree hours or heat unit to determine maturity and harvesting of pea. The number of degree hours above $4.4°C$ required to bring a certain cultivar to maturity is calculated each season. The average of few seasons helps to ascertain proper maturity stage of a cultivar on the basis of day-to-day temperature record. However, Reath and Wittwer (1952) suggested taking photoperiod also into account in addition to degree day hours for predicting maturity stage accurately. According to Pal (1976) the requirement varied from 1534 heat units for earliest cultivar to 3942 heat units for late cultivar.

For fresh market the pods are hand picked. Several pickings are needed, as all pods do not mature at the same time. As the pods attain marketing stage they turn light green and become well filled up. Usually 3-4 pickings are made at 7-10 days interval (Nath, 1976). Picking by hand is expensive and time consuming and increases cost of cultivation substantially. In Western countries, where labour is expensive, periodical picking is not possible, hence pea viner pulls up the whole plants and then the marke-table pods stripped. In hand picking well-graded pods are harvested, whereas in latter the harvested pods are separated into different maturity grades (Singh and Joshi, 1970). Peas for processing are separated into four grades depending on seed size and smallest size fetches highest price being very tender.

Harvesting of Peas

8.2.0 Yield

The early maturing cultivars produce 25-40 quintals of edible pods, medium maturing cultivars 65-80 quintals and late maturing cultivars 85-115 quintals/ha. The shelling percentage ranges between 35 and 45 per cent (Nath, 1976).

8.3.0 Storage

To retain post harvest quality, peas are harvested before physiological maturity is reached. Green pea has a tough pod which is separated from peas and discarded, prior to eating. Shortly after harvest, loss of sweetness and crispness, as well as degreening and the development of mealiness, may degrade the quality of snow pea pods (Pariasca *et al.*, 2001). Pea is a non-climacteric, highly respiring (>60 mL CO_2/kg/h at 5 °C) and perishable commodity that can be stored at temperatures near 0 °C and 85 to 90 per cent relative humidity to extend its shelf life for maximum 2 weeks (MacGillivray, 1961; Scetar *et al.*, 2010; Kader, 1992; Suslow and Cantwell, 1998). Most often green peas are frozen or processed in developed countries like U.S. (Basterrechea and Hicks, 1991) whereas; peas are harvested just before physiological maturity is attained and used mainly for culinary purpose, due to inadequate freezing and processing facilities in developing countries. Freezing converts liquid water into ice, which greatly reduces microbial and enzymatic activities, however oxidation and respiration are weakened effectively by low temperature (Haiying *et al.*, 2007). Freshly harvested vegetables continue their respiration and metabolic processes which are associated with deterioration of products during storage (Sharma and Gupta, 2006). The rate of respiration primarily decides the storage quality and shelf life under different storage environment.

Good quality peas should be uniformly bright green, fully turgid and free from decay and damage caused by black calyxes, freezing, insects, mechanical, mildew or other diseases for better market price. Freezing is an energy intensive preservation method below 0 °C and requires costly infrastructure with uninterrupted power supply. The main disadvantage of freezing is that it requires uninterrupted power supply; to maintain the frozen condition without temperature fluctuation for maintaining better quality during storage. Peas are graded primarily based on appearance and freezing temperature abuse adversely affects the appearance of peas, when peas are transferred from frozen condition to atmospheric condition due to thawing and wilting. Freezing itself will decrease food quality (Haiying *et al.*, 2007). Quick freezing is thought to better keep the shape, nutrition and taste of food (Lester, 1995). Even quick freezing of pre treated mushroom, green cauliflower, navy bean and pea pod would produce smaller ice crystals, which cause some irreversible damage to microstructures. The impacts of freezing on food quality are directly related with the growth of ice crystals, which can break cellular walls (Anzaldua-morales *et al.*, 1999). Modified atmosphere packaging (MAP) is of the new applied food packaging techniques offering a prolonged shelf life of respiring products (Day, 1996). Despite many advantages of MAP, a very little information is available on the MAP of shelled green pea. The shelf life of shelled peas packed in 25 µm thickness low and high density polyethylene (LDPE and HDPE) was 45, 17, 7 and 4 days when stored at the temperatures of "11, 5, 15 °C and room temperature, respectively (Sandhya and Singh, 2004). The study conducted by Anurag *et al.* (2016) revealed that shelled green peas can be stored in MAP with 3 perforations (0.4 mm dia) in the temperature range of 4 to 10 °C and 90–94 per cent RH to extend shelf life with marketable quality for 24 days.

Peas kept inside cold room conditions (0°C and 93 per cent RH, Ecofrost) with zip-lock bag packaging with 2 per cent ventilation helped in extending the storage-life up to 14 days under cold storage (Jadhav, 2018).

In Western countries, peas shipped to long distance are pre-cooled with ice cold water and transported in refrigerated trucks in order to retain their sweetness and quality.

9.0 Diseases and Pests

9.1.0 Diseases

The important diseases of pea and their control measures are described below.

Fungal Diseases

9.1.1 Powdery Mildew

The disease is caused by *Erysiphe polygoni*. It attacks leaves first producing faint, slightly discoloured specks from which greyish white powdery growth of mycelium and spores spread over leaf, stem and pod. The leaves turn yellow and die. The fruits do not either set or remain very small. It causes defoliation. The disease develops late in season and reaches to maximum intensity at the time of pod formation. The early cultivars escape the disease (Nath, 1976).

The disease is of economic importance due to yield reduction and reduced seed quality under severe conditions as 33 to 69 per cent of pea foliage has been reported to be infected with powdery mildew (Suneetha *et al.*, 2014). Yield losses up to 47 per cent have been reported due to powdery mildew (Munjal *et al.*, 1963; Nisar *et al.*, 2006).

Singh and Singh (1978) and Shukla *et al.* (1976) reported effective control of disease by formulations of wettable sulphur such as sulfex and thiovit at 3 kg/ha which confirmed the earlier finding of Kotasthane (1975). Rathi (1977) reported that elosal 8 WP three times at 10-day interval also controlled the disease effectively, whereas karathane (dinocap), dikar (manocap), morocide (binapacryl) and morestan (quinomethionate) were equally effective. Raut and Wangikar (1979) and Khatua *et al.* (1977) found 0.03 per cent calixin most effective followed by karathane (0.2 per cent) and bavistin (100 ppm). Singh and Naik (1977) also found similar results by two fortnightly sprays, which increased the yield by about 9 per cent. Systemic fungicides reduced the disease intensity of powdery mildew with propiconazole at 0.2 per cent giving the best result (Hooda and Parashar, 1985). Extensive research throughout the agrochemical industry expanded options for powdery mildew control in the 1980s through introduction of several triazoles (sterol demethylation inhibitors) and two additional members of the morpholine group, fenpropimorph

Diseases of Pea

Powdery mildew infection in pea leaves and pods

Rust infection in pea leaves   **Root rot infection in pea**

Pests of Pea

Pea aphid

Pea pod borer

Pea weevil

Pea leaf miner

and fenpropidin. These have proven very effective in controlling pea powdery mildew (Ransom *et al.*, 1991; Warkentin *et al.*, 1996). More control options are recently available with the broad-spectrum fungicides strobirulins and anilinopyrimidines and the powdery mildew specifics spiroxamine and quinoxyfen (Hollomon and Wheeler, 2002).

There is need to develop cultivars resistant to powdery mildew because the disease is of common occurrence. Only two recessive (*er1* and *er2*) and one dominant (*Er3*) genes for powdery mildew resistance have been described so far in *Pisum* germplasm (Harland, 1948; Heringa *et al.*, 1969; Fondevilla *et al.*, 2007). Gene *er1* provides from complete to moderate levels of resistance (Cousing, 1965; Heringa *et al.*, 1969; Tiwari *et al.*, 1997a, b; Fondevilla *et al.*, 2006) and is in widely used in pea breeding programmes. A recent study indicates that resistance provided by er1 is due to a loss of function of PsMLO1, a MLO (Mildew Resistance Locus O) gene (Humphry *et al.*, 2011). Gene *er2* (Heringa *et al.*, 1969) confers complete resistance that was effective in some locations but ineffective in others (Tiwari *et al.*, 1997a, b; Fondevilla *et al.*, 2006). Gene *Er3* was recently identified in *Pisum fulvum* and has been successfully introduced into adapted *Pisum sativum* material by sexual crossing (Fondevilla *et al.*, 2007a, 2010).

Tyagi *et al.* (1978) reported P 185 and 6578 immune to *E. polygoni* on the basis of natural and artificial inoculation test. Joi *et al.* (1975) reported Giant Sugar as resistant and several other lines (American Wonder, Sel.1, Mores-55) moderately resistant, whereas Naik *et al.* (1975) reported EC 33866 as moderately resistant with 25 per cent leaves infected. Sokhi *et al.* (1979) showed that resistance in P 185, 6585, 6587 and T 10 were governed by a recessive gene, whereas in P 388 and 6588 by two recessive genes. One of the genes in P 388 and 6588 appeared to be same as the gene in above cultivars. Singh *et al.* (2015) screened 101 pea accessions against this disease and found Alaska, ACTomour, Arka Ajit, Angoori, CHP-I C-96, C-778, DAP-2, HUVP-3, JP-15, JP-20, JP-141, JP-625, Punjab-89, PMR-4, PMR-62, PMVAR-1, VRP-22, VRPMR-9, VRPMR-11, and KTP-8 as highly resistant under Ludhiana condition, Punjab, India. Resistance was found in four lines APL-55, APL-69, APL-80 and Line 1-2SPS5 carrying *er2* gene tested at Mid Hill conditions of Himachal Pradesh, India (Rahman *et al.*, 2018). A total of 310 germplasm lines of pea were screened under artificial epiphytotic conditions in greenhouse for resistance against powdery mildew using the criteria of disease severity, apparent infection rate and area under disease progress curve (AUDPC) by Chaudhary and Banyal (2017). Out of which 31 lines *viz.*, HFPU, P-1797, P-1783, P-1052, HFP-7, HFP-8, P1808, P-1820, P-1813, P-1377, P-1422-1, P-1811, IPF-99-25, KMNR-400, LFP-566, LFP-569, LFP-552, LFP-573, JP-501-A/2, PMR21, KMNR-894, P-1280-4, P-1436-9, P-200-11, IPFD-99-13, HVDP-15, DPP-43-2, LFP-517, LFP-570, JP Ajjila and JP-15 showed highly resistant reaction against powdery mildew at Palampur, Himachal Pradesh, India. Screening of 15 pea entries/varieties against powdery mildew was conducted by Nag *et al.* (2018) at Raipur, India and they found Arka Ajeet, 2012/PMPM-5, 2011/PMPM-4, 2011/PMPM-3, 2012/PMPM-3, IP-3, 2011/PMPM-2, 2011/PMPM-5, and 2011/PMPM-1 showed minimum powdery mildew disease severity (< 10.00 per cent) and categorised as resistant types.

9.1.2 *Fusarium* Wilt and Near Wilt

The two diseases are similar in several respects and are caused by two distinct races of *Fusarium oxysporum* f. *pisi*. The near wilt has higher optimum temperature for development in the soil than wilt and is more damaging in warm season, on late plan-ting and late cultivars (Thompson and Kelly, 1957). The disease spreads through the infected soil and in severe cases whole plant wilts and stem shrivels. The affected seeds may also carry the disease to next season. The near wilt develops more slowly than wilt. An early infection may cause death of plant, whereas few poorly filled pods appear in late infection. The infection takes place from soil through roots and the organism follows water conducting tissue obstructing the movement of water. The vascular elements are brick red in colour. Most of the wilt resistant cultivars are susceptible to near wilt. The disease can considerably be reduced by following long crop rotations (Nath, 1976). Seed treatment with benlate and dexan controlled the disease as reported by Menzies and Wright (1976). Gangopadhyay and Kapoor (1976) found effective control with wet seed treatment for 30 minutes in 0.1 per cent difolatan + 0.2 per cent captan at 30°C followed by keeping for 3.5 hours at room temperature before sowing in inoculated soil. Sen and Kapoor (1975) found best results by soil drench with cerobin or bavistin, which gave effective control of disease up to 60 days after treatment. *In vitro* testing of fungicides revealed that Score, Topsin-M and Raxil ultra produced best inhibition of the pathogen (Rehman *et al.*, 2014; Khan *et al.*, 2016).

Six races, known to cause disease, have been described in the world. Race 1 was discovered in Wisconsin in 1924 (Linford, 1928) and was later found in Washington, Idaho and New York (Wade *et al.*, 1929). Soon after resistance to race 1 was incorporated into cultivars, a second race which overcame resistance to race 1 was discovered (Snyder and Walker, 1935). This second race, designated as race 2, was called near wilt since symptoms became noticeable later in the season than for race 1. Races 1 and 2 were the only economically important wilt races in the United States until race 5 appeared in north western Washington in 1963 (Haglund and Kraft, 1970). Races 1 and 2 are known to occur throughout the world, while races 5 and 6 are only important in western Washington State (Haglund and Kraft, 1979). Race 3 is present in Europe and England, while race 4 is found in Canada (Hagedorn, 1984). Genetic resistance to races 1, 2, 5, and 6 is conferred by different single dominant genes (Hagedorn, 1984) and is available in numerous germplasm releases (Haglund and Anderson, 1987; Kraft and Giles, 1978; Kraft and Tuck, 1986). Resistance to race 2 was first discovered in an adapted breeding line in 1945 by researchers in Wisconsin (Hare *et al.*, 1949). A selection from this line was later released as Delwiche Commando (Canner, 1945).

The development of resistant cultivars is the only effective means to control the disease. The cultivar Alaska has been reported resistant to wilt, whereas Delwiche Commando and New Era resistant to both (Choudhury, 1967). Singh and Joshi (1970) have reported Bonneville and Early Perfection to be resistant to near wilt disease whereas Ramphal and Choudhury (1978) reported Sel.17 and Sel.45 highly resistant and Alaska resistant to *Fusarium* wilt. Haglad and Kraft (1979) reported several races of *Fusarium oxysporum* f. *pisi*. The variety 'Kala Nagni' was immune to the New Delhi strain and 'Alaska' was resistant to this disease (Kalloo, 1998). To identify genetic sources of resistance against *Fusarium* wilt in pea, a study was conducted by Shubha *et al.* (2016) at IARI, New Delhi, India. Three genotypes (GP-6, GP-55 and GP-942) were found to be highly resistant and four genotypes (GP-17, GP-48, GP-473, GP-941) were resistant against this disease and identified as new donor source. They also found that all the popular cultivated varieties (Arkel, Pusa Pragati, AP-3, VRP-6, VL-7, VL-10 and Arka Ajit) were highly susceptible to wilt. Twenty pea varieties/lines were evaluated against *Fusarium* wilt caused by *Fusarium oxysporum* f. sp. *pisi* by sowing them in sick plot at Faisalabad, Pakistan by Husnain *et al.* (2019). They found only a single variety Garrow performed as moderately resistant by showing 21 per cent plant mortality in the field.

9.1.3 Root Rots

Several organisms like *Rhizoctonia, Fusarium, Pythium* and other soil-borne organisms cause the seed decay, damping off and root rot. These result in poor crop stand and lower yield.

Fusarium root rot can be responsible for 30–57 per cent of yield losses in pea based on commercial fields and artificially inoculated field plots in Canada (Basu *et al.*, 1976) and artificially inoculated plants in large field plots in Eastern Washington (Kraft and Berry, 1972). It is caused primarily by the pathogen *Fusarium solani* (Mart.) Sacc. f. sp. *pisi* (F. R. Jones) W. C. Snyder and H. N. Hans (*Fsp*) worldwide (Basu *et al.*, 1976; Persson *et al.*, 1997; Kraft and Pfleger, 2006; Etebu and Osborn, 2010; Hamid *et al.*, 2012). However, other *Fusarium* species can cause root rot (Clarkson, 1978; Persson *et al.*, 1997) and can be predominant in certain pea growing areas. For instance, *F. avenaceum* has been isolated more frequently than any other *Fusarium* species in field surveys of field pea conducted in North Dakota (Chittem *et al.*, 2010) and Canada (Feng *et al.*, 2010). *Fsp* infection initiates at or near the cotyledon-hypocotyl junction. As infections progress, regions immediately under the soil and upper region of the taproot also become infected (Kraft and Pfleger, 2006). Jung *et al.* (1999) observed that pea plants infected by artificial inoculation of *Fsp* initially show round or irregular light brown lesions that progress to dark black lesions on below ground stems. Although the optimum temperature for growth of *Fsp* is approximately 25–30 °C, infection can take place at temperatures as low as 18°C (Kraft and Pfleger, 2006).

Resistance to *Fusarium* sp. has been reported as a quantitative trait regulated by more than one gene (Grunwald *et al.*, 2003; Mukankusi *et al.*, 2011) and is the most economic means of managing *Fsp*. Quantitative trait loci have been identified previously for this pathogen (Weeden and Porter, 2008; Feng *et al.*, 2013), and pea genotypes with pigmented flowers and seed coats tend to have higher levels of partial resistance to *Fusarium* sp. than those with white flowers and green seed coats (Grunwald *et al.*, 2003).

The seed treatment with fungicides like ceresan, spergon, arasan and soil drenching with difolatan, captan and brassicol reduces the disease (Shukla *et al.*, 1977). Locke *et al.* (1979) obtained excellent control of seed rot caused by *Phythium ultimum* by seed treatment of 0.6 g/kg seed with metalaxyl. Robertson (1976) controlled damping off by carboxin, penaminosulf and terrazole treatment.

In Taiwan, pea wilt affected by root rot (*Fusarium solani* f. sp. *pisi*) was controlled by seed treatment with thiram + graphite (2 g/kg seed) or by soil amendment with calcium cyanamide at 300 kg/ha (Lin, 1991).

Lacicowa and Pieta (1994) and D'ella *et al.* (1998) reported the effectiveness of various isolates of *Trichoderma* in controlling the root rot of pea caused by *F. solani* f. sp. *pisi*. Similarly, Kapoor *et al.* (2006) also reported that *T. harzianum* had strong antagonistic activity against *Fusarium solani* f.sp *pisi*. Sufficient antagonism was expressed by bio-agents especially *Trichoderma* sp., *Gliocladium* sp. and *P. fluorescence* when applied against root rot pathogen of pea (Hamid *et al.*, 2012).

Seeds with seed-coat coloured by anthocyanins were found resistant to *Pythium ultimum*, whereas those with uncoloured coats were susceptible (Stasz *et al.*, 1980). Gangopadhyay and Kapoor (1975) reported Kelvedon Wonder, EC 8633 and EC 3833 resistant to *Sclerotinia* rot at Kullu valley. Dwarf Gray Sugar cultivar was reported resistant to pre-emergence and post-emergence damping off caused by *Rhizoctonia solani* (Amin *et al.*, 1981), whereas Okh *et al.* (1978) reported six lines resistant to damping off and root rot caused by *Pythium ultimum*.

Based on root disease severity (RDS) values, the most *Fsp*-resistant Austrian winter, green fresh, green dry, yellow dry, green winter and yellow winter pea were identified as PI 125673, 5003, 'Banner', 'Carneval', PS 05300234, and 'Whistler', respectively at USA (Bodah *et al.*, 2016).

Bacterial Disease

9.1.4 Bacterial Blight

The disease is caused by bacteria *Pseudomonas syringae* pv. *pisi*. The affected plants develop watery, olive-green blisters on stems and leaf bases and water-soaked oily spots upon pods and leaves. The disease is favoured by cold and humid weather. It is most severe after frost injury.

It is a seed-borne disease first recorded in the USA in 1915 (Sackett, 1916).The disease is also reported as soil-borne (MacGillivray, 1961). However, *P. syringae* pv. *syringae* has been reported to occur in pea crops in Australia (Wimalajeewa and Nancarrow, 1984; Clarke, 1990) and overseas (Taylor and Dye, 1972; Jindal and Bhardwaj, 1989; Lawyer and Chun, 2001), but has been regarded as less important than *P. syringae* pv. *pisi*, which is reported to cause disease over a wider range of environmental conditions (Taylor and Dye, 1972; Lawyer and Chun, 2001). Worldwide, seven races of *P. syringae* pv. *pisi* are currently recognized (Taylor *et al.*, 1989). The interaction of races and cultivars is controlled by a gene-for-gene relationship with avirulence genes in the pathogen matched by resistance genes in the host (Bevan *et al.*, 1995). Race-specific resistance is widespread in commercial pea cultivars, however, there are no cultivars known to be resistant to race 6. The most common resistance gene R3 was present in a worldwide *Pisum* germplasm collection (Taylor *et al.*, 1989; Elvira-Recuenco and Taylor, 2001).

There is little reported information regarding yield loss caused by *P. syringae* pv. *syringae* in pea. Jindal and Bhardwaj (1989) reported a severe outbreak in northern India attributed to *P. syringae* pv. *syringae*, but the extent of yield loss was not quantified. The average yield loss in the resistant cultivars in the presence of infected field pea stubble was 23 per cent, whereas in the susceptible cultivars the yield loss was 75 per cent (Richardson and Hollaway, 2011).

According to Taylor and Dye (1976) slurry treatment of pea seed with streptomycin sulphate (2.5 g/kg seed) reduced primary infection of bacterial blight by 90 per cent. Application of dust gave little control. Soaking seed in streptomycin solution for 2 hours gave better control than slurry treatment but was toxic to several cultivars.

Viral Diseases

Several viral diseases of pea have been reported and are becoming serious in some parts of the world. Pea seed-borne mosaic virus (PSbMV, genus Potyvirus, family Potyviridae) was the most widespread, but infections by Bean leaf roll virus (BLRV, Luteovirus, Luteoviridae), Beet western yellows virus (BWYV, Polerovirus, Luteoviridae), Bean yellow mosaic virus (BYMV, Potyvirus, Potyviridae), Cucumber mosaic virus (CMV, Cucumovirus, Bromoviridae) and Alfalfa mosaic virus (AMV, Alfamovirus, Bromoviridae) were reported as well (Latham and Jones, 2001; Freeman *et al.*, 2013). Natural infection of garden and canning peas by Soybean dwarf virus (SbDV, Luteovirus, Luteoviridae, syn. Subterranean clover red leaf virus, SCRLV) has been reported in Tasmania (Johnstone, 1978) and by Subterranean clover stunt virus (SCSV, Nanovirus, Nanoviridae) on the central tablelands of New South Wales (Grylls, 1972). Zimmer and Alikhan (1976) reported the symptoms of pea seed-borne mosaic virus disease from Canada. Hampton *et al.* (1976) reported the disease from the USA. It is reported to be spread by aphids and *Chenopodium amaranticolor, Vicia faba* and Perfection type peas are reported to be its indicator plants. Considerable loss in processing pea was reported if the disease infected 2 weeks after emergence (Kraft and Hampton, 1980). Pea aphid transmits pea enation mosaic virus and Perfected Freezer 60 has been reported to be its source of resistance (Hagedorn and Hampton, 1975). Provvidenti and Granett (1976) reported Plantago mottle virus from New York which developed leaf veinal chlorosis, mottle and necrosis at 15 to 25°C but at higher temperature (35°C) infection was reduced and the symptoms suppressed.

Among the different potential sources for Bean leaf roll virus (BLRV) resistance that Leur *et al.* (2013) evaluated, G-1000 deserves a special mention. This vegetable type pea, bred at the Cornell University Agricultural Station (Geneva, New York), was developed to combine resistance to a range of potyviruses (Provvidenti *et al.*, 1991). Its immunity to both Bean yellow mosaic virus (BYMV) and Pea seed-borne mosaic virus (PSbMV) was confirmed in field and greenhouse trials, but it also consistently showed a high level of BLRV resistance, both in Australia and in Syria (Leur *et al.*, 2013).

9.2.0 Pests

The important insect pests of peas are:

9.2.1 Pea Aphid (*Acyrthosiphon pisum*)

Pea aphid is one of the serious pests of pea crop. The pest feeds on a variety of agriculturally relevant crop plants in the Fabaceae, including pea (*Pisum sativum*), soybean (*Glycine max*), snap bean (*Phaseolus vulgaris*), and fava/broad bean (*Vicia faba*), (Swenson, 1954,1957; Nault *et al.*, 2004; Hampton *et al.*, 2015), and has been shown to vector bean yellow mosaic virus between these hosts (Hampton *et al.*, 2015; Swenson, 1954,1957). Several of these plants are also primary hosts for the highly aggressive bean pathogen *Pseudomonas syringae* pv. *syringae* B728a (PsyB728a), which establishes epiphytically and can invade the plant and cause disease (Beattie *et al.*, 1995, 1999; Hirano and Upper, 2000). Stavrinides *et al.* (2009) illustrated that aphids can also vector bacterial pathogens and that even seemingly host-restricted pathogens can have alternative host specificities and lifestyles. It attacks young vines sucking the juice from growing tip, later covering the whole plant (MacGillivray, 1961). It causes curling of the leaves and pods. The affected plants become stunted and the pods curl, have rough spots and fail to fill. The insect is a large type of plant lice, pale green in colour.

According to Barlow *et al.* (1977) pea plant infested with 50 aphids reduced the plant weight by 63.9 per cent in 11 days. The ETL of pea aphid is considered as 3-4 aphids/stem tip. It reduced the relative growth rate and efficiency of production of new tissue by 118 per cent (Barlow and Messmer, 1982). Single application of 1.5 kg/ha of mephosfolan + carbfuran or two applications of dimethoate were effective in checking pea aphid population (Mathur *et al.*, 1974). It can also be controlled by spray of 0.06 per cent nicotine sulphate or 0.25 per cent parathion (Nath, 1976). The cultivars Feltham First and Meteor have been reported to be resistant to pea aphid (Bintcliffe and Wratten, 1980).

Due to the known harmful effects of such conventional pesticides, there is a growing use of pesticide alternatives to reduce risks. Alternatives are currently being investigated and include the use of biorational compounds that are compatible with integrated pest management (Horowitz and Ishaaya, 2004). According to the US-Environmental Protection Agency biorational pesticides pose minimal risk to the environment, degrade quickly, leave minimal residue, are safe to handle, and relatively small quantities are required for effective control. Pesticides classified as biorational include various classes of insect growth regulators (IGRs), microbial products, synthetic molecules with novel modes of action and plant-derived compounds. Flonicamid and pymetrozine are two novel insecticides with selective activity against Homoptera, acting as feeding inhibitors with high mortality due to starvation (Harrewijn and Kayser, 1997; Denholm *et al.*, 1998; Morita *et al.*, 2007). Imidacloprid is the most important neonicotinoid insecticide with good systemic activity that acts as an agonist of the insect nicotinyl acetylcholine receptors, causing the insect to reduce or stop feeding and mobility. It is particularly effective against aphids, whiteflies and planthoppers (Boiteau and Osborn, 1997; Elbert *et al.*, 1998; Nauen *et al.*, 1998).

Flonicamid, a novel systemic insecticide with selective activity as feeding blocker against sucking insects, showed high toxicity over another feeding blocker, pymetrozine, and the neonicotinoid, imidacloprid against first-instar of *A. pisum* nymphs with an LC50 of 20.4 µg/ml after 24 h, and of 0.24 µg/ml after 72 h (Sadeghi *et al.*, 2009).

9.2.2 Pod Borer (*Etiella zinckenella* Tr.)

It is a serious pest of pea. The crop is attacked by many insect-pests among which pea pod borer and stem fly are serious pests in Uttar Pradesh (David and Ramamurthy, 2011). Bijjur and Verma (1997) reported 57 species of insects attacking pea crop with an annual monetary loss of 540 million Indian Rupees. Pea pod borer is a major pest of pea causing as high as 50.9 per cent pod infestation with 77.64 per cent seed damage resulting in 23.9 per cent loss in seed yield (Bachatly and Malak, 2001). Yadav and Chauhan (2000) observed that *Etiella zinckenella* caused 3.5 per cent to 30.8 per cent pod damage in pea crop in Uttar Pradesh alone. It is distributed throughout India with particular reference to Uttar Pradesh, Bihar, Madhya Pradesh and Punjab (Mathur and Upadhyay, 2006).

The young caterpillars first feed on the surface of the pods, bore into them and feed on the seeds. It can effectively be controlled if the egg masses, caterpillars and pupae are hand-picked in the beginning of the infestation.

Among the various insecticides evaluated against pod borer complex in pea, Neem soap 5 per cent + Rynaxypyr 20 SC @ 25 g a.i./ha treated plot showed 100 per cent reduction in larval population and gave the highest crop yield (Krishna *et al.*, 2019). Dhaka *et al.* (2011) found that the Indoxacarb had the comparable lower number of larvae as well as pod borer and seed infestation.

9.2.3 Pea Weevil (*Sitona lineatus* L.)

It is more severe in hilly areas. *S. lineatus* is native to Europe and North Africa where it is a well-known pest of field peas and faba beans. The elongate, yellow eggs are laid on small green pods in spring. In 15-18 days the eggs hatch and the larvae burrow through the pod into the seed (MacGillivray, 1961). They develop inside peas in 30-50 days, construct exit tunnel, come out and pupate. Havilckova (1980) reported a negative correlation between leaf thickness and feeding rate by weevils. A strong inhibition of feeding was induced by amino acid tyrosine. Among sugars, saccharose was most effective in stimulating feeding.

Soil nitrogen content has a pivotal role in determining the pest status of *S. lineatus*. Adding enough nitrogen to the soil at planting to meet plant requirements results in reduced nodulation (Vankosky *et al.*, 2011), thereby decreasing larval food resources, which in turns reduces weevil recruitment (Cárcamo *et al.*, 2015).

Methiocarb was effective at 0.05 per cent in controlling weevil population (Crowell, 1975). The insecticides used against pea aphid were found effective against pea weevil. Fumigation of dry peas with methyl bromide was necessary to kill at weevils in pea.

Foliar insecticide active ingredients that have been evaluated include phorate (King, 1981; Bardner *et al.*, 1983), cyhalothrin-lambda (also known as lambda-cyhalothrin; Steene *et al.*, 1999), permethrin (McEwan *et al.*, 1981; Bardner *et al.*, 1983; Griffiths *et al.*, 1986), and imidacloprid (Steene *et al.*, 1999). Cyhalothrin-lambda is registered in several jurisdictions in North America and also in Europe (Seidenglanz *et al.*, 2010). Cyhalothrin-lambda treatment reduced adult weevils by 56 per cent (Steene *et al.*, 1999). Application of permethrin (pyrethroid insecticide) decreased larval populations by approximately 50 per cent (Bardner *et al.*, 1983), likely due to mortality of adult females, as contact foliar insecticides have no direct impacts on eggs or larvae (Steene *et al.*, 1999). Seed coating with systemic insecticides for *S. lineatus* management in field peas has been studied in detail. There is consensus that systemic insecticides are more effective than foliar applications (Vankosky *et al.*, 2009, Seidenglaz *et al.*, 2010).

9.2.4 Leaf Miner [*Chromatomyia horticola* (Goweau)]

It is the most serious and regular pest (Shantibala and Singh, 2008) and more common in the Mediterranean area and occurs widely throughout Asia (Gencer, 2004). The pupae usually occur in the ground. The pest breeds in sugar beet fields and peas grown next to sugar beet may become heavily infested (MacGillivray, 1961) Bhalla *et al.* (1974) reported that fenitrothion and fenthion were most effective and persistent insecticides against the pest. Saito (2004) found than emamectin benzoate, fipronil, spinosad, chlorfenapyr, cyromazine and asoxation were very effective in controlling leaf miner. Desai *et al.* (2013) reported that fenvalerate 20 per cent EC was the most effective while neem oil 1 per cent and NSKE 5 per cent were relatively more effective.Yadav *et al.* (2016) observed profenofos 50 per cent EC @ 2ml/l was the most effective, followed by chlorantraniliprole 18.5 per cent SC @ 0.3 ml/l against the pest.

9.2.5 Nematode

The nematodes (*Meloidogyne* spp.) are also very serious affecting pea yield in some infested areas. Aldicarb 5 kg/ha gave good nematode control and improved crop productivity (Pailla and Lopez, 1979).

The pea crop is greatly damaged by birds, especially during germination stage. Seed treatment with 0.35 to 0.45 per cent by weight of Methiocarb as slurry before sowing was found very good bird repellent and reduced the bird damage considerably (Porter, 1977; Burgmans, 1979).

9.3 Physiological Disorders

9.3.1 Hail Damage

A symptom occurs where hail has hit the foliage white streaks and other markings appear on leaves and stems. Leaves are torn and pods broken open. Exposed seeds discolour rapidly.

9.3.2 Frost Injury

Plants vary in reaction and leaves are bleached and mottled. Flower buds are aborted. Growing points may become blind. Developed pods have white mottling on surface.

Disorders of Pea

Frost injury

Hail damage

Water congestion

9.3.3 Water Congestion

Leaf tips of newest leaves are pinched and necrotic following heavy rain during the summer when plants are growing rapidly. Brown discolouration occurs at leaf tip. Apical cells of developing leaflets are ruptured and tips of newly formed leaves die back. Peas on fertile soils are most likely to show damage. Damage is insignificant and usually affects only a single pair of leaves.

10.0 Seed Production

Pea is a self-pollinated crop. Hence two cultivars can be grown in adjacent plots without any danger of out-crossing. Care should be taken to avoid mechanical mixture. However, an isolation distance of 20 m is recommended for production of foundation seed. Pea crop produces off-type plants which multiply rapidly if not rogued out every year. Hence roguing of off-type plants at flowering and fruiting stage is necessary.

The seed of garden pea of almost all the cultivars can be produced efficiently where the table crop is raised. The plains of North India and hills are suitable for seed production. The method of seed production is also similar as in case of table crop except that the pods are harvested when mature and dried and seeds shelled out. Three field inspections shall be made, the first before flowering, the second at flowering and third at edible pod stage. The maximum off-type permitted percentage for foundation and certified seed is 0.10 and 0.50, respectively. Although a self pollinated crop, pea is well known for producing off type plants. Hence, rigorous rouging must be undertaken at flowering and fruiting stage. When almost 90 per cent pods on the plants mature and turn dry, the whole plants are uprooted and collected on the threshing floor. After about a week the seeds are separated out from the pods by threshing and winnowing. Threshing is done by a thresher and extreme care should be taken during threshing to prevent injury to the seed. The ripe and dry pods can also be picked up by hand and threshed on small scale. Usually the moisture content of seeds at this time is higher therefore the drying must be resorted to maintain the specified moisture content of 9 per cent for ordinary pack and 8 per cent for vapour proof pack (Table 3). The seeds maintain viability for two years under normal storage conditions.

Table 3: Seed Standards for Pea

Factor	Standards for each Class	
	Foundation	Certified
Pure seed (minimum)	98.0 per cent	98.0 per cent
Inert matter (maximum)	2.0 per cent	2.0 per cent
Other crop seeds (maximum)	None	5/kg
Weed seeds (maximum)	None	None
Other distinguishable varieties (maximum)	5/kg	10/kg
Germination including hard seeds (minimum)	75 per cent	75 per cent
Moisture (maximum)	9.0 per cent	9.0 per cent
For vapour-proof containers (maximum)	8.0 per cent	8.0 per cent

In Assam, India, Saharia (1986) observed that a delay in sowing from 10 November to 20 November, 30 November or 10 December decreased average seed yields from 1.42 to 1.24, 0.80 and 0.44 t/ha, respectively in 6 pea cultivars. He obtained highest yield in cv. KPSD 1 (2.91-2.98 t/ha) for crop sown on 10 and 20 November and in cv. Pusa 10 (1.4 t/ha) for crops sown on 10 December. Keeping the pods on plant too long after drying results in shattering of seeds. Rains at the time of seed harvest may result in germination of seed in the pod itself. Singh *et al.* (1962) have reported seed yield of 14 q/ha in NP 29 cultivar at Katrain.

11.0 Crop Improvement

11.1.0 Breeding Objectives

The main objectives in breeding pea crop and yield, regional adaptability, suitable plant type, lodging and shattering resistance, disease and insect resistance, environmental stress resistance, quality and effective nitrogen fixation.

11.1.1 Breeding for High Yield

The yield potential of pea for green pods as well as for the grains is at present very low and there exists a scope for its improvement. Continued yield improvement is necessary for pea to remain an attractive option compared to cereals and oilseeds in crop rotations. Improving yield involves addressing many biotic and abiotic stresses, using a large set of strategies including diverse germplasm as parents, making many crosses, selecting for major gene traits under conditions conducive to selection, and yield testing of a large number of breeding lines.

Pre-breeding activities as phenotypic and genetic assessment of germplasm collections are key functions of a breeding program to obtain basic information about the genetic relationships among accessions, inheritance patterns of some important traits, and to select parental lines for subsequent crossing cycles (Ranalli and Cubero, 1997). In this regard, the characterization of germplasm banks of pea crop worldwide has been crucial for the development of agriculture because they are the reservoirs of genetic diversity. A large number of *ex-situ* germplasm collections have been reported around the world in public domain as compiled by Ambrose (1995). In India, about 2000 pea germplasm accessions are conserved at NBGPGR, New Delhi, IIVR, Varanasi and IIPR, Kanpur, Besides, a few state agricultural universities rich in vegetable pea germplasm are G.B. Pant University of Agriculture and Technology, Pantnagar, Punjab Agricultural University, Ludhiana, Haryana Agriculture University, Hisar, JNKVV, Jabalpur and Indian Agricultural Research Institute, New Delhi, India.

Important achievements were obtained in pea cultivars through conventional breeding over the past 20 years. Yield gains of approximately 2 per cent per year have been achieved (Warkentin *et al.*, 2015).The environmental factors limiting yield should be taken into account while developing genotypes for yield. In garden pea number of green pods/plant, green pod weight, pod length and number of seeds/pod have been shown to be the major yield components affecting the green pod yield. These yield components usually do not show component compensation effect and therefore, simultaneous improvement for these characters should be possible.

Pod and grain characteristics, which directly influence the yield and its quality, should be given due consideration. Well-filled medium-sized pods having better shelling percentage and wrinkled type seeds are the better quality and yielding factors being considered in breeding pea. The cultivar Perfection New Line is very high yielding (Nath, 1976).

Singh (1991, 1995) has compiled extensive information on genetics and breeding of peas including listing of superior lines with multiple disease resistance in pea. Kalloo (1993) and Narsinghani and Tewari (1993) have also given detailed accounts of pea breeding.

A few examples of donors on peas are as follows:

☆ Earliness: Asauji, Lucknow, Bonia, Hans, EC 3

☆ More pods/plant: PL P 26, 50, 69, 179, 279, 496

☆ Long pods: EC 109171, 109176, 109190, 109195

☆ Bold pods: EC 4103, 6185, 95924

☆ Powdery mildew: EC 326, 42959, 109190, 109196, T 10, P 185, P 288, PC 6578, B 4048, P 6587, P 6588, BHU 159, EC 42959, IC 4604, JP 501, VP 7906

☆ Wilt: Early Perfection, Bonneville, PL 43, 124, 6101, Glacier

☆ Rust: PJ 207508, 222117, EC 109188, EC 42959, IC 4604, PJ 207508, JP Batri Brown 3, JP Batri Brown 4

☆ Pea mosaic: American Wonder, Perfection Canner's Gem, Dwarf White Sugar, Little Marvel

☆ Leaf miner: EC 16704, 21711, 25173

☆ Pea stem fly (tolerant): Bonneville, Asaugi, Boach Sel., GC 141, IP 3 (Pant Uphar), Dwarf Gray Sugar, T 10, T 163

11.1.2 Breeding for Greater Stability and Regional Adaptability

An important task of the breeders not only is to increase the productivity of the plant, but new variety should provide the highest yield in favorable conditions and at the same time have high adaptive capacity to form stable yield in different growing conditions. One of the most important indicators characterizing the resistance of plants to adverse environmental factors homeostasis them, which is a universal property of the body to minimize the adverse impact of the external environment and may interact with it. The criterion homeostatic variety can be considered as the ability to maintain a low variability of the trait (Kondratenko *et al.*, 2014; Kosev and Othman, 2017). Pea is generally grown as cool season crop throughout India and in many other parts of the world. The crop can thrive well under irrigated and rainfed conditions. Recently, stable, quality and reliable yields under various environmental conditions have consistently gained importance over solely increased yield in the development of new cultivars. For example, the cultivar Alaska, an introduction from the USA, is heat tolerant and can successfully be grown at places where temperature at maturity is high, thus has wider adaptability (MacGillivray, 1961). Genotype-environment interaction was analysed using linear regression techniques to determine the stability and adaptability of advanced pea mutant lines for plant height, number of pods, number of seeds and seed weight per plant (Vassilevska-Ivanova *et al.*, 2008). Stability analysis demonstrated that two newly developed lines 2/950 - fasciata and 2/1384 - apulvinic type are less responsive to changed environmental conditions, having more stability and adaptation and can be grown over range environments maintaining good mean values of the investigated characters. Zelenov *et al.* (2014) reported that selection interest shown on pea type pleophilla is characterized by high intensity of photosynthesis, with balanced amino acid composition and productivity of green mass superior type peas with ordinary type leaves. The same authors found low homeostatic of a production process and the instability of grain yield in genotypes with pea pleophilla leaves.

11.1.3 Breeding for Suitable Plant Type

At prime maturity the quality is maximum but prime quality is usually associated with reduction in yield. In indeterminate plant, the pod yield is higher but the maturity is not uniform. Therefore, picking at different intervals is required and such cultivars can cater to the fresh market needs. Economic consideration indicates that the crop grown for processing should be harvested at one time. Since only the seeds contribute towards yield, it seems advisable to breeding cultivars with a determinate habit, thereby minimizing wasteful distribution of assimilates to plant parts that not to add to economical yield. Uniform maturity can be expected from single podded cultivars as compared to multipodded ones. The interplant uniformity in maturity can be further increased with the use of genetically determined character of simultaneous flowering (Marx, 1972). However, multipod forms such as Monogoplodnyik K 5555 (many fruited K 5555 which forms 12 flowers/peduncle) are used to increase yield (Bazhina, 1985). Developing two pods per node increased yield over one pod per node, but in most trials triple-podded cultivars have not performed significantly better than the best double-podded cultivars (Gritton, 1986). Devi *et al.* (2018) reported the development of a garden pea genotype 'VRPM-901-5' producing five flowers per peduncle at multiple flowering nodes, by using single plant selection approach from a cross 'VL-8 × PC-531'. In addition, five other stable genetic stocks, namely VRPM-501, VRPM-502, VRPM-503, VRPM-901-3 and VRPSeL-1 producing three flowers per peduncle at multiple flowering nodes were also developed by them. Besides, biological efficiency of plant with favourable balance between the vegetative and reproductive development should also be given due weightage for selection programme in pea breeding. Some foliage mutants *af, st, tl, etc.*, can be mentioned for such a selection improvement programme (Marx, 1974). The afila gene (*af*) results in plants without leaflets but with a proliferation of tendrils called semi-leafless type. The yielding ability of these types has been reported 90 to 100 per cent that of normal (Gritton, 1986).

11.1.4 Breeding for Resistance to Lodging and Shattering

Lodging is due to the collapse of pea canopies which occurs when the stem and petioles cannot support the weight of the canopy. In pea crops, wind is a secondary cause of lodging (Holland, 1990). The tendency to lodging varies greatly in different cultivars and is also related to plant growth stage. However, stem stiffness and plant height are the two main factors affecting the lodging of pea and other crops

(Amelin *et al.*, 1991). Since lodging resistance is a quantitative trait and many genes are involved, selection efficiency has been limited in traditional breeding. In general, breeding for lodging resistance has focused on the selection of short plants with short, thick internodes while retaining the same internode number (Samarin and Samarina, 1981; Obraztsov and Amelin, 1990; Park *et al.*, 1998; Amelin and Parakhin, 2003). Linear stem density has been considered a useful marker in selecting for lodging resistance in pea (Samarin, 1975; Obraztsov and Amelin, 1990; Amelin *et al.*, 1991). The semi-leafless trait has been important in improving lodging resistance over the past 20 years. Studies using F_2 pea plants derived from half-diallel crosses indicated that the variance of general combining ability (GCA) for lodging at the end of flowering and plant height was significant, but the specific combining ability (SCA) was not significant (Boros and Sawicki, 2001).

There is greater need to develop cultivars that are determinate with erect plant habit. These cultivars should not lodge, and hold seeds until harvest without shattering. These characters are important for machine harvesting. In pea, determinate and indeterminate stem types have been well described. The determinate types are less likely to lodge than the indeterminate types. Lodging resistance is required in pea grown for seeds because in machine harvesting severe losses may result from shattering. In this case only erect habit for mechanical picking is important. Green peas for canning, quick freezing, dehydration and fresh market also decide the plant type and method of harvesting.

Several genes controlling the traits associated with lodging in pea have been mapped, and molecular markers linked to these traits have been reported. For example, *Fa/fa*, which controls normal or fasciated stem type (Blixt, 1974); *Det/det*, which controls indeterminate or determinate growth habit (Marx, 1986); *Tl/tl*, which controls tendril type (tendril/tendrilless) (Makasheva and Drozd, 1987); *rms2, rms3* and *rms4*, which control the amount of branching (ramous) (Poole *et al.*, 1993); and *p275* marker, which was associated with plant height (Dirlewanger *et al.*, 1994), have been mapped in pea. Furthermore, a sequence characterized amplified region (SCAR) marker linked to *rms3* was developed and can be scored as a codominant marker. It mapped 16.7 cM from the *Rm3* locus (Rameau *et al.*, 1998).

The inter wining of tendrils in plants homozygous for recessive *of* gene (leafless form) gave the plants improved standing ability (Gritton, 1986). Shevchenko (1989) reported the use of leafless (tendrilled) forms in producing non-lodging type, by combining leaflessness with strong stem and petioles. The cultivars Neosypayush-chusya-1 (non-shedding-1), Mass 1 and Mass 4 are used as donors of shattering resistance and leafless varieties like Us 5 (Tendril 5) and Australia as donors of lodging resistance (Bazhina, 1985). In work on breeding for shattering resistance, Vetrova (1987) produced single and complex hybrids involving a form with altered funicle morphology (ensuring firm attachment of the seeds to the pod), in Moldavia. The resistant cultivar VOMO 84 was bred. Türk *et al.* (2011) reported that pea half tendril varieties had significantly better habitat, increased productivity and resistance to lodging capacity compared to normal varieties with leaf type, but with fewer seeds in beans.

11.1.5 Breeding for Abiotic Stress (Drought, frost, salinity and environmental pollution) Resistance

There are large climatic variations between pea cropping areas, between years, and even within a cropping year (Annicchiarico and Iannucci, 2008). Drought is one of the most limiting factors in pea cropping in many parts of the world, affecting both quality and quantity of the yield (Boyer, 1982; Ali *et al.*, 1994). Although precipitation during flowering has been proven to be the most important factor in seed yield, peas can suffer from drought during both the vegetative phase or during reproductive development. Both early (Ali *et al.*, 1994) and terminal drought may be severe depending on cropping environments (Annicchiarico and Iannucci, 2008), thus, occurrence and distribution of rainfall can also affect the degree of losses (Stoddard *et al.*, 2006). Development of new varieties with wide adaptation ability including drought tolerance is the primary aim of pea breeding works (Abd-El Moneim *et al.*, 1990), and it is necessary to increase competitiveness of legumes (Dar and Gowda, 2010). The cultivar Early Badger reported tolerant to heat and drought (Choudhury, 1967) can be well exploited for drought resistance breeding. Dwarf type pea plants "Progress" No. 9 was found more drought resistant compared to tall phenotype "Alaska" (Iwaya-Inoue *et al.*, 2003). Water of dwarf control and no-elongation zone of GA1-treated "Progress" No. 9 grown under light condition was more sustained during 3-h air drying compared to those of "Alaska" (Iwaya-Inoue *et al.*, 2003). Experiment was conducted to compare the drought resistance of conventional leaf and semileafless cultivars in pea (Semere and Froud-Williams, 2001). They found that the leafy cultivar "Bohatyr" produced significantly more leaf area than semileafless "Grafila". Epicuticular wax also plays an important role in control the loss of water from the cuticle. Under drought conditions the wax load of pea cultivars increases significantly and it is accompanied by increased residual transpiration rate (Sanchez *et al.*, 2001). Under drought stress the roots of pea are grown deeper in the soil than those under irrigated conditions. About 34 per cent of the total pea roots are deeper than 0.23 m in the dry soil, whereas only about 20 per cent of roots rooted in this depth of soil profile under irrigated conditions (Benjamin and Nielsen, 2006). However, osmotic stress induced by PEG 6000 results in shortening of primary root and increases lateral root number (Kolbert *et al.*, 2008). Drought stress reduces the seed number in an intensity dependent manner and the distribution of seeds is also affected: more seeds develop on the basal phytomers of drought-stressed pea plants than on control plants (Guilioni *et al.*, 2003). Drought-tolerant pea genotypes have better turgor maintenance, which is significantly related to osmotic adjustment (Sanchez *et al.*, 1998). Increased levels of the metabolites in drought-stressed pea plants are detected including proline, valine, threonine, homoserine, myoinositol, c-aminobutyrate, and trigonelline (Charlton *et al.*, 2008).

Salinity is a major abiotic stress limiting germination, plant vigour and yield of leguminous crops especially in arid and semi-arid regions. Food legumes are relatively salt sensitive compared with cereal crops, thus farmers do not consider growing food legumes in salinized soils. The sensitivity in legumes may be due to salt affecting bacterial activity and nitrogen fixation (Toker and Mutlu, 2011; Egamberdieva and Lugtenberg, 2014). Salt stress led to reduction in shoot growth of pea plants (Delgado *et al.*, 1994). Pea is very sensitive to salinity problem in the areas of cultivation, for which suitable salinity tolerant cultivar can be developed through effective breeding procedures. Air pollution in the industrial areas affects the yield in peas.

Cold tolerance has been an important trait for improvement in crop adaptation in many countries (Materne *et al.*, 2011). Genetic variation has been identified in pea (Redden *et al.*, 2005), and the overall level of tolerance of this crop is greater than other legumes (Materne *et al.*, 2011). Cultivar Multifreezer can be a good source of frost tolerance and the cultivar Alderman which is also suitable for hilly regions (Yawalkar, 1969) can be used for transferring frost resistance in other suitable cultivars.

Air pollution in the industrial areas affects the yield in peas. Some of the varietal sources resistant to pollution can be explored for stress resistance breeding.

11.1.6 Breeding for Disease Resistance

Attempts have been made to breed disease resistant cultivars in pea. In addition to development of improved cultivars adapted to the different production areas and mechanical harvesting, *etc.*, breeders have also given attention to breeding pea cultivars for disease resistance. The important diseases in pea are *Fusarium* wilt, powdery and downy mildews and virus diseases namely, bean yellow mosaic, pea enation mosaic, and pea dwarf mosaic.

Probably the most unique source of resistant pea germplasm is Minnesota 494-All (King *et al.*, 1981). It has moderate resistance to *A. euteiches*, *F. solani* f. sp. *pisi* and *P. ultimum* and is resistant to races 1, 2 and 6 of *F. oxysporum* f. sp. *pisi*. However, this line has pigmented flowers and many other undesirable horticultural traits. Marx *et al.* (1972) reported that tolerance in PI 175227 to *A. euteiches* was associated with dominant, wild type alleles at 3 unlinked marker loci (Le, A, and PI). They further reported that substitution of recessive alleles which express horticulturally desirable traits at each of these loci resulted in a reduction in root rot tolerance. However, Kraft (1988) reported that resistance to *A. euteiches* was recovered in breeding lines with desirable horticultural traits. Lewis and Gritton (1988) reported that resistance to *A. euteiches* appears to be quantitatively inherited with low heritability. A recurrent selection program where disease pressure is intense was used to increase resistance to common root rot in horticulturally acceptable types. "Dark Skin Perfection", "Freezer 76110" and "Wando" are reported to have partial resistance to *R. solani* (Shehata *et al.*, 1981). Two breeding lines, VR-74-410-2 and VR-1492-1, were released which are resistant to PSbMV, race 1 and 2 of *F. oxysporum* f. sp.pisi and *Fusarium* root rot (Kraft and Giles, 1978). Kraft (1981) released two F8 resistant lines (792022 and 792024) which have combined resistance to *F. solani* f. sp. *pisi* and *F. oxysporum* f. sp. *pisi* races 1 and 2 (Kraft, 1981). Later, four F5 breeding lines (86-638, 86-2197, 86-2231 and 86-2236) were released (Kraft, 1988), which have resistance to *A. euteiches*, *F. solani* and races 1, 2, and 5 of *F. oxysporum* f. sp. *pisi*.

Kalia and Sharma (1988) observed that cultivars resistant to powdery mildew contained higher levels of phenolics and phenol-oxidizing enzymes than the susceptible cultivars. Munshi *et al.* (1987) observed greater quantities of orthodihydroxyphenols and polyphenols in cultivar resistant to powdery mildew. Early Badger is a reported source of *Fusarium* wilt resistance (Choudhury, 1967). The cultivar Delwiche Commando developed by University of Wisconsin from cross between Admiral and Pride is a resistant source of common wilt and near wilt diseases. Appreciable resistance to *Fusarium oxysporum* was shown by the hybrids IP2, Koroza, Rondo, Koroza and also by cv. Raro (Puzio-Idezkowska, 1990). Other source of wilt resistance has been mentioned earlier in the text. Some other cultivars showing resistance to root rot and damping off have also been mentioned. As mentioned earlier, there is a need to develop cultivar resistant to powdery mildew which is of common occurrence. The lines P 185, 6585, 6587, T.10, Sel.1, Mores-5, EC 3386 and cultivars Giant Sugar, American Wonder and PMR lines are good sources of powdery mildew resistance and these can be made use of in breeding programme (Tyagi *et al.*, 1978; Joi *et al.*, 1975; Naik *et al.*, 1975; Sokhi *et al.*, 1979). As regard virus diseases, sources of resistance to PSbMV include PI 193586 and PI 193835 which have the recessive gene *sbm 1* (Hagedorn and Gritton, 1973). Other sources of resistance to PSbMV include: X78123, X78126 and X78127 which are known to carry *sbm 1* (Khetarpal *et al.*, 1990), and PLP40, PLP350, EC15184, EC17451 and PLP564 (Khetarpal *et al.*, 1990). Seven breeding lines, OSU-547-29, OSU-559-6, OSU-564-3, OSU-584-16, OSU-589-12, OSU- 615-15 and OSU-620-1, resistant to PSbMV and pea enation mosaic virus (PEMV) were released in 1988 (Baggett and Kean, 1988). Hagedorn and Hampton (1975) also reported cultivar Perfected Freezer 60 as the source of resistance to pea enation mosaic; Horal and Perfection (Yen and Fry, 1956), Bonneville, Pride and Wisconsin Perfection (Johnson and Hagedorn, 1958) are the sources of bean yellow mosaic virus resistance; and PI 193586 and PI 193835 (Stevenson *et al.*, 1970) are the sources of resistance to pea seed-borne mosaic virus disease. All these sources provide ample scope for their exploitation and utilization in breeding programmes for disease resistance in pea.

11.1.7 Breeding for Insect Resistance

A number of insect species feed on pea. The important and most injurious insect pest is pea aphid, pea moth, pea pod borer, pea weevil and leaf miner. The three mechanisms: antixenosis, antibiosis and tolerance contribute resistance to insect pests of legume crops (Soundararajan *et al.*, 2013). Several physico-chemical characteristics contribute to insect resistance in legumes (Clement *et al.*, 1994). Presence of dense covering of hairs or trichomes on the leaves/pods confers resistance to many insect species. Allomones such as arcelins, L-canavanine, polyhydroxy alkaloids and saponins have been reported to confer resistance to insect pests in legumes (Dilawari and Dhaliwal, 1993).

Breeding for insect resistance is necessary for quality and high yield, which are the mainly affected components. Besides the above-mentioned insects, there are also the prime vectors of other pea diseases (Hagedorn, 1973). Some LMR lines are good sources of leaf miner resistance breeding. The cultivars Feltham First and Meteor are sources of resistance to pea aphid (Bintcliffe and Wratten, 1980). Sources of pea weevil and pod borer resistance can also be sought for breeding programme.

Kalloo (1993) and Narsinghani and Tewari (1993) have mentioned the following donor parents for biotic stress resistance in pea breeding.

☆ Powdery mildew: EC 326, 42959, 109190, 109196, T 10, P 185, P 288, PC 6578, B 4048, P 6587, P 6588, BHU 159, EC 42959, IC 4604, JP 501, VP 7906

☆ Wilt: Early Perfection, Bonneville, PL 43, 124, 6101, Glacier

☆ Rust: PJ 207508, 222117, EC 109188, EC 42959, IC 4604, PJ 207508, JP Batri Brown 3, JP Batri Brown 4

☆ Pea mosaic: American Wonder, Perfection Canner's Gem, Dwarf White Sugar, Little Marvel

☆ Leaf miner: EC 16704, 21711, 25173

☆ Pea stem fly (tolerant): Bonneville, Asaugi, Boach Sel., GC 141, IP 3 (Pant Uphar), Dwarf Gray Sugar, T 10, T 163

11.1.8 Breeding Cultivars for Higher Nutritive Quality

Plants can synthesise all 20 amino acids contained in proteins, but human beings can only make 11 of these, the non-essential amino acids. The other nine, known as the essential amino acids, must be obtained from the diet. With regard to dietary requirements, legume proteins tend to be low in the essential sulphur amino acids (SAAs) cysteine and methionine, and in tryptophan. Dietary guidelines suggest adults should consume 15 mg/kg of methionine and cysteine and 4 mg/kg of tryptophan per day (United Nations University, 2002). Both major globulin proteins in pea contain low concentrations of SAAs, although legumin contains slightly higher levels (1.01–1.78 per cent of protein) (Casey and Short, 1981) than vicilin proteins, which largely contain no SAAs (0–0.2 per cent) (Shewry *et al.*, 1995; Casey *et al.*, 1998). The importance of this lack of SAAs relates to whether or not pea is consumed as part of a balanced diet, as cereal crops provide SAAs lacking in pulses if consumed concurrently and pulses provide lysine, an amino acid which is lacking in cereals (Shewry and Tatham, 1999). However, if pea was to become more of a primary protein source, then increases in the levels of essential amino acids would vastly improve its nutritional value. Albumin proteins in pea have been shown to contain higher levels of SAAs than globulins; however, anti-nutritional properties have been reported for pea albumins, linked to lower digestibility (Vigeolas *et al.*, 2008; Clemente *et al.*, 2015; Vaz Patto *et al.*, 2015). Incorporation of high levels of pea into the diets of nonruminant livestock, such as pigs, has been shown to depress growth, likely due to the low digestibility of albumin proteins such as lectin (Le Gall *et al.*, 2007).

Anti-nutrients are compounds which can severely affect the bioavailability of nutrients in plant foods (Hurrell and Egli, 2010). Even though pea seeds are nutrient-rich, the presence of anti-nutrients in sufficient quantities can reduce nutrient bioavailability and diminish nutritional value. A highly abundant antinutrient in many legume seeds is phytic acid (or phytate). Phytic acid is the major storage form of phosphorus, representing 50–85 per cent of total plant phosphorus in seeds. Ruminants are able to break down phytate through the presence of phytase in their polygastric digestive systems; however, humans and monogastric livestock lack the required enzymes in sufficient quantities (Gupta *et al.*, 2015). Because phytate accumulates while seeds mature, immature peas, widely consumed as garden peas or petit pois, have a lower phytate concentration than mature peas but the same concentration of iron–ferritin (Moore *et al.*, 2018). It has been shown that the iron: phytate ratio in peas is positively correlated with iron bioavailability and is higher in immature than mature pea seeds. Seed protease inhibitors constitute another group of compounds that can reduce the value of pulse seeds within the food and feed industries (Clemente *et al.*, 2015).Trypsin/chymotrypsin inhibitors prevent the breakdown of nutritionally valuable proteins in the gut, and their presence means that seeds must be heattreated before use (Clemente *et al.*, 2015). The most abundant trypsin/chymotrypsin inhibitors in pea seeds are Bowman–Birk inhibitors, a specific class of protease inhibitors discovered more than 70 years ago (Birk, 1985). Polyphenols and tannins are well known to affect the bioavailability of micronutrients in food in general (Hurrell and Egli, 2010) and in pulse crops in particular (Wang *et al.*, 1998).

Raffinose oligosaccharides are sugar compounds associated with flatulence (Gaw³owska *et al.*, 2017). The predominant raffinose oligosaccharides in pea seeds are raffinose, stachyose and verbascose. These compounds are digested by bacterial microflora in the large intestine, sometimes causing flatulence and discomfort in humans, as well as diarrhoea and reduced performance in monogastric livestock. Domesticated pea species and advanced breeding lines have been shown to contain lower levels of raffinose oligosaccharides compared to other groups, suggesting that previous selection by breeders has already favoured lines with lower concentrations of these compounds (Gaw³owska *et al.*, 2017).

Anti-nutritional factors contribute to plant defence against pests and diseases (Rehman *et al.*, 2019), meaning that their removal could compromise plant performance in the field. Therefore, both the potential agronomic and health benefits of tannins and other anti-nutrients, including Bowman–Birk inhibitors (Kennedy *et al.*, 1996; Clemente and Olias, 2017), must be considered when attempting to reduce their levels through breeding programmes.

In comparison with other cultivated legumes, pea has a relatively low content of seed protein, and it usually lies within range of 15-35 per cent (Pate, 1975). Efforts have been made to improve protein content (Kalinina *et al.*, 1985; Vetrova and Petrushina, 1987; Varlakhov *et al.*, 1987) and to combine earliness with high protein and tryptophan content of the seeds (Kalinina *et al.*, 1985). Certain amount of variation in other seed constituents like vitamins, amino acids, sugar and minerals exists among cultivars. Cultivar Little Marvel is very sweet and cultivars Lincoln, Thomas Laxton, Alderman are having excellent canning and freezing quality. These characteristics can be transferred into new high yielding cultivars following systematic breeding programme.

Significant progress has been made in multiple areas that will assist with improving the nutritional value of pea, notably the alteration of pea seed protein composition (Domoney *et al.*, 2013), the identification of the *sbeI* gene, which increases resistant starch in pea (Bhattacharyya *et al.*, 1990), and the removal of anti-nutrients such as phytate (Warkentin *et al.*, 2012), protease inhibitors (Clemente *et al.*, 2015) and vicine-convicine (Duc *et al.*, 2004).

11.2.0 Methods of Breeding

Pea falls in the group of self-pollinated crops and hence the methods employed for the improvement of self-pollinated crops can very well be adopted for its improvement. The basic methods by which new cultivars can be developed are (i) introduction, (ii) selection, and (iii) hybridization. These methods tend to follow the systematic sequence of steps developed for utilization of their best advantage. The initial step is to assemble cultivars currently under cultivation and to identify the cultivars that are adopted to the local conditions. The selection is practised to isolate the highest yielding genotypes from the local materials. For example, cultivars Asauji and Lucknow Boniya are selections from local materials. Finally the important characteristics of genotype are established, which are combined following the hybridization procedure. The sequence of these methods is in chronological order.

The main breeding programmes at the Gorku Agricultural Experiment Station, (the then USSR) are intraspecific hybridization of geographically distant forms and directional selection (Bazhina, 1985). Hybridizing ecologically and geographically distant forms, chemical mutagenesis and selection of transgressive forms were considered effective breeding methods (Zhogina, 1986). Because non-shedding is controlled by recessive genes 2-3 cycles of intercepted backcrossing between non-shedding forms is recommended for producing high yielding, non-shedding lines (Khangildin, 1984).

11.2.1 Introduction

The most common interpretation of a plant introduction is a crop species or a cultivar or genetic strain of a crop species, introduced from one country into another. Most of the peas grown in India have been introduced from various European countries and the USA. These introduced cultivars have been either adapted as such and released for cultivation, or plants selected from the varietal population, multiplied and distributed as superior cultivar or used in hybridization programme owing to their desirable characters such as disease or insect resistance and some other important characteristics. Examples of using in hybrid are T.19 ´ Early Badger-Early December (Wakankar and Mahadik, 1961); GC 195-developed from T.19 ´ Little Marvel (Wakankar, 1978). The introduced cultivars are tested for their acclimatization to the local conditions of the environment and soil, *etc.* Pea is introduced from various international germplasm centres in the form of seeds. In the USA, the germplasm centre is the North East Regional Plant Introduction Station, Geneva, New York, which maintains about 2000 accessions. The Weibullsholm Pisum Genetica Association Collection, Landskrona, Sweden, maintains about 2000 lines. The pea collection of the Instirure fur Kulturpflanzenforschung, Gattersleben, Germany, contains about 1600 lines. The Laboratoro Del Germplasm, Bari, Italy includes a *Pisum* collection. In India, ICRISAT, Hyderabad, A.P., maintains the collection of field pea germplasm (Gritton, 1980). The pea accessions for various objectives are introduced from these sources directly by the breeders or their organisations. In India, the National Bureau of Plant Genetic Resources is the agency for introduction of pea germplasm from international sources. Account of introduced cultivars has been given in the text explaining varietal characteristics. Arkel, Meteor (Choudhury and Ramphal, 1975), Early Badger (Choudhury, 1967), Perfection New Line (Nath, 1976), Alderman (Yawalkar, 1969), Little Marvel and Superb (Singh and Joshi, 1970) are introductions.

11.2.2 Selection

Selection is the oldest breeding procedure employed by man since he began to cultivate crops. It is essentially a process by which individual plant or its groups are sorted out from the mixed populations. For effective use of this method the presence of genetic variability is a prerequisite. Generally two methods of selection are practised in pea. These are (i) mass selection and (ii) pure line or single plant selection. At Ekaterininsk, in Russia cvs. ObroztsovChiftik 6, Indo and Imperial were selected for combination of characters, notably, seed yield, quality and resistance to fungal diseases (Yankov, 1989). In India a number of varieties in pea have been developed through selection such as Asauji, Harbhajan (from exotic stock), NP 29, Pant Uphar, and Ooty-1.

11.2.3 Hybridization

In this method of breeding, pea cultivars are crossed and plants possessing the combined desirable features of the parental cultivars are selected from the segregating progenies. In addition to combining the phenotypic features of the parental cultivars it is also possible to select plants from a cross progeny superior to the parental lines in feature of the quantitative nature such as yield, seed weight, winter hardiness, *etc.*, where inheritance is determined by polygenes. In this method, parents are selected on the basis of desirable characters under transfer combination and are crossed artificially. In pea this crossing and selfing programme is relatively easy due to large floral parts and its self-pollinating nature. For artificial crossing the emasculation and pollination procedures are adopted. In a cross developed from pure line cultivars, the plants will be highly heterozygous but have similar genotypes. Genetic segregation will start in the F_2 generation and heterozygosity will reduce to half with each seeding self-generation.

In hybridization programme, in order to combine maximum of the desirable traits in a cultivar under development, the crosses can be attempted in the form of single cross, double cross, three-way cross and multivarietal cross. The selection is operated from the F_2 generation segregants.

Ochatt *et al.* (2004) confirmed the strong cross incompatibility existing between *P. sativum* and *Lathyrus sativus* as described by Campbell (1997), where the putative hybrid turned out to be pea and they also reported on the production of fertile sexual hybrids between *Pisum sativum* L. as the female and *P. fulvum* as the male parent, without any need for the use of a wild type *P. sativum* accession as bridging cross.

The cultivar Oregon 523 was derived by 7 generation of single plant selection from the cross OSUB 190-18-2 ´ Small Sieve Freezer (Baggett and Kean, 1987). Yankov (1989) noticed that leafless (tendrilled) forms were late, but line 21/87, selected in the F4 of Neosypayushchusya 1 (non-shedding-1) ´ K 7779 was promising for its earliness. Hybridization between Australian cv. Massey Gem and Local cultivar Harabona and subsequently by pureline selection the cultivar Mattar Ageta-6 has been developed (Dhillon *et al.*, 1989). Cultivar Vica was developed from Nike Express (Schmelcz, 1985).

Over the years many pea varieties developed in India were the results of either single cross (Pant Uphar × Arkel: VL-Ageti Matar-7; T19 × Early Badger: Jawahar Marar 3; T19 × Little Marvel: Jawahar Marar 4; Early Badger × IP3 (Pant Uphar): Pant Matar 2; Bonneville × P23: Hisar Harit; Massey Gem × Harabona: Matter Ageta 6; Old Sugar × Early Wrinkled Dwarf 2-2-1: VL- Matar-3; Pusa 02 × Morrasis-55: Punjab Harit; Bonneville × 6587: Azad Pea 2); double cross [(Arkel × JM5) × ('4bc' × JP 501): Jawahar Peas 54; (JMI × JP 829) × (46 c × JP 501): Jawahar Peas 83]; or triple cross [(JMI × R 98 B) × JP 501 A/2: Jawahar Peas 15] hybridization followed by selection.

11.2.4 Utilization of Hybrid Vigour

In self-pollinated crops, utilization of hybrid vigour is dependent upon a system for economically producing the F_1 seeds. Hybrid vigour in pea, as in other crops, is dependent on the parents used in hybrid combination. Hybrids from some parents exhibit no yield increase over the parents, while others had over 25 per cent increase. Therefore, individual hybrid is evaluated to select desirable hybrid combinations for cultivation. Except for a report of simple recessive type of genetic male sterility, there is non-availability of effective male sterility types and the incompatibility system (Gritton, 1980). Thus, the use of natural crossing in producing hybrids is almost restricted. However, even in pea, usually considered to be a self-pollinated crop, cross-pollination can be quite extensive with some genotypes and environments. Reported outcrossing percentage ranges from zero (White, 1971) to 60 (Harland, 1948) which could be made use of in producing hybrid seed commercially. Besides, a well defined array of genetic marker is available in *Pisum* (Yarnell, 1962; Blixt, 1972), from which the dominant alleles of *r, rb* and *i* expressed in cotyledonary stage of F_1 provides the identification of hybrids from such natural cross hybridization programme. However, leaf, seed and flower colour markers can also be exploited for development of hybrids. In spite of this, the self-pollinating behaviour due to cleistogamous flower structure prohibits the awaiting prospects of hybrid seed production and utilization of hybrid vigour in increasing yield potential in pea. At last, the less number of seeds produced in cross-pollination attempted and high cost of hybrid seed in comparison to that involved in seed production of pure line cultivars are also the other limiting factors to such a breeding programme.

Heterosis has been recognized in pea since the early studies of Mendel in 1866. Later reports on heterosis in pea have been documented by Krarup and Davis (1970), Gritton (1975), and Sarawat *et al.* (1994). Heterosis in predominantly self pollinating species is of interest, but difficult to apply in practice without sufficient means to achieve cross pollination. Studies of F_1 hybrids in pea have been conducted with the intent of capitalizing on heterosis in commercial seed production and for the potential of using data from F_1 hybrids to identify superior populations for subsequent selection and cultivar development. To date no known hybrids have been developed or commercialized.

Rao and Narsinghani (1987) observed highest heterosis in crosses Kinnauri ´ R 701, R 710 ´ EC 33866 and 46C ´ 251A. El-Murabaa *et al.* (1988) reported that cultivars Kelvedon Wonder and Little Marvel proved to be good general combiners for all characters and can be used in heterosis breeding programmes. Similarly, Singh *et al.* (1989) found that the best general combiners were T163, 6113 and P209 for seed yield and EC 33866 and BR 12 for protein content. The best specific combinations for yield and protein content were T 163 ´ EC 33866 and BR12 ´ P209, respectively. The cvs. Ruga, Amino and Miranda had good general combining ability (GCA) for yield (Csizmadia, 1990). Singh (1990) observed that Rachna was the best parent producing significant positive GCA effects, generally heterotic effects and additive gene effects. Recently many workers have reported heterotic response for pod yield and its component traits in pea (Sharma *et al.*, 1998; Sharma *et al.*, 2007). According to Karmaker and Singh (1990) VP-8005 was a good combiner for seeds per pod and Arkel for dwarf structure. Of the hybrids, Gloriosa × JP-169 was the best specific cross for yield and Arkel × VP-8005 was a promising cross for seeds per pod. Singh and Singh (1990a) evaluated a 12-parent diallel cross set of pea for yield related characters. Two F_1 hybrids T-163 × Bonneville and T-163 × Sel-2 were identified as most promising for yield per plant. Singh and Singh (1990b) derived the information on combining ability and reported that parental lines T163, PG-3, Selection 2 and Bonneville were the good general combiners for yield. Brar *et al.* (2012) observed heterotic response in crosses Arkel × C-96 for days taken to 50 per cent flowering and days taken to maturity, C-308 × C-400 for plant height, MA-6 × C-400 for pod length, Arkel × C-400 for number of pods per plant, JM-5 × C-96 and NDVP-10 × C-400 for shelling percentage. They found P-1, AP-1 and PMR19 were most promising donors for improvement of green pod yield per plant. The highest significant heterosis for yield was observed in the crosses DDR-23 × HUDP-15 and Arkel × CAUP-1 over better parent and standard check, respectively (Rebika, 2017).

12.0 Biotechnology

12.1.0 Micropropagation

Tissue culture techniques provide an opportunity to improve crop plants through the selection of clones with desirable characters. Progress in crop improvement through *in vitro* techniques has been hindered by the inability of tissues to regenerate plants with high frequency. Grain legumes including pea are known as recalcitrant plants *in vitro*. Previous works on pea have included shoot organogenesis from callus of shoot apices (Gamborg *et al.*, 1974); adventitious shoot formation from immature leaf explants (Marginiski and Kartha, 1981; Rubluo *et al.*, 1983); Callus production as well as shoot organogenesi (Hussy and Gunn,1984); regeneration from cotyledonary nodes (Jackson and Hobbs 1990) and from hypocotyl explants (Nielsen *et al.*, 1991); regeneration from nodal explants of pea seedlings (Nauerby *et al.*, 1990) and also from embryonic axis (Schroeder *et al.*, 1993).

Tissue culture work on pea for micropropagation and organogenesis involves use of various explants using leaflets, ovary, ovule and embryos besides axillary buds. Murashige and Skoog (MS) medium has been the widely used medium for pea tissue culture. Hussey and Gunn (1984) obtained vigorously growing calluses from cultivars Puget and Upton, but not from Maro, Melton or Vedette. Calluses were induced on the basis of plumules, excised from germinated seed, cultured on MS medium supplemented with 1 mg/l BA and 4 and 8 mg/l IBA. Shoot regeneration took place 2-4 weeks after transfer to the same medium with IBA reduced to 0.25 mg/l, and continued after subculturing to fresh media. Fujioka *et al.* (2000) established a cultural method for regenerating Japanese pea cultivars and reported differences in callus formation, shoot and root regeneration by *in vitro* culture of immature leaflets. Shoot regeneration occurred in 5 of 6 Japanese pea cultivars from primary calluses induced from immature leaflets cultured *in vitro*. More than 30 per cent of explants from young leaflets of 2- to 3-day-old seedlings regenerated shoots. Plant regeneration can be done sequentially by inducing shoot and root formation. Fujioka *et al.* (1999) found that among these 3 culture methods, ovary culture regenerated plants most efficiently. When an excised ovary was cultured before anthesis, the ovule failed to grow, whereas after anthesis, the older the ovary was, the faster the ovules grew and germinated. Ovules of cultivars Misasa, Oranda and Kishu-usui all grew and germinated in ovary culture. In ovary culture, the ovule germinated faster at 25°C than at 20°C. The addition of NAA at 0.5 or 1.0 mg/l or BA at 1.0 mg/l to the medium promoted ovule growth in ovary culture but the addition of 0.5 mg GA_3/l suppressed it.

Malik and Saxena (1992) established axenic seedling cultures from mature seeds on MS medium supplemented with thidiazuron (TDZ). Pea seedlings exhibited a unique pattern of shoot formation, which was accomplished in two distinct phases. Multiple shoots developed within a week, from the nodal regions of the primary epicotyl in a medium that contained 5-50 mM TDZ. Ozcan (1995) cultured embryos of the pea cv. Orb and Consort embryos on MS medium supplemented with 1, 2 or 4 mg BA + 0.02 or 0.2 mg NAA + 0 or 2 mg kinetin/l. For both cultivars, the largest numbers of shoots were produced on medium containing 4 mg BA + 0.02 mg NAA. Excised shoots showed 80-90 per cent rooting after 7-10 days in half-strength MS medium supplemented with 1 mg IBA/l, and grew into normal fertile plants.

The best result of shoot formation was achieved when hypocotyls explants were cultured on MS-medium supplemented with 2 mg/l BA+1 mg/l NAA (Ghanem *et al.*, 1996). However, immature cotyledon explants showed the highest frequency of shoot formation with 1 mg/l BA. Data of *in vitro* rooting showed that maximum root frequency occurred on culture medium containing half strength of MS salts, 40 g/l sucrose and 2 mg/l NAA.

Pea seeds were used as the source material for the establishment of mutagenic shoots and roots (Supe and Roymon, 2011). Seeds were treated with different radiations at various ranges (0.6-2.0KR). MS medium supplemented with BAP (3mg/l) in combination with Kn (1mg/l) or NAA (0.5mg/l) was found to be most effective in initiating multiple shoots. Microshoots rooted best *in vitro*, in half strength MS medium supplemented with IAA (0.1mg/l). Regenerated plantlets were successfully established in soil with a survival rate of 95 per cent.

Transformation in Pea

Agrobacterium rhizogenes-mediated transformation of Pisum sativum roots as a tool for studying the mycorrhizal and root nodule symbioses. (A) Pea seedlings were incubated in containers filled with Jensen medium placed within ventilated plastic box. (B) Incubation of pea seedlings in jars after transformation. (C) Callus formation on wounded sites of young cv. Frisson pea seedlings treated with A. rhizogenes. (D) General view of pea seedling after incubation on an antibiotic. The arrows indicate the callus formation (Leppyanen *et al.*, 2019).

Transformation in Pea

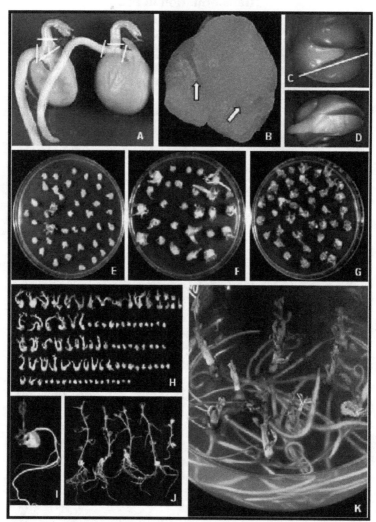

In vitro and *In vivo* **Agrobacterium mediated transformation in pea**

A - germinating etiolated pea seedlings at the stage (4-d-old) proper for isolation of cotyledonary node segments - lines mark cut areas. B - isolated cotyledonary node with axillary buds (arrows) used for cocultivation. C - imbibed pea seed with partially removed one cotyledon (D) ready for cocultivation. E, F, G effect of vacuum infiltration (E) and sonication (F) during cocultivation on regeneration of transformed cotyledonary nodes on selection Km + medium; control culture (G) - sonication, Km - medium. H - test of intrinsic kanamycin tolerance in pea cotyledonary nodes *in vitro*; from the top to bottom: 0 (control), 50, 100, 150 and 200 mg(Km) dm^{-3} (medium). I, J - development of *in vivo* infected trimmed seeds in perlite saturated with selection Km + medium. K - root induction on shoots isolated from cotyledonary nodes on selection Km + medium; second selection step (Svabova *et al.,* 2005).

12.2.0 Molecular Markers

12.2.1 Genetic Linkage

Genetic maps have proven to be useful to uncover the molecular bases of monogenic characters such as Mendel's characters (Ellis *et al.*, 2011) and also to decipher the determinism of complex agronomically-important traits. QTLs responsible for the genetic control of yield-related traits, seed protein content, aerial and root architecture, and biotic and abiotic stress resistance have been detected under multiple environmental conditions and located on different maps. In addition to QTL mapping analyses in bi-parental populations, association analyses have emerged as a complementary approach to dissect quantitative traits in pea by exploiting natural genetic diversity and ancestral recombination events characterizing germplasm collections. Diverse sets of cultivars with distinct geographic origins were used to determine associations between genetic markers and seed mineral nutrient concentration (Kwon *et al.*, 2012; Cheng *et al.*, 2015; Diapari *et al.*, 2015), seed low-carbohydrate content (Cheng *et al.*, 2015), seed lipid content (Ahmad *et al.*, 2015), yield-related traits (Kwon *et al.*, 2012), disease/pest resistance, and morphological traits such as flower color and seed coat color.

Specific markers linked to major genes were developed for use in breeding, especially for trypsin inhibitors in pea seeds (Page *et al.*, 2002; Duc *et al.*, 2004), flowering (Weller and Ortega, 2015), lodging resistance (Zhang *et al.*, 2006) and resistance to diseases such as powdery mildew (Ghafoor and McPhee, 2012; Reddy *et al.*, 2015), pea enation and seed borne mosaic virus (Frew *et al.*, 2002; Jain *et al.*, 2013), fusarium wilt (McClendon *et al.*, 2002), *Ascochyta* blight (Jha *et al.*, 2015), and rust (Barilli *et al.*, 2010). Some other resistance, flowering or seed composition genes were reviewed by Warkentin *et al.* (2015). Marker-assisted selection was conducted in early generation (F_2) breeding populations using markers linked in coupling to two major QTLs controlling lodging resistance and was demonstrated more efficient than phenotypic selection (Zhang *et al.*, 2006). Recently, marker-assisted backcrossing (MABC) was successfully used to introgress one to three of these seven main *Aphanomyces* root rot resistance QTLs (Hamon *et al.*, 2013) into several recipient agronomic lines (Lavaud *et al.*, 2015). Evaluation for resistance of the subsequent 157 BC5/6 Near Isogenic Lines (NILs) validated the effect of the major and some minor QTLs in controlled conditions and showed QTL × genetic background interactions. AMABC strategy was also used to introgress three frost tolerance QTLs among the main four QTLs identified by Lejeune-Hénaut *et al.* (2008). Field evaluations of 125 QTL-NILs validated the effect of these QTLs in the spring-type genetic background Eden (Hascoët *et al.*, 2014). So far, marker-assisted construction of QTL-NILs has mainly allowed QTL effects to be validated. The rational use of these genetic regions in breeding strategies can now be considered in order to combine favorable alleles at complementary QTLs to improve multiple stress resistance in agronomic material. In parallel to strategies considering the combination of individual QTLs, genomic selection seems a promising approach in pea, as first suggested by the prediction of the date of beginning of flowering and 1000 seed weight using a subset of 331 SNP markers genotyped in a reference collection of 372 pea accessions (Burstin *et al.*, 2015). Increasing the marker coverage of the genome by using the newly-developed GenoPea 13.2K SNP Array (Tayeh *et al.*, 2015a) further improved prediction accuracies (Tayeh *et al.*, 2015b).

In order to determine whether pea gene sequences contain enough polymorphism to be used as genetic markers, Pavy *et al.* (1998) examined the molecular variability at the DNA sequence level within different lines and wild ecotypes of pea. Each region was specifically amplified and polymorphism was identified by electrophoretic mobility and by direct sequencing of PCR products. The observed polymorphism illustrated the possibility of developing molecular markers, as all the analyzed loci were successfully localized. Polymorphism was detected either as DNA conformational polymorphism following non-denaturing polyacrylamide gel electrophoresis or as CAPS (cleaved amplified polymorphism sequence). These genetic markers were able to establish bridges between different existing pea genetic maps.

Polans *et al.* (1990) used RFLP markers to follow the transmission of ctDNA from parents to their F_1 offspring. Only maternal plastid markers were seen in the F_1 progeny of each cross examined, irrespective of the pollen ctDNA level of the male parent. The same result was obtained for F_1 progeny produced from crosses using pollen characterized by comparatively high ctDNA content, even when offspring were sampled at early developmental stages. Thus, there appeared to be little correspondence between pollen cytological data indicating potential paternal plastid transmission and data from molecular markers studies confirming that *P. sativum* generally follows a uniparental mode of plastid inheritance. Insufficient F_1 progeny was examined to exclude instances of trace biparentalism.

Brosche *et al.* (1999) identified sixteen ultraviolet-B (UV-B) radiation-regulated pea genes and divided functionally, the corresponding proteins into 4 groups, namely, Chloroplast-localized proteins, protein turnover enzymes, proteins involved intrace-llular signalling and phenyl propanoid or flavanoid biosynthesis. Efficient genome mapping was demonstrated through a combination of bulked segregant analysis (BSA) with DNA amplification fingerprinting (DAF) by Men *et al.* (1999). McCallum *et al.* (1997) studied developmental, environmental and genetic factors affecting seed colour in the progeny of a cross between 2 white-flowered (*aa*) green cotyledon peas using the pale large-seeded cultivar Primo and the greener small-seeded Prussian Blue OSU442-15 using QTL mapping.

Molecular markers have been successfully used to map genes controlling economically important traits including resistance to powdery mildew, *Aschochyta* blight, mosaic virus, *etc.* Tiwari *et al.* (1998) analyzed resistance to powdery mildew (*Erysiphe pisi*) and identified three RAPD markers linked to *er-1*, and 4 amplified fragment length polymorphism markers linked to *er-2*.

12.2.2 Identification of Pathogens and Genes of Resistance

Dirlewanger *et al.* (1994) analyzed the F_2 population of *P. sativum* consisting of 174 plants by RFLP and random amplified polymorphic DNA (RAPD). *Ascochyta pisi* race C resistance, plant height, earliness and number of nodes were measured in order to map the genes responsible for their variation. They identified molecular markers linked to each resistance gene were found: *Fusarium oxysporum* wilt (6 *cM* from *Fw*), powdery mildew (*Erysiphe polygoni*) (11 *cM* from *er*) and bean yellow mosaic potyvirus (15 *cM* from *mo*). QTLs (quantitative trait loci) for *Ascochytapisi* race C resistance which were mapped, with most of the variation explained by only three chromosomal regions. Timmerman *et al.* (1993) determined the location of *sbm-1* on the *P. sativum* genetic map was determined by linkage analysis with eight syntenic molecular markers. A strong association was found between one of these patterns and resistance to pea seed-borne mosaic potyvirus. Yu *et al.* (1995) identified two RAPD markers linked to *en*, the gene conferring resistance to pea enation mosaic virus of pea.

12.3.0 Genetic Transformation

Several methods have been developed for delivering foreign DNA into plant cells. Among others they comprise *Agrobacterium*-mediated gene transfer, and vectorless methods such as chemically induced DNA uptake, electroporation-induced DNA uptake and particle bombardment.

Pea is a natural host for *Agrobacterium* (De Cleene and De Ley, 1976), and both the bacterial strain and plant cultivar used have been shown to influence the tumour response in a number of inoculation studies (Bercetche *et al.*, 1987; Haws *et al.*, 1989; Hobbs *et al.*, 1989; Hussey *et al.*, 1989; Puonti-Kaerlas *et al.*, 1989; Robbs *et al.*, 1991). The *Agrobacterium* strains A281 and C58 are the most efficient for inducing tumour formation in pea, while the octopine strains appear to be either avirulent, or only slightly virulent (Hobbs *et al.*, 1989; Hussey *et al.*, 1989; Puonti-Kaerlas *et al.*, 1989; De Kathen and Jacobsen, 1990; Robbs *et al.*, 1991). Results of *in planta* inoculations, however, can seldom be used as reliable indicators of *in vitro* transformation, and thus screening for optimum combinations for production of transgenic plants should not be done solely by screening tumour induction frequency of wild-type strains (Puonti-Kaerlas *et al.*, 1989; 1990; Hobbs *et al.*, 1989).

A large amount of both basic and applied studies on genetic transformation has been reported in pea. *Agrobacterium*-mediated transformation has been used in most of the studies. Here, only certain applied results are presented. Readers are advised to refer the chapter by Modi *et al.* (2001) for additional information. Hussey *et al.* (1989) achieved transformation of the meristematic cell in the shoot apex of cultured pea shoots by *A. tumefaciens* and *A. rhizogenes*-mediated gene transfer. Jordan and Hobbs (1993) used a rapid regeneration system for studies of *Agrobacterium*-mediated transformation in pea using cotyledonary node explants. However, attempts to repeat this procedure were unsuccessful, probably due to a low number of explant cells competent for both regeneration and transformation by the *Agrobacterium* strains used. Paunti-Kaerlas *et al.* (1990) developed a static transformation system of pea using *Agrobacterium*-mediated gene transfer that allowed regeneration of fertile transgenic pea plants from transformed calli selected for antibiotic resistance.

Explants from axenic shoot cultures were cocultivated with disarmed *A. tumefaciens* carrying hygromycin phosphotransferase gene as selectable marker. DNA and RNA analyses of the calli, regenerated plants and their progenies confirmed their transgenic nature. Nadolska-Orczyk and Orczyk (2000) studied the factors influencing the efficiency of *Agrobacterium*-mediated transformation of pea using a highly efficient, direct regeneration system namely the virulence of *Agrobacterium* and the efficacy of selection agents.

Proteinase inhibitors have been used to increase resistance to insect pests in transge-nic plant (Charity *et al.*, 1999). Skot *et al.* (1994) used transgenic *Rhizobia* in a novel approach to the biological control of *Sitona*. Two transcriptional fusions were made in which the coding sequence of the insecticidal crystal protein gene (*cry iiia*) from *Bacillus thuringiensis* subsp. *tenebrionis* was fused to either the promoter of the rhizosphere enhanced *rhiA* gene from *R. leguminosarum* bv. *viciae*, or the promoter of the nodule specific *nifH* gene from *R. leguminosarum* bv. *trifolii*. The 2 chimeric genes were transferred to *R. leguminosarum* bv. *viciae* and *R. leguminosarum* bv. *trifolii*. But, they concluded that a toxin with much higher toxicity against larvae of *Sitona* is necessary for this approach to succeed as a biological control method.

Another use of transformation has been to induce resistance to viral diseases. Chowrira *et al.* (1998) tested transformed peas expressing the pea enation mosaic enamovirus (PEMV) coat protein gene (PEMV-CP) for their resistance of PEMV infection peas (var. Sparkle). The plants were transformed *in planta* by injection/electroporation of axillary meristems with a chimeric *pemv* coat protein gene construct. R1 progenies of these plants were shown to harbour the transgene. They found that PEMV coat protein mediated resistance can reduce virus replication, and may provide economic levels of protection against PEMV. Transgenic pea lines carrying the replicase (*nib*) gene of pea seed-borne mosaic potyvirus (PSbNV) were generated used in experiments to determine the effectiveness of induced resistance upon heterologous isolates (Jones *et al.*, 1998). Three pea lines showed inducible resistance in which an initial infection by the homologous isolate (PSbMV-DPDI) was followed by a highly resistant state. Resistance was observed in plants in either the homozygous or hemizygous condition and resulted in no overall yield loss despite the initial infection. Resistance was associated with a loss of both viral and transcriptional gene silencing.

13.0 References

Abd-Alla, I.M., Abdelfattah, M.A. and El-Hafez, A.A.A. (1972) *Bietragezur Tropischen and Subtropischen Land Wirtscheft and Topenveterinarmedizin*, **10**: 215-238.

Abd-El Moneim, A.M., Cocks, P.S. and Mawlawy, B. (1990) *Plant Breed.*, **104**: 231-240.

Ageev, V.V. and Demkin, V.I. (1987) *Agrokhimiya*, No. 7, pp. 61-67.

Ahmad, S., Kaur, S., Lamb-Palmer, N.D., Lefsrud, M. and Singh, J. (2015) *Crop J.*, **3**: 238-245.

Ali, M., Zaid, M.M. and Yahya, S.F. (2014) *Int. J. Chem. Env. Bio. Sci.*, **2**: 1.

Ali, S.M., Sharma, B. and Ambrose, M.J. (1994) *Euphytica*, **73**: 115-126.

Ambrose, M.J. (1995) *Diversity*, **11**: 118–119.

Amelin, A.V. and Parakhin, N.V. (2003) *Kormoproizvodstvo*, **2**: 20-25.

Amelin, A.V., Obraztsov, A.S., Lakhanov, A.P. and Uvarov, V.N. (1991) *Selektsiya-i-Semenovodstvo-Moskva*, **2**: 21-23.

Amin, K.S., Pal, A.B. and Brahmappa (1981) *Indian Phytopath.*, **34**: 77-79.

Annandale, J.G., Campbell, G.S., Olivier, F.C. and Jovanovic, N.Z. (1999) *Irrigation Sci.*, **19**: 65-72.

Annicchiarico, P. and Iannucci, A. (2008) *Field Crops Res.*, **108**: 133–142.

Anonymous (1989) *Bulletin of Taichung District Agricultural Improvement Station*, No. 22, pp. 1-2.

Anurag, R.K., Manjunatha, M., Jha, S.N. and Kumari, L. (2016) *J. Food Sci. Technol.*, **53**: 1640–1648.

Anzaldua-morales, A., Brusewitz, G.H. and Anderson, J.A. (1999) *J. Food Sci.*, **64**: 332–335.

Aziz-Ur-Rahman, R. Rathour, Viveka Katoch and Rana, S.S. (2018) *Int. J. Curr. Microbiol. App. Sci.*, **7**: 1441-1450.

Bachatly, M.A. and Malak, V.S.G.A. (2001) *Egypt. J. Agric. Res.*, **79**: 489-497.

Bagget, J.R. and Kean, D. (1987) *HortScience*, **22**: 330-331.

Banga, R.S., Yadav, A. and Malik, R.S. (1998) *Indian J. Weed Sci.*, **30**: 145-48.

Bardner, R., Fletcher, K.E. and Griffiths, D.C. (1983) *J. Agric. Sci.*, **101**: 71–80.

Barilli, E., Satovic, Z., Rubiales, D. and Torres, A.M. (2010) *Euphytica*, **175**: 151–159.

Barlow, C.A. and Messmer, I. (1982) *Econ. J. Ento.*, **75**: 765-768.

Barlow, C.A., Randolph, P.A. and Randolph, J.C. (1977) *Canadian Ento.*, **109**: 1491-1502.

Basterrechea, M. and Hicks, J.R. (1991) *Sci. Hort.*, **48**: 1–8.

Basu, P.K., Brown, N.J., Cre ˆte, R., Gourley, C.O., Johnston, H.W., Pepin, H.S. and Seaman, W.L. (1976) *Can. Plant Dis. Surv.*, **56**: 25–32.

Basuchaudhuri, P. (2016) *Indian J. Plant Sci.*, **5**: 25–38.

Baswana, K.S. and Legha, P.K. (1995) *Indian J. Agron.*, **40**: 139-140.

Bazhina, T.A, (1985) In: *SelektsiyaiSortorayaagrotekhnika Zernovykh Kul'turna Severo-Vostoke Nechernozemno zany RSFSR*, Kirov, USSR, pp. 27-30.

Beattie, G.A. and Lindow, S.E. (1995) *Phytopathol.*, **33**: 145–172.

Beattie, G.A. and Lindow, S.E. (1999) *Phytopathol.*, **89**: 353–359.

Behringer, F.J. (1991) *Dissertation Abst., Internatl. B, Sciences and Engineering*, **52**: 607B-608B.

Belford, R.K., Connell, R.Q., Thompson, R.J. and Dennis, C.W. (1980) *J. Sci. Food and Agri.*, U.K., **31**: 857-869.

Benjamin, J.G. and Nielsen, D.C. (2006) *Field Crops Res.*, **97**: 248–253.

Ben-Zeiev, N. and Zohary, D. (1973) *Israel J. Bot.*, **22**: 73-91.

Bercetche, S., Chirqui, D., Adam, S. and David, C. (1987) *Plant Sci.*, **52**: 195-210.

Bevan, J.R., Taylor, J.D., Crute, I.R., Hunter, P.J. and Vivian, A. (1995) *Plant Pathol.*, **44**: 98–108.

Bhalla, G.S., Singh, B. and Sandhu, G.S. (1974) *Haryana J. Hort. Sci.*, **3**: 207-209.

Bhattacharyya, M.K., Smith, A.M. and Ellis, T.H.N. (1990) *Cell*, **60**: 115–122.

Bijjur, S. and Verma, S. (1997) *Pesti. Res. J.*, **9**: 25-31.

Bintcliffe, E.J.B. and Wratten, S.D. (1980) *Tests of Agronomicals and Cultivars*, **1**: 52-53.

Birk, Y. (1985) *Int. J. Pept. Prot. Res.*, **25**: 113–131.

Bisen, A.L., Saraf, R.K. and Joshi, G.C. (1991) *Orissa J. Hort.*, **19**: 57-63.

Bisseling, T., Staveren, W.Van and Cammen, V.Van (1980) *Biochemical and Biophysical Research Communications*, **93**: 687-693.

Blackshaw, R.E. (1998) *Postemergence Weed Control in Pea (Pisum sativum) with Imazamox Weed Technology*, pp. 64-68.

Bland, B.F. (1971) *Crop Production: Cereals and Legumes*, Academic Press, London and New York.

Blixt, S. (1972) *Agri. Hort. Genet., Landskrona*, **30**: 1-293.

Blomfield, P.D. (1958) *New Zealand J. Sci. Techl.*, **36**: 46.

Bodah, E.T., Porter, L.D., Chaves, B. and Dhingra, A. (2016) *Euphytica*, **208**: 63–72.

Boiteau, G. and Osborn, W.P.L. (1997) *Canadian Entomologist*, **129**: 241-249.

Boros, L. and Sawicki, J. (2001) *Instytut Hodowli i Aklimatyzacji Roslin, Radzikowie*, **216**: 417-423.

Boyer, J.S. (1982) *Science*, **218**: 443–448.

Bradnock, W.T. and Mathews, S. (1970) *Hort. Res.*, **10**: 50-58.

Brar, P.S., Dhall, R.K. and Dinesh (2012) *Veg. Sci.*, **39**: 51-54.

Bressani, R. and Elias, L.G. (1988) In: Summerfield, R.J. (Ed.), *World Crops: Cool Season Food Legumes*, Kluwer Academic Publishers, Dordrecht, The Netherlands, pp. 381-404.

Brosche, M., Fant, C., Bergkvist, S.W., Strid, H., Svensk, A., Olsson, O. and Strid, A. (1999) *Biochimica et Biophysica Acta, Gene Structure and Expression*, **1447**: 185-198.

Bundy, J.W. (1971) In: Wareing, C.P.F and Cooper, J.C. (Eds.), *Potential Crop Production*, Heinamann, London.

Burstin, J., Salloignon, P., Chabert-Martinello, M., Magnin-Robert, J.B., Siol, M., Jacquin, F., Chauveau, A., Pont, C., Aubert, G., Delaitre, C., Truntzer, C. and Duc, G. (2015) *BMC Genomics*, **16**: 105.

Campbell, C.G. (1997) *Grass Pea, Lathyrus sativus L., Rome, Gatersleben/IPGRI.* p. 92.

Cannell, R., Gales, K., Snaydon, R. and Suhail, B. (2008) *Annals Appl. Biol.*, **93**: 327-335.

Canner. (1945) *Canner*, **101**: 11.

Cárcamo, H.A., Vankosky, M.A., Wijerathna, A., Olfert, O.O., Meers, S.B. and Evenden, M.L. (2018) *Ann. Entomol. Soc. Am.*, **111**: 144–153.

Casey, R. and Short, M.N. (1981) *Phytochemistry*, **20**: 2–23.

Casey, R., Domoney, C. and Forster, C. (1998) *J. Plant Physiol.*, **152**: 636–640.

Cebula, S., Wajtaszek, T. and Poniedzialek, M. (1987) *Zeszyty Naukowe Akademii Rolniczejim Hugona Kollataja W Krakowle, Ogrodnictwo*, **211**: 185-200.

Chamberland, E. (1982) *Canada J. Soil Sci.*, **62**: 663-672.

Chan, P.K. and Gresshoff, P.M. (2009) *Roles of Plant Hormones in Legume Nodulation. Encyclopedia of Life Support Systems (EOLSS) Biotechnology, Vol. VIII.*

Charity, J.A., Anderson, M.A., Bittisnich, D.J., Whitecross, M. and Higgins, T.J.V. (1999) *Mol. Breed.*, **5**: 357-365.

Charlton, A.J., Donarski, J.A., Harrison, M., Jones, S.A., Godward, J. and Oehlschlager, S. (2008) *Metabolomics*, **4**: 312–327.

Chaubey, C.N. (1977) *Indian J. Agric. Res.*, **11**: 119-121.

Chaudhary, J. and Banyal, D.K. (2017) *Indian Phytopath.*, **70**: 69-74.

Chauhan, D.V.S. (1968) *Vegetable Production in India*, Ram Parsad and Sons, Agra, India.

Chekrygin, P.M. (1987) *Selektsiya I Semenovodstvo, kiev*, **63**: 52-54.

Cheng, P., Holdsworth, W., Ma, Y., Coyne, C.J., Mazourek, M. and Grusak M.A. (2015) *Mol. Breed.*, **35**: 75.

Chittem, K.R., Porter, L., McPhee, K., Khan, M. and Goswami, R.S. (2010) *Phytopathology*, **100**: S25.

Choudhury, B. (1967) *Vegetables*, National Book Trust, India, New Delhi.

Choudhury, B. and Ramphal (1975) *Indian Hort.*, **19**: 25-27.

Chowrira, G.M., Cavileer, T.D., Gupta, S.K., Lurquin, P.F. and Berger, P.H. (1998) *Transgenic Res.*, **7**: 265-271.

Clarke, R. (1990) *Plant Protect.*, **5**: 160–161.

Clarkson, J.D.S. (1978) *Plant Pathol.*, **27**: 110–117.

Clement, S.L., Sharaf El-Din, N., Weigand, S. and Lateef, S.S. (1994) *Euphytica*, **73**: 41-50.

Clemente, A. and Olias, R. (2017) *Curr. Opin. Food Sci.*, 14: 32–36.

Clemente, A., Arques, M.C. and Dalmais, M. (2015) *PLoS One*, **10**: e0134634.

Clemente, A., Carmen Marín-Manzano, M., Jiménez, E., Carmen Arques, M. and Domoney, C. (2012) *Br. J. Nutr.*, **108**: 135–144.

Clemente, A., Gee, J.M., Johnson, I.T., Mackenzie, D.A. and Domoney, C. (2005) *J. Agric. Food Chem.*, **53**: 8979–8986.

Cousing, R. (1965) *Annales de l'Amélioration des Plantes*, **15**: 93–97.

Crowell, H.H. (1975) *J. Econ. Ento.*, **68**: 275-276.

Csizmadia, L. (1990) *ZoldsegtermesztesiKutatoIntezetBulletinje*, **23**: 79-87.

D'ella, A., Roberti, R., Brunelli, A. and Cisce, A. (1998) *Att. Gio Fitopat*, pp. 661-666.

Dar, W.D. and Gowda, C.L.L. (2010) *5th Abstract Book of the "International Food Legumes Research Conference and 7th European Conference on Grain Legumes," Antalya, April 26–30, 2010.*

Das, S.K. (2016) *J. Crop Weed*, **12**: 110-115.

David, B.V. and Ramamurthy, V.V. (2011) *Ele. Econ. Entomol.*, p. 190.

Davies, D.R., Berry, G.J., Health, M.C. and Dawkins, T.C.K. (1985) In: *Pea (Pisum sativum* L.) (R.J. Summerfield and E.H. Roberts eds), Williams Collins Sons and Co. Ltd., London, UK, pp. 147-198.

Day, A. (1996) *Postharvest News Inf.*, **7**: 31–34

De Candolle, A. (1896) *Origin of Cultivated Plants*, Hafner Publishing Co., New York.

De Cleene, M. and De Ley, J. (1976) The host range of crown gall. Bot. Rev. **42**: 389-466.

De Kathen, A. and Jacobsen, H.J. (1990) *Plant Cell Rep.*, **9**: 276-279.

Delgado, M.J., Ligero, F. and Lluch, C. (1994) *Soil Biol. Biochem.*, **26**: 371–376.

Denholm, I., Horowitz, M., Cahill, M. and Ishaaya, I. (1998) In: *Ishaaya, I. and Degheele, D. (Eds.) Insecticides with novel modes of action; mechanisms and application*, pp. 260-282.

Desai, A.R., Gonde, A.D., Raut, S.A. and Kumar, A. (2013) *Bioinfolet-A Quar. J. Life Sci.*, **10**: 945-946.

Devi, J., Mishra, G.P., Sanwal, S.K., Dubey, R.K., Singh, P.M. and Singh, B. (2018) *PLoS One*, **13**: e0201235.

Dhaka, S.S., Singh, G., Ali, N., Mittal, V. and Singh, D.V. (2011b) *Agriculture Research Information Center, Hisar, India*, **42**: 331-335.

Dhesi, N.S. and Namdpuri, K.S. (1965) *Vegetable Growing in Punjab*, Punjab Agric. Univ., Ludhiana.

Dhillon, G.S., Singh, T. and Singh, M. (1989) *J. Res., Punjab Agric. Univ.*, **26**: 733.

Diapari, M., Sindhu, A., Warkentin, T.D., Bett, K. and Tar'An, B. (2015) *Mol. Breed.*, **35**: 30.

Dilawari, V.K. and Dhaliwal, G.S. (1993) In: *Dhaliwal, G.S. and Dilawari, V.K. (Eds.) Advances in Host Plant Resistance to Insects. Kalyani Publishers, New Delhi, India*, pp.394- 421.

Dirlewanger, E., Isaac, P.G., Ranade, S., Belajouza, M., Cousin, R. and De-Vienne, D. (1994) *Theo. Appl. Genet.*, **88**: 17-27.

Dolgopolova, L.N. and Lakhanov, A.P. (1979) *Khimiya v Sol'skankhozyaistve*, **17**: 27-29.

Domoney, C., Knox, M. and Moreau, C. (2013) *Func. Plant Biol.*, **40**: 1261–1270.

Duc, G., Marget, P., Page, D. and Domoney, C. (2004) In: *Muzquiz, M., Hill, G.D., Cuadrado, C., Pedrosa, M.M. and Burbano, C. (Eds.), Recent Advances of Research in Antinutritional Factors in Legume Seeds and Oilseeds, Proceedings of the Fourth International Workshop on Antinutritional Factors in Legume Seeds and Oilseeds,* pp. 281–285.

Duke, J.A. (1981) *Hand Book of Legumes of World Economic Importance*, Plenum Press, New York, pp. 199-265.

Eastin, J.A. and Gritton, E.T. (1969) *Agron. J.*, **61**: 612-615.

Egamberdieva, D. and Lugtenberg, B. (2014) In: Miransari, M. (Ed.) *Use of Microbes for the Alleviation of Soil Stresses. Springer Science+Business Media, New York*, pp. 73–96.

El-Aassar, M.R., Hafez, E.E., El-Deeb, N.M. and Fouda, M.M. (2014) *Int. J. Biol. Macromol.*, **69**: 88–94.

El-Beheidi, M.A., El-Mansy, A.A. and Khalil, M.A.L. (1978) *Annals Agric. Sci.*, **9**: 169-186.

El-Behidi, M.A., Gad, A.A., El-Sawah, M.H. and Elhady, H.M. (1985) *Zoldsegtermesztesikutato Intezet Bulletinje*, **18**: 17-25.

Elbert, A., Nauen, R. and Leicht, W. (1998) In: *Ishaaya, I. and Degheele, D. (Eds.) Insecticides with novel modes of action; mechanisms and application*, pp. 50-73.

Ellis, T.H.N., Hofer, J.M.I., Timmerman-Vaughan, G.M., Coyne, C.J. and Hellens, R.P. (2011) *Trends Plant Sci.*, **16**: 590–596.

El-Murabaa, A.I., Waly, E.A., Abdel-Aal, S.A. and Zayed, G.A. (1988) *Assiut J. Agric. Sci.*, **19**: 223-233.

Elneklawy, A.S., El-Maksoud, H.K.A. and Selim, A.M. (1985) *Annals of Agric. Sci.*, **23**: 1365-1373.

El-Rokiek, K.G., El-Din, S.A.S. and El-Wakeel, M.A. (2019) *Bull. Natl. Res. Cent.* **43**: 193.

Elvira-Recuenco, M. and Taylor, J.D. (2001) *Euphytica*, **118**: 305–311.

Etebu, E. and Osborn, A.M. (2010) *Phytoparasitica*, **38**: 447–454.

Evans, L.S., Lewis, K.F. and Vella, F.A. (1980) *Plant Soil*, **56**: 71-80.

FAO. (2016) *Food and Agriculture Organization, Definition and classification of commodities. 4. Pulses and derived products.*

Farrington, P.A. (1974) *J. Aust. Inst. Agric. Sci.*, **40**: 99-108.

Feng, J., Hwang, R., Chang, K.F., Conner, R.L., Hwang, S.F., Strelkov, S.E., Gossen, B.D., McLaren, D.L. and Xue, A.G. (2013) *Can. J. Plant Sci.*, **91**: 199–204.

Feng, J., Hwang, R., Chang, K.F., Hwang, S.F., Strelkov, S.E., Gossen, B.D., Conner, R.L. and Turnbull, G.D. (2010) *Plant Pathol.*, **59**: 845–852.

Fernandez Pascual, M., Delernzoc, Pozuelo. J.M. and Defelipe, M.R. (1992) *J. Pl. Physiol.*, **140**: 385-390.

Fondevilla, S., Carver, T.L.W., Moreno, M.T. and Rubiales, D. (2007a) *Plant Breed.*, **126**: 113–119.

Fondevilla, S., Carver, T.L.W., Moreno, M.T. and Rubiales, D. (2006) *Eur. J. Plant Pathol.*, **115**: 309–321.

Fondevilla, S., Cubero, J.I. and Rubiales, D. (2010) *Plant Breed.*, doi: 10.1111/j.1439 0523.2010.01769.x.

Fougereux, J.A., Dore, T., Ladonne, F. and Fleury, A. (1997) *Crop Sci.*, **37**: 1247-1252.

Frank, P.A. and Grigs, B.H. (1957) *Weeds*, **5**: 206-217.

Freeman, A., Spackman, M., Aftab, M., McQueen, V., King, S., van Leur, J.A.G., Loh, M.H. and Rodoni, B. (2013) *Australas. Plant Path.*, **42** (in press).

Frew, T.J., Russell, A.C. and Timmerman-Vaughan, G.M. (2002) *Plant Breed.*, **121**: 512–516.

Fujioka, T., Fujita, M. and Iwamoto, K. (2000) *J. Japanese Soc. Hort. Sci.*, **69**: 656-658.

Fujioka, T., Fujita, M., Miyamoto, Y. (1999) *J. Japanese Soc. Hortic. Sci.*, **68**: 384-390.

Gamborg, O., Constabel, F. and Shyluk, J. (1974). *Physiologia Plantarum.*, **30**: 125-128.

Gangopadhyay, S. and Kapoor, K.S. (1975) *Veg. Sci.*, **2**: 28-30.

Gangopadhyay, S. and Kapoor, K.S. (1976) *Veg. Sci.*, **3**: 74-78.

Gangwar, M.S., Singh, H.N., Singh, S., Singh, K. and Gupta, R.A. (1998) *Ann. Agri. Res.*, **19**: 386-389.

Gaw³owska, M., Swie zcicki, W. and Lahuta, L. (2017) *Genet. Resour. Crop Evol.*, **64**: 569–578.

Gencer, L. (2004) *Turk. J. Zool.*, **28**: 119-122.

Ghafoor, A. and McPhee, K. (2012) *Euphytica*, **186**: 593–607.

Ghanem, S.A., El-Bahr, M.K., Saker, M.M. and Badr, A. (1996) *Giorn. Bot. Ital.*, 130: 695-705.

Gladish, D.K. and Rost, T.L. (1993) *Envir. Expt. Bot.*, **33**: 243-258.

Glaman, G., Bucurescu, I. and Vilceanu, G. (1989) *ProductiaVegetalaHortic.*, **38**: 9-13.

Gogoi, A.K., Kalita, H., Pathak, A.K. and Deka, J. (1991) *Indian J. Agron.*, **36**: 287-288.

Gonzalez, A., Gonzalez, C., Murua and Royuela, M. (1996) *Weed Sci.*, **44**: 31-37.

Goulden, D.S. and Scott, R.E. (1993) *New Zealand J. Crop and Hort. Sci.*, **21**: 265-266.

Govorov, L.I. (1928) *Bull. Appl. Bot. Genet. and Pl. Breed.*, **19**: 497-522.

Griffiths, D.C., Bardner, R. and Bater, J. (1986) *FABIS Newsl.*, **14**: 30–33.

Gritton, E.T. (1975) *Crop Sci.*, **15**: 453-457.

Gritton, E.T. (1980) *Hybridization Crop Plants* (W.R. Fehr and H.H. Hadley Eds.), *Am. Soc. Agron. and Crop Sci. Soc.*, Pub. Madison, Wisconsin, USA.

Gritton, E.T. (1980) In: *American Society of Agronomy*, (W.R. Fehr and H.H. Hadley eds.) Inc., and Crop Science Society of America, Inc., Wisconsin, USA, pp. 347-356.

Gritton, E.T. (1986) *Pea breeding*, In: Breeding Vegetable Crops (M.J. Bassett ed.), AVI Pub. Comp. Inc., Connecticut, USA.

Grunwald, N.J., Coffman, V.A. and Kraft, J.M. (2003) *Plant Dis.*, **87**: 1197–1200.

Grylls, N.E. (1972) *Australian J. Experi. Agri. Ani. Husb.*, **12**: 668-674.

Guilioni, L., We ´ry, J. and Lecoeur, J. (2003) *Funct. Plant Biol.*, **30**: 1151–1164.

Gukova, M.M. and Arbuzova, I.N. (1969) *IzvestiaTimirjazev. Cel'hoz. Akad*, No. 2, pp. 90-98.

Gupta, R.K., Gangoliya, S.S. and Singh, N.K. (2015) *J. Food Sci. Technol.*, 52: 676–684.

Gupta, Y.P. (1983) *Acta AgronomicaAcademiaeScientiarumHungarieae*, **32**: 180-182.

Hagedorn, D.J. (1973) *Breeding Plants for Disease Resistance* (Nelson ed.), Penn. State Univ. Press, USA.

Hagedorn, D.J. (1984). *Am. Phytopath. Soc., St. Paul, Minn.* p. 30.

Hagedorn, D.J. and Gritton, E.T. (1973) *Phytopathology*, **63**: 1130-1133.

Hagedorn, D.J. and Hampton, R.O. (1975) *Plant Dis. Reptr.*, **59**: 895-899.

Haglad, W.A. and Kraft, J.M. (1979) *Phytopathol.*, **69**: 818-820.

Haglund, W.A. and Anderson, W.C. (1987) *HortSci.*, **22**: 513-514.

Haglund, W.A. and Kraft, J.M. (1970) *Phytopath.*, **60**: 1861-1862.

Haiying, W.Z., Shaozhi, Z. and Guangming, C. (2007) *LWT Food Sci. Technol.*, **40**: 1112–1116.

Hall, B.J. (1962) *Agri. Gaz. N.S.W.*, **73**: 15-17.

Hamid, A., Bhat, N.A., Sofi, T.A., Bhat, K.A. and Asif, M. (2012) *Afr. J. Microbiol. Res.*, **6**: 7156-7161.

Hamon, C., Coyne, C.J., McGee, R.J., Lesné, A., Esnault, R. and Mangin, P. (2013) *BMC Plant Biol.*, **13**: 45.

Hampton, R.O., Jensen, A. and Hagel, G.T. (2005) *J. Econ. Entomol.*, **98**: 1816–1823.

Hampton, R.O., Mank, G.A., Hamilton, R.I., Draft, J.M. and Meuhlbauer, F. (1976) *Plant Dis. Reptr.*, **60**: 455-459.

Hanolo, W. and Pulung, M.A. (1994) *Acta Horti.*, **369**: 335-339.

Hardas, M.G. and Deshmukh, N.Y. (1963) *Indian J. Agron.*, **8**: 341-344.

Hare, W.W., Walker, J.C. and Delwiche, E.J. (1949) *J. Agr. Res.*, **78**: 239–251.

Harker, K.N. (2001). *Canadian J. Pl. Sci.*, **81**: 33942.

Harland, S.C. (1948) *Heredity*, **2**: 263-269.

Harrewijn, P. and Kayser, H. (1997) *Pesticide Sci.*, **49**: 130-140.

Hascoët, E., Jaminon, O., Devaux, C., Blassiau, C., Bahrman, N. and Bochard, A.M. (2014) *Towards fine mapping of frost tolerance QTLs in pea, in 2nd PeaMUST Annual Meeting (Dijon).*

Havilckova, H. (1980) *Entomologia Experimentalis et Applicata*, **27**: 287-292.

Hawes, M.C., Robbs, S.L. and Pueppke, S.G. (1989) *Plant Physiol.*, **90**: 180-184.

Hebblethwaite, P.D. and McGowan, M. (1980) *J. Sci. Food Agric.*, **31**: 1131-1142.

Hegedus, O. and Trnka, L. (1989) *Shornik UVTIZ, Zahradnictvi*, **16**: 147-150.

Heringa, R.J., Van Norel, A. and Tazelaar, M.F. (1969) *Euphytica*, **18**: 163–169.

Highkin, A,R. and Lang, A. (1966) *Planta*, **68**: 94-98.

Highkin, H.R. (1960) *Cold Spring Harb. Symp. Quant. Bio.*, **25**: 231-238.

Hillman, H.D. (1954) *Saatgutwirtschafe*, **6**: 13-14.

Hirano, S.S. and Upper, C.D. (2000) *Microbiol. Mol. Biol. Rev.*, **64**: 624–653.

Hobbs, S.L.A., Jackson, J.A. and Mahon, J.D. (1989) *Plant Cell Rep.*, **8**: 274-277.

Holland, M.R. (1990) *Diss. Abstr. Int. –B Sci. Eng.*, **50**: 4284B.

Hollomon, D.W. and Wheeler, I.E. (2002) In: Bélanger, R.R., Bushnell, W.R., Dik, A.J. and Carver, T.L.W. (Eds.) *The powdery mildews, a comprehensive treatise. APS Press, St. Paul*, pp. 249–255.

Hooda, I. and Parashar, R.D. (1985) *Veg. Sci.*, **12**: 42-44.

Horowitz, A.R. and Ishaaya, I. (2004). *Insect pest management: field and protected crops. Springer-Verlag.*

Huisman, J. and van der Poel, A.F.B. (1994) In: *Expanding the Production and Use of Cool Season Food Legumes* (F.J. Muehlbauer and W.J. Kaiser eds.), Kluwer Academic Publishers, Dordrecht, The Netherlands, pp. 53-76.

Hulse, J.H. (1994) In: *Expanding the Production and Use of Cool Season Food Legumes* (F.J. Muehlbauer and W.J. Kaiser eds.), Kluwer Academic Publishers, Dordrecht, The Netherlands, pp. 77-97.

Humphry, M., Reinstädler, A., Ivanov, S., Bisseling, T. and Panstruga, R. (2011) *Mol. Plant Pathol.*, doi: 10.1111/J.1364-3703.2011.00718.X

Hurrell, R. and Egli, I. (2010) *Am. J. Clin. Nutr.*, **91**: 1461S–7S.

Husnain, S.K., Khan, S.H., Atiq, M., Rajput, N.A., Abbas, W. and Mohsin, M. (2019) *Pak. J. Phytopathol.*, **31**: 89-96.

Hussain, A., Malukhera, A. and Habib, M. (1976) *J. Agric. Res.*, **14**: 92-93.

Hussey, G. and Gunn, H.V. (1984) *Plant Science Letters*, **37**: 143-148.

Hussey, G., Johnson, R.D. and Warren, V. (1989) *Protoplasma*, **148**: 101-105.

Hussy, G. and Gunn, V.H. (1984) *Plant Sci. Lett.*, **37**: 143-148.

Iwaya-Inoue, M., Motooka, K., Ishibashi, Y. and Fukuyama, M. (2003) *J. Facul. Agricult. Kyushu. Univ.*, **48**: 29–38.

Jackson, A.J. and Hobbs, A.S. (1990) *In vitro Cell Dev. Biol.*, **26**: 835-838.

Jadhav, P.B. (2018) *Int. J. Curr. Res.*, **10**: 66167-66170.

Jain, N.L. and Choudhury, B. (1963) *Indian J. Hort.*, **20**: 129-134.

Jain, S., Weeden, N.F., Porter, L.D., Eigenbrode, S.D. and McPhee, K. (2013) *Crop Sci.*, **53**: 2392–2399.

Jakobsen, I. (1985) *Physiol. Plantarum*, **64**: 190-196.

Jambunathan, R., Blain, H.L., Dhindsa, K.H., Hussein, L.A., Kogure, K., Li Juan, L. and Youssef (1994) In: *Expanding the Production and Use of Cool Season Food Legumes* (F.J. Muehlbauer and W.J. Kaiser eds.), Kluwer Academic Publishers, Dordrecht, The Netherlands, pp. 98-112.

Jana, S.K. and Paria, N.C. (1996) *Env. Ecol.*, **14**: 535-537.

Janick, J., Schery, R.W., Woods, F.W. and Rutten, V.W. (1969) *Plant Science, Freeman, San Fransisco.*

Jarak, M., Jelicic, Z., Milosevi c, N. and Govedarica, M. (1989) *Nauka u Praksi*, **19**: 117-124.

Jensen, E.S. (1989) *In Legumes in farming systems* (P. Plancquaert and Hagger eds.).

Jha, A.B., Tar'An, B., Diapari, M., Sindhu, A., Shunmugam, A. and Bett, K. (2015) *Euphytica*, **202**: 189–197.

Jindal, K. and Bhardwaj, S. (1989) *Plant Dis. Res.*, **4**: 165–166.

Johnson, K.W. and Hagedorn, D.J. (1958) *Phytopathology*, **48**: 451-453.

Johnstone, G.R. (1978) *Australian J. Agric. Res.*, **29**: 1003-1010.

Joi, M.B., Mote, U.N. and Sonone, H.N. (1975) *Res. J. Mahatma Phule Agric. Univ.*, **6**: 160-161.

Jones, A.L., Johanson, I.E., Bean, S.J., Bach, I. and Maule, A.J. (1998) *J. Gen. Virol.*, **79**: 1241-1255.

Jordan, M.C. and Hobbs, S.L.A. (1993) *In Vitro Cell. Dev. Biol. Plant.*, **29**: 77-82.

Jung, Y.S, Kim, Y.T., Yoo, S.J. and Kim, H.G. (1999) *Plant Pathol.*, **15**: 44–47.

Kader, A.A. (1992) In: *Kader, A.A. (ed.). Postharvest Technology of Horticultural Crops. Univ California Pub. 3311, USA*, pp. 85–92.

Kalia, P. and Sharma, S.K. (1988) *Theor. Appl. Genet.*,**76**: 795e799.

Kalinina, N.V., Veselkova, K.I., Vavilova, E.I. and Bushueva, R.A. (1985) In: *Selektsiya I sortovayaagrotekhuikaZernovykh Kultur naSevero-VostokeNecherno-ZemnoiZong RSFSR Kim*, USSR, pp. 30-37.

Kalloo, G. (1993) In: Kalloo, G. and Bergh, B.O. (Eds.), *Genetic Improvement of Vegetable Crops, Pergamon Press, Oxford and New York.* pp. 409-425.

Kalloo, G. (1998) *Vegetable Breeding, Vols. I-III (Combined ed.). Panima Edu. Book Agency.*

Kanaujia, S.P., Sharma, S.K. and Rastogi, K.B. (1998) *Ann. Agril. Res.*, **19**: 219-221.

Kapoor, A.S., Paul, Y.S. and Singh, A. (2006) *Indian Phytopathol.*, **59**: 467-474.

Karmaker, P.G. and Singh, R.P. (1990) *Veg. Sci.*, **17**: 95-98.

Kay, D. (1979) *TPI Crop and Product Digest*, No. 3, pp. 26-47.

Kellenbarger, S., Silveira, V., McCready, R.M., Owens, H.S. and Chapman, J.L. (1951) *Agron. J.*, **43**: 337-340.

Kennedy, A.R., Beazer-Barclay, Y. and Kinzler, K.W. (1996) *Cancer Res.*, **56**: 679–682.

Khan, S.A., Awais, A., Javed, N., Javaid, K., Moosa, A., Haq, I.U., Khan, N.A., Chattha, M.U. and Safdar, A. (2016) *Pak. J. Phytopathol.*, **28**: 127-131.

Khangildin, V. Kh. (1984) *In Genet. Selekts. issled. naUrale Inf. materialy, Sverdlork*, USSR, pp. 95-96.

Khatua, D.C., Maiti, S., Bandopadhyay, S. and Sen, C. (1977) *Indian J. Mycol. Pl. Path.*, **7**: 173-174.

Khetarpal, R.K., Maury, Y., Cousin, R., Burghofer, A. and Varma, A. (1990) *Annals Appl. Biol.*, **116**: 297-304.

King, J.M. (1981) *Proc. 1981 BCPC Pest and Diseases, 16–19 November 1981, Brighton, Englan*, pp. 327–331.

King, T.H., Davis, D.W., Shehata, M.A. and Pfleger, F.L. (1981) *HortScience*, **16**: 100.

Kof, E.M., Chuvasheva, E.S., Kefeli, V.I. and Kandykov, I.V. (1998) *Russian J. Plant Physiol.*, **45**:279-387.

Kohli, U.K., Thakur, I.K. and Shukla, Y.R. (1992) *Hortic. J.*, **5**: 59-61.

Kohli, U.K., Thakur, I.V. and Shukla, Y.R. (1992) *Ann. Agri. Res.*, **13**: 394-395.

Kolar, J.S. and Sandhu, K.S. (1989) *Indian Frng.*, **34**: 17-18.

Kolbert, Z., Bartha, B. and Erdei, L. (2008) *Physiol. Plant*, **133**: 406–416.

Kosev, V. and Othman, O.M. (2017) *Acad. J. Sci. Res.*, **5**: 685-691.

Kotasthane, S.R. (1975) *Sci. and Cult.*, **41**: 450-452.

Kraft, J.M. (1981) *Crop Sci.*, **21**: 352-353.

Kraft, J.M. (1988) *Crop Sci.*, **29**: 494-495.

Kraft, J.M. and Berry, J.W. (1972). *Plant Dis. Rept.*, **56**: 398-400.

Kraft, J.M. and Giles, R.A. (1978) *Crop Sci.*, **18**: 1098.

Kraft, J.M. and Hampton, R.O. (1980) *Plant Disease*, USA, **69**: 22-24.

Kraft, J.M. and Pfieger, F.L. (2006) *Compendium of pea diseases and pests, 2nd edn. APS Press, Saint Paul*.

Kraft, J.M. and Tuck, J.A. (1986) *Crop Sci.*, **26**: 1262–1263.

Krarup, A. and Davis, D.W. (1970) *J. Am. Soc. Hortic. Sci.*, **95**: 795-797.

Krarup, A. and Ross, E.O. (1979) *Agro. Sur.*, **7**: 84-88.

Krishna, H., Singh, A.K., Kumar, P. and Kumar, S. (2019) *J. Entomol. Zool. Std.*, **7**: 1250-1252.

Kudsk, P. (1992) *Shell. Agric.*, **13**: 29-30.

Kumar, A., Sharma, B.C., Nandan, B. and Sharma, P.K. (2009) *Indian J. Weed Sci.*, **41**: 23-26.

Kumar, K. and Rao, K.V.P. (1992) *Ann. agril. Res.*, **13**: 245-248.

Kumar, R., Gupta, K.R. and Singh, V.P. (1989) *Indian Fmg.*, **39**: 9-10.

Kuo, J.Y. (1988) *Bulletin of Taichung District Agricultural Improvement Station*, No. 20, pp. 49-60.

Kwon, S.J., Brown, A.F., Hu, J., McGee, R., Watt, C. and Kisha, T. (2012) *Genes Genomics*, **34**: 305–320.

Lacicowa, B. and Pieta, D. (1994) *Ann. Univ. Mar. Cu. Ekl. Sect. EEE Hort.*, **7**: 119-135.

Lamprecht, H. (1956) *Agri. Horti. Genet.*, *Landskrona*, **14**: 1-4.

Latham, L.J. and Jones, R.A.C. (2001a) *Australian J. Agric. Res.*, **52**: 397-413.

Lavaud, C., Lesné, A, Piriou, C., Le Roy, G., Boutet, G. and Moussart, A. (2015) *Theor. Appl. Genet.*, **128**: 2273–2288.

Lawson, H.N. and Rubens, T.G. (1970) *Proc. 10th Br. Weed Control Conf.*, pp. 638-645.

Lawyer, A. and Chun, W. (2001) In: *Kraft, J.M. and Pfleger, F.L. (Eds) Compendium of pea diseases and pests. The American Phytopathological Society, St. Paul*, pp. 22–23.

Le Gall, M., Quillien, L. and Seve, B. (2007) *J. Ani. Sci.*, **85**: 2972–2981.

Lejeune-Hénaut, I., Hanocq, E., Bethencourt, L., Fontaine, V., Delbreil, B. and Morin, J. (2008) *Theor. Appl. Genet.*, **116**: 1105–1116.

Leppyanen, I.V., Kirienko, A.N. and Dolgikh, E.A. (2019) *Peer J.*, **7**: e6552.

Lester, E.J. (1995) *Freezing effects on food quality. Marcel Dekker Inc, New York*.

Lewis, M.E. and Gritton, E.T. (1988) *Pisum Newsl.*, **20**: 20-21.

Lin, Y.S. (1991) *Plant Protection Bulletin* (Taipei), **33**: 34-36.

Linford, M.B. (1928) *Wisc. Agr. Expt. Sta. Res. Bull.*, **85**: 43.

Locke, J.C., Papavizas, G.C. and Lewis, J.A. (1979) *Plant Dis. Reptr.*, **63**: 725-788.

MacGillivray, J.J. (1961) *Vegetable Production*, McGraw-Hill Book Co., Inc., New York, Toronto, London.

Makasheva, R.K. and Drozd, A.M. (1987) *Pisum Newsl.*, **19**: 31–32.

Makasheva, R.Kh. (1983) *The Pea*, Oxonian Press Pvt. Ltd., New Delhi, India, p. 267.

Malik, K.A. and Saxena, P.K. (1992) *Australian J. Plant Physiol.*, **19**: 731-740.

Malik, R.S. and Bhandari, A.R. (1994) *Indian J. Agril. Sci.*, **64**: 847-849.

Malik, R.S. and Kumar, K. (1996) *J. Indian Soc. Soil Sci.*, **44**: 508-509.

Malik, R.S. and Kumar, K. (1997) *Agricultural Sci. Digest (Karnal)*, **17**: 129-131.

Mani, V.P. and Shridhar (1996) *J. Hill Res.*, **9**: 309-311.

Marginiski, L.K. and Kartha, K.K. (1981) *Plant Cell Rept.*, **1**: 64-66.

Marocchi, G. (1977) *InformatoreAgrario.*, **33**: 26101-26109.

Martyniuk, S., Oroñ, J. and Martyniuk, M. (2005) *Acta Soc. Bot. Polon.*, **74**: 83–86.

Marx, G.A. (1972) *Pisum News Letter*, **4**: 28-29.

Marx, G.A. (1974) *Pisum News Letter*, **6**: 60.

Marx, G.A. (1986) *Pisum Newsl.*, **18**: 45–48.

Massoud, M.B., Sakouhi, L. and Chaoui, A. (2019) *J. Plant Nutr.*, **42**: 1230–1242.

Materne, M., Leonforte, A., Hobson, K., Paull, J. and Gnanasambandam, A. (2011) In: *Pratap, A. and Kumar, J. (Eds.) Biology and Breeding of Food Legumes. CAB International*, pp. 49–62.

Mathur, A.C., Krishniah, K. and Tandon, P.L. (1974) *Indian J. Hort.*, **31**: 286-288.

Mathur, Y.K. and Upadhyay, K.D. (2006) A *Text Book of Entomology. Rama Publication, New Delhi*, pp. 189-191.

Matos, A.T., Carrijo, O.A., Guedes, A.C. and Ferreira, P.E. (1984) In: *Resumos, XXIV CongressoBrasileiro de Olericultura e I Reuniao Latino-Americana de Olericultra*, p. 158.

McCallum, J., Timmerman Vaughan, G., Frew, T. and Russell, A. (1997) *J. Amer. Soc. Hort. Sci.*, **122**: 218-225.

McClendon, M.T., Inglis, D.A., McPhee, K.E. and Coyne, C.J. (2002) *J. Am. Soc. Hortic. Sci.*, **127**: 602–607.

McCue, A.S. and Minotli, P.L. (1979) *Proc. Northeastern Weed Sci. Soc.*, **33**: 106.

McEwen, J., Bardner, R., Briggs, G.G., Bromilow, R.H., Cockbain, A.J., Day, J.M., Fletcher, K.E., Legg, B.J., Roughley, R.J. and Salt, G.A. (1981) *J. Agric. Sci.*, **96**: 129–150.

Meadley, J.T. and Milbourn, G.M. (1970) *J. Agric. Sci.*, **74**: 273-278.

Meagher, W.R., Johnson, C.M. and Stout, P.R. (1952) *Pl. Physiol.*, **27**: 223-230.

Men, A.E., Borisov, A.Y., Rozov, S.M. Ushakov, K.V., Tsyganov, V.E., Tikhonovich, I.a. and Gresshoff, P.M. (1999) *Theo. Appl. Genet.*, **98**: 929-936.

Menjkova, K.A. (1954) *Trud. Inst. Genet.*, **21**: 179-182.

Menzies, S.A. and Wright, M.J. (1976) *New Zealand Commercial Grower*, **31**: 11.

Mikanova, O., Kubat, J., Vorisek, K. and Randova, D. (1995) *RostlinnaVyroba*, **41**: 423-425.

Mishriky, J.F., El-Fadaly, K.A. and Badawi, M.A. (1990) *Bulletin, fac. Agric., Univ. of Cairo*, **41**:

785-797.

Modi, M.K., Barua, S.J.N. and Deka, P.C. (2001) *Leguminous vegetables*, (V.A. Parthasarathy, T.K. Bose and P.C. Deka eds.), Biotechnology of Horticultural Crops, Vol. 2, NayaProkash, Calcutta, pp. 284-325.

Moore, K.L., Rodrýguez-Ramiro, I. and Jones, E.R. (2018) *Sci. Rep.*, **8**: 6865.

Morita, M., Ueda, T., Moneda, T., Koyanagi, T. and Haga, T. (2007) *Pest Manag. Sci.*, **63**: 969-973.

Muehlbauer, F.J. (1987) *Crop Sci.*, **27**: 1089-1890.

Muehlbauer, F.J. and Tullu, A. (1997) *Pisum sativum,* New Crop Fact Sheet, Center for New crop and Plant Products, Purdue University, p. 12.

Mukankusi, C.M., Melis, R.J., Derera, J., Buruchara, R.A. and Mark, D. (2011) *Afr. J. Plant Sci.*, 5: 152–161.

Munjal, R.L., Chenulu, V.V. and Hora, T.S. (1963) *Indian Phytopath.*, 16: 268-270.

Munshi, G.D., Jhooty, J.S. and Bajaj, K.L.K. (1987) *Indian J. Mycol. Pl. Path.*, **17**: 280-283.

Muthuswami, S., Thangaraj, T. and Nanjan, K. (1980) *South Indian Hort.*, **28**: 63.

Nadolska-Orczyk, A. and Orczyk, W. (2000) *Mol. Breed.*, **6**: 185–194.

Nag, U.K., Khare, C.P., Markam, V. and Dewngan, M. (2018) *Pharma Innov. J.*, **7**: 11-15.

Naik, L.B. (1989) *Indian J. Hort.*, **46**: 234-239.

Naik, L.B. (1995) *Ann. Agril. Res.*, **16**: 108-110.

Naik, L.B., Sinha, M.N. and Rai, R.K. (1991) *J. Nuclear Agril. Biol.*, **20**: 21-24.

Naik, S.M.P. Singh, S.D. and Yadavendra, J.P. (1975) *Indian J. Mycol. Pl. Path.*, **5**: 200.

Narsinghani, V.G. and Tiwari, A. (1993) In: *Chadha, K.L. and Kalloo, G. (Eds.). Advances in Horticulture, Malhotra Publishing House, New Delhi, India,* pp. 217-233.

Nascimento, W.M., Giordano, L.B., Camara, F.L.A. and Leite, S.L.S. (1987) *HorticulturaBrassileira*, **5**: 34-36.

Nath, Prem (1976) *Vegetable for the tropical Region,* Indian Council of Agricultural Research, New Delhi.

Nauen, R., Hungenberg, H., Toloo, B., Tietjen, K. and Elbert, A. (1998) *Pesti. Sci.*, **53**: 133-140.

Nauerby, B., Madsen, M., Christiansen, J. and Wyndaele, R. (1990) *Plant Cell Rept.*, **9**: 676-679.

Nault, B.A., Shah, D.A., Dillard, H.R. and McFaul, A.C. (2004) *Environ. Entomol.*, **33**: 1593–1601.

Negi, S.C. (1992) *Indian J. Agron.*, **37**: 772-774.

Negm, A.Y., Abdel Samad, A.M., Abdel aziz, O. and Waly, A.F.A. (1997) *Egyptian J. Agril. Res.*, **75**: 855-867.

Nelson, D.C. and Nylund, R.E. (1962) *Weeds*, **10**: 224-229.

Nelson, L.M. and Edie, S.A. (1991) *Soil Biology and Biochemistry*, **23**: 681-688.

Newman, C.W., Newman, R.K. and Lockerman, R.H. (1988) In: *World Crops: Cool Season Food Legumes* (R.J. summerfield ed.), Kluwer Academic Publishers, Dordrecht, The Netherlands, pp. 405-411.

Niielsen, S.V.S., Poulsen, G.B. and Larsen, M.E. (1991) *Physiol. Plant.*, **82**: 99-102.

Nisar, M., Ghafoor, A., Khan, M.R. and Qureshi, A.S. (2006) *Acta. Biol. Cracov. Bot.,* **48**: 33-37.

Obraztsov, A.S. and Amelin, A.V. (1990) *Sel'skok hozyaistvennaya-Bologiy,* **1**: 83-89.

Ochatt, S.J., Benabdelmouna, A., Marget, P., Aubert, G., Moussy, F., Ponte´caille, C. and Jacas, L. (2004) Euphytica, **137**: 353–359.

Okh, S.H., King, T.H. and Kommedahl, T. (1978) *Phytopath.,* **68**: 1644-1649.

Oono, R. and Denison, R.F. (2010) *Plant Physiol.,* **154**: 1541–1548.

Ozcan, S. (1995) *Turkish J. Bot.,* **19**: 427-429.

Ozdemir, S., Karadavut, U. and Erdogan, C. (1999) *Turkish J. agril. For.,* **23**: 869-874.

Padrit, J., Hampton, J.G., Hill, M.J. and Watkin, B.R. (1996) *J. Appl. Seed Prodn.,* **14**: 41-45.

Page, D., Aubert, G., Duc, G., Welham, T. and Domoney, C. (2002) *Mol. Genet. Genomics,* **267**: 359–369.

Pailla, C.A. and Lopez, R. (1979) *AgronomiaCostarricense,* **3**: 83-95.

Pal, A.N. (1976) *Indian J. agric. Sci.,* **46**: 104-105.

Pal, B.P., Sikka, S.M. and Singh, H.B. (1956) *Indian J. Hort.,* **13**: 64-65.

Pandita, M.L., Vashistha, R.N. and Singh, K. (1977) *Haryana J. Hort. Sci.,* **6**: 166-169.

Pariasca, J.A.T., Miyazaki, T., Hisaka, H., Nakagawa, H. and Sato, T. (2001) *Postharvest Biol. Technol.,* **21**: 213–223.

Park, C.K., Jung, C.S., Baek, I.Y., Shin, D.C., Kwack, Y.H., Suh, H.S., Lee, S.K., Oh, Y.J., Son, C.K. and Choi, J.K. (1998) *RDA. J. Crop Sci.,* **39**: 136-140.

Pate, J.S. (1975) *Crop Physiology,* Some Histories, (T. Evans ed.), Cambridge Univ. Press.

Patel, A. (2014) *Biochem. Comp.,* **2**: 1–9.

Patel, T.S., Katare, D.S., Khosla, H.K. and Dubey, S. (1998) *Crop Res. (Hisar),* **15**: 54-56.

Patseva, M.A. and Kosenko, L.V. (1991) *Microbiology* (New York), **60**: 184-188.

Paul, Y., Sharma, B.N., Saimbhi, M.S. and Mann, G.S. (1976) *J. Res., Punjab Agric. Univ.,* Ludhiana, **13**: 290-293.

Paunti-Kaerlas, J., Eriksson, T. and Engsteom, P. (1990) *Theor. Appl. Genet.,* **80**: 209-252.

Pavy, N., Drouaud, J., Hoffman, B., Pelletier, G. and Brunel, D. (1998) *Agronomie,* **18**: 209-224.

Persson, L., Bbattu2dker, L. and Larsson-Wikstro, M. (1997) Plant Dis., **81**: 171–174.

Petkova, R. (1997) *Pochvoznanie, Agrokhimiya y Ekologiya,* **32**: 64-66.

Podlesna, A. and Wojcieska, U. (1996) *ZeszytyProblemowePostepowNaukRolniczych,* **434**: 13-17.

Polans, N.O., Carriveau, J.L. and Coleman, A.W. (1990) *Current Genet.,* **18**: 477-480.

Polhill, R.M. and van der Maesen, L.J.G. (1985) In: *Grain Legume Crops* (R.J. Summerfield and E.H. Roberts eds.), Collins, London, UK, pp. 3-36.

Popov, N., Petkov, N. and Miteva, N. (1996) *Rasteniev"dni-Nauki.*, **33**: 68-71.

Porter, R.E.R. (1977) *New Zealand Exp. Agril.*, **5**: 335-338.

Powell, A.A. and Mathews, S. (1979) *J. Exp. Bot.*, **30**: 193-197.

Powell, A.A. and Mathews, S. (1980) *J. Agric. Sci.*, **95**: 35-38.

Prasad, R.N. and Prasad, A. (1998) *Indian J. Hort.*, **55**: 164-167.

Prasad, R.N. and Prasad, A. (1999) *Scientific Hort.*, **6**: 133-135.

Provvidenti, R. and Granett, A.L. (1976) *Ann. Appl. Biol.*, **82**: 85-89.

Pulung, M.A. and Harjadi (1994) *Acta Hortic.*, **369**: 306-310.

Puonti-Kaerlas, J., Stabel, P. and Eriksson, T. (1989) *Plant Cell Rep.*, **8**: 321-324.

Purewall, S S. (1957) *Vegetable cultivation in North India. Farm Bulletin. No. 36. Indian Council of Agricultural Research. New Delhi, India.*

Purseglove, J.W. (1974) *Tropical Crops*, Dicotyledons, Longman, London.

Puzio-Idzkowska, M. (1990) *BiuletynInstytutuHodowli I Aklimatyzacji Roslin*, No. 173-174, pp. 113-115.

Radziszewski, J. and Rola, J. (1991) *Ochrona Roslin*, **35**: 18-19.

Raina, J.N., Thakur, B.C. and Bhandari, A.R. (1998) *J. Indian Soc. Soil Sci.*, **46**: 562-567.

Raman, R. and Krishnamoorthy, R. (2005) *Legume Res.*, **28**: 128-30.

Ramana Rao, K.V., Gangwar, S., Bajpai, A., Keshri, R., Chourasia, L. and Soni, K. (2017) *Legume Res.*, **40**: 559-561.

Rameau, C., Denou, D., Farval, F., Haurogne, K.M., Josserand, J., Laucou, V., Batge, S. and Murget, I.C. (1998) *Theor. Appl. Genet.*, **97**: 916-928.

Ramírez-Bahena, M.H., García-Fraile, P., Peix, A., Valverde, A., Rivas, R., Igual, J.M., Mateos, P.F., Martínez-Molina, E. and Velázquez, E. (2008) *Int. J. Syst. Evol. Microbiol.*, **58**: 2484– 2490.

Ramphal and Choudhury, B. (1978) *Indian J. Agric. Sci.*, **48**: 407-410.

Ranalli, P. and Cubero, J.I. (1997) *Field Crops Res.*, **53**: 69-82.

Randhawa, K.S. and Saimbhi, M.S. (1976) *Progressive Hort.*, **7**: 53-60.

Ransom, L.M., O'Brien, R.G. and Glass, R.J. (1991) *Australas. Plant Path.*, **20**: 16– 20.

Rao, V.S.N. and Narsinghani, V.G. (1987) *Indian J. Genet. Pl. Breed.*, **47**: 137-140.

Rathi, Y.P.S. (1977) *Pantnagar J. Res.*, **2**: 55-57.

Raut, B.T. and Wangikar, P.D. (1979) *Pesticides*, **13**: 21-23.

Reath, A.N. and Wittwer, S.H. (1952) *Proc. Amer. Soc. Hort. Sci.*, **60**: 301-314.

Rebika, T. (2017) *Int. J. Curr. Microbiol. App. Sci.*, **6**: 45-50.

Redden, B., Leonforte, A., Ford, R., Croser, J. and Slattery, J. (2005) In: Singh, R.J. and Jauhar, P.P. (Eds.), *Genetic Resources, Chromosome Engineering, and Crop Improvement: Grain Legumes, Vol. 1. CRC Press, Boca Raton, FL*, pp. 49–83.

Reddy, D.C.L., Preethi, B., Wani, M.A., Aghora, T.S., Aswath, C. and Mohan, N. (2015) *J. Hortic. Sci. Biotechnol.*, **90**: 78–82.

Rehman, A., Mehboob, S., Sohail, M., Gondal, A.S., Idrees, M. and Ali, M. (2014). *Pak. J. Phytopathol.*, **26**: 309313.

Rehman, H.M., Cooper, J.W. and Lam, H.M. (2019) *Plant, Cell Environ.*, **42**: 52–70.

Renfrew, J.M. (1973) *Palaeoethnobotany*, Methuen, London.

Reynolds, J.D. (1960) *Agriculture Land.*, **66**: 509-513.

Riah, N., Bena, G. Djekoun, A., Heulin, K., de Lajudie, P. and Laguerre, G. (2014) *Syst. Appl. Microbiol.*, **37**: 368-375.

Richardson, H.J. and Hollaway, G.J. (2011) *Australasian Plant Pathol.*, **40**: 260–268.

Robbs, S.L., Halves, M.C., Lin, H.J., Pueppke, S.G. and Smith, L.Y. (1991) *Plant Physiol.*, **95**: 52-57.

Roberts, E.H. and Roberts, D.L. (1972) *Viability of Seeds*, (E.N. Roberts ed.), Chapman and Hall, London.

Robertson, G.I. (1976) *New Zealand Commercial Grower*, **31**: 9.

Rodriguez Maribona, B., Tenorio, J.L., Conde, J.R. and Ayerbe, L. (1993) *InvestigacionAgraria, Produccion Vegetables*, **8**: 153-164.

Rubluo, A., Kartha, K.K., Mronginski, L.A. And Dyek, J. (1983) *J. Plant Physiol.*, 117: 119-130.

Sackett, W.G. (1916) *Bull. Colorado Agric. Experi. Sta.*, **218**: 3–43.

Sadeghi, A., Van Damme, E.M. and Smagghe, G. (2009) *J. Insect Sci.*, **9**: 65.

Saha, R. (2011) *Indian J. Agric. Sci.*, **81**: 633–636.

Saharia, R. (1986) *Indian J. Agron.*, **32**: 377-379.

Saimbhi, M.S., Prakash, J. and Singh, K. (1970) *Indian J. Weed Sci.*, **2**: 51-55.

Saimbhi, M.S., Sandhu, K.S., Singh, D., Kooner, K.S., Dhiman, J.S. and Dhillon, N.P.S. (1990) *Haryana J. Hort. Sci.*, **19**: 198-201.

Saini, J.P. and Thakur, S.R. (1996) *Indian J. Agril. Sci.*, **66**: 514-517.

Saito, T. (2004) *Appl. Entomol. Zool.*, **39**: 203-208.

Samarin, N.A. (1975) *Ordena Lenina-i-Ordena-Druzhby-Narodov-Instituta-Rastenievodstva-Imeni-N.-I.Vavilova*, **53**: 51-55.

Samarin, N.A. and Samarina, L.N. (1981) *Tr. Prikl. Bot. Genet. Sel.*, **70**: 36-39.

Sanchez, F.J., Manzanares, M., de Andres, E.F., Tenorio, J.L. and Ayerbe, L. (2001) *Eur. J. Agron.*, **15**: 57–70.

Sanchez, F.J., Manzanares, M., de Andres, E.F., Tenorio, J.L. and Ayerbe, L. (1998) *Field Crops Res.*, **59**: 225–235.

Sandhya and Singh, A.K. (2004) *J. Res. Punjab Agri. Uni.*, **41**: 110–118.

Saravanan, A. and Nambisan, K.M.P. (1994) *Madras Agril. J.*, **81**: 573-574.

Sarawat, P., Stoddard, F.L., Marshall, D.R. and Ali, S.M. (1994) *Euphytica*, **80**: 39-48.

Sarikova, D. (1995) *Agrochemia Bratislava*, **35**: 95-98.

Scetar, M., Kurek, M. and Galic, K. (2010) *Croatian J. Food Technol. Biotechnol. Nutri.*, **5**: 69–86.

Schmelcz, L. (1985) *Hort. Abstr.*, **55**: 542.

Schroeder, E.H., Schotz, A.H.T., Wardley-Richardson, B.S., Pencer and Higgins, T.J.V. (1993) *Plant Physiol.*, **101**: 75 1-757.

Scott, R.E. and Goulden, D.S. (1993) *New Zealand J. Crop and Hort. Sci.*, **21**: 263-264.

Seidenglanz, M.R.J., Smýkalová, I., Poslušná, J. and Kolabattu2ík, P. (2010) *Plant Protect. Sci.*, **46**: 19–27.

Sekhon, H.S., Singh, G. and Brar, J.S. (1993) *Proc. Indian Society of Weed Science International Symposium, Hisar, India, Indian Society of Weed Science, Vol. III, pp.141-146.*

Semere, T. and Froud-Williams, R.J. (2001). *J. Appl. Ecol.*, **38**: 137–145.

Sen, B. and Kapoor, I.J. (1975) *Veg. Sci.*, **2**: 76-78.

Sen, S. and Kavitkar, A.G. (1958) *Indian J. Agric. Sci.*, **28**: 31-42.

Shantibala, T. and Singh, T.K. (2008) *Ann. Plant Protec. Sci.*, **16**: 353-355.

Sharga, A.N. and Jauhari, O.S. (1970) *Madras Agric. J.*, **57**: 216-221.

Sharma, A., Sood, M., Rana, A. and Singh, Y. (2007) *Indian J. Hortic.*, **64**: 410-414.

Sharma, A.R. and Vats, O.P. (1986) *Indian J. Weed Sci.*, **18**: 250-253.

Sharma, A.R. and Vats; OP. (1986) *Indian J. Agron.*, **33**: 214-216.

Sharma, R.N., Mishra, R.K, Pandey, R.L. and Rastogi, N.K. (1998) *Annals Agric. Res.*, **19**: 58-60.

Sharma, S.R. and Gupta, A.K. (2006) *J. Res. Punjab Agri. Uni.*, **43**: 208–213.

Shehata, M.A., Davis, D.W. and Anderson, N.A. (1981) *Plant Dis.*, **65**: 417-419.

Shevchenko, A.M. (1987) *Selektsiya I Semenovodstvo*, USSR, No. 5, pp. 27-30.

Shevchenko, A.M. (1989) *Selektsiya I Semenovodstvo*, USSR, No. 5, pp. 20-22.

Shewry, P.R. and Tatham, A.S. (1999) In: *Shewry, P.R. and Casey, R. (Eds.), Seed Proteins*, pp. 11–33.

Shewry, P.R., Napier, J.A. and Tatham, A.S. (1995) *The Plant Cell*, **7**: 945–956.

Shubha, K., Dhar, S., Choudhary, H., Dubey, S.C. and Sharma, R.K. (2016) *Indian J. Hort.*, **73**: 356-361

Shukla, P., Singh, R.P. and Prasad, M. (1977) *Indian J. Mycol. Pl. Path.*, **7**: 164-168.

Shukla, P., Singh, R.R., Misra, A.N. and Singh, D.V. (1976) *Indian J. Mycol. Pl. Path.*, **6**: 203-204.

Shukla, Y.R., Kohli, U.K. and Sharma, S.K. (1993) *Hortic. J.,* **6**: 129-131.

Sidorova, K.K., Uzhintseva, L.P., Vavrenteva, L.I., Druganov, A.G. and Pentegova, V.A. (1986) *FiziologiyaiBiokhimiyaKul'turnykhRastenii,* **18**: 274-279.

Singh, B.P. (1991) *South Indian Hort.,* **39**: 381.

Singh, D.V. and Singh, R.R. (1978) *Pesticides,* **12**: 33-34.

Singh, G. (2003) *Agric. Rev.,* **24**: 217-222.

Singh, H. and Kumar, A. (1979) *J. Res., Punjab Agric. Univ.,* Ludhiana, **16**: 164-168.

Singh, H.B., Thakur, M.R. and Bhagchandani, P.M. (1962) *Indian J. Hort.,* **20**: 148-154.

Singh, J., Dhall, R.K. and Aujla, I.S. (2015) *SABRAO J. Breed.Genet.,* **47**: 384-393.

Singh, K. and Sharma, R.N. (1968) *Punjab Hort. J.,* **8**: 245-248.

Singh, K., Kumar, K. and Pandita, M.L. (1973) *Indian J. Weed Sci.,* **5**: 42-47.

Singh, K.N. (1990) *Indian J. Pulse Res.,* **3**: 19-24.

Singh, K.N., Santoshi, U.S. and Singh, H.G. (1989) *Indian J. Genet. Pl. Breed.,* **47**: 115-117.

Singh, M.N. and Singh, R.B. (1990a) *Crop Improv.,* **17**: 117-22.

Singh, M.N. and Singh, R.B. (l990b) *Indian J. Genet.,* **50**: 359-63.

Singh, O. and Raj, B. (1988) *Ann. Agric. Res.,* **9**: 13-19.

Singh, R. and Choudhury, S.L. (1970) *Indian J. Agron.,* **16**: 302.

Singh, S., Dahiya, B.S. and Sindu, P.S. (1980) *Genetica Agraria,* **34**: 289-298.

Singh, S., Dutt, O.M. and Singh, S. (1997) *Indian J. Plant Physiol.,* **2**: 90-92.

Singh, S.D. and Naik, S.M.P. (1977) *Indian J. Agric. Sci.,* **47**: 87-89.

Singh, S.K., Tomar, B.S., Anand, A., Kumari, S. and Prakash, K. (2018) *Indian J. Agric. Sci.,* **88**: 1730–1734.

Skot, L. (1983) *Physiol. Plant,* **59**: 585-589.

Skot, L., Tamms, E. and Mytton, L.R. (1994) *Plant and Soil,* **163**: 141-150.

Skrobakova, E. (1995) *Agrochemia Bratislava,* **35**: 61-62.

Slinkard, A.E., Bascur, G. and Hernandez-Bravo, G. (1994) In: *Expanding the Production and Use of Cool Season Food Legumes* (F.J. Muehlbauer and W.J. Kaiser eds.), Kluwer Academic Publishers, Dordrecht, The Netherlands, pp. 195-203.

Smart, J. (1990) *Grain Legumes: Evolution and genetic resources,* Cambridge University Press, Cambridge, UK, p. 200.

Smartt, J. (1976) *Tropical Pulses,* Tropical Agricultural Series, Longman Group Ltd., London.

Sneddon, J.L. and Squibbs, F.L. (1958) *J. Nat. Agric. Bot.,* **8**: 378-422.

Snyder, W.C. and Walker, J.C. (1935) *Zentralbl. Bakteriol. Parasitenkd.,* **91**: 355-78.

Sokhi, S.S., Jhooty, J.S. and Bains, S.S. (1979) *Indian Phytopath.,* **32**: 571-579.

Soundararajan, R.P., Chitra, N. and Geetha, S. (2013) *Agri. Reviews*, **34**: 176-187.

Srivastava, T.K., Ahlawat, I.P.S. and Panwar, J.D.S. (1998) *Indian J. Plant Physiol.*, 3: 237-239.

Stanisavljevic, N.S., Ilic, M.D., Matic, I.Z., Jovanovic, Ž.S., Cupic, T. and Dabic, D. (2016) *Nutr. Cancer.*, **68**: 988–1000.

Stasz, T.E., Harman, G.E. and Marx, G.A. (1980) *Phytopathology*, **70**: 730-733.

Stavrinides, J., McCloskey, J.K. and Ochman, H. (2009) *Appl. Environ. Microbiol.*, **75**: 2230–2235.

Steene, van de F., Vulsteke, G., de Proft, M. and Callewaert, D. (1999) *Z. Pflanzenkr. Pflanzenschutz.*, **106**: 633–637.

Stevenson, W.R., Hagedorn, D.J. and Gritton, E.T. (1970) *Phytopath.*, **60**: 1315.

Stoddard, F.L., Balko, C., Erskine, W., Khan, H.R., Link, W. and Sarker, A. (2006) *Euphytica*, **147**: 167–186.

Suneetha, T.B., Gopinath, S.M. and Naik, S.L. (2014) *Int. J. Innov. Res. Adv. Eng.*, **6**: 33-36.

Supe, U. and Roymon, M.G. (2011) *Indian J. Fund. Appl. Life Sci.*, **1**: 203-208.

Suslow, T.V. and Cantwell, M. (1998) *Peas: snow and snap pod peas. Perishables handling quarterly. Univ of calif. Davis CA*, **93**: 15–16.

Sutcliffe, J.F. and Pate, J.S. (1977) *The Physiology of Garden Pea*, Academic Press, London, New York.

Svabova, L., Smykal, P., Griga, M. and Ondrej, V. (2005) *Biologia Plantarum.*, **49**: 361-370.

Swenson, K.G. (1954) *J. Econ. Entomol.*, **47**: 1121–1123.

Swenson, K.G. (1957) *J. Econ. Entomol.*, **50**: 727–731.

Szukala, J., Maciejewski, T. and Sobiech, S. (1995) *Prace z ZakresuNaukRolniczych*, **79**: 119-125.

Tayeh, N., Aluome, C., Falque, M., Jacquin, F., Klein, A. and Chauveau, A. (2015a) *Plant J.*, doi: 10.1111/tpj.13070.

Tayeh, N., Klein, A., Le Paslier, M. C., Jacquin, F., Houtin, H. and Rond, C. (2015b) *Front. Plant Sci.*, **6**: 941.

Taylor, J.D. and Dye, D.W. (1976) *New Zealand J. Agric.*, **19**: 91-95.

Taylor, J.D., Bevan, J.R., Crute, I.R. and Reader, S.L. (1989) *Plant Pathol.*, **38**: 364–375.

Tewari, A.N., Husssain, N.K. and Singh, B. (1997) *Indian J. Weed Sci.*, **29**: 94-97.

Thompson, H.C. and Kelly, W.C. (1957) *Vegetable Crops*, McGraw-Hill Book Co., Inc., New York, Toronto, London.

Thompson, T., Patel, G.S., Pandya, K.S., Dabhi, J.S. and Pawar, Y. (2015) *Int. J. Farm Sci.*, **5**: 8–13.

Tian, Chang Fu, Wang, En Tao, Wu, Li Juan, Han, Tian Xu, Chen, Wen Feng, Gu, Chun Tao, Jin Gang Gu and Chen, Wen Xin (2008) *Int. J. Syst. Evol. Microbiol.,* **58**: 2871-2875.

Timmerman, G.M., Frew, T.J., Miller, A.L., Weeden, N.F. and Jermyn, W.A. (1993) *Theor. Appl. Genet.,* **85**: 609-615.

Tiwari, K.R., Penner, G.A., Warkentin, T.D., Rashid, K.Y. and Menzies, J.G. (1998) *3rd European conference on grain legumes, Opportunities for high quality, healthy and added value crops to meet European demands,* Valladolid, Spain, 14-19 November, 1998, pp. 120-121.

Tiwari, K.R., Warkentin, T.D., Penner, G.A. and Menzies, J.G. (1999a) *Can. J. Plant Pathol.,* **21**: 159–164.

Toker, C. and Mutlu, N. (2011) In: *Pratap, A. and Kumar, J. (Eds.), Biology and Breeding of Food Legumes. CAB International,* pp. 241–260.

Trebuchet, G., Chopinet, R. and Drowsy, J. (1953) *Ann. Amel. Plantes,* **3**: 147-251.

Tsyganok, N.S. (1990) *SelektsiyaiSemenovodstvo (Moskva),* No. 2, pp. 29-31.

Turk, M., Albayrak, S. and Yuksel, O. (2011) *Turk. J. Field Crops.,* **16**: 137-141.

Tyagi, R.N.S., Mathur, A. and Bhatnagar, L.G. (1978) *Indian J. Mycol. Pl. Path.,* **8**: 15.

United Nations University (2002) *WHO Tech. Rept. Series,* **935**: 1–265.

Unkovich, M., Herridge, D., Peoples, M., Cadisch, G., Boddey, B., Giller, K., Alves, B. and Chalk, P. (2008) *Measuring plant-associated nitrogen fixation in agricultural systems. Clarus design, Canberra.*

van Leur, J.A.G., Kumari, S.G., Aftab, M., Leonforte, A. and Moore, S. (2013) *New Zealand J. Crop Hortic. Sci.,* **41**: 86-101.

Vankosky, M., Dosdall L.M. and Cárcamo, H.A. (2009) *CAB Reviews: Perspectives in Agriculture, Veterinary Science, Nutrition and Natural Resources, Wallingford, UK.*

Vankosky, M.A., Cárcamo, H.A. and Dosdall, L.M. (2011a) *J. Econ. Entomol.,* **104**: 1550–1560.

Vassilevska-Ivanova, R., Naidenova, N. and Kraptchev, B. (2008) *Comptes rendus de l'Acadeµmie bulgare des sciences: sciences matheµmatiques et naturelles,* **61**: 955-962.

Vavilov, N.I. (1926) *TrudydyByuroPrikl. Bot.,* **16**: 139-148.

Vavilov, N.I. (1951) *Chronica Bot.,* **13**: 1-366.

Vaz Patto, M.C., Amarowicz, R. and Aryee, A.N.A. (2015) *Cri. Rev. Plant Sci.,* **34**: 105–143.

Ventorino, V., Chiurazzi, M., Aponte, M., Pepe, O. and Moschetti, G. (2005) *Curr. Microbiol.,* **55**: 512–517.

Verma, M.L. and Bhandari, A.R. (1998) *Hort. J.,* **11**: 55-65.

Verma, M.L., Bhandari, A.R. and Raina, J.N. (1997) *Internatl. J. Trop. Agril.,* **15**: 195-198.

Vetrova, *E.g.* (1987) In: *SelektsiyaiSemenovodstvopolevykhKul'tur v Moldavskoi*, SSR, pp. 64-69.

Vetrova, *E.g.* and Petrushina, N.V. (1987) In: *Urozhii Kachestvo Produktsii Osnovnykh Polevykh Kul'tur Moldavii, Kishinev*, Moldavian, SSR, pp. 78-90.

Vigeolas, H., Chinoy, C. and Zuther, E. (2008) *Plant Physiol.*, **146**: 74-82.

Wade, B.L. (1929) *Wis. Agr. Expt. Sta. Res. Bull.*, **97**: 32.

Wade, B.L. (1943) *Circ. U.S. Dept. Agric.*, No. 676, p. 12.

Wakankar, S.M. (1978) *Indian Hort.*, **22**: 11.

Wakankar, S.M. Mahadik, C.N. (1961) *Indian Fmg.*, **11**: 8-12.

Wang, X., Warkentin, T.D. and Briggs, C.J. (1998) *Euphytica*, **101**: 97-102

Warkentin, T.D., Delgerjav, O. and Arganosa, G. (2012) *Crop Sci.*, **52**: 74-78.

Warkentin, T.D., Rashid, K.Y. and Xue, A.G. (1996) *Can J. Plant Sci.*, **76**: 933-935.

Warkentin, T.D., Smykal P., Coyne, C.J., Weeden, N., Domoney, C. and Bing, D. (2015) In: *De Ron, A.M. (Eds.) Grain Legumes, Series Handbook of Plant Breeding*, pp. 37-83.

Weeden, N.F. and Porter, L. (2008) *Pisum Genet.*, **39**: 35-36.

Weller, J.L., Liew, L.C., Hecht, V.F.G., Rajandran, V., Laurie, R.E. and Ridge, S. (2012) *Proc. Natl. Acad. Sci., U.S.A.*, **109**: 21158-21163.

White, O. (1971) *Proc. Amer. Philos. Soc.*, **56**: 487-588.

Whyte, R.O., Leissner, G.N. and Tremble, H.C. (1953) *Legume in Agriculture*, FAO Agric. Stu., No. 21, FAO, Rome.

Wielbo, J., Marek-Kozaczuk, M., Mazur, A., Kubik-Komar, A. and Skorupska, A. (2011) *Brit. Micr. Res. J.*, **1**: 55–69.

Wielbo, Jerzy, PodleœnA, A., kiDaJ, D., Podleœny, J. and SkorupSka, A. (2015) *Microbes Environ.*, **30**: 254-261.

Wimalajeewa, D.L.S. and Nancarrow, R.J. (1984) *Aust. J. Exp. Agric. Ani. Husb.*, **24**: 450-452.

Yadav, J.L. and Chauhan, R. (2000) *Int. J. Trop. Agric.*, **18**: 169-172.

Yadav, L.M., Sharma, P.P. and Maurya, K.R. (1996) *Madras Agril. J.*, **83**: 142-147.

Yadav, S.K.M. and Agnihotri, R.S. (2016) *Indian J. Entomol.*, **78**: 373-391.

Yankov, I.I. (1989) *SelektsiyaiSemenovodstvo (Moskva)*, No. 5, pp. 31-33.

Yarnell, S.H. (1962) *Bot. Rev.*, **28**: 465-537.

Yawalkar, K.S. (1969) *Vegetable Crops of India*, Agri-Horticultural Pub. House, Nagpur, India.

Yen, D.E. and Fry, P.R. (1956) *Aust. J. Agric. Res.*, **7**: 272-280.

Yu, J., Gu, W.K., Providenti, R. and Weeden, N.F. (1995) *J. Amer. Soc. Hort. Sci.*, **120**: 730-733.

Zelenov, A.N., Shchetinin, V.Y. and Sobolev, D.V. (2008) *Agrarnaya nau-ka*, **2**: 19-20.

Zhang, C., Tar'An, B., Warkentin, T., Tullu, A., Bett, K.E. and Vandenberg, B. (2006) *Crop Sci.*, **46**: 321–329.

Zhogina, V.A. (1986) In: *Selektsiyai Semenovod, Korm. itekhn. Kultur, Krasnodar*, USSR, pp. 8-11.

Zimmer, R.C. and Alikhan, S.T. (1976) *Canada Agric.*, **21**: 6-8.

2

FRENCH BEAN

A. Chattopadhyay and V.A. Parthasarathy

1.0 Introduction

The French bean (syn. Kidney bean, haricot bean, snap bean, and navy bean) is one of the most important leguminous vegetables. It is the world's most important food legume. Farmers grow common beans in two forms, as dry beans (rajmah) and snap beans (the green pods are consumed as a vegetable).

The total worldwide cultivated area of green beans is 1,527,613 hectares, producing 21,720,588 tons, as reported by FAO (2017). China is the world's leading producer of green beans, with a total cultivated area of 635,385 hectares and a production of 17,031,702 tons. Latin America produces nearly half of the world's supply of dry beans. Brazil, Mexico, and Central America are the major producers in this continent. Africa is considered to be a secondary centre for bean genetic diversity. Bean consumption is particularly high in African countries-for example, per capita consumption of bean ranges from 50 to 60 kg per year in Rwanda, Kenya, and Uganda (Broughton *et al.*, 2003; Buruchara *et al.*, 2011). Dry beans account for 57 per cent of the world's food legume production, having twice the production and market value of chickpeas, the next leading food pulse. Over 80 per cent of bean production in developing countries is from subsistence farming of semi-arid regions and sub-humid to humid growing environments (Assefa *et al.*, 2019). In these areas, most producers are small-scale farmers who use unimproved bean cultivars.

2.0 Composition and Uses

2.1.0 Composition

It is a nutritious vegetable. Table 1 below depicts the nutritive value of its tender pod.

Table 1: Composition of French Bean Green Pod (per 100 g of edible portion)*

Moisture	91.4 g	Vitamin C	11 mg
Protein	1.7 g	Nicotinic acid	0.30 mg
Fat	0.1 g	Calcium	50 mg
Carbohydrates	4.5 g	Phosphorus	28 mg
Fibre	1.8 g	Iron	1.70 mg
Minerals	0.5 g	Potassium	129 mg
Vitamin A	221 I.U.	Sulphur	37 mg
Thiamine	0.08 mg	Sodium	4.30 mg
Riboflavin	0.06 mg	Copper	0.21 mg

* Aykroyd (1963).

2.2.0 Uses

In India, it is grown for tender vegetable, shelled green beans and dry beans (rajmah), while in the U.S.A. it is grown for processing in large quantities. The immature seeds are boiled or steamed, and used as a vegetable. The mature seeds are dried, and stored for future use. They must be thoroughly cooked before consumption, and are best if soaked in water for about 12 hours prior to cooking. They can be boiled, baked, pureed or ground into powder. The seeds can also be sprouted, and used in salads or cooked. Young leaves are consumed raw or cooked as a potherb.

3.0 Origin and Taxonomy

French bean belongs to the family Leguminosae. Initially Smartt (1976) has presented a comprehensive account of the origin and domestication of *Phaseolus vulgaris* L. During the process of domestication in common bean, several morphological changes have occurred. The cultivated French bean was an erect growing plant with determinate branching, whereas the wild type was indeterminate and profusely branched. The cultivated types have smaller number of nodes on the main axis, while the wild forms have more nodes. The internode length was relatively shorter in the cultivated types. The changes under domestication are typically loss of seed dormancy and pod dehiscence mechanism, a change from perennial to the annual life form and a great change in seed size correlated with modified shoot architecture. Stems tend to be thicker, leaves larger, branches fewer, the number of nodes may be reduced and internode length was shortened. This process culminates in evolution of self-supporting plants well adapted to mono-crop husbandry systems. This has also led to appearance of a vast variety of seed sizes, shapes and colour and selection for photoperiod insensitivity. But several evidences indicate that French bean was domesticated at least twice–from northern Andean and from Mesoamerican populations (Bitocchi *et al.*, 2013; Schmutz *et al.*, 2014; Ariani *et al.*, 2016; Cortés and Blair, 2017). These evolutionary and domestication histories arguably leave French bean with vulnerabilities to a wide range of biotic and abiotic stresses (Bitocchi *et al.*, 2013; Schmutz *et al.*, 2014), particularly as the crop has moved into new agroecological niches worldwide. According to Smartt (1976), *P. vulgaris* is the most successful American bean because of the presence of the optimum climate for its cultivation in America. Southern Mexico and Central America are considered to be the primary centre of origin, while Peruvian-Ecuadorian-Bolivian area to be the secondary centres.

Evidence for genetic diversity in cultivated common bean (*Phaseolus vulgaris*) is reviewed by Singh *et al.* (1991). Multivariate statistical analyses of morphological, agronomic, and molecular data, as well as other available information on Latin American landraces representing various geographical and ecological regions of their primary centers of domestications in the Americas, reveal the existence of two major groups of germplasm: Middle American and Andean South American, which could be further divided into six races. Three races originated in Middle America (races Durango, Jalisco, and Mesoamerica) and three in Andean South America (races Chile, Nueva Granada, and Peru).

Uses of French Bean

Fresh cut canned French bean

Baked French bean

Salad prepared from French bean   Canned seeds of French bean

A brief description is as follows:

A. Middle American Races

i) Mesoamerica

This race includes small-seeded (< 25 g/100 seed) land races of all seed colours and growth habits. The group is often characterised by an ovate, cordate or hastate terminal leaflet of trifoliate leaves and large, broad cordate or lanceolate bracteoles. Inflorescences are multinoded. Pods are 8-15 cm long, slender, fibrous or parchmented and easy to thresh. This race is distributed throughout the tropical low lands and intermediate altitudes of Mexico, Central America, Colombia, Venezuela and Brazil.

ii) Durango

These are predominantly of indeterminate, prostrate growth habit III, which is characterized by relatively small to medium ovate or cordate leaflets, thin stems and branches, short internodes, and fruiting commencing from and concentrated in basal nodes. These landraces often possess small ovate bracteoles with a pointed tip. The pods are medium sized (5-8 cm), flattened with 4-5 flattened rhombohedric seeds of medium size (25-40 g/100 seeds). This race is distributed in semiarid central and northern high lands of Mexico and South-western USA.

iii) Jalisco

This race is often characterized by indeterminate growth habit IV. Plant height can be over 3 m in its natural habitat. The terminal leaflet of trifoliolate leaves is hastate, ovate, or rhombohedric and sometimes relatively large. Stems and branches are weak and have medium-sized or long internodes. Most germplasm from this race possesses medium-sized, cordate, ovate, or lanceolate bracteoles. Fruiting is distributed either along the entire length of the plant or mostly in its upper part. Pods are 8-15 cm long and have five to eight medium-sized seeds, whose shape is round, oval, or slightly elongated and cylindrical or kidney-shaped. Their natural habitat is the humid high lands of central Mexico and Guatemala, where maximum diversity is found.

B. South American Races

i) Nueva Granada

Germplasm is mostly of growth habits I, II, and III with medium (25-40 g/100 seeds) and large seeds (< 40 g/100 seeds) of often kidney or cylindrical shapes which vary greatly in colour. Leaves are often large with hastate, ovate, or rhombohedric central trifoliolate leaflets and long, dense, straight hairs. Stem internodes are intermediate to long. Bracteoles are small or medium, and ovate, lanceolate, or triangular. Dry pods are fibrous, hard, medium to long (10-20 cm), and leathery, and possess four to six seeds. The pod beak often originates between the placental and ventral sutures. This race is distributed mostly at intermediate altitudes (< 2000 m) of the northern Andes in Colombia, Ecuador, and Peru, but it is also found in Argentina, Belize, Bolivia, Brazil, Chile, Panama, and some Caribbean countries, including the Dominican Republic, Haiti, and Cuba.

ii) Chile

Landraces are predominantly of indeterminate growth habit III. These are characterised by relatively small or medium hastate, rhombohedric, or ovate leaves; short internodes; small or medium, and narrowly triangular, spatulate, or ovate bracteoles; light pinkish or white flower; medium-sized (5-8 cm) pods, often with reduced fibre content; and round to oval seeds (three to five per pod).

Morphologically, these landraces largely resemble germplasm from race Durango, except that seeds of race Chile are round or oval, and fruiting is more sparse. In some of the landraces (*e.g.*, 'Coscorron' (G 4474) and 'Frutilla' (G 5852), pods exhibit an attractive anthocyanin striping, and in many countries these are harvested for green seeds (green shelled or "granados") before physiological maturity. This race is distributed in relatively drier regions at lower altitudes in the southern Andes (southern Peru, Bolivia, Chile, and Argentina).

iii) Peru

Key morphological characteristics of germplasm belonging to this race are the large hastate or lanceolate leaves (often basal) and long and weak internodes with either indeterminate or determinate type IV climbing growth habit. In its natural habitat, it is always grown in association with maize and other crops. Pods are often long (10-20 cm) and leathery. Fruiting is distributed either along the entire stem length or only in the upper part of the plants. Seeds are large and often round or oval but can also be elongated. This group is highly photoperiod sensitive and is adapted to moderately wet and cool temperatures often requiring more than 250 days to maturity. The race is distributed from the northern Colombian high lands (>2000 m) to Argentina.

According to Evans (1976), the domestication of *P. vulgaris* might have occurred in Brazil and Northern Argentina from a wild form of which *P. aborigineus* is a modern survivor. Yarnell's (1965) crossing experiments have confirmed *P. aborigineus* Burkart, to be a progenitor of *P. vulgaris*.

It is now generally agreed that *Phaseolus* originated in the new world and that old world species previously included in *Phaseolus* should be assigned to *Vigna* (Evans, 1976). Westphal (1974) reviewed the evidence for the revision of these two genera. There are four cultivated species in the New World:

- ☆ *Phaseolus vulgaris*: The common, haricot, navy, French or snap bean
- ☆ *P. coccineus*: The runner or scarlet runner bean
- ☆ *P. lunatus*: The Lima (large-seeded), sieva (small-seeded), butter or Madagascar bean
- ☆ *P. acutifolius* var. *latifolius*: The tepary bean.

All the species are diploid with 2n=2x=22. *P. coccineus* is generally cross-pollinated and the other three species are self-pollinating with only a small amount of cross-pollination.

4.0 Cultivars

The French bean cultivars are classified as follows, according to M.A.F.F. Bulletin No.87 (cf. Bland, 1971).

Types of French Bean

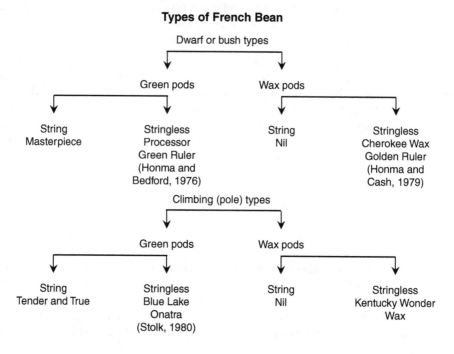

There are a large number of French bean cultivars. Thompson and Kelly (1957) classified cultivars as follows:

1. Snap beans – for vegetable pods.
2. Green-shell beans – used in the green-shelled condition.
3. Dry-shell beans – used in the dry state (field beans).

Each group is again divided into indeterminate climbing (pole) and determminate (bush) types. This growth habit classification divides beans into four groups: Type I (determinate bush), Type II (indeterminate bush), Type III (indeterminate semi climber), and Type IV (indeterminate climber) (Singh, 1991). Besides growth habit classification, beans are sometimes also classified by origin–specifically, by the major Andean and Mesoamerican gene pools and races within those pools (Singh *et al.*, 1991b; Beebe *et al.*, 2013).

Some of the green-shell bean cultivars are Low's Champion, French Horticultural, Dwarf Horticultural, Brilliant, Flash (Thompson and Kelly, 1957) and Green Lime Light (Kemp, 1977) which are bush type cultivars. Pole type cultivars are London Horticultural and Red Cranberry.

Growth Habits and Cultivation Technique of French Bean

Field view of bush and pole type cultivation of French bean

Bush and pole type French bean

Pole type French bean cultivation over trellises

Some of the important snap bean cultivars are as follows:

There are three types of bush beans, namely, flat, oval and round types.

The flat types are Bountiful, Plentiful, Green Ruler, Golden Ruler and Romano.

The oval types include Pusa Parvati, Contender, Spartan Arrow, Premier, Tendergreen, King Green, *etc.*

The pole types include Blue Lake, Kentucky Wonder and Phenomenal Long Podded. Most of the local types native to North-eastern Region of India are pole types.

Budanova (1985) suggested a classification of French bean cultivars based on seed shape. He classified them as *sphaericus, ellipticus, oblongus, subcompressus* and *compressus.*

Tendercrop and Cascade are popular processing cultivars. However, the latest con-cept in the U.K. is to go in for cultivars with short pods so that they can be processed whole (Bingham, 1978).

At IIHR, Bangalore, cultivars Selection 2, Selection 5 and Selection 9 have been developed. Selection 9 has been released under the name Arka Komal (Rajendran and Satyanarayana, 1988). The cultivar Rao Doce was developed in Colombia from the cross Carioca and Bat 76 and introduced to Brazil in 1984. It gives a mean yield of 1-2 tonnes/ha. Plants are erect with indeterminate growth habit and showed greater tolerance of angular leaf spot (Anon., 1987). In Hungary, Nemeskeri (1988) reported that cultivars Forum and Rovet were suitable for canning and showed drought resistance. A pencil-podded bean, Buvet, suitable for freezing, had good quality pods and was drought tolerant.

In Romania, Fana, Cascade and Amboy are high yielding cultivars (Glaman *et al.,* 1989). In Brazil, cultivar Alessa, developed from a cross between Green Isle and Blue Lake 274 can be grown without staking, has a growth period of 52-55 days and produces an average yield of 115 quentals per hectare (Leal, 1989). Some of the elite Chinese French bean cultivars are G0438, G0482 and Pingyum 2 (Duan and Ding, 1989).

Some of the resistant lines/cultivars include Wisconsin 17 and 28 for resistance to *Pseudomonas syringae* (Hagedorn and Rand, 1980a), Wisconsin 46 (Hagedorn and Rand, 1980b) for resistance to root rot complex (*Pythium, Fusarium* and *Rhizoctonia*), B 4000-3 for bean rust resistance (Wyatt *et al.,* 1977) and B 4175 for nematode resistance (Wyatt *et al.,* 1980). A French bean cv. IAPAR 8-Rio Negro is resistant to all races of *Colletotrichum lindemuthianum* and bean common mosaic virus. It is also moderately resistant to *Uromyces phaseoli*. It is indeterminate in growth habit, normally 70 cm in height and yields about 3 tonnes/ha (Alberini *et al.,* 1987). Scully *et al.* (2000) developed new F7 rust-resistant snap bean breeding lines Belt Glade RR-1, BeltGlade RR-2, and BeltGlade RR-3, with the *Ur-4* and *Ur-11* genes, using modified backcross and pedigree breeding methods. PI 181996 (containing the *Ur-11* resistance gene) was used as a pollen parent and crossed sequentially with a series of 7 female parent cultivars. Pod colour for the 3 lines are uniform light to medium green, and pods are straight to slightly curved, with a round transect and minimal interlocular constriction. Seeds are monochrome, off-white to ivory, semi-shiny to semi-dull with a small yellow hilar ring. Seed shape is cuboid to reniform with a round to off-round transect shape and a distinctive flat spot. Seed test weight averaged 28.5-30 g/100 seeds.

Mullins and Straw (2001) evaluated eleven filet snap bean cultivars Axel, Carlo, Dandy, Flevoro, Masai, Maxibel, Minuette, Nickel, Pluto, Rapier and Teseo in Tennessee, USA. Minuette, Pluto Carlo, Masai and Minuette were the most productive cultivars. Maxibel, Carlo, Dandy and Teseo were among cultivars that produced the highest percentage of pods with larger sieve size. Flevoro, Nickel and Pluto pods were firmer than pods of all cultivars. Pods of Minuette, and Rapier were darker in colour than pods of all cultivars except Axel and Teseo. Maxibel produced the longest pods, while Axel produced shorter pods than all cultivars except Masai and Rapier. Masai and Nickel produced the smoothest pods. Dandy and Maxibel pods had the most curvature.

A few of the latest cultivars released from different countries are listed below:

Haibushi

The cultivar has been developed in Japan by Nakano *et al.* (1997) by pure line selection from a heat tolerant Malaysian accession. The line developed showed higher heat tolerance and higher yield potential than commercial varieties in trials carried out in summer in several areas of southern Japan. The line showed tolerance of heat over both short and long periods of time. When the line was exposed to high temperatures for one month, the critical mean air temperature for pod setting ranged from 28 to 29.5°C.

Uel 1

Selected from the cross 78BP8 × NY76-2812-15 and developed by USDA (USA). This new tropical, erect *Phaseolus vulgaris* line has light-green, cylindrical pods (8.2 mm diameter × 12 mm length), which are ready for harvest 60 days after sowing. Flowers and mature seeds are white (Castiglioni *et al.*, 1993).

Macarrao Preferido AG-482

This green-podded, vegetable *P. vulgaris* cultivar was developed from a cross between Macarrao Favorito AG480 and a line developed from crosses involving BGF$_1$458 and Cornell 49AG480 which are sources of resistance to rust (*Uromyces phaseoli* var. *typica* [*U. appendiculatus*]), and Macarrao Rasteiro 274. It is indeterminate, has a growth period of 100 days and produces pods approximately 18 cm long and weighing 11 g. Macarrao Preferido AG482 is resistant to the main physiological races of *U. phaseoli* and also exhibits resistance to anthracnose (*Colletotrichum lindemuthianum*) (Carrijo, 1993).

Taichung 1

This cultivar was developed using Black Creaseback, which is susceptible to rust (*Uromyces phaseoli* var. *typica* [*U. appendiculatus*]), backcrossed 4 times to rust resistant 15R55BK, and released in 1988. It has an indeterminate growth habit and bears smooth, light green pods. Average days to first and last harvest were 53.5

and 92.5 for spring and 55.5 and 90.5 for autumn crops, respectively. Average pod yields for 1982 to 1984 were 224.0 and 195.0 q/ha for spring and autumn crops, respectively. These were 7.1 and 4.8 per cent higher than for Black Creaseback. Trials to screen for rust from 1987 to 1988 at 4 locations showed that the cultivar was highly resistant in mountain areas and resistant to highly resistant in lowland areas (Guu, 1989).

Alessa

This cultivar, obtained by the pedigree method from a cross between Green Isle and Bush Lake 274, has shown promise in trials throughout Rio de Janeiro State, combi-ning average bean yields of 115 q/ha with good resistance to rust diseases. It has a determinate habit and a growth period of 52-55 days, producing erect plants with white flowers and seeds. Pods are flat, approximately 16.5 cm long and show synchronous ripening and good postharvest storage quality (Leal and Bliss, 1990).

Andra

This cultivar, obtained by the pedigree method from a cross between Blush Blue Lake 274 and Cascade, has shown promise in trials throughout Rio de Janeiro State, combining average bean yields of 135 q/ha with good resistance to rust and anthracnose (*Colletotrichum lindemuthianum*). It has a determinate habit and an average growth period of 55 days, producing erect plants with white flowers and seeds. Pods are cylindrical, approximately 17 cm long and show synchronous ripening and good post-harvest storage quality. Andra is considered suitable for both fresh consumption and processing (Leal, 1990).

Cultivars of French bean can conveniently be classified according to growth habit, into two types namely, bush type and pole type.

Bush Type Cultivars

Contender

It is an introduced cultivar from USA. It is high yielding, early and stringless cultivar; flower light purple; pod light green, round, fleshy and thick and without fibres, 12-18 cm long and curved at tips. First picking 50-55 days after sowing. It is tolerant to powdery mildew and mosaic.

Giant Stringless

It is an introduced cultivar from USA. Early and bushy in habit; pods are green, medium long, slightly curved, tender, meaty and stringless. Seeds are glossy and yellowish brown. First picking 45-50 days after sowing. High yielder.

Bountiful

A heavy yielding, introduced and bush type cultivar. Clusters of beans are borne on the main stem. The tender green pods are ready for harvest within a week of flowering.

Premier

An introduced cultivar from USA. It is a high yielding and bush type cultivar. The fruits are neither round nor flat in cross section (oval).

Jampa

An introduced cultivar from Mexico. It is early and flower within 50 days of sowing. The pods are flat, smooth, non-stringy and pale green in colour. The seeds are black, smooth, small and flat. Very high yielder.

Pusa Parbati

This cultivar was developed at IARI, New Delhi, India through X-ray irradiation of the American cultivar "Wax Pod". It is an early, bush type with attractive, round, curved, meaty, stringless and light green pods of 18-20 cm in length. First picking can be done after 45 days of sowing. Average yield is 80-90 q/ha. It is tolerant to mosaic and powdery mildew.

VL-Baunil

This bush type cultivar was developed at Almora, India. Plants are dwarf (40 cm); stem and leaves are green; flowers are white with purple tinge. Pods are long (15-16 cm), round, fleshy, slightly curved, stringless and light green in colour. First picking can be done after 45 days of sowing. Moderately susceptible to foliar diseases. Average yield is 110-120 q/ha.

Arka Komal (Sel.9)

This erect and bushy cultivar was developed at IIHR, Bangalore, India by pure line selection from an Australian introduction. Plants with bright green stem and foliage. Pods are straight, flat, tender and green. It is very good for transport because of extended shelf-life. Seeds are large and light brown. Crop duration is 65- 70 days. Average yield is 180-200 q/ha.

Arka Sukomal

High yielding rust resistant pole bean cultivar. Plants are indeterminate which grow more than 2.0 m in height. Cultivar takes 60 days for 1st harvest. Pods stringless, oval, green and long (23 cm). Ten pod weight is 87 g. Yield potential is 24 q/ha in 100 days. Suitable for both *kharif* and *rabi* seasons.

Arka Anoop

Plants bushy, photo-insensitive. Resistance to both rust and bacterial blight. Pods long, flat and straight. Pod Yield is 20 q/ha in 70-75 days.

Arka Sharath

Plants are bushy and photo-insensitive. Round, string less, smooth pods suitable for steam cooking. Pods are crisp, fleshy with no parchment and perfectly round on cross section. Pod yield is 185 q/ha in 70 days.

Arka Arjun

Plants are bushy, vigorous and photo-insensitive. Pods are green, stringless with smooth surface. Suitable for both *rabi* and summer. Resistant to *MYMV* disease. Pod yield is 170 q/ha in 70 days.

Pant Anupama (UPF-191)

This cultivar was developed at GBPUAT, Pantnagar, India through selection. Plants are upright, 45 cm in height, bushy, foliage medium green and early bearing; pods are round, 12-15 cm long, smooth, tender, non-stringy and green with concentrate fruiting at mid height of the plant. First picking after 55-65 days of sowing. Resistant to angular leaf spot and bean common mosaic virus diseases.

YCD-1

Developed at Yercaud, India through pure line selection from a local type collected from Shevory hills. It is a dual purpose cultivar. Pods are slightly flat, 15 cm long with bold and attractive dark purple seeds. Field tolerant to root rot, rust, yellow mosaic and anthracnose. Average green pod yield 90 q/ha and seed yield 60 q/ha.

Phule Surekha

Developed at MPKV, Rahuri, India through selection from the introduced cultivar Jampa Improved. Pods are 9-10 cm long, flat and light green. Suitable for cultivation in kharif, rabi and spring-summer season. Tolerant to anthracnose, yellow mosaic and wilt diseases. Average yield is 150 q/ha.

Kashi Sampann (VRFBB-1)

Late maturing, bush type high yielding with round, light green pods, tolerant to GYMV and high yield 250-300 q/ha.

Kashi Rajhans (VRFBB-2)

Bushy cultivar has dark green, thin round slighty curved fleshy, free from parchment and smooth surface pod. Tolerant to French Bean Golden Yellow Mosaic Virus. Yield potential is 241.30 q/h.

Swarna Priya (CH-812)

This cultivar has been developed through pure line selection at CHES, Ranchi, India and identified through AICRP-VC in 2001. Pods are flat, green, fleshy having good cooking quality. Dried seeds are bold and maroon in colour, suitable for 'daal' preparation. First harvest can be take 50-55 days after planting. Suitable for mid September sowing and is tolerant to powdery and downy mildew and resistant to leaf minor under field conditions. It has been recommended for cultivation in Jharkhand, Sikkim, Meghalaya, Manipur, Nagaland, Mizoram, Tripura, Arunachal Pradesh, Madhya Pradesh, Maharashtra and Andaman and Nicobar Islands and has yield potential of 120-140 q/ha.

Pole Type Cultivars

Kentacky Wonder

An introduced and late cultivar from USA. Pods are green, fleshy, curved, large (22.5 cm), stringless and borne in clusters. Pods become stringy at later stage. Seeds are glossy and yellowish brown in colour. It is tolerant to wilt. Average yield is 100-120 q/ha.

Pusa Himlata

This pole type cultivar was developed at Katrain, India. Vine medium in height (2.5 m) with comparatively small internmode length. Each node bear 3-4 flowers. Pods are medium sized (14 cm), round, meaty, stringless and light green. First picking starts 60 days after sowing. Average yield is 260 q/ha.

TKD-1

Developed at Thadiyankudisai, India through advanced generation selection from the cross Selection-1 × PV118. It is suitable for growing in hills. The pods are green, long, flat with low fibre content. First harvest 60 days after sowing. Crop duration 90-100 days. Average yield is 50-60 q/ha.

KKL-1

Developed at Kodaikanal, India through pure line, selection from a local type of Kodaikanal. It is best suited for hilly areas of 1800-2400 m altitude. Pods are green ana long (28 cm) with bold milky white flat grains. Average pod yield is 70 q/ha and seed yield (if grown only for seed purpose) 30 q/ha.

Kashi Param (IVFB-l)

This cultivar has been developed through pureline selection at IIVR, Varanasi, India and notified by CVRC Notification no. 2035(E) dated 28.11.2006. Plants are determinate (70 cm long) with dark green leaves. Pods are fleshy, length 14.70 cm, round, dark green, seed colour dark brown to green. This cultivar has been recommended for cultivation in J&K, H.P. and Uttarakhand and has yield potential of 120-140 q/ha.

Arka Suvidha (IIHR-909)

This cultivar has been developed through hybridization (Blue Crop × Contender) followed by pedigree selection at IIHR, Bangalore and notified by CVRC Notification no. 2035(E) dated 28.11.2006. Plants are bushy and photo-Insensitive. Pods are straight, oval, light green, fleshy and stringless. It has been recommended for cultivation in J&K, H.P. and Uttarakhand and has yield potential of 180-195 q/ha in 70 days of crop duration.

Azad Rajmah-1

This cultivar has been developed at CSAUA&T, Kanpur, India and notified by CVRC Notification no. 597 (E) dated 25.4.2006. Plants are dwarf with dark green foliage. Pods are light green, smooth, tender stringless and length 12-15 cm. It has yield potential of 75-80 q/ha.

Phule Suyash (GK-7)

This cultivar has been developed through selection from Shahpur Local cultivar at MPKV, Rahurui, India and notified by CVRC Notification no. 597(E) dated 25.04.2006. Pods are attractive tender, straight to slightly curved, smooth and green. Plants are highly tolerant to wilt and bean mosaic virus. It has been recommended for the cultivation in Maharashtra and has yield potential of 230-245 q/ha in 70-80 days of crop duration.

Swarna Lata (CH-819)

This cultivar has been developed through pure line selection at CHES, Ranchi, India and identified for release through AICRP-VC in 2001. This is suitable for mid June and September sowing. Pods are medium long, fleshy and stringless with good cooking quality. First harvest can be taken 55-60 days after planting. It is tolerant to powdery and downy mildew and resistant to leaf minor under field conditions. This cultivar has been recommended for cultivation in Jharkhand, J&K, Himachal Pradesh and Uttarakhand and has yield potential of 120-140 q/ha.

5.0 Soil and Climate

5.1.0 Soil

It cannot grow well in extreme acid as well as in alkaline soil. The optimum soil pH range where this crop grows well is around 5.5-6.0. Liming is needed if soil pH is less than 5.5. Although beans can be grown on all types of soils, medium textured silt loams or clay loams are best for obtaining high yields (Schafer, 1979).

5.2.0 Climate

French beans are grown as winter crop in plains, while it can be grown all through the year except winter, in hills. It cannot withstand drought as well as very heavy rainfall and frost. Even though, many of the cultivars are photo-insensitive, certain cultivars develop floral buds only during short days but would abscise during long days (Morgan, 1976). For best growth and yield, the optimum soil temperature is 25-30°C (Wendt, 1978). High temperature (27-32°C) drastically decreased pod production even though branching and flowering were increased (Kigel $et al.$, 1991). For pole types the maximum and minimum temperatures for seed germination and growth are 25°C and 18-20°C, respectively, while temperatures below 13-14°C (minimum) and above 25°C (maximum) are limiting (Dauple $et al.$, 1979). However, in mid-hills of North-eastern region, particularly in Meghalaya, pole beans are grown from March to December when highest summer temperature reaches up to 32°C. Waters $et al.$ (1983) found that N-fixation rate and node fresh weight in $P.$ $vulgaris$ cultivars were over 10-fold higher at a cool, high rainfall mountain area (Papayan) than at a hot, medium rainfall valley (Palmira) in Colombia.

6.0 Cultivation

6.1.0 Seeds and Sowing

Seed quality is a major factor for proper germination. Transverse cotyledon cracking (TVC) has been attributed to be a major problem (MacCollum, 1953).

Promising Cultivars of French Bean

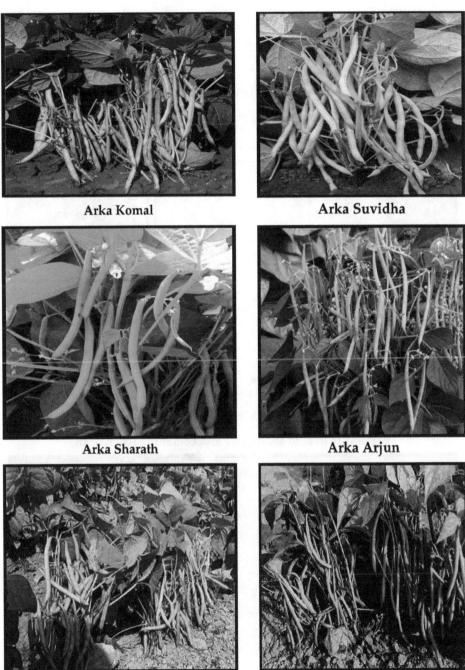

Arka Komal

Arka Suvidha

Arka Sharath

Arka Arjun

Kashi Sampann

Kashi Rajhans

Seed Types, Seed Sowing and Seed Germination in French Bean

Variability in seed shape, size and colour in French bean

Seed sowing (left) and germination (right) in French bean

Dickson *et al.* (1973) and Dickson and Boettger (1976, 1977) have studied this problem in detail. According to them, TVC varied from 0 to 95 per cent among bean cultivars and was enhanced by planting dry seed in wet soil. White seeded cultivars are more susceptible to this problem. McCollum (1953) found that the seed-coat permeability was associated with TVC because susceptible cultivars imbibe water rapidly. Morris *et al.* (1970) reported that structural weakness appeared to be in cell wall as rupture occurred along cell walls. Dickson *et al.* (1973) concluded that seed moisture content was responsible for TVC. Subsequent experiments of Dickson and Boettger (1976) proved that hard seed-coat is essential for resistance to TVC and seed coat shattering (SH). The cultivars resistant to such disorders showed a germination percentage of 60-80, while susceptible ones showed 20 per cent and less. Commercial seeds contain about 7.7-13.7 per cent moisture. Roos and Monolo (1976) found that seeds containing 12 per cent moisture had better germination percentage than low moisture containing seeds when soil temperature was below 10°C but they lost moisture in dry soil quickly. The viability of seed would be lost within 9 months in storage if moisture content were more than 6.7 per cent at storage temperature of 30°C. While with the same moisture content, if seeds were stored at 20°C, it would remain viable for 3 years with 85 per cent germination (Zink *et al.*, 1976). Another germination disorder, hypocotyl necrosis, was found to be associated with low calcium content in seeds (Shannon *et al.*, 1967). But Dickson *et al.* (1973) found that Ca can affect seed, if it was within limit but had generally less effect than Mg. But soils rich in calcium and magnesium can offset the problem.

Seed treatment with carboxin has been reported to significantly increase per cent stand (Keinath *et al.*, 2000). A matriconditioning procedure based on the matric properties of Micro-Cel E (a synthetic calcium silicate) and expanded vermiculite No. 5 has proved effective in improving seedling emergence in growth chambers (Khan *et al.*, 1992). Conditioning in Micro-Cel E, alone or in combination with GA_3, thus appears to be a viable alternative to conditioning seeds in liquid carriers (Khan *et al.*, 1992). Subjecting the French bean seeds to osmotic priming with PEG 8000 at –1.25 Mpa results in better germination (Parthasarathy *et al.*, 1993).

To combat the seed-borne fungi, which might affect the seed germination, seed treatment with benomyl, thiram, captan or carboxin was recommended (Ellis *et al.*, 1977). Thakur *et al.* (1991) found that pre-emergence damping off was best controlled by seed treatment with benomyl, benalaxyl + copper oxychloride and thiabendazole. Post-emergence damping off was least after treatment with carbendazim, bitertanol, benomyl and with thiabendazole.

The soil amendments with low C: N ratio increased the incidence of *Rhizoctonia solani,* in any decomposition level, whereas the amendment with high C: N ratio did not interfere on the incidence of the pathogen (Fenille and de Souza, 1999). The incidence of *R. solani* on *P. vulgaris* plants, in a soil amended with both castor-oil cake or sugarcane bagasse, was independent of soil moisture conditions (Fenille and de Souza, 1999). The AR3-2S is a soil amendment containing 10 per cent (W/W) cattle manure, 20 per cent rice chaff, 10 per cent crab shell meal, 6 per cent urea, 20 per cent calcium superphosphate, 4 per cent potassium chloride, and 30 per cent mineral ash (Hsieh *et al.*, 1999). In large-scale field trials, AR3-2S consistently showed its potential to suppress southern blight, both in the seedling and mature stages of snap bean. AR3-2S applied to infested soil at 20 g per sowing hole, 10 days before sowing achieved significant control (Hsieh *et al.*, 1999). Crucifers can be used as soil amendments to control soilborne plant pathogens, due to their content of glucosinolates. Soil amendment by the addition of dried, ground canola leaves and petioles to the furrow at sowing increased the number of nodules on bean roots in all years and damping-off was not affected by the addition of canola at sowing (Smith, 2000).

French beans are sown during September to November in South Indian plains, during February in North Indian plains, while it is grown from April to October in hills.

In Brazil, the number of days required for maturity was more with sowing in autumn/winter (May/June) than with sowing in spring/summer (rainy season) or in summer/autumn (dry season) (Vieira, 1991).

Sowing in the spring (24-26 April) gave higher yields than sowing in summer (20-24 July) but the later sowing produced better quality pods (Corokalo *et al.*, 1992). Use of plastic tunnels produced more vigorous plants and all the cultivars studied showed increased growth with higher quality pods with the sowing date of 29 February compared to the other sowing dates (Singer *et al.*, 1999).

6.2.0 Spacing

Increased plant population has been found to increase yields, which can be obtained by more plants/row and close spacing of rows. Row spacing affected yield and colour, *viz.*, narrow rows (22.9 cm) had less colour intensity and uniformity resulting in reduced sensory quality (Tompkins *et al.*, 1972). Drake and Silbernagel (1982) while confirming the earlier results, also found that narrow rows (22.9 cm) produced bean pods containing 20 per cent more ascorbic acid than wide row grown plants (55.9 cm). They found that row spacing had a major influence on all the quality attributes of frozen snap beans. More drip loss, less moisture and increased soluble solids were noticed from pods of narrow row grown plants. Mack (1983) reported a yield increase of 20-38 per cent at higher plant densities (15.2-30.5 cm, *i.e.*, and 43-65 plants/m²) than at lower densities (91.4-cm rows, 22-29 plants/m²). For U.P. hills, Pandey *et al.* (1974) suggested a spacing of 60 × 10 cm, while for Kodai hills in South, the spacing recommended is 25 × 25 cm (Prasad *et al.*, 1978). In the hilly region of Darjeeling district of West Bengal, the recommended spacing is 40 × 10 cm (Anon., 1982-83). Venkataratnam (1973) recommended 25 cm and Katyal (1977) recommended 45-60 cm × 10-15 cm. The spacing varies among cultivars as well as growing seasons. Bull (1977) found in New Zealand that seeds sown during December yielded well with close spacing, while during January no effect of spacing was noticed. Narrow spacing along with a spray of Ethrel at the rate of 0.25-kg a.i./ha at second true leaf stage gave higher yields (Wraight and Rogers, 1978). Narrow spacing showed higher yield by virtue of more number of pods/plant, while pod number and average pod weight were negatively associated (Duranti and Lanza, 1978). An increase of plant number from 6 to 9/30 cm would result in yield increase of about 15 per cent and further increase to 12 plants/30 cm would augment a further yield increase of 2.5 per cent which (12 plants/30 cm) seemed to be an optimum number (Tompkins *et al.*, 1979). Another advantage of narrow spacing would be that the pods would be borne higher up from the ground thus preventing infections to pods from soil.

In France, Bourillet (1989) observed that by reducing the inter-row spacing from 40 to 20 cm and increasing plant density from 30 to 40 plants/m^2 the yield potential of bean increased from 10.5 to 12.34 tonnes/ha and resulted in a net production gain of 8 per cent. Reducing inter-row spacing also decreased weed population, reduced lodging and facilitated mechanical harvesting.

Pole (climbing) beans are spaced more between rows compared to narrow spacing of bush beans. According to Thompson and Kelly (1957), pole beans are planted in hills of South-Eastern United States in a spacing of 3′ × 3′ (0.9 × 0.9 m), 3′ × 4′ (0.9 × 1.2 m) or 4′ × 4′ (1.2 × 1.2 m) with 4-6 seeds/hill and thinned to 3-4 plants, while in the Pacific coast they observed an intra-row spacing of 3″-5″ (7.6-12.7 cm) and inter-row spacing of 4′-7′ (1.2-2.1 m). Crespo and Torres (1979), however, recommended a closer spacing of 30-45 cm (between rows) and 8-23 cm (within rows) depending upon pole cultivars, the highest being for the cultivar Blue Lake 45 × 23 cm.

Marketable yields increased linearly as row spacing decreased from 92 to 31 cm (Tyson and Kostewicz, 1986). The amount of weed competition as measured by weed weight and number was not affected by row spacing where plant populations were constant. Weed competition generally declined linearly as the row spacing decreased from 92 to 31 cm when the number of seeds/row was constant (Tyson and Kostewicz, 1986). Snap bean shows sensitivity to single- and twin-drill configurations in yield response. Twin-drill plots of snap bean yielded over twice as much as single-drilled beans (Parish *et al.*, 1996).

Seed rate, hence, varies depending upon the cultivar and spacing. About 50-75 kg would be required for dwarf bean, while it would be about 25-30 kg for pole types/ha.

6.3.0 Manuring and Fertilization

Gabr (2000) suggested cattle manure combined with chicken manure could be used for snap bean production.

French bean is highly responsive to fertilizer and has a marked response to nitrogen. French bean lacks biological N fixation because of poor or no nodulation (Ghosal *et al.*, 2000; Usha *et al.*, 2019). Hence it needs liberal N fertilization (100-120 kg/ha) which is also an important constituent of chlorophyll and takes part in protein synthesis and vegetative growth.

The fertilizer requirement varies among cultivars. Mack (1983) found no significant interaction between plant density and doses of fertilizer. He found that higher rates of fertilizer application produced higher yield of pods. Stang *et al.* (1979), however, reported that plant density was directly related to the level of nitrogen fertilization up to the highest level of 100 kg/ha used by them. But in Himachal Pradesh, Sharma *et al.* (1976) found no influence due to N, while P at the rate of

60-90 kg/ha was found to increase yield. Pandey *et al.* (1974) recommended 125 kg/ha of phosphorus in the U.P. hills. Singh (1987) reported that optimum doses of nitrogen and phosphorus are 67.3 kg and 79.7 kg/ha, respectively. Cardoso *et al.* (1978) also found no significant effect of different sources of nitrogen on yield.

The poor response to nitrogen may be due to the inherent capacity of legumes to fix atmospheric nitrogen. However, a starter dressing of nitrogen in low quantum is needed. Riguad (1976) also observed in hydroponics cultures, that nitrogen fixation by bean nodules were strongly inhibited when calcium nitrate was present in the medium. The *Rhizobium* bacteria, in symbiosis, produced substantial quantities of cytokinins (Puppo and Riguad, 1978). Ruschel and Ruschel (1975) also found that plant weight and total nitrogen content was augmented by *Rhizobium* inoculation even without addition of nitrogen. Nodulation decreases with the increase in nitrogen application. Ssali and Keya (1985) found that French bean plants fixed 27.7 to 35.2 kg nitrogen/ha when 0-20 kg N/ha was applied and nitrogen fixation was reduced by 15-47 per cent by a nitrogen rate of 100 kg N/ha. Correa *et al.* (1990) reported that seed yield with Mo application was increased by soil inoculation with *Rhizobium phaseoli*. Neptune *et al.* (1978, 1979) recommended placement of 60 kg N+120 kg P/ha, 10 cm deep and 5 cm on each side of row after covering with a soil layer before sowing for best uptake of fertilizer but they found no influence of nitrogen on P utilization.

When *Rhizobium* inoculated French bean plants were grown under different N sources it was found that nitrate and ammonium were more inhibitory to nodulation than urea (Hine and Sprent, 1988). Ghosal *et al.* (2000) conducted a field experiment in Bihar, India to evaluate the effect of varying N rates (0, 40, 80, 120 and 160 kg/ha) and times of application (full basal, 1/2 as basal + 1/2 at branching, 1/2 as basal + 1/2 at 50 per cent pod formation, 1/3 as basal + 1/3 at branching + 1/3 at 50 per cent pod formation, and 1/4 as basal + 1/2 at branching + 1/4 at 50 per cent pod formation) on the growth and yield of cv. HUR-15. N at 160 kg/ha gave the highest values for number of pods per plant, weight of unfilled pods per plant, grains per pod, grain weight per plant, 100-grain weight and dry weight per plant. N at 160 kg/ha also gave the highest grain (24.88 q/ha) and straw yields (26.22 q/ha). Among the different timings of application, N applied 1/2 as basal and 1/2 at branching gave the highest values for all the yield components, and the highest grain (20.43 q/ha) and straw yield (20.65 q/ha).

Phosphorus application enhances nodulation (Graham and Rosas, 1979). The maximum requirement of P was found to be 50 kg/ha (Parodi *et al.*, 1977). They also observed that P application and high organic matter in soil aided in getting high yield did not affect relative and total N and K contents in plants. While P at the rate of 60-240 kg/ha enhanced yield, K, however, decreased production, the economic rate of P application being 55 kg/ha (Eira *et al.*, 1974). Phosphorus as single superphosphate at 120 kg/ha has been found to improve yield (Roy and Parthasarathy, 1999).

Passarinho and Ricardo (1988) viewed that earliness in flowering is positively correlated with efficiency of K utilization. From sand culture experiments, Borys and Mamys (1978) found that the largest number of pods per plant was obtained from plants treated with Ca, Mg, Na and S at the rate of 80, 100, 90 and 72 ppm, respectively. But Smith (1976) cautioned about over-fertilization. He recommended 25 lb each of N and P/acre and found that K though increased vigour did not increase the yield.

Venkataratnam (1973) reported use of 12-15 tonnes of farmyard manure and 50 kg superphosphate per hectare. Singh *et al.* (1975) recommended 125 kg, 40 kg and 100 kg of superphosphate, urea and muriate of potash, respectively for French beans in U.P. hills. According to Jasrotia and Sharma (1999) application of FYM along with phosphorus had a more marked effect on green pod yield and quality of French bean than phosphorus alone. French bean responds significantly to phosphorus application and the best response was noticed at 80 kg P_2O_5. Application of FYM at 20 t/ha along with 80 kg P_2O_5/ha further enhanced the green pod yield. Yield quality parameters like crude protein and total soluble solids also showed a substantial increase at the 80 kg P_2O_5/ha level in the presence of FYM. Chavan *et al.* (2000) conducted a field experiment during the winter season with the cultivars Contender, Arka Komal and Waghya supplied with 3 rates of N (0, 25 and 50 kg/ ha) and 3 rates of P (0, 25 and 75 kg/ha). A basal application of a half rate of N and a full rate of P and K at sowing, and a top dressing of a half rate of N after one month were applied. Seeds were evaluated of N, P and K contents, total dry matter and protein production. The highest P uptake (6.3 kg/ha) by seeds and straw was recorded in both Waghya and Arka-Komal. Waghya recorded the highest total dry matter (17.2 q/ha), seed protein production (128.0 kg/ha) and N and K uptake (31.7 and 12.0 q/ha, respectively). The highest total P uptake (8.5 kg/ha) was recorded from the highest N rate (50 kg/ha). Total P uptake increased linearly with increase in P rates.

Rana *et al.* (1998a) reported that dry matter production (seed and straw) increased significantly up to 120 kg N/ha. N content and uptake of N and P also exhibited a similar trend. Increase in seed dry matter, P content and uptake of N and P was significant up to 100 kg P_2O_5/ha. Straw dry matter increased significantly up to 50 kg P_2O_5/ha. Subsequently they (Rana and Singh, 1998b) found that the mean increase in seed yield with 120 kg N/ha compared with 0, 40 and 80 kg N/ha was 66.6, 21.7 and 7.0 per cent, respectively. Growth and yield parameters generally followed the same trend. Applied P also increased seed yield, and 100 kg P_2O_5/ ha gave 39.8 and 7.4 per cent more yields than 0 and 50 kg P_2O_5/ha, respectively. Arya *et al.* (1999) found that high doses of N and K increased the days to 50 per cent flowering (*e.g.*, 50 days with N 50 + P 75 + K 100), but high P and lower K induced early flowering (*e.g.*, 47 days with N 50 + P 100 + K 50). The minimum number of days to seed maturity (81.32 days) was achieved with N 25 + P 75 + K 50. Highest 100-seed weight (46.78 g) was achieved with N 50 + P 125 + K 50. The highest seed germination percentage (90.2 per cent) and seed vigour index (916.0) were recorded with N 25 + P 50 + K 50. Seed yield was highest (13.27 q/ha) with N 25 + P 75 + K 100. It is concluded that N promotes growth and delays the generative phase and that P positively affects seed yield. It is suggested that N 25 + P 75 + K 50 is the best combination in terms of economics and seed yield. Baboo *et al.* (1998) found that the seed yield was highest with 120 kg N and 100 kg P_2O_5. Deshpande *et al.* (1995)

reported that seed yield increased with rate of P application and with decrease in intra-row spacing. A combination of 10 cm intra-row spacing and application of 75 kg P_2O_5/ha gave the highest yield of 2.72 t/ha. Ahlawat (1996) reported that lower plant density of 222 000 plants/ha showed significant increase in pods/plant compared with that of 333 000 plants/ha. Phosphate application greatly improved the yield attributes (pods/plant and seeds/pod), seed yield and the N and P uptake. The response of applied P was linear up to 26.4 kg P/ha.

Jitendra *et al.* (2018) obtained relatively bolder seeds with more number of seeds per pod, and higher bearing capacity per plant with the application of 120 kg N/ha. Singh *et al.* (2018) also proved that nitrogen at the rate of 120 kg/ha was superior to 90, 60 and 30 kg/ha in respect of growth, yield attributes and seed yield of French bean Allahabad, India. Srinivas and Naik (1988) recommended the optimum fertilizer levels of 125.0 N/ha for French bean. Pod yield increased proportionately to N levels up to 160 kg/ha. Production increase in response to N fertilization increased weight and number of pods/plant and harvest index. Sharma *et al.* (1996) found that increasing level of N significantly increased seed yield, number and weight of pods/plant and number of seeds/pod up to 120 kg N/ha. Application of N in three equal splits gave higher seed yield and yield attributes of the crop. Tewari and Singh (2000) conducted a study to determine the optimum and economical dose of nitrogen (0, 40, 80, 120 or 160 kg/ha) and phosphorus (0, 20, 40 or 60 kg/ha) for better growth and seed yield of French bean and found that application of 120 kg N/ha produced significantly higher number of pods per plant, weight of seeds per plant, number of seeds per pod and seed yield. The differences in the plant growth characters may be due to the genetic variability within the genotype itself or may be due to the environmental effects. Prajapati *et al.* (2003) reported that application of 120 kg N/ha enhanced significantly higher growth and yield attributing parameters as well as seed yield of French bean. French bean is the main source of plant protein, of which nitrogen is the main constituent necessary for growth and chlorophyll synthesis and is a part of the chlorophyll. Nitrogen is essential constituent of amino acid and helps in protein synthesis, fruit formation. As compared to other leguminous crops, nodule formation in roots of French bean is very less or even absent in the plain land. Therefore, it is high nutrient demanding crop and due to this the crop responds sharply to high doses of nitrogen (Ghosal *et al.*, 2000). Application of vermicompost (10 t/ha) + N (50 kg/ha) + *Rhizobium* (15 g/kg seed) + PSB (15 g/kg) + P_2O_5 (80 kg/ha) + K_2O (80 kg/ha) has recorded highest growth, yield parameters and seed yield in French bean at Varanasi, India (Barcchiya and Kushwah, 2017). Usha *et al.* (2019) obtained the maximum seed yield (1.97 t/ha) from 120 kg N/ha +10 kg S/ha (N3S1). Hence, the nitrogen and sulphur fertilizer management is very important agronomic practice for higher and efficient production of French bean seeds.

Dwivedi (1995) tested acidulated products of Lalitpur rock phosphate with the mineral acids HNO_3, HCl, H_2SO_4 and H_3PO_4 at 25, 50, 75, and 100 per cent acidulation on an acid Inceptisol (pH 4.8) in India on French bean. Acidulation increased the solubility of rock phosphate, matching the degree of acidulation, irrespective of the acid used. Among different acidulants, H_2SO_4 was the best followed by H_3PO_4, HCl and HNO_3. Acidulation at 25 per cent with sulphuric acid gave the largest bean yield, maintained greatest available P status of soil throughout the growth period and recorded largest uptake of N and P. It also had a pronounced effect on the availability coefficient ratio, fertilizer use efficiency, P use efficiency and relative agronomic effectiveness (RAE). The RAE of partially acidulated rock phosphate products (PARP) followed the order: H_2SO_4-PARP > H_3PO_4-PARP > HCl-PARP > HNO_3-PARP > RP > SSP.

French beans are susceptible to various micronutrients. The symptoms of deficiency and correction are given in Table 3. Foliar application of B, Cu, Mo, Fe, Zn, Mn and Mg each at 0.1 per cent was found effective in increasing pod yield under polyhouse condition (temperature 10-25°C and RH 80-92 per cent) (Jana and Kabir, 1987). Higher green pod yield (189.89 q/ha) of French bean was obtained by soil application of $FeSO_4$ @ 15 kg/ha and $ZnSO_4$ @ 10 kg/ha along with GRDF which was closely followed by foliar application of each chelated iron and chelated zinc @ 0.2 per cent along with GRDF (Bhamare *et al.*, 2018).

Hsieh *et al.* (1999) studied the effect of a soil amendment (AR3-2S) treatment on the control of snap bean southern blight caused by *Sclerotium rolfsii* [*Corticium rolfsii*]. The AR3-2S amendment was modified from a previously described AR3-2 amendment designed for the control of lily southern blight [same causal organism]. The AR3-2S amendment included 10 per cent (W/W) cattle manure, 20 per cent rice chaff, 10 per cent crab shell meal, 6 per cent urea, 20 per cent calcium superphosphate, 4 per cent potassium chloride, and 30 per cent mineral ash. During field trials, the effect of AR3-2S, AR3-2 and calcium cyanamide treatments were evaluated. Two weeks before sowing, 2000 kg amendment/ha or 250 kg calcium cyanamide/ha were mixed mechanically into naturally infested soils. Each treatment was further divided into mulching and non-mulching treatments for two weeks. All treatments increased sclerotial mortality and significantly reduced the incidence of southern blight in both mulched and non-mulched treatments. In large-scale field trials, AR3-2S consistently showed its potential to suppress southern blight, both in the seedling and mature stages of snap bean. AR3-2S applied to infested soil at 20 g per sowing hole, 10 days before sowing achieved significant control.

Crucifers can be used as soil amendments to control soil borne plant pathogens, due to their content of glucosinolates (Smith, 2000). Emergence of snap beans in field soil was reduced by at least 64 per cent by the addition of dried, ground canola leaves and petioles to the furrow at sowing. Soil amendment with the tissue increased the number of nodules on bean roots in all years. In plots with reduced stand, leaf area was increased and yield on a per-plant basis was larger than in plots with a better stand. Frequency of isolation of fungi that cause damping-off was not affected by the addition of canola at sowing (Smith, 2000).

6.4.0 Irrigation

Water stress has a marked influence on pod yield, number of seeds, sieve size distribution, colour, firmness and sloughing (separation from epidermal cells during blan-ching) (Gonzalez and Williams, 1978; Sistrunk *et al.*, 1960). Drake and Silbernagal (1982) found that irrigation treatment significantly influenced the turbidity of the brine from the canned snap beans. They found that the method of irrigation had a strong influence on snap bean colour, ascorbic acid content, relative firmness, and turbidity and drained weight. According to them, sprinkle irrigated fresh and canned snap bean pods contained more ascorbic acid than rill irrigated ones, the latter produced a more intense dark colour, less turbid brine and less drained weight, thus indicating that rill irrigated beans contain less moisture. Byron (1948) found that lack of soil moisture affected chemical composition but due to higher dry matter from low soil moisture conditions the difference became negligible when calculated on dry weight basis. Bonanno and Mack (1983) found that the percentage of set pods, pod length and number of seed per pod were reduced by low irrigation, while Fibre content of pods and weight/seed increased by low irrigation. They opined that low irrigation, which increased the canopy temperature, was responsible for high fibre content. Kattan and Fleming (1956) found that drought injury and water consumption increased with the age of plants. The pods were malformed and poor in colour than beans from irrigated plots. Rolbiecki *et al.* (2000) reported from Poland that the average long-term effect of overhead irrigation of the crops (increases marketable yields caused by irrigation) was 119 per cent in French bean. Better irrigation effects were obtained on soils characterized by small water capacity. On loose sandy soils, sprinkler irrigation was the basic cropping factor for vegetables.

The yield is reduced when there is a stress during flowering (Dubetz and Mahalle, 1969) and post-bloom stage (Maurer *et al.*, 1969). Pre-bloom irrigation is not essential if moisture is present in soil but irrigation is necessary during bloom and pod development period (Gableman and Williams, 1960). The stress has no influence on harvest time and pod size but it affects the yield (Gonzalez and Williams, 1978).

The effect of water stress on dry beans can vary depending on the frequency, duration, timing and intensity of stress and is highly variable in response due to interacting factors (Nielsen and Nelson, 1998; Terán and Singh, 2002; Nuñez Barrios *et al.*, 2005). It is generally accepted that water stress adversely affects the physiological, morphological, and growth traits of dry bean (Loss and Siddique, 1997), however, the magnitude to which the reduction occurs varies from region to region and there is a lack of consistency in the reported response of dry bean to water stress. Ramirez and Kelly (1998) reported that the characteristics of many dry bean traits reduced under water stress conditions and resulted in reduced seed yield from 22 per cent to 71 per cent. In the upland central plateau of Mexico, Gallegos and Shibata (1989) reported many morphological traits were negatively affected by water stress, which resulted in a reduced number of leaves and reduced the seed yield by 34 per cent to 50 per cent. A water stress study conducted by Stoker (1974) on dwarf beans in the sub-tropical climate of New Zealand reported a 13 per cent to 24 per cent reduction in seed yield caused mainly by the abscission of flowers and young pods. In another study, Miller and Burke (1983) reported maximum dry bean seed yield and biomass when irrigated to replace 100 per cent of estimated crop evapotranspiration (1.0ETc) and observed no significant increase in seed yield or biomass above 1.0ETc. While, Webber *et al.* (2006), found no significant differences in aboveground biomass and seed yield for common bean for the three irrigation levels tested (moderate, recommended, and above recommended). In addition to this, water stress decreases root growth and results in decreased nutrient uptake (Mooney *et al.*, 1991), and also, decreases nitrogen fixation in root nodules (Ramos *et al.*, 1999; Serraj and Sinclair, 1998) by reducing the contact between root and soil and thereby, impacting the growth and seed yield.

About 6-7 irrigation during the growing season would be required at regular intervals as the plants are susceptible to water stress at all stages of growth and development (Miranda and Belmar, 1977). Short watering intervals (2-3 days) during vegetative stage augments marketable and total yield with no effect on pod length while watering intervals during main flowering stage has no effect on yield or quality (M'Ribus, 1985).

Table 2: Symptoms of Micronutrient Deficiency and Recommendation

Nutrient	Deficiency Symptoms	Nutrient Recommendations	Reference
Boron	Normally seedlings are affected. Thick roots and leathery leaves, seedlings are stunted. Symptoms akin to witches' broom	1 kg B/ha as borax (soil application)	Singh and Singh (1988)
Zinc	Interveinal chlorosis. Normally in soils of pH above 7.0	$ZnSO_4$ @ 25 kg/ha (soil application)	Thung (1991)
Iron	Due to mobile nature it affects young leaves. Chlorotic leaves occur under alkaline condition	Foliar application of polyphosphate of iron chelate	Mortvedt and Giordano (1971)
Molybdenum	Essential for nodulation. Collapse of tissues between main veins and along edges of leaflets. Stunting and defoliation occur	1 kg ammonium molybdate/ha	Zaumeyer and Thomas (1957) Franco and Day (1980)
Copper	Richitic with shoot internodes and young leaves turn blue green.	0.5 to 1.0 kg copper sulphate as soil application	Lucas and Knezek (1972) Thung (1991)
Manganese	Dwarfing of plants. Interveins become golden before sowing yellow, occur at pH higher than 6-7 in organic or highly limed soils	5 to 10 kg of manganese sulphate	Fitts *et al.* (1967)

Uncini *et al.* (1988) from Italy reported that in dwarf beans cv. Taylors Hort, a total water rate of 1800 m^3/ha (over the whole growing cycle) should satisfy the crop's requirements and give a shelled bean yield of 67 q/ha. Stone *et al.* (1988) in Brazil observed that irrigation of *Phaseolus vulgaris*, when soil water tension reached 125, 250, 375, 500, 625 or 750-m bar increased yield as the tension at which irrigation was applied increased. In field experiments in France, irrigating French bean cv. Faria after flowering at a rate equivalent to 120 per cent evapotranspiration (ETP) increased pod number and improved pod quality compared with irrigation at 100 per cent ETP (Le Delliou, 1989). In Fort Valley, USA, Singh (1989) suggested an irrigation schedule based on 80 per cent Epan (open pan evaporation) for high snap bean production.

Excess moisture also adversely affects the production. It has been reported that even 1 day of flooding reduces photosynthesis in snap bean and causes a decrease in dry weight of the plant (Singh et al., 1981). Singh et al. (1996) gave irrigation during winter seasons for the cv. HUR 15 at 0, 0.25, 0.50, 0.75 or 1.00 irrigation water: cumulative pan evaporation (IW: CPE) ratios. They had given no NPK fertilizers, 40: 30: 20 kg N: P_2O_5: K_2O/ha, or 2 or 3 times these application rates. Seed yield and net returns increased with up to 0.75 IW: CPE ratio and the highest NPK rate. Uptake of N, P and K increased with fertilizer rate and was highest at 0.75 IW: CPE. Dahatonde et al. (1992) found for the same cultivar that the seed yield increased from 0.38 to 0.92 t/ha with increase in number of irrigations and it increased with up to 90 kg N/ha. There was a significant interaction between irrigation and N application. There were no significant effects of treatments on test weight. Dahatonde and Nalamwar (1996) found for the cultivar HUR15, irrigation at an IW: CPE ratio of 1.2 (6 irrigations) resulted in the highest seed yields for both years (mean 0.76 t/ha) and a significant increase in seed yield to N fertilizer application was observed up to 90 kg N/ha. Application of 120 kg N/ha did not result in any further increase in yields compared with the 90 kg N/ha application. The highest water-use efficiency was observed with irrigation at 1.2 and 1.0 IW: CPE ratios during the first and second years, respectively, while irrigation at 1.2 IW: CPE ratio + 90 kg N/ha application resulted in the highest seed yield due to a positive significant interaction effect. Durge et al. (1997) reported that the yield of *Phaseolus vulgaris* cv. VL-63 was 626, 662 and 815 kg/ha with irrigation at irrigation water: cumulative pan evaporation ratios of 0.5, 0.75 and 1.0, respectively, and was highest (957 kg) with 150 kg N/ha. Singh (1993) found that dry weight per plant at maturity increased with the number of irrigation and with higher fertility level. Irrigation proved essential for late-sown crops. Contrary to expectations, there was no visible increase in virus attack in late-sown crops. Advantages of late sowing included improved pod quality owing to the wetter and cooler conditions late in the summer, a considerably extended harvest season and the use of late beans as catch crops (Reichel, 1992).

Nandan and Prasad (1998a) reported that seed yield in the first year was highest (1.31 t/ha) when given 3 irrigations at 25, 50 and 75 days after sowing, while in the second year the highest yield of 1.35 t was given when irrigating at a 0.8 IW: CPE [irrigation water: cumulative pan evaporation] ratio. Yield and water-use efficiency increased with increasing N rate in both years. Nandan and Prasad (1998b) later reported that leaf-area index and dry-matter production were maximum at 3 irrigations, 0.8 IW: CPE (IW, irrigation water; CPE, cumulative pan evaporation) ratio and 120 kg N/ha.

The actual crop evapo-transpiration (ETc) for green bean varies with environments. Based on a lysimeter study, Tarantino and Rubino (1989) calculated the seasonal water requirement for maximum bean yield as 235 L/m^2. According to Sazen et al. (2008), a significant linear relationship was found between supplied water and pod yield ranging from 276 to 472 L/m^2 for green bean cultivated in the open field. El-Noemani et al. (2010) recommended applying 371 L/m^2 for maximum pod yield of green beans cultivated in an open field. Although there are several studies demonstrating the effects of a water regime on green bean in open fields (Sazen *et al.*, 2008; Abd El-Aal *et al.*, 2011), the results from such studies are limited only to the examined area and mostly done using a single cultivar. Saleh *et al.* (2018) concluded that 80 per cent of ET was quite enough to achieve the maximum productivity for green beans. El-Noemani *et al.* (2010) also noted that increasing the irrigation amount up to 100 per cent of ET prompted the highest growth, although the maximum pod yield was achieved by 80 per cent of ET. Under stress conditions, water deficit had a significant impact on plant growth, leading to a decline in growth, leaf area development, and photosynthetic capacity (Bayuelo-Jimenez *et al.*, 2003). The reduction in bean productivity (number of pods per plant and seed biomass) due to heat stress was associated with reduced leaf water content (Omae *et al.*, 2004, 2005). Under scarce water in semi-arid areas, irrigation water management aims to provide sufficient water to replenish depleted soil water in time to avoid physiological water stress in growing plants, using modern irrigation technologies such as a drip irrigation system (Saleh *et al.*, 2012; Abdel-Mawgowd *et al.*, 2006).

Measurements of actual evapotranspiration of French bean grown in a ventilated and heated plastic tunnel-type greenhouse were closely correlated with values of evapotranspiration within and outside the greenhouse estimated by the Penman method (Martins and Gonzalez, 1995). Guimaraes *et al.* (1996b) found genotypic variation in water use and suggested breeding programmes. Guimaraes *et al.* (1996a) studied the effect of water absorption efficiency of the roots and linear root density on drought resistance in the field in 3 cultivars under 3 moisture treatments – irrigation, moderate water stress and severe water stress. Water absorption efficiency increased with soil depth, independently of water deficiency level or cultivar.

Velich (1993) found a good correlation with a number of characters, *e.g.*, with root length and with root + foliage weight in trials involving polyethylene glycol treatment, with root weight in irrigation trials, and with canopy temperature (25 and 35°C). He reported that the data obtained are thought useful for establishing values for a drought tolerance index which would allow the classification of lines and varieties for use in breeding.

A few workers as affected by salinity and irrigation method studied growth response of French bean (Scholberg and Locascio, 1999; Pascale *et al.*, 1997).

The combined effect of soil gel-conditioner and irrigation water quality and level on growth, productivity, and water-use efficiency of snap bean in sandy soils was studied by Al-Sheikh and Al-Darby (1996). Generally, the addition of the gel-conditioner promotes seed germination and increases plant height, leaf area index and dry matter of shoots and roots.

In a pot experiment in a polyhouse, Upreti and Murti (1999) exposed French beans (*Phaseolus vulgaris*) cv. Contender (water stress tolerant) and IIHR-909 (susceptible) to water stress for 0-12 days. Water stress decreased nodule number and mass, root length and content of zeatin riboside in nodules and roots. The tolerant cultivar showed better nodulation and higher levels of zeatin riboside in the nodules and roots with and without water stress.

A very recent comprehensive study was undertaken to quantify the effect of irrigation treatments on dry bean growth, seed yield, and yield components and to develop empirical models between dry bean growth indices and environmental conditions, seed yield, and yield components in response to water deficit in the arid to semi-arid climate of USA (Rai *et al.*, 2020). The models developed in this study are among the first investigated models developed using the measured data for dry beans. The results of this research showed that both dry bean growth and seed yield were strongly affected by water deficit. However, the degree to which the vegetative growth and seed yield were reduced was dependent on the weather conditions. Water deficit at the beginning of the vegetative growth dramatically reduced dry bean h, leaf area index (LAI), and normalized difference vegetative index (NDVI) and shortened the number of days to reach the maximum h, LAI, and NDVI. The growth models presented in this research for different irrigation treatments can act as useful tools to assess in-season crop growth and detection of stress. It was observed that an irrigation amount in the range of 250–270 mm promotes optimum growth (highest LAI, NDVI, and h) and result in high seed yield. Irrigation beyond that range does not lead to a significant increase in plant growth and seed yield. Reduction in 25 per cent of irrigation from full irrigation treatment (FIT) resulted in an average reduction of 30 per cent in dry bean seed yield. Much higher seed yield losses of 40–93 per cent should be expected as a result of severe water deficit. Inter-relationship among different growth parameters presented in this study would allow for easy parameter conversions, assessing plant growth without destructive sampling, help build more accurate whole-plant growth models with fewer required parameters, and improve simulations at a regional scale using remote sensing technology.

The combination of suitable cultivars, improvement of WUE, use of modern irrigation systems, and optimization of cultivation management practices can improve cost-effectiveness and minimize problems of water shortages in French bean with a particular emphasis on sustainable resource management and environmental protection.

6.5.0 Weed Control and Staking

Inadequate weed control reduces yield. However, smothering of weeds occurs with the full development of canopy. Teasdale and Frank (1980) observed that narrow spacing reduced weed competition and hoeing increased yield on an average by about 11 per cent. About two weedings may be needed till the plants can cover well and to prevent weed competition with the crops. Pre-emergence application of bentazone (basagran) at 2.0 kg/ha along with thiobencarb (bolero) at 0.5 kg/ha or bentazone at 0.5 kg/ha+methabenzthiazuron (tribunil) help in getting better weed control and yield (Martinez and Soto, 1978). Other pre-emergence herbicides recommended are linouron or monolinouron at the rate of 0.375 kg/ha along with methyl 1-N- (3, 4-dichlorophenyl) carbamate at the rate of 3 kg/ha (Stevenson, 1977) and surflan (oryzalin) at 0.75-3.0 kg/ha or phenoxylene at the rate of 2-8 1/ha (Aponte *et al.*, 1976) or dinitroaniline herbicides (Glaze and Phatak, 1978). Ethofumesate at 1.5 kg/ha is the safest post-emergence herbicide (Stevenson, 1977). Leela *et al.* (1972) recommended anachlor at the rate of 3-kg a.i./ha. Cox

(1979) recommended application of anachlor on soil surface after sowing without incorporating the herbicide in soil.

Weeds can be managed effectively in French bean by pre-planting application of fluchloralin @ 1.00 kg/ha and pre-emergence application of pendimethalin @ 1.00 kg/ha to get higher productivity and profitability at both locations Jammu and Kashmir, and Uttar Pradesh, India (Kumar *et al.*, 2014; Panotra and Kumar, 2016). Application of imazethapyr 100 g/ha at 20 DAS can be useful for effective and economical weed control in French bean at Maharashtra, India (Goud and Dikey, 2016). Increased growth and higher dry matter accumulation of kharif French bean was recorded with the weed free treatment closely followed by the pendimethalin 30 per cent EC @ 1.0 kg a.i./ha (PE) + one hoeing at 30 DAS and Quizalofop-p-ethyl 5 per cent EC @ 100 g a.i./ha at 20 DAS + one hoeing at 30DAS at Maharashtra, India (Chavan *et al.*, 2019)

In Bulgaria, the best herbicidal effect was achieved with a system involving the use of tobakron (33 per cent metolachlor +17 metobromuron) at 5 l/ha after sowing and basagran (50 bentazone) at 2.5 litres and iloxan (28 per cent diclofop methyl) at 3 litres at the 2nd to 3rd trifoliate leaf stage (Ivanov *et al.*, 1988).

Yield losses of 59 per cent have been attributed to weed interference in dry beans in Ontario which is substantially greater than other field crops such as winter wheat (3 per cent), spring cereals (12 per cent), soybean (38 per cent), and corn (52 per cent) (Anonymous, 2013). Trifluralin (600 g ai/ha), s-metolachlor (1050 g ai/ha), halosulfuron (35 g ai/ha), imazethapyr (45 g ai/ha) applied pre-plant incorporated (PPI) caused minimal and transient injury in French bean and provided excellent full season control of common lambs quarters, redroot pigweed, common ragweed, and wild mustard and fair control of green foxtail (Soltani *et al.*, 2014).

Bentazone has been found to be selective in controlling weeds like *Portulaca oleracea, Ambrosia artemisiifolia, Brassica kaber, Polygonum pensylvanicum, Solanum sarachoides, Xanthium pensylvanicum, Cyperus esculentus* and *Cirsium arvense* (Orr *et al.*, 1977), while Surflan or Phenoxylene controls *Echinochloa colonum* and *Rottboellia exalta*.

At the time of tendril development, the pole type varieties are trained on stakes. The vines are trained by erecting poles on either side of rows and connecting them by rope wires. Staking prevents rotting of pods, improves pod quality and facilitates intercultural operations such as spraying, weeding, harvesting *etc.*

6.6.0 Mulching, Cover Crops and Netting Technology

Mulching is a desirable cultural management practice which is reported to conserve moisture, control weeds, moderate soil temperature, improve soil physical conditions by enhancing biological activity of the soil fauna and increase soil fertility (Mann and Chakor, 1989; Lal, 1989). Different kinds of organic (straw, rice husk, water hyacinth *etc.*) and inorganic mulches (black polythene, transparent polythene, agri silver mulch *etc.*) play important role in conserving soil moisture.

Han *et al.* (1989) noticed that polyethylene film mulch increased soil temperature, decreased the number of days to emergence and to flowering and increased stem height, number of pods/plant and 100 seed weight for both green and dry seed. Kamal *et al.* (2010) conducted a field experiment in Bangladesh to study the effect of mulching on growth and yield of French bean and the results suggested that black polythene mulch produced the highest yield (15.01 t/ha) and profitability (B: C ratio,1.74) over organic mulches (rice straw and water hyacinth) in French bean. Sahariar *et al.* (2015) also recorded the maximum yield of French bean in Bangladesh under black polythene mulch with a crop spacing of 30 cm × 15 cm.

Singh *et al.* (2011) application of N, P_2O_5 and K_2O fertilizer @ 8-15: 13-25: 10-20 kg/ha, vermicompost @ 2.50-3.75 t/ha, 4 cm thick mulch of dried crop residues and 50 per cent irrigation is the most suitable and sustainable strategy to improve plant growth, pod formation and pod yield of French bean, and soil health of mild-tropical climate during dry season of Mizoram, India. Prasad *et al.* (2014) obtained that maximum plant height, leaf area index, dry matter accumulation, crop growth rate and relative growth rate under paddy straw mulch @ 10 t ha^{-1} in French bean at West Bengal, India. A study at AVRDC, Taiwan revealed that the inorganic and organic mulches of black polythene, clear polythene, grass improved the yields of green bean (*Phaseolus vulgaris*), leaf area, leaf area index, soil temperature, weed infestation suppression and soil moisture content (Kwambe *et al.*, 2015). They recommended that farmers should use black polythene and clear polythene mulches for green bean production. If polythene plastic is scarce they advised to use grass mulch for green bean production.

An investigation was carried out to assess the effect of two mulching materials (rice straw mulch (RSM) and farmyard manure mulch (FYM)), three irrigation treatments (I100 per cent = 100 per cent, I85 per cent = 85 per cent and I70 per cent = 70 per cent) of crop evapotranspiration (ETc) and four mulch layer thicknesses (MLT0, 3, 6 and 9 cm) on French bean yield, its components, water use efficiency and soil salinity under drip irrigation (Abd El- Wahed *et al.*, 2017). The maximum values of bean yield were obtained under FYM compared to RSM. The greatest values of dry seed yield of bean were obtained with the no-deficit treatment (I100 per cent). The average bean yield value of MLT9 was increased by 9.67, 25.28 and 45.80 per cent than those of MLT6, 3 and 0, respectively.

Weed suppression is one possible benefit of including cover crops in crop rotations. Cover crops have the potential to influence weed competition in dry beans in several ways, including direct competition, allelopathy, and alterations to the soil environment, which can influence weed seed persistence, weed emergence, and growth (Conklin *et al.*, 2002; Creamer *et al.*, 1996; Dyck and Liebman, 1994; Fisk *et al.*, 2001; Snapp *et al.*, 2005). Direct competition with weeds occurs from the time of cover-crop emergence through termination and may result in reduced inputs to the weed seed bank, and therefore fewer weeds in the dry bean crop (Gallandt, 2006; Ross *et al.*, 2001; Teasdale, 1998).

Dufault *et al.* (2000) conducted research to determine (i) if supplemental nitrogen (N) at 60 or 120 kg/ha following winter cover crops of wheat or crimson clover (*Trifolium incarnatum* L.) affect yield of tomato and snap bean grown in rotation;

and (ii) the distribution and retention of soil nitrates in the soil profile as affected by N fertilization and cover cropping. Fresh market tomatoes cv. Mountain Pride and snap beans cv. Strike were grown in rotation for four years. French bean marketable yield summed over all years was 60 per cent lower in clover plots compared with fallow. Total marketable snap bean yield increased with 60 kg N/ha in one out of three years but was unresponsive to N in two out of three years. Soil nitrates to 1.2 m depth were higher after clover and wheat than after fallow. Nitrate level was highest in soil with clover and 120 kg N/ha. In all cover crop of fallow plots, as fertilizer N application levels increased, the soil nitrates also increased. Cover crops or fertilizer N application did not increase the retention of residual nitrates in the 1.2 m soil profile depth after four years of cropping. The impact of cover crops on weed dynamics in an organic dry bean system was studied in detail by Hill *et al.* (2016).

A number of simple technologies have been tested in different parts of the world and proved successful in protecting crops against adverse weather conditions and insect pests. Netting technology has been used in agriculture to protect crops against environmental hazards like excessive solar radiation, wind, hail and flying insects and improve plant microclimate through reduction in heat/chill, drought stresses, and moderation of rapid climatic stresses leading to improved crop yield and quality (Shahak *et al.*, 2004). The use of net covers in crop production offers a cheaper and less energy consuming technology than greenhouses (Shahak, 2008). Coloured net technology is on the other hand, an emerging technology, which introduces additional benefits on top of the various protective functions of nettings. Photo-selective nets which include the coloured nets are unique in that they both spectrally-modify as well as scatter the transmitted light, absorbing spectral bands shorter or longer than the visible range. Spectral manipulation has a potential for promoting physiological responses in plants while the scattering of light improves penetration into the inner canopy, all of which contribute towards better crop performance (Rajapakse and Shahak, 2007). Different crops respond differently to the spectral manipulation induced by different colours of net covers. French bean grown under the different coloured net covers showed relatively better growth and crop performance marked by more pods and higher total yields and percentage of marketable yields under a white net cover compared to those grown under blue net, yellow net, tricolour net, grey net, and open field condition (Ngelenzi *et al.*, 2017). Growing French bean under net covers hastened the rate of pod maturation more-so under the light-coloured colour-nets. Findings of this study demonstrate the potential of coloured net covers in improving French bean pod yield and quality under tropical field conditions of Nairobi, Kenya.

6.7.0 Effect of Growth Substances

Application of growth substances improved the plant growth, flowering and yield of pod in French bean. GA_3 sprayed on the plants at 50-200 ppm proved very effective in increasing the plant height and number of leaves (Deka and Das, 1975). Several workers (Rackham and Vaughn, 1959; Veinbrants and Rowan, 1971; Chakraborty and Sharma, 1982) also recorded the beneficial effect of gibberellins on vine growth. See and Foy (1983) reported that a combined foliar application of 0.8 kg ethephon and 3.2 kg daminozide/ha inhibited the vegetative growth of *Phaseolus vulgaris.*

Some photosensitive bean plants are poorly adapted to areas with short growing seasons because of late flowering and maturation. A technique to promote early flowering in certain late-maturing bean types would, therefore, be useful. In Nebraska, Coyne (1969) studied the effect of CCC, SADH and GA7 on time of flowering of short-day field bean cvs. Great Northern Nebraska No.1 and Sel. 27, when grown under long and short photoperiods. Sprays of CCC at a concentration of 500 ppm, applied when the first trifoliate leaf was emerging, promoted earlier flowering under long-day conditions, the flowering occurred as early as that of control plants grown under short-day conditions. The plants sprayed with SADH at a concentration of 1500 ppm or GA_7 at a concentration of 100 ppm and grown under both long and short days flowered almost at the same time as the respective control. Rafiqueuddin (1984) recommended the application of 400 ppm of cycocel (CCC) in 2 or 4 foliar sprays for reduction in plant height and higher pod and seed yield. Leprince (1989) reported that SADH or baronet (triapenthenol) applied at 750 g/ha at the beginning of flowering increased yields by an average of 6 and 8 per cent, respectively. Similar results were obtained when amchem (640 ml/ha) at early flowering or 400 ml/ha at full bloom stage were applied (Hurduc *et al.,* 1982). GA_3 at 250 ppm or Cycocel at 100 and 250 ppm produced maximum number of pods per plant, seed yield, seed index and protein content in seeds of French bean cv. Azad P-1 and Aparna (Bora and Sarma, 2006). Application of Salicylic acid (SA) at 10 M-3 was found promising for increasing all yield parameters (pod number, pod fresh and dry weight) of French bean followed by Paclobutrazol (PP333) at 10 ppm (Hegazi and El-Shraiy, 2007).

Rackham and Vaughn (1959) also reported increase in pod set with the use of gibberellin. El-Tahawi *et al.* (1982a) recorded that GA_3 at 10, 25 or 50 ppm applied to 30-day-old plants of cv. Giza-3 increased chlorophyll a and b and the total carbohydrate content of the leaves and they also noted that total N content of leaves and stems was highest with the application of 25 ppm GA_3. Nitrate N and amino N concentration in seeds increased with higher concentration from 10 to 50 ppm, while soluble N, ammonia and amide-N contents were unaffected. Total free amino acid content in the seeds was highest with 50 ppm and in the leaves and stems with 25 and 50 ppm. Seed RNA and DNA contents were highest following treatment with 25 and 50 ppm, respectively (El-Tahawi *et al.,* 1982b). Rathod *et al.* (2015) observed that application of GA_3 at 100 ppm produced the maximum number of green pods, yield per plant and yield per hectare in French bean (cv. Arka Komal) at Maharashtra, India. GA_3 at 30 to 70 ppm gradually increased crop growth rate (CGR), net assimilation rate (NAR) and relative growth rate (RGR), number of dry pods/plant, number of seeds/pod, 1000 seed weight, fresh fodder, fresh pod, dry seed yield and harvest index at Dhaka, Bangladesh (Noor *et al.,* 2017).

Ku *et al.* (1996) found that uniconazole at all concentrations (0.005, 0.01, or 0.02 mg/pot) protected against SO_2 injury; higher uniconazole concentrations provided complete protection. Higher concentrations of uniconazole reduced leaf enlargement and stem elongation by 42 and 53 per cent, respectively, reduced whole plant

Field Growing and Flowering in French Bean

Field view of bush type French bean

Field view of pole type French bean

Flowering stage in French bean

Pod Bearing Stage and Intercropping of French Bean

Pod bearing stage of bush and pole type French bean

Green and purple pod bearing in bush French bean

Brinjal + French bean based intercropping system

transpiration, increased SOD activity and total chlorophyll concentration, decreased MDA accumulation and delayed flowering slightly (1-1.8 days), but had little or no effect on cv. or on number or DW of pods. The increase in SOD activity and the decrease in MDA in SO_2 fumigated plants treated with uniconazole were associated with a reduction in pollution injury. It was suggested that uniconazole-induced SO_2 tolerance might be attributed in part to increased antioxidant activity of plant tissue, thereby reducing cellular damage from oxidative stress. Since stomatal resistance was unaffected by uniconazole treatment, stomatal closure was not a contributing factor to increased SO_2 tolerance. In a field trial in the winter season of 1987-88 at Banswara, Rajasthan, Bhatnagar *et al.* (1992) treated *P. vulgaris* with 20, 40 or 60 kg N/ha and mixtalol [triacontanol-based] as a seed treatment (2 g/kg seed), foliar application (2 ppm) pre-sowing and 30 days after sowing or a combination of seed treatment and foliar spray. Seed yield and N uptake in seed increased and CP percentage decreased with increase in N application rate. All methods of mixtalol application increased seed yield with seed treatment + foliar spray giving the highest seed yield (880 kg/ha) compared with the untreated control (810 kg/ha). Mixtalol also increased CP content and N uptake in seed.

At Kenya, French bean farming requires water supply, especially around flowering and grain filling stages. Irrigating the crop at one-day interval, where there is plenty of water or two-day interval, where there is partial availability, and spraying of cytokinin @ 0.50 ml/litre in both conditions resulted the highest yield of beans (Kalawa *et al.*, 2018).

7.0 Harvesting and Yield

7.1.0 Harvesting

The French bean is ready for harvest from 40 days onwards after germination depending on cultivar. It takes about 7-12 days after flowering for the pods to be ready for picking. In India pods are hand-picked and tender ones are harvested. But in advanced countries, where manual labour is costly machine harvesting is done under once-over harvest system.

About 2-3 pickings are made for bush beans, while it may be 3-5 for pole beans. The quality varies among harvests. Loss of crispness during storage was attributed to the loss of water and an increase in water-soluble pectin (Sistrunk *et al.*, 1960). They also found a large increase of pectin in beans from last harvest and attributed the difference between first and last harvest to the development of more insoluble cell wall substances.

Cultivars differ in seed and fibre content as well as in other quality factors (Gould, 1951). Seed weight is a major indicator of green bean harvest maturity. Silbernagel and Drake (1978) proposed a seed index based upon the formula, seed index = [(seed wt/total pod wt) × 100] × 10-seed length. The seed index would help to do away with many hours of tedious fibre analysis. The seed index was positively correlated with the fibre development of large and medium sieved snap beans whether fresh or canned. However, they suggested the evaluation of other factors such as yield, days and/or heat units to harvest maturity, sieve size distribution and processed product quality at the point when at least 95 per cent of the harvested pods are still within seed index limits for Fancy Grade (Table 4), *i.e.,* no more than 5 per cent Extra Standard.

Most quality conscious processors prefer cultivars that have largest proportion of pods in Fancy and/or Extra Standard grades at the highest potential yield.

Dependence on sieve size development alone as an indicator of optimum harvest maturity can result in processed beans with excessive seed and/or fibre (Robinson *et al.,* 1964).

Table 3: Seed Index Guide in Estimated Snap Bean 'Grade'*

Grade	Maximum (per cent) Seed	Max. Per cent Seed Length Sieve Size			Max. Seed Index Value Sieve Size		
		4 sv	*5 sv*	*6 sv*	*4 sv*	*5 sv*	*6 sv*
Extra Fancy	4	80	90	100	320	360	400
Fancy	8	90	100	110	720	800	880
Extra Standard	12	100	110	120	1200	1320	1440
Top Standard	16	110	120	130	1760	1920	2080
Standard	24.9	120	130	140	2988	3237	3486

* Silbernagel and Drake (1978)

Culpepper (1936) noted a marked change in the composition of snap beans with maturity and suggested the per cent by weight of seed sewed as an index of maturity. Reddy and Sulladmath (1975) found the various parameters, *viz.,* seed size, per cent seed, dry matter content, AIS (alcohol insoluble solids) and distribution of pods according to sieve size, as reliable maturity indices. The maturity stage was reached from 9 to 15 days, depending upon the cultivar.

The prevalent practice in the canning and freezing industry is to judge quality by size distribution. With the mechanical harvesting, the timing of the snap bean harvest has become crucial, not only on the entire yield obtained in a single picking, but also the stripping of the plants in a single operation brings in much wider range in size and quality of raw beans than hand-picked ones (Robinson *et al.,* 1964).

7.2.0 Yield

The yield depends upon type and cultivars. On an average, 60-200 quintals of green pods/ha for bush varieties and 90-260 quintals of green pods/ha for pole varieties are obtained. The yield of dry bean or bean seeds varies from 12 to 18 quintals per ha.

8.0 Diseases and Pests

8.1.0 Diseases

Large number of pathogenic fungi, bacteria and viruses can infect French bean crop. The important diseases caused by these are the following.

Harvesting, Sorting, Grading, Packaging and Marketing of French Bean

Harvesting of French bean pods

Sorting and grading of French bean

Marketing of French bean

Packaging of French bean for long distance transportation

Diseases of French Bean

Sclerotinia blight

Fusarium root rot (left) and rust (right) infection in French bean

Anthracnose infection in French bean leaves and pods

Fungal Diseases

8.1.1 Bean Anthracnose

Anthracnose was first described from plant specimens obtained in Germany in 1875 (Walker, 1957). Since then, the disease has become one of the most important and widely distributed throughout the world. It has been reported in USA, European countries, Canada, Latin America. In Africa, it is particularly important in Uganda, Kenya, Tanzania, Rwanda, Burundi, Ethiopia and D.R. Congo. In Brazil more than 25 different *C. lindemuthianum* races have been identified (Thomazella *et al.*, 2002). In Tanzania yield losses remain very high (40 - 80 per cent) and are estimated to be worth $ 304 million per annum. In Uganda, anthracnose is the most important disease in the high altitude, low temperature areas (Opio *et al.*, 2006). In Sudan, field losses in these regions, due to seedling, leaf, stem and pod infections, are up to 90 per cent under climatic condition favourable to the disease. The infected seeds are the most important means of dissemination of this pathogen, which explains its worldwide distribution (Mudawi *et al.*, 2009).

Anthracnose is the most common disease of beans. The fungus *Colletotrichum lindemuthianum* can infect, besides French bean, other legume vegetables like scarlet bean, lima bean, cowpea, dolichos bean, *etc.* The disease is most severe in high rainfall subtropical to temperate areas than in tropical areas.

The most characteristic symptoms of the disease are black, sunken, crater like cankers on the pods. The lesions remain isolated by yellow-orange margins and under high humidity, dull salmon coloured ooze is formed in the centre. Similar spots are also found on the cotyledons and stem of young seedlings and when severe, can cause seedling mortality. The fungus is seed-borne. Use of healthy seeds, clean cultivation, avoidance of overhead irrigation is cultural management for control. Use of organic sulfur fungicides like thiram, dithane Z-78 and the systemic fungicides like benlate or bavistin give satisfactory control (Babu and Ciotea, 1976). Sequential applications of pyraclostrobin @ 0.1 kg active ingredient (a.i.)/ha at the early flowering (40 per cent bloom) and late flowering (80 per cent bloom) stages usually resulted in the lowest disease severity on all plant parts and in the highest yield (Conner *et al.*, 2004). Benlate (500 g a.i./kg WP) as a seed dressing at a rate of 2 g/kg seed, difenoconazole (250 ml a.i./EC) at a rate of 87.5 g a.i./ha as a foliar spray and Benlate (500 g a.i./kg WP) seed dressing at a rate of 2 g/kg seed followed by foliar spray of difenoconazole (250 ml a.i./EC) at a rate of 87.5 g a.i./ha effectively reduced anthracnose severity and incidence and increased the yield per plot and 100 g seed weight (Beshir, 2003). Mancozeb seed treatment at a rate of 3 g/kg seeds followed by carbendazim foliar spray at a rate of 0.5 kg/ha and carbendazim seed treatment at a rate of 2 g/kg seeds followed by carbendazim foliar spray at a rate of 0.5 kg/ha have been suggested to reduce anthracnose severity and incidence (Mohammed *et al.*, 2013).

Integrated disease management is considered the most effective approach to minimize the yield losses to anthracnose (Menezes and Dianese, 1988). The integration of soil solarization, mancozeb seed treatment at a rate of 3 g/kg seeds and carbendazim foliar spray at a rate of 0.5 kg/ha were found to be effective in reducing bean anthracnose epidemics (Mohammed *et al.*, 2013). Botanicals and biopesticides (10 per cent extracts of *Adenocalymma alliaceae, Azadirachta indica* and *Lawsonia inermis*, 0.4 per cent talc formulation of *T. viride* and *Pseudomonas fluorescens* along with fungicides carbendazim (0.2 per cent) and mancozeb (0.4 per cent) were evaluated in a greenhouse and field experiment in India and gave promising results (Ravi *et al.*, 2000).

8.1.2 Bean Rust

Bean rust caused by *Uromyces appendiculatus* is worldwide in distribution and is particularly severe in high humidity areas. The rust infects French bean, Lima bean, Dolichos bean, Vigna bean, scarlet runner bean and several other species of *Phaseolus*.

Rust pustules are formed on all above ground plant parts but are more frequent on underside of leaves. The rust pustules are small, round, powdery and brown coloured. Fewer dark brown telia are formed at later stages. Although sulfur fungicides are effective against the disease, disease avoidance by adjustment of sowing dates and varietal selection are recommended for control (Rangaswami, 1975).

Numerous fungicides, including chlorothalonil, tebuconazole, propiconazole and some dithiocarbamates (like mancozeb, maneb) are effective in controlling rust, but proper timing of fungicide application, which is essential to improve economic return, requires good disease monitoring and a weather forecasting system. Foliar sprays with triadimefon @ 500 ml/ha applied four times starting right after the appearance or onset of the disease and continued at ten days interval managed the disease better than the other fungicides (Azmeraw and Hussien, 2017).

8.1.3 *Cercospora* Leaf Spot

Leaf spot caused by *Cercospora cruenta* is a fairly widespread disease. Lower leaves are first affected. Water-soaked lesions develop on the leaf lamina which soon turn reddish brown to brown. The spots are somewhat irregular being bound by veins. At late stages the spots appear ashy-grey with purplish borders. Such lesions may also develop on the pods. Treat the seeds with thiram or captan @ 3 g/kg of seeds. Spraying with copper oxychloride or organic sulfur fungicides is recommended for control.

8.1.4 Angular Leaf Spot

Leaf spot caused by *Phaeoisariopsis griseola* is a serious disease under humid weather conditions. The symptoms appear as dark-brown to greyish angular spots with a distinct margin. The spots may develop greyish mold on underside of the leaves. The disease may cause defoliation under serious infection conditions. The lesions on pods show reddish-brown centres and black borders. The lesions on stem are dark and elongated.

The integrated disease management includes combination of seed treatment with *Trichoderma harzianum* @ 0.5 per cent followed by pre-sowing soil application of *T. harzianum* @ 2.5 kg/50 kg FYM/ha and two periodic sprays of mycobutanil

@ 0.2 per cent at fifteen days interval started with the first appearance of disease proved most effective in limiting the angular leaf spot disease and enhanced the green pod yield (Adikshita and Sharma, 2017).

8.1.5 Powdery Mildew

Powdery mildew caused by *Erysiphe polygoni* is a common disease of beans and other legume vegetables and pulses. The fungus has a very wide host range. White powdery growth occurs on leaves, spreading to cover the stem and other plant parts. In severe cases, the entire plant dries up. Dusting with powdered sulfur, spraying of wettable sulfur preparations or kelthane are recommended for control. Systemic fungicides like benlate or bavistin and calixin also give effective control of powdery mildew as seen in related crops (Singh and Naik, 1976). Effective control is possible with fungicides such as triadimefon and propiconazole (Singh *et al.*, 2016). Host plant resistance is an alternative to manage the disease effectively. Out of 52 genotypes of dry beans screened against diseases, only four genotypes EC150250, $BLF_1 01$, EC 565673A and GPR 203 portrayed resistance reaction against powdery mildew and stem rot over the years (Mishra *et al.*, 2019).

8.1.6 Root Rot and Wilt, and Ashy-Grey Stem Blight and Charcoal Rot

Dry root rot and wilt caused by *Rhizoctonia solani, Pythium* spp., and *Fusarium solani* f. sp. *phaseoli* (Beebe and Pastor-Corrales, 1991; PortaPlugia and Aragona, 1997), and ashy-grey stem blight and charcoal rot caused by *Macrophomina phaseolina* (PortaPlugia and Aragona, 1997) are of localized importance in certain areas. The incidence of root rots depends on temperature, soil moisture level, and their interactions (Tivoli *et al.*, 1997). The pathogens involved are soil-borne and have several important legume host plants. Infections caused by the former remain restricted to the base of plants including any or all the roots. Infections are more severe at flowering stage of plants. The taproot shows bright red discolouration and the plant wilts. *Macrophomina phaseolina* is severe in warm humid areas and has a very wide host range among other economically important crop plants. Symptoms include damping off and killing of young seedlings, black cankerous lesions on the cotyledons and stem, dry root rot which are blackened, stem blight of older plants, *etc.* Poor soil and crop management, nutrient deficiencies increase disease severity. Phytosanitary measures, use of organic manures, *etc.*, are recommended for control of these diseases.

Rapid seedling emergence can reduce root rot incidence (Phillips, 1989) thus the use of different sowing techniques can be directly related with the presence of root rot, especially those techniques that provide a favourable microclimate for their development (Barrera, 1997; Valenciano *et al.*, 2006). The use of fungicides increases seed germination and plant growth, reduces damping-off, and improves seedling emergence (Gupta *et al.*, 1999; Valenciano *et al.*, 2006).

In U.S.A., green house study indicated the possibility of using binucleate *Rhizoctonia* like fungi as biocontrol agent of root rot caused by *Rhizoctonia solani* (Cardoso and Echandi, 1987).

Soaking green bean seeds in chemical inducers in sorbic or benzoic acids 5.0 per cent (v/w) and coated with biocontrol agent *T. harzianumin* dividedly or in combination treatments significantly reduced root rot diseases and increased survivality of bean plants under either green house or field conditions (El-Mohamedy *et al.*, 2015).

Treating the seeds with captan or thiram @ 3 g/kg pod seeds or seed coated with biocontrol agent *T. viride* can reduce ashy-grey stem blight and charcoal rot disease in French bean. Long-term crop rotation with non-leguminous crops also reduces severity of the disease.

Bacterial Diseases

8.1.7 Bacterial Blight

Common bacterial blight caused by *Xanthomonas axonopodis* pv. *phaseoli* is the most important bacterial disease of French bean and other beans. The disease is characterized by irregular, sunken, red to brown leaf spots surrounded by a somewhat narrow yellowish halo. Several spots may coalesce to form irregular patches. Premature leaf fall occurs consequent to infection. Cankerous lesions may develop on the pods and under suitable environment entire plant may be blighted. Symptomatologically similar, Fuscous blight of bean is caused by a variety of the pathogen, *X. phaseoli fuscans* (Burk.) Dowson (Patel and Jindal, 1972). Dolichos bean and other *Phaseolus* spp. are alternative hosts of the pathogens.

Yield losses are into the range of 10 and 40 per cent, depending on the intensity of the disease, degree of bean susceptibility and environmental conditions that favour the progress of the disease (Saettler, 1989; Opio *et al.*, 1996).

Immunity to common bacterial blight was not detected in beans, although many lines have shown high resistance (Sherf and MacNab, 1986). Resistance to bacterial blight in bean has been described as a quantitative trait (QTL) with low to medium heritability (Silva *et al.*, 1989). Some sources of resistance to *Xap* have been reported in *Phaseolus coccineus* (Yu *et al.*, 1998), tepary bean *P. acutifolius* (Yu *et al.*, 1998; Singh and Munoz, 1999) and in common bean, *P. vulgaris* (Schuster *et al.*, 1983). The interspecies cross between *P. vulgaris* and either *P. acutifolius* or *P. coccineus* have frequently been used to transfer the resistance-related traits in common bean breeding programs (Taran *et al.*, 2001). Five weakly susceptible lines (HR 45, Oreol XAN-159, XAN -208, and XAN -273) were identified and recommended as possible sources of tolerance in breeding programme (Popovic *et al.*, 2012).

There is no effective chemical control measure to check this disease. Growing resistant varieties is the only practical approach. Use of disease free seed, deep ploughing, long term crop rotation and destruction of diseased plant debris helps to reduce the disease. Hot water seed treatment at 50°C for ten minutes followed by dipping in Streptocycline solution @ 1 g/10 litres of water is effective in reducing the seed infection.

Viral Diseases

Potyviruses can be separated into five basal lineages: the *Sugarcane mosaic virus* group; the *Bean yellow mosaic virus* (BYMV) group; the *Onion yellow dwarf virus* group, and the *Pea seedborne mosaic virus* group. In addition, there are two supergroups: the *Potato virus Y* (PVY) supergroup and the *Bean common mosaic virus* (BCMV) supergroup (Gibbs and Ohshima, 2010). The phylogeographic pattern of the species within the BCMV lineage suggests that this group originated in South and East Asia (Gibbs *et al.*, 2008a, 2008b). It is estimated that 3600 years ago, the BCMV lineage first emerged and now includes 19 virus species, including BCMV itself and the closely related Bean common mosaic necrosis virus (BCMNV) (Gibbs *et al.*, 2008b).

BCMV and BCMNV are the most common and most destructive viruses that infect common beans as well as a range of other cultivated and wild legumes (Morales, 2006). Yield losses due to BCMV and BCMNV can be as high as 100 per cent (Damayanti *et al.*, 2008; Li *et al.*, 2014; Saqib *et al.*, 2010; Singh and Schwartz, 2010; Verma and Gupta, 2010).

BCMV produces chlorotic, crinkled and stiff young leaves as primary symptoms. Leaves droop with shortened petioles. At later stages, secondary symptoms are produced on leaf blades of trifoliate leaves as general chlorosis and mottling with a definite pattern of dark and light green areas. The compound leaves show downward curling and rolling which differentiates the disease from BYMV. Shortening of petiole, stunting of plants and deformation of pods are other characteristic symptoms. In plants of susceptible bean lines, BCMV and BCMNV can have very similar symptoms that include dwarfing, mosaic, leaf curling, and chlorosis (Flores-Esteʹvez *et al.*, 2003). However, even "symptomless" infection of hosts in which these effects are not manifested can decrease crop yield to the extent of 50 per cent (Morales, 2006). The most striking difference in symptomology between BCMV and BCMNV relates to the hypersensitive necrotic reaction (black root disease) that occurs in bean plants containing the dominant I resistance gene (Ogliari and Castao, 1992). Early experiments showed that this gene conferred strong hypersensitive resistance to BCMV (Ali, 1950). The hypersensitive reaction is a form of resistance that involves localization of a pathogen to the initially infected cells or to a zone of cells in the immediate vicinity of the primary infection site and that is often associated with programmed death of cells in this zone (Loebenstein, 2009).

All known strains of BCMV and BCMNV can be transmitted through seed or spread by aphids. Thus, an outbreak can be triggered by the use of contaminated seed stock and amplified by aphid-mediated virus transmission to generate an epidemic (Hampton, 1975). Alternatively, the viruses may infect a healthy crop population when viruliferous aphids immigrate from infected wild plants (Sengooba *et al.*, 1997). Aphid vectors reported for BCMV and BCMNV include *Macrosiphum solanifolii, Macrosiphum pisi, Macrosiphum ambrosiae, Myzus persicae, Aphis rumicis, Aphis gossypii, Aphis medicaginis, Hyalopterus atriplicis,* and *Rhopalosiphum pseudobrassicae* (Zaumeyer and Meiners, 1975; Zettler and Wilkinson, 1966).

Use of disease-free seeds for BCMV and spraying the crop with dimethoate, monocrotophos or 1 per cent mineral oil for vector control can give effective control of these diseases (Singh *et al.*, 1981).

For the effective control of major insect and mite pests of French bean, an IPM module including seed treatment by captan @ 3 g/kg seed, weeding frequency (3 times at 15 days interval), yellow sticky trap @ 20 nos./ha, mechanical destruction (removing by hand picking and destroying them), intercrop with coriander (2: 1 ratio), release of *Coccinnela transversalis* for effective control of aphid and two spotted spider mites was found to be most effective at Assam, India (Sharmah and Rahman, 2017).

The cultivars Top Crop, Bush Lake 274, Cotid, Pinto UI-III, Provider and Scotio were rated resistant to BCMV infection while Contender was susceptible and Bountiful as highly susceptible (Al- Fadnil and Al-Ani, 1987).

Genetic markers linked to resistance genes for BCMV and BCMNV have been identified and could be used to underpin marker-assisted selection (Haley *et al.*, 1994)

8.2.0 Pests

8.2.1 Bean Aphid (*Aphis craccivora*)

Aphis craccivora infestation causes characteristic damage to the tender aerial parts of the plants and produces symptoms like curling of leaves, twisting of twigs and developing fruits, and sometimes shedding of flowers. The infestation may assume serious proportion at any stage of crop growth. It can easily be controlled by application of systemic organophosphatic insecticides, *viz.*, spraying with 0.05 per cent phosphamidon or 0.025 per cent methyldemeton. Imidacloprid (0.03 per cent) or acetamiprid (0.15 per cent). An IPM module includng seed treatment by captan @ 3 g/kg seed, weeding frequency (3 times at 15 days interval), yellow sticky trap @ 20 nos./ha, mechanical destruction (removing by hand picking and destroying them), intercrop with coriander (2: 1 ratio), release of *Coccinnela transversalis* for effective control of aphid was found to be most effective at Assam, India (Sharmah and Rahman, 2017). Another IPM module including deep ploughing + carbofuran 3G @ 25 kg/ha + Delta traps @ 100/ha + *Bacillus thuringiensis* var. *kurstaki*, 90-120 billion spores/g @ 500g per hectare + imidacloprid 17.8 SL @ 0.05 per cent was found effective against bean aphid at Jammu, India (Mondal *et al.*, 2017).

8.2.2 Jassids

Empoasca fabae and *E. kraemeri* have also been found to be serious pests which produce typical 'hopper burn' symptoms on crop when severely infested. The same control measures as for aphid are also applicable for the control of these insects. It has been indicated that density of hooked trichome on leaf may be negatively correlated with the survival of nymphs of *E. fabae* (Pillemer and Tingey, 1978). The control measure are the same as for aphid.

8.2.3 Lygaeid Bug

Chauliops fallax Sweet and Schaeffer causes discolouration of leaves and pods, which has been found to be active from July to October with peak of its incidence during late July to August in Himachal Pradesh (Lal, 1974; Bala and Kumar, 2018).

Pests of French Bean

Aphid infestation

Jassids

Bean lady bird beetle infestation

Stem fly infestation

Physiological Disorder

Sunscald

8.2.4 Hairy Caterpillars

Ascotis imparata and *Spilosoma obliqua* have been reported as important pests of this crop in Karnataka, India (Puttaswamy and Reddy, 1981). The larvae cause characteristic skeletonization of leaves during the early gregarious stage and later they completely denude the plants. The pests can easily be controlled by systematic collection of the larvae during the early gregarious stage or by spraying with endosulfan at 0.07 per cent.

The other lepidopteran pests that cause sporadic damage include *Amsacta lactinae* and *Euproctis scintilans* (Subba Rao *et al.*, 1974). Spryaing of chlorpyriphos @ 0.25 per cent can be effective against this pest.

8.2.5 Stem Fly

Ophiomyia phaseoli often causes drying of the plants as the maggot bores inside the stem. It can be controlled by spraying the crop with 0.03 per cent diazinon or 0.05 per cent quinalphos (Jagtap *et al.*, 1979) or by placement of granules of 3 per cent carbofuran in the planting row at 5g/3m length before sowing (Hussein, 1978). It has been found that the resistance of the crop to this pest is associated with high leaf hair density, thin stem and long internodes (Rogers, 1979). Soil application of granular insecticides like phorate or foliar application of endosulfan is effective against this pest (Krishna Moorthy and Tewari, 1987). Petiole-mining thresholds based on number of petioles with mining is appropriate. Use of potassium chloride sprays @ (KCl, Laboratory Grade) spray at 0.4 per cent wt/vol (2.8 kg KCl in 700 litres water/ha) on 10, 15, and 20 d after planting, for management of bean fly was found somewhat effective (Krishna Moorthy and Srinivasan, 1989).

8.2.6 Bean Lady Bird Beetle

Epilachna varivestis is commonly known as Mexican bean beetle or bean lady beetle and has been known to be a serious pest at times in Southern United States. Both larvae and adults feed on the leaves, usually on the lower surface, leaving the upper surface more or less intact but such scraped away areas may break through upon drying out. Spraying with malathion or carbaryl, taking care that the lower surface of the leaves get well coverage, will control the insect (Metcalf and Flint, 1962). Spryaing of chlorpyriphos @ 0.25 per cent can be effective against this pest.

8.2.7 Pea Leaf Weevil

Pea leaf weevil, *Sitona lineatus* and some other species of root curculio are important insects in Europe and America. Though it is more important on clovers and alfalfa, they doubtlessly cause damage to other beans. Both larvae and adults of these insects also damage the plant but in this case the larvae are soil dwellers and damage the subaerial parts of the plant. The roots are scored and furrowed on the outside with numerous burrows and are often nearly girdled. This results in wilting and drying of the affected plants, especially during dry weather. Soil treatment with gamma-BHC or dieldrin at 2.24-4.48 kg a.i./ha has been found to be quite effective to control the larvae (Bardner and Fletcher, 1979).

Application of cyhalothrin-lambda reduced adult populations by approximately 56 per cent compared to untreated controls (Steene *et al.*, 1999). Other foliar insecticides evaluated for *S. lineatus* control include imidacloprid, cypermethrin, aldicarb, permethrin, deltamethrin, parathion, cyfluthrin, benomyl and carbofuran (Bardner and Fletcher, 1979; McEwen *et al.*, 1981; Bardner *et al.*, 1983; Arnold *et al.*, 1984a, 1984b; Griffiths *et al.*, 1986; Ester and Jeuring, 1992; Doré and Meynard, 1995; Sache and Zadoks, 1996; Steene *et al.*, 1999). Foliar applications have no direct effects on larvae, but timely application can decrease adult pea leaf weevil populations, egg production and eventual larval populations (Steene *et al.*, 1999). Foliar sprays must be applied as soon as weevils are detected in order to control larval populations, as adult weevils mate and oviposit soon after arrival. If applied too late, egg production is not prevented and control efforts will not impact larval populations (King, 1981; Bardner *et al.*, 1983; Ester and Jeuring, 1992).

8.3 Physiological Disorders

8.3.1 Sunscald

The symptoms on your green bean leaves are due to sun burn (commonly called sunscald). The appearance of brown patches (*i.e.*, scorching of leaf tissue) are due to intense sun light. Remove the affected leaves and provide shade to plants during afternoon intense sunlight. Also water the plants regularly.

8.3.2 Transverse Cotyledon Cracking

Due to planting dry seeds in wet soil. White seeded cultivars are more prone. Seeds with 12 per cent moisture has better germination.

9.0 Seed Production

The seed production of French bean is akin to the production of dry beans (rajma). Most of the varieties are day neutral and hence the day length does not affect the seeding habit except some vine varieties which are short day type. Inoculation of seed with French bean nodule bacteria, *Rhizobium phaseoli* is recommended when it is sown for first time in the field or grown in a poor soil. The culture is mixed thoroughly @ 30 g jaggery solution per kg of seed, sufficient to moist the seed. The seed are dried in the shade for half an hour before sowing. For seed crop, it should be grown in such season so that they become ready for harvest during rain free period. When almost 90 per cent pods on the plants mature and turn dry, the whole plants are uprooted and collected pods on the threshing floor. After about a week seeds are separated out from the pods by threshing and winnowing. Delayed harvesting results in shattering of pods and consequently reduced seed yield. Threshing is done by a thresher and extreme care should be taken during threshing to prevent injury to the seed. Seeds harvested and threshed directly off the crop were of good quality provided the seed moisture content in the crop had fallen to less than 25 per cent (Siddique *et al.*, 1987). According to Demir *et al.* (1994) the optimum harvest time for seed production is when seed moisture content reaches 14 per cent and further delay results in physiological aging and seed loss through shedding. During seed ripening stage, irrigation is stopped. Thompson and Kelly (1957) recommended the use of desiccants or chemical defoliants like potassium cyanate and borate-chlorate mixtures. In cultivars where ripening is uniform the plants are pulled in the morning and threshed in threshing floor. Care is needed in threshing to prevent seed injury.

The cleistogamous nature of flowers ensures self-pollination. Quagliotti *et al.* (1980) reported natural crossing to be less than 1.1 per cent. An isolation distance of 10 m is advisable for foundation seed production and 5 m for certified seed production. Proper rouging of seed crop is essential in order to maintain the purity of seeds. Rouging is conducted for off types and diseased plants before flowering, during flowering and during pod edible stage. Seed yield between 1000 and 1500 kg/ha is usually obtained. The following standards have to be maintained for producing foundation and certified seeds (Table 4).

Table 4: Standards for Foundation and Certified Seed of French Bean

Factor	Foundation Seed	Certified Seed
Field standards		
Off Types (per cent)	1.0	0.50
Other crop plants (per cent)	Nil	Nil
Objectionable weed plants (per cent)	Nil	Nil
Diseased plants (per cent)	0.5	0.5
Seed standards		
Pure seed (min.) (per cent)	98.0	98.0
Inert matter (max.) (per cent)	2.0	2.0
Other crop seeds (per cent)	Nil	Nil
Total weed seeds (per cent)	Nil	Nil
Objectionable weed seeds (per cent)	Nil	Nil
Germination (per cent)	75	75
Moisture (per cent)	9.0	9.0

There exists a large genotypic variation for seed yield. A plant can yield 9 to 18 pods/plant with 4-6 seeds/pod, and 100-seed weight ranged from 25 to 41 g (Peixoto *et al.*, 1993). On an average, from a hectare of seed crop about 1000-1800 kg seeds can be obtained depending upon the variety (Ramandeep *et al.*, 2018).

10.0 Crop Improvement

In the U.S.A. the release of cv. Tendercrop in 1958 revolutionized the snap bean improvement programme, and since then a number of cultivars suitable for canning, frozen processing industry, high density planting, better pod quality, mechanical harvesting, fresh market or shipping and disease resistant have been evolved (Silbernagel, 1986).

The French bean improvement is being carried out in some Research Institutes and Agricultural Universities in India. This has led to the identification/evolution of cultivars like VL Boni (Almora); UPF-191, UPF-204, *etc.* (Pantnagar); SVM (an interspecific cross) from Solan; Sel. 2, Sel. 5 and Sel. 9 (Bangalore) and Pusa Parvati from Indian Agricultural Research Institute, New Delhi, India. All the research institutes excepting Indian Council of Agricultural Research Complex, Shillong are involved in evolving bush bean lines. The work in North-Eastern region of India is towards evolving pole beans cultivars. Because of high rainfall, farmers prefer pole types as the fruiting height is much above the ground level thus preventing the pods from being infected with pathogens from soil. Besides, there are many local pole types being grown which can be improved for specific traits like yield and quality.

The original comprehensive gene list was prepared by Yarnell (1965). An undated list was prepared by Dickson *et al.* (1982). Bassett (1989, 1993 and 1996) prepared extensive additions, corrections, revisions, and style changes in the genes of *P. vulgaris.* (http:/beangenes.cws.ndsu.nodak.edu/genes/genlist3.htm).

The traits for which breeding work is being carried out are as follows:

i) High yield

ii) Quality for fresh market and sloughing free cultivars for canning industry

iii) Short pods for whole pod processing

iv) Resistance to pests and diseases

v) Exploitation of hybrid vigour by induction of male sterility.

Recently, breeding for resistance/tolerance to biotic and abiotic stresses has been a major research objective (Singh, 1992; Beebe, 2012). Some other breeding objectives, derived from breeding priorities of the International Center for Tropical Agriculture (CIAT) and Pan-African Bean Research Alliance (PABRA, http://www. pabra-africa.org) includes improvement of traits such as higher content of minerals (iron and zinc), and fast cooking time (particularly for developing countries), canning quality, harvest index, and market class/seed color (Beebe *et al.*, 2013; Assefa *et al.*, 2015, 2017). In the USA, Canada, and European countries, most bean production is by commercial farmers, with much of that production being for export (*e.g.*, small white Navy beans for UK processing and small black beans for Cuba and Mexico) or for specialized markets (*e.g.* 'alubias' white beans for export largely to Argentina and Spain). In these areas, improvement efforts have particularly focused on resistance to major diseases, including white mold, bacterial blight, rust, halo blight, anthracnose, and bean common mosaic virus, and to insect pests such as bean leaf beetle, stinkbugs, and aphids.

Vasic (1990) stated that the main problems of French bean breeding in Yugoslavia are once-over harvesting, resistance to ecological conditions and resistance to plant diseases and pests. He also emphasized on breeding for improved habits (height, pod distribution), pod structure, yields components, photosynthesis, cold resistance and resistance to bacteria, viruses and insects.

Singh (1998) described the basic requirements and alternative selection methods for simultaneous improvement of maximum number of agronomic characters of common beans in the shortest time possible as follows: (1) determination of the importance and priorities of bean production problems in the region; (2) definite objectives and priorities of breeding; (3) identification and use of reliable parental sources considered to be donors of necessary genes; (4) development of multiple-parent crosses with large number of pollination to produce sufficient seeds of each cross; (5) evaluation and selection for currently available markers (*e.g.*, for Zabrotes [*Z. subfasciatus*], bean common mosaic potyvirus, bean golden mosaic bigeminivirus, rust, anthracnose, apion [*A. godmani*], common bacterial blight, *etc.*) from hybridization; (6) alternative selection methods such as gamete selection and single seed descendants, gamete and pedigree selection using markers (GS-PUM), gamete selection and agronomic evaluation and selection of families in early generations (GS-AEF), or a combination and integration of the GS-PUM and GS-AEF methods; and (7) seeking genuine and direct collaboration and integration in all activities of researchers from different disciplines, institutions, countries and the entire region at key sites for reliable evaluation of bean nurseries in order to carry out simultaneous selection for a maximum number of agronomic traits. Moya Lopez *et al.* (2000) suggested that the principal breeding objectives for snap beans in Cuba are centered on developing cultivars that are resistant to high temperature, rainfall and light intensity, drought and salinity, and tolerant of common pests and diseases. A low nutrient requirement for production is also desired. Traditional methods of crop improvement, which look for natural variability and more modern methods involving somaclonal variation, radiomutagenesis, and diverse imported and local germplasm have been used. The selection methods used have facilitated the development of single genotype with a series of resistance genes. Selections are grown under different environmental conditions where they are evaluated using practical criteria useful to growers. In addition to the environmental stresses inherent in the environment of each test locality, plants are grown under reduced irrigation or with unconventional cultural practices. Important results include selections of snap bean that are adapted to high temperature, high rainfall, reduced irrigation, use of organic plant nutrition, and non-conventional cover crops and crop associations. These selections also show tolerance to the most common pests and diseases in Cuba.

It would be beyond the purview of the chapter to review the entire work on breeding; however, a brief account on breeding is dealt with.

10.1.0 Resistance Breeding

With over 200 different bean diseases reported, the pathogens causing significant yield losses to beans include bacteria, virus, fungi, and plant parasitic nematodes. Many of these diseases and insect pests have co-evolved with common bean (Beebe, 2012; Beebe *et al.*, 2013). Some of the most significant bean diseases in the tropics include bean angular leaf spot (ALS, *Phaeoisariopsis griseola*), anthracnose (ANT, *Colletotrichum lindemuthianum*), common bacterial blight (CBB), and viral diseases bean golden mosaic virus (BGMV) and bean common mosaic virus (BCMV) (Beebe and Corrales, 1991; Duc *et al.*, 2015; Miklas *et al.*, 2017). In temperate regions, the most common diseases are CBB, halo blight, rust, and white mold (Duc *et al.*, 2015).

Several insect pests and nematodes attack aerial and underground parts of the common bean (Cardona, 1989; Kornegay and Cardona, 1991; Singh, 1992,

2001; Cardona and Kornegay, 1999). Leafhoppers *E. kraemeri* and *E. fabae* are the most widely distributed insect pests in common bean fields on the American continents, especially in relatively drier areas. Bean pod weevil (*Apion godmani* and *A. aurichalceum* Wagner) causes severe damage to developing pods and seeds in the high lands of Mexico, in Guatemala, El Salvador, Honduras, and Nicaragua. In thehigh lands of Mexico and in the western USA, Mexican bean beetles (*Epilachna varivestis* Mulsant) also cause severe leaf damage, especially in late-maturing cultivars. Bean weevil, *Zabrotes subfasciatus* Boheman, in warm tropical and subtropical environments, and *Acanthoscelides obtectus* (Say), in cool and temperate environments, cause severe post-harvest problems when common bean is not properly stored (Cardona, 1989). Key insect pests in other parts of the world, including Africa, are the leafhopper, thrips, weevils, whitefly, bean fly (*Ophiomyia phaseoli* Tryon), aphids (*e.g., Aphis fabae* Scopoli), chrysomelids (*Ootheca* species), pod borer [*Maruca testulalis* (Geyer)], and mites [*Tetranychus cinnabarinus* Boisd., *Polyphagotarsonemus latus* (Banks)] (Karel and Autrique, 1989; Schwartz and Peairs, 1999). Occurrence, distribution, and losses caused by nematodes are poorly documented. However, root-knot (*Meloidogyne* species), lesion (*Pratylenchus* species), and soybean cyst (*Heterodera glycines*) nematodes can pose major threats to common bean production. The population densities of these nematodes are exacerbated by the continual plantings of susceptible cultivars. Rotation with other susceptible crops such as tomato, pepper, and potato aggravates problems for the succeeding common bean crop (Schwartz and Pastor-Corrales, 1989; Abawi and Widmer, 2000; Schwartz *et al.*, 2005). In addition to causing direct damage to foliar and underground plant parts including pod and seed yield and quality, insects are important as vectors of numerous common bean viruses and viral diseases (Schwartz *et al.*, 2005).

Significant progress has been made in developing cultivars with resistance to various diseases using conventional breeding. Markers associated with established resistance loci can be used for more efficient breeding to develop resistant cultivars. Some early examples of marker-assisted selection for bean diseases include 23 RAPD markers and 5 SCAR markers associated to 15 different resistance genes, described by Kelly and Miklas (1998). Molecular markers and linkage mapping of rust resistance genes have been reviewed by Miklas *et al.* (2002). Kelly and Vallejo (2004) provided a summary of markers, MAS, map location, and breeding value for anthracnose resistance. Similarly, Miklas *et al.* (2006) reviewed MAS in breeding for resistance to anthracnose, angular leaf spot, common bacterial blight, halo bacterial blight, bean golden yellow mosaic virus, root rots, rust, and white mold.

Principal factors determining strategies and methods used for breeding for resistance to insect pests and nematodes and other desirable traits include (1) the genetic distance between the cultivar to be improved and resistant donor germplasm, (2) the direct and indirect screening methods available, (3) the genetics of resistance, and(4) the number of resistances and other traits to be improved. Given the diversity and genetic distance between the cultivars of specific market classes to be improved and the resistance donor germplasm, a two- or three-tiered integrated breeding approach is often used to broaden the genetic base and introgress and pyramid resistance genes and QTL from across market classes, races, gene pools, and related species into successful new common bean cultivars (White and Singh, 1991; Kelly *et al.*, 1998; Singh, 2001). Often, the pedigree, mass pedigree, and single seed descent breeding methods suffice for transferring major resistance alleles and QTL between cultivars and elite breeding lines within market classes. Some form of backcrossing, such as recurrent backcrossing, inbred-backcrossing, or congruity backcrossing (*i.e.*, backcrossing alternately with either parent), becomes essential as the genetic distance between the cultivar under improvement and the resistance donor germplasm increases. Thus, for most of the between-market class, inter-racial, and inter-gene pool crosses, and for introgressing resistance alleles and QTL from wild populations of common bean and *Phaseolus* species in the secondary and tertiary gene pools, recurrent or congruity backcrossing or modifications become the methods of choice in early stages of the program. Gamete selection using multiple-parent crosses (Singh, 1994; Singh *et al.*, 1998; Asensio *et al.*, 2005, 2006; Tera´n and Singh, 2009) and recurrent selection (Kelly and Adams, 1987; Singh *et al.*, 1999; Teran and Singh, 2010a, b), respectively, may be more effective. But, the use of these latter methods is not common in common bean because of the large number of pollinations required and other demands on resources.

Kolotilov *et al.* (1989) selected cultivars for earliness, yield, various yield components, optimum pod insertion height for mechanical harvesting, resistance to bacterial disease, bean common mosaic potyvirus and protein content in seeds and found that the best cultivars for a combination of traits were K11358 from Poland, K 11264 from Romania and K 13940 from Hungary.

Hagedorn and Rand (1980a) used Wisconsin Selection 133 (resistant to bacterial brown spot, *Pseudomonas syringae*) to cross with Slimgreen (processing cultivar) as follows to produce breeding line BBSR 17 and BBSR 28.

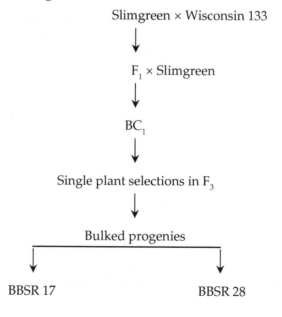

They (1980b) evolved RRR 46 for resistance to root rot complex caused by *Pythium* spp., *Fusarium solani* f. sp. *phaseoli* and *Rhizoctonia solani* and the origin is as follows:

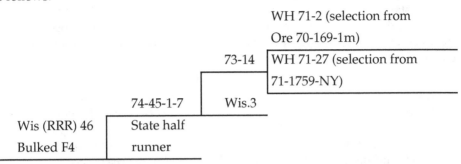

An F_{10} cross between breeding lines B 4137-1N (resistant to *Meloidogyne incognita*) and B 4124-2R was released as nematode resistant breeding line B 4175 (Wyatt *et al.*, 1980). One parent B 4137-1N is a selection derived from a complex parentage, *viz.*, Cascade, Bush Blue Lake 274, Bonus, Extender and P 165426; (Fassuliotis *et al.*, 1967; Wyatt *et al.*, 1980). The other parent has the pedigree of Provider, Wade, Commodore, Unrivaled Wax, Black Valentine and AsgrowStringless Greed pod. The breeding method involved in evolving B 4175 was bulk breeding method. Breeding work for resistance to anthracnose, rust, *etc.* has also been carried out. For mosaic resistance, the resistant line Wisconsin Refugee has been used to breed a resistant cultivar Cristal Blanco Fenix in Chile (Bascur and Cataki, 1977). Scully *et al.* (1990) developed a breeding line CU-M88 by using Midnight as the recurrent female parent for 5 generations and B21 as the donor providing the *By2* gene conferring resistance to bean yellow mosaic potyvirus (BYMV). Guu (1989) reported that the cultivar Taichung-1 was developed using Black Creaseback, which is susceptible to rust by backcrossing 4 times to rust resistant 15R55BK.

Resistance breeding in French bean is aimed at developing varieties that are resistant to infection by *Pythium*, rust (*Uromyces phaseoli* var. *typica*), anthracnose (*Colletotrichum lindemuthianum*), bean common mosaic potyvirus (BCMV), halo blight (*Pseudomonas savastanoi* pv. *phaseolicola*), angular leaf spot (*Phaeoisariopsis griseola*) and common bacterial blight (CBB; *Xanthomonas axonopodis* pv. *phaseoli*). Ginoux and Messiaen (1993) studied 20 *Pythium* species and strains and found that *P. ultimum* was the most aggressive, *P. sylvaticum* was moderately aggressive and palmate-coralloid strains were weakly aggressive. In tests of varietal reactions, black beans, some red, buff-coloured and mottled beans were more resistant than white ones, among which differences in susceptibility were differentiated using the weakly aggressive *Pythium* strains. French flageolet cultivars were more susceptible. The study of progeny from a cross between PI226895 (the most resistant black bean cultivar) and Elsa (a flageolet vert) demonstrated an increase in susceptibility conferred by 2 recessive genes inducing discolouration and flageolet quality. High levels of resistance were found in buff-coloured and black-seeded lines in progeny from this cross. At the biochemical level, susceptibility was associated with high levels of soluble substances (sugars, nitrogen compounds) in seed exudates during imbibition, low content of leucoanthocyanins and seed coat thickness. The pleiotropic influence of the gene *P* on *P. vulgaris* characteristics was demonstrated in studies of white mutants of the black-seeded cultivars PI226895 and Aiguillon. The best cultivar for these characteristics was Vernandon. It is suggested that the low resistance levels of white-seeded lines (which are preferred by the snap bean canning industry) can be improved with low levels of fungicides used for seed treatment. Breeding of light-buff-coloured lines, which give less unpleasant water colouration in cans, is recommended.

Carrijo (1993) developed MacarraoPreferido AG-482, a new French bean cultivar resistant to rust and anthracnose. This green-podded, vegetable cultivar was deve-loped from a cross between MacarraoFavorito AG480 and a line developed from crosses involving BGF_1458 and Cornell 49AG480, sources of resistance to rust (*Uromyces phaseoli* var. *typica* [*U. appendiculatus*]), and MacarraoRasteiro 274. It is indeterminate, has a growth period of 100 days and produces pods approximately 18 cm long and weighing 11 g. MacarraoPreferido AG482 is resistant to the main physiological races of *U. phaseoli* and also exhibits resistance to anthracnose.

Three virus-resistant French bean germplasm lines, USWA-64, USWA-67, and USWA-68 were released by Hang *et al.* (1999). These bean lines possess the dominant *I* gene conditioning resistance to bean common mosaic potyvirus (BCMV) and have complete resistance to curly top virus (CTV). The three lines produce a high yield of well-distributed green pods on an erect, bushy plant. USWA-64 is an F_12 bulk population from a single plant selection in the F5 generation from the cross (NVRS-196 × OSU-4091-3) × (5BP3 × Valliant). Resistant to all US endemic strains of BCMV including the NL8 strain, USWA-64 possesses the recessive *bc-1* gene, which, in combination with *I*, imparts partial protection to the hypersensitive necrosis syndrome due to strains of bean common mosaic necrosis potyvirus. The outstanding features of USWA-64 are early maturity, excellent quality blue lake flavoured pods and uniquely crinkled leaves. The dark-green pods are smooth, straight, round, 14.6 cm long and are stringless. USWA-67 is an F_25 bulk population from a single plant selection in the F_20 generation from the cross (OSU-1604 × 5BP-7) × RH-13. USWA-67 was derived from mass selection for small pod size during each generation. It also has resistance to anthracnose (Co-2; *Colletotrichum lindemuthianum*) and to all known strains of halo blight (*Pseudomonas savastanoi* pv. *phaseolicola*) in the USA. Pods of USWA-67 are medium to dark-green, smooth, straight, round to slightly heart-shaped, and 11.25 cm long. USWA-68 is an F_12 bulk population from a single plant selection in the F_5 generation from the cross BARC-1 × [(GN-31 × D-9) × OSU-10183]. USWA-68 has additional resistance to some strains of rust (*Uromyces appendiculatus*). Pods are medium-green, straight, round, and 12 cm long with a densely pubescent surface.

Paula *et al.* (1998) investigated the resistance of bean cultivars to anthracnose, angular leaf spot (*Phaeoisariopsis griseola*) and rust (*Uromyces appendiculatus*) in Minas Gerais, Brazil. Cultivar Manteiga Maravilha was resistant to races 64, 65, 69, 73 and 87 of *C. lindemuthianum*. Cultivar Novirex was resistant to races 64, 65, 69 and 87, and cultivar Argus to races 69 and 81. No cultivar was resistant to race 89 (the most common). Cultivar Manteiga Maravilha was also resistant to races 31.21, 31.23, 59.39 and 63.55 of *P. griseola*. Cultivar MacarraoPreferido and Novirex were resistant to races 31.21, 59.39 and 63.55. The other cultivars were susceptible to at least 4 inoculated races. All cultivars were susceptible to race 63.23 (one of the most common races). Cultivar MacarraoFavorito was immune to isolates Ua-1 and Ua-4 of *U. appendiculatus* and susceptible to other isolates. The best results were obtained with cultivar MimosoRasteiro (rust), cultivar Novirex (anthracnose and angular leaf spot) and cultivar Argus (anthracnose). Feng *et al.* (2000) investigated the factors influencing the pathogenicity of snap bean rust fungus and the resistance of the host. Urediniospore germination rate and penetration of the germ tubes were reduced in mature leaves, by high temperatures (above 28°C), short dew periods (shorter than 6 h) and continuous light during germ tube penetration. Host resistance changed with inoculum concentration. Infection frequency was correlated with the site of inoculation, *e.g.*, adaxial or abaxial leaf surfaces. *U. appendiculatus* urediniospores can be stored at –4°C and kept from 9 months without a reduction in pathogenicity. They reported a rapid, reliable and efficient method for screening snap bean resistance. A total of 260 accessions of snap bean were screened for rust resistance. A promising line 99-17-2-1-4-12-3 developed through crossing between Arka Bold × Arka Komal, with resistance to rust with high pod yield and good pod quality was selected and named Arka Anoop and released for commercial cultivation in India (Aghora *et al.*, 2007).

For improved resistance to common bacterial blight (CBB; *Xanthomonas axonopodis* pv. *phaseoli*) in snap beans, Rodrigues *et al.* (1998a) made diallel crosses (without reciprocals) using three snap bean cultivars (high yielding) and two dry bean lines (CBB resistant). Leaves were inoculated with 107 bacterial cells/ml using the razor blade procedure. A dissecting needle was used to inoculate pods. Leaves were inoculated 25 days after planting and pods when seeds were full developed. Leaf and pod disease reactions were determined eight days after inoculation. Leaf reaction was rated according to a scale ranging from 1 (resistant) to 5 (susceptible), and pod reaction was rated according to the length of water-soaking lesion from the point of inoculation. Significant differences were detected for parental effect for leaf reaction, indicating that additive genic effects are involved. Lines BAC-6 and A-794 were the genotypes that showed the best results. Significant differences were found in both parental and heterosis effect for pod reaction and hybrids Alessa × A-794, Alessa × BAC-6, HAB 52 × A-794 and HAB 52 × BAC-6 were the superior combinations. Subsequently, Rodrigues *et al.* (1999) determined the combining ability for disease resistance in three snap bean genotypes. Bac-6 and A794 were considered superior genotypes for leaf resistance. Non-additive effects were predominant in pod reactions, and Alessa × Bac-6, Alessa × A794 and Hab 52 × Bac-6 were the best combinations.

10.2.0 Improvement of Pod and Yield Characters

Pod yield and its components are quantitatively inherited and are highly influenced by environments (Singh, 1991), so understanding the relation between yield and its components is important to set effective selection criteria and breeding strategies. In several studies, a high correlation have been found between yield and plant height, flower per inflorescence, pods/plant, pod weight, pod length, pod width, seeds/pod, and 100 seed weight (Pandey *et al.*, 2013; Patil *et al.*, 2018;

Lyngdoh *et al.*, 2018). Hence, yield component traits have been used as selection criteria to improve pod yield and cultivar development.

The majority of efforts toward increasing seed yield under favorable environments has come from improvement in pods/plant, seeds/pod, and seed weight (Beebe *et al.*, 2013). However, under unfavorable conditions (*e.g.*, drought), other traits including biomass partitioning indices (pod partitioning index, harvest index, and pod harvest index) have been used as key traits to improve yield (Beebe *et al.*, 2013; Assefa *et al.*, 2013, 2017; Rao *et al.*, 2017).

Previous studies showed that hybridization of interracial bean varieties had higher yield, particularly in crosses between Mesoamerican with Durango or Jalisco races (Beebe, 2012). Increasing yield potential has also been achieved through breeding for abiotic stress tolerance. Beebe *et al.* (2008) reported that yield could increase under drought conditions through photosynthate remobilization and biomass translocation, implying that yield improvements can be made under drought condition.

Bassett (1993) observed a new gene for a flower colour pattern described as white banner (*wb*) in material derived from the cross Harvester snap bean (*P. vulgaris*) × PI273666 scarlet runner bean (*P. coccineus*). The WB character consists of a white banner petal and pale violet wings (veronica-violet 639/2). Inheritance of the mutation was studied in crosses involving *P. vulgaris* breeding line 5-593, which has bishops-violet (wild-type) flowers, and genetic stocks vBC25-593 (white flowers) and bluBC25-593 (blue flowers). Segregation in F_2 and F_3 progenies from the cross vBC25-593 × WB supported the hypothesis that a single recessive gene that is non-allelic with the V locus controls WB. An allelism test with bluBC25-593 provided evidence that WB is not allelic with the *blu* locus. The gene symbol *wb* is proposed for the gene producing the WB phenotype.

Rodrigues *et al.* (1998b) evaluated snap bean cultivars Alessa, Hab52 and Hab198 and dry bean breeding lines Bac-6 and A-794 and their diallel crosses for pod number, seed number, pod length, pod diameter, fibre content and plant height in 1996. GCA was significant for all characters evaluated. SCA was significant for pod diameter, fibre content and plant height. Bac-6, A-794 and Alessa were considered superior genotypes for pods/plant, seeds/plant, pod length and fibre content. Non-additive effects were predominant in pod reaction and plant height. The best combinations were Alessa × Bac-6, Alessa × A-794 and Hab 52 × Bac-6. De Carvalho *et al.* (1999) crossed bush snap bean cultivars Alessa, Andra, Cota and Cascade in a diallel mating design without reciprocals. Together with the F_1 hybrids were evaluated for 8 agronomic characters. Significant effects for general combining ability (GCA) were identified for pod number, pod weight, pod length, pod diameter, days from sowing to flowering, and plant height. Specific combining ability (SCA) was also significant for all characters, showing that dominance/epistatic effects were involved in the control of these characteristics. For all traits, Cota was considered the best parent, while the best combination was Cota × Cascade. Alessa × Cascade was also a good combination for yield. Likewise many studies on combining ability in French bean resulted identification of numerous good general- and specific-combiners which could be utilized in future breeding (Gonçalves Vidigal *et al.,* 2008; Arunga *et al.,* 2010; Iqbal *et al.,* 2012; Gonçalves *et al.,* 2015; Arruda *et al.,* 2019).

Mutation techniques are the best methods to enlarge the genetically conditioned variability of a species within a short time and have played a significant role in development of many crop varieties. Thus mutation breeding is a potent tool for creating variability, particularly in species where hybridization is difficult and therefore the French bean is readily suited for genetic improvement through induced mutation breeding as hybridization is difficult in that plant due to small flower size and cleistogamous flower structure.

The improvement in yield contributing characters of French bean is possible through induced mutation breeding programme (Mahamune, 2018). Mutants in French bean having improved grain yield have been obtained (Gotoh, 1968; Rubaihayo, 1975; Husseina and Disouki, 1979; Al-Rubeal, 1982). The mutants resistance to diseases (Fadal, 1983; Micke, 1983) and altered seed characteristics such as colour, brightness and size are also on record (Barbosa *et al.,* 1988). Highest mutation frequency for flower colour and seed characteristics was induced by gamma rays of 20 kR dose (Mahamune and Kothekar, 2011). They also obtained mutants with different flower colours and altered size, shape and coat colour of seeds. EMS @ 0.25 per cent treated M_2 phenotypic progeny showed different flower colour mutants like white, purple, blue, red, and yellow in French bean (Borkar and More, 2010). EMS @ 0.15 per cent can effectively be used for inducing maximum variability in yield component traits in French bean (Ramandeep *et al.,* 2018). The induced mutations are also useful in inducing biochemical diversity in French bean (Mahamune *et al.,* 2017). In India, only one variety 'Pusa Parvati' has been released through mutation breeding (Kharkwal and Shu, 2009) and this mutant was developed by mutation from wax podded accession "EC 1906" followed by selection.

Induced mutation by Ethyl methane sulfonate allowed isolation of several useful seed coat colour mutants in a shorter time with less labour and cost (Guimaraes *et al.,* 1989). Bassett and Shuh (1982) reported the possibility of using cytoplasmic male sterility in F_1 hybrid seed production where high bumble bee population is available.

10.3.0 Genetic Resources

Significant national collections of French bean are maintained at the USDA, in Pullman, Washington, USA (about 15,000 accessions), the Institute für Pflanzengenetik und Kulturpflanzenforschung, Germany (about 9000 accessions), in Brasilia, Brazil (CENARGEN/EMBRAPA, with about 6000 accession), in Beijing, China (CAAS, Institute of Genetic Resources with more than 5000 accessions), the National Center for Plant Genetic Resources in Alcala de Henares, Spain (with more than 5000 bean accessions) and NBPGR, New Delhi, India (with nearly 4000 bean accessions). The largest collection of French bean genetic resources is maintained under the auspices of the Food and Agriculture Organization (FAO) treaty, under International Treaty on Plant Genetic Resources for Food and Agriculture (ITPGRFA), at CIAT in Cali, Colombia (around 36,000 accessions), with a backup at the Svalbard Global Seed Vault in Norway, where more than 50,000 accessions are now held. In addition to French bean (*P. vulgaris*) and various wild *Phaseolus* species, these collections include four other domesticated *Phaseolus* species: year-long bean (*Phaseolus dumosus*), runner bean (*Phaseolus coccineus*), tepary bean (*Phaseolus acutifolius*), and lima bean (*Phaseolus lunatus*). Most of these collections were made from the centers of origin, mainly Andean and Mesoamerican regions.

Genetic diversity has been extensively studied in French bean using different types of markers, including seed protein (*e.g.*, phaseolin) (Gepts *et al.*, 1986; De La Fuente *et al.*, 2012) and isozyme analysis (Koenig and Gepts, 1989). Other molecular markers used for geneticdiversity in CB are DNA restriction fragment length polymorphism (RFLP) (Khairallah *et al.*, 1990, 1992), nuclear RFLP (Becerra Velasquez and Gepts, 1994), allozymes (Singh *et al.*, 1991a; Santalla *et al.*, 2002), and random amplified polymorphic DNA (RAPD) (Freyre *et al.*, 1998; Beebe *et al.*, 2000). Similar reports have also demonstrated genetic diversity through use of amplified fragment length polymorphism (AFLP) markers (Beebe *et al.*, 2001; Papa and Gepts, 2003; Zizumbo-Villarreal *et al.*, 2005), SSR markers (Gaitán- Solís *et al.*, 2002; Blair *et al.*, 2006a), DNA sequencing (Gepts *et al.*, 2008), and single nucleotide polymorphism (SNP) markers (Galeano *et al.*, 2009a, 2009b, 2012; Blair *et al.*, 2013). These tools help answer different questions related to evolution, domestication, and diversity of French bean, which is not possible to answer with the use of phenotypic methods alone (Arif *et al.*, 2010). For example, genes related to domestication from the Andean and Mesoamerican domestication events and evolutionary traits such as shattering have been identified (Bellucci *et al.*, 2013; Gaut, 2014) as have polymorphism in drought-related genes (Cortés *et al.*, 2012a, 2012b). Due to their cost-effectiveness, efficiency, and simplicity, SNP, SSR, and AFLP markers have been the most commonly used markers studies on French bean genetic diversity.

Phenotypic variation in a core collection of common bean (*Phaseolus vulgaris* L.) in the Netherlands was reported by Zeven *et al.* (1999). Forty accessions, forming a core collection of mainly bush type of the common bean germplasm in the Netherlands, were evaluated for 14 qualitative and quantitative traits. These and an additional 117 Dutch accessions, mainly collected in private home gardens, were also evaluated for phaseolin seed protein pattern, and morphological and agronomic traits at the International Center for Tropical Agriculture (CIAT, Spanish acronym), Cali, Columbia between 1987 and 1997. Multivariate and principal component analyses at both WAU and CIAT indicated existence of one large group with no discernable patterns among Dutch common bean collections of landraces, garden forms and cultivars. However, when phaseolin, an evolutionary, biochemical marker, was used as an initial classification criterion followed by use of morphological markers, the two major gene pools; Andean and Middle American with two races in each (Chile and Nueva Granada in Andean, and Durango and Mesoamerica in Middle American) were identified. The Andean gene pool was predominant (136 of 157 accessions), especially the race Nueva Granada (126 accessions) characterized by the bush determinate growth habit type T. phaseolin and I. Occurrence of a large number of recombinants strongly suggested considerable hybridization and gene exchange between Andean and Middle American gene pools, thus blurring the natural boundaries and forming a large single group of common bean germplasm in the Netherlands. The inter-gene-pool recombinants of both dry and French beans should be of special interest to breeders for use as bridging-parents for development of broad-based populations. Traka-Mavrona *et al.* (2000) described a functional breeding and maintaining programme of intra-selection in a traditional snap bean cultivar. The programme was applied in three stages. The first thing examined was the existing genetic variability of source material for earliness and pod yield potential. Single-plant frequency distributions with positive skewness for earliness showed that the frequency of unfavourable alleles was high. For total pod yield, distribution was found normal. Thus, the end-target should be selection for early maturity, keeping quality, and stabilizing high yield. The seed shape uniformity was added as third criterion of selection. Secondly, combined pedigree intraselection, based on widely spaced single-plant performance, for the prementioned traits was applied for three successive generations. The evaluation of the third-generation families revealed progenies with high yield, earliness and stability of performance. Thirdly, the end product of the programme applied was to restore or even improve the cultivar. The evaluation of improved selections of fourth-generation families and of the source material, at dense stand, showed that all families were the only ones producing high and stable early fresh pod harvest, even 53 days after planting (53.25-80 g/plant, compared with 0 g/plant of the control). The total pod yield of all the families was 219 to 276 per cent superior compared with source material. Conclusively, the widely spaced single-plant combined pedigree intras-election was proved reliable and effective in restoring or even improving the local cultivar of snap bean according to update demands.

10.4.0 Improvement of Quality Characters

French bean is an important dietary component for many people worldwide. It is a major source of protein in the Americas and in parts of Asia and Africa where animal products are either scarce or too expensive for widespread consumption. Despite its present extensive cultivation, it could contribute more to the world's food reserves. Grain yields remain low, nitrogen fixation is low and variable, and seed protein nutritional qualities are less than optimum for man and other animals. If the potential of the common bean is to be fully realized, improvements in these characteristics must be made.

Although French beans are rich in essential amino acids, they fall short in sulphur containing amino acids, requiring the consumption of larger amount of protein to meet the nutritional requirements for these essential amino acids (Geil and Anderson, 1994). Therefore, it has become essential to develop varieties with increased amount of proteins with higher levels of sulphur containing amino acids. Along with many nutritional components, French bean contains considerable amount of different antinutritional factors like protease inhibitors, lectins, condensed tannins, α-amylase inhibitors and flatulence sugars. These factors cause flatulence (Olson *et al.*, 1981) and low protein digestibility (Nielsen, 1991). Protease inhibitors affect the digestibility of proteins. Trypsin and chymotrypsin inhibitors are reported in French bean. Five different types of trypsin inhibitors have been isolated from French bean (Tsukamoto *et al.*, 1983). Whitaker and Sgarbieri (1981) purified and studied trypsin and chymotrypsin inhibitors from French bean seeds. Lectins also called as haemagglutinins, are special kinds of proteins that effectively bind with carbohydrate and are distinct from enzyme and antibodies. They are extremely toxic when supplied through improperly cooked seeds (Grant and Driessche, 1993 and Rodhouse *et al.*, 1990). Phytohaemagglutinins and the lectin related proteins present in bean seeds are toxic to monogastric animals and hence lower the nutritional value of French bean. Lectins derived from French bean resulted in impairment in transport of nutrients across the intestinal wall, increased catabolism of liver, lowering of blood insulin level and inhibition of brush border hydrolases (Pusztai, 1987). Condensed tannins cause a decrease in digestibility of proteins and carbohydrates as a result of formation of insoluble enzyme resistant complex with tannins (Reddy *et al.*, 1985). Tannin activity was studied by Weder *et al.* (1997) in French bean. The α-amylase inhibitors interfere with the starch digestion in animals by inhibiting pancreatic amylase. á-amylase inhibitor Phaseolamin was studied by Marshall and Lauda (1975) which was found specific to only animal α—amylases and inactive against plant, bacterial and fungal enzymes. Different molecular weight α-amylase inhibitors have been isolated and characterised from French bean (Power and Whitaker, 1977; Pick and Wober, 1978 and Frels and Rupnow, 1984). Various raffinose oligosaccharides (*e.g.* raffinose, stachyose and verbascose) present in beans have been found to be contributing to flatulence (Reddy *et al.*, 1989; Guzman-Maldonado and Paredes-Lopez, 1998). All such anti-nutritional factors not only affect the bioavailability of nutrients but also cause acceptability problems and create physiological disorders after their consumption. It is, therefore, essential to improve existing varieties to minimize the anti-nutritional factors.

Organoleptic traits also determine the niche value of crops like dry bean (rajmah). In many parts of India, the rajmah enjoys its niche status mainly on account of its characteristic organoleptic qualities that have shaped its persistence as an important component of farming systems despite the crop having become less competitive due to relegation to low input farming systems. Superior organoleptic characteristics of some traditional common bean landraces have been pointed to as the reason for their persistence in cultivation, despite the advance of new commercial cultivars. Obtaining reliable information on sensory traits is not easy due to the need for panels of tasters and the weak relationship between sensory and chemical traits found thus far.

The various quality characters such as mechanical resistance of seeds, calcium concentration and storage life have been studied in detail in French bean. Bay *et al.* (1995) studied mechanical damage resistance in three genotypes of snap bean seeds; a white-seeded commercial cultivar and two near isogenic lines (white and dark-seeded) selected for mechanical damage resistance. The effect of impact, at a velocity of 10 m/s, was investigated on eleven specific locations on the seed surface. Averaged over genotype, impacts at the top, front or side of the embryonic axis lobe reduced germination more than corresponding impacts at the chalazal lobe, while impacts on the back of either lobe had little detrimental effect. Impacts on the front or top of the axis had a greater detrimental effect on the commercial variety than the dark-seeded breeding line. Differential susceptibility of the genotypes to impact was studied at the morphological level. Coverage/protection of the embryonic axis by the cotyledons was allocated to categories by visual examination after removal of the seed coat, and the embryonic axis exposed area and length was measured by an image analysis system. The commercial variety had a larger exposed axis area than the breeding lines. One mechanism responsible for the greater resistance to mechanical damage in the breeding lines compared with the commercial variety was better embryonic axis protection by the cotyledons. Mekwatanakarn *et al.* (1997) packed green snap beans (cultivars 91G, Derby, Bronco, Hialeah and Prosperity) in Cryovac PD 941 polyolefin film and stored at 5°C. Four of the 5 cultivars can be kept for 3 weeks under these conditions. Cultivar 91G was the exception and developed chilling injury symptoms 7 days after packing. Gas compositions were similar within all packages and equilibrated at about 4 per cent CO_2 and 5 per cent O_2. There were significant differences in colour measurements between cultivars after storage. Prosperity and Hialeah showed better general appearance compared with Bronco and Derby. Weight loss was low (<3 per cent after 21 days) with only slight cultivar differences. Initial quality assessments were favourable.

Variation in calcium concentration among sixty S1 families and four cultivars of snap bean was studied by Quintana *et al.* (1996). Pod yield and Ca concentration of pods and foliage were determined for a snap bean population, which included 60 S1 families plus 4 commercial cultivars. The experimental design was an 8 × 8 double lattice, repeated at 2 locations (Arlington and Hancock, Wisconsin). Snap beans were planted in June 1993 and machine harvested in August 1993. Calcium analyses were made using an atomic absorption spectrophotometer. Significant differences were detected in pod Ca concentration and yield among the S1 families. Pod size and Ca concentration were inversely correlated ($R2 = 0.88$). Distinct differences between the locations were not observed, and higher Ca genotypes remained high regardless of location or pod size. Low correlation ($R2 = 0.21$) between pod and leaf Ca concentrations was found. Pods of certain genotypes appeared to have the ability to import Ca more efficiently than others, but this factor was not related to yield. Subsequently, they (Quintana *et al.*, 1999a) designed a study to compare snap and dry beans for pod Ca concentration, and to identify genetic resources that might be useful in breeding programmes directed at increasing Ca concentration in bean pods. Pods from 8 snap bean and 8 dry bean cultivars were evaluated for Ca concentration for two years. Snap beans (4.6 ± 0.7 mg/g dry weight) had significantly higher pod Ca concentration than did dry beans (4.2 ± 0.6 mg/g dry weight). Within snap beans, Checkmate had the highest pod Ca concentration (5.5±0.3 mg/g dry weight). Within dry beans, GO122 had the highest (5.1±0.4 mg/g dry weight) pod Ca concentration (3.6±0.3 mg/g dry weight). Six cultivars had podded Ca concentrations significantly higher than the overall mean (4.4±0.3 mg/g dry weight).

Quintana *et al.* (1999b) suggested that genetic variability in pod Ca concentration is caused mainly by differences in flow rate, rather than differences in sap Ca concentration. Hystyle showed a 1.6-fold greater flow rate, 1.5-fold greater pod Ca concentration and 1.7-fold greater Ca absorbed than Labrador. Flow rate correlated positively with Ca absorbed and with Ca concentration in pods of size no. 4 and total pods. Plant maturity influenced sap Ca concentration and Ca translocated increased as plants matured. These results provide evidence that flow rate differences may cause variability for pod Ca concentration in snap beans. To understand the genetics that control pod Ca concentration in snap beans, Quintana *et al.* (1999c) evaluated two snap bean populations consisting of 60 genotypes, plus 4 commercial cultivars used as checks. These populations were CA2 ('Evergreen' × 'Top Crop') and CA3 ('Evergreen' × 'Slimgreen'). The experimental design was an 8 × 8 double lattice repeated each year. No Ca was added to the plants grown in a sandy loam soil with 1 per cent organic matter and an average of 540 ppm Ca. To ensure proper comparison for pod Ca concentration among cultivars, only commercial sieve size no. 4 pods (a premium grade, 8.3 to 9.5 mm in diameter) were sampled and used for Ca extractions. After Ca was extracted, readings for Ca concentration were done via atomic absorption spectrophotometry. In both populations, genotypes and years differed for pod Ca concentration (P = 0.001). Several snap bean genotypes showed pod Ca concentrations higher than the best of the checks. Overall mean pod Ca concentration ranged from a low of 3.82 to a high of 6.80 mg/g dry weight. No differences were detected between the populations. Significant year × genotype interaction was observed in CA2 (P = 0.1), but was not present in CA3. Population variances proved to be homogeneous. Heritability for pod Ca concentration ranged from 0.48 (CA2) to 0.50 (CA3). Evidently, enhancement of pod Ca concentration in beans can successfully be accomplished through plant breeding.

Mencinicopschi and Popa (1999) performed physico-chemical and sensory analyses were performed after harvesting, after freezing at –30°C with forced air circulation, and after 5 or 9 months' storage at –20°C without exposure to light. In snap beans, organoleptic scores were best when the cultivar had a high carbohydrate content, a pH of 8-8.5 and an alcohol insoluble solids content <8.5-9 per cent. The most suitable cultivars for freezing were Nerina, Fina Verde, Echo, Crest, Atlantic, Aurelia, Vidra 9, Lena and Lavinia.

Among different traits, seed traits have been found as the most predominant one which is most important in common bean and major determinants of commercial acceptability of varieties (Saba *et al.*, 2016). They found marked variation in seed coat colour, coat pattern, shape and size and cooking quality that are in favour of need of common bean growers of Kashmir valley of India. Seed traits have also been considered highly heritable traits, therefore important in breeding programmes (Blair *et al.*, 2010).

Texture is a very important indicator of green bean product quality for consumers and it is the parameter that is most obviously changed when pods are canned or frozen. A recent study aimed to establish the correlation between instrumental (using Stable Micro System Texture Analyzer) and sensory texture analyses assays in snap bean pods Pevicharova *et al.*, 2015). The rupture force correlated significantly and negatively with the sensory traits: parchment layer free, crispness, and stringlessness. Firmness measurements of the raw pods by the texture analyzer indirectly gave enough information to use it as a tool for sensory texture analyses of the processed beans. This could allow breeders to evaluate textural quality of cooked pods in the early stages of the breeding and selection programme, thus saving time and costs in doing sensory texture analysis. It is possible to screen cultivars for better sensory quality combined with multiple disease resistance, indirectly based on the consumer's preference.

10.5.0 Abiotic Resistance

At the International Center for Tropical Agriculture (CIAT), abiotic stress breeding has received attention for more than 20 years, in particular for resistance to drought, tolerance to low soil P and nitrogen (N) availability, and resistance to high Al in acid soils (Lynch and Beebe, 1995; Thung and Rao, 1999; Rao, 2002; Beebe *et al.*, 2008). One advantage in the breeding of beans is the wide genetic variability, both within the species, and in sister species with which it can be crossed. Wild common bean presents four gene pools (Tohme *et al.*, 1996), two of which are amply represented in cultivated bean (Singh *et al.*, 1991), and a third of which there is evidence of incipient domestication (Islam *et al.*, 2001).

Breeding for drought resistance has a long history, not only in CIAT but also in the Mexican national bean program of INIFAP (Acosta-Gallegos *et al.*, 1999), and in the program of EMBRAPA in Brazil (Guimaraes *et al.*, 1996). Improved drought resistance has resulted from combining germplasm adapted to the dry high lands of Mexico with small seeded types from lowland Central America, through recurrent selection within each genepool. In the case of drought, roots have long been recognized as playing an important role. A drought resistant line, BAT 477, presented deep rooting under drought stress, permitting access to soil moisture at greater depths (Sponchiado *et al.*, 1989; White and Castillo, 1992). However, deep rooting alone does not assure drought resistance. Data on root density at various levels of the soil profile suggest that deep rooting genotypes are not always the best yielding materials (CIAT, 2007; CIAT, 2008). Data on stomatal conductance and canopy temperature depression confirm that these genotypes are accessing water, but this is not translated into greater yield (CIAT, 2008). In some cases and in some environmental conditions, it appears that partitioning to roots can be at the expense of grain production. Rather, drought resistance appears to result from a combination of mechanisms including a deeper root system, stomatal control and improved photosynthate remobilization under stress (CIAT, 2007; CIAT, 2008). In particular, remobilization of photosynthate from vegetative shoot structures to pods, and from pod wall to grain, is an important mechanism of drought resistance (Rao *et al.*, 2007). Sources of drought resistance have been found in the Durango race and in tepary bean (Beebe, 2012; Rao *et al.*, 2013; Asfaw and Blair, 2014; Mukeshimana *et al.*, 2014). Several drought-resistant lines have also been identified in Africa (Asfaw *et al.*, 2012; Mukeshimana *et al.*, 2014). Breeders and physiologists should focus on improving the traits related to photosynthate mobilization [(pod harvest index (PHI), pod partitioning index (PPI), and harvest index (HI)] from vegetative parts of the plant to the pod walls and seeds under drought conditions (Beebe *et al.*, 2013; Rao *et al.*, 2017). However breeding for improved adaptation to drought is complex because several traits are involved in resistance mechanisms, and the traits are quantitatively inherited and highly affected by environments (Mir *et al.*, 2012). Use of MAS for improving drought resistance was explored by Schneider *et al.* (1997), who identified QTLs for drought using Random Amplified Polymorphic DNA (RAPD) markers. In this study, yield was improved by 11 per cent under drought and 8 per cent under normal conditions by using five RAPD markers (Schneider *et al.*, 1997). Genotype by environment interactions affecting drought QTL are reported by Chavarro and Blair (2010), Asfaw *et al.* (2012), Asfaw and Blair (2012), Blair *et al.* (2012), and Mukeshimana *et al.* (2014). The QTLs identified in all those studies could be important tools for MAS in bean breeding programs to select indirectly for drought tolerance traits that are difficult to screen in large populations.

Common bean is considered to be relatively sensitive to Al compared to other crops (Thung and Rao, 1999). The major site of Al perception and response is the root apex (Ryan *et al.*, 1993), and the distal part of the transition zone (1-2 mm) is the most Al-sensitive apical root zone (Kollmeier *et al.*, 2000). Common bean is also known to differ from cereals through a lag phase in expression of Al resistance mechanisms after exposure to Al (Rangel *et al.*, 2007). Common bean exhibits a typical pattern II type of response to Al treatment, characterized by a delay of several hours in an Al-induced exudation of organic acids, particularly citrate (Stass *et al.*, 2007). Rangel *et al.* (2009) showed that the inhibition of root elongation is induced by apoplastic Al and that the induced and sustained recovery from the initial Al stress in Al-resistant common bean genotype is mediated by reducing the stable-bound Al in the apoplast thus allowing cell elongation and division to resume.

High temperature (HT) stress is a major bean production constraint (Rainey and Griffiths, 2005; De Ron *et al.*, 2016). HT (greater than 30 °C day and/or greater than 20 °C at night) causes significant reduction in yield and quality and limits environmental adaptation. The major effect of high temperature is shown as inhibition of pollen fertility that results in blossom drop. This causes reduced seed number and quality of the seed. Researchers have identified heat-tolerant bean genotypes from diverse gene pools (Porch and Jahn, 2001; Porch, 2006; Porch *et al.*, 2013). Development of HT varieties under adverse environments would also increase resilience for the future global climate change threats (Porch *et al.*, 2013; Gaur *et al.*, 2015).

Bean is sensitive to low temperatures, which can limit production in the early part of the season. Differences among genotypes for tolerance to suboptimal temperatures were reported by Dickson and Boettger (1984). The unifoliate and the first trifoliolate leaf stages were the most sensitive to freezing temperatures in bean (Meyer and Badaruddin, 2001). Their estimated temperature to cause 50 per cent mortality was " 3.25 °C, although regrowth after survival was limited, meaning few plants made it to maturity. Interspecific introgression of portions of the tepary bean genome into French bean is a promising method for increasing tolerance to extreme temperatures (Souter *et al.*, 2017). Rodino *et al.* (2007) reported seven cultivars of *P. coccineus* that showed ability to germinate, emerge, and grow under cold temperature–thus showing potential source of cold-tolerant genes in interspecific hybridization with French bean.

Low soil phosphorus (P) availability causes significant bean yield loss in the tropics (Ramaekers *et al.*, 2010; Beebe, 2012). About 50 per cent of bean growing regions worldwide are affected by low soil P (Nielsen *et al.*, 2001; Beebe, 2012). Progress has been made in developing tolerant cultivars with better P acquisition efficiency, involving higher total root length, root surface, and shallow root angle under low P (Ochoa *et al.*, 2006; Beebe, 2012; Rao *et al.*, 2016). One of the key mechanisms identified to increase access to P is greater topsoil foraging resulting from root architectural, morphological, and anatomical traits (Lynch, 2011). Shallower root growth angle of axial or seminal roots increases the topsoil foraging and thereby contributes to greater acquisition efficiency of P from low P soils. Root QTLs associated with P acquisition in low soil P environments were reported in Beebe *et al.* (2006), which are linked with root parameters such as total and specific root length. QTLs associated with P use efficiency were also reported by Cichy *et al.* (2009a).

Breeding for resistance to drought and Al toxicity may illustrate a dilemma in the breeding for better root systems. Given the importance of roots in confronting abiotic stress, one expects that root vigor should favor better yield. However, by stimulating more partitioning of photosynthates to roots, better yield may not necessarily result. A breeder may be altering the overall pattern of partitioning between vegetative and reproductive structures to the detriment of yield. Instead of more root biomass, we may need to breed for more efficient roots that use the same biomass to best advantage, for example, through longer root hairs, through thinner roots and greater specific root length, through greater organic acid exudation, *etc.* (Liao *et al.*, 2004; Yan *et al.*, 2004; Beebe *et al.*, 2006; Nord and Lynch, 2009), while maintaining or improving partitioning to grain. These are important research issues that must form part of the strategy for fitting the right root system to each production environment.

Nakano *et al.* (1994) evaluated common bean germplasm collected from Malaysia and Thailand. The first mission collected 8 accessions of black-seeded snap bean germplasm in Malaysia while the second mission sent to Thailand collected 28 accessions. In Thailand, local varieties of field bean were not found, but black-seeded snap bean varieties were distributed throughout the country and were presumed to be local cultivars introduced many years ago. It was shown that the black-seeded snap bean varieties from Malaysia and Thailand might have higher potential for heat tolerance than varieties previously reported to be tolerant of high temperatures. For evaluation of heat resistance, seeds of collected accessions, and several heat tolerant lines, were sown at the Okinawa Subtropical Station, Japan (24.3°N, 124.1°E) during the summers (mean daily temperatures 28-28.5°C). Several of the Malaysian accessions and some from southern Thailand had high pod yields and were judged heat tolerant. Nakano *et al.* (1997) evaluated germplasm of snap bean from a variety of sources under the subtropical conditions of Japan. The germplasm comprised 323 breeding lines from CIAT in Colombia, 20 entries from the Philippines, Sri Lanka and Indonesia, 6 reputedly heat tolerant varieties from the USA and 37 accessions collected in Malaysia and Thailand. One of the Malaysian accessions produced the most heat tolerant plants. Pure line selection was undertaken using one of these plants. The line developed showed higher heat tolerance and higher yield potential than commercial varieties in trials carried out in summer in several areas of southern Japan. The line showed tolerance of heat over both short and long periods of time. When the line was exposed to high temperatures for one month, the critical mean air temperature for pod setting ranged from 28 to 29.5°C. The line was registered under the name Haibushi.

10.6.0 Demand-led Breeding

Common bean has steadily evolved from primarily a smallholder subsistence crop (Katungi *et al.*, 2009) to market oriented production (Buruchara *et al.*, 2011). This shift in focus has necessitated a revision in the varietal development process and seed system. The hands-on nature of participatory variety selection (PVS) has evolved from more contractual and consultative to demand-led breeding (DLB) (Persley and Anthony, 2017) where multidisciplinary researchers work closely with bean value chain actors to develop bean varieties that meet the needs of farmers and others in the value chain. This paradigm shift in bean breeding has been toward a value chain focused approach, with relatively less emphasis placed on the farmer-focused approach of a few farmers engaged in selecting varieties for ecological suitability. Through DLB, bean breeders can also enhance varietal diversity through involvement of actors throughout the value chain, as well as minimizing effort that might be invested in developing varieties that are unacceptable to farmers and local communities, and traders/processors. DLB is also able to exploit genotype by environment interaction by taking advantage of specific adaptations to particular locations and growing conditions, such as periodic drought or soil mineral toxicity.

11.0 Biotechnology

11.1.0 Micropropagation

Production of multiple buds from apical and axillary explants of 4 diverse genotypes was achieved on Murashige and Skoog (MS) basal medium plus benzyladenine (BA) after 28 days of culture by Allavena and Rossetti (1986). Genotypes differ in their responses to culture and to added growth regulators (BA and NAA). Elongation *in vitro* was achieved with buds from axillary (but not apical) explants of these genotypes. Protocols for micropropagation (plant regeneration) suitable for routine utilization for breeding and other purposes, have been reported by many workers (Mohamed, 1990; Mohamed *et al.*, 1992; Kalantidis and Griga, 1993).

11.2.0 Molecular Markers

Various molecular markers are being used successfully for linkage map, identification of genes for resistance to various diseases and agronomic traits. A brief review is presented below.

Markers associated with established resistance loci can be used for more efficient breeding to develop resistant cultivars. Some early examples of marker-assisted selection for bean diseases include 23 RAPD markers and 5 SCAR markers associated to 15 different resistance genes, described by Kelly and Miklas (1998). Molecular markers and linkage mapping of rust resistance genes have been reviewed by Miklas *et al.* (2002). Kelly and Vallejo (2004) provided a summary of markers, MAS, map location, and breeding value for anthracnose resistance. Similarly, Miklas *et al.* (2006) reviewed MAS in breeding for resistance to anthracnose, angular leaf spot, common bacterial blight, halo bacterial blight, bean golden yellow mosaic virus, root rots, rust, and white mold.

Bassett and Gepts (1988) presented a comprehensive review describing the linkage mapping of marker genes in *P. vulgaris* and a descriptive list of the 31 genes appearing on the map are also given. Linkages involving genes for disease resistance and protein (arcelin and lectin) synthesis are mentioned. The mapping and utilization of molecular markers (isoenzymes and RFLPs) was also included.

11.2.1 Genetic Linkage and Identification of Genotypes

Linkage mapping enables identification of associations between traits and markers, for both simple Mendelian traits and quantitatively inherited traits (QTLs). The first widely used genetic map in bean was developed from a backcross (BC) mapping population between Mesoamerican line 'XR-235-1-1' and 'Calima' (Andean cultivar (Vallejos *et al.*, 1992). This linkage map included 9 seed proteins, 9 isozymes, 224 RFLP, and seed and flower color markers. These molecular markers were placed on 11 linkage groups, spanning 960 centimorgans (cM). The second genetic map was developed using RFLP markers, spanning 827 cM. These markers were placed on an F_2 mapping population (cross of BAT93 by Jalo EEP558), with 142 markers being assigned to 15 linkage groups (Nodari *et al.*, 1993). A third genetic map was developed by Adam-Blondon *et al.* (1994) from the cross between Ms8EO$_2$ and Core. This map contained 51 RFLPs, 100 RAPDs, and two sequence-characterized amplified region (SCAR) loci and spanned 567.5 cM across 12 linkage groups. These three maps were mainly based on RFLPs, though few seed protein and isozyme markers were also included (McClean *et al.*, 2004). A consensus map was then developed utilizing these linkage maps on BAT93 × Jalo EEP558 (BJ) as a core map (McClean *et al.*, 2004). The creation of this consensus map has provided bean breeders with the means for combining all the genetic information from multiple populations developed from diverse genetic background. It also provided the opportunity to map more loci than from single cross populations and also increased important markers over different genetic backgrounds (Rami *et al.*, 2009). Numerous subsequent maps have been generated, using a succession of marker types (González *et al.*, 2018).

SSR markers (also called microsatellites), which are also typically though not exclusively PCR-based, have been extensively used in bean genetic studies. SSR markers were first reported in bean by Yu *et al.* (1999, 2000), with 15 different microsatellite markers included in a molecular linkage map constructed primarily using RAPD and RFLP markers. Blair *et al.* (2003) integrated 100 SSR markers in two linkage maps along with RFLP, AFLP, and RAPD markers. Much more saturated SSR-based maps were reported by Córdoba *et al.* (2010) and Blair *et al.* (2014). Since then, several bean genetic studies have been implemented using SSR markers and have further been employed for map comparison and integration.

In the last decade, SNP assay methods have become far more efficient. In common bean, SNP frequency is relatively high, with approximately one SNP per 88 bp across a genome of ~ 588 Mbp–implying more than six million SNPs are expected in the genome (Gaitán-Solís *et al.*, 2008; Schmutz *et al.*, 2014; Blair *et al.*, 2018). An important recent SNP map is the high resolution Mesoamerican × Andean cross of Stampede × Red Hawk produced by Song *et al.* (2015), which utilized 7276 SNP markers in an F_2 mapping population of 267 RILs. This was used to anchor sequence scaffolds into pseudomolecules in the first reference genome assembly for common bean (Schmutz *et al.*, 2014). Many bean SNPs have been discovered through sequencing and genotyping by sequencing (Bhakta *et al.*, 2015; Ariani *et al.*, 2016; Schröder *et al.*, 2016), and some older markers have been converted to 'Kompetitive Allele Specific PCR' (KASP)-based SNP assays (Cortés *et al.*, 2011). Improved molecular marker technologies may enable bean breeders and geneticists to speed up cultivar development.

Nodari *et al.* (1992) established two genomic libraries to provide markers to develop an integrated map combining molecular markers and genes for qualitative and quantitative morpho-agronomic traits in common bean. Llaca *et al.* (1994) investigated the differences in the chloroplast DNA (ctDNA) of taxa belonging to the *P. vulgaris* complex with a view to clarifying relationships among species of this complex.

Many workers have used RAPD-PCR technique to study the polymorphism in snap bean (Skroch and Nienhnis, 1995; Vasconcelos *et al.*, 1996; Graham *et al.*, 1994; Alzate Marin *et al.*, 1996; Beebe *et al.*, 1995). Haley *et al.* (1994) evaluated the degree of RAPD marker variability between and within commercially productive market classes representative of the Andean and Middle American gene-pools of common bean. Beebe *et al.* (1995) studied the relative degree of genetic diversity among 76 common bean breeding lines developed for Central America using RAPD. They also included six black seeded cultivars from Brazil and Argentina for comparison and found that the red-seeded and black-seeded beans formed distinct clusters with no overlaps, confirming them to be distinct populations.

According to Gepts *et al.* (2000), in the last decade increased attention has been devoted to genetic resources of wild relative of crop plants, in general, and wild legumes, in particular. Wild relatives have two major functions in genetic resources studies: (1) they provide a geographic framework of reference to elucidate patterns of genetic diversity and domestication; and (2) they constitute an increasingly important source of diversity for a wide range of traits. Examples from their (Gepts *et al.*, 2000) research are supplied on the phylogeography of *P. vulgaris*; (3) the inheritance of the domestication syndrome in *P. vulgaris*; and (4) an assessment of gene flow between wild and cultivated *P. vulgaris*. In all these studies, molecular markers, including DNA sequences, have provided a wealth of data, which, together with phenotypic and ecological data have significantly increased understanding of the genetic resources of these legumes. Freyre *et al.* (1996) found that analysis of phaseolin and molecular markers (isozymes and RFLPs) indicate that this gene pool consists of two major groups, Mesoamerican and Andean, and a third intermediate group found in north-western South America. RAPD analysis of genetic diversity correlated well with genetic diversity obtained with other markers. Moreover, the ease of analysis allowed a large number of bands to be obtained which was conducive to greater sensitivity and identification of geographic subgroups and accessions of hybrid origin.

11.2.2 Mapping for other Characters

Five genes, namely *t*, *z,l*, *Bip*, and *j* control the expression of partly coloured seed coat patterns in common bean. Since these genes have complex interactions, giving various seed coat patterns, it is often difficult to assign specific effects to a particular gene. The problem is compounded by multiple alleles at T, Z, and L. Molecular markers linked to those pattern genes have the potential to assist in identifying the genes involved in specific interactions. Brady *et al.* (1998) initiated a series of experiments to identify useful molecular markers linked to these genes. A group of 3 parental genotypes and 5 backcross lines were screened by RAPD to identify putative markers. Potential markers were then mapped in appropriate BC3-F_2 populations. To date, recombination has not been detected between RAPD markers OAM131350 and OM19400 and the *T* gene. A single recombinant was detected between RAPD marker OAM10560 and the Z gene, and the marker mapped at a distance of 1.4 cM.

11.2.3 Mapping Resistance Genes

11.2.3.1 Anthracnose Resistance

Anthracnose, caused by the fungus *Colletotrichum lindemuthianum*, is a severe disease of common bean controlled, in Europe, by a single dominant gene, *Are*. Adam Blondon *et al.* (1994) constructed four pairs of near-isogenic lines (NILs) in which the *Are* gene was introgressed into different genetic backgrounds (Coco, Early-wax, Processor and Slender-White). These pairs of NILs were used to search for DNA markers linked to the resistance gene. Nine molecular markers, five RAPDs (random amplified polymorphic DNA) and four RFLPs, discriminated between the resistant and the susceptible members of these NILs. A backcross progeny of 120 individuals was analysed to map these markers in relation to the *Are* locus. Five out of the nine markers were linked to the *Are* gene within a distance of 12.0 cM. The most tightly linked, a RAPD marker, was used to generate a pair of primers that specifically amplified this RAPD (sequence characterized amplified region, SCAR).

A RAPD marker designated OQ41440, generated by a 5′-AGTGCGCTGA-3′ deca-mer primer, was found tightly linked in coupling with the *Are* gene (Young and Kelly, 1996). The bracketing molecular markers allowed tagging of the *Are* allele with a selection fidelity of 99 per cent. Use of the OQ41440 and B3551000 RAPD markers for marker-based selection will afford the opportunity to retain the *Are* anthracnose resistance gene in bean germplasm, as other epistatic resistance genes are characte-rized, and incorporated into contemporary bean cultivars (Young and Kelly, 1996). Subsequently, they (Young *et al.*, 1998) identified two independently assorting dominant genes conditioning resistance to bean anthracnose in an F_2 population derived from the highly resistant *P. vulgaris* differential cultivar, G2333. One gene was allelic to the *Co-4* gene in the differential cultivar TO and was named Co-42, whereas the second gene was assigned the temporary name Co-7 until a complete characterization with other known resistance genes can be conducted.

Pyramiding major resistance genes using marker-assisted selection can develop new cultivars of the common bean with durable resistance to anthracnose. To this end, it is necessary to identify sources of resistance and molecular markers tightly linked to the resistance genes. Geffroy *et al.* (1998) transformed a RAPD marker, ROH20450, linked to the Mesoamerican *Co-2* anthracnose resistance gene, into a SCAR (sequence characterized amplified region) marker, SCH20. Since this SCAR marker was found to be useful mainly in the Andean gene pool, a new PCR-based marker (SCAreoli) was identified for indirect scoring of the presence of the *Co-2* gene.

The SCAreoli SCAR marker is polymorphic in the Mesoamerican as well as in the Andean gene pool and should be useful in MAS. De Arruda *et al.* (2000) studied the inheritance of resistance to anthracnose in the cultivar TO (carrying the *Co-4* gene), to identify random amplified polymorphic DNA (RAPD) markers linked to Co-4, and to introgress this gene in the cultivar Ruda. Populations F_1, F_2, F_2; 3, BC_1s, and BC_1r from the cross Ruda × TO were inoculated with race 65 of *C. lindemuthianum*.

11.2.3.2 Other Diseases

Molecular markers have been used for studying the resistance to various and construct partial linkage map. The diseases for which such studies have been reported are as follows:

CBB (Common bean blight, Xanthomonas compestris pv. phaseoli)

(Jung *et al.*, 1996, 1997; Bai *et al.*, 1997; Tar-an *et al.*, 1998; Boscariol *et al.*, 1998; Park *et al.*, 1999a; Souza *et al.*, 2000).

Bean rust (Uromyces appendiculatus var. appendiculatum)

(Jung *et al.*, 1998; Park *et al.*, 1999b).

Bean mosaic

(Kelly *et al.*, 1995).

Macrophomina phaseolina

(Olaya *et al.*, 1996).

11.3.0 Genetic Transformation

Modi *et al.* (2001) in the recent review on various aspects of biotechnology of leguminous vegetables summarised the advancement in genetic transformation in *Phaseolus vulgaris*. Different research on gene transfer in French bean is presented in Table 5.

Table 5: Summary of Research on Gene Transfer in French Bean

Gene	Method	References
Lindane resistance gene and GUS (*uid A*)	Agrobacterium	Eissa *et al.* (2000)
α -amylase inhibitors (αA1-1 and αA1-2)	-	Morton *et al.* (2000)
Methionine-rich storage albumin	Bolistic transformation	Aragao *et al.* (1999)
Viral antisense RNAs	Bolistic	Aragao *et al.* (1998)
Bar (bilanafos resistance)	Electroporation	Saker and Kuhne (1998)
Endochitinase class IV chitinase gene (*AtchitIV*)	-	Gerhardt (1997)
gfp gene	Conjugal plasmid transfer	Normander *et al.* (1998)

Gene	Method	References
Phenylalanine ammonialyase (*pal*) promoter/ β-glucuronidase gene fusion	-	Seguin *et al.* (1997)
Antisense chalcone synthase [naringenin-chalcone synthase] (*chs*) gene	*A. rhizogens*	Colliver *et al.* (1997)
Chalcone synthase [naringenin- chalcone synthase] (*chs*) gene	TMV	Faktor *et al.* (1997)
GUS gene under control of the CaMV 35S promoter	*A. tumifaciens*	Barros *et al.* (1997)
Intron-containing the gus gene (*uidA*)	*A. tumifaciens*	Zhang *et al.* (1997)
Canavalin promoter-β-glucuronidase gene fusion	Bolistic	Kim and Minamikawa (1997)
npt II and *uidA* (pMP90)	*A. tumefaciens* C58CIRifR	Dillen *et al.* (1997)
β-phaseolin gene (phas)	-	Geest and Hall (1996)
Glutamine synthetase *gln-\α\gene*	*A. tumefaciens*	Watson and Cullimore (1996)
gus transformation using micro-projectile bombardment	*Agrobacterium*-mediated	Brasileiro *et al.* (1996)
gus	-	Kim and Minamikawa (1996)
gus	Electroporation	Dillen *et al.* (1995)
Isopentenyl transferase gene (*ipt*)	*A. tumefaciens* C58	Song *et al.* (1995)
gus	Particle bombardment	Aragao *et al.* (1993)
gus, *bar* (encoding phosphinothricin acetyltransferase) and bean golden mosaic bigemini-virus coat protein genes	Electric-discharge mediated particle acceleration	Russell *et al.* (1993)
Bean chalcone [naringeninchalcone] synthase promoter-GUS fusion 2S albumin from Brazil nut (*Bertholletia excelsa*) and β-glucuronidase, both under the control of the 35S CaMV promoter	*Agrobacterium rhizogenes*	Franklin *et al.* (1993)
npt II	-	Aragao *et al.* (1992)
CaMV 35S promoter fused to the luciferase included the RNA leader sequence omega (omega), derived from tobacco mosaic tobamovirus, and intron 1 from the maize alcohol dehydrogenase gene	*A. tumifaciens* and *rhizogenes*	McClean *et al.* (1991)

Genetic transformation in French bean has been reported using *Agrobacterium*, particle bombardment and electroporation. Genga *et al.* (1990 a,b) reported genetic transformation of bean by using *Agrobacterium*-mediated gene transfer system in *Phaseolus* spp. Allavena and Bernacchia (1991), attempted genetic transformation of commom bean by using high velocity particle bombardment. Hoyos *et al.* (1992) achieved transient expression in sexual embryos of common bean. Genga *et al.* (1992) reported successful transformation of *Phaseolus* species by high velocity micro-projectile bombardment.

Genetic Transformation in French bean

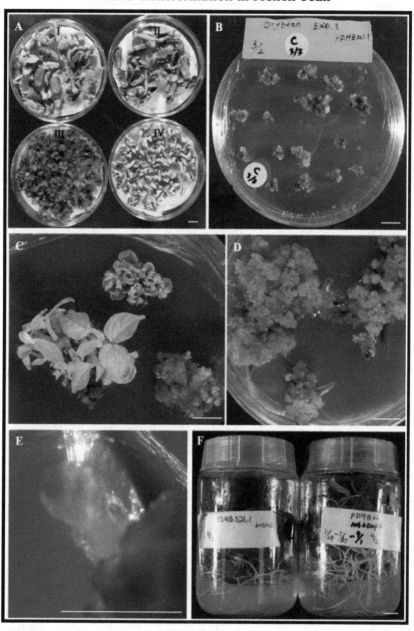

Agrobacterium tumefaciens-mediated transformation of 'Olathe' pinto bean. (A) Explants after 4-day co-cultivation in the dark at 25 °C; (I, II) Half-seeds explants; (III) Precultured embryo axes explants; (IV) Embryo axes explants; (B) Precultured embryo axes explants after 3-week selection on selection RM1; (C) Embryo axes explants after 6-week selection on selection RM3; (D, E) GS-resistant calli and green buds induced from the precultured embryo axes explants after 6-week selection on selection RM3; (F) Rooting of the GS-resistant plants produced from the precultured embryo axes explants.

Aragao *et al.* (1992) transformed mature bean (*P. vulgaris*) embryos using biolistic method with a plasmid containing the sulphur-rich 2S-albumin gene from Brazil nut and achieved transient expression whereas Franklin *et al.* (1993) could get transformed green bean (*P. vulgaris*) callus using both leaf disc and hypocotyl explants in cocultivation with *A. tumefaciens*. Aragao *et al.* (1996) also studied the parameters influencing transient gene expressions in bean (*P. vulgaris*) in response to transformation by electrical particle acceleration device. A calcium spermidine procedure for coating gold particles with DNA was reported to be better than the standard Ca-PO4 procedure. Russell *et al.* (1993) recovered stably transformed transgenic of *P. vulgaris* plants using electrical discharge-mediated particle acceleration method to transfer *gus*, *bar* and bean golden mosaic virus (BGMV) coat protein genes. The stability of the transformed characters was shown across 5 generations of selfing without loss of expression as well as in the crosses with non-transformed plants.

Aragao *et al.* (1996) generated stable transgenic plants of *P. vulgaris* exploiting the biolistic method to transfer two plasmids harbouring different genes. One of the plasmids contained the *2S* albumin gene of Brazil nut and the antisense sequence of *ac1*, *ac2*, *ac3* and *bc1* genes from the bean golden mosaic virus (BGMV). The inheritance pattern of the foreign genes in the progenies was shown to be Mendelian in nature. Aragao and Rech (1997) also discussed the morphological factors and other parameters influencing recovery of transgenic bean.

Stable transformation of common bean (*Phaseolus vulgaris* L.) has been successful, to date, only using biolistic-mediated transformation and shoot regeneration from meristem containing embryo axes. In a study conducted by Song *et al.* (2020), using precultured embryo axes, and optimal co-cultivation conditions resulted in a successful transformation of the common bean cultivar Ola the using *Agrobacterium tumefaciens* strain EHA105. Plant regeneration through somatic embryogenesis was attained through the preculture of embryo axes for 12 weeks using induced competent cells for *A. tumefaciens*-mediated gene delivery. Using *A. tumefaciens* at a low optical density (OD) of 0.1 at a wavelength of 600 nm for infection and 4-day co-cultivation, compared to OD 600 of 0.5, increased the survival rate of the inoculated explants from 23 per cent to 45 per cent. Selection using 0.5 mg L-1 glufosinate (GS) was effective to identify transformed cells when the *bialaphos resistance* (*bar*) gene under the constitutive 35S promoter was used as a selectable marker. After an 18-week selection period, 1.5 per cent -2.5 per cent inoculated explants, in three experiments with a total of 600 explants, produced GS-resistant plants through somatic embryogenesis.

12.0 References

Abd El-Aal, H., El-Hwat, N., El-Hefnawy, N. and Medany, M. (2011) *Am. Eurasian J. Agric. Environ. Sci.*, **11**: 79–86.

Abd El-Wahed, M.H., Baker, G.A., Ali, M.M. and Abd El-Fattah, Fatma A. (2017) *Sci. Hort.*, **225**: 235-242.

Abdel-Mawgowd, A.M. (2006) *Aust. J. Basic Appl. Sci.*, **2**: 443–450.

Adam Blondon, A.F., Sevignac, M., Bannerot, H. and Dron, M. (1994) *Theor. Appl. Gen.*, **88**: 865-870.

Adikshita and Sharma, M. (2017) *Int. J. Environ. Sci. Te.*, **6**: 1555-1559.

Aghora, T.S., Mohan, N., Somkuwar R.G. and Ganeshan, G. (2007) *J. Hort. Sci.*, **2**: 104-107.

Ahlawat, I.P.S. (1996) *Indian J. Agril. Sci.*, **66**: 338-342.

Alberini, J.L., Mohan, S.K., Menezes, J.R. De, Silva, W.R. Da, Oliari, L. and Meyer, R.C. (1987) *PesquisaAgropecuariaBrasileira*, **22**: 995-998.

Al-Fadnil, F.H. and Al-Ani, R.A. (1987) *J. Agric. Water Res. Res., Plant Prod.*, **6**: 99-114.

Ali, M.A. (1950) *Phytopathology*, **40**: 69-79.

Allavena, A. and Rossetti, L. (1986) *Sci. Hortic.*, **30**: 37-46.

Allavena, A. and Bermacchia, G. (1991) *Ann. Rep. Bean Improv. Coop.*, **34**: 137-138.

Al-Sheikh, A.A. and Al-Darby, A.M. (1996) *Arab Gulf J. Scientific Res.*, **14**: 767-793.

Alzate Marin, A.L., Baia, G.S., Martins Filho, S., Paula Junior, T.J. de., Sediyama, C.S., Barros, *E.g.* de., Moreira, M.A., De Paula, Junior, T.J., De Barros, *E.g.* (1996) *Brazilian J. Gen.* **19**: 621-623.

Anonymous (1982-83) *Annual report*, Research Project for Improvement in the Production of Horticultural Crops in the Hilly Region of Darjeeling District, Bidhan Chandra Krishi Viswavidyalaya, West Bengal.

Anonymous (1987) *Documentos -EMCAPA,* No. 40, p. 6.

Aponte, A., Moro, V.S., Soto, J.M. and Jimenez, J. (1976) *CIARCO,* **6**: 19-24.

Aragao, F.J.L, Barros, L.M.G, Brasileiro, A.C.M., Ribeiro, S.G., Smith, F.D., Sanford, J.C., Faria, J.C. and Rech, E.L. (1996) *Theor. Appl. Genetic.*, **93**: 142-150.

Aragao, F.J.L. and Rech, E.L (1997) *Intern. J. Plant Sci.*, **158**: 157-163.

Aragao, F.J.L., Barros, L.M.G., Sousa, M.V. de., Sa, M.F.G. de., Almeida, E.R.P., Gander, E.S., Rech, E.L. de, Sousa, M.V. de and Sa, M.F.G. (1999) *Gen. Mol. Biol.*, **22**: 445-449.

Aragao, F.J.L., Grossi, de, Sa, M.F., Almeida, E.R., Gander, E.S. and Rech, E.L. (1992) *Plant Mol Biol.*, **20**: 357-359.

Aragao, F.J.L., Grossi, de, Sa, M.F., Davey, M.R., Brasileiro, A.C.M., Faria, J.C. and Rech, E.L.(1993) *Plant Cell Rep.*, **12**: 483-490.

Aragao, F.J.L., Ribeiro, S.G., Barros, L.M.G., Brasileiro, A.C.M., Maxwell, D.P., Rech, E.L. and Faria, J.C. (1998) *Mol. Breed.*, **4**: 491-499.

Ariani, A., Teran, Y., Berny, J.C. and Gepts, P. (2016) *Mol. Breed.*, **36**: 87.

Arnold, A.J., Cayley, G.R., Dunne, Y., Etheridge, P., Greenway, A.R., Griffiths, D.C., Phillips, F.T., Pye, B.J., Rawlinson, C.J. and Scott, G.C. (1984b) *Ann. Appl. Biol.*, **105**: 369-377.

Arnold, A.J., Cayley, G.R., Dunne, Y., Etheridge, P., Griffiths, D.C., Phillips, F.T., Pye, B.J., Scott, G.C. and Vojvodic, P.R. (1984a) *Ann. Appl. Biol.*, **105**: 353-359.

Arruda, I.M., Moda-Cirino, V., Koltun, A., Zeffa, D.M., Nagashima, G.T. and Gonçalves, L.S.A. (2019) *Agron.*, **9**: 371.

Arunga, E., Esther, R., Henk A.V. and Owuoche, J.O. (2010) *Afr. J Agric. Res.*, **5**: 1951-1957.

Arya, P.S., Sagar, V. and Singh, S.R. (1999) *Scientific Hortic.*, **6**: 137-139.

Assefa, T., Mahama, A.A., Brown, A.V., Cannon, E.K.S., Rubyogo, J.C., Rao, I.M., Blair, M.W. and Cannon, S.B. (2019) *Mol. Breed.*, **39**: 20.

Assefa, T., Rao, I.M., Cannon, S.B., Wu,J., Gutema, Z., Blair, M.W., Paul, O., Alemayehu, A. and Dagne, B. (2017) *Plant Breed.*, **136**: 548-561.

Assefa, T., Wu, J., Beebe, S.E., Rao, M.I., Marcomin, D. and Rubyogo, J.C. (2015) *Euphytica*, **203**: 477-489.

Aykroyd, W.R. (1963) ICMR Special Rept. Series, 42.

Azmeraw, Y. and Hussien, T. (2017) *Adv. Crop Sci. Tech.*, **5**: 314.

Baboo, R., Rana, N.S. and Pantola, P. (1998) *Ann. Agric. Res.*, **19**: 81-82.

Babu, V. and Ciotea, V. (1976) *BiologieVegetela*, **28**: 67-71.

Bai, Y., Michaels, T.E., Pauls, K.P. (1997) *Genome*, **40**: 544-551.

Bala, K. and Kumar. S. (2018) *J. Entomol. Zool. Stud.*, 6: 1514-1518.

Barcchiya, J. and Kushwah S.S. (2017) *Legume Res.*, **40**: 920-923.

Bardner, R. and Fletcher, K.E. (1979) *J. Agric. Sci.*, **92**: 109-112.

Bardner, R., Fletcher, K.E. and Griffiths, D.C. (1983) *J. Agric. Sci.*, **101**: 71-80.

Barrera, J. (1997) *Escuela Nacional de Agricultura, Chapingo, México.*

Barros, L.M.G., Gama, M.I.C.S., Goncalves, C.H.R. deP., Barreto, C.C., santana, E.F. and Carneiro, V.T. (1997) *Pesquisa Agropecuaria Brasileira.*, **32**: 267-275.

Bascur, B.G. and Cataki, K.C. (1977) *Investigacion y Progreso Agricola*,9: 52-53.

Bassett, M.J. (1989) *Annu. Rep. Bean Improv. Coop.*, **32**: 1-15.

Bassett, M.J. (1993) *Annu. Rep. Bean Improv. Coop.*, **36**: 6-23.

Bassett, M.J. (1993) *J. Amer. Soc. Hort. Sci.*, **118**: 878-880.

Bassett, M.J. (1996) *Annu. Rep. Bean Improv. Coop.*, **39**: 1-19.

Bassett, M.J. and Gepts, P. (1988) In: *Genetic Resources of Phaseolus beans*, pp. 329-353.

Bassett, M.J. and Shuh, D.M. (1982) *J. Amer. Soc. Hort. Sci.*, **107**: 791-793.

Bay, A.P.M., Taylor, A.G. and Paine, D.H. (1995) *Plant Varieties and Seeds*,8: 151-159.

Bayuelo-Jimenez, J.S., Debouck, D.G. and Lynch, J.P. (2003) *Field Crops Res.*, **80**: 207–222.

Beebe, S.E. (2012) *Plant Breed. Rev.*, **36**: 357-426.

Beebe, S.E. and Pastor-Corrales, M.A. (1991) In: Schoonhoven, A van and Voysest, O. (Ed.), *Common beans: research for crop improvement. Wallingford-Cali, CAB International and CIAT.* pp. 561-618.

Beebe, S.E., Ochoa, I., Skroch, P., Nienhuis, J. and Tivang, J. (1995) *Crop Sci.,* **35:** 1178-1183.

Beebe, S.E., Rao, I.M., Blair, M.W. and Acosta-Gallegos, J.A. (2013) *Front Physiol.,* p. 4.

Beshir, T. (2003) *Biology and Control of Bean Anthracnose in Ethiopia. A Ph.D. Thesis submitted to the Faculty of Natural and Agricultural Sciences, University of Free State. Bloemfontein, South Africa.*

Bhamare, R.S., Sawale, D.D., Jagtap, P.B., Tamboli, B.D. and Kadam, M. (2018) *Int. J. Chem. St.,* **6:** 3397-3399.

Bhatnagar, G.S., Porwal, M.K. and Nanawati, G.C. (1992) *Indian J. Agric. Sci.,* **62:** 280-281.

Bingham, R.J. (1978) *Horticulture Industry* (September), **33:** 36.

Bitocchi, E., Bellucci, E., Giardini, A., Rau, D., Rodriguez, M., Biagetti, E., Santilocchi, R., Spagnoletti, Zeuli P., Gioia, T., Logozzo, G., Attene, G., Nanni, L. and Papa, R. (2013) *New Phytol.,* **197:** 300–331.

Bland, B.F. (1971) *Crop Production: Cereals and Legumes,* Academic Press London and New York, pp. 303-326.

Bonanno, A.R. and Mack, H.J. (1983) *J. Amer. Soc. Hort. Sci.,* **108:** 832-837.

Bora, R.K. and Sarma, C.M. (2006) *Asian J. Plant Sci.,* **5:** 324-330.

Borkar, AT. and More, A.D. (2010) *Adv. Biores.,* **1:** 22-28.

Borys, M.W. and Mamys, I. (1978) *RocznikiAkademiiRoljiczej w PoznaniuOgrodnictwo,* **98:** 3-12.

Boscariol, R.L., Souza, A.A., Tsai, S.M. and Camargo, L.E.A. (1998) *FitopatologiaBrasileira,* **23:** 132-138.

Bourillet, D. (1989) *UNILEC Informations, Union NationaleInterprofessionnelle des Legumes de Conserve,* No. 63, pp. 16-17.

Brady, L., Bassett, M.J. and Mc Clean, P.E. (1998) *Crop Sci.,* **38:** 1073-1075.

Brasileiro, A.C.M., Aragao, F.J.L., Rossi, S., Dusi, D.M.A., Barros, L.M.G. and Rech, E.I. (1996) *J. Amer. Soc. Hort. Sci.,* **121:** 810-815.

Broughton, W.J., Hernandez, G., Blair, M.W., Beebe, S.E., Gepts, P. and Vanderleyden, J. (2003) *Plant Soil,* **252:** 55–128.

Budanova, V.I. (1985) *Genetike I Selektsii,* **91:** 55-64.

Bull, P.B. (1977) *New Zealand Commericial Grower,* **32:** 23.

Buruchara, R., Chirwa, R., Sperling, L., Mukankusi, C., Rubyogo, J.C. and Mutonhi, R. (2011) *Afr.Crop Sci. J.,* **19:** 227–245.

Byron, E.J. (1948) *Proc. Amer. Soc. Hort. Sci.,* **51:** 457-462.

Cardona, C. (1989). In: Schwartz, H.F. and Pastor-Corrales, M.A. (Eds.), *Bean production problems in the tropics, CIAT, Cali, Colombia*, pp. 505-570.

Cardona, C. and Kornegay, J. (1999) In: Clement, S.L. and Quisenberry, S.S. (Eds.), *Global Plant Genetic Resources for Insect-Resistant Crops. CRC Press, Boca Raton, FL.*

Cardoso, A.A., Fontes, L.A.N. and Vioerira, C. (1978) *Revista Ceres*, **25**: 292-295.

Cardoso, J.E. and Echandi, E. (1987) *Plant Dis.*, **71**: 167-170.

Carrijo, I.V. (1993) *HorticulturaBrasileira*, 1993, **11**: 56.

Castiglioni, V.B.R., Takahashi, L.S.A., Athanazio, J.C., De Menezes, J.R., Fonseca, M.A.R., and Castilho, S.R. (1993) *Horticultura-Brasileira*, **11:** 164.

Chakraborty, P. and Sharma, C.M. (1982) *Indian J. Plant Physiol.*, **25**: 292-295.

Chavan, K.A., Suryavanshi, V.P. and Patil, A.A. (2019) *J. Pharmacog. Phytochem.*, **8**: 113-116.

Chavan, M.G., Ramteke, J.R., Patil, B.P., Chavan, S.A. and Shaikh, M.S.I. (2000) *J. Maharashtra Agril. Univ.*, **25**: 95-96.

Colliver, S.P., Morris, P. and Robbin, M.P. (1997) *Plant Mol. Biol.*, **35**: 509-522.

Conklin, A.E., Erich, M.S., Liebman, M., Lambert, D., Gallandt, E.R. and Halteman, W.A. (2002) *Plant Soil*, **238:** 245–256.

Conner, R.L., McAndrew, D.W., Kiehn, F.A., Chapman S.R. and Froese N.T. (2004) *Can. J. Plant Pathol.*, **26**: 299-303.

Corokalo, D., Miladinovic, Z., Zdravkovic, M. and Brkic, S. (1992) *SavremenaPoljoprivreda*, **40**: 115-120.

Correa, J.R.V., Netto, A.J., Rezende, P.M. De and Andrade, L.A. De. B. (1990) *PesquisaAgropecuariaBrasileira*, **25**: 513-519.

Cortés, A.J., Blair, M.W. (2017) In: Grillo, O. (Ed.) *Rediscovery of landraces as are source for the future, ISBN978-95351-5806-6.*

Cox, T.I. (1979) New Zealand Commercial Grower, **34**: 21.

Coyne, D.P. (1969) *HortScience*, **4**: 100, 117.

Creamer, N.G., Bennett, M.A., Stinner, B.R. and Cardina, J. (1996) *J. Am. Soc. Hortic. Sci.*, **121**: 559–568.

Crespo, G.M. and Torres, C.J. (1979) *J. Agric. Univ. Puerto Rico*, **63**: 465-468.

Culpepper, C.W. (1936) *Fd. Res.*,1: 357-376.

Dahatonde, B.N. and Nalamwar, R.V. (1996) *Indian J. Agron.*, **41:** 265-268.

Dahatonde, B.N., Turkhede, A.B. and Kale, M.R. (1992) *Indian J. Agron.*, **37:** 835-837.

Damayanti, T., Susilo, D., Nurlaelah, S., Sartiami, D., Okuno, T. and Mise, K. (2008) *Journal J. Gen. Plant Pathol.*, **74**: 438-442.

Dauple, P., Gumbard, C. and Peyriere, J. (1979) *Pepinieristes Horticulteurs Maraichers*, No. 193, pp. 37-41.

De Arruda, M.C.C., Alzate Marin, A.L., Chagas, J.M., Moreira, M.A., de Barros, *E.g.* (2000) *Phytopathology*, **90:** 758-761.

De Carvalho, A.C.P.P., Leal, N.R., Rodrigues, R. and Costa, F.A. (1999) *HorticulturaBrasileira*, 17: 102-105.

Deka, D. and Das, N. (1975) *J. Assam Sci. Soc.*, **18:** 67-70.

Demir, I., Yanmaz, R. and Gunay, A. (1994) *Bahce*, **23:** 59-65.

Deshpande, S.B., Jadhav, A.S. and Deokar, A.B. (1995) *J. Maharashtra Agril. Univ.*, **20:** 423-425.

Dickson, M.H. and Boettger, M.A. (1976) *J. Amer. Soc. Hort. Sci.*, **101:** 541-544.

Dickson, M.H. and Boettger, M.A. (1977) *J. Amer. Soc. Hort. Sci.*, **102:** 498-501.

Dickson, M.H., Duczmal, K. and Shannon, S. (1973) *J. Amer. Soc. Hort. Sci.*, **98:** 509-513.

Dickson, M.H., Hunter, J.E., Boettger, M.A. and Cigna, J.A. (1982) *J. Amer. Soc. Hort. Sci.*, **107:** 231-234.

Dillen, W., Clercq, J. de., Goossens, A., Montagu, M. van., Angenon, G., De, Clercq, J. and Van, Montagu, M. (1997) *Theor. Appl. Genet.*, **94:** 151-158.

Dillen, W., Engler, G., Montagu, M. van., Angenon, G. and Van, Montagu, M. (1995) *Plant Cell Rep.*, **15:** 119-124.

Doré, T. and Meynard, J.M. (1995) *J. Appl. Entomol.*, **119:** 49-54.

Drake, S.R. and Silbernagel, M.J. (1982) *J. Amer. Soc. Hort. Sci.*, **107:** 239-242.

Duan, X. and Ding, G. (1989) *ZuownPinzhongZiyuan* No. 2, p. 45.

Dubetz, S. and Mahalle, P.S. (1969) *J. Amer. Soc. Hort. Sci.*, **94:** 479-481.

Duc, G., Agrama, H., Bao, S., Berger, J., Bourion, V. and De Ron A.M. (2015) *Crit. Rev. Plant. Sci.*, **34:** 381-411.

Dufault, R.J., Decoteau, D.R. Garrett, J.T., Batal, K.D., Granberry, D., Davis, J.M., Hoyt, G. and Sanders, D. (2000) *J. Veg. Crop Prodn.*, **6:** 13-25.

Duranti, A., and Lanza, A.M.R. (1978) *Annalidella Facolta di Scienze Agrariedellauniversitadegli Studi di Napoli Portici*, **12:** 52-61.

Durge, V.W., Khan, I.A., Dahatonde, B.N. and Vayas, J.S. (1997) *Annals Plant Physiol.*, **11:** 223-225.

Dwivedi, G.K. (1995) *J. Indian Soc. Soil Sci.*, **43:** 231-236.

Dyck, E. and Liebman, M. (1994) *Plant Soil*, **167:** 227–237.

Eira, P.A. Da., Pessanha, G.G., Britto, D.P.P. De S. and Carbajal, A.R. (1974) *PesquisaAgropecuasiaBrasileira*, **9:** 121-124.

Eissa, A.E., Bisztray, G. and Velich, I. (2000) *Intl. J. Hort. Sci.*, **6:** 32-35.

Ellis, M.A., Galvez, E.G.E. and Sinclair, J.B. (1977) *Turrialba*, **27:** 37-40.

El-Mohamedy, R.S.R., Shafeek, M.R. and Fatma, A.R. (2015) *J. Agric. Technol.*, **11:** 1219-1234.

El-Noemani, A.A., El-Zeiny, H.A., El-Gindy, A.M., El-Sahhar, E.A. and El-Shawadfy, M.A. (2010) *Aust. J. Basic Appl. Sci.*, **4**: 6185–6196.

El-Tahawi, B.S., Diab, M.A., El-Hadidi, A.Z., Habib, M.A. and Draz, S.N. (1982a) *Minufiya J. Agric.*, **6**: 289-301.

El-Tahawi, B.S., El-Hadidi, A.Z., Diab, M.A., Habib, M.A. and Draz, S.N. (1982b) *Res. Bull. Fac. Ain Shams Univ.*, No. 1904, p. 16.

Ester, A. and Jeuring, G. (1992) *FABIS Newsl.*, **30**: 32-41.

Evans, A.M. (1976) *Evolution of Crop Plants*, (Ed. N.W. Simmonds), Longman, London pp. 168-172.

Faktor, O., Kooter, J.M., Loake, G.J., Dixone, R.A. and Lamb, C.J. (1997) *Plant Sci.*, **124**: 175-182.

Fassuliotis, G., Hoffman, J.C. and Deakin, J.R. (1967) *Nematologia,*13: 141.

Feng, D.X., Zhu, G.R. and Li, B.D. (2000) *Acta PhtophylocicaSinica*, **27**: 249-254.

Fenille, R.C. and de Souza, N.L. (1999) *PesquisaAgropecuariaBrasileira*, **34**: 1959-1967.

Fisk, J.W., Hesterman, O.B., Shrestha, A., Kells, J.J., Harwood, R.R., Squire, J.M. and Sheaffer, C.C. (2001) *Agron J.*, **93**: 319–325.

Flores-Este´vez, N., Acosta-Gallegos, J. and Silva-Rosales, L. (2003) *Plant Dis.*, **87**: 21-25.

Franco, A.A. and Day, J.M. (1980) *Turrialba*, **30**: 99-105.

Franklin, C.I., Trieu, T.N., Cassidy, B.G., Dixon, R.A. and Nelson, R.S (1993) *Plant Cell Rep.*, **12**: 74-79.

Freyre, R., Rios, R., Guzman, L., Debouck, D.G. and Gepts, P. (1996) *Econ. Bot.*, **50**: 195-215.

Gableman, W.H. and Wiliams, D.F. (1960) *Wisconsin Agric. Exp. Sta. Res, Bull.*, 221.

Gabr, S.M. (2000) *Alexandria J. Agril. Res.*, **45**: 201-212.

Gallandt, E.R. (2006) *Weed Sci.*, **54**: 588–596.

Gallegos, J.A.A. and Shibata, J.K. (1989) *Field Crops Res.*, **20**: 81–93.

Geest, A.H.M. van der., Hall, T.C. and Van der Geest, A.H.M. (1996) *Plant Mol. Biol.*, **32**: 579-588.

Geffroy, V., Creusot, F., Falquet, J., Sevignac, M., Adam Blondon, A.F., Bannerot, H., Gepts, P. and Dron, M. (1998) *Theor. Appl. Gen.*,**96**: 494-502.

Genga, A., Allavena, A., Ceriotti, A. and Bollini, R. (1990a) *Acta Horti.*, **280**: 527-536.

Genga, A., Ceriotti, A., Boelini, R. and Allavena, A. (1990b) *Ann. Rep. Bean Improv. Coop.*, **33**: 75.

Genga, A.M., Allavena, A., Ceriotti, A. and Bollini, R. (1992) *Acta Hort.*, **300**: 309-313.

Gepts, P., Papa, R., Coulibaly, S., Mejia, A.G., Pasquet, R. and Oono, K. (2000) *Proc. International Workshop on Genetic Resources*, Ibaraki, Japan, 13-15 October, 1999: Part 1, wild legumes, 2000, pp. 19-31.

Gerhardt, L.B.de A., Sachetto, Martins, G., Contarini, M.G., sandroni, M., Ferreira, R. de P., Lima, V.M. de., Cordeiro, M.C., Oliveira, D.E. de., Margis, Pinheiro, M., De, Lima, V.M. and De Oliveira, D.E. (1997) *FEBS Letters*, **419**: 69-74.

Ghosal, S., Singh, O.N. and Singh, R.P. (2000) *Legume Res.*, **23**: 110-113.

Gibbs, A.J. and Ohshima, K. (2010) *Annu. Rev. Phytopathol.*, **48**: 205-223.

Gibbs, A.J., Ohshima, K., Phillips, M.J. and Gibbs, M.J. (2008a) *PLoS One*, **3**: 2523.

Gibbs, A.J., Trueman, J. and Gibbs, M.J. (2008b) *Arch. Virol.*, **153**: 2177-2187.

Ginoux, J.P. and Messiaen, C.M. (1993) *Agronomie*, **13**: 283-292.

Glaman, G., Bucurescu, I. and Valceanu, G. (1989) *Productia Vegetala Horticultura*, **38**: 6-8.

Glaze, N.C. and Phatak, S.C. (1978) *Proc. 31st Annual Meet. South Weed Sci. Soc.*, ISI, USDA, Tifton, Georgia, p. 151.

Gonçalves, J.G.R., Chiorato A. Fernando, da Silva, D.A., Esteves, J.A.F. and Bosetti, F. (2015) *Sérgio Augusto Morais Carbonell*, **1**: 149-155.

Gonçalves Vidigal, M.C., Silvério, L., Elias, H.T., Filho, P.S.V., Kvitschal, M.V., Retuci, V.S. and da Silva, C.R. (2008) *Pesq. agropec. bras., Brasília*, **43**: 1143-1150.

Gonzalez, A.R. and Williams, J.W. (1978) *Arkans. Fm. Res.*, **27**: 3.

Goud, V.V. and Dikey, H.S. (2016) *Indian J. Weed Sci.*, **48**: 191–194.

Gould, W. (1951) *Ohio Agric. Exp. Sta. Res. Bull.*, 701.

Graham, G.C., Henry, R.J., Redden, R.J. (1994) *Aust. J. Exp. Agric.*, **34**: 1173-1176.

Graham, P.H. and Rosas, J.C. (1979) *Agron. J.*, **71**: 925-926.

Griffiths, D.C., Bardner, R. and Bater, J. (1986) *FABIS Newsl.*, **14**: 30-33.

Guimaraes, C.M., Brunini, O. and Stone, L.F. (1996a) *Pesquisa Agropecuaria Brasileira.* **31**: 393-399.

Guimaraes, C.M., Stone, L.F. and Brunini, O. (1996b) *Pesquisa Agropecuaria Brasileira.* **31**: 481-488.

Guimaraes, M.A., Barbosa, H.M., Vieira, C. and Sediyama, C.S. (1989) *RevistaBarsileira de Genetica*, **12**: 93-101.

Guu, J.W. (1989) *Bulletin of Taichung District Agril. Improvement Station*, No. 22, p.13-25.

Hagedorn, D.J. and Rand, R.E. (1980a) *HortScience*, **15**: 208-209.

Hagedorn, D.J. and Rand, R.E. (1980b) *HortScience*, **15**: 529-530.

Haley, S.D., Miklas, P.N., Afander, L. and Kelley, J.D. (1994) *Ann. Rep. Bean Improv. Coop.*, **36**: 12-13.

Hampton, R.O. (1975) *Phytopathology*, **65**: 1342-1346.

Han, K.Y., Lee, K.S., Suh, J.K. and Lee, Y.S. (1989) *Reports of the Rural Development Administration*, Horticulture, Korea Rep., **31**: 18-22.

Hang, A.N., Silbernagel, M.J. and Miklas, P.N. (1999) *HortScience*,**34**: 338.

Hegazi, A.M. and El-Shraiy, A.M. (2007) *Australian J. Basic Appl. Sci.*, **1**: 834-840.

Hegde, D.M. and Srinivas, K. (1989) *Progress. Hort.*, **21**: 100-105.

Hill, E.C., Renner, K.A., Sprague, C.L. and Davis, A.S. (2016) *Weed Sci.*, **64**: 261–275.

Hine, J.C. and Sprent, J.I. (1988) *J. Expt. Bot.*, **39**: 1505-1512.

Honma, S. and Bedford, C.L. (1976) *Res. Rept.*, Michigan State Univ. Agric. Exp. Sta., East Lansing, **306**.

Honma, S. and Cash, J.N. (1979) *Res. Rept.*, Michigan State Univ. Agric. Exp. Sta., East Lansing, **382**.

Hoyos, R.A., Roca, W., Hosfield, G.H. and Meyer, J. (1993) *Ann. Rep. Bean Improv. Coop.*, **35**: 68-69.

Hsieh, T.F., Kuo, C.H. and Wang, K.M. (1999) *Plant Pathol.Bull.*, **8**: 157-162.

Hurduc, N., Parjol-Savulescu, L. and Popa, G.F. (1982) *AnaleleInstitutului de CercetaripentruCereale Si Plante TechniceFundulea*, **49**: 303-313.

Hussein, M.Y.B. (1978) *Pertanika*,**1**: 36-39.

Iqbal1, A.M., Nehvi, F.A., Wani, S.A., Dar, Z.A., Lone, A.A. and Qadri, H. (2012) *SAARC J. Agri.*, **10**: 61-69.

Ivanov, L., Velev, B., Rankov, V., Manuelyan, K., Poryazov, I., Benevski, M., Petrova, R. and Lambrev, S. (1988) *RastenievdniNauki*, **25**: 95-100.

Jagtap, A.B., Awate, B.G. and Nair, L.M. (1979) *J. Maharashtra Agric. Univ.*, **4**: 83-84.

Jana, B.K. and Kabir, J. (1987) *Veg. Sci.*, **14**: 124-127.

Jasrotia, R.S. and Sharma, C.M. (1999) *Fert. News*, **44**: 59, 61.

Jitendra, M., Chamola, B.P., Rana, D.K. and Singh, K.K. (2018) *Int. J. Curr. Microbiol. App. Sci.*, **7**: 676-681.

Jung Geun Hwa., Skroch, P.W., Coyne, D.P., Nienhuis, J., Arnaud Santana, E., Ariyarathen, H.M., Kaeppler, S.M., Bassett, M.J. and Jung, G.H. (1997) *J. Amer. Soc. Hort. Sci.* **122**: 329-337.

Jung, G., Coyne, D.P., Bokosi, J., Steadman, J.R. and Nienhuis, J. (1998) *J. Amer. Soc. Hort. Sci.*, **123**: 859-863.

Jung, G., Coyne, D.P., Skroch, P.W., Nienhuis, J., Arnaud Santana, E., Bokosi, J., Ariyarathne, H.M., Steadman, J.R., Beaver, J.S. and Kaeppler, S.M. (1996) *J. Amer. Soc. Hort. Sci.*, **121**: 794-803.

Kalantidis, K. and Griga, M. (1993) *RostlinnaVyroba.*, **39**: 115-128

Kalawa, I.L., Wafula, W.N., Korir, N.K. and Gweyi-Onyango, J. (2018) *Asian J. Res. Crop Sci.*, **1**: 1-8.

Kamarr, S.M.A.H.M., Isramr, M.K., Kawochars M.A., Mahfuz', M.S. and Savem, M.A. (2010) *Bangladesh J. Environ. Sci.*, **19**: 63-66.

Karel, A.K. and Autrique, A. (1989) In: Schwartz, H.F. and Pastor-Corrales, M.A. (Eds.), *Bean production problems in the tropics. CIAT, Cali, Colombia*, pp. 455-504.

Kattan, A.A. and Fleming, J. W. (1956) *Proc. Amer. Soc. Hort. Sci.*, **68**: 329-342.

Katyal, S.L. (1977) *Vegetable Growing in India*, Oxford and IBH Publishing Co., pp. 60-68.

Keinath, A.P., Batson, W.E. Jr., Caceres, J., Elliott, M.L., Sumner, D.R., Brannen, P.M., Rothrock, C.S., Huber, D.M., Benson, D.M., Conway, K.E., Schneider, R.N., Motsenbocker, C.E., Cubeta, M.A., Ownley, B.H., Canaday, C.H., Adams, P.D., Backman, P.A. and Fajardo, J. (2000) *Crop Protection*, **19**: 501-509.

Kelly, J., Saettler, A., and Morales, M. (1983) *Annu. Rep. Bean Improv. Coop.*, **27**: 38-39.

Kelly, J.D., Afanador, L., Haley, S.D. (1995) *Euphytica*, **82**: 207-212.

Kemp, G.A. (1977) *Canada J. Plant Sci.*, **57**: 1013-4.

Khan, A.A., Maguire, J.D., Abawi, G.S. and Ilyas, S. (1992) *J. Amer. Soc. Hort. Sci.*, **117**: 41-47.

Kigel, J., Konsens, I. and Ofir, M. (1991) *Canada J. Pl. Sci.* **71**: 1233-1242.

Kim, J.W. and Minamikawa, T. (1996) *Plant Sci.*, **117**: 131-138.

Kim, J.W., Minamikawa, T. and Kim, J.W. (1997) *Plant Cell Physiol.*, **38**: 70-75.

King, J.M. (1981) *Proc. BCPC Pest Dis.*, pp. 327-331.

Kolotilov, V.V., Buravtseva, T.V. and Kolotilova, A.S. (1989) *Selektsiyai Semenovodsto* (Moskava) No. 5, pp. 33-34.

Kornegay, J. and Cardona, C. (1991) In: Schoonhoven, A. and van, O. Voysest (Eds.), *Common Beans: Res.Crop Improv.*, pp. 619–648.

Krishna Moorthy, P. N. and Srinivasan, K. (1989) *J. Econ. Entomol.*, **82**: 246-250.

Ku, J.H., Drizek, D.T. and Mirecki, R.M. (1996) *J. Korean Soc. Hort. Sci.*, **37**: 767-772.

Kwambe, X.M., Masarirambi, M.T., Wahome, P.K. and Oseni, T.O. (2015) *Agric. Biol. J. N. Am.*, **6**: 81-89.

Lal, O.P. (1974) *Indian J. Ent.*, **36**: 67-68.

Lal, R. (1989) *Adv. Agron.*, **42**: 147-151.

Le Delliou, B. (1989) *UNILEC Informations*, Union NationaleInterprofessionelle des Legumes de Conserve, No. 64, 15-17.

Leal, N.R. (1990) *HorticulturaBrasileira*, 1990, **8**: 29.

Leal, N.R. (1989) *Serie Documentos*, Empresa de PesquisaAgropecuaria do Estodo do Rio de Janeiro, No. 16, p. 6.

Leal, N.R. and Bliss, F. (1980) *HorticulturaBrasileira*, 1980, **8**: 29-30.

Leela, D., Dhuria, H.S. and Shukla, V. (1972) *3rd Int. Symp. Subtrop. Trop. Hort.*, Bangalore pp. 71-72.

Leprince, S. (1989) *UNILEC Informations*, Union NationaleInterprofessionelle des Legumes de Conserve No. 64, pp. 18-19.

Li, Y.Q., Liu, Z.P., Yang, Y.S., Zhao, B., Fan, Z.F. and Wan, P. (2014) *Plant Dis.*, **98**: 1017.

Liu, S.S.G., Liu, M., Ji, Y., He, H. and Gruda, N. (2018) *Horticulturae*, **4**: 3.

Llaca, V., Delgado Salinas, A. and Gepts, P. (1994) *Theor. Appl. Gen.*, **88**: 646-652.

Loebenstein, G. (2009) *Adv. Virus Res.*, **75**: 73-117.

Loss, S.P. and Siddique, K.H.M. (1997) *Field Crops Res.*, **52** ; 17–28.

Lyngdoh, Y.A., Thapa, U., Shadap, A., Singh, J. and Tomar, B.S. (2018) *Legume Res.*, **41**: 810-815

M'Ribus, K. (1985) *Acta Hortic.*, **153**: 145-149.

MacCollum, J.P. (1953) *Plant Physiol.*, **18**: 267-274.

Mack, H.J. 1983) *J. Amer. Soc. Hort. Sci.*, **108**: 574-758.

Mahamune, S.E. and Kothekar, V.S. (2011) *Rec. Res. Sci. Technol.*, **3**: 33-35.

Mann, J.S. and Chakor, J.S. (1989) *Indian J. Agron.*, **39**: 219-282.

Martineaz, A.O. and Soto, A.A. (1978) BoletinTecnico, *Faculted de Agronomia, Universidad de Costa Rica*, **11**: 14.

Martins, S.R. and Gonzalez, J.F. (1995) *RevistaBrasileira de Agrometeorologia*, **3**: 31-37.

Maurer, A.R., Omrod, D.P. and Scott, N.J. (1969) *Cand. J. Plant Sci.*, **49**: 271-278.

McClean, P., Chee, P., Held, B., Simental, J., Drong, R.F. and Slightom, J. (1991) *Plant Cell Tissue Organ Cult.*, **24**: 131-138.

McEwen, J., Bardner, R., Briggs, G.G., Bromilow, R.H., Cockbain, A.J., Day, J.M., Fletcher, K.E., Legg, B.J., Roughley, R.J., Salt, G.A., Simpson, H.R., Webb, R.M., Witty, J.F. and Yeoman, D.P. (1981) *J. Agric. Sci.*, **96**: 129-150.

Mekwatanakarn, W., Richardson, D.G. and Saltveit, M.E. (1997) *Postharvest Hort.-Series* -Department-of-Pomology,-University-of-California. **18**: 59-65.

Mencinicopschi, G. and Popa, M. (1999) *Agri-Food Quality II: quality management of fruits and vegetables - from field to table*, Turku, Finland, 22-25 April, 1998. pp. 161-163.

Menezes, J.R. and Dianese, J.C. (1988) *Phytopathology*, **78**: 650-655.

Metacalf, C.L.l. and Flint, W.B. (1962) Destructive and Useful Insects, their Habits and Contol, (4th edn.), McGraw-Hill Book Co. Inc., New York, London, p. 1087.

Miklas, P.N., Fourie, D., Chaves, B. and Chirembe, C. (2017) *Crop Sci.*, **57**: 802-811.

Miller, D.E. and Burke, D.W. (1983) *Agron. J.*, **75**: 775–778.

Miranda, N. O. and Belmar, N. C. (1977) *Agricultura Tecnica*, **37**: 111-117.

Mishra, R.K., Parihar, A.K., Basvaraj, T. and Kumar, K. (2019) *Legume Res.*, **42**: 430-433.

Modi, M.K., Barua, S.J.N. and Deka, P.C. (2001) *Leguminous Vegetables* (V.A. Parthasarathy, T.K. Bose and P.C. Deka eds.), In: *Biotechnology of Horticultural Crops, Vol. 2, NayaProkash, Calcutta*, pp. 284-325.

Mohamed, M.F., Read, P.E., Coyne, D.P. (1992) *J. Amer. Soc. Hort. Sci.*, **117**: 332-336.

Mohamed, M.F. (1990). *Assiut J. Agric. Sci.*, **21**: 361-371.

Mohammed, A. (2013) *J. Plant Pathol. Microb.*, **4**: 193.

Mohammed, A., Ayalew, A. and Dechassa, N. (2013) *J. Plant Pathol. Microb.*, **4**: 182.

Mondal, A., Shankar, U., Abrol, D.P., Singh, I. and Norboo, T. (2017) *Int. J. Curr. Microbiol. App. Sci.*, **6**: 1441-1448.

Mooney, H.A., Winner, W.E. and Pell, E.J. (1991) *Response of Plants to Multiple Stresses*. Academic Press; Elsevier: Amsterdam, The Netherlands.

Morales, F.J. (2006) In: G. Loebenstein and J. P. Carr (Eds.), *Natural resistance mechanisms of plants to viruses*, The Netherlands: Springer, pp. 367-382.

Morgan, D.G. (1976) *J. Sci Food Agri.*, **27**: 793-794.

Morris, J.L.; Campbell, W.F. and Polard, L.H. (1970) *J. Amer. Soc. Hort. Sci.*, **95**: 541-543.

Morton, R.L., Schroeder, H.E., Bateman, K.S., Chrispeels, M.J., Armstrong, E. and Higgins, T.J.V. (2000) *Proc. Natl. Acad. Sci.*, USA, **97**: 3820-3825.

Moya Lopez, C., Alvarez Gil, M., Ponce Brito, M., Bertoli Herrera, M., Gonzalez Cepero, M.C., Dell' Amico Rodriguez, J., Morales Arvero, C. and Crane, J.H. (2000) *Proc. Interamerican Soc. Trop. Hortic.*, Barquisimeto, Venezuela, 27 Sept-2 Oct., 1998, No. 42 pp. 327-334.

Mudawi, H.I., Idris, M.O. and El Balla, M.A. (2009) *Korean J. Agric. Sci.*, **17**: 118-130.

Mukeshimana, G., Paneda, A., Rodriguez-Suarez, C., Ferreira, J.J., Giraldez, R. and Kelly, J.D. (2005) *Euphytica*, **144**: 291-299.

Mullins, C.A. and Straw, R.A. (2001) *HortTech.*, **11**: 124-127.

Nakano, H., Boonmalison, D., Egawa,Y., Vanichwattanarumruk, N., Chotechuen, S., Hanada, T. and Momonoki, T. (1994) *Japanese J. Trop. Agrl.*, **38**: 239-245.

Nakano, H., Momonoki, T., Miyashige, T., Otsuka, H., Hanada, T., Sugimoto, A., Nakagawa, H., Matsuoka, M., Terauchi, T., Kobayashi, M., Oshiro, M., Yasuda, K., Vanichwattanarumruk, N., Chotechuen, S. and Boonmalison, D. (1997) *JIRCAS J.*, **5**: 1-12.

Nandan, R. and Prasad, U.K. (1998b) *Indian J. Agric. Sci.*, **68**: 75-80.

Nandan, R. and Prasad, U.K. (1998a) *Indian J. Agron.*, **43**: 550-554.

Nemeskeri, E. (1988) *Acta Hortic.*, **220**: 493-498.

Neptune, A.M.L., Muraoka, T. and Lourenco, S. (1978) *Turrialba*, **28**: 171-202.

Neptune, A.M.L., Muraoka, T. and Stewart, J.W.B. (1979) *Turrialba*, **29**: 29-34.

Ngelenzi, M.J., Mwanarusi, S. and Otieno, O.J. (2017) *Sust. Agric. Res.*, **6**: 1.

Nielsen, D.C. and Nelson, N.O. (1998) *Crop Sci.*, **38**: 422–427.

Nodari, R.O., Koinange, E.M.K., Kelly, J.D. and Gepts, P. (1992) *Theor. Appl. Genet.*, **84**: 186-192.

Noor, F., Hossain, F. and Ara, U. (2017) *J. Asiat. Soc. Bangladesh Sci.*, **43**: 49-60.

Normander, B., Christensen, B.B., Molin, S. and Kroer, N. (1998) *Appl. env. Microbiol.*, **64**: 1902-1909.

Nuñez Barrios, A., Hoogenboom, G. and Nesmith, D.S. (2005) *Sci. Agric.*, **62**: 18–22.

Ogliari, J. and Castao, M. (1992) *Pesqui. Agropecu. Bras.*, **27**: 1043-1047.

Olaya, G., Abawi, G.S. and Weeden, N.F. (1996) *Phytopathology*, **86**: 674-679.

Omae, H., Kumar, A., Egawa, Y., Kashiwaba, K. and Shono, M. (2004) *Jpn. J. Trop. Agric.*, **48**: 5–6.

Omae, H., Kumar, A., Egawa, Y., Kashiwaba, K. and Shono, M. (2005) *Jpn. J. Trop. Agric.*, **49**: 1–7.

Opio, A.F., Allen, D.J. and Teri, J.M. (1996) *Plant Pathol.*, **45**: 1126-1133.

Opio, A.F., Mugagga-Mawejje, D. and Nkalubo, S. (2006) *Progress report on bean anthracnose research in Uganda. MUARIC bulletin.* p.16.

Orr, J.E., Carter, C.W. and Kukas, R.D. (1977) *Proc. Western Soc. Weed Sci.*, **30**: 42-43.

Pandey, G.K., Seth, J.N., Phogat, K.P.S. and Tewari, S.N. (1974) *Progressive Hort.*, **6**: 41-47.

Pandey, V., Singh, V.K. and Upadhyay, D.K. (2013) *Int. J. Plant Res.*, **26**: 438.

Panotra, N. and Kumar, A. (2016) *Int. J. Appl. Sci.*, **4**: 275-283.

Parish, R.L., LaBorde, C.M. and Raiford, T.J. (1996) *J. Vegetable Crop Prodn.*, **2**: 65-75.

Park, S.O., Coyne, D.P., Mutlu, N., Jung, G. and Steadman, J.R. (1999a) *J. Amer. Soc. Hort. Sci.*, **124**: 519-526.

Park, S.O., Coyne, D.P., Bokosi, J.M. and Steadman, J.R. (1999b) *Euphytica*, **105**: 133-141.

Parodi, A. E., Opazo, A. J.D. and Mosjidis, CH.J. (1977) *Agricultura Tenica*, **37**: 12-18.

Parthasarathy, V.A., Parthasarathy, U. and Ashwath, C. (1993) *Ann. Plant Physiol.*, **7**: 206-210.

Pascale, S.de., Barbieri, G., Ruggiero, C. and Chatzoulakis, K.S. (1997) *Acta Hortic.*, **449**: 649-655.

Pasev, G., Kostova, D. and Sofkova, S. (2014) *J. Phytopathol.*, **162**: 19-25.

Passarinho, Jap and Ricardo, C.P.P. (1988) *In I journadasportuguesas de prateaginosas*, 225-236.

Patel, P.N. and Jindal, J.K. (1972) *Indian Phytopath.*, **27**: 387-391.

Patil, H.E., Akalade, S., Patel, B.K. and Sarkar, M. (2018) *Int. J. Genet.*, **10**: 514-517.

Paula, T.J. Jr., Pinto, C.M.F., Da Silva, M.B., Nietsche, S., de Carvalho, G.A. and Faleiro, F.G. (1998) *Revista Ceres*. **45**: 171-181.

Peixoto, N., Oliveira e Silva, L., Thung, M.D.T. and Santos, G. (1993) *HorticulturaBrassileira*, **11**: 151-152.

Pevicharova, G., Sofkova-Bobcheva, S. and Zsivanovits, G. (2015) *Int. J. Food Prop.*, **18**: 1169-1180.

Phillips, A.J.L. (1989) *Canadian J. Microbiol.*, **35**: 1132-1140.

Pillemer, E.A. and Tingey, W.M. (1978) *Ento. Exp. App.*, **24**: 83-94.

PoPovic, T., STarovic, M., alekSic1, G., Zivkovic1, S., JoSic, D., iGnJaTov M. and Milovanovic, P. (2012) *Bulg. J. Agric. Sci.*, **18**: 701-707.

Porta-Plugia, A. and Aragona, M. (1997) *Field Crops Res.*, **53**: 17-30.

Prajapati, M.P., Patidar, L.R. and Patel, B.M. (2003) *Legume Res.*, **26**: 79-84.

Prasad, B.V.G., Chakravorty, S., Saren, B.K. and Panda, D. (2014) *HortFlora Res. Spect.*, **3**:: 162-165.

Prasad, R.D., Singh, K.N. and Bondale, K.V. (1978) *Mysore J. Agric. Sci.*, **12**: 221-223.

Puppo, A. and Riguad, J. (1978) *Physiol. Plant*, **42**: 202-206.

Puttaswamy and Reddy, D.N.R. (1981) *Curr. Res.*, **10**: 39-41.

Quagliotti, L. Lepori, G. and Raldi, A. (1980) In: *Seed Production*, (P.D. Hebblethwaite ed.), Butterworths, London, Boston, pp. 569-584.

Quintana, J.M., Harrison, H.C., Nienhuis, J., Palta, J.P. and Grusak, M.A. (1996) *J. Amer. Soc. Hort. Sci.*, **121**: 789-793.

Quintana, J.M., Harrison, H.C., Nienhuis, J., Palta, J.P. and Kmiecik, K. (1999a) *HortScience*, **34**: 932-934.

Quintana, J.M., Harrison, H.C., Nienhuis, J., Palta, J.P., Kmiecik, K. and Miglioranza, E. (1999c) *J. Amer. Soc. Hort. Sci.*, **124**: 273-276.

Quintana, J.M., Harrison, H.C., Palta, J.P., Nienhuis, J., Kmiecik, K. and Miglioranza, E. (1999b). *J. Amer. Soc. Hort. Sci.*, **124**: 488-491.

Rackham, R.L. and Vaughn, J.R. (1959) *Plant Dis. Reptr.*, **43**: 1023-1026.

Rafiqueuddin, M. (1984) *Legume Res.*, **7**: 43-47.

Rai, A., Sharma, V. and Heitholt, J. (2020) *Dry Bean [Phaseolus vulgaris L.] Growth and Yield Response to Variable Irrigation in the Arid to Semi-Arid Climate Sustainability*, **12**: 3851.

Rajapakse, N.C. and Shahak, Y. (2007). In: Whitelam, G. and Halliday, K. (Eds.), *Light quality manipulation by horticulture industry. Blackwell Publishing, UK.*, pp. 290-312.

Rajendran, R. and Satyanarayana, A. (1988) *Indian Hort.*, **33**: 6-7.

Ramandeep, Dhillon, T.S., Dhall, R.K. and Gill, B.S. (2018) *Agric. Res. J.*, **55**: 219-223.

Ramirez-Vallejo, P. and Kelly, J.D. (1998) *Euphytica*, **99**: 127–136.

Ramos, M., Gordon, A.J., Minchin, F.R., Sprent, J.L. and Parsons, R. (1999) *Ann. Bot.*, **83**: 57–63.

Rana, N.S. and Singh, R. (1998b) *Indian J. Agron.*, **43**: 367-370.

Rana, N.S., Singh, R. and Ahlawat, I.P.S. (1998a) *Indian J. Agron.*, **43**: 114-117.

Rangaswami, G. (1975) Diseases of Crop Plants in India (2nd edn.), Prentice Hall of India Pvt. Ltd., New Delhi, p. 250.

Rathod, R.R., Gore, R.V. and Bothikar, P.A. (2015) *J. Agric Vet. Sci.*, **8**: 36-39.

Ravi, S., Sabitha, D., Valluvaparidasan, V. and Jayalakshmi, C. (2000) *Legume Res.*, **23**: 170-173.

Reddy, R. and Sulladmath, U.V. (1975) *Haryana J. Hort. Sci.*, **4**: 186-196.

Reichel, S. (1992) *GartenbauMagazin.*, **1**: 91-93.

Riguad, J. (1976) *Physiologie Vegetable*, **14**: 297-308.

Robinson, W.B., Wilson, D.E., Moyer, J.C. and Atkins, J.D. (1964) *Proc. Amer. Soc. Hort. Sci.*, **84**: 339-347.

Rodrigues, R., Leal, N.R. and Lam Sanchez, A. (1998b) *HorticulturaBrasileira*, **16**: 61-64.

Rodrigues, R., Leal, N.R. and Pereira, M.G. (1998a) *Bragantia.*, **57**: 241-250.

Rodrigues, R., Leal, N.R., Pereira, M.G. and Lam Sanchez, A. (1999) *Genet. Mol. Biol.*, **22**: 571-575.

Rogers, D.J. (1979) *J. Aust. Ent. Soc.*, **18**: 245-250.

Rolbiecki, S., Rolbiecki, R., Rzekanowski, C. and Zarski, J. (2000) *Acta Hortic.*, **537**: 871-877.

Roos, E.E. and Monolo, J.R. (1976) *J. Amer. Soc. Hort. Sci.*, **101**: 321-324.

Ross, S.M., Ling, J.R., Izaurralde, C. and O'Donovan, J.T. (2001) *Agron J.*, **93**: 820–827.

Roy, N.R. and Parthasarathy, V.A. (1999) *Indian J. Hort.*, **56**: 317-320.

Ruschel, A.P. and Ruschel, R. (1975) *PesquisaAgropecuariaBrasileira*, **10**: 11-17.

Russell, D.R., Wallace, K.M., Bathe, J.H., Martinell, B.J. and McCabe, D.E. (1993) *Plant Cell Rep.*, **12**: 165-169.

Saba, I., Sofi, P.A., Zeerak, N.A., Bhat, M.A. and Mir, R.R. (2016) *SABRAO J. Breed. Genet.*, **48**: 359-376.

Sache, I. and Zadoks, J.C. (1996) *Eur. J. Plant Pathol.*, **102**: 51-60.

Saettler, A.W. (1989) In: H. Schwartz and M.a. Pastor-corrales (eds.). *Bean Production Problems in the Tropics.* Cali, Columbia, pp. 261-283.

Sahariar, M.S., Karim, M.R., Nahar, M.A., Rahman, M. and Islam, M.U. (2015) *Prog. Agric.*, **26**: 129-135.

Saker, M.M. and Kuhne, T. (1998) *Biologia Plantarum*, **40**: 507-514.

Saleh, S.A., El-Shal, Z.S., Fawzy, Z.S. and El-Bassiony, A.M. (2012) *Aust. J. Basic Appl. Sci.*, **6**: 54–61.

Saqib, M., Nouri, S., Cayford, B., Jones, R.A.C. and Jones, M.G.K. (2010) *Australas. Plant. Path.*, **39**: 184-191.

Sazen, S.M., Yazar, A., Akyildiz, A., Dasgan, H.Y. and Gencel, B. (2008) *Sci. Hortic.*, **117**: 95–102.

Schafer (1979) *The Manuring of Pulses*, Verlagsgesellschaft fur Ackerbau, mbH Hannover, p. 68.

Schmutz, J., McClean, P.E., Mamidi, S., Wu, G.A., Cannon, S.B. and Grimwood, J. (2014) *Nat Genet.*, **46**: 707-13.

Scholberg, J.M.S. and Locascio, S.J. (1999) *HortScience*, **34**: 259-264.

Schuster, M.L. and Coyne, D.P. (1981) *Hortic. Rev.*, **3**: 28-58.

Schwartz, H.F. and Peairs, F.B. (1999) In: Singh, S.P. (Ed.), *Common bean improvement in the twenty-first century.* Kluwer Academic Press, Dordrecht, the Netherlands. pp. 371-388.

Scully, B., Provvidenti, R., Halseth, D.E. and Wallace, D.H. (1990) *HortScience*, **25**: 1314-1315.

Scully, B.T., Beiriger, R.L., Olezyk, T. and Stavely, J.R. (2000) *HortScience*, **35**: 1180-1182.

See, R.M. and Foy, C.F. (1983) *J. Plant Growth Regulation*, pp. 9-17.

Seguin, A., Laible, G., Leyva, A., Dixon, R.A. and Lamb, C.J. (1997) *Plant Mol. Biol.*, **35**: 281-291.

Sengooba, T.N., Spence, N.J., Walkey, D.G.A., Allen, D.J. and Femi Lana, A. (1997) *Plant Pathol.*, **46**: 95-103.

Serraj, R. and Sinclair, T.R. (1998) *Plant Soil,* **202**: 159–166.

Shahak, Y. (2008) *Acta Hortic.*, **770**: 161-168.

Shahak, Y., Gussakovsky E.E., Cohen, Y. and Lurie, S. (2004) *Acta Hortic.*, **636**: 609-616.

Shannon, S., Natti, J.J. and Atkin, J.D. (1967) *Proc. Amer. Soc. Hort. Sci.*, **90**: 180-190.

Sharma, H.M., Singh, R.N.P., Singh, H. and Sharma, R.P.R. (1996) *Indian J. Pulses Res.*, **9**: 25-30.

Sharma, R.K., Sengupta, K. and Pacharui, D.C. (1976) *Progress. Hort.*, **8**: 65-68.

Sharmah, D. and Rahman, S. (2017) *J. App. Nat. Sci.*, **9**: 674-679.

Sherf, A. F. and MacNab, A. A. (1986) *Vegetable Diseases and Their Control. Wiley, New York.*

Siddique, M.A., Somereset, G. and Goodwin, P.B. (1987) *Australian J. Expt. Agric.*, **27**: 179-187.

Silbernagel, M.J. (1986) In: *Breeding Vegetable Crops,* (ed. M.J. Bassett), AVI Pub. Comp. Inc. Connecticut, USA.

Silbernagel, M.J. and Drake, S.R. (1978) *J. Amer. Soc. Hort. Sci.*, **103**: 257-260.

Silva, L. O., Singh, S. P. and Pastor-Corrales, M. A. (1989) *Theor. Appl. Genet.*, **78**: 619-624.

Singer, S.M., Helmy, Y.I., Sawan, O.M. and El Abd. (1999) *Acta Hortic.*, **491**: 221-228.

Singh, S.P., Gepts, P. and Debouck, D.G. (1991) *Eco. Bot.*, **45**: 379–396.

Singh, S.P., Nodari, R. and Gepts, P. (1991a) *Crop Sci.*, 31: 19–23.

Singh, B. (1987) *Indian J. Agron.*, **32**: 223-225.

Singh, B.K., Pathak, K.A., Verma, A.K., Verma, V.K. and Deka, B.C. (2011) *Veg. Crops Res. Bull.*, **74**: 153-165.

Singh, B.P. (1989) *HortScience*, **24**: 69-70.

Singh, B.P. and Singh, B. (1990) *J. Indian Soc. Soil Sci.*, **38**: 769-771.

Singh, K., Singh, U.N., Singh, R.N., Bohra, J.S. and Singh, K. (1996) *Fert. News*, **41**: 39-42.

Singh, M.K. (1993/95) *J. Appl. Biol.*, **3**: 112-115.

Singh, R.D., Seth, J.N., Pande, G.K. and Kuksal, R.P. (1975) *Indian Hort.*, **20**: 17-19.

Singh, S. P. and Munoz, C. G. (1999) *Crop Sci.*, **39**: 80-89.

Singh, S.D. and Naik, S.M.P. (1976) *Indian J. Mycol. Plant Path.*, **6**: 99.

Singh, S.J., Sastry, K.S. and Sastry, K.S.M. (1981) *Gartenbauwissenchaft*, **46**: 88-91.

Singh, S.P. and Schwartz, H.F. (2010) *Crop Sci.*, **50**: 2199-2223.

Singh, S.P. (1991) In: Schoonhoven, A. and Voysest, O. (Eds.), *Common beans: research for crop improvement.* CAB Int, CIAT, Cali, Wallingford, pp. 383–443.

Singh, S.P. (1992) *Plant Breed. Rev.*, **10**: 199-269.

Singh, S.P. (1998) *AgronomiaMesoamericana*, **9**: 1-9.

Singh, S.P. (2001) *Crop Sci.*, **41**: 1659-1675.

Singh, S.P., Gepts, P. and Debouck, D.G. (1991b) *Econ. Bot.*, **45**: 79–396.

Singh, V.K., Kumar, C., Kumar, M., Nirala, D.P. and Singh, R.K. (2018) *J. Pharmacog. Phytochem.*, **1**: 1138-1141.

Singh, V.K., Mathuria, R.C., Gogoi, R. and Aggarwal, R. (2016). *Indian Phytopath.*, **69**: 357–362.

Sistrunk, W.A., Frazier, W.A., Clarkson, V.A. and Cain, R.F. (1960) *Proc. Amer. Soc. Hort. Sci.*, **76**: 389-396.

Skroch, P.W. and Nienhuis, J. (1995) *Theor. Appl. Genet.*, **91**: 1078-1085.

Smartt, J. (1976) *Tropical Pulses*, Longman, London., p. 348.

Smith, C.B. (1976) *Res. Rept., Penn. State Univ. Coll. Agri.*, No. 353, p. 7.

Smith, V.L. (2000) *HortScience*, **35**: 92-94.

Snapp, S.S., Swinton, S.M., Labarta, R., Mutch, D., Black, J.R., Leep, R., Nyiraneza, J. and O'Neil, K. (2005) *Agron. J.*, **97**: 322–332.

Soltani, N., Nurse, R.E., Shropshire C. and Sikkema, P.H. (2014) *Adv. Agric.*, p. 7.

Song, G., Han, X., Wiersma, A.T., Zong, X., Awale, H.E., Kelly, J.D. (2020) *PLoS ONE* **15**(3): e0229909. https://doi.org/10.1371/journal.pone.0229909.

Song, J.Y., Choi, E.Y., Lee, H.S., Choi, D.W., Oh, M.H. and Kim, S.G. (1995) *J. Plant Physiol.*, **146**: 148-154.

Souza, A.A., Boscariol, R.L., Moon, D.H., Camargo, L.E.A. and Tsai, S.M. (2000) *Gen. Mol. Biol.*, **23**: 155-161.

Srinivas, K. and Naik (1990) *Haryana J. Hort. Sci.*, **19**: 160-167.

Srinivas, K. and Naik, L.B. (1988) *Indian J. Agric. Sci.*, **58**: 707-708.

Ssali, H. and Keya, S.O. (1985) *East African Agril. For. J.*, **45**: 277-283.

Stalin, P., Shammugan, K., Thamburaj, S. and S. Ram Das (1989) *South Indian Hort.*, **37**: 28-33.

Stang, J.R., Mack, H.J. and Rowe, K.E. (1979) *J. Amer. Soc. Hort. Sci.*, **104**: 873-875.

Steene, F., van de, G., Vulsteke, M., de Proft, D. and Callewaert. (1999) *Zeitschrift für Pflanzenkrankheiten und Pflanzenschutz*, **106**: 633-637.

Stevenson, M.R. (1977) *New Zealand Commercial Grower*, **32**: 20.

Stoker, R. (1974) *New Zealand J. Exp. Agric.*, **2**: 13–15.

Stolk, J.H. (1980) *Groentenen Fruit*, Netherlands, **35**: 28-29.

Stone, L.F., Moreira, J.A.A., Silva, S.C.Da (1988) *PesquisaAgropecuariaBrasileira*, **23**: 161-167.

Subba Rao, P.V., Rangarajan, A.V. and Azeez Besha, A. (1974) *Indian J. Ent.*, **36**: 227-228.

Suryanarayana, V. and Kumar, J.P. (1981) *Veg. Sci.*, **8**: 130-134.

Tar' an, B., Michaels, T.E. and Pauls, K.P. (1998) *Pl. Breed.*, **117**: 553-558.

Tar'an, B., Michaels, T. E. and Pauls, K. P. (2001) *Genome*, **44**: 1046-1056.

Tarantino, E. and Rubino, P. (1989) *Cent. Di Studio Sull Ortic. Ind.*, **36**: 228–234.

Teasdale, J.R. (1998) In: Hatfield, J.L., Buhler, D.D. and Stewart, B.A. (Eds.), *Integrated Weed and Soil Management. Chelsea, MI: Ann Arbor Press*, pp. 247–270

Teasdale, J.R. and Frank, J.R. (1980) *Proc. Northeastern Weed Sci. Soc.*, **34**: 109.

Terán, H. and Singh, S.P. (2002) *Crop Sci.*, **42**: 64–70.

Tewari, J.K. and Singh, S.S. (2000) *Veg. Sci.*, **27**: 172-175.

Thakur, R.S., Sugha, S.K. and Singh, B.M. (1991) *Indian J. Agric. Sci.*, **61**: 230-232.

Thomazella, C., Gonçalves-Vidigal, M.C., Vidigal Filho, P.S., Nunes, W.M.C. and Vida, J.B. (2002) *Crop Breed. Appl. Biot.*, **2**: 55-60.

Thompson, H.C. and Kelly, W.C. (1957) *Vegetable Crops*, McGraw-Hill Book Co. Inc., New York.

Thung, M. (1991) In: *Common beans: research for crop improvement* (Schoonhoven, A. van and Voysest, O. eds), CAB International, Wallingford, UK, pp. 737-834.

Tivoli, B., Halila, H. and Porta-Puglia, A. (1997) In: Tivoli, B. and Caubel, G. (ed.), *Les légumineuses alimentaires méditerranéennes. Les Colloques, Institut National de la Recherche Agronomique.* pp. 81-101.

Tompkins, F.D., Guin, R.S. and Mullins, C.A. (1979) *Tennessee Farm and Home Sci. Prog. Rept.* No. 110, pp. 41-44.

Tompkins, F.D., Sistrunk, W.A. and Horton, R.D. (1972) *Arkans. Fm. Res.*, **21**: 4.

Traka-Mavrona, E., Georgakis, D., Koutsika-Sotiriou, M. and Pritsa, T. (2000) *Agron. J.*, **92**: 1020-1026.

Tyson, R.V. and Kostewicz, S.R. (1986) *Proc. Fla. Sta. Hort. Sci.*, **99**: 358-362.

Uncini, L., Brandozzi, C. and Ferrari, V. (1988) *InformatoreAgrario*, **44**: 75-80.

Upreti, K.K. and Murti, G.S.R. (1999) *J. Plant Biol.*, **26**: 187-190.

Usha, S.A., Uddin, F.M.J., Rahman, M.R. and Islam, A.M.R. (2019) *J. Pharmacog. Phytochem.*, **8**: 1218-1223.

Valenciano, J.B., Casquero, P.A., Boto, J.A. and Guerra, M. (2006) *Field Crops Res.*, **96**: 2-12.

Vasconcelos, M.J.V., Barros, *E.g.* de., Moreira, M.A., Vieira, C. and De Barros, *E.g.* (1996) *Brazilian J. Gen.*, **19**: 447-451.

Vasic, M. (1990) *Zadaciimogucnostiselekciji u resavanju problem asemenarst vaboranije semenarstvo*, **7**: 17-21.

Veinbrants, N. and Rowan, K.S. (1971) *Aust. J. Biol. Sci.*, **24**: 1347-1349.

Velich, I. (1993) *ZoldsegtermesztesiKutatoIntezetBulletinje*, **25**: 103-115.

Venkataratnam, L. (1973) *Beans in India*, Directorate of Extension, Ministry of Agriculture, New Delhi, p. 64.

Verma, P. and Gupta, U. (2010) *Indian J. Microbiol.*, **50**: 263-265.

Vieira, C. (1991) *Revista Ceres*, **38**: 438-443.

Walker, J.C. (1957) *Plant Pathology. McGraw-Hill, New York, USA.*

Waters, I. Jr., Graham, P.H., Breen, P.J., Mack, H.J. and Rosas, J.C. (1983) *J. Agric. Sci.*, **100**: 153-158.

Watson, A.T. and Cullimore, J.V. (1996) *Plant Sci.*, **120**: 139-151.

Webber, H.A., Madramootoo, C.A., Bourgault, M., Horst, M.G., Stulina, G. and Smith, D.L. (2006) *Agric. Water Manag.*, **86**: 259–268.

Wendt, T. (1978) *Gemuse*, **14**: 288-289.

Westphal, E. (1974) *Agric. Res. Rept., Wageningen*, **815**: 129-176.

Wraight, M. and Rogers, B. (1978) *New Zealand Commercial Grower*, **33**: 27.

Wyatt, J.E., Fassuliotis, G., Johnson, A.W., Hoffman, J.C. and Deakin, J.R. (1980) *HortScience*, **15**: 530.

Wyatt, J.E., Hoffman, J.C. and Deakin, J.R. (1977) *HortScience*, **12**: 505.

Yarnell, S.H. (1965) *Bot. Rev.*, **31**: 247-330

Young, R.A. and Kelly, J.D. (1996) *J. Amer. Soc. Hort. Sci.*, **121**: 37-41.

Young, R.A., Melotto, M., Nodari, R.O. and Kelly, J.D. (1998) *Theor. Appl. Gen.*, **96**: 87-94.

Yu, Z.H., Stall, R.E. and Vallejos, C. E. (1998) *Crop Sci.*, **38**: 1290-1296.

Zaumeyer, W. and Meiners, J. (1975) *Annu. Rev. Phytopathol.*, **13**: 313-334.

Zaumeyer, W.J. and Thomas, H.R. (1957) *Phytopathology*, **47**: 454.

Zettler, F.W. and Wilkinson, R.E. (1966) *Phytopathology*, **56**: 1079-1082.

Zeven, A.C., Waninge, J., VanHintum, T. and Singh, S.P. (1999) *Euphytica*, **109**: 93-106.

Zhang, Zhan Yuan, Coyne, D.P., Mitra, A. and Zhang, Z.Y. (1997) *J. Amer. Soc. Hort. Sci.*, **122**: 300-305.

Zink, E., Almedia D'A.L. and Lago, A.A.D.O. (1976) *Bragantia*,**35**: 443-451.

③

COWPEA

A. Chattopadhyay, V. A. Parthasaraty,
A. K. Chakraborti and P. Hazra

1.0 Introduction

Cowpea (*Vigna unguiculata*) is grown throughout India for its long green pods as vegetable, seeds as pulse and foliage as vegetable, and as fodder. When grown for dry seeds, it is also known as black-eye pea, Kaffir pea, China pea, southern bean. The cultivars grown for their immature pods are variously known as asparagus bean, snake bean, yard-long bean. Cowpea originated in Africa and is widely grown in Africa, Latin America, Southeast Asia and in the southern United States. The history of cowpea dates to ancient West African cereal farming, 5 to 6 thousand years ago, where it was closely associated with the cultivation of sorghum and pearl millet (Davis *et al.*, 2000).

2.0 Composition and Uses

2.1.0 Composition

It is rich in nutritive value which is given in Table 1.

Table 1: Composition of Cowpea Green Pod (per 100g of edible portion)*

Moisture	84.6 g	Phosphorus	74 mg
Protein	4.3 g	Iron	2.5 mg
Fat	0.2 g	Vitamin A	941 I.U.
Minerals	0.9 g	Riboflavin	0.09 mg
Fibre	2.0 g	Thiamine	0.07 mg
Carbohydrates	8.0 g	Nicotinic acid	0.9 mg
Calcium	80 mg	Vitamin C	13.0 mg

* Aykroyd (1963).

2.2.0 Uses

It is used as food at both the green shell and dry stage. It is also grown for hay, silage, pasture for all types of stock and as a source of protein, especially of lysine in the staple cereal diets of subsistence and farming communities. Cowpea seed is a nutritious component in the human diet, as well as a nutritious livestock feed. Nutrient content of cowpea seed is summarized in Table 2.

The protein in cowpea seed is rich in amino acids, lysine and tryptophan, compared to cereal grains; however, it is deficient in methionine and cystine when compared to animal proteins. Therefore, cowpea seed is valued as a nutritional supplement to cereals and an extender of animal proteins. Cowpea can be used at all stages of growth as a vegetable crop. The tender green leaves are an important food source in Africa and are prepared as a pot herb, like spinach. Immature snapped pods are used in the same way as snap beans, often being mixed with other foods. Green cowpea seeds are boiled as a fresh vegetable, or may be canned or frozen. Dry mature seeds are also suitable for boiling and canning (Davis *et al.*,

2000). In India, cowpea cultivars grown for tender pods as vegetable are either true Sequipedalis forms (sub-species *sesquipedalis*) originally introduced from southeast Asia and integrades (intermediate types) of Unguiculata (sub-species *unguiculata*) and true Sequipedalis forms, while the cultivars grown for dry seed are either true Unguiculata forms or integrades of Unguiculata and Biflora (sub-species *cylindrica*) forms and the cultivars grown for fodder are either ture Biflora (sub-species *cylindrica*) or integrades of Unguiculata and Biflora forms (Hazra *et al.*, 2011).

Table 2: Nutrient Content of Mature Cowpea Seed*

Protein	24.8 per cent
Fat	1.9 per cent
Fiber	6.3 per cent
Carbohydrate	63.6 per cent
Thiamine	0.00074 per cent
Riboflavin	0.00042 per cent
Niacin	0.00281 per cent

* Davis *et al.* (2000).

3.0 Origin and Taxonomy

Cowpea is in cultivation from very ancient times in the tropics of old world. Its country of origin is uncertain. Vavilov (1939) considered India as the main centre of origin. Some believe that it is of tropical or Central African in origin where wild races are found even now. In India it has been known since the Vedic times. However, it is generally agreed that the cowpea is of African origin as conspecific wild forms are found in Africa but are absent in Asia.

Cowpea belongs to the family Leguminosae, subfamily Fabaceae and genus *Vigna*. *Vigna* is a pantropical genus of about 170 species, 120 being in Africa, 22 in India and South-East Asia and a few in America and Australia (Faris, 1965). Cowpea is a warm-season, annual, herbaceous legume. Plant types are often categorized as erect, semi-erect, prostrate (trailing), or climbing. There is much variability within the species. Growth habit ranges from indeterminate to fairly determinate with the non-vining types tending to be more determinate. Cowpea generally is strongly taprooted. Root depth has been measured at 95 inches 8 weeks after seeding (Davis *et al.*, 2000).The seed coat can be either smooth or wrinkled and of various colors including white, cream, green, buff, red, brown, holstein, and black. Seed may also be speckled, mottled, or blotchy. Many are also referred to as 'eyed' (blackeye, pinkeye purple hull, *etc.*) where the white colored hilum is surrounded by another color (Davis *et al.*, 2000). Cowpea is a day neutral crop. Flowers are borne in multiple racemes on 8 to 20 inches flower stalks (peduncles) that arise from the leaf axil. Two or three pods per peduncle are common and often four or more pods are carried on a single peduncle. The presence of these long peduncles is a distinguishing feature of cowpea and this characteristic also facilitates harvest. The open display of flowers

Cowpea Production in the World

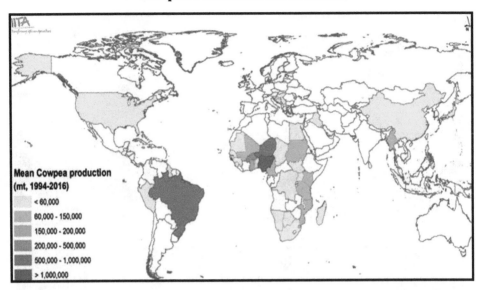

Uses of Cowpea

Cowpea as a fodder (left), Deep fried cowpea fritters (middle) and Cooked black eyed cowpea (right)

Soup (left) and salad (middle) and curry (right) prepared from black eyed cowpea

Growth Habits and Cultigroups of Cowpea

Bush type cowpea

Pole type cowpea

Vigna unguiculata cv-gr. Biflora (left), Unguiculata (middle) and Sesquipedalis (right)

above the foliage and the presence of floral nectaries contribute to the attraction of insects. Cowpea primarily is self pollinating (Davis *et al.,* 2000).

The species is a herbaceous annual and exists in different forms and cultivars. Based on variations in growth habit, size and position of pods, shape and colour of seed, these are broadly classified under three groups. Some authors had given specific status to these groups earlier (Hector, 1936). But many authors are now convinced that there is no sharp distinction between these species. They not only merge with one another but also cross freely to produce intermediate types. Thus all of them are best considered under a single variable complex, *Vigna unguiculata* and different variants, wild and cultivated, are classified as subspecies or varieties (Sampson, 1936; Sellschop, 1962; Verdcourt, 1970). According to Verdcourt (1970), there are five subspecies of *V. unguiculata.* Two subspecies are wild: subsp. *dekindtiana* in the African savanna zone and Ethiopia, and subsp. *mensensis* in forests. Other three subspecies are cultivated and widespread in India, and the Far East (Steele, 1976). A brief description of the three subspecies are given below:

(a) subsp. *unguiculata,* syn. *V. sinensis* (common cowpea): plants viny, sometimes erect with short pendant pods, 20-30 cm long and small seeds. This includes a large number of varieties distributed mostly in Africa as compared to those in tropical and subtropical Asia (Sampson, 1936).

(b) subsp. *cylindrica* syn. *V. catjang, V. cylindrica* (catjang bean): plants are more erect than those of *V. unguiculata,* pods small, 7-13 cm long, more or less erect. Seeds small, 5-6 mm long, kidney-shaped, nearly as thick as broad bean (Sampson, 1936).

(c) subsp. *sesquipedalis* (syn. *V. sesquipedalis, V. sinensis* var. *sesquipedalis* (asparagus bean or yard-long bean): plants are trailing or climbing, pods pendant, 30-90 cm long, fleshy and inflated tending to shrink when dry. Seeds elongated kidney-shaped, 8-12 mm long. Cultivars belonging to this subsp. are found in Indonesia, Philippines and Sri Lanka. They are also known in India (Sampson, 1936).

Classification of cowpeas was a matter of debate as they were ranked as species, varieties or sub-species. Some authorities however, do not consider the cultivated sub-species as distinct and lumped them under one sub-species *V. unguiculata,* sub-species *unguiculata* and differentiate them by intraspecific category "cultigroup". The sub-species *unguiculata, sesquipedalis, cylindrica* and *textilis* are renamed as cultigroup Unguiculata, Sesquipedalis, Biflora and Textilis. Similarly the wild sub-species *dekindtiana* and *mensensis* were put under single sub-species *dekindtiana* and distinguished them by varietal category.

The cytotaxonomy of *Vigna* is relatively simple (2n = 2x = 22) and with apparently little genetic and no chromosomal divergence of the cultivars from their putative ancestor (Steele, 1976). The five subspecies of *V. unguiculata* are interfertile, but all attempts to hybridize cultivars with other *Vigna* species, notable *V. luteola, V. marina* and *V. nilotica,* proposed as wild progenitors have failed (Faris, 1965).

4.0 Cultivars

Different cultivars respond differently to photoperiod (Hazra *et al.*, 2011). Cultivars of viny and vegetable type under *V. unguiculata* sub-species *sesquipedalis* (cultigroup Sesquipedalis) are photosensitive in nature and basically short day crop.

Integrades or intermediate types of the sub-species *unguiculata* (cultigroup Unguiculata) and *sesquipedalis* (cultigroup Sesquipedalis) mainly cultivated as vegetable cowpea show both photo sensitive and photo non-sensitive characters. Vegetable cowpea cultivars developed through combination breeding involving the cultivars sub-species *unguiculata* (cultigroup Unguiculata) and *sesquipedalis* (cultigroup Sesquipedalis) generally show photo non-sensitive character.

Cultivars under sub-species *unguiculata* (cultigroup Unguiculata) and sub-species *cylindrica* (cultigroup Biflora) generally show photo non-sensitive character.

The following cultivars of vegetable cowpea have been developed by Agricultural Research Institutes and SAUs in India.

Pusa Phalguni

It has been selected from a Canadian cultivar (Pal *et al.*, 1956). It is a bushy, dwarf cultivar suitable for sowing in February-March in Northern India. Pods are dark green, erect 10-12 cm long and appear in two flushes. Pods get ready for harvest in 60 days and the yield of green pod is about 70-75 q/ha.

Pusa Barsati

It has been selected from a collection from the Philippines (Pal *et al.*, 1956). It is an early cultivar suitable for growing during the rainy season. It flowers in 35 days and pods become ready in 45 days. Pods are light green, pendant, 26-28 cm long. Yield is about 70-75 q/ha.

Pusa Dofasli

It has been evolved by crossing Pusa Phalguni with a cultivar from the Philippines (Singh *et al.*, 1968). It is a bushy cultivar suitable for sowing in both spring-summer and rainy seasons. It flowers in 40 days. Pods are light green and 18 cm long. It produces about 75-80 quintals of green pods per hectare.

Pusa Komal

It is widely adaptable, photo insensitive, indeterminate, bushy cultivar developed at IARI. It flowers in 40-45 days, produces 20-22 cm long pod with synchronous ~ bearing habit and good quality pod. The cultivar is highly resistant to bacterial blight and produces about 100 quintals of green pods per hectare in 2-3 flushes (Anon., 1987).

IT82D-716

The cultivar IT82D-716 resistant to diseases, thrips and bruchids and yields about 20.0 q seed/ha (Nwabudike and Gast, 1996).

'Ebony PR' syn. Line 4A

Derived from the cross Bechuana White × Blackeye 5, this diploid (2n = 22), spreading, forage type cowpea cultivar is submitted for the registration of plant cultivar rights in Australia. Ebony PR was evaluated together with Red Caloona and Meringa in trials conducted at Lawes, Queensland in 1996. Ebony PR is resistant to *Phytophthora vignae* races 3 and 4 while both Red Caloona and Meringa are susceptible. Number of seeds per pod and 1000-seed weight averaged 16.6 and 129.9 g, respectively, in Ebony PR; the corresponding values were 12.6 and 67.8 in Red Caloona and 15.8 and 109.1 in Meringa (Anon., 1996).

Philippine Early

This cultivar was introduced from Philippines by IARI, New Delhi, India. It was used extensively for fresh vegetable purpose in early periods (1950s).

CO-2

Developed at Coimbatore, India through advanced generation selection from the cross C 521 (grain type) × C 419 (vegetable type). Plants are semi spreading in habit. Pods are long (26 cm), green and having less fibre content. Crop duration is 85 days. Average yield is 90 q/ha.

Vamban-2

Developed through pure line selection from IT81-D-1228-10. Pods are long and green. Average pod yield is 90-100q/ha.

Cowpea 263

A dual season and dwarf cultivar developed at Ludhiana through selection from Bangalore local, a germplasm collection from IIHR, Bangalore, India. Pods are green, thick, meaty, tender, medium long (20 cm) and contain 3.57 per cent crude protein on fresh weight basis. It is resistant to mosaic and golden mosaic virus. Average yield is 100-120 q/ha.

Pusa Rituraj

A photo and thermo insensitive cultivar developed at IARI, New Delhi, India through advanced generation selection from the cross Pusa Dofasli × EC 26410. It is profuse bearing and all season cultivar. Pods are 22-24 cm long, thin, less fibrous with brown seeds. This cultivar can be used for both green pods as vegetable and dry seed as pulse.

Arka Garima

Developed at IIHR, Bangalore, India through back cross pedigree method. of breeding from the cross TUV-762 × *Vigna unguiculata* subsp. *sesquipedalis* (cultiroup Sesquipedalis). Plants are indeterminate but less viny, vigorous, bushy and spreading with light green leaves and purple flowers. Pods are light green, thick, long (22-24 cm), fleshy and stringless. Average yield is 80-100 q/ha in 90-100 days crop duration.

Sel 2-1

Developed at NDUAT, Faizabad, India through selection. Plants are 70-75 cm in height and semi-viny in nature. Pods are green and medium long (25-30 cm) with black seeds. It is susceptible leaf spot and virus disease.

Arka Samrudhi (IIHR Sel-16)

An early maturing cultivar developed at IIHR, Bangalore, India through pedigree selection from the cross Arka Garima × Pusa Komal. Plants are erect, bushy, 70-75 cm tall and photoinsensitive in nature. Plants are erect, bushy, 70-75 cm tall and photo-insensitive. Pods are green, medium thick and long (15-18 cm), tender, fleshy without parchment, good cooking quality. It has been recommended for cultivation in Rajasthan, Gujarat, Haryana, Delhi, M.P. and Maharashtra. Pod Yield is 190 q/ha in 70-75 days.

Arka Mangala

Developed at IIHR, Bangalore, India. Plants tall (3-4 m), pods are very long (80 cm), light green, stringless, round, tender with crisp texture and matures in 60 days. Suitable for kharif and rabi. Pod yield is 250 q/ha in 100 days.

Bidhan Barbati-1

Developed at BCKV, West Bengal, India through modified back cross pedigree method of breeding from the cross EC 243954 (Unguiculata) × EC 305827 (Sesquipedatis). Plants are compact and determinate with dark green foliage and white flower having yellow colour at the base of the standard petal. First pod picking 55-60 days after sowing. Pods are green, tender, thick, fleshy and medium long (25.2 cm) and borne on long and stout peduncle. Seeds are bold, thick, slightly reinform with Holstein coloured with buff and brown. Pod protein is 3.87 per cent. Resistant to cowpea mosaic and golden mosaic virus. Average yield is 130-135 q/ha.

Bidhan Barbati-2

Developed at BCKV, West Bengal, India through modified back cross-pedigree method of breeding from the cross V-70 (Biflora) × Sel TM-3 (Sesquidalis). Plants are semi-determinate and loose framed with yellowish green foliage and light purple-violet flowers. Pods are light green, tender, thick, solid, fleshy and medium long (25.8 cm) and borne on short and soft peduncle. Seeds are flat, reinform and holstein coloured with buff and dark tan. Pod protein is 4.0 per cent. Average yield is 155-160 q/ha.

Kashi Shyamal

This cultivar has been developed at IIVR, Varanasi, India. Plants are dwarf and bushy type (70-75 cm height), suitable for both *Kharif* and *Zaid* seasons, having 3-4 branches, early in flowering (40 days after sowing) with first harvesting after 48 days of sowing producing 35-40 green pods per plant with pod length of 25-30 cm. It is tolerant to Golden mosaic virus. It has been recommended for cultivation in Punjab U.P., Bihar and Jharkhand and has yield potential of 80-90 q/ha.

Kashi Gauri (VRCP-2)

This cultivar has been developed at IIVR, Varanasi, India. Plants are bush type, dwarf, photo-insensitive and early cultivar suitable for sowing in both spring summer and rainy season. It flowers in 35-38 days and pods get ready for first harvest in 45-48 days. Pods are 25-30 cm long, light green as compared to Kashi Shyamal, soft, fleshy and free from parchment layer. Plants are resistant to golden mosaic virus and *Psedocercospora cruenta*. It has been recommended for cultivation in Karnataka, Tamil Nadu and Kerala and has yield potential of 110-130 q/ha.

Arka Suman

This cultivar has been developed through hybridization (Pusa Komal × Arka Garima) followed by pedigree selection at IIHR, Bangalore, India. Plants are erect, bushy, photo-insensitive with pods above the canopy. Pods are medium long, tender, fleshy, crisp, without parchment good cooking quality. It has been recommended for cultivation in Karnataka. This cultivar has yield potential of 170-190 q/ha.

Anaswara

This cultivar has been developed at KAU, Vellanikkara, India. Vine is semi-determinate type, bears purple flowers and suitable for cultivation as Thadapayar. Pods medium long (28.13 cm) light green, have bold creamy seeds, weight 12.5 g and 19 seeds per pod. This cultivar has been recommended for cultivation in Kerala and has yield potential of 100-125 q/ha.

Swarna Suphala (CHCP-2)

This cultivar has been developed through introduction at HARP, Ranchi, India. Pods are straight light green, 30-35 cm long with bulged appearance of seeds. Mature dried seeds are bicoloured (mottling of brown and sandalwood). It is suitable for cultivation in summer and rainy seasons. First harvest can be taken 50-55 days after planting. It is resistant to rust and cowpea golden mosaic virus and tolerant to pod borer during summer under field conditions. It has been recommended for cultivation in Jharkhand. Bihar, Karnataka and Kerala and has yield potential of 210-250 q/ha.

Swarna Harita

Developed through pure line selection. Pods are dark green, very long (50-60 cm), straight, round and fleshy having excellent cooking quality. Mature dried seeds are light brown, elongated and kidney shaped. Tolerant to mosaic viruses and rust under field condition. First harvest 50-55 days after sowing. Average fresh pod yield is 300-350 q/ha.

Swarna Sweta

Developed through pure line selection. Pods are medium long (30-35 cm), white, straight, round and fleshy pods having very good cooking quality. Resistant to mosaic viruses and rust and tolerant to pod borer under field condition. First harvest 50-55 days after sowing. Average fresh pod yield is 250-300 q/ha.

Swarna Mukut

Developed through hybridization followed by pedigree selection. Pods are straight round, light green pod (20-25 cm). Mature dried seeds are kidney shaped and yellowish brown. Under field conditions, the cultivar is least infected by cowpea mosaic viral disease and pod borer infestation. First harvest 45-50 days after sowing. Average fresh pod yield is 120-150 q/ha.

Sel-263

This cultivar has been developed through pedigree selection at PAU, Ludhiana, India. Plants are dwarf and can be grown in both spring and rainy season. Pods are green, thick, fleshy, tender and medium (20 cm long). Plants are resistant to mosaic and golden mosaic viruses. This cultivar has been recommended for cultivation in states of Punjab, U.P. and Bihar has yield potential of 200-220 q/ha.

Narendra Lobia-2 (NDCP-13)

This cultivar has been developed through hybridization (Sel-2-1 × Red Seeded) followed by pedigree selection at NDUA and T, Faizabad, India. Plants are bushy, early flowering, dark green and 28 cm long. First picking starts after 50 days of seed sowing and suitable for *kharif* and *zaid* cultivation. It has been recommended for cultivation in West Bengal, Assam, North East States, Andmand and Nicobar Islands, Punjab, U.P., Bihar, Jharkhand, Karnataka, Tamil Nadu and Kerala and has yield potential of 75-100 q/ha.

Kashi Kanchan (IIVRCP-4)

This cultivar has been developed at IIVR, Varanasi, India. It is a bush type and dwarf (50-60 cm height), photo-insensitive and early cultivar suitable for sowing in both spring-summer and rainy seasons. It flowers in 40-45 days and pods get ready for harvest in 50-55 days. It produces 40-450 pods per plant of 30-35 cm long. The pods are dark green, tender, pulpy with less fibre and free from parchment layer. The cultivar is resistant to golden mosaic virus and *Pseudocercospora cruenta*. It has been recommended for cultivation in Uttar Pradesh, Chhattisgarh, Madhya Pradesh and Maharashtra and has green pod yield potential of 150-200 q/ha.

VR-5

This cultivar has been developed at IIVR, Varanasi, India. It is a bush type and early maturing, suitable for spring summer and rainy seasons sowing. Flowering starts 35-40 days after sowing and produces 40-45 pods per plant. It has been recommended for cultivation in Uttar Pradesh, Chhattisgarh, Madhya Pradesh and Maharashtra and has green pod yield potential of 150-175 q/ha.

Bidhan Sadabahar (BCCP-3)

Developed at BCKV, West Bengal, India through modified back cross-pedigree method of breeding from the cross the cross between Unguiculata and Sesquidalis cultigroups. Plants are determinate with green foliage and light purple-violet flowers, very early maturing, suitable for spring summer and rainy seasons sowing. Flowering starts 30-35 days after sowing. Pods are green, tender, thick, solid, fleshy and medium long (25.8 cm) and borne on long peduncle. Seeds are flat, reinform and biscuit coloured. Pod protein content is 4.2 per cent. Average yield is 160-180 q/ha. It has been recommended for cultivation in West Bengal, Uttar Pradesh, Chhattisgarh, Madhya Pradesh and Maharashtra.

Lola

Developed at KAU, Vellanikkara, India through pureline selection from Kerala Local. Plants viny, pods green smooth, long (30.35 cm) and tender. Yield potential is 200 q/ha.

Vyjayanthi

Developed at KAU, Vellanikkara, India through pureline selection from Perumpadavam Local (VS 21-1). Plants viny, pods green smooth, long (30.35 cm) and tender. Yield potential is 200 q/ha.

KMV-1

Developed at KAU, Vellanikkara, India through pureline selection from Manjeri Red Plain.

Malika

Developed at KAU, Vellanikkara, India through Single plant selection from Thiruvananthapuram.

Vellayani Jyothika

Developed at KAU, Vellanikkara, India through selection from Sreekaryam Local. Pole type, high yielding (190 q/ha) with long light green pods.

Sharika

Developed at KAU, Vellanikkara, India through selection from Valiyavila Local (SPS).

Kashi Nidhi

Plants are dwarf, erect and bushy, with 20-25 peduncle per plant. Fruits are green, 25-30 cm long. Seed colour is reddish brown. Golden mosaic virus and *Pseudocercospora cruenta* tolerant with an average pod yield of 140-150 q/ha. Better yield and keeping quality suitable for distant marketing. Recommended for release and cultivation in the states of Uttar Pradesh, Bihar, Haryana, Punjab and Jharkhand. Vide gazette notification number S.O. 2363(E), 04.10.2012.

Kashi Unnati

This is a photo-insensitive cultivar. Plants of this cultivar are dwarf and bushy, height 40-50 cm, branches 4-5 per plant, early flowering (30-35 days after sowing), first harvesting at 40-45 days after sowing, produces 40-45 pods per plant. Pods are 30-35 cm long, light green, soft, fleshy and free from parchment. The cultivar is resistant to golden mosaic virus and *Pseudocercospora cruenta*, and gives green pod yield of about 125-150 q/ha. This has been identified by Horticultural Seed

Sub-Committee and Notified during XIII meeting of Central Sub-Committee on Crop Standard Notification and Release of Varieties for Horticultural Crops for the cultivation in Punjab, U.P., and Jharkhand.

Kashi Sudha

Golden mosaic virus and *Pseudocercospora cruenta* tolerant, Identified for Uttar Pradesh, Bihar, Jharkhand, Bihar, Andhra Pradesh, Odisha, Chhattisgarh, Madhya Pradesh and Maharashtra by AICRP-VC.

Konkan Wali

Pods are Light green with 35 to 40 cm length, tender pods suitable for vegetable. It is mature in 145-160 days with yield potential of 64 q/ha.

Pant Lobia-1

Developed by GBPUAT, Pantnagar, India. It has 45-50 cm plant height, Resistant to YMV and drought tolerant, 27 per cent protein content. It is mature in 65-70 days with yield potential of 150-200 q/ha.

Pant Lobia-2

Developed by GBPUAT, Pantnagar, India. It has tolerance to major bacterial and viral diseases like yellow mosaic, photo insensitive and drought tolerant, adaptable to zaid season. It is mature in 75-80 days with yield potential of 140-180 q/ha.

Pant Lobia-4

Developed by GBPUAT, Pantnagar, India. It has tolerance to major bacterial and viral diseases like yellow mosaic, photo insensitive and drought tolerant, adaptable to zaid season. It is mature in 60-65 days with yield potential of 140-180 q/ha.

Pant Lobia-3

Developed by GBPUAT, Pantnagar, India. It has 50-55 cm tall. Resistant to YMV and bacterial blight, bush type, seeds are kidney to oval shape and brown in colour. It has 27 per cent protein. It is mature in 65-70 days with yield potential of 140-180 q/ha.

Tender Cream

Tender Cream was derived from an initial cross of Floricream and Ala. 963.8, a line resistant to cowpea curculio (*Chalcodermus aeneus*). In a breeding programme involving pedigree, line and recurrent selection other lines were selected included *Cercospora* leaf spot resistant CR17-1-34. The Tender Cream phenotype is similar to that of Carolina Cream and Bettergreen. Pods are slightly curved and 14-16 cm long containing 12-14 seeds. Seeds are small (20 g/100 seeds), cream coloured and ovate to reniform with smooth testas. Canned samples of Tender Cream scored well in quality tests. In yield trials Tender Cream has the same yield potential as California Cream and out yielded Early Acre and White Acre in the 1992, 1993 and 1994 Regional Southernpea Cooperative Trials. In addition to resistance to cowpea curculio, ELISA tests indicated that Tender Cream is resistant to *Cercospora cruenta*

Promising Cowpea Cultivars

Kashi Unnati

Kashi Kanchan

Ankur Gomti

Bidhan Barbati-2

Bidhan Barbati-1

Bidhan Sadabahar

Types of Seed, Seed Germination and Crop Establishment

Variability in seed shape, size and colour in cowpea

Seed germination (left) and vegetative stage (right) in cowpea

[*Mycosphaerella cruenta*], southern blight (*Corticium rolfsii*), rust (*Uromyces phaseoli* [*U. appendiculatus*]) and powdery mildew (*Erysiphe polygoni*). In tests it showed a high level of resistance to the root-knot nematode *Meloidogyne incognita* (Fery and Dukes, 1996). In Maharashtra, India Rajput *et al.* (1991) noticed that selection 61B is an early maturing cultivar suitable for vegetable purpose. An early maturing cultivar IT 82 E32 with red seeds and fairly good resistance to disease have been reported from Ghana (Hossain and Asafo-Adjei, 1987). In Indonesia, Soedomo and Sunaryono (1988) observed that cultivars BSH 1 and BSH 3 were most promising for seed production and cultivars EGH2 and BS7 were suitable for fresh pod yields.

Timsina (1988) from Nepal reported that for vegetable purpose the cultivars IT81 D1228-15 and IT81 D1228-10 performed better and produced 90.51 and 90.63 quintals green pods/ha respectively. A cultivar BR 10-Piaui from Brazil is immune for cowpea severe mosaic, cowpea rugose mosaic and cowpea severe mottle viruses and very resistant to cowpea golden mosaic virus and gives 72 per cent higher yield than that of traditional cultivars (Santos *et al.*, 1987). In Netherlands, the best yielding cultivars under glasshouse condition were recorded to be 84087, NZ and Guirlande (Heij, 1989).

5.0 Soil and Climate

Cowpea can be grown in all types of well drained soil with pH 5.5 to 6.5. It is a warm-season crop well adapted to many areas of the humid tropics and subtropical zones, and tolerates heat and dry conditions, but intolerant to frost (Davis *et al.*, 2000). Cowpea thrives best between 21 and 35°C. It can be grown successfully in spring-summer, summer-rainy and rainy-autumn and early autumn seasons in the plains. It cannot withstand heavy rainfall and water-logging. Different cultivars respond differently to temperature and day length and thus there are distinct cultivars for spring-summer and rainy seasons.

6.0 Cultivation

6.1.0 Season and Sowing

Cowpea can be sown during in spring-summer and rainy seasons. In mild climate, it can be grown almost throughout the year. But in that case photo-insensitive cultivars are to be grown. It is usually sown in February/March in the northern plains in India and in December/January in the southern plains for spring-summer crop. For summer-rainy and rainy season crops, sowing is done in May-June and July-August, respectively all over the plains of India. According to Davis *et al.* (2000) cowpea should not be planted until soil temperatures are consistently above 65°F (18.3°C) and soil moisture is adequate for germination and growth. Seeds will decay in cool, wet soils.

Sowing is done in a well-prepared field by broadcasting or line sowing. Usually 20-25 kg/ha of seed is required for bush type cultivars grown in spring-summer season, 12-15 kg/ha of seed for bush type cultivars grown in rainy season, and 8-10 kg seeds/ha for indeterminate, viny cultivars grown in rainy-autumn and early autumn season. Seed inoculation with *Rhizobium* culture (*cowpea miscellany* group) before sowing is required for quick nodulation on roots which fix atmospheric nitrogen. About 375-500 g *Rhizobium* culture is sufficient for treating the seeds required for one hectare. Sowing in line facilitates better intercultural operations and aftercare. Line sowing can be done with a seed drill operated by a tractor, bullock or manual labour. Spacing between rows should be 45-60 cm and between plants within a row should be 25-30 cm for bushy cultivars and 75×60 cm for Indeterminate, viny type cultivars. In the U.S.A. high level of pod production was achieved using extremely high densities of planting from 100 to 400 thousand plants/ha with cultivars VCR 193, VCR 206 and Snapea (Kwapata and Hall, 1990). According to Davis *et al.* (2000) traditionally, cowpea in the United States are sown in rows, 75 cm apart with seeds spaced 5 to 10 cm in the row. Higher plant populations achieved by using narrow rows 30-50 cm has been used in commercial plantings. They suggested that highly determinate types may be planted 5-8 cm apart. Vines of indeterminate types require more space, and a final stand with 20-25 cm between plants and 75 cm rows is considered to be a minimally acceptable population (Davis *et al.*, 2000). Stacking with jute stick is required to support the plants of indetermite, viny cultivars.

6.2.0 Manuring and Fertilization

Being a leguminous crop it does not require heavy fertilization. If cowpea is sown for the first time in a field, seeds should be inoculated with *Rhizobium* culture before sowing. This helps in quick nodulation of the roots for symbiotic nitrogen fixation. Although cowpea *Rhizobium* is normally widespread, seed inoculation with *Rhizobium* specific to cowpea would be beneficial in areas where it is not present (Davis *et al.*, 2000). Mandal *et al.* (1999) identified several *Rhizobium* strains for cowpea but local rhizobia invariably out-populated the introduced strains. Until recently, it was assumed that indigenous *Bradyrhizobium* spp. that effectively nodulate cowpea was abundantly present in tropical soils (Caldwell and Vest, 1968; Singleton *et al.*, 1992; Kimiti and Odee, 2010) and therefore inoculation was not necessary. However Chidebe *et al.* (2018) recorded a group of highly diverse and adapted cowpea-nodulating microsymbionts which included *Bradyrhizobium pachyrhizi*, *Bradyrhizobium arachidis*, *Bradyrhizobium yuanmingense*, and a novel *Bradyrhizobium* sp., as well as *Rhizobium tropici*, *Rhizobium pusense*, and *Neorhizobium galegae* in Mozambican soils. In modern agriculture systems, cowpea can contribute with 70–350 kg nitrogen per ha through biological nitrogen fixation (BNF) (Quin, 1997).

Excess nitrogen (N) promotes lush vegetative growth, delays maturity, may reduce seed yield and may suppress nitrogen fixation. The plant will perform well under low N conditions due to a high capacity for N fixation. A starter N rate of around 12 kg/acre is sometimes required for early plant development on low-N soils (Davis *et al.*, 2000). Singh and Sikka (1955) recommended an application of 17 kg nitrogen and 28 kg phosphorus per hectare before sowing. Mehta (1959) reported that about 22-45 kg nitrogen per hectare was sufficient for growing cowpea. Application of 10-20 kg nitrogen, 50-70 kg phosphorus and 50-70 kg potash per hectare has been recommended by Chauhan (1972). Kumar and Singh (1990) observed that green pod yields increased with increasing levels of P_2O_5 and K_2O up to 80 kg/ha. A soil test is the best way to determine soil nutrient levels. In general, at least 12 kg P/acre and 18 kg K/acre are recommended on soils of medium fertility but individual soils will vary in fertilizer requirements. Band fertilizer 7-10 cm deep and 5-7 cm away from the seed, or broadcast, including nitrogen, before planting is also practised (Davis *et al.*, 2000). Chattopadhyay and Dutta (2003) observed that application of phosphorus fertilization up to 80 kg/ha significantly increased pod yield and nodulation of cowpea. They also recorded dual inoculation with *Rhizobium* and Phosphate-solubilizing bacteria (PSB) resulted in conspicuous increase in nodule number and weight, leg-haemoglobin content and Nitrogenase activity of nodules and pod yield over no inoculation. Khan *et al.* (2015) also recorded that application of vermicompost @ 6.0 t/ha and combined seed inoculation with the *Rhizobium* + PSB significantly increased the seed yield, straw yield, biological yield, protein content in seed, total root nodules, effective root nodule, leghaemoglobin content, nitrogen capacity and net returns over control and other treatments. Chatterjee and Bandyopadhyay (2015) also recorded that combined use of seed treatment with molybdenum (0.5 g/kg seed) and biofertilizers (*Rhizobium* + PSB) along with foliar spray of boron at 4 weeks of planting significantly enhanced the growth and yield attributes of cowpea and registered 42 per cent and 54 per cent improvement in number of pod and pod yield/plant, respectively over control, Cowpea is sensitive to zinc deficiency. In zinc deficient soils, application of Zinc Sulphate @ 10-15 kg/ha is beneficial to the crop.

Most legumes have the capacity to generate symbiotic interactions with bacteria called rhizobia that fix atmospheric nitrogen (N) benefiting the plant, which in turn delivers carbon to the bacteria. This symbiosis reduces the production costs and the risk of environmental pollution due to the use of synthetic N fertilizer. However the symbiotic capacity of cowpea is often impaired in the presence of fertilizer N. One approach to overcoming this "fertilizer-N" barrier would be to identify cowpea genotypes that are tolerant of fertilizer N. Harper and Gibson (1984) while investigating nitrate tolerance in legume species, concluded that efforts to overcome nitrate inhibition of nodule formation and function should focus also on uptake of nitrate and its metabolism through nitrate reductase (NR) activity. The effectiveness of symbioses of cowpea genotypes that have little or no nitrate-induced NR activity would probably be, at worst, only mildly susceptible to fertilizer N. Nitrite is also a product of NR activity. Nitrite, and the products of its reduction by nitrite reductase, may interfere with N_2 fixation (Trinchant and Rigaud, 1984). Not all plants accumulate nitrite (Naik *et al.*, 1982) and others accumulate it without any apparent deleterious effects on the plant (Nair *et al.*, 1988) Notwithstanding, cowpea lines with low nitrite contents or from which nitrite is absent may well have more effective symbioses than nitrite accumulators (Singh and Usha, 2003).

6.3.0 Irrigation and Interculture

Cowpea is a shallow-rooted crop and requires less moisture for its growth. It is sensitive to water-logging. Therefore, light irrigation should be given. Irrigation prior to flowering helps in pod setting and another irrigation should be given after the pods have set. After harvesting of green pods of the first flush, one more irrigation improves yield. This will help in initiation of second flush of flower. Irrigation given at IW: CPE 0.3 through sprinkler form at 60±1.5 cm water table depth favours the higher grain yield and nutrient uptake by cowpea whereas flooded irrigation with deep water table condition accelerated nutrient leaching (Dasila *et al.*, 2006). There have been several irrigation and crop water use studies made on cowpea (Shouse *et al.*, 1981; Saunders *et al.*, 1985; Ziska and Hall, 1983; Fapohunda *et al.*, 1984), but only a few have related the water use to a reference ET and presented a crop coefficient function. Turk and Hall (1980) found that, when well-watered and with complete ground cover on a sandy soil, the cultivar 'California Blackeye No. 5' (CB-5) had a crop coefficient for use with class A pan evaporation (Kcp) that averaged 0.94, using a pan located in a field of cowpea. In a field experiment, Rao and Singh (2004) measured the water use by cowpea in an arid climate with a maximum use of 6.8 mm/day and a Kcm = 1.19 during the vegetative stage of growth. Andrade *et al.* (1993) found a crop coefficient for use with the Penman reference ET (Kcn) of 1.16 at 42 days-after-planting (DAP) for a determinate variety. Souza *et al.* (2005), in a 69-day season using lysimeters, found the average Kcm = 1.27 at the flowering stage of cowpea. The Kcm increased steadily from the beginning up to flowering and peaked at 1.35 on 50 DAP; it then decreased rapidly until harvest time. There was no obvious mid-season plateau as given by FAO-56. Total water use for the season was 337 mm. Souza *et al.* (2005) also used class A pan evaporation as a reference ET but the resulting Kcp-values seem strangely high, with one peak at 1.52 and another at 1.46. Aguiar *et al.* (1992), with one full 69-day season of data from a sandy soil in a very humid climate, showed much lower values with Kcn of 1.10 and 1.04 for the flowering and fruiting stages, respectively, and an average mid-season value of 1.05. The total water use was 306 mm. The reference ET they used was 85 per cent of class A pan evaporation, which was considered to be equivalent to the Penman ET. DeTar (2009) recorded the average value for the crop coefficient during the mid-season plateau was 0.986 for the coefficient used with pan evaporation, and it was 1.211 for the coefficient used with a modified Penman equation for ET0 from the California Irrigation Management and Information System (CIMIS).

One hoeing should be done about 4 weeks after sowing for controlling weeds and helping in root aeration. Spraying of maleic hydrazide at 50-200 ppm just before flowering has been reported to increase the yield of pod (Choudhury and Ramphal, 1960). Nabi *et al.* (2016) recorded that the application of GA_3 up to 33.33 ppm would be the most suitable for obtaining the greater yield of cowpea at Dhaka, Bangladesh.

7.0 Harvesting and Yield

As in other beans, pods should be harvested when they are tender. Harvesting should be done at short interval before the pods become fibrous and unfit for marketing. Marketable pods are available in about 45 days in case of an early cultivar and may continue up to 100 days in flushes. The pod length and pod weight reaches maximum at 12 days after anthesis, which is considered the optimum harvest stage for vegetable use. It produces about 50-80 quintals of green pod per hectare. Cowpea can be harvested at three different stages of maturity: green snaps, green-mature, and dry. According to Davis *et al.* (2000) depending on temperature, fresh-market (green-mature) pods are ready for harvest 16 to 17 days after bloom (60 to 90 days after planting). Harvest date for green snap pods is normally specified by the processor. Mechanical harvest requires the use of a snap bean or green pea

Intercropping, Irrigation and Mulching in Cowpea

Intercropping cowpea with cashew

Intercropping cowpea with maize

Furrow irrigation (left) and mulching with crop residue (right) at initial growth stages in cowpea

Pod Bearing and Harvesting of Cowpea

Pod bearing stage of cowpea

Harvesting of cowpea pods

Cowpea pods in bundles ready for marketing

harvester. Most domestic cowpea production in U.S.A. is mechanically harvested, however, hand harvested cowpeas suffer less damage and the harvest season may continue over a 1- to 3-week period (Davis *et al.*, 2000). Mature green cowpeas are normally harvested mechanically by some type of mobile viner in U.S.A.

8.0 Diseases and Pests

8.1.0 Diseases

Major diseases of cowpea and their control measures are described below:

Fungal Diseases

8.1.1 Anthracnose

This is a fungal disease caused by *Colletotrichum lindemuthianum*. It attacks stems, leaves and pods. Infected portions show dark brown, sunken spots with raised reddish or yellowish margins. Tissues of necrotic spots on leaves collapse, become thin and papery and produce abundant pinkish slimy mass in humid weather. Infected stems crack and rot. It is seed transmissible. Control measure consists of use of healthy seeds, seed treatment with 0.125 per cent solution of ceresan for half an hour (Choudhury, 1972) or dry dressing of seed with captan at the rate of one teaspoonful per kg of seed (Reddy, 1968) and crop rotation. Spray the crop with 0.1 per cent carbendazim or hexaconazole or 0.2 per cent chlorothalonil at 10 days interval gives good control (Hazra *et al.*, 2011).

8.1.2 Die-back

It is caused by *Colletotrichum capsici*. Twigs and branches dry up from tip downwards. Small, black, dot-like structures appear on dried up portions. Infected pods shrivel. Spray the crop with 0.1 per cent carbendazim or hexaconazole or 0.2 per cent chlorothalonil at 10 days interval gives good control of disease (Hazra *et al.*, 2011).

8.1.3 Ashy Stem Blight

This is a fungal disease caused by *Macrophomina phaseolina*. It causes brown lesions at the collar portion. The lesions spread rapidly covering the entire stem portion and killing the growing point. Vascular portion of the roots turns brown, rootlets rot, causing the plants to dry up. The disease is seed-borne. Prophylactic measures consist of use of disease free seed and seed dressing with captan or thiram at 2-3 g per kg of seed before sowing. Spraying the crop with 0.1 per cent carbendazim/carbendazim (1 g) + mancozeb (2 g)/l or Captan @ 2.5 g/l along with sticker checks the spread of the infection effectively (Hazra *et al.*, 2011).

8.1.4 Powdery Mildew

The fungus *Erysiphe polygoni* causing powdery mildew, attacks almost all parts of the plants. White powdery patches first appear on leaves and then spread to stems and green pods. In severe cases defoliation occurs. The disease generally appears late in the season. It can be controlled by spraying 0.5 per cent wettable sulphur as soon as the disease appears. Growing of resistant variety Pusa Komal and spraying of benlate or bavistin or tridemorph at the rate of 0.15 per cent thrice at weekly interval has also been reported to be effective.

Bacterial Disease

8.1.5 Bacterial Blight

It is caused by *Xanthomonas vegnicola*. The primary damage is high mortality of the seedlings, especially when the seeds used for sowing are from a severely diseased crop (Singh and Patel, 1977). Infected leaves show light yellow, irregular to circular spots with necrotic brown centre, later changing to straw colour. Dark green, water-soaked spots of variable shape and size appear on pods which later become yellow and dry. Affected leaves often fall off early. Control measures consist of use of disease free seed and growing of resistant cultivars. Three cultivars of cowpea, 779, 868 and 1552 resistant to bacterial blight have been developed by Singh and Patel (1983) at I.A.R.I., New Delhi, India. Hazra *et al.* (2011) advocated some control measures which include growing of resistant variety Pusa Komal, use of disease free healthy seeds, seed treatment with 0.01 per cent streptocycline solution for 30 minutes before sowing and spraying of crop with 0.01 per cent streptocycline + 0.2 per cent copper oxychloride.

Viral Disease

8.1.6 Mosaic

It is a viral disease. The affected leaves develop a typical mosaic of broad and raised dark-green patches along with chlorotic streaks or spots. The diseased plant remains stunted with reduced and malformed leaves. The disease is seed-borne and trans-mitted by aphids. Cowpea cultivars Early Sugarcrowder, Sowanee and Taylor have been reported to be resistant (Capoor and Varma, 1956). Cultivar RBIO-Piaui is multiple resistant to viruses (Santos *et al.*, 1987). Three varieties Bidhan Barbati-1, Bidhan Barbati-2 and Bidhan Sadabahar (BCCP-3), resistant to Cowpea Mosaic Virus and Golden Mosaic Virus, have been developed from BCKV, West Bengal, India, (Hazra *et al.*, 2001; Chattopadhyay *et al.*, 2014). Use of disease free seed, removal of diseased plants and spraying with some systemic insecticides, *e.g.*, imidacloprid (0.03 per cent) or acetamiprid (0.15 per cent) can keep the disease under check.

8.2.0 Pests

Major insect pests of cowpea and their control measures are described below:

8.2.1 Galerucid Beetle (*Madurasia obscurella*)

It causes damage to the foliage, resulting in small sieve like holes on the leaves. The insect is active during evening, night and early morning hours. During hot sun it hides under debris and loose soil. The tiny grub feeds on the root hairs and nodules causing considerable damage. It can be controlled by soil application of systemic granular insecticide, *e.g.*, phorate or aldicarb 10 G at the rate of 10-15 kg per hectare or carbofuran 3 G at the rate of 30-33 kg per hectare at the time of sowing. This will protect the crop up to 30-40 days. Spraying of chlorpyriphos at the rate of 2.5 ml per litre of water can also effectively control the pest.

Diseases of Cowpea

Web blight (left) and Root rot (right) infection in cowpea

Bacterial leaf blight (left) and Mosaic (right) infection in cowpea

Cercospora leaf spot (left) and Anthracnose (right) infection in cowpea

Brown rust (left) and Charcoal rot (right) infection in cowpea

Pests of Cowpea

Pod borer infestation in cowpea

Aphid infestation in cowpea

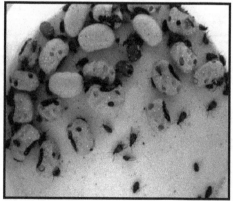

Pod sucking bug (left) and pulse beetle (right) infestation in cowpea

8.2.2 Aphid (*Aphis* sp.)

The tiny insects, are grey or black in colour which suck the cell sap of the tender parts of the plant, mostly the leaves. When the insects are in abundance, they attack the developing pods causing reduction in growth and yield. They also spread cowpea mosaic virus disease (Capoor, 1967). If the attack is at early stage when there is no pod, spraying of imidacloprid (0.03 per cent) or acetamiprid (0.15 per cent) can also control the pest. If the crop is in pod, all edible pods should be harvested and the crop should be sprayed with Malathion 50 E.C. at the rate of 2 ml per litre of water. Hendawy *et al.* (2018) observed that primiphosmethyl @ 375 ml/100 l proved to be the most effective compound against cowpea aphid.

8.2.3 Jassid (*Amrasca kerri*)

The adult insect is wedge-shaped, about 2 mm long and green in colour. The nymphs (youngs) are wingless and found in abundance on the lower surface of the leaves. The nymphs and adults pierce the plant tissues and suck the cell sap. Initial damage is yellowing of the leaf margins followed by curling up. Spraying of imidacloprid (0.03 per cent) or acetamiprid (0.15 per cent) can control the pest.

8.2.4 Pod Borer (*Adisura* sp., *Heliothis armigera*)

Sometimes it becomes a serious pest of beans, especially of *Dolichos* bean. They first feed on the surface of the pods, bore into them and feed on the seeds. Spraying of endosulfan at the rate of 2 ml per litre of water can control this pest effectively. Singh and Singh (2015) observed least per cent pod damage, highest pod yield and benefit: cost ratio by application of *Bacillus thuringiensis* 5 WG (0.025 per cent) and fipronil 5 SC (0.015 per cent)

8.2.5 Bean Weevil (*Bruchus* sp.)

This pest infests seeds in the store, and damages the quality both for consumption and sowing purposes. It can be controlled by fumigating the material under airtight condition with phosphine gas available in the form of tablets, *e.g.*, celphos, phosfume *etc.* It should be applied at the rate of 1-2 tablets per tonne of material or per cubic metre of space. Visarathanonth *et al.* (1990) recorded that pirimiphos-methyl at 10 ppm was the most effective in controlling the insect up to 36 weeks in store while chlorpyrifos-methyl, fenitrothion and methacrifos at both rates and 5 ppm pirimiphos-methyl were effective from 20 to 28 weeks.

9.0 Seed Production

There is no difference between the methods of raising crop for production of green pods and seeds. In case of seed crop, land in which one cultivar of cowpea was grown in the previous year should not be used for growing another cultivar in the following year to avoid contamination with the self sown plants from the previous crop. Cowpea is a self-pollinated crop. Maximum numbers of flowers open between 7 and 9 am (Krishnaswamy *et al.*, 1945). An isolation distance of 50 m for the production of foundation seed and 25 m for the production of certified seed should be provided between fields of two cultivars (Anon., 1971). A minimum of two inspections shall be made, the first before flowering and the second at flowering and pod formation stage. Ripe and dry pods are harvested by hand picking or by cutting the plants in case of last flush. To avoid shattering of seeds, harvesting should be done when half to two-thirds of pods have matured. Some cultivars shatter more than others. This problem is not encountered with the cultivars having fleshy, inflated pods. Threshing is done by beating with a stick or by a thresher. When cowpea is grown for seed, extreme care should be taken during threshing to prevent injury to the seed. Seeds maintain viability for two years under normal storage conditions. Vegetable cultivars of cowpea are less seed yielder than pulse types. Seed yield varies from 10 to 12 q/ha depending on the cultivar. Field and seed standards developed by the Central Seed Certification Board, Government of India for cowpea is given in Table 3.

Table 3: Standards for Foundation and Certified Seed of Cowpea

Contaminants/Factors	Foundation Seed	Certified Seed
Field standards		
Field of other varieties (m)	10.0	5.0
Fields of the same variety not conforming to varietal purity requirements for certification (m)	10.0	5.0
Off types (per cent)	0.10	0.20
Plants affected by seed borne diseases (per cent)	0.10	0.20
Seed standards		
Pure seed (min.) (per cent)	98.0	98.0
Inert matter (max.) (per cent)	2.0	2.0
Other crop seeds (max.) (per cent)	Nil	10.0/kg
Total weed seeds (max.) (per cent)	Nil	10.0/kg
Other distinguish varieties (per cent)	5.0/kg	10.0/kg
Germination including hard seeds (min.) (per cent)	75.0	75.0
Moisture (max.) (per cent)	9.0	9.0
For vapour-proof containers (maximum) (per cent)	8.0	8.0

10.0 Crop Improvement

Different National and International Research Organizations notably the IITA has been actively developing improved cowpea cultivars with high yields, early maturity, pest and disease resistance (Boukar *et al.*, 2018). The needs for agronomic traits differ from one region to another. For example, in the tropical regions, earliness, erect growth habit, resistance to biotic stresses (insect pests, pathogens, weeds), drought tolerance, high and stable seed yield, high harvest index, and appropriate seed quality were outlined as required cowpea agronomic traits (Abadassi, 2015; Boukar *et al.*, 2018). Detailed characterization of subspecies *unguiculata, sesquipedalis* and *cylindrica* of *V. unguiculata* was done employing wide array of genotypes based on different growth, leaf, root, nodule, flower, pod, seed and stomatal characters in several set of experiments (Hazra, 1991; Hazra *et al.*, 1993, 1996a; Chattopadhyay, 1995). According to Singh *et al.* (2016), pure Line, pedigree and recurrent selection could be followed in cowpea improvement using different crosses for utilizing additive, dominance, additive × additive, additive × dominance and dominance × dominance effects for different traits. Most breeding programmes use conventional and molecular breeding tools to harness cowpea genetic variation for breeding. Hall (2012) discussed the need for breeding for water-use efficiency, deeper rooting, and heat tolerance. One of the most efficient ways to study and breed cowpea for adaptation to drought is by optimal cycle lengths determination. This is suitable for cowpeas growing under rain-fed conditions in semi-arid zones through a hydrologic budget analysis method outlined by Hall (2012).

The International Atomic Energy Agency (IAEA) has been supporting member states in genetic improvement of various crops including cowpea through the use of artificial mutagenesis such as gamma rays, X-rays, and ethyl methanesulphonate (EMS) (Jain, 2005; Maluszynski *et al.*, 2000). This has led to the development and release of improved cowpea cultivars in Africa, Asia, and Latin America with different agronomic traits of interest. The application of induced mutation breeding techniques in cowpea has increased in most countries across Africa as a faster way to enhance genetic variation (Goyal and Khan, 2010; Singh *et al.*, 2013). Induced mutation fit well in cowpea genetic enhancement as most cowpea breeding initiatives aim at broadening the genetic bases of the crop to adapt to various cropping systems and agro-ecologies, and also in the development of consumer-preferred varieties (Lima *et al.*, 2011; Singh *et al.*, 2003). Despite all the previous efforts, there is still a big gap in cowpea improvement in order to increase productivity. Therefore, it is recommended for future and current cowpea breeding efforts to be geared towards traits such as earliness, erect growth habit, resistance to insect pests, pathogens and weeds, drought tolerance, high and stable pod and seed yield, high harvest index, and appropriate seed quality (Abadassi, 2015).

To alleviate the devastations caused by numerous cowpea production constraints, breeding programs across the globe are implementing both molecular and conventional breeding to develop improved lines with high grain yield potential, resistance to biotic stresses, tolerance to abiotic factors, adaptation to major production agro-ecologies, and traits preferred by consumers and producers. Sources of genes for several of these traits have been identified through screening of the germplasm available in different countries. The International Institute of Tropical Agriculture (IITA) is maintaining in its genetic resources center about 15,000 accessions of cultivated cowpea and more than 2000 wild relatives. Mining these resources has resulted in the identification of several sources of resistance to biotic and abiotic stresses. Several authors have reported on those germplasm lines that are important sources of resistance for use in breeding programs (Ferry and Singh, 1997; Singh, 2002; Boukar *et al.*, 2015). These genetic sources of desirable traits have been used in hybridization programmes to generate several segregating populations, which were used to select plants with good combinations of target traits (high yield potential, resistance to biotic and abiotic stresses, and consumer preferences). Different breeding methods applicable to self-pollinated crops are employed in cowpea genetic improvement including mass selection and pure line breeding, pedigree selection, single seed descent, bulk selection, backcrossing, mutation breeding, and farmer-participatory varietal selection. Generally, combinations or modifications of these breeding methods are also adopted as necessary.

Cowpea is a major source of protein, minerals, and vitamins in the daily diets of the rural and urban masses in the tropics, particularly in West and Central Africa where it complements the starchy food prepared from cassava, yam, sorghum, millet, and maize. Systematic efforts have begun at IITA and a few other institutions to develop improved cowpea varieties with enhanced levels of protein and minerals combined with faster cooking and acceptable taste. Singh *et al.* (2006) screened 52 improved and local cowpea varieties to estimate the extent of genetic variability for protein, fat, and minerals. On a fresh weight basis (about 10 per cent moisture), the protein content ranged from 20 to 26 per cent, fat content from 0.36 to 3.34 per cent, iron content from 56 to 95.8 ppm, and manganese content from 5 to 18 ppm. The improved cowpea varieties IT89KD-245, IT89KD-288, and IT97K-499-35 had the highest protein content (26 per cent), whereas the local varieties like Kanannado, Bauchi early, and Bausse Local had the lowest protein content (21 to 22 per cent). Appropriate crosses have been made to study the inheritance of protein, fat, and iron contents and to initiate a breeding program for improving these quality traits. There have been earlier reports on the extent of genetic variability for quality traits in cowpea. Hannah *et al.* (1976) reported high methionine content in Tvu 2093 and Bush Sitao (3.24-3.4 mg/g) dry seeds compared to 2.75- 2.88 mg/g seeds of the check variety, G-81-1.

Cowpea is highly self-pollinated in most environments, the result of a cleistogamous flower structure and simultaneous pollen shed and stigma receptivity. Like any other legume pollination and getting good set is very difficult in cowpea which hinders the breeding programmes. High relative humidity improved pod production following cross pollination using a technique involving pollen deposition with dehisced anthers between 08.00 and 12.00 h. The most successful pollination treatment involved the application of pollen using a small brush at dawn between 05.45 and 07.00 h following emasculation the previous afternoon (Venter, 1996). An alternative method is to collect open flowers of the male parent in the early morning, refrigerate them at 4 to 10°C until late afternoon. At this time, mature but unopened floral buds are chosen, emasculated, and then pollinated with the male flowers (Ehlers and Hall, 1996).

The extent of genetic variability and diversity in a crop is of paramount importance to plant breeders for developing new and improved cultivars with desirable characteristics which includes both farmers' and breeders' preferred traits. It is reported that genetic diversity in cowpea has declined over the past years due to various biotic and abiotic factors (Fang *et al.*, 2007). Some farmers reported loss of their local varieties overtime due to frequent droughts, damage by insect pests both in the field and in storage (Horn *et al.*, 2015; Stejskal *et al.*, 2006). Gbaguidi *et al.* (2013) reported loss of genetic diversity in African cowpea at an increasing rate of 28 to 60 per cent in some agro-ecologies. It is postulated that artificial selection for better performing varieties could have accelerated the loss of genetic diversity because of negative selection against poor performing types from a narrow genetic base. In addition, genetic variation also is restricted within specific breeding programmes in the absence of complementary pre-breeding programmes (Gbaguidi *et al.*, 2013). Studies on germplasm collected from North America, Asia and Africa revealed a narrow genetic base of cowpea (Fang *et al.*, 2007). The same studies further have shown a strong genetic relatedness among germplasm collections of USA and Asia with that of African cowpea collections. Well-characterized germplasm is useful to incorporate economic traits through designed crosses. Genetic diversity is routinely assessed using agro-morphological or phenotypic markers. In cowpea breeding, both quantitative and qualitative phenotypic characters are extensively used in germplasm characterization, classification and selection (Hazra *et al.*, 1993, 1996a; Chattopadhyay *et al.*, 1996; Molosiwa *et al.*, 2016). Quantitative traits include number of branches per plant, days to 50 per cent flowering, days to 50 per cent maturity, number of pods per plant, pod length, pod width, seed weight, number of seeds per pod and seed yield (Molosiwa *et al.*, 2016). Use of phenotypic characteristics is a common approach because they form the most direct measure of the phenotype, readily available and relatively cheaper requiring simple equipment. However, phenotypic markers are subject to environmental influences in the field that may mask the real genetic variation among genotypes. However, molecular marker techniques are regarded as powerful in determining the genetic diversity and finger-printing of germplasm. Boukar *et al.* (2016) reported on linkage maps used to identify quantitative trait locus (QTLs) for desirable traits in cowpea. QTLs are important in identifying molecular markers (such as SNPs or AFLPs) that correlate with an observed trait in breeding. Once the genetic markers that define the QTL have been identified, breeders can use them to select individuals with desired QTL. Some advanced breeding programme in Africa makes use of Effective Field-based High-throughput Phenotyping Platforms (HTPPs) which are robust plant breeding selection programmes (Araus and Cairns, 2014).

In cowpea, green pod yield is the most important character to be considered in its improvement. Pod yield of vegetable cowpea is complex, highly variable and is associated with a number of component characters (Ullah *et al.*, 2011).

Many workers reported that the characters like primary leaf area, leaf dry weight, number of pods per plant, number of branches per plant, pod length, pod width, number of seeds per pod and 100 seed weight should be considered by a plant breeder while selecting a plant type for getting maximum vegetable pod yield (Jana *et al.*,1982; Hazra, 1991; Chattopadhyay *et al.*, 1997; Lal *et al.*, 2007; Nath *et al.*, 2009; Udensi *et al.*, 2012; Chattopadhyay *et al.*, 2014). Out of these characters studied by them number of pods per plant, pod yield per plant, pod weight, number of seeds per pod, 100 seed weight and pod length attained the highest broad sense heritability and genetic advance values. Significant variation in both pod and seed protein contents among the genotypes belonging to three cultigroups indicated the possibility of improving protein contents for both pods and seeds (Hazra *et al.*, 1996b). However, protein content of green pods did not show significant correlation with pod yield in their studies. Thus, for improving protein yield/plant, selection of high pod yielding genotypes with appreciable protein content from a wide gene pool is feasible. The genotypes could be screened for protein content in the early stages of pod development. Walle *et al.* (2018) concluded that seed yield in cowpea can be improved by focusing on traits pod length, seed length, seed thickness, seed width, biomass and harvest index.

Genotype by environment interaction (G × E) is a differential response of genotypes when grown across environments. Multi-environmental trials (METs) are required to quantify the magnitude of G × E interaction and to recommend varieties with narrow or broader adaption (Ramburan *et al., 2012*). G × E interaction has the advantage to crop improvement that targets broad adaptation, but it can also represent opportunities to genetic improvement for specific sites (Annicchiarico *et al.*, 2010). Chattopadhyay *et al.* (2001) and Shiringani and Shimelis (2011) determined the relative yield response and stability among selected improved cowpea genotypes to make recommendation for wide or specific adaptation through G × E analysis and results showed significant interactions among genotypes, planting dates and locations for green pod and seed yield, respectively. Research on the occurrence and molecular mechanisms of phenotypic plasticity and G × E in plant populations was carried out by Des Marais *et al.* (2013), and revealed that G × E was often caused by changes in the magnitude of genetic effects in response to the environment, and associated with diverse genetic factors and molecular variants.

Through selection from exotic collections and hybridization, some useful cultivars have been developed in India. The cultivar Pusa Phalguni was selected from the Canadian cultivar Dolique Du Tonkin, whereas the cultivar Pusa Barsati was selected from a collection from the Philippines (Pal *et al.*, 1956). The cultivar Pusa Dofasli was developed from a cross Pusa Phalguni × Philippines selection at I.A.R.I New Delhi (Singh *et al.*, 1968). Photoinsensitive cv. Pusa Komal, also developed at I.A.R.I., is a product of cowpea strains P-85-2 and P426 (Singh and Patel, 1990). Two heterotic combinations of *unguiculata × sesquipedalis* and *cylindrica × sesquipedalis* have been advanced up to late generation following modified back cross-pedigree method to develop two vegetable cowpea varieties, 'Bidhan Barbati-1' and 'Bidhan Barbati-2' which were released from West Bengal, India (Hazra *et al.*, 2001). Recently a bushy cultivar Bidhan Sadabahar (BCCP-3) was developed from a cross Bidhan Barbati-1 × Kashi Kanchan followed by pedigree selection at BCKV, West Bengal, India (Anon., 2018). In Brazil, direct and reciprocal crosses were made between the cultivars CNC 0434 and T Vu 612 and seedlings in the Fa, F and F5 were inoculated in the field with a mixture of cowpea viruses. Subsequently, a line BRlO-Piaui immune to multiple resistance to viruses with high yield was evolved by Santos *et al.* (1987).

In a series of cross combinations, manifestation of considerable heterosis up to 72 per cent over respective better parents could be realized particularly for pod yield and pod number/plant (Hazra, 1991; Chattopadhyay *et al.*, 2003). Genetic divergence of the parents did not all the time reflect the manifestation of heterosis (Hazra, 1991).

The success of crossing was generally low, in the range 0-45 per cent (Hazra, 1991; Chattopadyay, 1995) and many reasons for poor crossing success are proposed from histological and cytological points of view (Ojomo, 1971). High pollen fertility was recorded in the cultigroups but pollen germination was generally poor, indicating some restriction at pre-fertilization stage which may be one of the reasons of poor success (Hazra *et al.,* 1990). Less fluctuating and more equitable temperatures coupled with intermediate relative humidity conditions in the spring-summer season in West Bengal, India gave somewhat greater success than in winter or in the rainy season; genotypes in the three cultivar groups, Biflora, Unguiculata and Sesquipedalis were crossable under these conditions (Hazra, 1991; Chattopadyay, 1995). Nevertheless, success varied widely between genotypes within groups. Wide genetic divergence of the parents also affected crossability adversely. There was evidence for unilateral incongruity between particular members of the Sesquipedalis cultigroup and of the other two cultigroups (Hazra *et al.,* 2000).

Several studies indicated predominance of additive gene effects for both pod length and weight and these characters could be improved by simple breeding scheme such as pedigree method (Hazra *et al.,* 1994). However, partial dominance of short and light pod of Unguiculata and Biflora genotypes over long and heavy pods of Sesquipedalis genotypes indicated that there would be less likelihood of recovering Sesquipedalis-like long podded segregates in the advanced generations (Hazra *et al.,* 1993). Additive genetic variance was more important for both pod and seed protein contents (Hazra *et al.,* 1996b).

There is a scope for its improvement, especially in respect of plant ideotype. Breeder may produce determinate cultivars which are highly productive utilizing determinate growth from subsp. *cylindrica* and desirable seed types from subsp. *unguiculata* (Steele, 1976). The inclusion of subsp. *dekindtiana* may be utilised in breeding for resistance to different pathogens and insect pests (Steele, 1976). He further reported that the protein content of seeds can be increased (the range reported is 22-35 per cent) and its nutritive value could be improved by increasing the proportion of sulphur containing amino acids. Combination of erect, determinate, and early maturity habits of the genotypes of cultigroup Unguiculata (ssp. *unguiculata*) or Biflora (ssp. *cylindrica*) with pod length, succulence and fleshiness of the genotypes of cultigroup Sesquipedalis (ssp. *sesquipedalis*) to develop relatively short, non-viny types with earliness, synchrony in pod bearing and medium-long fleshy pods would be an ideal proposition for vegetable cowpea breeding as proposed by Hazra (1991). Hazra (1991) and Chattopadhyay (1995) also observed that repeated back crossings with the recurrent Sesquipedalis parent did not prove worthwhile as the dominant viny character was transferred to the progenies. On the other hand, pedigree selection from inter-cultigroup cross combinations could not recover non-viny segregates with medium-long pods of good quality. Hazra (1991), Som and Hazra (1993) and Chattopadhyay (1995) proposed advancement of generations from highly heterotic inter-cultigroup cross combinations (Unguiculata × Sesquipedalis or Biflora × Sesquipedalis) by modified backcross-pedigree method. In this breeding method the F_1 of such cross combination was backcrossed once with Sesquipedalis parent followed by subsequent pedigree selection up to seven generations (Hazra, 1991; Som and Hazra, 1993).

In the strategy for improving protein yield, selection of high-yielding genotypes with appreciable protein content in pods and seeds should be encouraged and for effective selection, a wide gene pool needs to be developed. Maternal inheritance is widely reported for protein content of different legumes so high protein parent should be kept as female in hybridization program. In the proposed modified backcross-pedigree method of breeding (Hazra, 1991), the segregates need to be tested for plant frame, pod yield and protein content to strike a proper balance between pod yield and protein content.

10.1.0 Physiological Traits

Cowpeas are rarely insensitive to photoperiod; they are typically quantitative short-day plants in which floral bud (f) is delayed when photoperiod (P) is longer than the critical photoperiod (Pc). Therefore, in order to quantify genotypic variation in temperature sensitivity, genotype was regressed against the mean trial f in circumstances where P < Pc (*i.e.* approximately 13 h d-1) and mean temperature (T) was between 19 and 28°C. Correspondingly, in order to assess genotypic variation in photoperiod sensitivity, trials in which T was near optimal (25-28°C) and P ranged from 10 to 14.5 h d-1 were used, (Craufurd *et al.*, 1996a). Ehlers and Hall (1996) classified the genotypes into 11 groups based on photoperiod response (change in position of the first reproductive node or period to appearance of floral buds in long compared with short days), juvenility (minimum period for appearance of floral buds when grown under short days), and suppression of floral bud development and pod set under hot, long days. This classification system will aid breeders and agronomists in their understanding of the genetic variation for these characteristics and in choosing genotypes with appropriate juvenility, photoperiod response, and heat tolerance for breeding and agronomy programmes serving tropical and subtropical production environments.

The stability of twenty one genotypes of cowpea, comprising landraces and varieties, grown in 22 photothermal environments in Nigeria and Niger, West Africa, indicated no significant differences between genotypes in temperature sensitivity, but revealed significant differences in photoperiod sensitivity. Regression coefficients from the stability analysis were strongly correlated (r = 0.94, 19df) with a photoperiod sensitivity constant, 'c', determined from a photothermal flowering model (Craufurd *et al.*, 1996b). Ohler and Mitchell (1996) manipulated photoperiod and harvesting of cowpea (*Vigna unguiculata*) canopies were manipulated to optimize productivity for use in future controlled ecological life-support systems. They measured productivity by edible yield rate (EYR; edible shoot DW per unit area per day), shoot harvest index (SHI; edible shoot DW as a percentage of total shoot DW) and yield-efficiency rate (YER; EYR/non-edible shoot DW). Breeding lines IT84S-2246 (S-2246) and IT82D-889 (D-889) were grown in a greenhouse under 8-, 12- or 24-h photoperiods. S-2246 was short-day and D-889 was day-neutral for flowering. Photoperiod did not affect EYR of either breeding line for any harvest treatment. However, photoperiod had no effect on SHI or YER for D-889 for any harvest scenario. Breeding for selection for heat need not require very sophisticated growth chambers. Selecting heat-tolerant cowpea genotypes for high pod yields under field conditions are as effective as using growth chambers with controlled environments (Marfo, 1996).

Studies on variability in stomatal frequency, stomatal length and breadth on the adaxial and abaxial leaf surfaces of 7, 12 and 6 genotypes of the *V. unguiculata* cultivar groups *Unguiculata*, *Biflora* and *Sesquipedalis*, repectively revealed that the cowpea genotypes were significantly different for all the stomatal characters. Association of high heritability and moderately high genetic advance for adaxial stomatal frequency suggested its probable control by additive gene action (Hazra *et al.*, 1996b).

Breeding for drought adaptation by selecting carbon isotope discrimination (DELTA) may be a useful selection criterion because of its correlation with transpiration efficiency. However, effectiveness of indirect selection will depend on the realized heritability of DELTA and genetic correlations with other traits contributing to yield (Menendez and Hall, 1996). Selection for low DELTA would be more efficient in families in advanced generations rather than in single F_2 plants, and could result in some indirect selection for low HI (Menendez and Hall, 1996). Through molecular techniques, researcher identified RNA sequences for drought-sensitive and drought-tolerant effects with two cowpea genotypes (CB46, drought sensitive, and IT93K503-1, drought-tolerant) (Barrera-Figueroa *et al.*, 2011).

10.2.0 Disease Resistance

More than 40 species of fungi have been reported to cause diseases in cowpea (Bailey *et al.*, 1990). The most destructive fungal disease of cowpea includes leaf smut (false smut or black spot), caused by *Protomycopsis phaseoli* (Bailey *et al.*, 1990). Fungal diseases cause leave smut, stem rot, as well as root rot (Bailey *et al.*, 1990). Yield losses due to serious epidemics were reported in Nigeria, the Sudan savanna and Sahel (Adejumo *et al.*, 2001). The losses ranged from 20 to 100 per cent (Mbeyagala *et al.*, 2014). Sources of resistance to fungal pathogens have been identified, and screening techniques developed for: anthracnose (Adebitan *et al.*, 1992), cercospora leaf spot (Fery and Dukes, 1977), septoria leaf spot, rust (*Uromyces appendiculatus* (Pers.: Pers.) Unger), and brown blotch (Abadassi *et al.*, 1987; Adebitan *et al.*, 1992). Ashy stem blight is important in Africa and India but to-date no genetic resistance to this disease has been identified. Bacterial blight is an important disease in the southeastern US, and sources of resistance are known, but little progress has been made in incorporating resistance to this disease into commercial cultivars (Patel, 1985).

10.2.1 Rust Resistance

Breeding for resistance to rust has been attempted seriously by many workers. Based on logistic and Gompertz growth rates and area under disease progress curve (AUDPC) values several genotypes including V38, APC813, APC83 and V17 were identified as possessing favourable slow rusting behaviour (Cherian *et al.*, 1996a). In a subsequent study they found that the varieties IT84D-449, IT86D-364, IT86D-373, IT86D-498, IT86D-1038, IT87D-1827, IT87S-1390, IT87S-1393, IT87S-1459 and IT845-2246 possessed good slow rusting resistance (Cherian *et al.*, 1996b). Ryerson and Heath (1996) studied the nature of the cowpea rust resistance genes, present in resistant cultivars, by macroscopic and microscopic examination of the interaction between race 1 of *Uromyces vignae* [*U. appendiculatus*] and the progeny of a cross between a resistant and a susceptible cowpea cultivar. The monokaryotic and dikaryotic forms of the fungus were investigated on all plants. The different levels and inheritance of resistance patterns, shown by the F_2 generation and subsequent progeny, suggested the presence of multiple genes and the presence of dominant and recessive resistance components. Some of the resistance elements acted preferentially toward the monokaryon. Full elucidation of the different plant phenotypes is possible after microscopic examination, as some cytologically different interactions with the fungus appeared identical macroscopically (Ryerson and Heath, 1996).

10.2.2 Resistance to other Diseases

Leina *et al.* (1996) observed a distinct host-specific interaction among the different species of *Pseudocercospora*. Cultivar specific interactions were most pronounced between cowpeas and *P. cruenta*. A direct correlation occurred between the variation in peroxidase activity in the soluble fraction of inoculated leaves and resistance to infection in okra and cowpea cultivars. The soluble fraction of inoculated leaves had higher peroxidase activity than either mitochondrial or chloroplast extracts.

Susceptibility of cowpea genotypes to dry root rot caused by *Macrophomina phaseolina* increased with increasing soil moisture stress. Burman and Lodha (1996) studied water relation parameters of the healthy and diseased plants of cowpea genotypes at the time of mild and severe moisture stress (20-40 days after sowing). Shoot water potential decreased significantly in the healthy and diseased plants of susceptible genotype ARS Durgapura compared with resistant V-265. Higher susceptibility of ARS Durgapura to *Macrophomina* infection was associated with impairment of its water uptake processes.

Adebitan and Olufajo (1998) reported that the variety IAR7/180-4-5 showed multiple disease resistance to scab (*Elsinoe phaseoli*), anthracnose (*Colletotrichum lindemuthianum*) and bacterial blight (*Xanthomonas campestris* pv. *phaseoli*). New cultivars developed by the International Institute of Tropical Agriculture, Nigeria is resistant to anthracnose (*Colletotrichum lindemuthianum*), web blight (*Rhizoctonia solani*), brown blotch (*Colletotrichum capsici*), Cercospora leaf spots (*Cercospora cruenta* (*Mycosphaerella cruenta*) and *Cercospora canescens*), Septoria leaf spot (*Septoria vignae*), bacterial blight (*Xanthomonas campestris* pv. *vignicola*, Cowpea yellow mosaic virus, Blackeye cowpea mosaic virus, Southern bean mosaic virus and Cowpea aphid-borne mosaic virus are 'Vuli-1' (Mligo and Singh, 2007), 'NGVU-05-25' (Singh *et al.*, 2006), Ayiyi (AsafoAdjei and Singh, 2005), Korobalen (Toure and Singh, 2005). In India, Dhiman *et al.* (1989) found appreciable multiple resistance to bacterial blight (*Xanthomonas campestris* pv. *vignicola*), cowpea mosaic comovirus and anthracnose (*Colletotrichum lindemuthianum*) in the variety Sel-263. The International Institute of Tropical Agriculture has also developed improved cowpea germplasm lines, namely IT90K-59, IT97K-205-8 and IT97K-499-35 with combined resistance to the parasitic plants *Striga gesnerioides* and *Alectra vogelii* and also resistant to major diseases (Singh *et al.*, 2006). Some sources of resistance to different fungal and bacterial diseases of cowpea identified by several workers are given in Table 4.

Table 4: Sources of Resistance to Major Diseases of Cowpea*

Disease	Source of Resistance	Reference
Anthracnose (*Colletotrichum spp.*)	TV × 3236TE 97-411	Latunde-Dada *et al.* (1999); Barreto *et al.* (2007)
Cercospora	IT89KD-288, IT97K-1021-15, IT97K-463-7, IT97K-478-10, IT97K-1069-8, IT97K-556-4TVx 3236	Singh (1998), and Singh (1999); Latunde-Dada *et al.* (1999)
Macrophomina	L-198 and CN × 377-1EV-265	Rodriguez *et al.* (1996); Uday and Lodha (1996)
Aspergillus flavus	IT82E-16 and IT81D-1032	Zohri (1993)
Smut	IT97K-556-4, IT95K-1090-12, IT95K-1091-3, IT95K-1106-6IAR-48, IT97K-506-6	Singh (1998), and Singh (1999)
Rust	IT97K-1042-8, IT97K-569-9 (Uromyces), IT97K-556-4, IT97K-1069-8, IT95K-238-3, IT97K-819-118, IT90K-277-2, IT97K-1021-15, IT96D-610, IT86D-719	Singh (1998), and Singh (1999)
Septoria	Tvu 12349, Tvu11761, IT95K-398-14, IT90K284-2, IT95K-1090-12, IT97K-1021-15, IT98K-205-8, IT98K-476-8, IT97K-819-118, IT95K-193-12, Tvu 1234, IT95K-1090-12	Singh (1998), and Singh (1999)
Scab	IT98K-476-8, IT97K-1069-8, TVx 3236, IT95K-398-14, IT97K-1021-15, IT95K-1133-6Kvu 46, Kvu 39, and Kvu 454	Singh (1998), and Singh (1999), Nakawuka and Adipala (1997)
Ascochyta	Tvu 11761	Singh (1998), and Singh (1999)
Powdery mildew	C 7, C 200, C 265, C 347 and C 402	Mishra *et al.* (2005)
Bacterial blight	IT95K-398-14, IT95K-193-12, IT81D-1228-14, IT95K-1133-6, IT97K-556-4, IT97K-1069-8, IT90K-284-2, IT91K-93-1, IT91K-118-20, IT90K-284-2, IT91K-93-10, and IT91K-118-20	Singh (1998), and Singh (1999), Wydra and Singh (1998)
Multiple resistance	Vuli-1NGVU-05-25AyiyiIT90K-59, IT97K-205-8 and IT97K-499-35IAR7/180-4-5	Mligo and Singh (2007), Singh *et al.* (2006), AsafoAdjei and Singh (2005); Singh *et al.* (2006)

*Adebitan and Olufajo (1998).

10.2.3 Virus Resistance

Cucumber mosaic virus (CMV) and blackeye cowpea mosaic virus (BlCMV) interact synergistically in dually infected plants of cowpea (*Vigna unguiculata* subsp. *unguiculata*) to cause cowpea stunt disease, the most damaging viral disease of this crop in the USA (Gillaspie, 2001). The most important factors that constrain cowpea production in the northeastern region of Brazil are the virus diseases, caused mainly by cowpea severe mosaic virus (CSMV) of the group Comovirus, cowpea aphid borne mosaic virus (CABMV) of the group Potyvirus, cucumber mosaic virus

(CMV) of the group Cucumovirus, and cowpea golden mosaic virus (CGMV) of the group Geminivirus (Lima and Santos, 1988). Cucumber mosaic virus (CMV) and blackeye cowpea mosaic virus (BlCMV), are reported as the major viruses occurring in asparagus beans (*Vigna unguiculata* subsp. *sesquipedalis*) in Taiwan (Chang *et al.*, 2002). From Pakistan 7 viruses known to be seedborne in cowpea: blackeye cowpea mosaic potyvirus (BlCMV), cowpea aphid-borne mosaic potyvirus (CABMV), cucumber mosaic cucumovirus (CMV), cowpea mosaic comovirus (CPMV), cowpea severe mosaic comovirus (CSMV), cowpea mottle virus (CPMoV) and southern bean mosaic sobemovirus (SBMV) have been identified (Bashir and Hampton, 1993). Shoyinka *et al.* (1997) could identify six viruses, cowpea aphid-borne mosaic potyvirus (CAMV), blackeye cowpea mosaic potyvirus (BlCMV), bean southern mosaic sobemovirus (SBMV), cowpea mottle carmovirus (CMoV), cowpea (yellow) mosaic comovirus [cowpea mosaic comovirus] (CpMV) and the cowpea strain of cucumber mosaic cucumovirus (CMV-CS) throughout all agroecological zones in Nigeria. In Sri Lanka, blackeye cowpea mosaic potyvirus (BlCMV) the seedborne, aphid-transmitted virus cause severe and prevalent disease of vegetable cowpea (Jeyanandarajah and Brunt, 1996).

Bashir and Hampton (1996) tested cowpea cultivars and lines (51) by mechanical inoculation against 7 geographically and pathogenically diverse isolates of blackeye cowpea mosaic potyvirus (BlCMV), to identify genetic resources with comprehensive BlCMV resistance. The diversity among BlCMV isolates was illustrated by the range of responses to inoculation among cowpea genotypes, many of which were either immune to or tolerant of individual BlCMV isolates.

Most cowpea cultivars are resistant to infection by cucumber mosaic cucumovirus strain Y (CMV-Y) that causes a hypersensitive reaction followed by the development of necrotic local lesions (Nasu *et al.*, 1996). It has been anticipated that resistance to CMV-Y depends upon the existence of a resistance (R) gene in cowpea. Cowpeas were screened to isolate susceptible cultivars. Of 38 cultivars tested, PI 189375 was identified as susceptible based on symptoms and the presence of progeny virions on noninoculated upper leaves. Protoplasts prepared from PI 189375 and the resistant cultivar Kurodane Sanjaku supported virus multiplication. As there was no difference between the capacities for virus multiplication in the 2 cultivars at a single-cell level, it is suggested that the specific R gene is expressed intact in Kurodane Sanjaku plants. F_1 and F_2 populations were then made by reciprocal crosses between Kurodane Sanjaku and PI 189375. Reactions of F_1 and F_2 populations to CMV-Y inoculation demonstrated that the R gene was inherited as a single dominant gene. The resistance locus was designated as *Cry* (cowpea R gene to CMV-Y).

Resistance to CSMV, CABMV, and CGMV has already been incorporated in some of the released varieties like BR 10-Piaui (Santos *et al.*, 1987), BR 12-Canindé (Cardoso *et al.*, 1988), BR 14-Mulato (Cardoso *et al.*, 1990), BR 17-Gurguéia (Freire Filho *et al.*, 1994), EPACE 10 (Barreto *et al.*, 1988), Setentão (Paiva *et al.*, 1988), BR 16-Chapeo-de-couro (Fernandes *et al.*, 1990) and Bidhan Barabati-1, Bidhan Barbati-2 and Bidhan Sadabahar (Hazra *et al.*, 2001).

10.3.0 Insect Resistance

Considerable progress has been made in developing cowpea varieties resistant to several insects. Pandey *et al.* (1995) reported TVu 908 to be resistant to leaf beetles and the IC20533 from Madhya Pradesh showed least damage by semilooper (*Plusia nigrisigna*). Prasad *et al.* (1996) selected six cowpea genotypes resistant to various insect pests (TVX7 against *Heliothis armigera* [*Helicoverpa armigera*], GC82-7 against *Maruca testulalis* [*M. vitrata*], DLPC216 against *Psidia tikora* [*Cydia tychora*] and C152, C190 and GC82-7 against bruchids). Singh *et al.* (1996) reported several improved cowpea varieties with combined resistance to aphid, thrips, and bruchid. Of these, IT90K-76, IT90K-59, and IT90K 277-2 are already popular varieties in several countries. Among the new varieties IT97K-207-15, IT95K-398-14, and 98K- 506-1 have a high level of bruchid resistance (Singh, 1999). From Iran, cowpea cultivar 'Kamran' showed the lowest (1.25 per cent) infestation level for *Callosobruchus maculates* under field condition (Ghadiri and Sohrabi, 1999). In India, two cowpea mutants, ICV11 and ICV12, resistant to *A. craccivora*, were developed from the M_3 generation of ICV1 seeds irradiated with 20 kR gamma rays (Pathak, 1988). Nkansah and Hodgeson (1995) confirmed resistance of TVu 801 and TVu 3000 to the Nigerian aphid strain but found that the two lines were susceptible to aphids from the Philippines indicating the existence of different aphid strains. The mechanism of resistance in the resistant cowpea varieties TVu62, TVu408, TVu2740, TVu3273, TVu3509 and TVu9944 to *Aphis craccivora* was investigated with artificial infestations in screened cages and it was found to include antibiosis, manifested as high mortality of nymphs, reduced weights, shortened lifespan and low fecundity of adults (Ofuya, 1988).

Only low levels of resistance have been observed for *Maruca* pod borer and pod bugs, which cause severe damage and yield reduction in cowpea. Selection and breeding for cowpea accessions, to combine such resistance with morphological, biochemical and biophysical traits could enhance the low levels of resistance and ultimately lead to the effective management of this pest. Jagginavan *et al.* (1995) observed cowpea lines P120 and C11 to be least damaged by *Maruca*. Pubescence (trichomes) in wild and cultivated cowpeas (*Vigna vexillata* and *V. unguiculata*) adversely affected oviposition, mobility, and food consumption and utilization by the pyralid *Maruca testulalis*. Oghiakhe (1996) suggested that it would be advantageous to use TVnu 72 (*Vigna vexillata*, highly resistant and highly pubescent), in a breeding programme to incorporate pubescence into high-yielding commercial cultivars for resistance to *M. testulalis* and possibly other major pests. Jackai *et al.* (1996) evaluated a large number of accessions belonging to a few wild *Vigna* species (*V. unguiculata* subsp. *dekindtiana*), *V. oblongifolia* and *V. vexillata* were evaluated using choice (DCAT) and no-choice (NCFT) laboratory feeding bioassays to determine their resistance to *Maruca vitrata*. The most resistant accessions belonged to *V. vexillata*, followed by those from *V. oblongifolia*, with a few outstanding exceptions from *V. unguiculata* subsp. *dekindtiana*. It is suggested that both antibiosis (post-ingestive effects) and antixenosis (deterrence to boring into the pods to feed) mechanisms of resistance are involved. The results are discussed further in relation to the origin, domestication and use of these accessions in cowpea improvement. Veeranna and Hussain (1997) found TVX- 7 to be most resistant to *Maruca* and has a high density of trichomes (21.41/mm²). Veerappa (1998) screened 45 cowpea lines for resistance to *Maruca* pod borer and observed that the tolerant lines 29 had higher phenol and tannin contents compared to susceptible lines. This is in line with the general observation that cowpea varieties with pigmented calyx, petioles, pods, and pod tips suffer less damage due to *Maruca*. However, Oghiakhe *et al.* (1993) earlier observed that despite the differences in phenol concentration between cultivars, correlation showed that phenol does not play any significant role in cowpea resistance to *M. testulalis*. Tayo (1989) examined the anatomy of damaged and undamaged young stems, peduncles and young pods of 3 varieties of cowpea from transverse sections in order to evaluate an anatomical basis for resistance to the pod borer, *Maruca testulalis*. It was suggested that the smaller diameter and the relative abundance of strengthening tissues in the stem and peduncle of TVu 946, by limiting total ingestible biomass, might be restricting the damage to these organs by the pod borer. As indicated earlier, a distant wild relative of cowpea *Vigna vexillata* has shown high levels of resistance to pod borer (*Maruca vitrata*) and bruchid but all the efforts made at IITA to transfer *Maruca* resistance genes from *Vigna vexillata* to cowpea have not so far been successful.

The genetics of resistance to the aphid (*Aphis craccivora*) was investigated utilizing resistant cowpea lines Vs 350, Vs 438 and Vs 452 and it was found that resistance to the aphid was governed by a single dominant gene (Joseph and Peter, 2003). Githiri *et al.* (1996) studied the inheritance of aphid resistance and allelic relationships among sources of resistance in the parents, F_1, F_2, F_3, and backcross populations of cowpea (*Vigna unguiculata*) crosses using 8 resistant and one susceptible (Tvu 946) cultivars as parents. Each 4-day-old seedling was infested with five fourth-instar aphids. Seedling reaction was recorded 14-16 days after infestation when the susceptible check was killed. The segregation data from eight crosses between resistant and susceptible cowpea cultivars indicated that aphid resistance was inherited as a monogenic dominant trait. Segregation data from crosses among eight resistant cultivars indicated that one or two loci and modifier(s) were involved in the expression of resistance to aphids.

10.4.0 Nematode Resistance

Root-knot nematodes (*Meloidogyne* spp.) are serious pests of cowpea and many other crops worldwide. Several species of *Meloidogyne* are pathogenic to cowpea. Several sources of resistance to nematodes were identified including some of the improved breeding lines with high yield potential (Rodriguez *et al.*, 1996; Roberts *et al.*, 1996, 1997; Fery and Dukes, 1996; Ehlers *et al.*, 2000; and Singh, 1998). Some of the varieties with high yield and nematode resistance are IITA3, Habana-82, Incarita-1, IT86D364, IT87D14638, Vinales 144, P902 and IITA7, IT849-2049, IT89KD-288, IT86D-634, IT87D- 1463, IT95K-398-14, IT96D-772, IT96D-748, IT95K-222-5, IT96D-610, IT87K-818-18, and IT97K-556-4. Among these varieties, IT89KD-288 was found to be resistant to four strains of *Meloidogyne incognita* in USA (Ehlers *et al.*, 2000). Singh *et al.* (1996) found IT89KD-288 to be high yielding and highly resistant to nematodes in the trials conducted at Kano (Nigeria), where nematode attack is very severe in the dry season planting with irrigation. Screening of cowpea germplasm for additional resistance to the root-knot nematodes, *M.incognita* and *M.*

javanica, revealed an accession (IT84S-2049) from Africa with resistance to diverse populations of both root-knot species (Roberts *et al.*, 1996). Choudhury *et al.* (2005) could identify 19 cultivars as resistant out of 149 cultivars tested. Hemeng (1989) identified GH 438-76-1 as immune and IT 84S-2137 as highly resistant cultivars against *M. incognita*. From Pakistan, Khan and Husain (1989) reported the line IC-503 as moderately resistant to *Meloidogyne incognita*. Tender Cream is resistant to *Cercospora cruenta* (*Mycosphaerella cruenta*), southern blight (*Corticium rolfsii*), rust (*Uromyces phaseoli* [*U. appendiculatus*]) and powdery mildew (*Erysiphe polygoni*). In tests it showed a high level of resistance to the root-knot nematode *Meloidogyne incognita* (Fery and Dukes, 1996). California Blackeye 27 (CB27) has been developed by the University of California, Riverside for its heat tolerance and broad-based resistance to *Fusarium* wilt and root-knot nematodes (Ehlers *et al.*, 2000).

The non-viny vegetable cowpea cultivar Bidhan Barbati-2 showed resistance against *Meloidogyne incognita* under Rajasthan condition (Yadav and Ramesh Chand, 2003). Adegbite *et al.* (2006) found that root-gall levels were correlated negatively with number of pods and leaves and the cultivar IT84S2246-4 seemed to be the most resistant since it had a gall index of 1.5 and a reproduction factor of 0.45. Segregation of resistance to gene Rk-virulent *M. incognita* in progenies from IT84S-2049 × CB3 showed that resistance in IT84S-2049 is governed by one dominant nuclear gene 'Rk' (Roberts *et al.*, 1996). Resistance in IT84S-2049 is conferred by an additional dominant allele of the *Rk* locus, or by another gene locus very tightly linked to *Rk* within 0.17 map units. The symbol *Rk2* is proposed to designate this new resistance factor in IT84S-2049. Thus, Rk may be a complex nematode resistance locus, analogous to those reported for other plant pathogen-host combinations. Ehlers *et al.* (2000) found that additional resistance in the line, H8-8R is conferred by a single recessive gene which was independent of gene Rk, conferred partial resistance when expressed alone, and has an additive effect of increasing resistance in the presence of gene Rk. Line TVu4554 in the pedigree of H8-8R was identified as the probable donor parent of the recessive gene, for which we propose the gene symbol rk3. Sirohi and Dasgupta (1993) provided the evidence for the early induction of phenylalanine ammonia lyase, the first key enzyme of the phenypropanoid pathway, and it's relationship with resistance expression in cowpea cultivar C- 152 inoculated with *Meloidogyne incognita* race 1.

11.0 Biotechnology

11.1.0 Tissue Culture

In vitro techniques such as micropropagation have aided plant breeders, and proved useful for the propagation of sexually sterile hybrids and isolation of solid mutants. Shoot tip multiplication offers a direct approach for micropropagation. This process circumvents the callus stage, thus reducing the possibility of somaclonal variations. Shoot tips are a suitable explant source for cowpea micropropagation, and can be used for callus induction. Brar *et al.* (1997) induced shoot multiplication in cowpea cv. Georgia-21 using shoot tip explants. Shoot tips of 5 mm were isolated from *in vitro*-grown seedlings, and were cultured on Murashige and Skoog medium containing BA at 1.0, 2.5 or 5.0 mg/l or kinetin at 1.0, 2.5 or 5.0 mg/l combined

In-Vitro Propagation of Cowpea

**Multiple Shoot Induction and Plant Regeneration from
Cotyledonary Node Explants of Cowpea.**

a. Explant at the time of culture (Bar 1 mm), b. Shoot induction from axils of explant within 1
week, c. Shoot proliferation from mother explants in subsequent reculture, d. Proliferation of
multiple shoots within 4 weeks of culture, e. A rooted shoot after 2 weeks of culture, f. plant
acclimatized in greenhouse (Bakshi *et al.,* 2012).

with 2,4-D at 0.01, 0.1 or 0.5 mg/l or NAA at 0.01, 0.1 or 0.5 mg/l. Cultures were maintained under a 12-h photoperiod (40 μmol m-2 s-1) at 23 ± 2°C. Treatments with BA induced more shoot proliferation than those with kinetin, the highest number of shoots being produced with 5 mg BA/l in combination with 0.01 mg NAA or 2,4-D/l. Callus proliferated from the basal ends of shoot pieces in all treatments. The cultures also formed roots in the presence of kinetin, but not on BA-containing medium. To produce whole plants, the shoots were separated and rooted on 0.1 mg NAA/l. The resulting plants grew normally under greenhouse conditions. Multiple shoots from shoot meristems of 3 to 5-day-old *in vitro* grown seedlings of Turkish cowpea cv. Akkiz was obtained in MS supplemented with 0.50 mg/l BAP - 0, 0.10, 0.30 and 0.50 mg/l NAA (Aasim *et al.*, 2008). Maximum mean number of 2.60 shoots per explant was obtained on MS without NAA. Regenerated shoots were rooted on MS containing 0.50 mg/l IBA where up to seven adventitious secondary shoots arose from the base of mother shoot were also recorded. These shoots could also be rooted easily on the same rooting medium. Rooted plants were adapted at room temperature in soil mix in pots. All plants flowered and set seeds in the growth room after three months. Anand *et al.* (2000) induced embryogenic callus from primary leaves of *V. unguiculata* cv. P152 in MS medium containing 2,4-D. Greenish white, friable embryogenic calluses were used to establish suspension cultures. A shaking speed of 90 rpm and 0.4 ml packed cell volume per 25 ml medium were found to be optimal for maintaining suspension cultures. Globular, heart-shaped and torpedo-shaped embryos were developed in suspension culture containing 4.52 μM 2,4-D. Maturation of cotyledonary stage somatic embryos was achieved on 0.05 μM 2,4-D, 5 μM abscisic acid and 3 per cent mannitol. Twenty two percent of the embryos were converted into plants and survived; survival in the field was 8-10 per cent.

The production of whole plant of cowpea from calli is an efficient, reliable and rapid strategy (Odutayo *et al.*, 2005). They observed that root development in the cells was promoted by the action of the auxin, naphthalene acetic acid (NAA), at low cytokinin concentration. After five weeks, multiple shoots ranging from two to four developed from the calli cultures after being subcultured on media with high concentration of cytokinin, benzylaminopurine (BAP). Percentages shoot production from calli grown on media with 1 μM concentration of BAP was 45.5 per cent while calli subcultured on 4 μM BAP produced 87.5 per cent shoot production. Percentage rate of survival was between 21-26 per cent in the hardened transplanted plantlets.

Genotype proved to be one of the significant factors that affect regeneration capacity in cowpea. The importance of genotype in cowpea regeneration was first demonstrated by Brar *et al.* (1999). They regenerated less than 50 per cent of the 36 US cowpea genotypes they evaluated by initiating cotyledons on induction medium containing 1/3 MS fortified with 66.6 BA and subsequent shoots regeneration on MS supplemented with 4.44 μM/L BA. In addition to the differences in regeneration capacity among the genotypes, a great variation in regeneration frequencies (1 to11 per cent) and number of shoots per explant (2 to 14) were also observed among the genotypes. Monoharan (2008), Aasim *et al.* (2008), Bakshi *et al.* (2012) and Sawardekar *et al.* (2013) reported significant influence of genotype on *In vitro* regeneration of cowpea. In all the reports, cowpea genotypes not only differed in their regeneration potentials, but also in the degree to which they responded to prevailing culture conditions. The variations might be due to both genetic and physiological differences among the genotypes, which underscore the need for genotype-dependent regeneration protocol in cowpea. Recently, Sani *et al.* (2015) gave a summary of cowpea regeneration work carried out so far and discussed approaches employed as well as challenges of developing efficient regeneration systems in cowpea.

11.2.0 Molecular Markers

Various molecular markers like RAPD, SSR, AFLP, RFLP and biochemical markers have been used in cowpea breeding, such as genetic diversity analysis, genetic linkage map construction, QTL mapping, *etc.*

11.2.1 Analysis of Genetic Diversity

For cowpea breeding, the genetic diversity information is extremely important, which is the basis of breeding and genetic research. RAPD is widely used in cowpea genetic analysis because it is simple and little DNA is required. The RAPD technology was proved to be a useful tool in the characterization of the genetic diversity among cowpea cultivars by Nkongolo (2003), Ba *et al.* (2004), Zannou *et al.* (2008), Malviya *et al.* (2012). SSR is the most frequently used marker in the genetic diversity analysis of cowpea. The earliest cowpea SSR research was conducted by Li *et al.* (2001) and 27 SSR primers have been developed. After that, SSR research on cowpea from different areas, mainly Africa and Asia, has been carried out by many workers (Ogunkanmi *et al.*, 2008; Asare *et al.*, 2010; Sawadogo *et al.*, 2010; Xu *et al.*, 2010; Badiane *et al.*, 2012). Similarly AFLP was recognized as one of the most efficient molecular markers to assess the genetic diversity of cowpea (Coulibaly *et al.*, 2002; Fang *et al.*, 2007). The advantages of combination markers proved to be more effective for assessing diversity in cowpea. Many workers used different combination of markers such as, RAPD and SSR (Diouf *et al.*, 2005), AFLP and SAMPL (Tosti and Negri, 2008), AFLP and SSR (Gillaspie *et al.*, 2005), SSR and ISSR (Tantasawat *et al.*, 2011); ISSR and a combined RAPD-ISSR (Ghalmi *et al.*, 2010).

11.2.2 Construction of Genetic Linkage Map

Genetic linkage map refers to chromosomal linear linkage map which uses chromosome recombinant exchange rate as relative length units and mainly consists of genetic markers. It can be used to locate and mark the target gene to promote the application of marker-assisted breeding in practice. At the same time, it reveals the genetic basis of traits controlled by multiple genes and provides an important tool for map-based cloning. Therefore, building a high-density genetic linkage map is of great significance. The cowpea genetic linkage map is mainly constructed by a cross between a wild species or a cultivated species in the wild type and a cultivar because of its relatively narrow genetic background. There are not many current cowpea genetic maps which are usually constructed with RIL (recombinant inbred lines), the most commonly used mapping population. The first map to be constructed was based mainly on the segregation of RFLP markers in the progeny of a cross between an improved cultivar and a putative wild progenitor type (*Vigna unguiculata* subsp. *dekindtiana*). The map consisted of 92 markers placed in eight linkage groups that spanned a total genetic distance of 684 cm. After that, Menendez *et al.* (1997) constructed a genetic linkage map within the cultivated gene pool of cowpea. The map consisted of 181 loci, comprising 133 RAPDs, 19 RFLPs, 25 AFLPs, three morphological markers, and a biochemical marker (dehydrin). These markers identified 12 linkage groups spanning 972 cm with an average distance of 6.4 cm between markers. On the basis of the two maps above, Ouedraogo *et al.* (2002) constructed an improved genetic linkage map, which was based on the segregation of various molecular markers and biological resistance traits. The new genetic map of cowpea consists of 11 LGs (linkage groups) spanning a total of 2670 cm, with an average distance of 6.43 cm between markers. They also discovered a large, contiguous portion of LG1 that had been undetected in previous mapping work. This region, spanning about 580 cm, was composed entirely of AFLP markers.

Shim *et al.* (2001) screened five hundred and twenty random RAPD primers for parental polymorphism. There are six linkage groups of 40 cM or more, and five smaller linkage groups range from 4.9 to 24.8 cM. The average linkage distance between pairs of markers among all linkage groups was 6.87 cM. The number of markers per linkage group ranged from 2 to 32. The longest group 1 spans 190.6 cM, while the length of shortest group 11 is 4.9 cM. They felt that this map needs to be saturated further with various markers such as RFLP, AFLP, SSR and various populations and primers. In addition, morphological markers and biochemical markers should be united to construct a comprehensive linkage map.

Ouedraogo *et al.* (2002) developed an improved genetic linkage map for cowpea combining AFLP, RFLP, RAPD, biochemical markers, and biological resistance traits. They constructed an improved genetic linkage map based on the segregation of various molecular markers and biological resistance traits in a population of 94 recombinant inbred lines (RILs) derived from the cross between the breeding lines IT84S-2049 and 524B. A set of 242 molecular markers, mostly amplified fragment length polymorphism (AFLP), linked to 17 biological resistance traits, resistance genes and resistance gene analogues (RGAs) were scored for segregation within the parental and recombinant inbred lines. These data were used in conjunction with the 181 random amplified polymorphic DNA (RAPD), restriction fragment length polymorphism (RFLP), AFLP and biochemical markers previously mapped to construct an integrated linkage map for cowpea. The new genetic map of cowpea consists of 11 linkage groups (LGs) spanning a total of 2670 cM, with an average distance of 6.43 cM between markers. Astonishingly, a large, contiguous portion of LG1 that had been undetected in previous mapping work was discovered. This region, spanning approximately 580 cM, is composed entirely of AFLP markers (54 in total). In addition to the construction of a new map, molecular markers associated with various biologi-cal resistance and (or) tolerance traits, resistance genes and RGAs were also placed on the map, including markers for resistance to *Striga gesnerioides* races 1 and 3, cowpea mosaic virus, cowpea severe mosaic virus, blackeye cowpea mosaic poty-virus, southern bean mosaic virus, *Fusarium* wilt and root-knot nematodes. These markers will be useful for the development of tools for marker-assisted selection in cowpea breeding, as well as for subsequent map-based cloning of various resistance genes. Ouedraogo *et al.* (2001) used AFLP and bulked segregant analysis to identify molecular markers linked to resistance of cowpea to parasitism by *Striga gesnerioides*. The identification of AFLP markers linked to *Striga* resistance provides a stepping stone for a marker-assisted selection programme and the eventual cloning and characterization of the gene(s) encoding resistance to this noxious parasitic weed.

The construction of current cowpea genetic map is mainly based on efficient molecular markers such as AFLP, SSR and SNP (Kongjaimun *et al.*, 2012; Andargie *et al.*, 2011; Xu *et al.*, 2011; Muchero *et al.*, 2009). RAPD markers are generally not used to construct genetic maps due to the poor reproducibility. High-density genetic map provides a powerful tool for analysing the heredity of target gene, monitoring specific genes or genomic regions transmitted from parent to next generation, as well as map-based cloning. Therefore, more high-density genetic map of cowpea should be developed by taking advantages of molecular markers.

11.2.3 Molecular Markers Linked to Biotic Resistance

In breeding program, using molecular markers to select the target trait is called MAS (marker-assisted selection), which is the main application of molecular markers. In Africa the parasitic weed *Striga gesnerioides* is the main biotic factor restricting yield of cowpea. Growing cultivars that have resistance to the parasitic weeds is the best way. Searching for more molecular markers tightly linked to the resistance traits against parasitic of cowpea will greatly improve breeding efficiency. Ouedraogo *et al.* (2001) identified three AFLP markers and seven AFLP markers that were linked to Rsg2-1, a single dominant gene controlling resistance to *S. gesnerioides* race 1, and Rsg4-3, a single dominant gene controlling resistance to *S. gesnerioides* race 3, respectively. Both of them were located within linkage group 1 of the cowpea genetic map. Boukar *et al.* (2004) identified four AFLP markers, and mapped 3.2, 4.8, 13.5 and 23.0 cm, respectively, from Rsg1, a gene in IT93K-693-2 that gives resistance to race 3 of *S. gesnerioides*. The AFLP fragment from marker combination E-ACT/M-CAC, which was linked in coupling with Rsg1, was cloned, sequenced, and converted into a SCAR (sequence characterized amplified region) marker named SEACTMCAC83/85, which was co-dominant and useful in breeding programs. Rust disease, incited by the fungus *Uromyces vignae*, is one of the major diseases in cowpea production. Li *et al.* (2001) determined that rust resistance was controlled by a single dominant gene designated Rr1. An AFLP marker (E-AAG/M-CTG) was converted to a SCAR marker, named ABRSAAG/CTG98, and the genetic distance between the marker and the Rr1 gene was estimated to be 5.4 cm. Aphid not only hinders growth, transmits virus, but also causes abnormal of flower, leaf and bud. Yield losses of up to 35 per cent and 40 per cent have been attributed to aphid infestation in Africa and Asia, respectively (Singh and Allen, 1980). Myers *et al.* (1996) found one RFLP marker, bg4D9b, to be tightly linked to the aphid resistance gene (Rac1). The close association of Rac1 and RFLP bg4D9b presented a real potential for cloning this insect resistance gene.

11.2.4 QTL Mapping

The location of genes controlling quantitative traits in the genome is known as QTL (quantitative trait loci). QTL could be detected by employing molecular markers in genetic linkage analysis, *i.e.* QTL mapping. With the help of molecular markers linked to QTL, the heredity of some related QTL could be tracked and the ability of genetic manipulation to QTL is greatly enhanced, thus improving the accuracy and predictability to select genotypes with superior quantitative trait.

Kongjaimun *et al.* (2012) developed a genetic linkage map of yardlong bean using 226 SSR makers from related *Vigna* species and to identify QTLs for pod length. One major and six minor QTLs were identified for pod length variation between yardlong bean and wild cowpea. Andargie *et al.* (2011) identified the QTLs of cowpea agronomic traits related to domestication (seed weight, pod shattering) by SSR markers. Six QTL for seed size were revealed with the phenotypic variation ranging from 8.9 per cent -19.1 per cent. Four QTL for pod shattering were identified with the phenotypic variation ranging from 6.4 per cent -17.2 per cent. The QTL for seed size and pod shattering mainly clustered in two areas of LGs 1 and 10. Fatokun *et al.* (1992) developed genomic maps for cowpea based on RFLP markers. Using these maps, major QTLs for seed weight had been identified. Muchero *et al.* (2009) reported the mapping of 12 QTL associated with seedling drought tolerance and maturity in a cowpea recombinant inbred (RIL) population. Regions harbouring drought-related QTL were observed on linkage groups 1, 2, 3, 5, 6, 7, 9, and 10 accounting for between 4.7 per cent and 24.2 per cent of the phenotypic variance. Further, two QTL for maturity were mapped on linkage groups 7 and 8 separately from drought-related QTL. Some QTL of resistance to disease and insects have also been identified. CoBB (cowpea bacterial blight), caused by *Xanthomonas axonopodis* pv. vignicola (*Xav*), is a worldwide major disease of cowpea. Agbicodo *et al.* (2010) used a SNP (single nucleotide polymorphism) genetic map with 282 SNP markers constructed from the RIL population to perform QTL analysis. Three QTLs, CoBB-1, CoBB-2 and CoBB-3 were identified on linkage group LG3, LG5 and LG9, respectively. Besides, Muchero *et al.* (2011) identified the QTL for *Macrophomina phaseolina* resistance and maturity in cowpea with SNP markers. Muchero *et al.* (2010) also identified three QTL for resistance to *Thrips tabaci* and *Frankliniella schultzei* based on an AFLP genetic linkage map. These QTLs were located on linkage groups 5 and 7 accounting for between 9.1 per cent and 32.1 per cent of the phenotypic variance.

11.2.5 Isozyme Analysis

Isozymes have been used in cowpea to study the relationships as well as for characterization of germplasm. Sonnante *et al.* (1996) scored isoenzyme variation in 25 accessions of wild and cultivated *Vigna unguiculata*, 49 accessions of 7 wild species belonging to section Vigna, and 11 accessions of *V. vexillata* (subgenus Plectrotropis) was scored at 17 putative loci to assess genetic relationships within and among species. The wild species selected for this study are among those which carry important agronomical traits useful in cowpea (*V. unguiculata*) breeding programmes. Low levels of intraspecific variation were observed for *V. heterophylla*, *V. luteola* and *V. racemosa*, whereas the other species showed a higher polymorphism. *V. unguiculata* possessed intraspecific genetic distances comparable to those previously found by other authors. Most of the isoenzyme variation was apportioned among species. Odeigah and Osanyinpeju (1996) investigated the total seed protein, globulin and albumin fractions of 20 cowpea accessions from IITA gene bank by SDS-PAGE. While there was no correlation between seed colour and total seed protein banding pattern, 6 insect-resistant cultivars were characterized by the presence of the 39 and 20 kD subunits. The globulins were the predominant class of the total seed proteins and consisted mainly of 64, 58, 56 and 14 kD subunits which make up CP1 and CP2, the major globulins. The albumins in all accessions were a heterogeneous protein fraction consisting of both high and low molecular weight subunits. It was suggested that the insect-resistant cultivars may be genetically related and that the 39 and 20 kD subunits may be involved in the insect resistance mechanism. Reis and Frederico (2001) used isozymes to evaluate accessions of cultivated cowpea (*V. unguiculata* subsp. *unguiculata*, *sesquipedalis* and *biflora*) from different countries for variability. Comparative analysis of esterase zymograms made it possible to identify some of the accessions studied. However, the cultivated groups biflora and *sesquipedalis* could not be distinguished from one another or from *unguiculata*.

11.3.0 Genetic Transformation

The interest in cowpea for genetic transformation has been due to the discovery of cowpea trypsin inhibitor gene (CpTI) for insecticidal activity. Murdock (1992) suggest the focus of studies on genetic transfer in cowpea in the following four areas, namely, (i) developing improved bioassay systems to use in finding and testing specific insect resistance genes; (ii) identifying specific genes that confer resistance to specific post-flowering pests; (iii) attempting to make interspecific crosses between wild, insect-resistant *Vigna* species and cultivated *V. unguiculata*; and (iv) the genetic transformation of cowpea, using particle-mediated and *Agrobacterium*-mediated gene transfer.

One of the early attempts in genetic transformation study was that of Garcia *et al.* (1986) using leaf discs inoculated with an *Agrobacterium tumefaciens* strain harbouring a Ti-plasmid-derived vector that contained 2 copies of a chimaeric kanamycin resistance gene. By culturing the leaf discs in selective medium, kanamycin resistant callus was obtained. Transformation of this callus was confirmed by the detection of nopaline synthase activity and by Southern blot hybridization, revealing the integration of the kanamycin resistance gene in the plant DNA. Later, a number of researchers have reported success with *Agrobacterium* (Penza *et al.*, 1991; Filippone, 1990; Gnanam *et al.*, 1995; Muthukumar *et al.*, 1996; Sahoo *et al.*, 2000). Other methods used include electroporation (Penza *et al.*, 1992; Akella and Lurquin, 1993; Chowrira *et al.*, 1995) and by particle gun (Matsuoka *et al.*, 1997). Yamaguchi *et al.* (1997) analysed 40 independent genes that are responsive to drought in cowpea (*V. unguiculata*), and studied the structure of their gene products.

The transfer of insect resistance genes to cowpea, by genetic transformation, has the potential to address some of these problems and could have a major impact on food security on the African continent (Machuka, 2000). Garcia *et al.* (1986, 1987) were among the first to attempt transformation experiments in cowpeas and kanamycin-resistant callus was obtained but no plants could be regenerated. Penza *et al.* (1991) and Muthukumar *et al.* (1996) used longitudinal mature embryo slices and mature de-embryonated cotyledons, respectively. After explant co-

Genetic Transformation in Cowpea

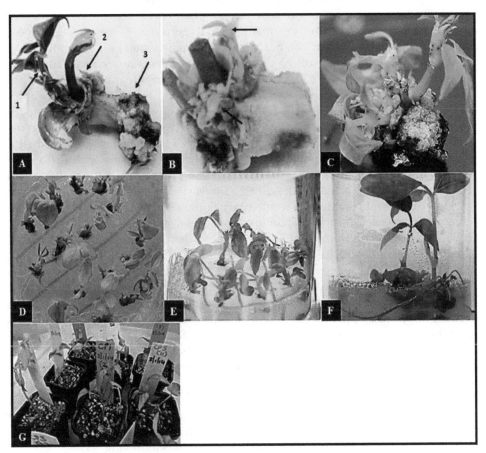

In vitro Regeneration of Cowpea Explants following Co-cultivation with *Agrobacterium*.

(A) Cowpea explant with cotyledon and primary shoots (arrow 1) on shoot induction medium (SIM) with selection at 2 weeks after co-cultivation, regenerated small buds (arrow 2) and callus (arrow 3) are visible (B) explant with cotyledon and primary shoot removed at 4 weeks leaving a clump with shoot buds (arrows) (C) multiple shoots formed on callus (D) multiple shoots separated onto SIM with 30 mg/L geneticin at 8 weeks (E) individual shoots grown with 30 mg/L geneticin at 10 weeks, (F) individual shoots rooting on elongation and rooting medium at 14 weeks and (G) rooted plantlets in soil at 16 weeks (Bett *et al.,* 2019).

cultivation on media containing BAP, there was no evidence of stable integration of the selectable marker or the reporter genes (Penza *et al.*, 1991). Muthukumar *et al.* (1996) obtained four cowpea plants after selection on hygromycin. Southern blot hybridization analysis of one transgenic plant confirmed the integration of the hpt gene, but the seeds of this plant failed to germinate. Ikea *et al.* (2003) also observed transformation in cowpea. However, the transgenes were transmitted to only a small proportion of the progeny and there was no evidence for stable integration. It appears there was an unstable transgene integration, which could not be fixed in a homozygous line. Popelka *et al.* (2006) described a protocol for Agrobacterium-mediated genetic transformation of cowpea and demonstrate for the first time stable transmission and expression of two co-integrated genes in the progeny of transgenic plants. This technology paved the way for biotechnological complementation of existing conventional breeding programs with the aim of developing cowpea germplasm with increased insect resistance, among other useful characteristics. Ivo *et al.* (2008) described a novel system of exploiting the biolistic process to generate stable transgenic cowpea plants. The system was based on combining the use of the herbicide imazapyr to select transformed meristematic cells after physical introduction of the mutated *ahas* gene (coding for a mutated acetohydroxyacid synthase, under control of the ahas 50 regulatory sequence) and a simple tissue culture protocol. The *gus* gene (under control of the *act*2 promoter) was used as a reporter gene. The transformation frequency (defined as the total number of putative transgenic plants divided by the total number of embryonic axes bombarded) was 0.90 per cent. Southern analyses showed the presence of both *ahas* and *gus* expression cassettes in all primary transgenic plants, and demonstrated one to three integrated copies of the transgenes into the genome. The progenies (first and second generations) of all self-fertilized transgenic lines revealed the presence of the transgenes (*gus* and *ahas*) co-segregated in a Mendelian fashion. Western blot analysis revealed that the GUS protein expressed in the transgenic plants had the same mass and isoelectric point as the bacterial native protein. This was the first report of biolisticmediated cowpea transformation in which fertile transgenic plants transferred the foreign genes to next generations following Mendelian laws.

12.0 References

Aasim, M., Khawar, K. M. and Özcan, S. (2008) *Bangladesh J. Bot.*, **37**: 149-154.

Abadassi, J. (2015) *Int. J. Pure App. Biosci.*, **3**: 158-165.

Adebitan, S.A. and Olufajo, O.O. (1998) *Indian J. Agric. Sci.*, **68**: 152–154.

Adebitan, S.A., Ikotun, T., Dashiell, K.E. and Singh, S.R. (1992) *Plant Sci.*, **76**: 1025–1028.

Adegbite, A.A., Amusa, N.A., Agbaje, G.O. and Taiwo, L.B. (2006) *Veg. Sci.*, **12**: 21-27.

Adejumo, T.O., Florini, D.A. and Ikotun, T. (2001) *Crop Prot.*, **20**: 303-309.

Aguiar, J.V.J., Lea˜ o, M.C.S. and Saunders, L.C.U. (1992) *Ciên. Agron., Fortaleza*, **23**: 33–37.

Akella, V. and Lurquin, P.F. (1993) *Plant Cell Rep.*, **12**: 110-117.

Anand, R.P., Ganapathi, A., Anbazhagan, V.R., Vengadesan, G. and Selvaraj, N. (2000) *In Vitro Cell Dev. Biol-Plant*, **36**: 475–480.

Andrade, C.L.T., Silva, A.A.G., Souza, I.R.P. and Conceicao, M.A.F. (1993) *EMPRAPA-CNPAI, Teresina*, p.6.

Annicchiarico, P., Harzic, N. and Carroni, A.M. (2010) *Field Crop Res.*, **119**: 114-124.

Anonymous (1971) *Indian Minimum Seed Certification Standards*, Central Seed Committee, Min. Food Agric., Comm. Dev. Co-op., New Delhi, p. 102.

Anonymous (1987) *Recommended Varieties and Package and Practices, Directorate of Vegetable Research, ICAR, New Delhi*, p. 22.

Anonymous (1996) *Plant Varieties J.*, **9**: 4, 25.

Anonymous (2018). *Proc. Central Variety Release Committee, Ministry of Agriculture and Farmers' Welfare, Government of India, New Delhi.*

Araus, J.L. and Cairns, J.E. (2014) *Trends Plant Sci.*, **19**: 52-61.

Asafo-Adjei, B. and Singh, B.B. (2005) *Crop Sci.*, **45**: 2650-2650.

Asare, A.T., Gowda, B.S., Galyuon, I.K.A., Aboagye, L.L., Takrama, J.F. and Timko, M.P. (2010) *Plant Genet. Resour.*, **8**: 142–150.

Aykroyd, W. R. (1963) *ICMR Special Rept.*, Series, No. 42.

Ba, F.S., Pasquet, R.S. and Gepts, P. (2004) *Genet. Resour. Crop Evol.*, **51**: 539-550.

Bailey, J.A., Nash, C., O'Connell, R.J. and Skipp, R.A. (1990) *Mycol. Res.*, **94**: 810-814.

Bakshi, S., Sahoo, B., Roy, N.K., Mishra, S., Panda, S.K. and Sahoo, L. (2012) *Plant Cell Rep.*, **31**: 1093-1103.

Barrera-Figueroa, B.E., Gao, L., Diop, N.N., Wu, Z., Ehlers, J.D., Roberts, P.A., Close, T.J., Zhu, Jian-Kang and Liu, R. (2011) *BMC Plant Biol.*, **11**: 127.

Barreto, A.L.H., Vasconcelos, I.M., Grangeiro, T.B., Melo, V.M.M., Matos, T.E., Eloy, Y.R.G., Fernandes, C.F., Torres, Davi Coe, Freire, F.C.O. and Oliveira, J.T.A. (2007) *Int. J. Plant Sci.*, **168**: 193-203.

Barreto, D.P.D., Santos, A.A. dos, Quindere, M.A.W., Vidal, J.C., Araujo, J.P.P., Walt, E.E., Rios, G.P. e and Neves, B.P. (1988) *Epace-10: Nova Cultivar DE Caupi PARA O CEARÁ. Fortaleza: EPACE.*

Bashir, M. and Hampton, R.O. (1993) *Plant Disease*, **77**: 948-951.

Bashir, M. and Hampton, R.O. (1996) *European J. Plant Pathol.*, **102**: 411-419.

Bett, B., Gollasch, S., Moore, A., Harding, R. and Higgins, T.J.V. (2019) *Front. Plant Sci.*, **10**: 219. doi: 10.3389/fpls.2019.00219.

Bliss, F.A., Barker, L.N., Franckowiak, S. D. and Hall, T. C. (1973) *Crop Sci.*, **13**: 656-660.

Boukar O, Fatokun CA, Roberts PA, Abberton M, Huynh BL, Close TJ, (2015) In: Ron, D. (Ed.). *Handbook of Plant Breeding*. Springer, New York, pp. 219-250.

Boukar, O., Belko, N., Chamarthi, S., Togola, A., Batieno, J., Owusu, E., Haruna, M., Diallo, S., Umar, M.L. and Olufajo, O. (2018) *Plant Breed.*, **138**: 415–424.

Boukar, O., Fatokun, C.A., Huynh, B.L., Roberts, P.A. and Close, T.J. (2016) *Front. Plant Sci.*, **7**: 757. https://doi.org/10.3389/fpls.2016.00757.

Boukar, O., Kong, L., Singh, B.B., Murdock, L. and Ohm, H.W. (2004) *Crop Sci.*, **44**: 1259-1264.

Brar, M.S., Al Khayri, J.M., Shamblin, C.E., McNew, R.W., Morelock, T.E. and Anderson, E.J. (1997) *In Vitro Cellular and Dev. Biol. Plant.*, **33**: 114-118.

Brar, M.S., Al-Khayri, J.M., Morelock, T.E. and Anderson, E.J. (1999) *In Vitro Cell Dev Biol-Plant*, **35**: 8–12.

Burman, U. and Lodha, S. (1996) *Indian Phytopathol.*, **49**: 254-259.

Caldwell, B.E. and Vest, G. (1968) *Crop Sci.*, **10**: 19–21.

Capoor, S.P. (1967) *Important Virus Diseases of Field and Garden Crops in India and their Control*, ICAR, New Delhi, p. 37

Capoor, S.P. and Varma, P. M. (1956) *Indian J. Agric. Sci.*, **26**: 95-104.

Cardoso, M.J., Freire Filho, F.R. e and Athayde Sobrinho. C. (1990) *Teresina: Embrapa-Uepae de Teresina*, p. 4.

Cardoso, M.J., Santos, A.S.A. dos, Freire Filho, F.R. e and Frota, A.B. (1988) *Teresina: Embrapa-Uepae de Teresina*, p. 3.

Chang, C.A., Chen, C.C., Yang, T.T. and Tsan, T.M. (2002) *Plant Path. Bull.*, **11**: 107-111.

Chatterjee, R. and Bandyopadhyay, S. (2015) *J. Saudi Soc. Agric. Sci.*, DOI: 10.1016/j.jssas.2015.11.001.

Chattopadhyay, A. (1995) *Ph.D. thesis, BCKV, Mohanpur, Nadia, West Bengal.*

Chattopadhyay, A. and Dutta, D. (2003) *Legume Res.*, **26**: 196-199.

Chattopadhyay, A., Dasgupta T., Hazra P. and Som, M.G. (2001) *Indian Agric.*, **45**: 141-146.

Chattopadhyay, A., Dasgupta, T., Hazra, P. and Som, M.G. (1997) *Indian Agric.*, **41**: 49-53.

Chattopadhyay, A., Dasgupta, T., Som, M.G. and Hazra, P. (1996) *Hort. J.*, **9**: 71-75.

Chattopadhyay, A., Hazra, P., Dasgupta T. and Nath, S. (2003) *The Hortic. J.*, **16**: 49-54.

Chattopadhyay, A., Pandiarana, N., Seth, T., Das, S., Chatterjee, S. and Dutta, S. (2014) *Legume Res.*, **37**: 19-25.

Chauhan, D.V.S. (1972) *Vegetable Production in India*, Ram Prasad and Sons, Agra, p. 392.

Cherian, S., Anilkumar, T.B. and Sulladmath, V.V. (1996a) *Mysore J. Agric. Sci.*, **30**: 374-379.

Cherian, S., Anilkumar, T.B. and Sulladmath, V.V. (1996b) *Mysore J. Agric. Sci.*, **30**: 153-158.

Chidebe, I.N., Jaiswal, S.K. and Dakora, F.D. (2018) *Appl. Environ. Microbiol.*, **84**: e01712-17.

Choudhury, B. (1972) *Vegetables, National Book Trust*, India, New Delhi, p. 220.

Choudhury, B. and Ramphal (1960) *Indian J. Hort.*, **17**: 129-132.

Choudhury, B.N., Rahman, M.F. and Bora, A. (2005) *Indian J. Nematol.*, **35**: 219.

Chowrira, G.M., Akella, V. and Lurquin, P.F. (1995) *Mol.Biotechnol.*, **3**: 17-23.

Coulibaly, S., Pasquet, R.S., Papa, R. and Gepts, P. (2002) *Theor. Appl. Genet.*, **104**: 358–366.

Craufurd, P.Q., Qi, A., Summerfield, R.J., Ellis, R.H. and Roberts, E.H. (1996b) *Expt. Agril.*, **32**: 29-40.

Craufurd, P.Q., Roberts, E.H., Ellis, R.H., Summerfield, R.J. (1996a) *Euphytica,***88**: 77-84.

Dasila, B., Singh, V., Kushwaha H.S., Srivastava, A. and Ram, S. (2006) *SAARC J. Agri.*, **14**: 46-55.

Davis, D.W., Oelke, E.A. Oplinger, E.S., Doll, J.D., Hanson, C.V. and Putnam, D.H. (2000) *Alternative field crops manual.* http://www. hort.purdue.edu/newcrop/afcm/cowpea.html.

Des Marais, D.L., Hernandez, K.M. and Juenger, T.E. (2013) *Annu. Rev. Ecol. Evol. Syst.*, **44**: 5– 29.

DeTar, W.R. (2009) *Agric. Water Manag.*, **96**: 53–66.

Dhiman, J.S., Lal, T. and Singh, S. (1989) *Trop. Agric.*, **66**: 17-20.

Diouf, D. and Hilu, K.W. (2005) *Genet. Resour. Crop Evol.*, **52**: 1057–1067.

Ehlers, J.D. and Hall, A.E. (1996) *Crop Sci.*, **36**: 673-679.

Ehlers, J.D., Matthews, W.C., Hall, A.E. and Roberts, P.A. (2000) *Crop Sci.*, **40**: 611–618.

Fang, J., Chao, C.C.T., Roberts, P.A. and Ehlers, J.D. (2007) *Genet. Resour. Crop Evol.*, **54**: 1197–1209.

Fapohunda, H.O., Aina, P.O. and Hossain, M.M. (1984) *Agric. Water Manage.*, **9**: 219–224.

Faris, D.G. (1965) *Canadian J. Genet, Cytol.*, **7**: 433-452.

Fatokun, C.A., Menancio-Hautea, D.I., Danesh, D. and Young, N.D. (1992) *Genetics*, **132**: 841–846.

Fernandes, J.B., Holanda, J.S. de, Simplicio, A.A., Bezerra Neto, F., Torres, J.e and Rego Neto, J. (1990) *Pesquisa Agropecuária Brasileira*, **25**: 1555–1560.

Ferry, R. L. and Singh, B.B. (1997) In: Singh, B.B., Mohan-Raj, D.R., Dashiell, K.E. and Jackai, L.E.N. (Eds.) *Advances in cowpea research*, pp. 13–29.

Fery, R.L. and Dukes, P.D. (1977) *J. Hort. Sci.*, **12**: 454–456.

Fery, R.L. and Dukes, P.D. (1996) *HortScience*, **31**: 1250-1251.

Filippone, E. (1990) In: *Cowpea Genetic Resources* (Eds. N.Q. Ng and L.M. Monti), pp. 175-181.

Freire Filho, F.R, Santos, A.A. dos, Araujo, A.G. de, Cardoso, M.J., da Silva, P.H.S., and Ribeiro, V.Q. (1994) *Teresina: Embrapa-Cpamn, Embrapa-Cpamn.* p. 6.

Garcia, J.A., Hille, J. and Goldbach, R. (1986) *Plant Sci., Irish Republic,* **44:** 37-46.

Garcia, J.A., Hille, J., Vos, P. and Goldbach, R. (1987) *Plant Sci.,* **48:** 89–98.

Gbaguidi A.A., Dansi A., Loko L.Y., Dansi M. and Sanni A. (2013) *Soil Sci.,* **3:** 121-133.

Ghadiri,V. and Sohrabi, M. (1999) *Seed and Plant,* **14:** 1-5.

Ghalmi, N., Malice, M., Jacquemin, J.M., Ounane, S.M., Mekliche, L. and Baudoin, J.P. (2010) *Genet. Resour. Crop Evol.,* **57:** 371–386.

Gillaspie, A.G., Hopkins, M.S. and Dean, R.E. (2005) *Genet. Resour. Crop Evol.,* **52:** 245–247.

Gillaspie, A.G.Jr. (2001) *Plant Disease,* **85:** 1004-1005.

Githiri, S.M., Ampong Nyarko, K., Osir, E.O. and Kimani, P.M. (1996) *Euphytica,* **89:** 371-376.

Githiri, S.M., Kiman, P.M. and Pathak, R.S. (1996) *Euphytica,* **92:** 307-311.

Gnanam, A., Muthukumar, B., Mammen, M., Veluthambi, K. and Mathis, P. (1995) *Photosynthesis: from light to biosphere.* Volume III. Proc. Xth Internatl. Photosynthesis Cong., Montpellier, France, 20-25 August 1995, pp. 707-710.

Goyal, S. and Khan, S. (2010) *Int. J. Bot.,* **6:** 194-206.

Hall, A. (2012) *Front. Physiol.,* **3:** 155.

Hazra, P. (1991) *Ph.D. Thesis, Bidhan Chandra Krishi Viswavidyalaya,* Mohanpur, West Bengal, India.

Hazra, P., Bhattacharya, C., Mandal, A.R. and Ghosal, K.K. (1990) *Indian Biol.,* **22:** 8-11.

Hazra, P., Chattopadhyay, A., Dasgupta, T. and Som, M.G. (2001) *Indian Hort.,* **46:** 13-15.

Hazra, P., Chattopadhyay, A., Karmakar, K. and Dutta, S. (2011) *Modern Technology in Vegetable Production,* New India Publishing Agency, New Delhi.

Hazra, P., Chattopadhyay, A., Som M.G. and Das P.K. (2000) *SABRAO J. Breed. Genet.,* **32** (1): 23-29.

Hazra, P., Das, P.K. and Som, M.G. (1994) *Indian J. Genet.,* **54:** 175-178.

Hazra, P., Das, P.K. and Som, M.G. (1996a) *Indian J. Genet.,* **56:** 553-555.

Hazra, P., Das, P.K. and Som, M.G. (1996b) *Crop Res., Hisar,* **11:** 78-83.

Hazra, P., Som, M.G. and Das, P.K. (1993) *Indian J. Hort.,* **50:** 358-363.

Hector, J. M. (1936) *Introduction to the Botany of Field Crops, II Non-Cereals,* Central News Agency Ltd., South Africa, p. 639:

Heij, G. (1989) *Acta Hortic.,* **242:** 305-311.

Hemeng, O.B. (1989) *Int. Nematol. Net. Newsl.*, **6**: 6-7.

Hendawy, M.A., Saleh, A.A.A., Jabbar, A.S. and El-Hadary, A.S.N. (2018) *Zagazig J. Agric. Res.*, **45**: 2367-2375.

Horn, L., Shimelis, H. and Laing, M. (2015) *Legume Res.*, **38**: 691-700.

Hossain, M. A. and Asafo-Adjei, B. (1987) *IITA Research Briefs*, International Institute of Tropical Agriculture, **8**: 7.

Ikea, J., Ingelbrecht, I., Uwaifo, A. and Thottappilly, G. (2003) *Afr. J. Biotechnol.*, **2**: 211–218.

Ivo, Nayche L., Nascimento, Cristina P., Vieira, Lý´via S., Campos, Francisco A.P. and Araga˜o, Francisco J.L. (2008) *Plant Cell Rep.*, **27**: 1475–1483.

Jackai, L.E.N., Padulosi, S. and Ng, Q. (1996) *Crop Protn.*, **15**: 753-761.

Jagginavan, S.B., Kulkarni, K.A., and Lingappa, S. (1995) *Karnataka J. Agric. Sci.*, **89**: 90–93.

Jain, S.M. (2005) *Plant Cell Tiss. Org. Cult.*, 82: 113–123.

Jana, Sakrajit., Som, M.G. and Das, N.D. (1982) *Veg. Sci.*, **9**: 96-106.

Jana, Sakrajit., Som, M.G. and Das, N.D. (1983) *Haryana J. Hort. Sci.*, **12**: 224-227.

Jeyanandarajah, P. and Brunt, A.A. (1996) *Trop. Sci.*, **36**: 129-137.

Joseph, S. and Peter, K.V. (2003) *Legume Res.*, **26**: 57-59.

Khan, T.A. and Husain, S.I. (1989) *Pakistan J. Nematol.*, **7**: 91-96.

Khan, V.M., Manohar, R.S. and Verma, H.P. (2015) *Asian J Biosci.*, **10**: 113-115.

Kimiti, J.M. and Odee, D.W. (2010) *Appl. Soil Ecol.*, **45**: 304–309.

Kongjaimun, A., Kaga, A., Tomooka, N., Somta, P., Vaughan, D.A. and Srinives, P. (2012) *Ann. Bot.*, **109**: 1185–1200.

Krishnaswamy, N., Nambiar, K. K. and Mariakulandai, A. (1945) *Madras Agric. J.*, **32**: 145-160.

Kumar, P. and Singh, N. P. (1990) *Haryana J. Hort. Sci.*, **19**: 210-212.

Kwapata, M. B. and Hall, A. E. (1990) *Field Crop Res.*, **24**: 1-10.

Lal, H., Rai, M., Shiv Karan, Verma, A. and Ram, D. (2007) *Acta Hort.*, **752**: 413-416.

Latunde-Dada A.O., O'Connell, R.J., Nash, C. and Lucas, J.A. (1999) *Plant Pathol.*, 48: 777-785.

Lee, J.R., Back, H.J., Yoon, M.S., Park, S.K., Cho, Y.H. and Kim, C.Y. (2009) *Korean J. Breed. Sci.*, **41**: 369-376.

Leina, M.J., Tan, T.K. and Wong, S.M. (1996) *Annals Appl. Biol.*, **129**: 197-206.

Li, C.D., Fatokun, C.A., Ubi, B., Singh, B.B. and Scoles, G. (2001) *Crop Sci.*, **41**: 189-197.

Lima, J.A. de A. and Santos, A.A. (1988) In: Araújo, J.P.P de. and Watt, E.E. (Eds.) *O Caupi no Brasil. Goiânia, EMBRAPA-CNPAF/Ibadan, IITA*, pp. 507–545.

Lima, K.d.S.C., Souza, L.B.E., Godoy, R.L.D.O., França, T.C.C. and Lima, A.L.D.S. (2011) *Radiat. Phys. Chem.*, **80**: 983-989.

Machuka, J. (2000) *Proc. World Cowpea Conference III, IITA, Ibadan,Nigeria*, pp. 213–222.

Maluszynski, K.N., Zanten, L.N. and Ahloowalia, B.S. (2000) *Mut. Breed. Rev.*, **12**: 1-12.

Malviya, N., Sarangi, B.K., Yadav, M.K. and Yadav, D. (2012) *Plant Syst. Evol.*, **298**: 523-526.

Mandal, J., Chattpadhyaya, A., Hizra, P., Dasgupta, T. and Som, M.G. (1999) *Crop Res. (Hissar)*. **18**: 222-225.

Manoharan, M., Khan, S. and James, O.G. (2008) *J. App. Hortic.*, **10**: 40-43.

Marfo, K.O. (1996) *Tropical Agric.*, **73**: 192-195.

Matsuoka, M., Kai, Y., Yoshida, T. (1997) *Bull. Chugoku Natl. Agril. Expt. Stn.*, **18**: 31-39.

Mayers, G.O., Fatokun, C.A. and Yound, N.D. (1996) *Euphytica*, **91**: 181-187.

Mehta, Y.R. (1959) *Vegetable Growing in Uttar Pradesh*, Bureau of Agric. Information, Lucknow.

Menendez, C.M. and Hall, A.E. (1996) *Crop Sci.*, **36**: 233-238.

Menendez, C.M., Hall, A.E. and Gepts, P. (1997) *Theor. Appl. Genet.*, **95**: 1210–1217.

Mishra, S.K., Singh, B.B. and Hegde, V. (2005) *Indian J. Genet. Plant Breed.*, **65**: 193-195.

Mligo, J.K. and Singh, B.B. (2007) *Crop Sci.*, **47**: 437–438.

Molosiwa, O.O., Gwafila, C., Makore, J. and Chite, S.M. (2016) *Int. J. Biodivers. Conserv.*, **8**: 153-163.

Muchero, W., Ehlers, J.D., Close, T.J. and Roberts, P.A. (2009). *Theor. Appl. Genet.*, **118**: 849–863.

Muchero, W., Ehlers, J.D., Close, T.J. and Roberts, P.A. (2011). *BMC Genomics*, **12**: 8.

Murdock, L.L. (1992) In: Thottappilly, G., Monti, L.M., Mohan Raj, D.R. and Moore, A.W. (Eds.) *Biotechnology: enhancing research on tropical crops in Africa*. pp. 313-320.

Muthukumar, B., Mariamma, M., Veluthambi, K. and Gnanam, A. (1996) *Plant Cell Rep.*, **15**: 980-985.

Myers, G.O., Fatokun, C.A. and Young, N.D. (1996) *Euphytica*, **91**: 181-187.

Nabi, A.J.M.N., Hawlader, M.H.K., Hasan, M.M., Haque, M.Z. and Rahaman, M.L. (2016) *Progres. Agric.*, **27**: 94-100.

Naik, M.S., Abrol, Y.P., Nair, T.V.R. and Ramarao, C.S. (1982) *Phytochemistry*, **21**: 495–504.

Nair, T.V.R., Kaim, M.S., Pathak, J.B. and Mythili, J. (1988) *Proc. Int. Cong. Plant Physiologists; IARI: New Delhi, India*, pp. 1064–1067.

Nakawuka, C. K. and Adipala, E. (1997) *Plant Dis.*, **81**: 1395-1399.

Nasu, Y., Karasawa, A., Hase, S. and Ehara, Y. (1996) *Phytopathol.*, **86:** 946-951.

Nath, V., Lal, H., Rai, M., Rai, N. and Ram, D. (2009) *Indian J. Plant Genet. Resour.*, **22:** 22-25.

Nkansah, P.K. and Hodgson, C.J. (1995) *Int. J. Pest Manag.*, **41:** 161–165.

Nkongolo, K.K. (2003) *Euphytica*, **129:** 219–228.

Nwabudike, F.O. and Gast, H. (1996) *Organisation and management of national seed programmes*. Proc. of a follow-up seminar-workshop held from 12 to 24 November 1994 in Aleppo, Syria, pp.121-123.

Odeigah, P.G.C. and Osanyinpeju, A.O. (1996) *Genet. Resour. Crop Evol.*, **43:** 485-491.

Odutayo, O.I., Akinrimisi, F.B., Ogunbosoye, I. and Oso, R.T. (2005) *African J. Biotech.*, **4:** 1214-1216.

Ofuya, T.I. (1988) *Int. Pest Cont.*, **30:** 68-69.

Oghiakhe, S. (1996) *J. Appl. Entomol.*, **120:** 549-553.

Oghiakhe, S., Makanjuola, W.A. and Jackai, L.E.N. (1993) *Int. J. Pest Manag.*, **39:** 261-264.

Ogunkanmi, L A., Ogundipe, O.T., Ng, N.Q. and Fatokun, C.A. (2008) *J. Food Agric. Environ.*, **6:** 253-268.

Ohler, T.A. and Mitchell, C.A. (1996) *J. Amer. Soc. Hort. Sci.*, **121:** 576-581.

Ojomo, O.A. (1971) *Trop. Agric. (Trinidad)*, **48:** 277–282.

Ouedraogo, J.T., Maheshwari, V., Berner, D.K., St Pierre, C.A., Belzile, F. and Timko, M.P. (2001) *Theo. Appl. Genet.*, **102:** 1029-1036.

Ouedraogo, J.T., Gowda, B.S., Jean, M., Close, T.J., Ehlers, J.D., Hall, A.E., Gillaspie, A.G., Roberts, P.A., Ismail, A.M., Bruening, G., Gepts, P., Timko, M.P. and Belzile, F.J. (2002) *Genome*, **45:** 175-188.

Paiva, J.B., Teófi, E.M. lo, Santos, J.H.R., Lima, J.A.A. dos, Gonçalves, M.F.B. e and Silveira, L. de F.S. (1988) *"Setentão": nova cultivar de feijão-de-corda para o estado do Ceará. Fortaleza: UFC*.

Pal, B. P., Sikka, S. M. and Singh, H. B. (1956) *Indian J. Hort.*, **13:** 64-73.

Pandey, K.C., Hasan, N., Bhaskar, R.B., Ahmed, S.T. and Kohli, K.S. (1995) *Indian J. Genet.*, **55:** 198–203.

Patel, P.N. (1985) In: Singh, S.R. and Rachie, K.O. (Eds.), *Cowpea Research, Production and Utilization. Wiley, New York*, pp. 205-213.

Pathak, R.S. (1988) *Proc. workshop improvement of grain legume production using induced mutations*, Pullman, Washington, USA, pp. 279-291.

Penza, R., Akella, V. and Lurquin, P.F. (1992) *BioTechniques*, **13:** 576, 578, 580.

Penza, R., Lurquin, P.F. and Filippone, E. (1991) *J. Plant Physiol.*, **138:** 39-43.

Popelka, J.C., Gollasch, S., Moore, A., Molvig, L. and Huggins, T.J.V. (2006) *Plant Cell Rep.*, **25:** 304-312.

Prasad, D.T., Umpathy, N.S. and Veeranna, R. (1996) *J. Plant Biochem. Biotechnol.,* **5**: 47-49.

Quin, F.M. (1997) In: Singh, B.B., Mohan Raj, D.R., Dashiell, KE, Jackai, L.E.N. (Eds.). *Advances in cowpea research,* pp. 9-15.

Rajput, J.C., Thorat, S.T., Shinde, P.P. and Palve, S.B. (1991) *Ann. Agric. Res.,* **11**: 325-326.

Ramburan, S., Zhou, M. and Labuschagne, M.T. (2012) *Field Crop Res.,* **129**: 71-80.

Rao, A.S. and Singh, R.S. (2004) *J. Agrometeorol.,* **6**: 39–46.

Reddy, D.B. (1968) *Plant Protection in India,* Allied Publishers, Bombay.

Reis, C.M. and Frederico, A.M. (2001) *Acta Hort.,* **546**: 497-501.

Roberts, P.A., Ehlers, J.D., Hall, A.E. and Matthews, W.C. (1997) In: Singh, B.B. and Raj, M. (Eds.), *Advances in Cowpea Research.*

Roberts, P.A., Matthews, W.C. and Ehlers, J.D. (1996) *Crop Sci.,* **36**: 889-894.

Rodriguez, I., Rodriguez, M.G., Sanchez, L. and Iglesias, A. (1996) *Revista de Proteccion Vegetal,* **11**: 63-65.

Ryerson, D.E. and Heath, M.C. (1996) *Canadian J. Plant Pathol.,* **18**: 384-391.

Sahoo, L., Sushma., Sugla, T., Singh, N.D., Jaiwal, P.K. and Sahoo, J. (2000) *Plant Cell Biotechnol. Mol. Biol.,* **1**: 47-54.

Sampson (1936) *Kew Bull. Addl.Ser.,* 12, p. 234.

Sani, L.A., Usman, I. S., Faguji, M.I. and Bugaje, S.M. (2015) *British Biotechnol. J.,* **7**: 174-182.

Santos, A. A., Freire Filho, F. R. and Cardoso, M. J. (1987) *Fitopatologia Brasileira,* **12**: 400-402.

Saunders, L.C.U., Castro, P.T., Bezerra, F.M.L. and Pereira, A.L.C. (1985) *Cieˆn. Agron., Fortaleza,* **16**: 75–81.

Sawadogo, M., Ouédraogo, J.T., Gowda, B.S. and Timko, M.P. (2010) *Afr. J. Biotechnol.,* **9**: 8146–8153.

Sawardekar, S.V., Jagdale, V.K., Bhave, S.G., Gokhale, N.B., Sawardekar, S.V. and Lipne, K.A. (2013) *Int. J. Appl. Biosci.,* **1**: 1-8.

Sellschop (1962) *Wealth of India,* vol. X, Council of Scientific and Industrial Research, New Delhi,

Shim, J.H., Chung, J.I. and Go, M.S. (2001) *Korean J. Crop Sci.,* **46**: 341-343.

Shiringani, R.P. and Shimelis, H.A. (2011) *Afr. J. Agric. Res.,* **6**: 3259-3263.

Shouse, P., Dasberg, S., Jury, W.A. and Stolzy, L.H. (1981) *Agron. J.,* **73**: 333–336.

Shoyinka, S.A., Thottappilly, G., Adebayo, G.G. and Anno Nyako, F.O. (1997) *Int. J. Pest Manag.,* **43**: 127-132.

Singh, A., Singh, Y.V., Sharma, A., Visen A., Singh, M.K. and Singh, S. (2016) *Legume Res.,* pp. 1-8.

Singh, B. and Usha, K. (2003) *J. Plant Nutr.*, **26**: 463–473.

Singh, B.B. (1998) *IITA Annual Rept.*, pp. 24-27.

Singh, B.B. (1999) *IITA Annual Rept.*, p. 26.

Singh, B.B. (2002) In: Fatokun, C.A., Tarawali, S.A., Singh, B.B., Kormawa, P.M. and Tamo, M. (Eds.) *Challenges and Opportunities for Enhancing Sustainable Cowpea Production*, pp. 3-13.

Singh, B.B. (2005) In: Singh, R.J. and Jauhar, P.P. (Eds.), *Genetic Resources, Chromosome Engineering and Crop Improvement*. Florida, USA. pp. 117-162.

Singh, B.B., Ajeigbe, H.A., Tarawali, S.A., Fernandez-Rivera, S. and Abubakar, M. (2003) *Field Crops Res.*, **84**: 169–177.

Singh, B.B., Asnate, S.K., Jackai, L.E.N. and d Hughes, J. (1996) In: *IITA Annual Report. International Institute for Tropical Agriculture, Ibadan, Nigeria.*

Singh, C. and Singh, N.N. (2015) *Indian J. Agric. Res.*, **49**: 358-362.

Singh, D and Patel, R.N. (1990) *Indian Hort.* 3S: Oct-Dec., pp. 23-25.

Singh, D. and Patel, P.N. (1977) *Indian Phytopath.*, **30**: 99-102.

Singh, D.P., Sharma, S.P., Lal, M., Ranwah, B.R. and Sharma, V. (2013) *Legume Res.*, **36**: 10–14.

Singh, H.B. and Sikka, S.M. (1955) *Indian Fmg.*, **4**: 163.

Singh, H.B., Mittal, S.P. and Kazim, M. (1968) *Indian Hort.*, **12**: 13.

Singh, K.B. and Mehndiratta, P. D. (1969) *Indian J. Genet.*, **29**: 104-109.

Singh, S.R. and Allen, O.J. (1980) *Agric. Fish Food.*, p. 667.

Singh, V., Ramkrishna, K. and Arya, R.K. (2006) *Indian J. Genet. Plant Breed.*, **66**: 312–315.

Singleton, P.W., Bohlool, B.B. and Nakao, P.L. (1992) In: Lal, R. and Sanchez, P.A. (Eds.). *Myths and Science of Soils of the Tropics*, pp. 135–155.

Sirohi, A. and Dasgupta, D.R. (1993) *Indian J. Nematol.*, **23**: 31-41.

Soedomo, R. P. and Sunaryono, H. (1988) *Buletin Penlitian Hortikultura,* **16**: 53-56.

Som, M.G. and Hazra, P. (1993) In: Kalloo, G. and Bergh, B. O. (Eds.) *Genetic Improvement of Vegetable crops, Pergamon Press, Oxford*, pp. 339–354.

Sonnante, G., Piergiovanni, A.R., Ng, Q.N. and Perrino, P. (1996) *Genet. Resour. Crop Evol.*, **43**: 157-165.

Souza, M.S.M., Bizerra, F.M.L. and Teo´filo, E.M. (2005) *Irriga. Botucatu*, **10**: 241–248.

Steele, W.M. (1976) *Evolution of Crop Plants*, Ed. N. W. Simmonds, Longman, London, New York.

Stejskal, V., Kosina, P. and Kanyomeka, L. (2006) *J. Pest. Sci.*, **79**: 51-55.

Tantasawat, P., Trongchuen, J., Prajongjai, T., Jenweerawat, S. and Chaowiset, W. (2011) *Aus. J. Crop Sci.*, **5**: 283–290.

Tayo, T.O. (1989) *Insect Sci. Appl.*, **10**: 631-638.

Timsina, J. (1988) *Institute of Agriculture and Animal Science*, Nepal **9**: 21-27.

Tosti, N. and Negri, V. (2008) *Genome*, **45**: 656-660.

Touré, M.A. and Singh, B.B. (2005) *Crop Sci.*, **45**: 2648–2648.

Trinchant, J.C. and Rigaud, J. (1984) *J. Plant Physiol.*, **116**: 209–217.

Turk, K.J. and Hall, A.E. (1980) *Agron. J.*, **72**: 434–439.

Uday, B. and Lodha, S. (1996) *Indian Phytopathol.*, **49**: 254-259.

Udensi, O., Arong, G., Obu, J., Ikpeme, E. and Ojobe, T. (2012) *Am. J. Exp. Agric.*, **2**: 320–335.

Ullah, M.Z., Hasan, M.J., Rahman, A.H.M.A. and Saki, A.I. (2011) *SAARC J. Agric.*, **9**: 9-16.

Vavilov, N. I. (1939) *Chromosome Atlas of Cultivated Plants*, George Allen Unwin Ltd., London.

Veeranna, R. and Hussain, M.A. (1997) *Insect Environ.*, 3: 15.

Veerappa, R. (1998) *Insect Environ.*, **4**: 5–6.

Venter, H.M. (1996) *The biodiversity of African plants. Proc.of the XIVth AETFAT Congress*, Wageningen, Netherlands, 22-27 August 1994, pp. 656-660.

Verdcourt, B. (1970) *Kew Bull.*, **24**: 507-569.

Visarathanonth, P., Khumlekasing, M. and Sukprakarn, C. (1990) In: Fujii, K., Gatehouse, A.M.R., Johnson, C.D., Mitchel, R. and Yoshida, T. (Eds.) *Bruchids and legumes: economics, ecology and coevolution. Series Entomologica*, pp.101-104.

Walle, Tesfaye, Mekbib, Firew, Amsalu, Berhanu and Gedil, Melaku (2018) *Am. J. Plant Sci.*, **9**: 2794-2812.

Wydra, K. and Singh, B.B. (1998) *IITA Annual Rept.*, pp. 25-27.

Xu, P., Hu, T., Yang, Y., Wu, X., Wang, B. and Liu, Y. (2011) *HortScience*, **46**: 1102–1104.

Xu, P., Wu, X., Wang, B., Liu, Y., Quin, D. and Ehlers, J.D. (2010) *Mol. Breed.*, **25**: 675–684.

Yadav, B.D. and Ramesh Chand. (2003) *Ann. Plant Protection Sci.*, **11**: 349-351.

Yamaguchi Shinozaki, K. and Shinozaki, K. (1997) *JIRCAS Internatl. Symp. Series*, **5**: 243-252.

Zannou, A., Kossou, D.K., Ahanchédé, A., Zoundjihékpon, J., Agbicodo, E. and Struik, P.C. (2008) *Afr. J. Biotechnol.*, **7**: 4407–4414.

Ziska, L.H. and Hall, A.E. (1983) *Irrig. Sci.*, **3**: 247–257.

Zohri, A. A. (1993) *Qatar Univ. Sci. J.*, **13**: 57-62.

CLUSTER BEAN

A. Chattopadhyay, V.A. Parthasarathy
and P. K. Maurya

1.0 Introduction

Cluster bean (*Cyamopsis tetragonoloba* (L.) Taub.), a drought hardy and salt tolerant leguminous crop, is grown for feed, fodder, vegetable, green manure and gum production. The crop is popularly known as Guar, Thupi, Urahi, Koth Avarai, Gavar, Gor Chikudu, Gorikaya, Kothavara, Guvar and Matki (sprouted seed) in different parts of India. It occupies an important place in dry land agriculture of mainly arid and semi-arid regions of Northwestern states of Rajasthan, Haryana, Gujarat and parts of Punjab, Uttar Pradesh and Madhya Pradesh (Arora *et al.*, 1985). The crop has been grown in Burma, Sri Lanka, Pakistan and Texas and Arizona states of the USA since long back (Prabhakara, 2011). Among the beans grown in India, cluster bean or guar is cultivated to a lesser extent.

2.0 Composition and Uses

2.1.0 Composition

The green pods of cluster bean are as rich in food value as that of French bean. The nutritive value of cluster bean has been shown in Table 1.

Table 1: Composition of Cluster Bean in Green Pod (per 100 g of edible portion)*

Moisture	81.0 g	Thiamine	0.09 mg
Carbohydrates	10.8 g	Riboflavin	0.09 mg
Protein	3.2 g	Vitamin C	47 mg
Fat	0.4 g	Vitamin A	316 I.U.
Minerals	1.4 g		

* Aykroyd (1963).

2.2.0 Uses

The tender pods are used as vegetable and in the southern parts of India they are dehydrated and stored for use. It is a nutritious fodder for livestock and the seeds are also fed to the cattle. Besides, the crop can be used for soil improvement and as a medicine. Its boiled seeds are used for the treatment of plague, enlarged livers, head swellings and swellings on broken bones. Importance of cluster bean as potential hypoglycaemic agents for the treatment of diabetes has been emphasized.

Matured seeds have tremendous industrial value due to presence of galactomannan polysaccharide popularly known as guar gum or mucilage (mannogalacton) in its endosperm. Cluster bean seeds are thus widely being used for extraction of gum. The gum content in seed varies from 19.2 to 47.9 per cent depending on the genotype (Vijay and Leela, 1989). The gum is composed of D-galactopyranose and D-mannopyranose units. Three principle enzymes namely, a-D-galactosidase, b-D-mannanase and b-D-mannosidase are responsible for the degradation of galactomannan in germinating seeds (Sehgal *et al.*, 1973). The gum is mainly used in about 25 major industries like, textiles, cosmetics, explosives, paper, mining, oil, food processing *etc.* apart from being used as adhesives on postage stamps, to impart smoothness and stability to bakery products and as a foam stabiliser in beer (Smith, 1976). Gum is also added in storage tanks particularly in sprinkler irrigation system (Singh *et al.*, 1994). The gum is also used as a stabilizer and thickener agent in food products *viz.*, ice cream, bakery mixes and salad dressings (Singh, 1989). In food it acts as bulk laxative and cholesterol controlling agent. Guar gum is exported to different countries like, USA, Japan, UK, France, Italy and Netherlands *etc.* and America by far is the major importer while India and Pakistan are major exporters.

Guar now accounts for around 18 percent of India's total agricultural exports. USA is the major importing country followed by Germany. Both of these countries account for more than 50 per cent of India's exports of guar products. The other important importing countries from India are Netherlands, U.K., Japan and Italy. The gum is mainly exported as: (i) gum treated and pulverized – constitutes ~83 per cent of cluster bean gum trade (ii) gum refined split – share percentage is ~16 per cent (iii) cluster bean meal [guar meal] – accounts for a ~1 per cent (USDA, 2014). The export value of treated/pulverized gum, refined split and meal during 2012-13 was INR 177.6, 33.9 and 1.4 billion, respectively.

Guar-meal after separation of gum is a potentially valuable source of protein (45 per cent) for animal feed. However, growth and egg production of chick was adversely affected due to 10 per cent saponin content in guar-meal, limiting its use to low levels in poultry feed.

The gum extraction factories utilize hardly about 40 per cent of the total seed production and most of the remaining seeds are used as cattle feed by the farmers. Cluster bean is also grown as a forage and green manure crop for enhancing the soil fertility through fixation of atmospheric nitrogen (50-60 kg/ha) and also for incorporation of organic matters in soil (Lal, 1985).

3.0 Origin, Taxonomy and Morphology

3.1.0 Origin

Diverse opinions exist regarding the origin of this crop. Guar was probably domesticated in a dry region of West Africa having been introduced from other part of Africa. Arab traders are thought to have introduced it to Asia via South India (Smith, 1976), while some opine the crop to be native to India even though no wild forms occur in this country (Anon., 1950; Venkataratnam, 1973). Guar is a native to India where it is grown principally for its green fodder and for the pods that are used for food and feed. Guar was introduced into the United States from India in 1903. Production in the United States is centered in Texas, Oklahoma, and Arizona, but it is also adapted to locations with more tropical climates, such as in Florida and Puerto Rico (Stephens, 1994). The crop is grown in India, Myanmar, Sri Lanka and Pakistan. Even in arid zones of the USA like Texas and Arizona it is being grown (Venkataratnam, 1973).

Uses of Cluster Bean

Fresh cluster bean pods used as vegetable (A) and dried pods (B) for gum extraction purposes

The cluster bean: (A) seed, (B) splits,(C) powdered form of gum

Guar gum powder

Cluster bean frozen product

Different dishes of cluster bean

Traits Variation and Flowering in Cluster bean

Trait variation in cluster bean. Branching: (**A**) Unbranched; (**B**) Sparse branching; (**C**) Heavy branching. Flower color: (**D**) White; (**E**) Pink; (**F**) Purple. Leaf shape: (**G**) Ovate; (**H**) Deltoid. Leaf margin: (**I**) Serrated; (**J**) Smooth.

Flowering in cluster bean

3.2.0 Taxonomy

Cluster bean or guar, *Cyamopsis tetragonolobus* (L.) Taub. (syn. *C. psoralioides* DC.) belongs to the tribe Galegae (Indigoferae) of leguminosae family. The genus consists of three species, out of which, *C. tetragonolobus* (2n=2x=14) is the only member of economic importance. The African species *C. senegalensis* seems to be the immediate ancestor to guar which is not found anywhere in the wild state (Smith, 1976).

3.3.0 Morphology

Cluster bean is an erect annual growing to a height of 3 m with stiff erect branches. Stems are angled, leaves trifoliate, ovate and serrate. The white or pink coloured flowers are small and borne on axillary raceme. Androecium is monodelphous (10 stamens). Pods are compressed, linear, erect and clustered, double ridge on dorsal side, single ridge below, length 4-10 cm, beaked; seeds 5-12 per pod, white to grey or black in colour, 5 mm long with an average weight of 0.06 g (Smartt, 1976).

4.0 Cultivars

The cultivars grown in South India are mostly vegetable types, while it is grown for seed in Northwest. There are two well-defined forms, namely, Giant and Dwarf (Anon., 1950). The types grown in Gujarat are Giant types while those of Punjab and UP are Dwarf. There are two Dwarf forms, namely, smooth (vegetable type) and hairy (fodder type).

Some of the cultivars of cluster bean in India are listed below.

Pusa Mausami

It is a selection from local cultivar of North India. It comes for the first picking in about 80 days after sowing. The pods are attractive, 10-12 cm in length. This cultivar is suitable for rainy season.

Pusa Sadabahar

It is a selection from a local cultivar 'Jaipuri' of Rajasthan. It is a non-branching type suitable for both summer and rainy seasons. The first picking can be done 55 days after sowing. The disadvantage with this cultivar is poor quality of pods (Dabas *et al.*, 1981).

Pusa Navbahar

It combines the good traits of both the above cultivars. The pods are longer (15 cm) and of better quality, the disadvantages being single-stemmed growth habit and susceptibility to bacterial blight and lodging (Dabas *et al.*, 1981).

Sharad Bahar

It is evolved on the basis of single plant selection from IC 11704 (a local collection from Maharashtra). It is a branching type producing 12 to 14 branches with 133 pods on an average per plant.

Pardeshi

It is a vegetable type cultivar, produces long pubescent pods about 12-15 cm long. It is grown mostly in Gujarat.

Besides the above cultivars, the National Bureau of Plant Genetic Resources has identified the following vegetable cultivars (Mittal *et al.*, 1977).

P 28-1-1

This cultivar was developed from a cross between Pusa Navbahar and IC 11521; it is photo insensitive, high branching and can be grown both during summer and rainy seasons. It yields about 140 to 150 q/ha.

Other cultivars evolved at different institutes are CP 78 from Coimbatore, 160-1, S 299-7 and S 279 from Anand and Punjab, Guar No. 1, 2, 3, 4 and 5 from Sirsa (Haryana).

Santa Cruz

It is a full season, sparsely branching, indeterminate, glabrous cultivar of USA, adapted for cultivation to high altitude (Ray and Stafford, 1985).

Cultivar, HG 75 gave the highest yields of 1.40-1.46 tonnes seeds, 0.45-0.56 tonnes proteins and 0.40 tonnes gum/ha (Jain *et al.*, 1987).

Dabas *et al.* (1981) suggested that a good vegetable type should have the traits like good yield, ideal plant type, disease tolerance, and smooth, long and succulent pods with less number of seeds.

Bundel Guar-1

This cultivar was selected from native accession B5/54. It gives green fodder yields of up to 350 q/ha (mean 257 q/ha), dry matter yields up to 65 q/ha (mean 50.82 q/ha) and yields of crude protein of up to 10.15 q/ha (mean 0.96 t/ha). The mean seed production is 10.49 q/ha. Reaching a height of 95-115 cm, Bundel Guar 1 is moderately resistant to *Xanthomonas campestris* pv. *cyamopsidis* and has a growth period of 130-135 days (Singh and Dixit, 1994).

Gaug 34

This cultivar released for cultivation in the cluster bean growing areas of India in 1991-92, has a yield potential of 10.5-20 q/ha, producing 0.3-11.2 per cent higher yields than other cultivars (RGC 967, IGFRI 212-1, HG 258 and HG 75). Gaug 34 has seeds with higher gum content than HG75 and RGC967 (30.9 per cent vs. 29.5 and 29.8 per cent, respectively). Bacterial blight [*Xanthomonas campestris* pv. *cyamopsidis*], *Alternaria* leaf spots and powdery mildew [*Leveillula taurica*] are major diseases of this crop. When tested under natural and artificial conditions, this cultivar was moderately resistant to these diseases (Bharodia *et al.*, 1993).

Maru Guar

This is a selection from IARI (India) collection for its superior performance in arid regions. The seed contains 32.1 per cent gum, higher than in other promising varieties. The protein content (31.7 per cent) is also higher than in other cultivars,

Types of Cluster Bean

Indian cluster bean   Twisted cluster bean

Promising Cultivar

Pusa Navbahar

except HG182 (33.5 per cent). The straight pods contain round, bold, greyish seeds (1000-seed weight = 33 g). Under normal rainfall, this cultivar shows resistance to *Alternaria* leaf-spot and bacterial blight. It matures in 97-100 days and in advanced trials (1982-85) at Jodhpur gave mean seed yields of 7.9 q/ha (Henry *et al.*, 1992).

Goma Manjari

This cultivar was released in 1998 from CHES, Godhra, India through selection from local germplasm. It is erect single stem with yield potential 88-103 q/ha. It is resistant to bacterial blight [*Xanthomonas campestris* pv. *cyamopsidis*], *Alternaria* leaf spots and powdery mildew [*Leveillula taurica*] disease.

IC 11388

Developed at NBPGR, New Delhi, India by single plant selection from a local collection of Sukhpur in Kutch. Average yield is 160 q/ha.

SelChes 13-7

Developed at Godhra, India through selection from local strain. It is tall, single stemmed, non-branching type and produce up to 35 clusters with 8-10 pods per cluster. Crop duration is 90-100 days. Average yield is 50-60 q/ha.

IC 11704

Developed at NBPGR, New Delhi, India by single plant selection. It is a branched type producing 12-14 branches per plant in crop duration of 120 days. Its average yield is 190-200 q/ha.

Other cultivars lines developed are CP78 (Coimbatore), S-299-7, S-279 (Ludhiana), Guar No.1, 2, 3, 4 and 5 (Hisar), *etc.*

5.0 Soil and Climate

5.1.0 Soil

It is a shallow rooted crop with surface feeding nature. The crop prefers well-drained sandy loam soil. It can tolerate saline and moderately alkaline soils with pH ranging between 7.5 and 8.0 and in heavy soils bacterial nodulation is inhibited (Venkataratnam, 1973). Studies on the effect of salinity and inoculation on growth, ion uptake and nitrogen fixation by guar indicated that a soil electrical conductivity (ECe) level of 8.3 dS/m caused a 60 per cent decline in dry matter and grain yield of guar plants whereas a soil ECe level of 10.0 dS/m was almost toxic. In contrast most of the studied strains of *Rhizobium* were salt tolerant. Nevertheless, nodulation, nitrogen fixation and total nitrogen concentration in the plant was markedly affected at high salt concentration. A noticeable decline in acetylene reduction activity occurred when the salinity level increased to 8.3 dS/m (El-Sayed, 1999a). Selvaraj and Kumari (1999) incubated root nodule segments of cluster beans in solutions containing nitrate, ammonium, tungsten or molybdenum. Amide-synthesizing indeterminate cluster bean nodules generally recorded very low enzyme activity compared with the ureide-synthesizing determinate nodules of the other species. Nitrate up to about 10 mM increased nitrate reductase activity (NRA), but higher concentrations mostly inhibited NRA. Cluster bean nodules were not affected by <20 mM NRA. Nitrate + ammonium generally increased NRA. Molybdenum increased NRA 3-fold in cluster beans. Tungsten decreased NRA in cluster beans at >50 µmol. Role of phosphatase-producing fungi on the growth and nutrition of cluster bean was studied by Tarafdar *et al.* (1995) in arid soils (loamy sand). A significant increase in phosphatase (acid and alkaline), dehydrogenase and nitrogenase activities was observed. Nodulation and nitrogenase activity was most affected upon inoculation. Inoculation significantly improved DM production, seed yield, number and DW of nodules and decreased shoot: root ratio. In general, there was a significant enhancement in the concentration of N, P, Ca, Mg, and Fe and Zn but no effect on Mn with the inoculation of any PPF. A significant improvement in K concentration due to inoculation *of Aspergillus fumigatus* and Cu concentration with the inoculation of *Aspergillus rugulosus* and *Aspergillus terreus* was noted. *A. rugulosus* was the best PPF tested, followed by *A. fumigatus.* They concluded that PPF have a significant effect on growth and nutrient uptake in guar and indicated the importance of these fungi production in an arid soil.

5.2.0 Climate

Cluster bean is a typical tropical vegetable crop preferring warm climate even though it can be grown in the subtropics during summer. The extent to which the crop can stand the heat is substantiated by the fact that it is grown during summer in arid zones of Rajasthan and Haryana, India where the temperature may go up to 44°C, average range being 30-40°C. However, the *kharif* crop yields better because irrigation has been found to enhance bud production (Rogers and Stafford, 1976). It prefers long-day conditions for growth and short-day conditions for induction of flowering.

6.0 Cultivation

6.1.0 Soil Preparation and Sowing

The soil is thoroughly prepared by ploughing and pulverizing it. The seeds are either broadcast or line sown at a distance of 25 to 30 cm (Venkataratnam, 1973). Dabas *et al.* (1981) recommended a spacing of 60 cm between rows and 12-15 cm within rows for Sharad Bahar cultivar. A spacing of 45 cm between rows gave higher yield than 30 or 60 cm apart (Jain *et al.,* 1987). Bains and Dhillon (1977) suggested a spacing of 45 × 15 cm. Rathore *et al.* (1990) obtained highest seed yield at closest spacing of 60 × 10 cm along with application of cycocel at 2000 ppm four weeks after sowing. Depending upon spacing and method of sowing, *i.e.,* broadcasting or line sowing, the seed rate varies from 15 kg to 40 kg/ha. According to Singh *et al.* (1994), 10-15 kg/ha seeds are sufficient for vegetable and grain crop and 35-40 kg/ha seeds for raising a fodder crop.

Because of more or less uniform climate prevailing all through the year in South India, it is grown round the year. Venkataratnam (1973) suggested growing cluster bean during May-June to augment supply of vegetable during the lean period. Singh *et al.* (1979) recommended sowing during June-July in Haryana, while for Punjab, Bains and Dhillon (1977) found better production by sowing during July-August.

In Hissar, India, cluster bean cultivars sown on 10 July gave higher seed, protein and gum yields than those sown on 25 June, 25 July or 10 August (Jain *et al.*, 1987). Early planting favours more vegetative growth, resulting lodging of the crop and low yields. The crop for fodder purpose can be sown from April to mid July.

Both flat bed and ridge and furrow methods are followed in raising the crop. Seeds are usually inoculated with *Rhizobium* culture (*Cowpea miscellany* group) @ 10 g per kilogram of seed for ensuring good nodulation and efficient nitrogen fixation (Singh *et al.*, 1994). Inoculated seeds should be kept under the shade to avoid high temperature and desiccation, so that inoculum remains viable till sowing. It is better to place the seeds at a depth of 5-7 cm in the lines. Yield of cluster bean at Parbhani during *Kharif* season was found highest by sowing in the centre of furrows and modifying the ridges about a month later in such a manner that the plant stems were placed in the centre of the ridges and such system may also ensure limited plant moisture loss. Studies conducted at Jodhpur has established that sowing in paired rows gives about 9 per cent (in normal season) and 46 per cent (in dry season) higher yields over traditional method of sowing (Singh *et al.*, 1994). Better weed control and microclimate in between two rows of a pair are the main reasons for giving higher yield in this system. The highest seed yield of cv. Pusa Sadabahar was obtained at the closest spacing (10 cm within the row) and spraying the crop with Cycocel @ 2000 ppm four weeks after sowing (Rathore *et al.*, 1990). In a greenhouse trail, obtained good seed and pod yield by treating the plants with 2000 ppm Chlormequat.

It is grown as main crop or mixed with other crops like cucurbits, cotton and sugarcane or as a border crop around the main crop line. However, cluster bean is best fitted in intercropping system particularly with pearl millet. Pearl millet and cluster bean inter-cropping proved advantageous because of difference in their growth and feeding habits. Inter-cropping also helps to stabilize farmer's income under dry land conditions. Combination of cluster bean (cv. DSE/16J) and pearl millet (cv. BJ 104) in 3: 1 ratio resulted the best performance at Jodhpur. In another trial, inter-cropping cluster bean variety Malosan or HG75 with Pearl millet at a row ratio of 2: 2 was superior in respect of equivalent yield, net returns, CBR and LER in Gujarat. In Jobner, Rajasthan pearl millet in paired row inter-cropped with 2 rows of cluster bean gave good response with regard to yield and net profit per unit area (Sharma *et al.*, 1988).

6.2.0 Manuring and Fertilization

The crop has the ability to fix atmospheric nitrogen, being a leguminous crop through symbiotic association with *Rhizobium*. The amount of fertilizer to be applied depends on soil type, availability of irrigation facilities, and weather condition. Most of the beans give good response to organic matter (Anonymous, 1950). The early crop sown under irrigated conditions is greatly benefited if five cartloads of FYM are applied to the land before sowing the crop (Yawalkar, 1992). When cluster bean is sown on poor soils after an exhaustive crop application of about 100-125 quintals of FYM or compost about one month before sowing is highly beneficial for the crop (Singh *et al.*, 2004).

Generally cluster bean shows good response to phosphorus and potash and poor response to nitrogen. So, a basal dose of 10-20 kg nitrogen/ha can be given to boost up the initial growth along with 50-70 kg each of phosphorus and potash per hectare (Anonymous, 1961). The entire quantity of FYM or other organic manure is generally added at land preparation and mixed with the soil before sowing while the full dose of NPK fertilizers are applied at the time of sowing. However, Singh *et al.* (1987) obtained good yield response by application of 40 kg N (half as basal and rest half 20 days after sowing), 60 kg P (as basal) and 500 ppm Cycocel (as foliar spray 25 days after sowing) in cv. Pusa Naubahar under rain fed condition.

It has been observed that deficiency of N greatly reduced pod weight and total plant weight. Weight of upper leaves, lower leaves and stems was markedly reduced due to the deficiency of Mg, K and N, respectively. Reductions in leaf concentration of elements under deficiency conditions were most severe in lower leaves for K, S and B, and in upper leaves for P, Ca, Mg and Zn (Haag *et al.*, 1990). Sathiyamoorthy and Vivekanandan (1995) sprayed cluster bean plants with 1.5 per cent ammonium nitrate, 2 per cent calcium nitrate, 2 per cent sodium nitrate or 1 per cent potassium nitrate on 4 dates at 10-day intervals and the effects were studied on vigour of the seeds produced. Seed germination was increased by 8-26 per cent and seedling vigour index by 32-83 per cent in plants given foliar N compared with water sprayed controls. Seedling vigour and seed germination was highest with potassium nitrate.

Singh *et al.* (1987) reported that under rainfed conditions, best results with regard to green pod and various yield components were obtained with 40 kg N (half as a basal dressing and half 20 days after sowing), 60 kg P (as a basal dressing) and 500 ppm cycocel (as foliar application 25 days after sowing) in cv. Pusa Navbahar. Seed inoculation with *Rhizobium japonicum* + 20 kg N/ha was most effective in increa-sing yield and uptake of N, P and K in seeds and straw also increased (Singh and Singh, 1990). Application of 20 kg P_2O_5/ha not only increased seed yield but also the gum and protein contents in seeds (Meena *et al.*, 1991). Seed inoculation or 40-60 kg P_2O_5 increased the protein contents but decreased the gum contents (Jain *et al.*, 1988).

Phosphorus nutrition has been found vital for cluster bean. Tarafdar and Rao (1990) found significant correlation between plant P uptake, dry matter yield and phosphatase activity in the rhizosphere of cluster bean. Higher shoot and root weight, N and P uptake has been found with *Glomus fasciculate* inoculation followed by application of phosphorus (Champawat and Somani, 1990). In sandy loam soil of Rajasthan, application of 20 kg P/ha was found optimum for getting the highest yield of cluster bean (Meena *et al.*, 1991). Single super phosphate has been found better than diammonium phosphate as the source of phosphorus fertilizer (Tiwana and Tiwana, 1994). Application of the fertilizer mixture 5-10 cm below the soil surface in furrows at the time of sowing results better utilization of plant nutrients.

Most of the studies implicated positive role of micronutrients however, Meena *et al.* (1991) could not record any effect of the trace elements on yield and yield components. Ghonisikar and Saxena (1973) found that two sprays of sodium molybdate (0.15 per cent) 15 and 30 days after seedling emergence gave better yield.

Significant role of zinc on growth and yield has been established from different studies. Incorporation of zinc sulphate @ 25 kg/ha along with recommended doses of fertilizers has been suggested for various soil types (Singh *et al.*, 1994). Application of zinc significantly influenced number of nodules, the nitrogenase activity of the nodules and carbohydrate and protein contents of the leaves (Nandwal *et al.*, 1990). El Sayed (1999b) found that the proportion of the total soil content of each element taken up by the guar, including that in roots, ranged from 0.97 to 2.40 per cent for Cu, from 0.90 to 3.08 per cent for Zn and from from 0.11 to 1.88 per cent for Mn. Shoot dry matter yield was significantly increased with the addition of Zn, either individually or in combination with Cu, Fe and Mn (Singh *et al.*, 2001). Other Zn combinations like, of Mn + Zn + B, Zn + B or Mn alone on non-calcic brown soil also proved better (Singh and Kothari, 1982). Response of molybdenum was also found quite encouraging and two spray of sodium molybdate @ 0.15 per cent at 15 and 30 days after seedling emergence gave better yield (Ghonsikar and Saxena, 1973). Role of cobalt and phosphorus in nutrition of leguminous crops is well established. Application of 2 mg cobalt and 40 mg P/kg of soil increased the green foliage and dry matter yield of cluster bean in a pot trial. Cobalt concentration significantly decreased with increasing levels of P and significantly increased with increasing Co levels (Singh and Singh, 1994).

6.3.0 Interculture

Cluster bean seeds take a week to germinate depending on the soil moisture status in the field. Thinning of crop stand should be done 10-15 days after sowing to maintain the desired plant population. Crop weed competition for moisture, nutrients, sunlight and space is intense under dry land condition and yield reduction as high as 70-90 per cent has been observed due to severe weed infestation in this growing condition. So it is necessary to keep the fields free from weeds mechanically by periodical weddings and hoeing. Generally two to three hand weeding are sufficient to keep the field weed free. However, the crop should be kept free from weeds for at least first 30 days after sowing to obtain high seed yield (Bhadoria *et al.*, 1992). Under rain fed condition, weeding between 30 and 45 days after sowing is sufficient (Kumar *et al.*, 1996) however, removing of weeds at 20 or 30 days after sowing was also found to increase pods per plant, water use efficiency and seed yield (Yadav, 1998).

Chemical weed control has also been advocated in cluster bean generally through pre-sowing application of different herbicides. Application of 2,4-D or Disodium Methane Arsenate (DSMA) @ 2.0 kg a.i./ha was found effective against *Parthenium hysterophorus* (Dutta *et al.*, 1976). Pre-sowing application of Fluchloralin @ 2 litres/ha checks weed growth up to 20-25 days and of Basalin @ 1 kg a.i./ha in upper 10 cm soil is effective against annual grasses and broad leaved weeds (Singh *et al.*, 2004). Another herbicide, Trifluralin @ 1.5 kg/ha was found effective when sprayed all over the field and incorporated in the upper 5 cm soil prior to sowing (Singh *et al.*, 1994). Application of Alachlor, Trifluraline and Nitrofen also proved effective in controlling the weeds and grain yield equal to that obtained from the hand weeded plots has been realized (Daulay and Singh, 1982). The weeds were kept under control at the later stages of crop growth due to the thick canopy of the crop.

6.4.0 Irrigation

Cluster bean is generally grown as a rainfed crop that does not need irrigation. However, one irrigation at 60 days after sowing increases the seed yield and harvest index of the crop (Meena *et al.*, 1991). During long duration of moisture stress, one or two supplementary irrigations, if available, prove effective. In summer, 2 to 3 irrigations at 10-12 days interval are sufficient to meet the evapo-transpiration requirement of the crop (Singh *et al.*, 1994).

Sharma and Kuhad (1999) studied effect of K on physiological processes under water stress using the cvs. HG-72 and HG-75 grown in pots under greenhouse conditions. Control plants were maintained at 12 per cent soil moisture content (SMC). Water stress at 5.5 per cent SMC created by withholding irrigation at flowering (55 DAS) stage and at 50 per cent pod formation (75 DAS) resulted in a significant decrease in relative water content (RWC), water potential (psiw) and osmotic potential (psis). Increased con-centration of K brought an improvement in these parameters under water stress. Improved water status with K in terms of RWC and psiw significantly increased the seed yield. Flowering was the most sensitive stage. HG-75, a drought-sensitive cultivar responded best to applied potassium.

Cluster bean–wheat sequence on sandy loam soil of Rajasthan gave an economic yield of 44.33 quintal/ha and water use efficiency of 5.67 kg/ha/mm (Singh *et al.*, 1998). Grass mulching is useful to prevent moisture loss from the soil and increase root growth, nodulation, shoot and plant growth and water use efficiency (Gupta and Gupta, 1983). However, proper drainage is more important during kharif season as the crop cannot tolerate water logging.

7.0 Harvesting, Yield and Storage

Green pods are harvested by twisting or cutting from 40-45 days onwards depending on the variety and generally continued up to 120 days. All the pods at the leaf axil mature simultaneously which facilitate the harvest. Dry beans are harvested when maximum pods are fully ripe and turned yellow until the lower pods become dry enough for shattering.

Grains usually mature within 90 to 100 days depending upon the variety, soil type, rainfall and its distribution (Singh *et al.*, 1994). In light soils of arid region, crop matures 10-15 days early. After harvesting, the stalks are left for drying for 1 to 2 weeks and then trampled over by bullocks or tractor to separate the seeds from the pod.

The crop for green forage should be harvested between flowering and pod formation stage (Singh, 1989) and it should not be allowed to over mature as it will change into woody and fibrous substances resulting loss of digestive nutrients and palatability (Sharma, 1992).

The yield varies from 5 to 8 t/ha of tender green pods and 6 to 10 q/ha of seeds.

Tender green pods are largely used as vegetable however, the pods are also dehydrated and stored. Pods are generally harvested early in the morning and generally sent to the market by packing in gunny bags or bamboo baskets. Young pods do not have longer shelf life due to high respiration rate and common deteriorative symptoms are shrivelling, chlorosis and over-maturity. Tender pods can be kept for two days with frequent sprinkling of water under ambient condition but can be stored for 15-20 days at 0 °C with 85-90 per cent relative humidity in cold storage (Singh *et al.*, 2004). Vitamin C content was found significantly higher in the tender pod without the stalk after subjecting to different processing techniques such as cutting, washing, washing before cutting and cooking with the use of lid (Tapadia *et al.*, 1995).

The pods contain many small seeds which when ripe become black, white or grey and are cooked as vegetable. The seeds show orthodox type of storage behaviour and remain viable for a short period under ambient condition. The dehydrogenase enzyme activity was found less in the seeds stored at ambient condition than in cold storage (Doijode, 1989). However, keeping dry seeds in sealed or laminated bags extends the seed viability for 36 months (Doijode, 1986).

8.0 Diseases and pests

8.1.0 Diseases

The common diseases of cluster bean are wilt (*Fusarium* sp.), bacterial blight (*Xanthomonas cyamopsidis*), powdery mildew (*Leveillula taurica*) and anthracnose (*Colletotrichum* sp.).

Bacterial Blight

It is the most serious seed-borne and bacterial disease usually occurs in kharif season throughout cluster bean growing areas of this country. The disease appears as small, circular, water-soaked spots on the dorsal surface of the leaf. These spots coalesced to form bigger spots surrounded by chlorotic area on older leaves. The pathogen invades vascular tissues and causes flaccidity of the affected portion. The flaccid spots become necrotic and turn brown. Finally defoliation and black longitudinal streak in the stem occur. The conditions favouring the disease spread are spattering rains, high humidity and warm temperature (28-30°C).

Seed treatment was found more effective than foliar sprays in controlling the disease (Sindhan *et al.*, 1999). Seed treatment with hot water at 56°C for 10 minutes with Streptocycline (0.025 per cent) has been found very effective (Gupta, 1977). Rathore (2000) recommended seed dipping for 3 hours in aqueous solution of Streptocycline (0.02 per cent). However, could control the incidence of blight by two sprays of Streptocycline (250 mg/l water) in the standing crop. Growing of tolerant varieties like RGC 471, HG 75, HG 182 and GAUG 63. Plant debris should be removed before sowing.

Alternaria Leaf Spot

This disease is mainly prevalent in Haryana, Tamil Nadu, and parts of Rajasthan. The disease appear as dark brown round to irregular spots varying from 2 to 10 mm in diameter on leaf blades. These spots spread rapidly to form circular lesions in humid weather condition. In severe cases, concentric rings of dark-brown conidiophores are developed and the leaflets become chlorotic and usually drop off.

Field Growing and Harvesting of Cluster Bean

Pod bearing stage of cluster bean

Field view of cluster bean cultivation

Cluster bean harvesting

Diseases of Cluster Bean

Alternaria leaf spot infection in cluster bean

Bacterial leaf blight infection in cluster bean

Two sprays of Zineb (0.2 per cent) at an internal of 15 days reduce the disease incidence. Application of Iprodione, which appeared to enhance the activities of oxidative enzymes, was found most effective against this disease. Application of trace elements was also effective to control the disease. The variety HG 182 has been found tolerant to this disease in both pot and field trials at Mysore, Karnataka (Shivanna and Shetty, 1991).

Myrothecium Leaf Spot

This disease appears as small to minute oil-soaked spots of 1-2 mm in diameter. Later, these spots turn brown in colour, coalesce with other and cover fairly large area of leaf. High temperature conditions favour the spread of this disease.

Spraying of Diathane Z-78 (0.2 per cent) 5 weeks after sowing is effective in reducing this disease.

Anthracnose

The disease appears as brown to black spots on leaves, petioles, and stems during rainy season. It can be controlled by spraying of 0.2 per cent Diathane Z – 78. Khan and Verma (1990) reported that when aqueous leaf extract of *Pseuderanthemum biocolor* is sprayed or rubbed on *C. tetragonolobus* stimulated the synthesis of a virus inhibitory agent (VIA) which is associated with the induction of systemic resistance against virus infection.

Powdery Mildew

The disease is mostly prevalent in Karnataka, Gujarat, and parts of Rajasthan. The disease appears as white mycelial patches dotted with the fruiting bodies on the abaxial leaf surface, stem and even on pod also. Severely infected plants are defoliated and weakened by premature drying up and death of infected leaves.

Solanki and Singh (1976) recommended spraying of N-tridecyl-2, 6-dimethyl morpholine at 0.05 per cent at fortnightly interval. Spraying of Benomyl (0.025 per cent) or Dinocap (0.1 per cent) is highly effective against this disease. Growing of tolerant variety GAUG 63 (Pandey *et al.*, 1993).

Dry Root Rot

This seed and soil borne pathogen is most common in sandy soils of all cluster bean growing regions. Warm temperature (28-35°C) and moisture stress condition during post flowering period favour its occurrence. Initially reddish-brown discolouration on the stem is noticed. In advance stage, the tissue becomes darkened and numerous small black sclerotia are formed both inside and outside of the tissue and also on the roots. Very few pods are formed on the infected plants.

Crop rotation with non-host or less susceptible crops like moth bean, pearl millet, *etc.* should be practiced. Seed treatment with Bavistin (0.2 per cent) reduces seed-borne infection, improves germination and consequently leading to the highest seed and also fodder yield (Rathore, 2000). Growing of tolerant varieties like, Kutch-8 and RGC 471 (Singh *et al.*, 1994). Mulching with crop stubbles (3.5 t/ha) has been found effective in reducing plant mortality (Singh *et al.*, 1994). Soil drenching with Quintozone 4 days prior to sowing of seeds ensures high of 73 per cent disease control.

8.2.0 Pests

The important insect pests infesting the crop are aphids, jassids, Bihar hairy caterpillar and pod borer. Sometimes, root-knot nematode (*Meloidogyne incognita*); white fly (*Acaudaleyrodes citri*) and Scarabeid (*Protaetia terrosa*) have been reported to cause serious damage to this crop. Short descriptions of some important insect-pests are given below:

Aphid (*Aphis craccivora*)

This tiny insect suck the sap from the tender aerial parts of the plant and shows some characteristics symptoms like curling of leaves, twisting of twigs and developing fruits and sometimes shedding of flowers. Seed development is highly affected and consequently seed yield is severely reduced.

Spraying the crop with 0.03 per cent imidacloprid or 0.15 per cent acetamiprid. Application of granular insecticide like Carbofuran (Furadan 3G @ 1.0 Kg a.i./ha).

Hairy Caterpillar (*Diacrisia obliqua*)

The larvae first feed on the foliage and under severe attack whole plant is destroyed showing characteristic skeletonization of leaves. Spraying of 0.25 per cent chlorpyriphos is very much effective against this pest.

Jassid (*Empoasca fabae*)

Both nymphs and adults suck cell sap usually from ventral leaf surface and the affected leaves show typical hopper burn symptoms in case of severe infestation. Spraying the crop with 0.03 per cent imidacloprid or 0.15 per cent acetamiprid.

Pod Borer (*Helicoverpa armigera*)

This caterpillar bores into the developing pods and feeds on the seeds thein as a result of which both pods and seeds are unfit for consumption.

Removal and destruction of infested pods at the initial stage of attack. Spray the crop with indoxacarb 0.12 ml/l or Spinosad 0.3 ml/l is effective.

In Northern Gujarat, massive white fly infestation for the first time in cluster bean.

Cluster bean genotypes like, JG-2, HGV-75 and GAUG-26 showed moderate resistance to the root knot nematode (Yadav *et al.*, 2000).

Dusting with 2 per cent Methyl parathion at the time of insect appearance and soil drenching with chlorpyriphos @ 2.5 litres/ha at 25 days after germination gave some protection against Scarabeid.

9.0 Seed Production

The crop is raised for its seed as indicated earlier and for its industrial uses. The seed crop of the vegetable type is not normally raised on commercial basis because of limited area under this crop. The agronomic practices for seed crop are more or less same as that of the crop raised for vegetable purposes. The seed rate for seed production is about 10-25 kg/ha and seed of about 1 t/ha can be obtained under dryland conditions. The plants are pulled out for threshing. But care must be exercised in threshing as indiscriminate threshing affects the seed quality.

Though beans are self-pollinated crops, yet some out-crossings have been reported. In cluster bean, 2 per cent out-crossing occurs (Anon., 1961b). It is, therefore, recommended that two cultivars of cluster bean should be sown 50 m apart in order to get pure seed (Anon., 1971).

10.0 Crop Improvement

It is a highly self-pollinated crop with very limited out crossing. Limited genetic diversity for the different traits coupled with yield losses caused by different biotic and abiotic stresses has constrained intensive breeding efforts of this crop. Despite these limitations, progress has been achieved by exploiting the available diversity and different breeding methods like pedigree selection in the cluster bean improvement program with good results. Hybridization has not been successful due to the small, delicate flower resulting in a low percentage of hybrid seed setting. Mutation breeding has helped to generate some genetic variability in certain important traits, but it has had limited impact on cluster-bean breeding. DNA-based molecular markers are being extensively used in genetic diversity analysis and phylogenetic studies. Efforts are also underway to develop genomic resources (SSR and SNP markers) that could be used in cluster-bean breeding. Limited transcriptomic and mi-RNA studies have also been recently reported. With the availability of advanced molecular tools it would be feasible to develop resources to aid in high-yielding, cluster-bean varieties with moderate to high gum content for a much preferred export commodity.

In India, cluster bean improvement program was started in 1950s. The initial breeding emphasis was on vegetable type genotypes but later, due to realization of cluster bean as industrial crop, research on seed type genotypes for gum content was intensified. Therefore, in 1961, systematic development of genetic resources was initiated by Division of Plant Introduction, Indian Agricultural Research Institute (Now NBPGR), New Delhi by collecting diverse guar germplasm sources from prime areas (Gujarat, Rajasthan and Punjab) of variability (Thomas and Dabas, 1982). Real momentum was picked up during 1965–70 by launching a scheme entitled: Collection and isolation of superior genotypes for gum purposes from, under Public Law 480 in 1954 (Agricultural Trade Development and Assistance Act). Consequently, two catalogues were published in succession containing information on 1150 accessions in 1981(Dabas *et al.*, 1981) and 3580 accessions in 1988 (Dabas *et al.*, 1989). In addition to this, independent collections were made and maintained at various research centres of state agriculture universities especially in the North-western Indian states (Rajasthan, Gujarat, Haryana and Punjab).

According to Dabas *et al.* (1981), a good vegetable type of cluster bean should have the traits like high yield with good quality, ideal plant type, smooth, long and succulent pods with less number of seeds and disease tolerance. Breeding objectives for cluster bean should be aimed at these characters. The improvement work on cluster bean is limited only to a few Research Stations. The cultivars Pusa Mausami and Pusa Sadabahar have been evolved from local selection at the Indian Agricultural Research Institute (IARI), New Delhi. Pusa Navbahar cultivar, also evolved at IARI, combines the good traits of the above mentioned cultivars. In Maharashtra, Sharad Bahar has been evolved from local selection. Besides these, the National Bureau of Plant Genetic Resources has identified a promising line P 28-1-1 (selected from a cross between Pusa Navbahar and IC 11521). It is photo-insensitive, high branching and can be grown during both summer and rainy seasons.

10.1.0 Genetic Diversity and Germplasm Resources

The availability of genetic diversity and its successful collection, maintenance, utilization and conservation is pre-requisite for crop improvement program. The genetic variability existing in cluster bean germplasm has been evaluated by using various morphological and biochemical traits. Conspicuous morphological variations are displayed for branching (branched/unbranched), pubescence (hairy/smooth), pod shape(straight/sickle), growth habit (determinate to indeterminate), pod bearing pattern (regular/irregular) (Saini _et al._, 1981). In India, the exhaustive catalogue by Dabas _et al._ (1989) has provided information on fifteen morpho-physiological and yield traits along with place of collection for all 3580 accessions. Many important traits like plant height (46.75–239 cm), clusters per plant (1.75–64.5), pods per plant (2.25–262.35), pod length (1.85–19.3 cm), seeds per pod (4.15–13.0) days to maturity (128–185 d), seed yield (0.95–59.7) and 100 seed weight (1.9–4.75 g) showed a high degree of diversity. Though, there was paucity of early flowering and maturing genotypes in this collection but was later on enriched by many collections evaluated at Regional Station, National Bureau of Plant Genetic Resources (NBPGR), Jodhpur, India where certain genotypes flowered as early as 28 days and matured within 70 days. Many other studies involving comparatively limited genotypes though reported considerable diversity towards desirable direction but with reduced range (Mishra _et al._, 2009; Pathak _et al._, 2011a, 2011c). Plant height and branches per plant on a higher side are important as they bear more clusters and branches (Dabas _et al._, 1989). High level of diversity for yield traits _viz._ number of pods, clusters _etc._ have also been reported by many studies alongwith other morphological traits (Mittal _et al._, 1977; Dabas _et al._,1982; Henry and Mathur, 2005; Mahla and Kumar, 2006; Pathak _et al._, 2009; Pathak _et al._, 2010a; Kumar _et al._, 2014; Girish _et al._, 2012; Sultan _et al._, 2012; Manivannan and Anandakumar, 2013; Manivannan _et al._, 2015; Kumar and Ram, 2015). Morris (2010) characterized 73 accessions collected from India, Pakistan and USA and reported enough genetic variability for pod length (32–110 mm), 100 seed weight (2.3–4.8 g) and number of days to 50 per cent maturity (96–185 days).

Thirty genotypes of cluster bean were assessed for genetic divergence using Mahalanobis D^2 technique (Kumar _et al._, 2014). Based on D^2 values, 30 genotypes were classified into five clusters. The percent contribution towards genetic diversity was highest from days to maturity (37.87 per cent) followed by pod yield (q/ha) (22.32 per cent) and seed yield/plant (g). On the basis of inter-cluster distances and _per se_ performance observed in the present study, a hybridization programme involving genotypes IC- 258087, IC-258092, IC-369789, IC-369868, IC-370490, IC-373427, IC-373480, IC-402293, IC-370478, IC-415137, IC-415142, IC-415157, IC-415159, IC-421242, IC-421798 and IC-421806 can be utilized as parents in the hybridization programme for the improvement of pod yield and quality.

A study was undertaken to evaluate the physiological variability for drought tolerance and genetic diversity among nineteen advanced breeding lines by using SSR marker (Rashmi and Mohan Kumar, 2018). Nineteen advanced breeding lines including the check variety Pusa Navbahar were raised in pots containing soil, sand and fertilizers in the ratio 1: 2: 1 and screened for relative water content (RWC), which is the most appropriate estimate of plant water status in terms of physiological consequence of cellular water deficit 35 DAS under drought. Maximum per cent of RWC was recorded in Pusa Navbahar (83.34 per cent) followed by the line COHBCB 41 (85.12 per cent). Molecular markers are being widely used in various areas of plant breeding as an important tool for evaluating genetic diversity and determining cultivars identity. A dendrogram constructed based on unweighed pair group method of arithmetic means (UPGMA) using data derived from 22 SSRs grouped the 19 cluster bean advanced breeding lines into two main clusters with 0.62 similarity. The ranges of dissimilarity among the cluster beans of advanced breeding lines were varied from 0.62 to 1.00.

Jukanti *et al.* (2014) evaluated 140 germplasm lines of cluster bean for agronomic and yield. Wide variation was observed among agronomic and yield-related traits among the accessions. High heritability (~85 per cent) coupled with high genetic advance (~30 per cent) was observed for total yield, pods per cluster and cluster on the main branch. Seed number and total yield exhibited significant positive correlation with pod length (0.55) and number of pods (0.85) respectively. Principal component analysis revealed significant variation among the characters with the first four principal components explaining about 70.8 per cent of the total variation. Projecting the germplasm accession onto the first two principal components revealed two groups: (i) accessions showing high pods per cluster, total pods and total yield and (ii) accessions having high seed number. The gum content and endosperm content ranged between 26.95 and 31.68 per cent and 36.4 to 42.5 per cent, respectively. RAPD analysis indicated polymorphism in banding pattern among the selected germplasm lines. Among the 140 germplasm lines evaluated, IC-421815 was the best performing line (yield - 31.53 g/plant; pod number/plant - 139.8; and pods per cluster - 8.6) compared to five check varieties.

Presently National Gene Bank at NBPGR, India is maintain about 5000 accessions of cluster bean along with *C. serrata* and *C. senegalensis* + two close wild relatives of guar. Cluster bean improvement work started at Durgapura (Rajasthan), Hisar (Haryana) and CAZRI (Rajasthan) with collection and assessment of germplasm which resulted selection/identification of a number of high yielding genotypes (Paroda and Saini, 1978; Henry *et al.*, 1992; Mahla *et al.*, 2011).

A number of studies have indicated significant correlations among economically important traits resulting in simultaneous improvement of various traits (Henry and Mathur, 2008; Mahla and Kumar, 2006; Shekhawat and Singhania, 2005; Singh *et al.*, 2005a, 2005b). Pod number which is the main component of green pod yield should be given maximum emphasis in selection programme of cluster bean (Vijay, 1988). Many economically important traits like pods per plant, pod length, clusters per plant, number of branches and hundred seed weight were found positively associated with seed yield (Sidhu *et al.*, 1982; Vijay, 1988; Gresta *et al.*, 2016). Positive correlations among grain yield and gum content were not followed for seed weight and gum percentage (Menon *et al.*, 1970; Mittal *et al.*, 1971). However, percent endosperm was positively correlated withgum percentage (Menon *et al.*, 1970; Lal and Gupta, 1977).

10.2.0 Induced Variation through Mutation

The utilization of artificial and/or induced mutation to create new elite alleles is a powerful tool in crops like cluster bean where variability at genetic level is low. However, in comparison to induced mutation, frequency of desired natural mutation is very low for accelerated plant breeding. Hence, artificial mutation using physical and chemical mutagens is the best way to expand genetic variability in short time period (Auti, 2012). Though, the success of mutation either natural or induced depends on efficiency to create desirable changes with least undesirable changes (Harten, 1998; Pathak, 2015). Artificial mutagenesis could be better option for enriching variation in guar having small cleistogamous flowers difficult for emasculation and further artificial crossing (Arora and Pahuja, 2008). Induced mutation by gamma rays in cluster bean was for the first time carried out by Vig (1965) where 6 °C was a source of radiation. He reported low fertility mutant in a gamma (200 Gray) irradiated population of cluster bean cv. Punjab G2. Semi-sterile variants reported by him were outcome of reciprocal translocations and compared with fertile plants; these variants were taller and low yielder and showed more racemes perplant with prolonged vegetative period. Study of Vig (1969) also pointed out that the guar chromosomes are radio-resistance as a dose as 30,000 kR as showed no detectable effect on the chromosomes. This might be a delayed effect of radiation due to small chromosomes size. Adverse effects of irradiation doses (10, 20, 40, 60, 80, 100,150 and 200 kR) on morphology of guar was recorded by Lather and Chowdhury (1972) where germination rate, seedling mortality rate and pollen fertility were inversely proportional to radiation dosage. Though, mutants generated at low doses (2, 5, 10, 15 and 20 kR) of gamma rays showed increments for number of seed per plant, protein and galactomannan in seeds of M_2 generation (Chaudhary *et al.*, 1973). A high yielding stable early-flowering mutant of Pusa Navbahar was developed by Rao and Rao (1982) in population created by irradiation from 6°Co source of X-rays (100 Gray). Mahla *et al.* (2005, 2010) and Velu *et al.* (2007, 2012) carried out physical (gamma-rays) and chemical (EMS) mutagenesis and detected a gradual decrement in plant height as well as many other agronomic traits with the increased mutagen dose in cluster bean cultivars. Similar results were also obtained previously by Bhosale and Kothekar (2010) in the cluster bean. They recorded increase in the mutation frequency with increased doses of gamma rays and chemical mutagens EMS and sodium azide (SA). Polyploidizationwas also explored as means to increase the variability by Vig (1963) and Bewal *et al.* (2009) however, no significant achievement was reported.

10.3.0 Distant Hybridization

Distant hybridization though might have potential in genetic improvement, has so far remained unexploited and less successful in cluster bean, due to incompatibility in species combinations probably because of fragile pollen germination, abnormal pollen tube development or incompatible interaction of pollen with stigma and style. A complete failure of conventional wide crossing between cluster bean and *C. serrata* has been revealed by Sandhu (1988) and this might be due to the unreceptiveness of stigma for foreign pollens. To make stigma receptive Sandhu (1988) also deployed other approaches like bud pollination, stigma/style amputation, use of organic solvents but was unsuccessful to get the stigma-pollen compatibility. Thus, in cluster bean, interspecific hybridization is limited because of pre-and/or post-fertilization barriers (Ahlawat *et al.*, 2013a, b). Mathiyazhagan (2007) reported failure in achieving interspecific hybridization between cultivated guar genotypes and wild species.

10.4.0 Male Sterility and Heterosis Breeding

Male sterility phenomenon was not exploited much in cluster bean (Arora and Pahuja, 2008). This is either due to non-availability of natural outcrossing system, or an efficient male-sterility system or both. A number of natural mutations have been reported in guar for male sterility (Stafford, 1989; Mittal *et al.*, 1968; Vig, 1965). Spontaneous, complete, as well as partial male sterility in cluster bean was observed by Mittal *et al.* (1968). He also mentioned that this trait was inherited monogenically where pollen fertility was recorded dominant over sterility. During induced mutagenesis, Vig (1965) and Kinnmann *et al.* (1969) found that reciprocal translocations phenomenon caused semi-sterility in guar and was responsible for setting of fewer seeds in pod, more racemes per plant and extra plant height than fertile plants. Later on in South Africa, Stafford (1989) recorded a partially sterile mutant in cluster bean having rosette-type inflorescence (raceme). During the study, Stafford (1989) also described that partial mate-sterility in cluster bean is governed by two genes, of which one is dominant epistatic while another is incomplete dominant. As per Stafford (1989) this partial male-sterility (pms) was totally unique from genetic male-sterility (ms) observed by Mittal *et al.* (1968).

Several workers described an extensive range of heterosis in cluster bean (Chaudhary *et al.*, 1981; Saini *et al.*, 1990; Arora *et al.*, 1998). The best overall combination of desirable characters was found in the cross HFG 516 × HFG 590 and therefore, recommended for use in cluster bean breeding programmes (Hooda *et al.*, 1990). However, for commercial exploitation of heterosis, stable male sterility along with complete fertility restoration is a pre-requisite. Due to absence of stable source of male sterility a limited work on heterosis breeding has been reported in cluster bean. The opportunity of heterosis breeding is practically low in the absence of methods for economic production of large quantities of hybrid seed (Pathak, 2015). Owing to unavailability of stable male sterility, efforts have been made to develop stable and operational male sterility by male gametocides or chemical hybridizing agents (CHAs) and physical agents. Nisha and Chauhan (2006) found benzotriazole, maleic hydrazide (MH) and ethrel (ethephon) to be useful in induction of male sterility in cluster bean. Various concentration and treatments of all three chemicals induced pollen sterility between 92 and 100 per cent. Although ethrel (0.3 per cent) was more sensitive and induce complete sterility but associated significantly in yield reduction. Chauhan and Nisha (2006) reported that foliar spray of benzotriazole induced 93–100 per cent pollen sterility in cluster bean. A comparative light and transmission electron microscopic study showed abnormal behaviour of tapetal mitochondria in pollen caused pollen abortion. Shinde and More (2010) analysed the effect of different mutagenic treatments on the pollen sterility while working on gamma rays mutagenesis. Gamma rays at 400 gray (40 kR) dosage were found to be most effective as induced maximum sterility (20.83 per cent). Lower doses of different mutagens single or in combination induce higher variability than compared to higher doses (Kumar *et al.*, 2013).

10.5.0 Biotechnology

Despite variability at morpho-agronomic level, isozymes variation study through electrophoretic technique in guar germplasm has not been employed in depth for genetic diversity analysis. Mauria (2000) made primary attempt in cluster bean to understand domestication through isozyme diversity in primitive landrace accessions from India along with released cultivars from USA, and two wild relatives (*C. serrata* and *C. senegalensis*). Later on, Brahmi *et al.* (2004) made attempt to resolve the diversity in cluster bean germplasm with allozyme markers and reported greater inter-population diversity as compared with the overall genetic diversity in guar. Because of environmental sensitivity, inconsistency, fewer loci and low polymorphism of protein markers (isozymes andallozymes), marker analysis inclined towards DNA-based marker system. Previous studies indicated that employing DNA markers in crop improvement can economize both time and resources (Shah *et al.*, 2015a). In the last decade or so, studies have been conducted for assessing genetic diversity and phylogenetic of cluster bean using DNA markers. Among various DNA based marker system, Random Amplification of Polymorphic DNA (RAPD) has been more frquently used in cluster bean for diversity analysis and to study genetic relationships (Weixin *et al.*, 2009; Punia *et al.*, 2009; Pathak *et al.*, 2011d; Rodge *et al.*, 2012; Sharma *et al.*, 2013, 2014a, b; Sharma and Sharma, 2013; Kuravadi *et al.*, 2013; Ajit and Priyadarshani, 2013; Kumar *et al.*, 2013; Kalaskar *et al.*, 2014; Patel *et al.*, 2014). Though, RAPD markers show low reproducibility and not repeatable but have been extensively used for assessment of genetic diversity, germplasm characterization, cultivar identification, genetic purity testing, and gene tagging. Sharma *et al.* (2014b) reported that Inter-Simple Sequence Repeat (ISSR) is more powerful than RAPD due to their higher capacity to reveal polymorphisms and greater potential to determine intra and inter-genomic diversity. As compared to other legumes, sequence based DNA markers especially microsatellites or simple sequence repeats (SSRs) are inadequate in cluster bean (Kumar *et al.*, 2015) consequently the molecular breeding efforts in cluster bean are not implemented swiftly (Kuravadi *et al.*, 2014).

10.5.1 Tissue Culture, Regeneration Protocols and Genetic Transformation

Success of *in-vitro* regeneration of any plant species depends on several factors like explants and explant source, medium com-position, type of hormones, media

In-Vitro Regenaration of Cluster bean

Direct shoot-bud differentiation from cluster bean cotyledonary node explants

(A) Explant types: type I cotyledonary node with main shoot and type II cotyledonary node after removal of main shoot. (B) Type II explant showing multiple shoot initials on MS + 5.0 µM TDZ in a 4 week-old culture. (C) Multiple shoot buds on a cotyledonary node explant on 5.0 µM BA +0.5 µM IAA, initially exposed to 5.0 µM TDZ. (D) Ex vitro rooted shoot, after 15 min treatment with 300 µM IBA. (E) Elongated shoots on 5.0 µM BA +0.5 µM IAA. (F) An acclimatised plantlet in an 8-cm diameter cup (Ahmad and Anis, 2007).

composition, culture condition *etc*. In context to in-vitro regeneration and tissue culture, the work on cluster bean was started by Ramulu and Rao (1987). Both these researchers reported many *in vitro* regeneration protocols on MS and B5 media for establishment and enhancement of callus in cluster bean (Ramulu and Rao, 1989, 1991, 1993, 1996). However, attempts for callus induction were not just limited to unorganized mass, but reports are available on direct differentiation from cotyledonary nodes using cytokinins (Prem *et al.*, 2003) and shoot organogenesis in cluster bean via callus culture (Prem *et al.*, 2005). To induce endosperm callus from embryo orcotyledon seed explants, Bhansali (2011) reported that MS media supplemented with 2, 4-D, IAA, NAA in combination with BAP is the most suitable. Till 2013, the work was carried out on cultivated species but Ahlawat *et al.* (2013a) successfully induced callusing in cultivated as well as two wild species *viz. C. serrata* and *C. senegalensis* using cotyledon explant. Whilst, most of the research proved 2,4-D and BAP as excellent plant growth regulators to induce callus and regeneration, Gargi *et al.* (2012) showed multiple shoot induction with gibberellins along with 2,4-D and BAP as PGRs whereas, Meghwal *et al.* (2014) showed that the combination of BA, kinetin and gibberellins was most suitable for multiple shoot generation. In order achieve to prompt shoot multiplication from cotyledonary node explants, Ahmad and Anis (2007) and Ahmad *et al.* (2012) used thidiazuron (TDZ-phenyl-urea derivatives) instead of conventional cytokinin activity of BAP. Verma *et al.* (2013) found regeneration with the same explant placed on medium supplemented with various combinations of indole-3-butyric acid (IBA), BAP and gibberellin and they also observed rooting on the same medium after 10 days of shooting. They transferred these rooted shoot plantlets for the acclimatization on coco peat mixture and showed successful hardening. Progress was also seen in the establishment of the regeneration protocol where Mathiyazhagan *et al.* (2013) reported somatic embryogenesis on MS medium fortified with 2 mg/l NAA, 0.5 mg/l and 3g/l charcoal using mature embryo as an explant in cultivated species of cluster bean. They also reported direct shoot regeneration from cotyledonary node placed on MS medium containing 1 mg/l Kn, 0.5 mg/l and 1 mg/l Zeatin on both cultivated and wild species. The information on somatic embryogenesis in cluster bean is still absent. The race for the establishment of simple and efficient regeneration protocol was started from the work done by Prem *et al.* (2003) who tried to regenerate plant from cotyledonary node explant and showed successful plantlet formation and subsequent hardening. And still the race is going on to establish "inefficient" regeneration protocol (Sheikh *et al.*, 2015).

An important work regarding promoter, which enhances tissue specific expression of transgene, was carried out by Rasmussen and Donaldson (2006). They isolated promoter of sucrose synthase gene from rice termed *rsus 3* and evaluated luciferase and GUS activity, under its regulation, in guar endosperm. They found the promoter very strong for endosperm tissue. Considering this work for the enhancement of galactomannan content through over expression of respective gene in endosperm, one can have an option to manipulate this promoter influencing over-expression of down-stream transgene. Similar promoter from guar (mannan synthase promoter) was also isolated by Naoumkina and Dixon (2011).

Ray and Stafford (1985) summarized the genetical studies done in cluster bean and had given the gene symbols for various qualitative characters in Table 2.

Table 2: A Summary of the Qualitative Genetic Studies in Guar and the Symbols Assigned*

Character	Reported Descriptive Name and Mode of Inheritance	Reported symbol
Plant type	Normal dominant	B
	Bushy: recessive	b
Branching habit	Non-branched: dominant	NA
	Branched: recessive	NA
	Branched: dominant	Br
	Non-branched: recessive	br
	Non-branched: dominant	B
	Branched: recessive	b
	Branched: dominant	Br1.Br2
	Non-branched: recessive	br1.br1 Br2-
	(Duplicate recessive epistasis)	Br1-br2br2
		br1br1br2br2
Raceme clustering habit	Clusters not at each node dominant	Cl
	Clusters at each node: recessive	cl
	Alternate bearing: dominant	NA
	All node bearing: recessive	NA
Growth habit	Indeterminate: dominant	De
	Determinate: recessive	de
Pubescent	Pubescent: dominant	G
	Glabrous: recessive	g
	Hairy: dominant	H
	Non-hairy: recessive	h
	Hairy: dominant	NA
	Non-hairy: recessive	NA
	Hairy: dominant	Hr
	Non-hairy: recessive	hr
Leaf size	Narrow leaflet: dominant	Ls
	Broad leaflet: recessive	ls
Male sterility	Male fertile: dominant	Ms
	Male sterile: recessive	ms
Leaf shape	Narrow leaflet: dominant	N
	Normal leaflet: recessive	n
Seed shape	Flat: dominant	R
	Round: recessive	r

Character	Reported Descriptive Name and Mode of Inheritance	Reported symbol
	Flat: dominant	NA
	Round: recessive	NA
Leaf colour	Green: dominant	Y
	Yellow: recessive	y
Asynapsis	Synapsis: dominant	NA
	Asynapsis: recessive	NA
Foliage colour	Dark green: dominant	NA
	White: recessive	NA
Flower colour	Purple: dominant	NA
	White: recessive	NA
Leaf margin	Serrated: dominant	NA
	Smooth: recessive	NA
Pod shape	Straight pod: dominant	NA
	Crescent pod: recessive	NA
Seed size	Intermediate: dominant	NA
	Small: recessive	NA

NA: No symbol assigned. *Ray and Stafford (1985).

11.0 References

Ahlawat, A., Dhingra, H.R. and Dhankar, J.S. (2013a) *J. Krishi Vigyan.*, **1**: 48–55.

Ahlawat, A., Dhingra, H.R. and Pahuja S.K. (2013b) *Afr. J. Biotechnol.*, **12**: 4813–4818.

Ahmad, N. and Anis, M. (2007) *J. Hort. Sci. Biotech.*, **82**: 585–589.

Ahmad, N., Faisal, M. and Anis, M. (2012) *Rend. Fis. Acc. Lincei.*, **24**: 7–12.

Ajit, P. and Priyadarshani, Y. (2013) *Int. J. Adv. Biotechnol. Res.*, **4**: 1021–1029.

Anonymous (1950) *Wealth of India*, Council of Scientific and Industrial Research, New Delhi, Vol. II, p. 407-408.

Anonymous (1961a) *Hand Book of Agriculture*, Indian Council of Agricultural Research, New Delhi.

Anonymous (1961b) *Agricultural and Horticultural Seeds*, FAO, United Nations, Rome, p. 531.

Anonymous (1971) *Indian Minimum Seed Certification Standards*, Central Seed Committee, Min. Food Agric., Comm. Dev. Co-op., New Delhi.

Arora, R.N. and Pahuja, S.K. (2008) *Plant Mutat. Rep.*, **2**: 7–9.

Arora, R.N., Ram, H., Tyagi, C.S. and Singh, J. (1998) *Forage Res.*, **24**: 159–162.

Arora, S.K., Joshi, U.N. and Jain, V. (1985) *Guar Res. Ann.*, **4**: 9-10.

Auti, S.J. (2012) *Bioremed. Biodiv. Bioavail.*, **6**: 27–39.

Aykroyd (1963) *ICMR Special Rept. Series*, p. 42.

Bains, D.S. and Dhillon, A.S. (1977) *J. Res., Punjab Agric. Univ., Ludhiana*, **14**: 57-61.

Bewal, S., Purohit, J., Kumar, A., Khedasana, R. and Rao, S.R. (2009) *Czech. J. Genet. Plant Breed.*, **45**: 143–154.

Bhadoria, R.B.S. Chauhan, G.S., Kushwaha, H.S. and Singh, V.N. (1992) *Indian J. Agron.*, **37**: 436-439.

Bhansali, R.R. (2011) *J. Arid Legumes.*, **8**: 77–82.

Bharodia, P.S., Zaveri, P.P., Kher, H.R., Patel, M.P. and Chaudhari, D.N. (1993) *Indian Farm.*, **43:** 31-33.

Bhosale, S.S. and Kothekar, V.S. (2010) *J. Phytol.*, **2**: 21–27.

Brahmi, P., Bhat, K.V. and Bhatnagar, A.K. (2004) *Genet. Resour. Crop Evol.*, **51**: 735–746.

Champawat, R.S. and Somani, L.L. (1990) *Transac. Indian Soc. Desert Technol.*, **15**: 17-22.

Chaudhary, B.S., Lodhi, G.P. and Arora, N.D. (1981) *Indian J. Agric. Sci.*, **51**: 638–642.

Chaudhary, M.S., Ram, H., Hooda, R.S. and Dhindsa, K.S. (1973) *Ann. Arid Zone*, **12**: 19–22.

Chauhan, S.V.S. and Nisha (2006) *J. Phytol. Res.*, **19**: 165–169.

Dabas, B.S., Mital, S.P. and Arunachalam, V. (1982) *Indian J. Genet.*, **42**: 56-59.

Dabas, B.S., Thomas, T.A. and Mehra, K.L. (1981) Catalogue on guar [*Cyamopis tetragonoloba* (L.) Taub.] germplasm. NBPGR, New Delhi, p. 146.

Dabas, B.S., Thomas, T.A., Mittal, S.P. and Chopra, D.P. (1981) *Indian Hort.*, **25**: 17-18.

Dabas, B.S., Thomas, T.A., Nagpal, R. and Chopra, D.P. (1989) Catalogue on guar [*Cyamopsis tetragonoloba* (L.) Taub.]. NBPGR, New Delhi, p. 131.

Daulay, H.S. and Singh, K.C. (1982) *Indian J. Agric. Sci.*, **52**: 758-763.

Doijode, S.D. (1986) *Prog. Hort.*, **18**: 218-221.

Doijode, S.D. (1989) *Indian J. Pl. Genet. Resour.*, **2**: 41-43.

Dutta, T.R., Gupta, J.N. and Gupta, S.R. (1976) *Sci. and Cult.*, **43**: 179-181.

El-Sayed, S.A.M. (1999b) *Egyptian J. Soil Sci.*, **39**: 85-96.

El-Sayed, S.A.M. (1999a) *Egyptian J. Soil Sci.*, **39**: 237-250.

Gargi, T., Acharya, S., Patel, J.B. and Sharma, S.C. (2012) *A.G.R.E.S.*, **1**: 1–7.

Ghonisikar, C.P. and Saxena, S.N. (1973) *Indian J. Agric. Sci.*, **43**: 938-941.

Girish, M.H., Gasti, V.D., Mastiholi, A.B., Thammaiah, N., Shantappa, T., Mulge, R. and Kerutagi, M.G. (2012) *Karnataka J. Agric. Sci.*, **25**: 498–502.

Gresta, F., Mercati, F., Santonoceto, C., Abenavoli, M.R., Ceravolo, G., Araniti, F., Anastasi, U. and Sunseri, F. (2016) *Ind. Crops Prod.*, **86**: 23–30.

Gupta, D.K. (1977) *Veg. Sci.*, **4**: 25-27.

Gupta, J.P. and Gupta, G.N. (1983) *Agric. Water Manag.,* **6**: 375–383.

Haag, H.P., Campora, P. and Forti, L.H.S.P. (1990) *Anais da Escola Superior de Agricultura Luiz de Queiroz,* **47**: 251-260.

Harten, A.M. (1998) *Mutation Breeding Theory and Practical Applications. Cambridge University Press,* New York.

Henry, A. and Mathur, B.K. (2005) *J. Arid Legumes,* **2**: 145–148.

Henry, A., Daulay, H.S. and Bhati, T.K. (1992) *Indian Farm.,* **42**: 24-25.

Hooda, J.S., Saini, M.L. and Singh, J.V. (1990) *Haryana Agric. Univ. J. Res.,* **20**: 28-34.

Jain, V., Joshi, U.N. and Taneja, K.D. (1988) *Agricultural Sci. Digest, India,* **8**: 9-11.

Jain, V., Yadav, B.D., Sharma, B.D. and Taneja, K.D. (1987) *Indian J. Agron.,* **32**: 378-382.

Jukanti, A., Bhatt, R.K., Sharma, R. and Kalia, R.K. (2014) *Diversity analysis (genetic, molecular and gum content) of cluster bean (Cyamopsis tetragonoloba L.) - An emerging industrial crop. 3rd International Conference on Agriculture and Horticulture on October 27-29, 2014 Hyderabad International Convention Centre, India.*

Jukanti, A.K., Bhatt, R.K., Sharma, R. and Kalia, R.K. (2015) *J. Crop Sci. Biotech.,* **18**: 83-88.

Kalaskar, S.R., Acharya, S., Patel, J.B., Sheikh, W.A., Rathod, A.H. and Shinde, A.S. (2014) *J. Food Legumes,* **27**: 92–94.

Khan, A.A.M.M. and Verma, H.N. (1990) *Ann. Appl. Biol.,* **117**: 617-623.

Kinnmann, M.L., Bashaw, E.C. and Brooks, L.E. (1969) *Crop Sci.,* **9**: 570.

Kumar, S., Joshi, U.N., Singh, V., Singh, J.V. and Saini, M.L. (2013) *Genet. Resour. Crop Evol.,* **60**: 2017–2032.

Kumar, S., Parekh, M.J., Patel, C.B., Zala, H.N., Sharma, R., Kulkarni, K.S., Fougat, R.S., Bhatt, R.K. and Sakure, A.A. (2015) *J. Plant Biochem. Biotechnol.,* **25**: 263–269.

Kumar, V. and Ram, R.B. (2015) *Int. J. Pure Appl. Biosci.,* **3**: 143–149.

Kumar, V., Ram, R.B. and Yadav, R.K. (2014) *Indian J. Sci. Technol.,* **7**: 1144–1148.

Kumar, V., Yadav, B.D. and Agarwal, S.K. (1996) *Haryana J. Agron.,* **12**: 43-46.

Kuravadi, A.N., Tiwari, P.B., Choudhary, M., Tripathi, S.K. and Dhugga, K.S. (2013) *Int. J. Adv. Biotechnol. Res.,* **4**: 460–471.

Kuravadi, A.N., Tiwari, P.B., Tanwar, U.K., Tripathi, S.K., Dhugga, K.S., Gill, K.S. and Randhawa, G.S. (2014) *Crop Sci.,* **54**: 1097–1102.

Lal, B.M. and Gupta, O.P. (1977) *Studies on galactomanans in guar and somecorrelations for selecting genotypes rich in gum content. Paper Presented in Workshop on Guar Research and Production Held on January 11-12 at CAZRI, Jodhpur,* pp. 124–130.

Lal, S. (1985) *Grow multipurpose use crop – Guar Kheti,* **38**: 24-27.

Lather, B.P.S. and Chowdhury, J.B. (1972) *Nucleus,* **15**: 16–22.

Lawande, K.E Subedi, P.P., Bhattarai, S. and Dhakal, J. (1990) *J. Maharashtra Agric. Univ.*, **16**: 112-113.

Mahla, H.R. and Kumar, D. (2006) *J. Arid Legumes*, **3**: 75–78.

Mahla, H.R., Kumar, D. and Shekhawat, A. (2010) *Indian J. Agric. Sci.*, **80**: 1033–1037.

Mahla, H.R., Kumar, D., Henry, A., Acharya, S. and Pahuja, S.K. (2011) *J. Arid Legumes*, **8**: 133–137.

Mahla, H.R., Shekhawat, A., Kumar, D. and Bhati, P.S. (2005) *J. Arid Legumes*, **2**: 282–286.

Manivannan, A. and Anandakumar, C.R. (2013) *Indian J. Sci. Technol.*, **6**: 5337–5341.

Manivannan, A., Anandakumar, C.R., Ushakumari, R. and Dahiya, G.S. (2015) *Bangladesh J. Bot.*, **44**: 59–65.

Mathiyazhagan, S. (2007) *Interspecific hybridization and plant regeneration in guar [Cyamopsis tetragonoloba (L.) Taub] M.Sc. Thesis. CCS Haryana Agricultural University*, Hisar, India.

Mathiyazhagan, S., Pahuja, S.K. and Ahlawat, A. (2013) *Legume Res.*, **36**: 180–187.

Mauria, S. (2000. *Indian J. Plant Genet. Resour.*, **13**: 1–10.

Meena, K.C., Singh, G.D. and Mundra, S.L (1991) *Indian J. Agron.*, **36**: 272-274.

Meghwal, M.K., Kalaskar, S.R., Rathod, A.H., Tikka, S.B.S. and Acharya, S. (2014) *J. Cell Tiss. Res.*, **14**: 4647–4652.

Menon, U., Dubey, M.M. and Bhargava, P.D. (1970) *Indian J. Hered.*, **2**: 55–58.

Mishra, S.K., Singh, N. and Sharma, S. K. (2009) In: Kumar, D. and Henry, A. (Eds.), *Perspective research activities of arid legumes in India, Indian Arid Legumes Society, CAZRI, Jodhpur, India*, pp. 23-30.

Mittal, S.P., Dabas, B.S. and Thomas, T.A. (1968) *Curr. Sci.*, **37**: 357.

Mittal, S.P., Dabas, B.S., Thomas, T.A. and Chopra, D.P. (1977) *Indian Hort.*, **22**: 15-17.

Mittal, S.P., Thomas, T.A., Dabas, B.S. and Lal, S.M. (1971) *Indian J. Genet. Plant Breed.*, **31**: 228–232.

Morris, J.B. (2010) *Genet. Resour. Crop Evol.*, **57**: 985–993.

Nandwal, A.S., Dabas, S., Bhati, S. and Yadav, B.D. (1990) *Ann. Arid Zone*, **29**: 99–103.

Naoumkina, M. and Dixon, R.A. (2011) *Plant Cell Rep.*, **30**: 997–1006.

Nisha, Chauhan, S.V.S. (2006) *J. Physiol. Res.*, **19**: 191–195.

Paroda, R.S. and Saini, M.L. (1978) *Forage Res.*, **4**: 9–39.

Patel, I.S. and Dodia, D.A. (1999). *Pestology*, **13**: 22-23.

Patel, K.A., Patel, B.T., Shinde, A.S., Sheikh, W., Parmar, L.D., Sharma, S.C., Parmar, R.G., Ravindrababu, Y. and Acharya, S. (2014) *Can. J. Plant Breed.*, **2**: 43–46.

Pathak, R. (2015) *Cluster Bean: Physiology, Genetics and Cultivation. Springer*, Singapore.

Pathak, R., Singh, M. and Henry, A. (2009) *Indian J. Agric. Sci.*, **79**: 559–561.

Pathak, R., Singh, M. and Henry, A. (2011a) *Indian J. Agric. Sci.*, **81**: 309–313.

Pathak, R., Singh, M. and Henry, A. (2011c) *Indian J. Agric. Sci.*, **81**: 402–406.

Pathak, R., Singh, S.K. and Singh, M. (2011d) *J. Food Legumes*, **24**: 180–183.

Pathak, R., Singh, S.K., Singh, M. and Henry, A. (2010a) *Indian J. Dryland Agric. Res. Dev.*, **25**: 87–92.

Prabaharan, M. (2011) *Int. J. Biol. Macromol.*, **49**: 117–124.

Prem, D., Singh, S., Gupta, P.P., Singh, J. and Kadyan, S.P.S. (2005) *Plant Cell Tiss. Org. Cult.*, **80**: 209–214.

Prem, D., Singh, S., Gupta, P.P., Singh, J. and Yadav, G. (2003) *In Vitro Cell Dev. Biol. Plant*, **39**: 384–387.

Punia, A., Yadav, R., Arora, P. and Chaudhary, A. (2009) *J. Crop Sci. Biotechnol.*, **12**: 143–148.

Ramulu, C.A. and Rao, D. (1987) *J. Swamy Bot. Cl.*, **4**: 29–31.

Ramulu, C.A. and Rao, D. (1989) *Vegetos*, **2**: 30–32.

Ramulu, C.A. and Rao, D. (1991) *J. Phytol. Res.*, **4**: 183–185.

Ramulu, C.A. and Rao, D. (1993) *Geobios*, **20**: 7–9.

Ramulu, C.A. and Rao, D. (1996) *J. Environ. Biol.*, **17**: 257–260.

Rao, S. and Rao, D. (1982) *Proc. Indian Natl. Sci. Acad.*, **48**: 410–415.

Rashmi K. and Mohan Kumar S. (2018) *Int. J. Agric. Sci.*, **6**: 5548-5550.

Rasmussen, T.B. and Donaldson, I.A. (2006) *Plant Cell Rep.*, **25**: 1035–1042.

Rathore, B.S. (2000) *Plant Dis. Res.*, **15**: 89-92.

Rathore, S.V.S., Kumar, P. and Singh, D.R. (1990) *Haryana J. Hort. Sci.*, **19**: 219-221.

Ray, D.T. and Stafford, R. E. (1985) *Crop Sci.*, **25**: 177-179.

Rodge, A., Jadkar, R., Machewad, G. and Ghatge, G. (2012) *Res. Plant Biol.*, **2**: 23–31.

Rogers, C.E. and Stafford, R.E. (1976) *Agron. J.*, **68**: 496-499.

Saini, M.L., Arora, R.N. and Paroda, R.S. (1981) *Guar Newsl.*, **2**: 7-11.

Saini, M.L., Singh, J.V. and Jhorar, B.S. (1990) *Guar. Agric. Sci. Digest.*, **10**: 113–116.

Sandhu, H.S. (1988) Interspecific hybridization studies in genus *Cyamopsis*. *Ph.D. Thesis. CCS Haryana Agricultural University, Hisar*, India.

Sathiyamoorthy, P. and Vivekanandan, M. (1995) *Legume Res.*, **18**: 50-52.

Sehgal, K., Nainawatee, H.S. and Lal, B.M. (1973) *Biochemie Und Physiologie Der Pflanzen*, **164**: 423–428.

Selvaraj, K. and Kumari, E.V.N. (1999) *Indian J. Plant Physiol.*, **4**: 271-276.

Shah, S.M., Shabir, G., Aslam, K., Sabar, M. and Arif, M. (2015a) *Environ. Plant Syst.*, **1**: 04–15.

Sharma, A. and Sharma, P. (2013) *Res. J. Recent Sci.*, **2**: 1–9.

Sharma, A., Mishra, S. and Garg, G. (2013) *Res. J. Pharma. Biol. Chem. Sci.*, **4**: 8–17.

Sharma, B.D., Taneja, K.D., Kairon, M.S. and Jain, V. (1988) *Indian J. Agron.*, **29**: 557–558.

Sharma, H.C. (1992) In: *Proceedings of Symposium on Resource Management for Sustainable Crop Production, RAU, Bikaner*, pp. 88–101.

Sharma, K.D. and Kuhad, M.S. (1999) In: *Recent advances in management of arid ecosystem. Proceedings of a symposium* (Faroda, A.S., Joshi, N.L., Kathju and Amal Kar (Eds.), held in India, March 1997, pp. 267-270.

Sharma, P., Kumar, V., Raman, K.V. and Tiwari, K. (2014a) *Adv. Biosci. Biotechnol.*, **5**: 131–141.

Sharma, P., Sharma, V. and Kumar, V. (2014b) *J. Agric. Sci. Technol.*, **16**: 433–443.

Sheikh, W.A., Dedhrotiya, A.T., Khan, N., Gargi, T., Patel, J.B. and Acharya, S. (2015) *Curr. Trends Biotechnol. Pharm.*, **9**: 175–181.

Shekhawat, S.S. and Singhania, D.L. (2005) *Forage Res.*, **30**: 196–199.

Shinde, M.S. and More, A.D. (2010) *Asian J. Exp. Biol. Sci.*, **1**: 31–34.

Shivanna, M.B. and Shetty, H.S. (1991). *Indian J. Agric. Sci.*, **61**: 856-859.

Sidhu, A.S., Pandita, M.L., Arora, S.K. and Vashistha, R.N. (1982) *Haryana Agric. Univ. J. Res.*, **12**: 225–230.

Sindhan, G.S., Hooda, I. and Parashar R.D. (1999) *J. Mycol. Plant Pathol.*, **29**: 110–111.

Singh, A., Kharub, A.S. and Singh, A. (2001) *Fert. Market. News*, **32**: 3-5.

Singh, H.B. and Singh, V. (1994) *Indian J. Plant Physiol.*, **37**: 221-223.

Singh, J.V., Chander, S. and Sharma, S. (2004) *J. Plant Improv.*, **6**: 128-129.

Singh, K., Kumar, S. and Taneja, K.D. (1979) *Haryana Agric. University, J. Res.*, **9**: 315-319.

Singh, K.C. (1989) *Ann. Arid Zone.*, **21**: 275–278.

Singh, N.P., Singh, R.V., Chaudhary, S.P.S. and Singh, J. (2005a) *J. Arid Legumes*, **2**: 97–101.

Singh, R.V. and Singh, R.R. (1990) *Ann. Agril. Res.*, **11**: 329-332.

Singh, R.V., Chaudhary, S.P.S., Singh, J. and Singh, N.P. (2005b) *J. Arid Legumes*, **2**: 102–105.

Singh, S.D., Gupta, J.P. and Singh, P. (1998) *Agron. J.*, **70**: 948–951.

Singh, S.J.P., Rajput, C.B.S. and Singh, K.P. (1987) *Gujarat Agric. Univ. Res. J.*, **13**: 1-6.

Singh, U.P. and Dixit, O.P. (1994) *Indian Farm.*, **43**: 31-32.

Singh, V. and Kothari, S.K. (1982) *J. Agric. Sci. Camb.*, **108**: 691- 693.

Singh, Y.P., Dahiya, D.J., Kumar, V. and Singh, M. (1994) *Crop Res.*, **6**: 394–400.

Smartt, J. (1976) *Tropical Pulses*, Longman, London, p. 348.

Smith, P.M. (1976) In: Simmonds, N.W. (Ed.), *Evolution of Crop Plants*, Longman, London, and New York, pp. 311-312.

Solanki, J.S. and Singh, R.R. (1976) *Indian J. Agric. Sci.*, **46**: 241-243.

Stafford, R.E. (1989) *Plant Breed.*, **103**: 43–46.

Stephens, J.M. (1994) *Fact Sheet HS-608, a series of the Horticultural Sciences Department*, Florida Cooperative Extension Service, Institute of Food and Agricultural Sciences, University of Florida.

Sultan, M., Rabbani, M.A., Zabta, K.S., Masood and M.S. (2012) *Pak. J. Bot.*, **44**: 203–210.

Tapadia, S.B., Arya, A.B. and Rohini Devi, P. (1995) *J. Food Sci. Technol.*, **32**: 513–515.

Tarafdar, J.C. and Rao, A.V. (1990) *J. Arid Env.*, **10**: 31–37.

Tarafdar, J.C., Rao, A.V. and Kumar, P. (1995) *J. Arid Env.*, **29**: 331-337.

Thomas, T.A. and Dabas, B.S. (1982) *Plant Genet. Resour. Newsl.*, **52**: 16–18.

Tiwana, V.S. and Tiwana, M.S. (1994) *Indian J. Ecol.*, **21**: 117-121.

USDA [United States Department of Agriculture] (2014) *An analysis of guar crop in India. Prepared by CCS National Institute of Agriculture Marketing, Jaipur for USDA, USA.*

Velu, S., Mullainathan, L. and Arulbalachandran, D. (2012) *Int. J. Curr. Trends Res.*, **1**: 48–55.

Velu, S., Mullainathan, L., Arulbalachandran, D., Dhanavel, D. and Poonguzhali, R. (2007) *Crop Res.*, **34**: 249–251.

Venkataratnam, L. (1973) *Beans in India*, Directorate of Extension, Ministry of Agriculture, New Delhi. p. 64.

Verma, S., Gill, K.S., Pruthi, V., Dhugga, K.S. and Randhawa, G.S. (2013) *Indian J. Exp. Biol.*, **51**: 1120–1124.

Vig, B.K. (1963) *Curr. Sci.*, **32**: 375–376.

Vig, B.K. (1965) *Sci. Cult.*, **31**: 531–533.

Vig, B.K. (1969) *Ohio J. Sci.*, **69**: 18.

Vijay, O.P. (1988) *Indian J. Hortic.*, **45**: 127–131.

Vijay, O.P. and Leela, N.K. (1989) *Indian J. Hort.*, **46**: 59-65.

Weixin, L., Anfu, H. and Peffley, E.B. (2009) *Chin. Agric. Sci. Bull.*, **25**: 133–138.

Yadav, B.D., Aggarwal, S.K. and Arora, S.K. (2000) *Guar Res. Ann.*, **5**: 24–27.

Yadav, R.S. (1998) *J. Agron. Crop Sci.*, **181**: 209–214.

Yawalkar, K.S. (1992) *Vegetable Crops of India*, Agri-Horticultural Publishing House, Nagpur, India.

5

LABLAB BEAN

A. Chattopadhyay, A.K. Chakraborti,
V.A. Parthasarathy and P.K. Maurya

1.0 Introduction

Lablab bean, Dolichos bean, hyacinth bean (*sem*) or Indian bean is grown throughout the country. It is a multi-utility and multi-beneficial leguminous crop grown for green pods as vegetable, dry seeds as pulse, fodder, green manure, cover crop, medicine and ornamental purpose (Ayyangar and Nambiar, 1935). It is one of the oldest legume crop known to be cultivated in dry and semi-arid regions of Asia, Africa and America. In India, it is grown as a field crop in Madhya Pradesh, Maharashtra, Andhra Pradesh and Tamil Nadu. It is also known by several other names such as butter bean, helmet bean, Egyptian kidney bean, lubia bean, to name a few. This crop has received little attention due to competition with other cash crops and availability of a limited number of cultivated varieties (Chattopadhyay and Dutta, 2010; Das *et al.*, 2015), although a broad spectrum of the genetic base is available in nature. It is grown either in pure stand or intercropped with cereals like finger millet, pearl millet, corn and sorghum, and with other crops like groundnut, castor in rainfed ecosystems.

2.0 Composition and Uses

2.1.0 Composition

This is an underexploited tropical legume that is valued for its nutritional and sensory attributes. According to Venkatachalam *et al.* (2002), it contains 30 per cent protein on a dry weight basis. *D. lablab* [*Lablab purpureus*] var. *lignosus* has more protein (20.9-29.2 per cent) than var. *typicus* (Saimbhi, 1992). The mean seed protein content of Lablab bean is 26.51 per cent with a highly significant negative correlation (r = -0.28) between seed protein and methionine content (Rangasamy *et al.*, 1993). Albumin, globulin, prolamin and glutelin accounted for approximately 20, 48, 1 and 31 per cent of the total seed proteins, respectively.

It is rich in nutritive value and the composition of green pod is given in Table 1.

Table 1: Composition of Lablab Bean Green Pod (per 100 g of edible portion)*

Moisture	86.1 g	Sodium	55.4 mg
Carbohydrates	6.7 g	Iron	1.7 mg
Protein	3.8 g	Potassium	74.0 mg
Fat	0.7 g	Sulphur	40.0 mg
Fibre	1.8 g	Vitamin A	312 I.U.
Minerals	0.9 g	Riboflavin	0.06 mg
Magnesium	34.0 mg	Thiamine	0.1 mg
Calcium	210.0 mg	Nicotinic acid	0.7 mg
Phosphorus	68.0 mg	Vitamin C	9.0 mg

*Aykroyd (1963).

2.2.0 Uses

It is primarily grown for green pods, which are cooked as vegetable like other beans. The dry seeds are also used for various vegetable preparations. The foliage of the crop provides hay, silage and green manure. Among the legumes, dolichos bean constitutes an important source of therapeutic agents used in the modern as well as traditional systems of medicine (Morris, 2003, 2009). The seeds are used as a laxative, diuretic, anthelmintic, anti-spasmodic, aphrodisiac, anaphrodisiac, digestive, carminative, febrifuge and stomachic (Chopra *et al.*, 1986; Kirtikar and Basu, 1995). The bean contains the potential breast cancer fighting a flavonoid known as kievitone (Hoffman, 1995). The flavonoid, genistein found in this bean may play a role in the prevention of cancer (Kobayashi *et al.*, 2002) and as a chemotherapeutic and/or chemopreventive agent for head and neck cancer (Alhasan *et al.*, 2001). Tyrosinase (polyphenol oxidase) is present in plant tissue and is important in fruit and vegetable processing as well as storage of processed foods. The bean contains tyrosinase, which has potential for the treatment of hypertension in humans (Naeem *et al.*, 2009a).

3.0 Origin and Taxonomy

Lablab bean presumably had its origin in tropical Asia, probably in India and from there it was introduced to China, Sudan and Egypt. Since antiquity it has been cultivated extensively in India. Some assign it to the American Centre of diversity. It has spread over the entire tropics and is cultivated throughout the warmer parts of the world from sea level to 2000 m. Lablab bean (2n=2x=22, 24) belongs to family *Leguminosae*. Verdcourt (1970) placed hyacinth bean in a separate genus *Lablab* from *Dolichos* and designated as *Lablab purpureus* (L.) Sweet (syn. *Dolichos lablab, D. purpureus*) that is now widely accepted. He classified *L. purpureus* into the following subspecies:

L. purpureus subsp. *purpureus* – It comprises of common cultivated races of the hyacinth bean.

L. purpureus subsp. *uncinatus* – It has more slender inflorescence with smaller pods, about 4 cm long, 1.5 cm wide.

L. purpureus subsp. *bengalensis* – Pods are linear, similar to kidney bean in appea-rance and quite dissimilar to those of other hyacinth beans. These races interbreed freely.

The species is diverse (Duke *et al.*, 1981) with the subspecies *typicus*, a perennial garden type, and *lignosus*, an annual field type. It is a potential perennial plant but cultivated as an annual or biennial with bushy erect and climbing races. It is extremely variable in all aspects. The distinguishing features of two botanical or cultivated types of dolichos bean are given in Table 2.

4.0 Cultivars

Though this crop has originated in India, very little work has been done to study the varietal characters and to improve the quality of pods and yield of the local strains available. Marked variation exists in the plant and pod characters among

Uses of Dolichos Bean

Soup (left) and fries (right) of dolichos bean

Curry of dolichos bean

Growth Habits and Types of Pods in Dolichos bean

Bush and Pole type dolichos bean

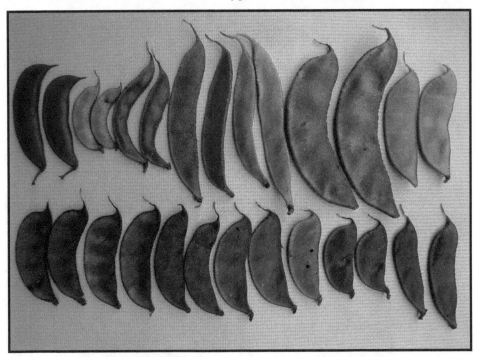

Variability in pod colour and shape of dolichos bean

the cultivars grown all over the country. Genetic studies of some of the economic characters have been made by Cruz and Ponnaiya (1969), Chikkadevaiah *et al.* (1979), Kabir and Sen (1991), Kabir *et al.* (1993), Das *et al.* (2018), and Nayek *et al.* (2018). On the basis of growth habit, cultivars have been classified into two groups, namely, the bushy field cultivar and twining pole garden cultivar. Pusa Early Prolific is one of the earliest cultivars recommended by the Indian Agricultural Research Institute, New Delhi. After that many public sector research institutes of India like ICAR-Indian Institute of Horticultural Research (IIHR), Bengaluru; KKV, Dapoli; CSAUA&T, Kanpur; MPKV, Rahuri; JNKVV, Jabalpur, UAS, Bangalore; TNAU, coimbatore and BCKV, West Bengal have developed many photo-insensitive pole and bush type dolichos varieties for vegetable purpose (Raghu *et al.*, 2018).

Table 2: Distinguishing Features of Two Botanical or Cultivated Types of Dolichos Bean

Features	*Lablab purpureus var. typicus*	*Lablab purpureus var. lignosus*
Common names	Garden bean, Hyacinth bean	Field bean, Indian bean, Lablab bean
Growth type	Indeterminate or semi-indeterminate	Determinate
Growth habit	Perennial twining herb usually trained on trellis (Poletype)	Perennial bush often grown as annual
Pigmentation	More pigmentation on stem, leaves and pods	Less pigmentation on stem, leaves and pods
Flowering	Thermo-photoperiod sensitive (short day)	Thermo-photoperiod insensitive
Flowering duration	60-90 days after sowing	40-50 days after sowing
Pod traits	Pods are Longer, flat and tapering. Long axis of seeds isparallel to suture of the pod	Pods are shorter in length and axis of seeds is perpendicular to suture of the pod
Parchment on pod wall	Pods are relatively less fibrous, soft, meaty, and whole pod is edible	Pods are firm-walled and fibrous, not suitable for wholepod consumption
Harvesting stage	Green immature pods and green seeds are harvested	Fully matured dry pods are harvested for dry seeds; However, pods are harvested for green seeds
Other traits	No oily substances and characteristic fragrance	It exude oily substances that emit characteristic fragrance
Yield potential	High	Less

During the last two decade many cultivars of Lablab bean have been developed and released in India. The cultivars are described under two broad classes *viz.*, pole type or pandal type and bush type.

A. Pole Type or Pandal Type

Pusa Early Prolific

It bears early long thin pods in bunches. It is suitable for sowing in early autumn and early spring in the northern plains (Choudhury, 1972).

CO- 1

Plant height of the cultivar is 60-70 cm, pods flat, green when tender and tan coloured at maturity and the duration is 140 days. It is a pure line selection from local strains, released in 1980 Tamil Nadu, India from the Agricultural University at Coimbatore, India.

CO- 2

Plant height is 60 cm, photoinsensitive, erect, bushy with 5-6 branches. Pods flat, green and tetra-seeded. The cultivar has been developed from cross Co. 8 × Co. 1 and released in Tamil Nadu in 1984.

Deepaliwal

It has been developed at Punjabrao Krishi Vidyapeeth, Maharashtra, India and identified in 1990 by the Directorate of Vegetable Research of the ICAR for release in agroclimatic zone No. 5 (sub humid to humid eastern and south eastern unlands) and 7 (semi arid lava plateau and central high lands). It is a high yielding, pole type cultivar maturing in 91 days.

Arka Adarsh

Pole type and photo-insensitive and early cultivar. Pods are borne in clusters and dark green coloured. Suitable for Karnataka, India. Developed by pedigree method of selection from F_7 generation involving (IIHR 178 × Arka Swagath). Pod yield is 300 q/ha in 120 days.

Arka Krishna

Pole type and photo-insensitive and early cultivar. Pods are borne in clusters and dark green coloured. Suitable for Karnataka, India. Developed by pedigree method of selection from F_7 generation involving (IIHR 178 × Arka Swagath). Pod yield is 300 q/ha in 120 days.

Arka Pradhan

Pole type and photo-insensitive cultivar. Pods are green in colour, smooth and shiny with undulating surface. Suitable for cultivation in Maharashtra, India. Developed by pedigree method of selection from F_7 generation involving (IC 556824 IPS-2 × Arka Swagath). Pod yield is 350 q/ha in 120 days.

Arka Visthar

Pole type and photo-insensitive cultivar. Pods are long, thick, very broad and dark green coloured. Suitable for cultivation in Tamil Nadu and North Eastern states.Developed by Pedigree method of selection from F_7 generation involving (IIHR 178 × Arka Swagath). Pod yield is 370 q/ha in 120 days.

Arka Bhavani

Pole type and photo-insensitive cultivar. Pods are slender, wavy and dark green coloured. Suitable for cultivation in Andhra Pradesh, India. Developed by pedigree method of selection from F_7 generation involving (IIHR 178 × Arka Swagath). Pod yield is 320 q/ha in 120 days.

Arka Prasidhi

Pole type and photo-insensitive cultivar. Pods are dark green, long, flat and slightly curved. Resistant to rust. Suitable for south Indian states. Developed by pedigree method of selection from F_7 generation involving (IC 556824 IPS-2 × Arka Swagath). Pod yield is 370 q/ha in 120 days.

Arka Swagath

Pole type and photo-insensitive cultivar and suitable for round the year cultivation. Pods are light green, medium long and suitable for Karnataka, India. Developed by pureline selection from IC 556736. Pod yield is 260 q/ha in 120 days.

Arka Amogh

Plants are medium tall and photo-insensitive. Pods are wavy, green, medium long and ready for harvest in 55 days. Suitable for Maharashtra, India. Developed by pedigree method of selection from F_7 generation involving (Arka Jay × Arka Vijay) × Konkan Bhushan). Pod yield is 190-200 q/ha in 75 days.

Arka Sambhram

Plants are medium tall and photo-insensitive. Pods are flat, light green, medium long, medium width and ready for harvest in 55 days. Suitable for Tamil Nadu, India. Developed by pedigree method of selection from F_7 generation involving (Arka Jay × Arka Vijay) × Konkan Bhushan). Pod yield is 190-200 q/ha in 75 days.

Arka Soumya

Plants are medium tall and photo-insensitive. Pods are slender, wavy, medium long and ready for harvest in 55 days. Suitable for Andhra Pardesh India. Developed by pedigree method of selection from F_7 generation involving (Arka Jay × Arka Vijay) × Konkan Bhushan). Pod yield is 190 q/ha in 75 days

JDL-79

This cultivar was developed at JNKVV, Jabalpur, India through selection from local types. Pods are thick, tender, long (11.8 cm), broad (3.6 cm), green with blackish-pink border along the sutures. It is tolerant to pod borer. Crop duration is 200 days. Average yield is 50-60 q/ha.

CO-4

Developed at Coimbatore, India by pure line selection from a local collection named "Sivappa avarai". The plants are purple throughout with green leaves having purple veins and flower. The pods are deep purple throughout, 10.2 cm long, 3.3 cm width, septate, fleshy and weighing 7.5 g with black seeds. Crop duration is 215-220 days. Average yield is 120-130 q/ha.

CO-5

Developed at Coimbatore, India through pure line selection from a local collection called "Kozhikkal avarai". The pods are long (13.4 cm), narrow (1.5 cm width), weighing 5.2 g, light green to white, tubular and curved with slightly serrated margin. Seeds are chocolate brown in colour. Crop duration is 230 days. Average yield is 60-70 q/ha.

Rajni

Developed at Kanpur, India through selection from local type. Pods are narrow (10.4 cm long and 1.2 cm width), oval in cross section like that of French bean, shinning green and weighing about 2 g. Crop duration is 200-210 days. Average yield is 70-80 q/ha.

Dasarawal

Developed at PDKV, Akola, India through selection from local collection. Pods are 7.8 cm long, 2.0 cm width, dirty green with purple tinge on both borders and weighing 3.2 g. Average yield is 70-80 q/ha.

JDL 53

Developed at JNKVV, Jabalpur, India through selection from local types. Pods are medium long (7.2 cm), thin, narrow (1.8 cm), flat, dull white with purple tinge along the sutures. It is less infested by pod borer. Crop duration is 200-220 days. Average yield is 100-120 q/ha.

125-36

This cultivar was developed at Gujarat, India through selection from local germplasm. Plant height 1 m., stem and foliage green; pods are short, medium broad and borne in clusters with 3-4 milky white seeds/pod. 80-100 pods are produced in a plant. Duration is 110-120 days.

KDB 403

Developed at Kalyanpur, India through selection from local collection. The pods are long (12.9 cm) and narrow (1.2 cm), shiny green and weighing 2 g. Crop duration is 180-210 days. Average yield is 50-60 q/ha.

KDB 405

Developed at Kalyanpur, India through selection from local collection. Pods are medium long (9.6 cm), very narrow (1.3 cm), round in cross section as that of French bean, dark green in the middle with light green borders and weighing 1.1 g. Crop duration is 180-200 days. Average yield is 30-40 q/ha.

Pusa Sem-2

Developed at IARI, New Delhi, India through selection from local collection: Plants with dark green foliage. Pods are dark green, 15-17 cm long, very tender, stringless, fleshy and borne in cluster of 11-13 pods in the spike above the plant canopy with 5-6 seeds. It is highly tolerant to anthracnose and yellow mosaic diseases and tolerant to aphids, jassids and pod borers. Crop duration is 230 days. Average yield is 150-180 q/ha.

Pusa Sem-3

Developed at IARI, New Delhi, India through selection from local collection. Plants with dark green foliage. The pods are flat, 15 cm long, green, very tender, stringless, meaty and borne in clusters of 10-12 pods. It is tolerant to anthracnose, bean yellow mosaic virus disease, aphid, jassid and pod borers. Average yield is 170-180 q/ha.

Kashi Sheetal (VRSEM-11)

Semi-pole type, tolerant to low temperature and DYMV. It can give yield of 180-200 q/ha. Cultivar is rich in protein 590.8mg/g and low in total sugar 0.621g/100g. Recommended for cultivation in Uttar Pradesh, India vide gazette notification number S.O. 692(E), dated 05.02.2019.

Kashi Harittima

Suitable for sowing in Kharif season. High yielding and good pod quality (green, tender and parchment free). It is moderately resistant to Dolichos Yellow Mosaic Virus diseases under field condition. Moderately tolerant to jassid, aphid and pod borer under field condition.

Phule Gouri (RHRWL-1)

This cultivar has been developed through selection from a germplasms at MPKV, Rahuri, India. Pods are whitish green, tender, flat and slightly curved, and pod length 7-9 cm. Fruits are smooth and attractive green with white strips at apical end and bigger in size. Plants are tolerant to pod borer, anthracnose, leaf spot and leaf minor under field condition. This cultivar has yield potential of 330-345 q/ha in 180 days of crop duration.

Swarna Utkrisht

This cultivar has been developed through pure line selection at HARP, Ranchi, India. Pods are straight, flat, green and fleshy pods (10-12 cm) having very good cooking quality. Mature seeds are light brown. Plants are tolerant to anthracnose, mosaic viruses and pod borer under field condition. First harvest can be taken 110-120 day after planting. It has been recommended for December-January sowing in Uttar Pradesh, Jharkhand, Bihar and Punjab and has yield potential of 375-400 q/ha.

Swarna Rituvar

Suitable for off-season cultivation in rainy season. Developed through pure line selection. Suitable for off-season cultivation in rainy season. Pod are flat, creamy white and fleshy pods (8-10 cm) having very good cooking quality. Mature seeds are black in colour. Tolerant to anthracnose and pod borer under field condition. First harvest 75-80 days after sowing. Average fresh pod yield is 125- 150 q/ha.

BCDB-1

This cultivar has been developed through pure line selection at BCKV, West Bengal, India. Pods are straight, flat, deep green with purple suture and fleshy pods having very good cooking quality. Mature seeds are black with white suture. Plants are tolerant to anthracnose, mosaic viruses and pod borer under field condition. First harvest can be taken 100-110 day after planting. It has been recommended for August-September sowing in West Bengal, U.P., Jharkhand, Bihar and Madhya Pradesh and has yield potential of 400 q/ha.

B. Bush Type

These cultivars are short duration with determinate and bushy growth habit and photoinsensitive nature. They have basically been developed from advanced generation selection from the cross between the genotypes belonging to *D. purpureus* var. *typicus* and *D. purpureus* var. *lignosus*. Pods of these cultivars, however, are not as long and tender as that of pole type vegetable hyacinth bean, but their photoinsensitive nature enable them to grow in any season of the year.

Hebbel Avare-1

This cultivar was developed at Bangalore, India through selection from the cross Local Avare × Red Typicus. It is photoinsensitive and is not a season bound cultivar. Pods are small, soft and has good cooking quality. Crop duration is 90-100 days. Average yield is 80 q/ha.

Hebbel Avare-3

Developed at Bangalore, India through selections from the segregating materials of the cross Hebbal Avare 1 × US 67-31 (USA). Plant height 65-75 cm, erect with determinate growth habit, photoinsensitive, flowers white, pods green, 2-3 seeded; seeds are brown, round and small; crop duration is 90-100 days. Average yield is 80-100 q/ha.

Hebbal Avare-4

Developed at Bangalore, India through advanced generation selections from the cross Hebbal Avare × CO 80. Plants are compact and first harvest within 60 days after sowing. Pods are soft and harvested in 5 pickings. Average yield is 60 q/ha.

CO-6

Developed at Coimbatore, India through advanced generation selections from the cross DL 3169 × CO-5 (pole type). The pods are slightly curved and bloated. Crop duration is 140 days. Average yield is 100-120 q/ha.

CO-7

Developed at Coimbatore, India through another selection from the segregating generation of the cross DL 3169 × CO-5 (pole type). Pods are long, flat, broad, succulent and greenish white in colour with bigger seeds. Crop duration is 140 days. Average yield is 100-120 q/ha.

CO-8

Developed at Coimbatore, India through selection from the segregating progenies the cross CO-5 × DL 3169. The pods are green, tubular and fleshy. Crop duration is 120 days. Average yield is 60-80 q/ha.

CO-9

Developed at Coimbatore, India from a spontaneous mutant with bushy habit. It is a dual purpose cultivar for green pods as well as dry seeds. It is tolerant to mosaic virus. Crop duration is 120 days. Average yield is 70-80 q/ha.

CO-10

Developed at Coimbatore, India through mutation breeding by irradiating 24 Krads gamma radiation to CO-6. Plants with purple pigmentation and pinkish-purple flower. It flowers 45 days after sowing. Pods are tubular, greenish white and curved. Crop duration is 120 days. Average yield is 50-60 q/ha.

CO-11

Developed at Coimbatore, India through advanced generation selection from the hybrid CO-9 × White Yanai Kathu (pole type). Plants are compact and it flower 40-45 days after sowing. Pods are light green, flat with purple margin. Average yield is 90-100 q/ha.

CO-13

Developed at Coimbatore, India through advanced generation selections from the cross CO-9 × Florikifield (pole type). Plants with white flowers and long green pods. Crop duration is 110-120 days. Average yield is 100 q/ha.

Wal Konkan-1

This cultivar was developed at Dr. BSSKKV, Dapoli, India through selection from the progenies of the cross Wal-2-K2 × WaL 125-136. Plant height 40-45 cm, bushy with long tendrils; leaves are dark green and smooth; flower in 60 days. Seeds are bold with brown coat colour. It is resistant to yellow mosaic virus; crop duration is 110-115 days. Average yield is 90-100 q/ha.

Arka Jay

Developed at IIHR, Bangalore, India through advanced generation selection from the cross Hebbel Avare-3 × IHR-99 (pole type). It is a dwarf cultivar, which start flowering in 44 days. Pods are medium long, thin, light green and slightly curved. Crop duration is 120 days. Average yield is 120 q/ha.

Arka Vijay

Developed at IIHR, Bangalore, India through advanced generation selection from the cross Hebbel, Avare-3 × Pusa Early Prolific (pole type). It starts flowering after 46 day of sowing. Pods are short, thin, dark green with characteristic aroma. Crop duration is 120 days. Average yield is 120 q/ha.

Konkan Bhushan

A dwarf cultivar developed at Dr. BSSKKV, Dapoli, India through hybridization of Hebbel Avare 3 × Wall and subsequent selections from the segregating, generations. The plant produces 125-180 pods. The pod is 13 cm long and 1.5 cm wide each weighing 3-4 g, tubular, tender, green and stringless. Crop duration is 100-110 days. Average yield is 80-100 q/ha.

CO-12

Developed at Coimbatore, India through advanced generation selection from the cross CO-9 × CO-4 (pole type). Pods are deep purple. Crop duration is 110 days. Average yield is 100-120 q/ha.

In China, an extremely early cultivar of Lablab bean Xiangbiandou 1 was developed by Peng *et al.* (2001). Dense planting of this new cultivar produces high yields. The yield is about 42000 kg/hm². It is resistant to disease and tolerant to cold.

5.0 Soil and Climate

Hyacinth bean grows on almost all types of soil of average fertility as in case of other beans (Nath, 1976) with pH 5.0 to 7.8. It cannot stand waterlogging.

It is sensitive to photoperiods and both short day and long day types are available (Anon., 1961). It prefers comparatively cool season, and moreover majority of traditional cultivars are temperature-and photoperiod-sensitive and requires short days for flowering. It is adapted to tropical and subtropical regions. Most cultivars are adapted to temperatures ranging between 18° and 30°C. High temperature does not affect its development. Frost damages leaves but light frost does not kill the plant. There are strains available which are drought resistant and are grown as a dry land crop in regions with 630 to 890 mm rainfall.

6.0 Cultivation

Like other beans, it also requires a good land preparation and tilth for sowing. It is sown in July-August with the onset monsoon. In South and Central India, it is usually grown as a mixed crop with ragi or sorghum. It is drilled at a spacing of about one metre in between ragi or sorghum. Ragi or sorghum earheads are harvested first leaving the stalks for giving support to the vines of Lablab bean. The vines grow on them profusely and flower in November-December giving a continuous crop of green pods as well as dry seeds throughout the winter and spring. The vines, when cut with sorghum straw give a mixed fodder with high nutritive value. If it is grown as a pure crop pole type varieties can be sown at a spacing of 1 × 0.75 m. In the Gangetic plains of West Bengal, India a spacing of 1.5 m both ways is optimum for pole type varieties (Chattopadhyay and Dutta, 2010). For dwarf, bushy cultivars, a spacing of 60 × 30 cm is adequate. Three to four seeds are sown per hill and later thinning is done to allow one plant to grow per hill. About 20-30 kg seed is required for sowing one hectare. In kitchen garden plants can be retained for 2-3 years.

Promising Cultivars of Dolichos bean

Arka Soumya

Arka Vijay

Arka Jay

Arka Swagath

Arka Krishna

Kashi Harittima

Variation in seed traits and Cultivation Practices of Dolichos bean

Variability for seed size, shape and color among dolichos bean germplasm collection

Field views of dolichos bean cultivation

Being a leguminous crop, dolichos bean is highly responsive to nitrogenous fertilizer application especially in early stage. Similarly, application of phosphorus influences symbiotic nitrogen fixation and serves dual purpose in legume by increasing the yield and quality of green pods of current as well as succeeding crop of the dolichos bean (Turuko and Mohammed, 2014). Potassium also plays a crucial role in legumes by enhancing the production of starch and sugar that benefit the symbiotic bacteria and thus enhances the fixation of nitrogen (Rustamani *et al.*, 1999). Lablab bean when grown with ragi is fertilized with 90 kg ammonium sulphate and 40 kg superphosphate per hectare after first weeding (Anon., 1961). Vijay and Vani (1990) suggested application of 10 to 15 tonnes of farmyard manure and 65 kg urea, 313 kg single superphosphate and 65 kg muriate of potash per hectare. Singh *et al.* (1993) reported that the cultivars JDL 53 and HD 60 require 44.4 and 47.7 kg P, respectively for best yield. In bush type beans Sel.1 and Sel. 2, Singh *et al.* (1992) standardised the optimum rates of P at 37.13 kg and 24.5 kg/ha, respectively. Application of inorganic fertilizers @ N: P: K 30: 60: 50 kg/ha was found optimum to raise a good crop in the Gangetic plains of West Bengal, India (Das *et al.*, 2015). At Karnataka, India green pod yield per plant and pod yield per hectare as well as higher total N, P and K uptake were recorded maximum in 25: 75: 50 kg NPK per hectare (Dalai *et al.*, 2019). Adoption of the best performing technology, maize/dolichos intercrop with combined application of 5 t/ha FYM and 60 kg/ha TSP+Urea, ought therefore to be tapered (in the short run) with prudent nutrient management strategies for system sustainability at Kenya (Sitienei *et al.*, 2017).

Pole type varieties require trellising usually made of bamboo poles and tie with nylon strings. Inter-cultivation is done to control weeds until vines spread between rows. Mathukia *et al.* (2018) recorded that pendimethalin 30 per cent EC 900 g a.i./ha as pre-emergence followed by one hand weeding at 45 days after sowing was found effective in controlling weeds in dolichos bean at Gujarat, India.

As the crop cannot stand water-logging, frequent irrigation should be avoided. Drip irrigation scheduling at 100 per cent Epan with fertigation level of 100: 100 N and K_2O kg/ha recorded maximum green pod yield and higher N, P and K uptake (Saileela *et al.*, 2015).

Shedding of flowers has been reported in this crop and some cultivars were observed to have a relatively higher pod set than others (Sheriff *et al.*, 1969). Only 10-20 per cent flowers develop into mature pods. Application of calcium chloride (0.1 per cent) + NAA (100 ppm) when the first inflorescence appears to flowering increases fruit set and productivity (Shivashankar and Kulkarni, 1989). Foliar spraying of NAA @ 40 ppm recorded higher growth and yield attributing parameters in dolichos bean (Pramoda and Sajjan, 2018). Higher plant height, number of leaves, chlorophyll content, leaf area and early flowering of dolichos bean were obtained by application of INM + foliar spray of Seaweed extract @ 5 per cent at 30, 60 and 75 days after sowing in coastal region of Tamil Nadu, India (Jaisankar and Manivannan, 2018).

7.0 Harvesting and Yield

Pods are harvested when they are green and succulent and have not become fibrous.

Pole type cultivar produces, on an average, 150 to 400 quintals of green pods per hectare and bush type cultivars can yield 40 to 60 quintals of green pods per hectare. When grown for consumption of seeds as pulse, it produces on an average about 10-15 quintals dry seeds per hectare.

8.0 Diseases and Pests

8.1.0 Diseases

Important diseases of lablab bean and their control measures are given below:

Fungal Diseases

8.1.1 Leaf Spot

It is a fungal disease caused by *Cercospora dolichi*. Circular to angular spots with grey centre and reddish border appear on leaves. They gradually cover the entire leaf surface as more number of spots appears. Spraying with 0.25 per cent mancozeb or zineb at 12-15 days interval commencing from first appearance of the disease can control the disease (Hazra *et al.*, 2011).

8.1.2 Ashy Stem Blight

This disease described under French bean and the symptoms and control measures have been discussed there.

8.1.3 Powdery Mildew

It is caused by the fungus *Leveillula taurica* var. *macrospora*. Description of the symptoms and control measures are the same as in French bean.

Viral Diseases

8.1.4 Yellow Mosaic

The leaves of the diseased plants develop bright yellow patches interspersed with green area. There is no dwarfing effect on vines or malformation of leaves, which are comparatively smaller than those on healthy vines. It is transmitted by whitefly (*Bemisia tabaci*). This disease is difficult to control. However, removal of diseased plants as soon as they are noticed and spraying of systemic insecticide imidacloprid (0.03 per cent) or acetamiprid (0.15 per cent) at regular interval can prevent the spread of the disease by controlling the vector.

8.2.0 Pests

Major insect pests like aphids and pod borer attacking Lablab bean are nearly the same as those described under French bean. The nature of damage and control measures is same as those of French bean.

9.0 Seed Production

Lablab beans are usually self-pollinated and partially cross-pollinated by insects.

Harvesting and Packaging of Dolichos Bean

Field views of dolichos bean cultivation

Packaging of dolichos bean pods for marketing

Diseases of Dolichos Bean

Anthracnose infection in dolichos bean leaves and pods

Mosaic infection in dolichos bean leaves

Pests of Dolichos Bean

Pod borer infestation in dolichos bean

Aphid infestation in dolichos bean

Different cultivars should be grown 50 m apart in case of foundation seed and 25 m apart in case of certified seed (Anon., 1971). The seed crop is inspected before flowering, during flowering and at pod maturity to rogue out off-type and diseased plants. For seed production it should be grown as a pure crop with staking of the vines in case of pole types. Ripe mature pods are handpicked from the standing crop. Threshing is done by beating pods with stick or by a roller. Seeds should be thoroughly cleaned and dried before bagging. Average seed yield is 6 to 8 quintals per hectare.

10.0 Crop Improvement

In spite of the fact that this crop can be grown in the plains almost throughout India, only during the last decade some attention has been paid to this crop for its improvement.

Genetic resources are the wealth/treasure of any country for continuous genetic improvement of economically important crops to cater the needs of present and future generations. The efforts to collect, conserve, characterize, evaluate and catalogue dolichos bean genetic resources are far from satisfactory given its multiple economic uses and ability to resist biotic and abiotic stresses. Shivashankar *et al.* (1971, 1977); Viswanath and Manjunath (1971); Viswanath *et al.* (1972); Chikkadevaiah *et al.* (1981) at the University of Agricultural Sciences (UAS), Bengaluru, India have collected, evaluated and catalogued dolichos bean germplasm on a limited scale. Wang *et al.* (1991) collected 385 selections belonging to 14 legume species including *Lablab purpureus* in Hainan Island of China during 1987-89 and evaluated them for agronomic characters. Xu-Xiang-shang *et al.* (1996) collected 32 dolichos bean accessions from Qinling-Bashan mountain region, Sichuan of China. They investigated their area of distribution, cultivation, morphological characteristics and recommended four elite cultivars for commercial production. Pujari (2000) and Shanmugam (2000) have established dolichos bean germplasm consisting of 60 accessions, which included 22 improved varieties and 38 local landraces collected from different districts of Odisha, West Bengal and Andhra Pradesh states of India. Thus, largest collection of dolichos bean genetic resources (650 accessions) is held at the University of Agricultural Sciences (UAS), Bengaluru, India. These accessions were characterized and evaluated for vegetative, inflorescence, pod and seed traits. A set of 70 descriptors based on 16 vegetative, 14 inflorescence, 20 pod and 20 seed traits were developed considering the spectrum of variability for these traits following the guidelines of Bioversity International (Byregowda *et al.*, 2015). Evaluation and characterization of both pole and bush type dolichos bean has been done in concerted manner at BCKV, West Bengal, India and the available germplasm showed wide diversity for different growth, pod and quality characters (Hazra, 1982; Kabir and Sen, 1986, 1987; Chattopadhyay and Dutta, 2010; Das *et al.*, 2015; Nayek *et al.*, 2017). A few of these are based on easily field assayable and simply inherited (single/oligogenic) descriptors such as growth habit, pod curvature, flower colour (Raut and Patil, 1985; Rao, 1987; Girish and Byregowda, 2009; Chattopadhyay and Dutta, 2010; Keerthi *et al.*, 2014a; Keerthi *et al.*, 2016) and seed traits (Ayyangar and Nambiar, 1936 a and b; Ayyangar and Nambiar, 1941; Patil and Chavan, 1961; D' cruz and Ponnaiya, 1968). These descriptors could be used as diagnostic markers of germplasm accessions for maintaining their identity and purity. They help minimize duplication and avoid mistakes in labeling the germplasm accessions and thereby enable their easy retrieval from the collection. They are also useful in conducting Distinctiveness (D), Uniformity (U) and Stability (S) test, a mandatory requirement for protecting varieties under Protection of Plant Varieties and Farmers' Rights (PPA and FR) Act of India and such other similar acts that are vogue in other countries (Byregowda *et al.*, 2015).

Considering that the genetic resources held at UAS, Bengaluru, India is unwieldy for precise characterization and evaluation and possibility of occurrence of duplicates due to repeated sampling of same accession and/or assigning different names/identity to the same accession, a core set consisting of 64 accessions was developed at UAS, Bengaluru (Vaijayanthi *et al.*, 2015 b and c) using PowerCore (v. 1.0) software, a program that applies advanced M-strategy with a heuristic search (Kim *et al.*, 2007). The core set retained more than 90 per cent of quantitative traits variability and polymorphism of qualitative traits in the base collection of 644 accessions. In similar efforts to reduce size and possible duplicates, Bruce and Maass (2001) in Ethiopia and Islam *et al.* (2014) in Bangladesh also developed core sets of 47 and 36 accessions from the base collections of 251and 484 accessions, respectively. Based on two years (2012 and 2014) of evaluation of core set, promising traits-specific accessions and those promising for multi-traits have been identified at UAS, Bengaluru (Vaijayanthi *et al.*, 2016a). The accessions promising for multi-traits were evaluated in multi-locations representing eastern, southern and central dry zones of Karnataka during 2015 to identify those widely/specifically adaptable to the three agro-climatic zones. The accessions such as GL 250, FPB 35 and Kadalavare were found widely adaptable to the three agro-climatic zones of Karnataka with relatively high fresh pod yield (Vaijayanthi *et al.*, 2016 b). These accessions are suggested for preferential use in breeding dolichos bean varieties widely adaptable to the three agro-climatic zones of Karnataka.

Genetic variability studies of yield components indicated the existence of wide genetic base among the various genotypes. Moreover, high heritability, coupled with high genetic gain for most of the characters showed the presence of appropriate genetic background for further selection with a view to improve yield and some of its component characters (Joshi, 1971; Singh *et al.*, 1979; Arunachala, 1979; Baswana *et al.*, 1980a; Nayar, 1982; Hazra, 1982; Kabir and Sen, 1986, 1987; Chattopadhyay and Dutta, 2010; Das *et al.*, 2015). Basu *et al.* (1999) studied the genetics of embryo weight, cotyledon weight and seed protein in Lablab bean. High values of GCV coupled with high heritability and high genetic gain indicated the scope of improvement in these characters through direct selection.

Path coefficient analysis revealed that pods per plant, pod length and pod width had positive direct contribution towards yield indicating that these are the important criteria for improvement by selection (Arunachala, 1979; Singh *et al.*, 1979; Pandey *et al.*, 1980; Baswana *et al.*, 1980b; Nayar, 1982; Kabir and Sen, 1989).

Dolichos bean have cleistogamous flower structure, so they are highly self-pollinated. In the young immature bud, the anthers are far behind the level of the stigma. As the bud grows the stamina filament elongate more rapidly than the style and the anther dehisce when the long stamen comes above or on a level with the stigma. Flower opens generally two days after anther dehiscence in the daytime. So, the flowers are to be emasculated and pollinated immediately before the anther dehiscence are found (Das *et al.*, 2014). No crossability barrier among the genotypes belonging to two growth habits (pole and bush types) was observed (Das *et al.*, 2014). Crossing success slightly varied between genotypes within groups. Wide genetic divergence of the parents also affected crossability adversely. The average crossing success was generally low ranging between 12 per cent (Bush × Bush type) and 30 per cent (Pole × Pole type). It is a self pollinated crop where degree of heterosis was theoretically considered less, but cross combinations showing heterotic vigour can be used for developing high yielding pure lines. Heterosis manifestation in dolichos bean is in the form of earliness in maturity, increased productivity, and better quality attributes (Chikkadevaiah, 1981; Hazra, 1982; Kabir and Sen, 1989; Singh *et al.*, 1986; Valu *et al.*, 2006; Patil *et al.*, 2011; Das *et al.*, 2014; Nayek *et al.*, 2018). However, only few reports have been made available in respect to heterosis for yield in this crop. Chikkadevaiah (1981) observed 43.96 per cent to 289.45 per cent heterobeltiosis for pod number per plant and for seed yield per plant in field bean, respectively. Hazra (1982) from West Bengal reported 142.42 per cent heterobeltiosis for pod yield per plant in a cross between JDL-85 and *Lal Sem* (local), while Kabir and Sen (1989) noted heterobeltiosis of 198 per cent for pod number and 144 per cent for pod yield in cross combination JDL 53 × Hebbal. Das *et al.* (2014) exhibited heterobeltiosis for pod yield per plant to the extent of 20.00 per cent and 9.20 per cent in crosses Arka Jay × DOLB VAR 2 and DOLP VAR 10 × DOLB VAR 2, respectively. Recently, Nayek *et al.* (2018) recorded heterobeltiosis for pod yield/plant to the extent of 114.26 and 86.63 per cent in crosses BCDB-2 × BCDB-15 and BCDB-10 × BCDB-5, respectively. They also been found that closely related parents (indeterminate × indeterminate) exhibited low heterosis, but crosses between diverse parents (indeterminate × determinate or indeterminate × semi-indeterminate) tended to show higher heterosis for pod yield and other important traits. Cultivars Kalianpur Type 2, 6802 and Rajani (Singh *et al.*, 1986); HD 18 (Kabir and Sen, 1990); BCDB 1, DOLP VAR 10, and DOLP VAR 5 (Das *et al.*, 2014); BCDB 10 and BCDB-2 (Nayek *et al.*, 2018) have been reported to be good general combiner in desirable directions for pod yield and its component characters.

Genetic analysis provides a guide line for the assessment of relative breeding potential of the parents or identifies best combiners in crops which could be utilized either to exploit heterosis in F_1 or the accumulation of fixable genes to evolve variety. Such studies not only provide necessary information regarding the choice of parents but also simultaneously illustrate the nature and magnitude of gene action involved in the expression of desirable traits. The type of gene action involved for the conditioning of different quantitative traits of dolichos bean was studied in detail by many workers. Number of inflorescences per plant, number of nodes per inflorescence, number of flower buds per inflorescence and seed protein content of pod were conditioned by additive gene action (Basu *et al.*, 1999; Das *et al.*,, 2014); length of inflorescence, number of pods per inflorescence, pod length and number of seeds per pod were controlled by non-additive gene effects (Das *et al.*, 2014), while days to first flowering, days to 50 per cent flowering, number of pods per plant, pod width, pod weight, shelling percentage of fresh pod and pod yield per plant were conditioned by both additive and non-additive gene effects (Das *et al.,* 2014; Nayek *et al.,* 2018). It is suggested that biparental mating between the selected recombinant and recurrent selection may prove very effective with a view to exploiting the additive and non-additive component of variation (Singh and Singh, 1981; Kabir and Sen, 1990; Das *et al.,* 2014; Nayek *et al.,* 2018). The inheritance of qualitative traits (raceme emergence, flower colour and pod curvature) in dolichos bean has been studied in detail by (Hanumantha Rao, 1987; Prasanthi, 2005; Keerthi *et al.,* 2014). Recent study of Keerthi *et al.* (2016) revealed biallelic monogenic control of photoperiod-induced sensitivity to flowering time and flower colour in F_2 and F_3 generations. While, growth habit and pod curvature were controlled by two genes that exhibit classical complementary epistasis, raceme emergence was controlled by two genes that displayed classical inhibitory epistasis. They observed that the dominant alleles, at two different unlinked pairs of genes are necessary for plants to exhibit indeterminate growth habit and bear straight pods, while any other combination of alleles at the two pairs of genes result in plants displaying determinate growth habit and bearing curved pods. They recorded the genes controlling growth habit, PSFT and raceme emergence were linked. Those controlling flower colour and pod curvature were segregated independent of each other.

In various parts of India, the farmers are popularly growing their own bred varieties of dolichos passed from many generations known by different vernacular names, collectively called farmers varieties. They known to possess unique traits and have economic importance. However, there is no internationally or nationally accepted DUS test guidelines available in dolichos; which is a mandatory requirement to register and protect farmer's varieties under Protection of Plant Varieties and Farmers' Right Authority (PPV and FRA), Government of India, New Delhi.

A good number of local strains are available in India. Some of the outstanding strains have been purified and selected as commercial cultivars. The JDL series of Jawaharlal Nehru Krishi Viswavidyalaya at Jabalpur and T series of Chandra Shekhar Azad University of Agriculture and Technology, Kalyanpur, Uttar Pradesh, are some of the examples of promising pole type cultivars in lablab bean. A pole type cultivar Pusa Early Prolific has been developed at Indian Agricultural Research Institute, New Delhi.

Some bushy dwarf types in the name of series Hebbal are available at the University of Agricultural Sciences, Bangalore, Karnataka. Though the green pods of this series cannot be consumed as vegetable, its bushy and dwarf plant characteristics has been used for incorporation in the commercial pole type cultivars so that it does not require any support for its growth. The photo-insensitive cultivars US 67-43 and US 67-44 from USA has also been utilized in the improvement programme at Bangalore. Hebbal Avare 1 which has been evolved from segregating materials of Local Avare × Red Typicus can be grown throughout the year. While Hebbal Avare 3 and Hebbal avare 4 developed from cross Hebbal Avare 1 × US 67-31 and Hebbal Avare × Co. 8, respectively were not only high yielding but compact and photo-insensitive. The ICAR-Indian Institute of Horticultural Research (IIHR), Bengaluru,

India can able to successfully introgress photo-insensitivity and determinate traits from *Lablab purpureus* var. *lignosus* (Hebbal Avare 3, a pulse type dolichos as a donor) into genetic background of *Lablab purpureus* var. *typicus* (Kanupu Chikudu, a most priced local garden bean), and developed two bush type vegetable dolichos varieties namely, Arka Jay and Arka Vijay suitable for round the year cultivation (Satyanarayana, 1985).

Research work at Tamil Nadu Agricultural University, Coimbatore has led to the release of dual purpose lines Co series (for vegetable and seed).

At Bidhan Chandra Krishi Viswavidyalaya, West Bengal, India lines selected in advance generations with combined desirabilities for dwarfness, photo-insensitive by incroporating Hebbal in hybridization programme also showed the potentiality of improved protein content and quality (Kabir *et al.*, 1992). Das *et al.* (2014) also incorporated dwarf and photo-insensitive genes from Arka Jay into viny, photosentive pole varieties to develop semi determinate lines with improved yield and seed protein content. Besides breeding of dwarf and bushy plant types with quality pod characters, earliness, breeding for resistance to bean mosaic virus, improved storage and higher seed protein content require special attention for effective and all-round improvement in this crop.

Breeding for resistance to insect pests is currently limited to screening and identification of resistance sources in germplasm and breeding lines of dolichos bean. Chakravarthy and Lingappa (1986) identified two stable sources of resistance to larval boring and to pod borers in field and laboratory conditions based on screening of 111 dolichos accessions. The two sources of resistance to pod borers exhibited significant degree of antibiosis as demonstrated by reduced larval survival, larval and pupal weights, prolonged larval duration and altered sex ratio (Chakravarthy and Lingappa, 1988). Combined effect pod colour, pubescence and fragrance appeared to be associated with resistance response. However, none of pod colour, pubescence and fragrance per se imparted resistance to pod borers. The accessions highly resistant to pod borers were highly susceptible to aphid infestation (Chakravarthy and Lingappa, 1988). Jagadeesh Babu *et al.* (2008) identified germplasm accessions such as GL 1, GL 24, GL 61, GL 69, GL 82, GL 89, GL 196, GL 121, GL 135, GL 412, and GL 413 with < 10 per cent insect damage as resistant to pod borers (*Heliothis armigera* and *Adisura atkinsoni*) and bruchids (*Collosobruchus chinensis*) based on two years (2004 and 2005) of field sowing of 133 germplasm accessions. Based on laboratory screening of 28 selected germplasm accessions under no choice conditions, Rajendra Prasad *et al.* (2013) identified resistant accessions, GL 77, GL 233 and GL 63 with least seed damages of 13.4 per cent, 14.69 per cent and 18.34 per cent, respectively (Rajendra Prasad *et al.*, 2013). During 2007 and 2008 at UAS, Bengaluru, another set of 132 germplasm accessions were screened for responses to infestation by *Helicoverpa armigera* and *Adisura atkinsoni* and bruchids. Pod damage due to *Helicoverpa* ranged from 1.8 (GL 187) to 36.8 per cent (GL 9), while *Adisura* infestation varied from 0.5 (GL 55) to 7.9 per cent (GL 14) and Bruchids infestation ranged from 0.6 (GL 187) to 40 per cent (GL 37). The average infestation of germpalsm accessions by *Helicoverpa* was 18 per cent followed by bruchids with 12.1 per cent and by Adisura with 3.1 per cent. The infestation by *Helicoverpa* was < 10 per cent in 13 accessions, while, that by bruchids was < 10 per cent in 42 accessions and by *Adisura* was < 10 per cent in 132 accessions. Another study at UAS, Bengaluru, indicated the role of both antixenosis and antibiosis mechanisms of resistance to damage by *Helicoverpa* in the germplasm accessions GL 233, GL 426, GL 357 and GL 187 which were found moderately tolerant (Rajendra Prasad *et al.*, 2015). Twenty eight dolichos bean genotypes were screened against pulse beetle, *Callosobruchus theobromae* and observed for ovipositional preference, number of eggs laid, emergence of adult, per cent adult emergence, number of seeds damaged and per cent seed damage at Bengaluru, Karnataka (Rajendra Prasad *et al.*, 2013). The genotype GL 77 recorded the lowest seed damage (13.4 per cent) followed by GL 233 (14.69 per cent) and GL 63 (18.34 per cent), and these entries were grouped as least susceptible. Lowest seasonal mean populations (16.11 and 17.22 insects/10 cm twig) of *Aphis craccivora* were recorded on genotypes 2013/DOLVAR-4 and 2013/DOLVAR-3, respectively hence they could be used in breeding programmes for developing resistant lines (Chouragade *et al.*, 2018).

As is true with insect pests, breeding dolichos bean for *Dolichos yellow mosaic virus* (DYMV) resistance is confined to identification of resistance sources. Singh *et al.* (2012) identified VRSEM 894, VRSEM 887 and VRSEM 860 as resistance to DYMV among 300 germplasm accessions based on initial field screening under natural infection followed by screening of 34 symptomless accessions using sap inoculation in field condition (Rai *et al.*, 2015). Rajesha *et al.* (2010) screened 195 genotypes of dolichos bean against anthracnose disease caused by *Colletotrichum lindemuthianum* under field conditions. They identified nine genotypes, such as GLB 3, GLB 4, GLB 8, GLB 9, GLB 11, GLB 19, GLB 60, GLB 166 and GLB 167 that were immune; 48 genotypes resistant and 83 genotypes moderately resistant to the disease. Deshmukh *et al.* (2012) screened 44 genotypes which included varieties and germplasm accessions of dolichos bean for responses to anthracnose infection under conditions and identified three varieties and two germplasm accessions resistant to the disease. Some of the useful resistant/tolerant sources against dolichos yellow mosaic virus (BCDB-1; Gomchi Green; Pusa Early Prolific; BCDB-2) and *Cercospora* leaf spot (BCDB-1; SEMVAR-8) have been identified in the eastern gangetic plains of India (Chattopadhyay and Dutta, 2010).

Dolichos bean has better inherent capacity to withstand moisture stress than the comparable legumes such as cowpea, horse gram, *etc.* (Nworgu and Ajayi, 2005; Ewansiha and Singh, 2006; Maass *et al.*, 2010) and adapt to acidic (Mugwira and Haque, 1993) and saline soils (Murphy and Colucci, 1999). With its deep root system, dolichos bean is not only drought tolerant (Kay, 1979; Cameron, 1988; Hendricksen and Minson, 1985), but, also has the ability to harvest soil minerals which otherwise not available for annual crops (Schaaffhausen, 1963 a, b). In the event of imminent extremities of abiotic stresses driven by climate change (IPCC, 2007), dolichos bean would be a better alternative to popular legumes. Thus, breeding and enhancing the economic value of dolichos bean would provide competitive edge to dolichos bean producers and enable preparing for unfriendly agriculture production climate. The reported research on breeding dolichos bean for resistance to abiotic stresses is limited. In the light of deficient knowledge of biochemical and physiological basis of its response to abiotic stresses, a study was initiated to evaluate the effect of drought on dolichos bean (D'souza and Devaraj, 2011). Effect of drought on dolichos (HA-4 cultivar) was evaluated in 10-d-old seedlings for 8 d after withholding water. The stress reduced dry and fresh weight, leaf number, surface area, root and shoot length, total chlorophyll and relative water content. Oxidative stress markers, H_2O_2, glutathione, malondialdehyde, proline, ascorbic acid, total phenols, and total soluble sugars were significantly elevated. Drought enhanced antioxidant enzymes, peroxidase and glutathione reductase, and reduced catalase in a time dependent manner in the leaves. POX and CAT in roots showed inverse relationship with the duration of stress, whereas GR exhibited increased activity. The metabolic activity of enzymes α-amylase and acid phosphatase increased temporally in leaves and roots. The plant showed ability to rehydrate and grow upon rewatering, and levels of antioxidant components correlated with drought tolerance of the plant.

11.0 Biotechnology

Research on biotechnology of *Dolichos* is very limited. Previous investigation on *in vitro* study of Lablab bean revealed that establishment of callus culture from cotyledon segments was possible on tobacco high salt basal medium supplemented with 2,4-D, kinetin and the adjuvent casein hydrolysate (Kumar, 1974). Shina *et al.* (1983) found good callus establishment from leaf explants of *Dolichos biflorus* using Blaydes basal medium with hormone supplement and picloram added to culture medium as adjuvent. Kabir and Sen (1990) reported that yellow cotyledon base taken aseptically from pre-soaked seed of Lablab bean should be the ideal source of *in vitro* induction of callus in MS basic medium with 2,4-D alone (4 mg/l) or in combination with kinetin (1 mg/l). MS medium with 130.2 mg/l Na2Fe EDTA supplemented with 2,4-D and kinetin was suitable for subsequent maintenance of callus. A few reports are available on use of protoplast culture and on the characterization of proteins.

Wei and Xu (1993) isolated protoplasts from immature cotyledons of *D. lablab* [syn. *Lablab purpureus*] and initially cultured in K8p liquid medium supplemented with 0.2 mg 2,4-D, 1 mg NAA and 0.5 mg BA/l. Sustained divisions resulted in mass production of cell colonies and small calluses after 6 weeks. Shoot formation was initiated on MSB medium with 0.1-0.25 mg/l IAA and 0.5-1.0 mg/l of both BA and zeatin. The regenerated potential of shoot tips, which can be exploited to get virus-free clones through micropropagation in dolichos bean (Sounder Raj *et al.*, 1989). *In vitro* regenerability is dependent on genotypic characters as well as growth regulator combinations (MS + 1.0 mg/l IBA + 2 per cent sucrose) (Kshirsagar *et al.*, 2018). The genotype Konkan Bhushan and G9 (No. 66) were proved better for regeneration and could be further taken for genetic manipulation studies. *Lablab purpureus* is a perfect candidate for embryogenesis and organogenesis (Tidke *et al.*, 2019). They have taken cotyledon as an explant and were cultured on MS media containing 2, 4-D [4.5-9.5µM]. The concentration of 2, 4-D at 4.5 and 5.5 µM showed the well defined globular, heart, torpedo shapes. The root tips were used for callus induction on MS supplemented with 2, 4-D, BAP, NAA, IBA *etc.* The combination of IAA (3.0 mg/l) + BAP (0.5 mg/l) and IAA (5.0 mg/l) + BAP (0.5 mg/l) showed compact, hard, greenish-white callus. Suspension culture was carried out using induced callus and inoculated on liquid media with 2, 4 D (2-4) + BAP (0.5) mg/l.

Genetic Transformation in Dolichos bean

In situ Bioassay in the T2 Generation Dolichos Bean Plants against *H. armigera.*

Performance of the wild-type (WT-1 to WT-3) and transgenic plants (T-1 to T-9) in the *in situ* bioassay against *H. armigera* larvae (Keshamma *et al.*, 2012).

The growth was observed in 2, 4-D (2) + BAP (0.5) mg/l. For indirect organogenesis pre-established callus of *in vitro* grown seedling used as an explant on MS media with different concentration of plant growth regulator. The shooting and rooting was observed.

Kabir *et al.* (1992) showed variations in seed protein content and protein profile through SDS-PAGE in the intervarietal segregants indicating further scope of improvement of seed protein in the desirable plant type.

Genomic tools such as DNA markers in dolichos bean breeding in still in infancy because they are not readily available. Nevertheless, sequence independent marker systems such as RAPD and AFLP have been used to detect and characterize genetic variation among germplasm accessions and breeding lines. Reported literature on the use of DNA markers in analysis of genetic variability in dolichos bean has been documented by many workers (Liu, 1996; Konduri *et al.*, 2000 a,b; Maass *et al.*, 2005; Patil *et al.*, 2009; Rai *et al.*, 2010; Kinmani *et al.*, 2012). Most of the studies on genetic diversity analysis are based on RAPD and AFLP. However, the information obtained from these markers is not reliable due to their poor reproducibility. Hence, sequence dependent simple sequence repeat (SSR) and single nucleotide polymorphism (SNP) are highly preferred by researchers owing to their simple inheritance and amenability for automation and high reproducibility. The use of cross species/genera SSR markers in crops where they are not available is considered as the most cost effective strategy as *de novo* development of SSR markers is both expensive and time consuming. Taking cue from several successful examples of cross transferability of SSR markers, Yao *et al.* (2012) demonstrated that all tested EST-SSR markers from soybean were cross transferable to dolichos bean although only 16 per cent of them could differentiate the genotypes. At UAS, Bengaluru, transferability of SSR markers from cowpea, soybean, *Medicago truncatula*, greengram and chickpea to dolichos bean was examined (Shivakumar and Ramesh, 2015; Uday Kumar *et al.*, 2016). Transferable SSR markers help enrich marker resources for various applications in dolichos bean genetics and breeding research such as (1) characterize and assess genetic variability in working germplasm and/or breeding lines, (2) fingerprint to identify duplicate germplasm accessions, (3) fingerprint varieties for protecting IPR and (4) Select genetically diverse genotype for effecting crosses to generate variability to identify genotypes with best combination of traits.

Ignacimuthu *et al.* (1997) analysed RFLP on the rRNA genes of populations of *Lablab purpureus, Dolichos trilobus* [*Vigna aconitifolia*], *V. bournii, V. grahmiana, V. unguiculata* and *V. wightii*. Polymorphisms between species were higher than those within species or between genera. Rouge and Risler (1990) recognized three stretches of homologous amono acid sequences along the complete amino acid sequences of 2-chain and single-chain lectins of *Dolichos* and their studies suggested that genes coding for lectins probably arose by duplication of a common ancestral gene followed by limited divergence.

The amenability and reproducibility of a tissue culture-independent *Agrobacterium tumefaciens*-mediated transformation strategy was analyzed in field bean and the stability of the transgenes was examined (Keshamma *et al.*, 2012). The protocol involves in planta inoculation of embryo axes of germinating seeds and allowing them to grow into seedlings *ex vitro*. Transformants were raised using a chimeric Bt gene, *cry1AcF*, and putative transformants were analyzed by PCR for both *cry1AcF* as well as the *nptII* genes. Bioassays against *Helicoverpa armigera*, the major pod borer, showed that several T1 plants performed well with 17 per cent of T1 plants harboring the transgene. Further, enzyme-linked immunosorbent assay (ELISA) and quick dip strip test confirmed the expression of the chimeric Bt toxin. The stability of the transgenes was checked in three generations for integration, expression, and efficacy against the two insects, *H. armigera* and *Spodoptera litura*. Southern blot analysis of 10 high expressing plants confirmed the integration of the transgene, whereas single copy integration of the T-DNA in 5 events was also evident.

12.0 References

Alhasan, S.A., Aranha, O. and Sarkar, F.H. (2001) *Clin. Cancer Res.*, **7**: 4174-4181.

Anonymous (1961) *Agricultural and Horticultural Seeds*, FAO, United Nations, Rome, p. 531.

Anonymous (1971) *Indian Minimum Seed Certification Standards*, Central Seed Committee, M. Food Agric., Comm. Dev. Co-op., New Delhi, p. 102.

Arunachala, A.S. (1979) *Mysore J. Agric. Sci.*, **13**: 369.

Aykroyd, W.R. (1963) *ICMR Special Rept. Series*, No. 42.

Ayyangar, G.N.R. and Nambiar, K.K.K. (1935) In: *The First Proc. Indian Acad. Sci.*, **1**: 57-867.

Ayyangar, G.N.R. and Nambiar, K.K.K. (1936a) *Proc. Indian Acad. Sci.*, **2**: 74-79.

Ayyangar, G.N.R. and Nambiar, K.K.K. (1936b) *Indian Acad. Sci.*, **4**: 411-433.

Ayyangar, G.N.R. and Nambiar, K.K.K. (1941) *Proc. Indian Acad. Sci.*, **15**: 95-113.

Basu A.K., Pal, D., Sasmal, S.C. and Samanta, S.K. (1999) *Veg. Sci.*, **26**: 37-40.

Baswana, K.S., Pandita, M.L., Pratap, P.S. and Dhankhar, B.S. (1980a) *Haryana Agric. Univ. J. Res.*, **9**: 52-55.

Baswana, K.S., Pandita, M.L., Pratap, P.S. and Dhankhar, B.S. (1980b) *Haryana Agric. Univ. J. Res.*, **10**: 485-489.

Bruce, C. and Maass, B.L. (2001) *Genet. Resour. Crop Evol.*, **48**: 261-272.

Byregowda, M., Gireesh, G., Ramesh, S., Mahadevu, P. and Keerthi, C.M. (2015) *J. Food Legumes*, **28**: 203-214.

Cameron, D.G. (1988) *Queensland Agric. J.* (March-April), pp. 110-113.

Chakravarthy, A.K. and Lingappa, S. (1986) *Colemania*, **2**: 43-51.

Chakravarthy, A.K. and Lingappa, S. (1988) *Insect Sci. Appl.*, **9**: 441-452.

Chattopadhyay, A. and Dutta, S. (2010) *Veg. Crops Res. Bull.*, **73**: 33-45.

Chikkadevaiah, G. (1981) *Indian J. Genet.*, **41**: 366-367.

Chikkadevaiah, Gj., Hiremath, S.R. and Shivashankar, G. (1979) *Experientia*, **35**: 171-172.

Chopra, R.N., Nayar, S.L. and Chopra, I.C. (1986) *Glossary of Indian Medicinal Plants.* Council of Scientific and Industrial Research, *New Delhi*.

Choudhury, B. (1972) *Vegetables*, National Book Trust, India, New Delhi, p. 220.

Choudhury, B. (1972) *Vegetables, National Book Trust,* India, New Delhi, p. 220.

Chouragade, V., Shukla, A., Sharma, A. and Bijewar, A.K. (2018) *J. Entomol. and Zool. Studies*, **6**: 1968-1971.

Claussen, W. (2005) *Plant Sci.*, **168**: 241–248.

Cruz, R.D. and Pannaiya, V.T.S. (1969) *Indian J. Genet.*, **29**: 139-141.

D'souza, M.R. and Devaraj, V.R. (2011) *Indian J. Biotechnol.*, **10**: 130-139.

Dalai, S., Evoor, S., Hanchinamani, C.N., Mulge, R., Mastiholi, A.B., Kukanoor, L. and Kantharaju, V. (2019) *Int. J. Curr. Microbiol. App. Sci.*, **8**: 187-195.

Das, I., Seth, T., Shende, V., Dutta, S., Chattopadhyay, A. and Singh, B. (2014) *SABRAO J. Breed. Genet.* **46**: 293-304.

Das, I., Shende, V.D., Seth, T. and Chattopadhyay, A. (2015) *J. Crop Weed*, **11**: 72-77.

Duke, J.A., Kretschmer, A.E., Jr., Reed, C.F. and Weder, J.K.P. (1981) In: *Handbook of legumes of world economic importance. Plenum Press, New York*, pp.102-106.

Ewansiha, S.S. and Singh, B.B. (2006) *J. Food Agri. Environ.*, **4**: 188-190.

Girish, G. and Byregowda, M.B. (2009) *Environ. Ecol.*, **27**: 571-580.

Hanumantha Rao, C. (1987) *Indian J. Genet.* **47**: 347–350.

Hazra, P. (1982) *M. Sc. (Ag.) Thesis*, Bidhan Chandra Krishi Viswavidyalaya, West Bengal, India.

Hazra, P., Chattopadhyay, A., Karmakar, K. and Dutta, S. (2011) *Modern Technology in Vegetable Production. New India Publishing Agency, New Delhi, India*, p. 480.

Hendricksen, R.E. and Minson, D.J. (1985) *Herb. Abstract*, **55**: 215-228.

Hoffman, R. (1995) *Biochem. Biophys. Res. Commu.*, **211**: 600-606.

Ignacimuthu, S., Schumann, K., Zink, D. and Nagl, W. (1997) *Curr. Sci.*, **72**: 624-626.

IPCC (Intergovernmental Panel on Climate Change). (2007) In: *"Climate Change 2007" – the IPCC fourth assessment report (AR4). [http://www.ipcc.ch/press/index.htm]*.

Islam, N., Rahman, M.Z., Ali, R., Azad, A.K. and Suttan, M.K. (2014) *Pakistan J. Agric. Res.*, **27**: 99-109.

Jagadeesh Babu, C.S., Byregowda, M., Girish, G. and Gowda, T.K.S. (2008) *Env. Eco.*, **26**: 2288- 2290.

Jaisankar, P. and Manivannan, K. (2018) *Int. J. Manag. Technol. Eng.*, **8**: 2780-2782.

Jaisankar, P. and Manivannan, K. (2018) *Plant Arch.*, **18**: 2194-2198.

Joshi, S.N. (1971) *Madras Agric. J.*, **58**: 367-371.

Kabir, J. and Sen, S. (1986) In: *Perspectives in Cytology and Genetics* (Eds. Manna, G.K. and Sinha, U.), **5**: 765-768.

Kabir, J. and Sen, S. (1987) *Ann. Agric. Res.*, **8**: 141-144.

Kabir, J. and Sen, S. (1989) *Trop. Agric.* (Trinidad), **66**: 281-283.

Kabir, J. and Sen, S. (1990) *Trop. Agric.* (Trinidad), **67**: 123-126.

Kabir, J. and Sen, S. (1991) *Legume Res.*, **14**: 120-124.

Kabir, J., Das, A. and Samanta, S.K. (1992) *Crop. Res.*, **5**: 512-516.

Kabir, J., Samanta, S.K. and Sen, S. (1993) *Crop Res.*, **6**: 270-275.

Kay, D.E. (1979) *Hyacinth bean-Food legume. Crop and Product Digest No. 3. Tropical Products Institute*, **16**: 184–196.

Keerthi, C. M., Ramesh, S., Byregowda, M., Mohan Rao, A., Rajendra Prasad, B.S. and Vaijayanthi, P. V. (2016) *J. Genet.*, **95**: 89-98.

Keerthi, C.M., Ramesh, S., Byregowda, M., Mohan Rao, A., Rajendra Prasad, B.S. and Vaijayanthi, P.V. (2014) *J. Genet.*, **93**: 203-206.

Keshamma, E., Sreevathsa, R., Manoj Kumar, A., Reddy, K.N., Manjulatha, M., Shanmugam, N.B., Kumar, V.A.R. and Udayakumar, M. (2012) *Plant Mol. Biol. Rep.*, **30**: 67–78.

Kim, K.W., Chung, H.K., Cho, G.T., Ma, K.H., Chandrabalan, D., Gwag, J.G., Kim, T.S., Cho, E.g. and Park, Y.J. (2007) *Bioinformatics*, **23**: 2155-2162.

Kinmani, E.N., Wachira, F.N. and Kinyua, M.G. (2012) *Amer. J. Plant. Sci.*, **3**: 3-32.

Kirtikar, K.R. and Basu, B.D. (1995) *Indian Medicinal Plants. Vol. I. 3rd Edn., Sri Satguru Publications, New Delhi.*

Kobayashi, T., Nakata, T. and Kuzumaki, T. (2002) *Cancer Lett.*, **176**: 17-23.

Konduri, V., Godwin, I.D. and Lin, C.J. (2000a) *Biomed. Life Sci.*, **100**: 866-871.

Konduri, V., Godwin, I.D. and Lin, C.J. (2000b) *Theor. Appl. Genet.*, **100**: 856-861.

Kshirsagar, J.K., Sawardekar, S.V., Sawant, G.B., Devmore, J.P. and Jadhav, S.M. (2018) *J. Pharmacog. Phytochem.*, **7**: 2782-2789.

Kumar, A. (1974) *Indian J. Expt. Biol.*, **12**: 595-596.

Liu, C.J. (1996) *Euphytica*, **90**: 115-119.

Maass, B. L., Jamnadass, R. H., Hanson, J. and Pengelly, B.C. (2005) *Genet. Resour. Crop Evol.*, **52**: 683 - 695.

Maass, B.L., Knox, M.R., Venkatesh, S.C., Angessa, T.T., Ramme, S. and Pengelly, B.C. (2010) *Tropical Plant Biol.*, **3**: 123.

Mathukia, R.K., Sagarka, B.K. and Gohil, B.S. (2018) *Res. and Rev.: J. Crop Sci. Technol.*, **7**: 39–43.

Morris, J.B. (2009) *J. Diet. Suppl.*, **6**: 263-279.

Mugwira, L.M. and Haque, I. (1993) *J. Plant Nutrition*, **16**: 37 - 50.

Murphy, A.M. and Colucci, P.E. (1999) *Livestock Res. Rural Dev.*, **11**: 96-113.

Myrene, R D'souza and Devaraj, V.R. (2011). *Ind. J. Biotechnol.*, **10**: 130-139.

Naeem, M., Khan, M.M.A. and Moinuddin Siddiqui M.H. (2009) *Sci. Hortic.*, **121**: 389-396.

Nath, P. (1976) *Vegetables for the Tropical Region*, ICAR, New Delhi.

Nayak, N.J., Maurya, P.K., Maji, A., Chatterjee, S., Mandal, A.R. and Chattopadhyay, A. (2017) *Int. J. Curr. Microbiol. App. Sci.*, **6**: 381-395.

Nayak, N.J., Maurya, P.K., Maji, A., Mandal, A.R. and Chattopadhyay, A. (2018) *Int. J. Veg. Sci.*, **24**: 390–403.

Nayar, K.M. (1982) *Mysore J. Agric. Sci.*, **16**: 486.

Nworgu, F.C. and Ajayi, F.T. (2005) *Livestock Research for Rural Development, [Online 14.11.2007 from: http://cipav.org.co/lrrd/lrrd17/11/nwor1712.htm]*.

Pandey, R.P., Assawa, B.M. and Assawa, R.K. (1980) *Indian J. Agric. Sci.*, **50**: 481-484.

Patil, A.B., Desai, D.T. and Patil, S.S. (2011) *Legume Res.*, **35**: 18-22.

Patil, G.D. and Chavan, V.M. (1961) *Indian J. Genet.*, **21**: 142-145.

Patil, P., Venkatesha, Ashok, T.H., Gowda, T.K.S. and Byregowda, M. (2009) *J. Food Legumes*, **22**: 18-22.

Peng, Y.L., Wang, X.M., Li, M. and Tang, C.W. (2001) *Acta Hortic. Sinica*, **28**: 480.

Pramoda and Sajjan, A.S. (2018) *Int. J. Pure Appl. Biosci.*, **6**: 423-430.

Prasanthi, L. (2005) *Legume Res.*, **28**: 233–234.

Pujari, I. (2000) *Elucidation on genetic variability in country bean with conventional and molecular strategies. MSc (Agri.) Thesis, OUAT, Bhubaneswar.*

Rai, N., Kumar, A., Singh, P. K., Singh, M., Datta, D. and Rai, M. (2010) *Afr. J. Biotech.*, **9**: 137-144.

Rai, N., Rai, K.K., Venkataramanappa, V. and Saha, S. (2015) *Appl. Biochem. Biotechnol.*, pp.1-15.

Rajendra Prasad, B.S., Jagadeesh Babu, C.S. and Byregowda, M. (2013) *Current Biotica*, **7**: 153-160.

Rajesha, G., Mantur, S.G., Ravishankar, M., Shadakshari, T.V. and Boranayaka, M.B. (2010) *Int. J. Plant Prot.*, **3**: 135-136.

Rangasamy, P., Kasthuri, R. Gomathinayagam, P., Khan, A.K.F., Murugarajendran, C. and Nagarajan, P. (1993) *Madras Agril. J.*, **80**: 709-710.

Rao, C.H. (1987) *Indian J. Genet.*, **47**: 347-350.

Raut, V.M. and Patil, V.P. (1985) *Maharashtra Agric. Univ.*, **10**: 292-293.

Rouge, P. and Risler, J.L. (1990) *Biochemical Systematics and Ecol.*, **18**: 29-37.

Rustamani, M.A., Memon, N., Leghari, M.H., Dhaunroo, M.H. and Sheikh, S.A. (1999) *Pak. J. Zoology*, **31**: 323-326.

Saileela, K., Raji Reddy, D., Devender Reddy, M. and Uma Devi, M. (2015) *Glob. J. Res. Anal.*, p. 4.

Saimbhi, M.S. (1992) *Agric. Rev.*, **13**: 209-218.

Satyanarayana, A. (1985) *Annual Report, IIHR, Hessuraghatta, Bangalore.*

Schaaffhausen, R.V. (1963a) *Economic Bot.*, **17**: 146-153.

Schaaffhausen, R.V. (1963b) *Turrialba*, **13**: 172-178.

Shanmugam, A. (2000) *Characterization of a country bean germplasm: Morphological, biochemical and molecular approaches. M.Sc Thesis, OUAT, Bhubaneswar.*

Sheriff, M.N., Rajagopalan, C.K. and Venugopal, K. (1969) *Madras Agric. J.*, **56**: 661-662.

Shina, R.R., Das, K. and Sen, S.K. (1983) *Indian J. Expt. Biol.*, **21**: 113-119.

Shivakumar, M.S. and Ramesh, S. (2015) *Mysore J. Agric. Sci.*, **49**: 263-265.

Shivashankar, G. and Kulkarni, R.S. (1989) *Indian Hort.*, **33 and 34**: 24-27.

Shivashankar, G., Chikkadevaiah and Hiremath, S.R. (1977) *Indian J. Genet.*, **37**: 353-371.

Shivashankar, G., Viswanathan, S.R., Manjunath, A. and Chandrappa, H.M. (1971) *Proc. Int. Symp. Sub-tropical and Tropical Horticulture, Bangalore, India.*

Singh, B., Medhi, R.P. and Parthasarathy, V.A. (1992) *Indian J. Agric. Sci.*, **62**: 692-694.

Singh, B., Medhi, R.P. and Pathasarathy, V.A. (1993) *Indian J. Hort.*, **50**: 364-369.

Singh, S.P. and Singh, H.N. (1981) *Z. Pfianzenziichtg*, **87**: 240-247.

Singh, S.P., Singh, H.N. and Srivastava, J.P. (1986) *Indian Agric.*, **30**: 147-152.

Singh, S.P., Singh, H.N., Singh, N.P. and Srivastava, J.P. (1979) *Indian J. Agric. Sci.*, **49**: 579-582.

Sitienei, R.C., Onwonga, R.N., Lelei, J.J. and Kamoni, P. (2017) *Agric. Sci. Res. J.*, **7**: 47-61.

Sivasubramanium, P., annappan, R.S., Iyemperumal, S., Rangaswamy, S.R.S. and Pandiyan, M. (1989) *Madras Agric. J.*, **76**: 411-412.

Smith, P.M. (1976) In: *Simmonds, Evolution of Crop Plant*, New York, p. 312.

Sounder Raj, V., Tejavathi, D.H. and Nijalingappa, B.H.M. (1989) *Curr. Sci.*, **58**: 1385-1388.

Tidke, S.D., Manore, S.G. and Tagad, S.N. (2019) *Trends Biosci.*, **12**: 919-924.

Turuko, M. and Mohammed, A. (2014) *World J. Agric. Res.*, **2**: 88-92.

Uday Kumar, H.R., Byre Gowda, M. and Ramesh S. (2016) *Mysore J. Agric. Sci.*, **50**: 372-375.

Vaijayanthi, P.V., Ramesh, S., Byregowda, M., Mohan Rao, A. and Keerthi, C.M. (2015C) *J. Crop Improv.*, **29**: 405-419.

Vaijayanthi, P.V., Ramesh, S., Byregowda, M., Mohan Rao, A. and Keerthi C. M. (2016a) *J. Crop Improv.*, **30**: 244-257.

Vaijayanthi, P.V., Ramesh, S., Byregowda, M., Mohan Rao, A., Keerthi, C.M. and Marry Reena, G.A. (2015a) *J. Food Legumes*, **28**: 5-10.

Vaijayanthi, P.V., Ramesh, S., Byregowda, M., Mohan Rao, A., Ramppa, H.K. and Chinnamade Gowda. (2016b) *Mysore J. Agric. Sci.*, **50**: 376-380.

Venkatachalam, M., Kshirsagar, H.H., Tiwari, R. and Sathe, S.K. (2002) *Annual Meeting and Food Expo* - Anaheim, California, Session 3ºC, Food Chemistry: Proteins.

Verdcourt, B. (1970) *Kew Bull.*, **24**: 379-448.

Vijay, O.P. and Vani, A. (1990) *Indian Hort.*, **35**: 14-17.

Viswanath, S.R., Shivashankar, G. and Manjunath, A. (1971) *Curr. Sci.*, **40**: 667-688.

Viswanath, S.R., Siddarammappa, R., Sivashankar, G. and Suresh, P. (1972) *Mysore J. Agric. Sci.*, **6**: 56-58.

Wang, F.J., Wang, T.Y. and Wang, F.N. (1991) *Crop Genet. Resour.*, **1**: 7-10.

Wei, Z.M. and Xu, Z.H. (1993) In: *Current Plant Science and Biotechnology in Agriculture*, **15**: 387-390.

Xu-Xiangshang, Tehg You De, Chen Xue Qun, Teng, Y.D. and Chen, X.Q. (1996) *Crop Genet. Resour.*, **3**: 20-21.

Yao, L.M., Zhang, L.D., Hu, Y.L., Wang, B. and Wu, T.L. (2012) *Indian J. Genet.*, **72**: 46-53.

⑥

WINGED BEAN

A. Chattopadhyay, V A. Parthasarathy
and P.K. Maurya

1.0 Introduction

Winged bean (Goa bean, Manila bean, princess pea, asparagus pea and four-angled bean) is gaining popularity as a tropical vegetable legume because of protein-rich pods, seeds, tender leaves, inflorescence and tubers, all of which are edible (Claydon, 1975; Ekpenyong and Borchers, 1978). The synonyms 'soya's rival' 'God-sent vegetable' have been attributed to this wonder vegetable. Its use has been more directed towards tender pods as vegetable and seeds as grain legume. But Stephenson *et al.* (1982) opined that the use of tuber should be more exploited since it is a high protein supplement for the predominantly carbohydrate diet. Studies carried out at Vietnam indicated that (1) the winged bean can be grown in all parts of Vietnam, (2) it gave higher yields than cowpea, (3) it could be harvested at times of the year when other vegetables were scarce, (4) it could be harvested continuously for 12 weeks compared with only 6 weeks for cowpea, and (5) pods, shoots, flower buds, flowers and tubers could be consumed (Hlava and Michl, 1990).

2.0 Composition and Uses

2.1.0 Composition

The nutritional composition of the edible parts of this fascinating crop is presented in Table 1.

Table 1: Proximate Composition of different Edible Parts of Winged Bean (g/100 g fresh weight)

Part	Water	Protein	Oil	Carbo-hydrate	Fibre	Ash	Author
Immature							
Pod	71.2	2.4	0.2-0.3*	3.1-3.8*	0.8-2.6*	0.4-1.9	Martin and Delpin (1978); *Newell and Hymowitz (1979)
Seeds	-	31.8-34.7	14.38-15.09		8.98-11.17		Rosario *et al.* (1981)
Tubers	-	8.2- 31.1 per cent	-	-	-	-	Hildebrand *et al.* (1982)
Leaves	77.7	5.7	1.1	3.0	-	-	Newell and
Flowers	84.2	5.6	0.9	3.0			Hymowitz (1979)

The globulin protein fraction of seeds is high in lysine, arginine and leucine, while the albumin fraction is high in cystine and methionine (Rosario *et al.*, 1981). Hildebrand *et al.* (1981) reported that seeds contain 2.8 to 4.8 per cent starch (mean 3.4 per cent) and 6.4 to 8.9 per cent soluble sugars. The winged bean oil contains fatty acids in the range of 53.8 to 68.5 per cent (Garcia *et al.*, 1979). Pospisil *et al.*

(1971) stated that winged bean oil is rich in unsaturated fatty acids. Eleven flavonoid glycosides were isolated and identified in the seed coats in winged bean (Hung *et al.*, 1986). The oil content and fatty acid composition of 19 genetically improved cultivars of winged bean seeds were estimated by Singh *et al.* (1995) as follows:

- ☆ Oil content - 14.5 to 22.7 per cent
- ☆ Oleic acid - 41.4±1.5 per cent
- ☆ Linoleic acid - 29.7±1.1 per cent
- ☆ Total unsaturated fatty acid - 66.8 to 78.9 per cent (73.9±0.9 per cent)
- ☆ Saturated fatty acids:
- ☆ Palmitic - 10.9±0.6 per cent
- ☆ Behenic - 6.7±0.4 per cent
- ☆ Stearic - 3.5±0.3 per cent

Parinaric acid, a toxic fatty acid was absent in seed oil of all varieties. The oil can be stored for longer periods, being low in linoleic acid (0.9±0.1 per cent). Ravindran *et al.* (1989) studied the seed polysaccharide composition of winged bean varieties. The non-starch polysaccharide fractions contained 63.2 to 67.5 per cent carbohydrates, 12.8 to 15.2 per cent protein, 10.4 to 12.1 per cent lignin and 3.2 to 4.8 per cent ash. Galactose (44.0 to 54.9 per cent) was the main sugar in all 5 varieties. Non-starch polysaccharides also contained 5.1 to 6.5 per cent arabinose, 4.3 to 6.5 per cent xylose, small amounts of glucose, rhamnose and fucose, 12.8 to 14.3 per cent uronic acid and 17.7 to 23.8 per cent cellulose. The results suggest that winged bean polysaccharides are essentially non-starch polysaccharide and are heterogeneous and insoluble.

With all the positive nutrition offered by winged bean, anti-nutritive factors (ANFs) also exist and have been extensively studied, including trypsin inhibitors, chymotrypsin inhibitors and hemagglutinins (Kortt, 1979,1984; Kortt *et al.*, 1989).

2.2.0 Uses

Winged bean has been a minor food for diverse ethnic groups in Asia and Africa for centuries (Sri Kantha and Erdman, 1984). In India, immature pods are eaten as a raw vegetable (The Wealth of India, 1969) or pickled (Verdcourt and Halliday, 1978). Elsewhere, the unripe seeds are used in soups and curries (Sri Kantha and Erdman, 1984, McKee, 1928), mature seeds are roasted and eaten like a peanut (McKee, 1928), and flowers are eaten and used to color dishes (National Academy of Sciences, 1975). In Ghana, Burma and Papua New Guinea tubers are used in a range of culinary preparations (National Academy of Sciences, 1975). It is also an ingredient for traditional Indonesian delicacies like tempeh kecipir, as well as in the preparation of snacks in Thailand (Erskine, 1979). Leaves and flowers are used as fried or boiled vegetable in Papua New Guinea (Khan *et al.*, 1977), eaten raw or cooked as stews in Indonesia (Sastrapradza and Lubis, 1975), eaten raw or cooked in Malaysia and as livestock feed in Bangladesh (Haq and Smartt, 1973). In high lands of Papua New Guinea (Khan *et al.*, 1977) and Myanmar (Purseglove, 1968) the tubers are delicious and prized. Ripe seeds are used after roasting in many countries. It is

Uses of Winged Bean

Curry of winged bean

Winged bean pickle

an excellent substitute for soybean and better than groundnuts (Pospisil *et al.*, 1971). Newell and Hymowitz (1979) presented a summarized review of the medicinal uses of winged bean. According to their review, the leaf infusion is used for treating eye and ear infections in Sumatra and Java, dyspepsia in East Java, leaf paste for boils, while seeds have a use in treatment of veneral diseases. One of the good uses of pods is as a slimming diet and in Sri Lanka it is used as a diabetic diet. In Malaccas and Sulawesi, the pods are considered to be good for blood. In Malaysia, the leaves form a component of an herbal preparation to treat smallpox (McKee, 1928). In Myanmar, the tubers are valued as poultice to treat vertigo (Burkill, 1935). Winged bean extracts show radical scavenging, antimicrobial and antioxidant activities (Nazri *et al.*, 2011; Yoga Latha *et al.*, 2007; Khalil *et al.*, 2013).

3.0 Origin, Distribution and Taxonomy

Diverse opinions exist regarding the origin of winged bean. A review by Zeven and De Wet (1982) suggests four areas as probable centers of origin: (a) Indo-Malayan - due to a long cultivation history in eastern Assam, India (Anon., 1969); (b) Asiatic origin - which suggests winged bean was domesticated there from an unknown and now extinct endemic Asian progenitor; (c) Papua New Guinea - based on the large genetic variation found there (Erskine and Khan, 1981; Matejka, 1987); and (d) African center - due to similarities with African species (Burkill, 1935). The African origin hypothesis has received much support due to observations that the morphology of *P. grandiflorus*, an African species, closely resembles winged bean (Smartt, 1980). Cytological and plant pathological evidence supports this (Harder and Smartt, 1992). It is possible that the progenitor of winged bean arose on the African side of the Indian Ocean and was carried east as a wild plant and then modified by human cultivation. Alternatively, it is possible that winged bean had a wider geographical distribution and was first domesticated in the Indian center or islands of Southeast Asia or Melanesia, as it is essentially a crop of Asia and the Western Pacific with considerable antiquity in these regions.

Masefield (1973), quoting Burkill, stated that Myanmar was the only country where winged bean was a field crop, while in South India also it was grown. The crop is grown in northeast Indian states of Manipur, Tripura and Mizoram where it could have been introduced from Myanmar by virtue of its nearness. Haq and Smartt (1973) stated that the winged bean had been growing in Bangladesh for the past two centuries. Presently, it is grown in India, Myanmar, Thailand, Vietnam, Malaysia, Indonesia (Lubis, 1978), Ghana (Karikari, 1978), Nigeria, Sri Lanka (Herath and Fernandez, 1978) and in many countries of the New World. In India, it is grown in Tamil Nadu, Kerala, Karnataka, Goa, Orissa, Maharashtra, West Bengal and entire N. E. states of India.

Psophocarpus tetragonolobus (L.) DC. belongs to family Leguminosae. Another species, *P. palmettorum* (2n=20) which is sometimes grown for its edible pods, is wild in tropical Africa.

Reports of many chromosomal counts resulted in confusion. Khan (1976) and Tixier (1965), Newell and Hymowitz (1979) and Harder (1992) reported a basic chromosome number of $\times = 9$, while Thuan (1975) reported 11. She *et al.* (2004)

examined finer karyotypic details of winged bean chromosomes using banding patterns and fluorescent *in situ* hybridization with the aid of 45s and 5s ribosomal RNA gene probes, illustrating the utility of these methods for species identification and differentiation.

4.0 Morphology and Nodulation

4.1.0 Morphology

Winged bean is predominantly a self-pollinated crop. However, cross-pollination to the extent of 7.6 per cent has been reported (Erskine, 1979). Even though winged bean is a perennial twining herb attains a height of 5 m, it is mostly grown as an annual with a wiry stem having a tendency for twining. Depending upon the cultivar, stems are green to purple in colour. Leaves are trifoliate and they range in shape from oval to ovate-lanceolate (Martin and Delpin, 1978) with an entire margin. The leaves are alternately borne subtended by a stipule. The inflorescence occurs in axils. The plant requires a day length shorter than 12 hours with a temperature of 27°/22°C for reproduction (Eagleton *et al.*, 1978). The inflorescence is a raceme bearing many flowers. The plant is 3–4 m tall and bears blue, bluish-white or purple flowers (National Academy of Sciences, 1975; Newell and Hymowitz, 1979), borne singly or in threes on pseudoracemes. The pods show four sides with wings protruding from angles. Pods vary in length (30–40 cm) and shape, containing 5–21 seeds. The seeds, varying from 5 to 40/pod, are spherical and vary in colour from white to tan to dark brown to almost black, the colour being influenced by anthocyanin and tannin contents, besides environment (Martin and Delpin, 1978). The pollen of winged bean is highly distinctive within the genus (Poole, 1979). Pollen grains are spheroidal, with sizes ranging from 42.3–51.6 μm (polar axis) and 43.4–49.9 μm (equatorial axis). Some varieties produce starchy underground irregular, spindle shaped tubers that are 2–4 min diameter, 8–12 cm long and around 50 g weight (National Academy of Sciences, 1975; Martin and Delpin, 1978).

4.2.0 Nodulation

The use of winged bean as a soil-improving crop is due to its ability to fix atmospheric nitrogen by means of its exceptionally large and numerous nodules, which were first quantified by Masefield (1952, 1957 and 1961; Motier *et al.*, 1998). Individual plant produced up to 440 nodules with their fresh weight reaching up to 700 lb/acre in Malaysia (Masefield, 1957), while in Myanmar a nodule was found to weigh up to 600 mg (Claydon, 1978). The nodules are semiglobose, oblate, pale brown, and smooth surfaced (Lim and Ng, 1977) and are capable of fixing 120-247 kg/ha of nitrogen (Drilion and Obordo, 1978). The nodulation commences two weeks after the emergence of seedlings, bacteroid stage appears after 3 weeks, while considerable number of bacteroid stage nodules appear after 4 weeks with the nodule distribution following a uniform pattern (Iruthayathas and Herath, 1981). It belongs to the cowpea cross inoculation group, nodulating with cowpea *Rhizobium* strains (EL, 32HI, LB756, *etc.*) in tropical and temperate environments (Elmes, 1976), even in soils with no previous legume cropping (Harding *et al.*, 1978). Studies carried out at different places have led to identification of high nodulating strains of *Rhizhobia*, *viz.*, JCv44 in West Bengal (Poi and Kabi, 1982), TAL 223 at Hawaii (Woomer *et al.*, 1978), RRIM 56 in Malaysia (Ikram and Broughton, 1978) and LBNC 3 in Sri Lanka (Iruthayathas and Herath, 1981). Deokar and Ruikar (1987) observed that seed inoculation with 3 *Rhizobium* strains increased seed yields of Nigerian winged bean by 6.2-23.6 per cent. The Nigerian strain was more effective than either Papua New Guinea or Sri Lankan strain. The difference observed in nodule producing capacity among winged bean cultivars is due to their genetic constitution (Iruthayathas and Herath, 1981). Besides, the nodule formation is affected by environmental factors. Water stress delays the nodule formation and nodule from stress plants would show more bacteroid containing cells per unit area up to 4 weeks (Rathfelder, 1981). Nodulation varies between soil types. On lateritic soil, nodule number was 20-54/plant (Bagchi *et al.*, 1991). As with other legumes, analysis of nitrogen and ä15N values in nodules, roots and shoots showed that N accretion in winged bean critically depends on the *Rhizobium* strains colonized in root nodules (Yoneyama *et al.*, 1986). Like other phaseoloid species, it is reported to be an ureide transporting plant, transporting fixed nitrogen from nodules as allantoin and allantoic acid (Yoneyama *et al.*, 1986). In an assessment of ã-radiation M4 mutants from UPS122 and cv. Kade 6/16, it was reported that lighter seed coat colour was associated with increased nodules per plant, likely due to the alterations in falvonoid biosynthetic pathways that could influence nod genes expression during symbiotic initiation between plant and rhizobia bacteria (Klu and Kumaga, 1999).

5.0 Cultivars

Because of occurrence of different ecotypes and land races, numerous cultivars are available. Martin and Delpin (1978) listed the following characters as the principal variations: (i) seed colour, (ii) seed size, (iii) ring around hilum, (iv) hard seed (germination with/without scarification), (v) vigour of vegetative growth, (vi) dwarf or normal stature, (vii) response to photoperiod, (viii) tuberous root production, (ix) degree of nodulation, (x) anthocyanin of stem, (xi) size of leaves and leaflets, (xii) flower colour, (xiii) flower colour distribution, (xiv) pod colour, (xv) pod colour distribution, (xvi) pod length, (xvii) pod width, (xviii) smoothness of pod surface, (xix) number of seeds/pod, (xx) cross-section of pod, (xxi) development of wings, (xxii) undulation of wings, (xxiii) total yield, (xxiv) perennial versus facultative annual growth habit, (xxv) drought resistance, (xxvi) hardness of seeds under cooking, (xxvii) bitterness of cooked seeds, (xxviii) protein content, and (xxix) oil content.

Martin and Delpin (1978) have compiled lists of cultivars with their characters. Besides, many cultivars have been identified in different places by various workers as given in Table 2.

Solanki and Saxena (1989) reported that cultivars UPS122 and LBNC3 produced the highest seed yield/plant (0.997 and 0.930 kg, respectively). The best varieties are EC38224, UPS62, UPS122 and LBNC3. The cultivar GRWB11 from USA is reported to be well adapted for both young pod and seed production in Japan (Nakanishi *et al.*, 1987). At Hisar, India winged bean cultivars gave pod yields of 0.41-10.20 t/

ha and root yields of 0.32-1.84 t/ha. Cultivar GRWB-24 gave the highest pod yield followed by cv. GRWB-10B with 10.11 t/ha while GRWB-25 gave the highest root yield followed by IC14981 with 1.74 t/ha (Khurana *et al.*, 1990). The most suitable cultivar for local conditions of Vietnam was Binh Mih (Hlava and Michl, 1990). Valicek *et al.* (1988) evaluated cultivars from South and S.E. Asia in Vietnam. Cultivars SLS 47 (Sri Lanka), 28-01 (Thailand) and UPM 207 (Malaysia) and to a lesser extent other Thai and Malaysian cultivars were recommended on account of their long growing season and high yields (7.1, 6.0 and 5.5 t seeds/ha (theoretical yields) for the cultivars as named, respectively). The local cultivar Binh Minh gave 4.3 t/ha.

Table 2: Summary of Results of Cultivar Evaluation in Winged Bean

Name of Place	Cultivars	Author
Venezuela	Tpt 3 (home gardens), Tpt 1, Tpt 14	Boscan Odor (1978)
Ghana	1/26, 4/24, 5/12, 6/16 (green pods); 24/3, 2/10, 15/16 and 11/6 (seed); 27/4, 26/6 and 10/6 (cover crop); 10/6, 27/4, 6/16 and 1/2 (tuber production)	Karikari (1978)
Sri Lanka	Puttalam (tubers); Colombo (tubers and seeds); Karaliyadde (seeds)	Herath and Fernandez (1978)
Hawaii	TAL 223	Woomer *et al.* (1978)
Papua New Guinea	UPS 122 (seed and tubers); PS 121, UPS 62 and UPS 47 (tubers)	Stephenson *et al.* (1982)
India	IIHR selections; 21, 60, 71 (Bangalore); WBC-2 (Meghalaya)	
Japan	UPS-31, Tpt 2, Urizun	Nakamura and Abe (1989)
India	EC38821B, EC27886, and IC95227	Ray *et al.* (2017)
USA	Local Green, Iriarte and Shikaku Mame	Tuquero and Takai (2018)
Malaysia	S319	Raai *et al.* (2020)

Long *et al.* (1993) identified a natural dwarf mutant, Gui Ai, of the line KUS101, which has climbing habit. Gui Ai forms many root nodules and combines vigorous branching with determinate habit (the apical buds differentiate into flower buds after the main stem ceases growth at the stage of 11-13 true leaves). The variety gives a pod yield of 22.95 t/ha and a seed yield of 3.7 t/ha. It can be sown between June and August in the Guangxi autonomous region of China. The sowing to flowering period is 63 days. The green pods contain 21.5 per cent protein and 5.1 per cent carbohydrates. The mature seeds contain 35.4 per cent protein. At Zamibia, seed yields/plant of winged bean cvs. Bogor, TPT-2, UPS-53 and GRWB-26 (day-neutral) were 263, 315, 65 and 7.4 g, respectively. 100-seed wt were 35.7, 29.1, 27.6 and 23.1 g. The period from sowing to 50 per cent flowering ranged from 14 weeks for GRWB-26 to 17 weeks for TPT-2 (Lesseps, 1988). The highest seed and root yields of 120 and 60 g/plant, respectively were given by cultivars ICA-66 while cultivar UPS-140 had the highest crude protein content in leaves (25.4 per cent), pods (17.9 per cent) and seeds (35.0 per cent) in Cauca valley of Columbia (Hernandez 'O *et al.*, 1987).

6.0 Soil and Climate

6.1.0 Soil

The crop does not appear to be very demanding in soil requirement (Masefield, 1973) but good soil texture favours germination of seeds as well as plant growth. Because of the ability of the crop to withstand moisture to a reasonable extent, it can also grow in heavy soils with poor drainage. Sandy and well-drained soils are good for tuber production.

6.2.0 Climate

The crop is typically tropical in nature and grows well in subtropics also. The crop is sensitive to photoperiod, humidity, rainfall, *etc.* The information on influence of various weather parameters except that of light is scarce. The plants require short-day conditions for flowering. In Puerto Rico (18° lat.), Martin and Delpin (1978) found that winged bean could be planted anytime when the day length varied between 11 and 13 hours, as they would develop enough vegetative phase. Planting during short days would result in flowering within 8 weeks. Uemoto *et al.* (1982) found the critical photoperiod for flower induction to be 12 hours with flowers being produced commonly in 8-11 hour photoperiod. Eagleton *et al.* (1980) were also of the same opinion for subtropical and Mediterranean summers. The optimum temperature suggested by them for flowering was 27°/22°C, while Uemoto *et al.* (1982) found 20°C to be optimum under Japanese conditions. Wong and Schwabe (1979) and Wong (1983) inferred from the Malaysian accession (M14/4), under a thorough controlled environment examination, that the 'critical' day length for flower induction was between 11 h 15 min and 12 h 15 min. A reduction in light intensity, during part of the photoperiod, lowered this critical day length. At short, potentially inductive day lengths, the optimal day temperature was found to be 26°C, whereas 32°C or 18 °C inhibited flowering.

For tuber production optimum thermo period was 24°/13°C while higher temperature (30°/22°C) promoted vegetative growth affecting tuber development. Tuberous root formation was induced by short days, and at 20°C under long days, but not at higher or lower temperatures while photo-insensitive cultivars formed tuberous roots under long and short days at 20 and 25°, and at 30° under long days (Okubo *et al.*, 1992). Day length has a significant and overriding influence on the pattern of dry matter partitioning and, therefore, yields (Harder, 1991). The composition and nutritional quality of the edible parts changed during the ontogeny of the plants and were modified by day length and flower removal (Harder, 1991).

Rainfall, detrimental to many legumes, was good for winged bean even though a short period of drought normally did not affect the crop (Martin and Delpin, 1978). Rathfelder (1981) also reported the positive influence of water in winged bean. Nangju and Baudoin (1979) found that winged bean was more suitable to a humid than sub-humid environment. However, water-logging decreased yield. In Japan, yields of all cultivars tended to decrease if sown after May (Nakamura and Abe, 1989) and certain cultivars like Urizun and Ishigaki flowered whenever air temperature was >20°C, while cultivars UPS-31 and Tpt-2 did not flower until October (Nakamura and Abe, 1989).

In the tropics, the crop may be grown at high altitude, corresponding to the conditions found in subtropical humid or subtropical monsoon climates which suggested that the crop may be introduced to the subtropics (Matejka, 1987).

7.0 Cultivation

7.1.0 Soil Preparation and Seed Sowing

The soil is ploughed and pulverized for sowing. Soil should be of good tilth so that tubers develop normally. The seeds pre-soaked for 2 days before planting would give an emergence of 83.4 per cent. They normally take about 10 days to germinate. The pre-soaked seeds germinate at 25°C in 5-6 days (Nangju and Baudoin, 1979). Low doses of gamma rays (–5 kR) stimulated germination and high doses (>10 kR) inhi-bited germination percentage, lengthened the period taken for germination and reduced survival percentage compared with untreated controls (Suma Bai and Sunil, 1993). The seeds are sown in flat beds. Ridges are recommended in places where flooding is a problem (Martin and Delpin, 1978). The spacing varies depending on the cultivar and location. The recommended spacings between rows and plants for winged bean are given in Table 3.

Table 3: Spacing Recommendations for Winged Bean

Between Rows	Between Plants	Author
4′ (121 cm)	2′ (60 cm)	MacMillan (1949)
4′ (121 cm)	2′ (60 cm)	Purseglove (1968)
24″-30″ (60-76 cm)	18″-24″ (45-60 cm)	Tindall (1968)
2′ (60 cm)	2′ (60 cm)	Pospisil *et al.* (1971)
150 cm	25 cm	Nangju and Baudoin (1979)
1.5 m	1 m	Martin and Delpin (1978)
1.5 m	60 cm	Sanchez *et al.* (1983)
75 cm	50 cm	Lee (1988)

Lee (1988a) recommended optimum spacing of 50 or 75 cm within the row for high yields of both green pods and dry seed of cv. Chimbu. Veeraragavathatham *et al.* (1988) reported that optimum plant density for high yields of good quality green pods (7.3-8.1 tonnes/ha) and a dry seed (1.9-2.0 tonnes/ha) was 20,000 plants/ha.

As drought would destroy the seedlings, it is important to irrigate, if sufficient soil moisture is not available. The plots are to be kept free from weeds.

7.2.0 Support

The seedlings show signs of twining a few weeks after germination. Support to the vines is essential. Normally, individual supports are given with local wood materials. In N.E. India where bamboo is available in plenty, bamboo supports are given. The use of 'subabool' (*Leucaena leucocephala*) has been found to improve plant growth and soil fertility in Nigeria (Nangju and Baudoin, 1979). Besides, 'subabool' reduced soil temperature. They found that support increased yield twice than that of control and suggested a height of 1 m for the supports. Wire supports are also used. Martin and Delpin (1978) suggested a Y-shaped trellis, connected by wires between the extremes and with strings between wires, which permitted a row of plants to grow at an angle following the strings and the pods which hang could easily be harvested. Valle *et al.* (1980) found that supports significantly enhanced yield in one cultivar (WB 21-8) but did not show any effect on yield of another cultivar (WB 10-3). They felt that supports, even though, had no effect on protein content, prevented pod rot by holding the pods from being affected by soil moisture.

Trials in Vietnam showed that plants grown with 'T' or 'A' shaped supports flowered 10 days earlier and produced higher pod number, leaves, branches and seed quantity. Highest seed yield was obtained with the 2 m high A-shaped frames. However, tuber yield was more in unsupported plants (Michl and Van Phu, 1985).

Motior *et al.* (1997) found that of the three support systems (wire trellis) at 0, 1 or 2 m height tried in Malaysia, seed yields from these 3 treatments were 0.31, 1.39 and 2.13 t/ha, respectively. They further studied the nitrogen accumulation and partitioning by winged bean in response to support systems (Motior *et al.*, 1998). Total nitrogen accumulation and nitrogen partitioning were determined throughout the growing season by measuring nitrogen content and concentration in above-ground plant tissues (leaves, stems, petioles and pods). Support heights of 1 and 2 m significantly increased total nitrogen accumulation in component parts of the plant, nitrogenase activity, nodulation, total dry matter accumulation and seed yield compared with the control. Plants grown on supports accumulated significantly higher leaf nitrogen at the vegetative stages and the contribution of seed nitrogen was also significantly higher compared with unsupported plants. Nitrogenase activity increased with onset of flowering but declined during the pod formation stage in plants grown on a support system. Unsupported plants showed higher nitrogenase activity prior to flowering possibly due to lack of photosynthate and, consequently, early senescence of leaves.

In the high lands of Papua New Guinea, crops of winged beans, grown for tubers, are traditionally pruned (Stephenson *et al.*, 1982). Herath and Fernandez (1978) reported that flower removal and pruning increased tuber yield. Vines for green pod yield are not pruned. Stephenson *et al.* (1982) found that tubers from pruned and unpruned plants bulked slowly up to 16 and 17 weeks, respectively, but tubers from pruned plants bulked over the following five weeks, whereas those from unpruned plants did not change significantly, the reason being higher mean leaf area of pruned plants than unpruned one at bulking time. Mulching with leaves and plant matter controls weeds and reduces moisture loss. In case of crop meant for tuber production mulching gave significantly higher yield.

7.3.0 Manuring and Fertilization

Information on nutritional requirement for winged bean is meagre. Islam *et al.* (2016) found the highest yield in vermicompost treatment compared to other treatments. Zusevics (1981) found that dry matter yield increased with phosphatic

Cultivation Techniques of Winged Bean

Seeds, Seed germination, flowering and fruiting of winged bean

Use of plastic mulch in winged bean cultivation

Use of plastic mulch in winged bean cultivation

Pod Bearing, Harvesting and Selling of Winged Bean

Pod bearing stage of winged bean

Harvesting of winged bean pods and tubers

Selling of winged bean tuber in the market

fertilizers but nitrogen and potassium had a negative influence. Besides, nitrogen also affected nodulation. Inoculation with strain Suchdol-3 increased nodule numbers 2.5 fold. Nitrogen application to the inoculated plants however, reduced nodulation (Gregr and Singh, 1978). Hikam *et al.* (1991) found that winged beans had a lower N concentration in mixed cropping system than in monoculture. The relative contribution of N fixation (determined by 15N isotope dilution) to total N accumulation in monocultures of winged beans was significantly greater at low N. As N availability decreased due to N level and intercropping, increases were observed in nodule DW and N fixation.

Parthipan and Kulasooriya (1989), on the other hand, observed that nodule number and dry weight were decreased by N application from 0-40 kg/ha in cultivars SLS-40 and SLS-44. Nodule number was highest in 45-day-old plants and nodule dry weight in 75-day-old plants. K application stimulated early root growth but had no effect on final root biomass. Under field condition winged bean seed inoculated with *Rhizobium* strain JCV-44 increased nodule number, plant dry weight, plant nitrogen content as P rate increased from 0 to 125 kg P_2O_5/ha. Seed yield also increased from 1.55 t without P to 2.75 t/ha at 125 kg P_2O_5/ha (Poi and Ghosh, 1986). Martin and Delpin (1978) recommended a mixed fertilizer application of 250 kg/ha.

Anugroho *et al.* (2010) examined biomass dry matter and nutrient uptake of live plant parts, leaf area index, and litter of winged bean (*Psophocarpus tetragonolobus*) at 12, 18, 24 and 30 weeks after sowing (WAS). P uptake was significantly higher in the leaf at 30 WAS and in the stem + petiole at all harvesting times. The total biomass of the leaf, stem + petiole and litter of winged bean was 317–561 g DM/m^2, and their N content was 12.3–17.7 g/m^2. The total biomass of live parts and litter of winged bean might be sufficient to suppress weeds and increase soil N.

At Nigeria, Ndegwe and Ikpe (2000) observed that potassium application did not increase seed and tuber yields in unstaked beans without flowers. In potassium-fertilized beans with flowers, staking markedly increased seed and tuber yields by 162.3 and 412.1 per cent, respectively, compared with unstaked, potassium-fertilized beans with flowers. Staked, potassium-fertilized and deflowered beans gave the highest tuber yield of 124.31 g/plant.

The highest amount of oleic (monounsaturated fatty acid) and linolenic acid were recorded with application of 40 kg N/ha + no inoculation (38.50 per cent) and no nitrogen + Rhizobium + PGPR inoculation (38.50 per cent), respectively (Kumari *et al.*, 2018). Maximum saturated fatty acid content (29.51 per cent) was recorded in treatment of 40 kg N/ha + PGPR, however minimum value (22.61 per cent) was obtained with application of 40 kg Nha[-1] and without inoculation of either PGPR or Rhizobium.

7.4.0 Effect of Growth Substances

Lee (1988b) reported that application of 2, 3, 5-triiodobenzoic acid (TIBA) at 30 or 45 ppm; b-naphthoxyacetic acid (NOA) at 50 or 75 ppm or (2-chloroethyl) trimethyl ammonium chloride (CCC) at 50 or 75 ppm to the foliage at 2- 4- and 5-leaf stage significantly increased number of fresh pods and yield of fresh pods.

8.0 Harvesting and Yield

The green pods to be used as vegetable are ready for harvesting 10 weeks after sowing (Tindall, 1968) and extends indefinitely (Masefield, 1973). Bagchi *et al.* (1989) suggested that for green bean the pods should be harvested 21-35 days after anthesis. Beyond that stage, winged bean should be utilized as a grain legume. The yield figures for leaves, flowers and green seeds are not available. Each vine yields about 25 pods every 5 or 6 days (Anon., 1969). The yield figures vary between countries and other cultivation conditions (Table 4).

In Choo, as quoted by Newell and Hymowitz (1979), reported green pod yield, seed yield and tuber yield up to 35526 kg, 4590 kg and 4980 kg/ha, respectively, in Malaysia. The yield varies with population densities. The optimum spacing for high yields (7.3-8.1 t/ha) of good quality green pods and for the highest yields of dry seeds (1.9-2.0 t/ha) was 20,000 plants/ha (Veeraragavathatham *et al.*, 1988). Winged bean plant grown at spacings of 30, 60, 90 or 120 cm in rows 1 m apart gave seed yields/plant of 140.7, 152.8, 105.9 and 123.3 g, respectively (Sarnaik and Baghel, 1988). Winged bean cv. Chimbu plants were spaced within the row at 12.5, 25, 50 and 75 cm with a between-row spacing of 1.22 m (equivalent to about 30 000, 40 000, 20 000 and 13 333 plants/ha, respectively) yielded 5.14, 5.29, 6.46 and 6.47 t/ha, respectively of green pods (Lee 1987). In West Bengal, India, pod yield was 2.00-3.01 t/ha in lateritic soil while on alluvial soil, pod yield averaged 2.55 t/ha with stakes and 1.38 t without stakes. Pod yield was positively correlated with pod number but not with nodule weight/plant (Bagchi *et al.*, 1991).

Table 4: Yield of Winged Bean as Reported in Literature

Place	Yield	Author
Puerto Rico	4 t/ha (green pods)	Martin and Delpin (1978)
	0.7-2.2 t/ha (seeds)	
Nigeria	12-26 t/ha (pods)	Nangju and Baudoin (1979)
	1-1.65 t/ha (seeds)	
Papua New Guinea	1.7-3.9 kg/1.5 sq.m. (green pod yield)	Kesavan and Erskine (1978)
	11 t/ha (tubers)	Stephenson *et al.* (1982)
Myanmar	2.5-6 t/ha (tubers)	Anon. (1969)
India	7.3-8.1 t/ha (green pods)	Veeraragavathatham *et al.* (1988)
India	1.9-2.0t/ha (seeds)	Veeraragavathatham *et al.* (1988)

9.0 Diseases and Pests

Winged bean is comparatively free from major diseases and pests (Masefield, 1973; Martin and Delpin, 1978). Nangju and Baudoin (1979) stated that winged bean is highly resistant to field weathering due to hard pod wall which is valuable in high rainfall areas and also protects against pod borers. They also recorded, for the first time, a minor disease characterized by necrosis and mosaic on young leaves transmitted by aphids (*Aphis craccivora*). The disease does not cause any yield

Disease of Winged Bean

Rust pustules on leaf and stem of winged bean plant

Pests of Winged Bean

Larva and moth of winged bean pod borer, *Maruca testulalis*

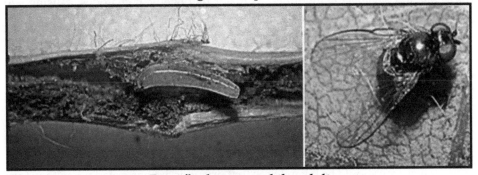

Bean fly damage and the adult

reduction. Martin and Delpin (1978) summarized Khan's report on a few diseases and pests as follows:

9.1.0 Diseases

Five major fungal diseases have been reported from winged bean: false rust (*Synchytrium psophocarpi*), dark leaf spot (*Pseudocercosa psophocarpi*), powdery mildew (*Oidium sp.; Erysiphe cichoracearum*), collar rot (*Macrophomina phaseolina, Fusarium semitectum, F. equiseti, F. moniliforme, Rhizoctonia solani*) and *Choanephora* blight (*Choanephora cucurbitarum*) (Erskine,1979; Fortuner *et al.*, 1979; Price, 1980). Detailed symptoms and management of some of these pests and diseases are well described (Reddy, 2015) but in the absence of comprehensive solutions, it is vital to screen germplasm through a uniform evaluation process to identify insect and disease resistant lines and enable marker assisted breeding programs to develop resistant varieties and cultivars (Erskine, 1979). Major viral diseases identified include necrotic mosaic virus which affects 9 per cent of the total young plants in the field, ring spot mosaic virus which causes approximately 10–20 per cent yield loss and leaf curl disease which has appeared only at Kpouebo, Ivory Coast (Erskine, 1979; Khan, 1982, Reddy, 2015; Fortuner, 1979). A new endornavirus, provisionally named Winged bean endornavirus 1, was recently discovered (Okada *et al.*, 2017).

9.2.0 Pests

Like most plants, winged bean is affected by insect species of the Orders *Hemiptera, Thysanoptera, Diptera, Coleoptera, Lepidoptera and Orthoptera* are reported as pests on winged bean. *Maruca testulalis* (bean pod borer) and *Helicoverpa armigera* (cotton bollworm) target pods and flowers. *Leucoptera psophocarpella* (winged-bean blotch miner) extensively damages leaves, whereas *Mylabris afzelii* and *M. pustulata* damage flowers (Khan, 1982). Caterpillars of *Aphis craccivora* (black bean aphid), *Henosepilachna signatipennis* (ladybird), *Ophiomyia phaseoli* (bean fly), *Lampides boeticus* (pea blue butterfly), *Nezara viridula* (southern green stink bug), *Podalia spp., Polyphagotarsonemus latus* and *Tetranychus urticae* are other pests that reportedly damage shoots, leaves and flowers (Reddy, 2015). Infestations of root-knot nematodes on winged bean have been shown, with *Meloidogyne javanica, M. incognita* and *M. arenaria* identified as major pathogenic nematodes, causing ~70 per cent tuberous root loss (Reddy, 2015).

Routine control measures for such diseases and pests in other legumes may be followed.

10.0 Seed Production

The seed crop of winged bean is cultivated in the same way as done for green pods. In many countries the crop is also grown for its seeds. Mature seeds production with support systems could also be increased by ratooning, in order to maximise cumulative yield from a single planting (Rahman, 1998). The pods on drying shatter.

Hence, care should be taken to see that dry pods are harvested when they turn brown. Unless the area under the crop increases sizeably and demand for seed would be less. As it is still grown mostly as a backyard crop, the home gardeners can leave a few pods from their crop for seed. About 1 to 1.5 tonnes of seeds can be produced in one hectare. Haq (1982) suggested that seed yield improvement could be achieved by pureline selection or mass selection of indigenous and exotic genetic resources and by hybridizing with selection within hybrid lines and the success would depend on the selection of parent and their adaptation to any given environment. Even when grown for seed purposes commercially, there is very little effect of spacing on crossing and the effects of distance on out-crossing were not significant (Parthipan and Senanayake, 1989b). Nangju and Baudoin (1979) found that winged bean (TPt2) seed yield was almost double that of cowpea, but lower than elite cultivars of soybean, pigeon pea, and jack bean when grown at Nigeria with annual rainfall of 1200 mm on string vertical support.

11.0 Crop Improvement

Considering the potentiality of winged bean, as a backyard crop in home gardens and as a commercial crop, breeding programmes have been initiated in several countries. Various characters that need to be introduced to improve this crop are (i) photo-insensitivity, (ii) earliness, (iii) home garden lines with extended harvest, (iv) heterosis breeding, (v) ideal plants having single leader with 40-50 nodes, and (vi) suitable cultivars for tubers, seeds and green pods (Satyanarayana *et al.*, 1978). Martin and Delpin (1978) suggested that breeding for selection from existing germplasm for better seed lines that contain less tannins is essential for promoting winged bean as a seed legume as they are hard to cook.

Thus development of early maturing, high yielding, dwarf, erect, and non-shattering pod bearing plants with reduced anti-nutritional factors constitute major breeding objectives in winged bean. Unlocking the genetic diversity housed in global germplasm collections is therefore critically important. Morphometric evaluations showed good diversity for a number of traits in germplasm collections, including variation in leaf size and shape and pod color in Asiatic germplasm (Khan, 1976). Selection for pods per plant, pod weight and roots per plant was recommended for improvement of yield (Pandita *et al.*, 1989). Shelling percent was positively and significantly correlated with seed yield, an association thought potentially valuable for selection purposes (Silva and Omran, 1987). Path coefficient analysis revealed that days to fruiting and days to harvest had a significant direct effect on yield (Solanki *et al.*, 1989). Satyanarayana *et al.* (1978) found that yield in winged bean may be improved by (i) increasing pod number/plant since this trait had the highest direct effects on yield through pod number and (ii) increasing pod weight and length because of their respective direct and indirect effects on yield. Kesavan and Erskine (1978), from their experiments in Papua New Guinea, reported general combining ability of the genotypes for green pod yield, yield related characters and flowering, thus proving additive genetic control. They opined that selection for earliness and high yield could prove successful. They suggested the following two approaches to selection:

    (i) In market gardens, where there is a premium on early yield, genotypes could be selected for earliness.

(ii) Selection could be made for extended periods of harvest to suit subsistence agriculture and home garden.

They also suggested the use of heterosis for commercial exploitation and frequent harvests to avoid stringiness and to maintain cooking quality. Genetic studies revealed dominance of purple over green for stem color, calyx color and pod wing color and rectangular over flat pod shape [(Erskine and Khan, 1977)]. Eagleton *et al.* (1978) showed variation in flowering habit. Good variability for nutritional and biochemical traits such as protein and oil content have also been shown (Harding *et al.*, 1978). Stephenson *et al.* (1982) presented the results of the investigations on the genetics of tuber production. They found that the tuber yield is controlled by polygenes. The additives genetic variances were higher than either dominance or environmental variance. They concluded that correlated responses to selection for tuber yield were possible, as phenotypic correlations between tuber yield and number of tubers/plant and haulm yield were significant. An encouraging trend in the genetical investigations on this crop is the recent interest taken by some scientists to work on the somatic cell genetics, *i.e.*, isolating mesophyll protoplasts and plant regeneration (Zakri, 1983). The identification of promising somaclones would aid in production of better cell lines.

Considerable effort has gone into developing superior winged bean varieties via mutation-breeding. In the absence of any naturally occurring dwarf/bush type winged bean, mutation breeding may be a valuable tool. Kesavan and Khan (1978) found gamma rays (10 to 20 krad) or ethyl methane sulphonate (EMS) at lower concentrations (0.05 per cent to 0.2 per cent) with longer treatment (1-12 hour) combined with 24 hours washing to be effective. Successful experiments resulted in a single, erect stem mutant, a multiple branched bushy mutant, and a mutant with long pods (Klu *et al.*, 1989). By gamma irradiation, three bushy mutants (Jugran *et al.*, 1985), an early-flowering mutation (Jugran *et al.*, 2001) and chlorophyll variants (Nath *et al.*, 1986) were recovered, as well as seed coat color, low tannin-producing, and nodulation mutants (Klu, 1996). Hakande (1992) successfully induced a spectrum of mutations and identified about 22 desirable morphological and chlorophyll mutants. Unfortunately, the majority of mutant lines recovered so far have been reported to be sterile. Induced mutations in winged bean for low tannin content were attempted by Klu *et al.* (1997). They obtained four mutants of winged bean with altered tannin content among the M_3 seeds present on M_2 plants following mutagenic treatment of seeds of cultivars UPS 122 and Kade 6/16 with 150-250 Gy gamma radiation. Mutants were selected from the most chimaeric parts of the M_1 plant, which had earlier been identified to be the first mature pods on the M_1 plant and the earliest formed M_2 seeds in the M_1 pods. The indirect selection of tannin mutants was based on seed coat colour changes in M_3 seeds. Mutants 3/1-10-12 and X 22 were selected from 1958 and 1883 M_2 plants of UPS122, respectively. Mutants 3/9-0-12 and 3/4-10-7 were selected from 1442 and 1011 M_2 plants of Kade 3/16, respectively. Mutant 3/4-10-7 had 25 per cent of the level of tannin of the wild type Kade 6/16. The other 3 mutants had similar or increased tannin levels. In Ghana, a mutant isolated in the M_2 of gamma irradiation of line UPS 122 did not flower throughout its growth period of 5 months but developed an underground tuber weighing about 100 g (Klu *et al.*, 1989). Effect of seed irradiation on some plant characteristics of winged bean was reported by Veeresh *et al.* (1995). Dry seeds of winged bean cv. Chimbu were exposed to different doses of gamma rays (10, 15, 20, 25, 30 and 35 krad) to study the effect on 6 plant characteristics. The results showed that there was a greater reduction at higher doses compared to lower doses for all the characters. There was no definite trend in the effect on vigour index (calculated by multiplying normal germination percentage with hypocotyl length), but there was a greater reduction at higher doses. High levels of radiation only significantly affected dry weight and shoot and root length. Treatment with 15 krad had a stimulatory effect on germination and fresh weight. The results indicate that this crop is very sensitive to gamma ray treatment at the early growth stages.

The breeding work on winged bean is yet to gain the proper place resulting in suitable cultivars for the protein hungry world.

12.0 Biotechnology

Among the vegetable legumes, research on winged bean is well reported. This crop could serve well as model system for vegetable biotechnology.

12.1.0 Tissue Culture

Considerable success has been done in developing micropropagation techniques in winged bean. Somatic embryogenesis has been well studied in winged bean and various explants were used even though leaf explants were found to be the best. Venketeswaran *et al.* (1990) established callus and suspension cultures from hypocotyl, cotyledon and epicotyl explants of several cultivars. Best callus growth occurred on MS medium supplemented with 1 mg/l of each of either 2,4-D + NAA or NAA + kinetin or benzy-ladenine (BA). Shoots were induced on callus upon transfer to medium supplemented with 1 mg BA/l with or without kinetin, and rooting occurred on medium supplemented with 1 mg IAA/l. Somatic embryogenesis occurred when fresh callus was maintained on medium with added 2,4-D and NAA, alone or in combination, and then transferred to full or half-strength MS medium. Cultivars Tpt1, Tpt2 and UPS122 formed somatic embryos or embryo-like structures in liquid suspension cultures. Later reports indicated that leaves were better materials for inducing embryogenesis. Ahmed *et al.* (1996) obtained somatic embryos from callus cultures derived from leaf explants. Initiation and development of the somatic embryos occurred with a two-step culture method. Callus cultures were initiated on MS medium with NAA and BAP, and upon transfer to a new medium with IAA and BAP produced somatic embryos. Maximum embryogenesis (60 per cent) was obtained on induction medium with 0.5 mg NAA + 1.0 mg BAP/l followed by transfer to a secondary medium with 0.1 mg IAA + 2.0 mg BAP/l. Optimal embryo germination and plantlet development was achieved on MS medium with 0.2 mg BAP + 0.1 mg IBA/l. The regenerated plants were successfully transferred to greenhouse conditions. Later, Gupta *et al.* (1997) found that seedling leaf segments of winged bean underwent direct somatic embryogenesis under appropriate incubation conditions. Initiation and development of the somatic embryos occurred using a two-step culture method. The culture procedure involved incubation for 28 days on MS basal medium supplemented with

0.1-0.5 mg NAA and 1.0-2.0 mg BA/l (induction medium) before transfer to MS medium supplemented with 0.1 mg IAA and 2.0 mg BA/l (embryo development medium). The initial exposure to low levels of NAA coincident with high levels of BA in the induction medium was essential for embryogenic induction. Maximum embryogenesis (43.3 per cent) was obtained with 0.2 mg NAA and 2.0 mg BA/l, and at least 14 days on the induction medium were required prior to transfer to the embryo development medium. The con-version frequency of cotyledonary embryos was 53.3 per cent upon culture on MS medium containing 0.1 mg ABA/l for 7 days followed by transfer to MS medium supplemented with 0.1 mg IBA and 0.2 mg BA/l. Following conversion, the regenerated plantlets were transferred to soil and showed normal morphological characteristics.

Micropropagation using protoplast culture with 2- mercaptoethanol supplemented enzyme solution was successful (Wilson _et al._, 1985). Genetic stability of _in vitro_ regenerated plants was tested by Koshy _et al._ (2013) using RAPD markers. They found that callus-derived plants showed less fidelity vis-à-vis plants regenerated from shoot tip culture, axillary bud proliferation and cotyledon culture. Induction of callus and shoot proliferation from leaf, petiole, node and epicotyl explants has been demonstrated (Naik _et al._, 2015). In an attempt to standardize gene transfer, Gill (1990) integrated plasmid encoded neomycin phosphotransferase (NPT-II) into the plant genome via epicotyl-derived protoplasts, but transformed calli could not be induced for shoot regeneration.

12.2.0 Protoplast Culture

It is suggested that winged bean can be a model system for the investigation of genetic modification in legumes. It was possible to isolate protoplasts from young, rapidly proliferating callus tissue, but not from leaf mesophyll. Brunel _et al._ (1981) found that polarity in callogenic behaviour was not marked in stem explants of winged bean. The external morphology of the callus strains isolated depended on the particular auxin used in the medium and no organogenesis was observed. Callus regeneration occurred from the anther walls or the filament and not from the pollen. Cuddihy and Bottino (1982) treated protoplasts from cells with an enzyme mixture were cultured on modified B5 medium at a plating density of 1-2 ? 105 protoplasts/ ml. After 24 h, the protoplasts synthesized cell walls. After 3 weeks, cell colonies were transferred to modified MS medium, on which callus tissue developed. Zakri (1984) isolated protoplasts from one-week-old suspension cultures with an enzyme mixture at pH 5.5.

Protoplasts were then cultured in modified B5 (Gamborg B5) medium. Cell division occurred within 3-4 days in protoplast populations of 150-300 ? 106/l. Cell colonies formed callus when transferred to MS medium containing 30 g/l sucrose and NAA and BA. Regeneration of plantlets was achieved by manipulating the various combinations of the growth regulators. Wilson _et al._ (1985) isolated protoplasts from suspension culture using an enzyme solution supplemented with 2-mercaptoethanol. Purified protoplasts suspended in supplemented Uchimiya and Murashige (UM) medium developed cell walls within 2 days and began dividing after 3-4 days to form colonies. Protoplast colonies formed callus on either UM or

In-Vitro Propagation of Winged bean

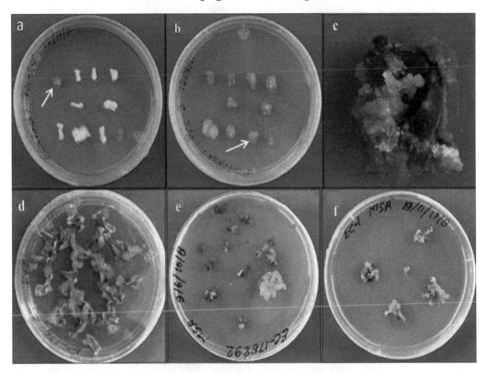

Callus Mediated Shoot Bud Induction and Shoot Regeneration from Hypocotyls and Cotyledon Explants of Winged Bean on Media Containing BAP and NAA or IAA.

(a) and (b). Callus induction from hypocotyl explants on media containing various concentrations of plant growth regulators, callus with regeneration potential was indicated with arrow markings; (c) Callus produced from cotyledon explants; (d) Callus induction from leaf discs incubated BAP and NAA containing medium; (e) Shoot buds and shoots produced from the callus which was obtained from hypocotyls explants; (f) Incubation of shoots for further proliferation on shoot induction medium (Singh *et al.,* 2014).

In-Vitro Propagation of Winged bean

***In vitro* Shoot Proliferation in Shoot Tip Explants Cultured on MS Medium Supplemented with different Concentrations of Kn, BAP & TDZ, in *P. tetragonolobus*.**

(A) Axillary shoot proliferation in shoot tip explants cultured on MS+Kn (1.5 mg/L), (B) Axillary shoot proliferation in shoot tip explants cultured on MS+BAP (1.5 mg/L), (C) Axillary shoot proliferation in shoot tip explants cultured on MS+TDZ (0.25 mg/L), (D) Shoot elongation cultured on MS+GA3 (1.0mg/L)+Charcoal (10 per cent), (E) Regenerated plantlets cultured on MS+IBA (0.75 mg/L) + Charcoal (5 per cent), (F) Hardening in earthen pot filled with garden soil (Naik *et al.*, 2015).

supplemented MS medium. On subculture on MS medium with various growth regulators shoots developed which were then rooted and grown to maturity.

12.3.0 Molecular Markers and Genomics

Winged bean has received little attention in terms of molecular breeding and genomic research, with few studies having used DNA markers for genetic analyses. Mohanty *et al.* (2013) analyzed 24 winged bean accessions and determined that ISSR markers were superior to RAPD markers. Chen *et al.* (2015) used ISSR markers to assess genetic diversity among 45 winged bean accessions and reported a narrow genetic base in their germplasm collection. Chapman (2015) sequenced a seedling transcriptome from the winged bean genotype Ibadan Local-1 and reported mining of ~1900 microsatellite and ~1800 conserved orthologous set loci. In a parallel effort to stimulate genomics assisted breeding programs in winged bean, Vatanparast *et al.* (2016) sequenced transcriptomes of multiple tissues from two Sri Lankan winged bean genotypes from United States Department of Agriculture germplasm accessions PI 639033 and PI 491423 and reported large-scale marker development, identifying over 5,000 single nucleotide polymorphisms and ~13,000 SSR markers between their Sri Lankan and Chapman's Nigerian genotypes. Most recently, Wong *et al.* (2017) developed 9,682 genic SSR markers via transcriptome sequencing from Malaysian accessions and validated 18 SSRs across 9 accessions. Later on, another study developed genic SSR markers from transcriptome data generated from a Malaysian accession. From this, the authors went on to validate 18 SSR markers upon nine genotypes originated from five countries (Malaysia, Bangladesh, Sri Lanka, Indonesia, and PNG) (Wong *et al.*, 2017). Yang *et al.* (2018) used five of these markers to screen 53 accessions from IITA, U.S. Department of Agriculture (USDA), and the National Agriculture and Food Research Organization (NARO, Japan) gene banks.

12.4.0 Genetic Transformation

Molecular biological studies in winged bean have been reported mainly on a winged bean Kunitz-type chymotrypsin inhibitor (WCI) expressed in seeds and tuberous roots and small amounts of the WCI protein and mRNA can also be detected in stems. In seeds, the expression of WCI is restricted to the period between the mid- and late-maturation stage. To understand the mechanisms that regulate the expression of WCI genes, the promoter activity of the upstream region of the WCI-3b gene, which encodes a major WCI protein, Sakata *et al.* (1997) analyzed in transgenic tobacco plants. By using a series of constructs with 5′ deletions in the upstream sequences, the region between -882 and -623, relative to the transcription start site, was shown to contain multiple sequences which are responsible for high level expression in mid-maturation stage seeds. However, when this region was fused to the CaMV 35S core promoter in both orientations, the chimaeric promoters showed only a weak transcription activity in transgenic tobacco plants. Further analyses using internal deletion constructs revealed that the region between -882 and -174 is required for the transcription activation. Disruption of the RY sequence at -517, which is conserved in many seed protein genes, resulted in a drastic reduction of the transcription activity in seeds. These results suggest that sequences necessary for high level induction of the WCI-3b gene transcription in developing seeds are dispersed in the region between - 882 and -174, and that the RY sequence is one of these sequences.

Habu *et al.* (1996) investigated the localization of WCI protein in stems of winged bean. The results demonstrated that the WCI protein was localized in sieve tubes. Furthermore, the 5' region of the *WCI-3b* gene, which exhibited strong transcriptional activity in developing seeds, also promoted transcription of a reporter gene in the phloem of stems of transgenic tobacco. Sakata *et al.* (1994) constructed sequential deletions of the promoter region of the *WCI-3b* gene, which encodes the major chymotrypsin inhibitor of winged bean and their expression was analyzed in transgenic tobacco plants and in bombarded winged bean seeds. In transgenic tobacco plants, a critical promoter region, which is important for high levels of expression in seeds, was identified, but deletion of this region had essentially no effect when bombarded into winged bean seeds. To analyze the regulation of gene expression in an heterologous plant, Peyachoknagul *et al.* (1994) constructed a chimaeric gene containing the 5' flanking region (about 1 kb) of the *WCI-2* gene fused with a coding region of a *gus* gene which was used as a reporter gene and introduced into tomato plants by *Agrobacterium*-mediated transformation. Some of the transformed tomato plants exhibited an irregular pattern of *gus* gene expression, which might have resulted from a chimaeric origin of the regenerated plants. However, a few transformed tomato plants showed expression of the *gus* gene in immature seeds as in winged bean. This suggests that a 1 kb 5' flanking region of WCI-2 is involved in organ specific and temporally regulated gene expression and can be recognized by the transacting factor(s) from tomato plant.

Winged bean was thought to be an important legume for the malnutrition in the third world countries. However, in spite of its importance, work during the last decade has been poor. Despite the latest phylogenetic analysis by Yang *et al.* (2018), the centre of origin of winged bean remains a matter that requires further investigations. Research efforts in this direction, based on larger collections of the cultivated *P. tetragonolobus* and related *Psophocarpus* species, could yield a clearer picture of the taxonomy, origin and evolution of winged bean, and allow for a strategic exploitation of the resources for future improvement. Winged bean is amenable for biotechnological studies would serve as a model system for other vegetable legumes. Lot of works particularly on chymotrypsin inhibitors using modern molecular techniques have been done. This should lead to the development of low tannin varieties. The identification of a dwarf mutant Gui Ai should lead to the development of non climbing, bushy type winged bean. The low tannin content in stems should be used to develop varieties which could be used as leafy vegetable.

13.0 References

Ahmed, R., Gupta, S.D. and De, D.N. (1996) *Plant Cell Rep.*, **15**: 531-535.

Anonymous (1969) *The Wealth of India, Vol. VIII*, Publications and Information Directorate, New Delhi, pp. 294-295.

Anugroho, F., Kitou, M., Kinjo, K. and Kobashigawa, N. (2010) *Plant Prod. Sci.*, **13**: 360–366.

Anugroho, F., Kitou, M., Kinjo, K. and Kobashigawa, N. (2010) *Agron. Crop Ecol.*, **13**: 360-366.

Bagchi, D.K., Banerjee, A. and Sasmal, B.C. (1989) *Trop. Agric., Trinidad*, **66:** 240-242.

Bagchi, D.K., Banerjee, A. and Bhattacharya, B. (1991) *Indian Agriculturist*, **35:** 1-7.

Boscan Odor, D.R. (1978) *Proc. Int. Symp. Winged bean*, Manila, pp. 211-214.

Brunel, A., Landre, C., Chardard, R. and Kovoor, A. (1981) In: *Tissue culture of economically important plants. Proc. internatl. symp.* (Ed, Rao, A.N.), Botany Department, National University of Singapore, Singapore, 28-30 April, 1981. 1982, 63-65.

Burkill, I.I. (1935) *Dictionary of Economic Products of Malay Peninsula*, University Press, Oxford Vol. 2, p. 2402.

Chapman, M.A. (2015) *Appl. Plant Sci.*, **3**: 1400111.

Chen, D., Yi, X., Yang, H., Zhou, H., Yu, Y., Tian, Y. and Lu, X. (2015) *Genet. Resour. Crop Evol.*, **62**: 823–828.

Claydon, A. (1975) *Science in New Guinea*, **3:** 103.

Claydon, A. (1978) *Workshop Rep. Dev. Potentiality of Winged bean*, Las Banos, Philippines, pp. 377-391.

Cuddihy, A.E. and Bottino, P.J. (1982) *Plant Cell Tissue Organ Cult.*, **1:** 201-209.

Deokar, C.D. and Ruikar, S.K. (1987) *J. Maharashtra Agril. Univ.*, **12:** 133-334.

Drilion, J.D. and Obordo, R.A. (1978) *Workshop Rep. Dev. Potentiality of Winged bean*, Las Banos, Philippines.

Eagleton, G.E.; Halin, A.H. and Chai, N.F. (1980) *Proc. Legumes in the Tropics*, Faculty of Agri., Univ. Pertanian Malaysia, Malaysia, pp. 133-144.

Eagleton, G.E.; Thurling, N. and Khan, T.N. (1978) *Workshop Rep. Dev. Potentiality of Winged bean*, Las Banos, Philippines, pp. 110-120.

Ekpenyong, T.E. and Borchers, R.L. (1978) *Workshop Rep. Dev. Potential of Winged bean*, Las Banos, Philippines.

Elmes, R.P.T. (1976) *Papua New Guinea Agr. J.*, **27**: 53–57.

Erskine, W. and Khan, T.N. (1977) *Euphytica*, **26**: 829–831.

Erskine, W. and Khan, T.N. (1981) *Field Crops Res.*, **3**: 359-364.

Fortuner, R., Fauquet, C. and Lourd, M. (1979) *Plant Dis. Rep.*, **63**: 194–199.

Garcia, V.A., Palmer, J.K. and Young, R.W. (1979) *J. Amer. Oil Chem. Soc.*, **56:** 931-932.

Gill, R. (1990) *Ann. Bot.*, **66**: 31–39.

Gregr, V. and Singh, (1978) *Agricultura Tropica et Subtropica*, **20:** 35-38.

Gupta, S.D., Ahmed, R. and De, D.N. (1997) *Plant Cell Rep.*, **16**: 628-631.

Habu, Y., Fukushima, H., Sakata, Y., Abe, H. and Funada, R. (1996) *Plant Mol.Biol.*, **32:** 1209-1213.

Hakande, T.P. (1992) *Cytogenetical studies in Psophocarpus tetragonolobus (L.) DC. Dissertation, BAM University.*

Haq, N. (1982) *Z. Pflanzenzuchtung,* **88:** 1-12.

Haq, N. and Smartt, J. (1973) *Trop. Grain Legume Bull.,* **15:** 35-38.

Harder, D. and Smartt, J. (1992) *Econ. Bot.,* **46:** 187–191.

Harder, D.K. (1991) *Dissertation Abst.Internatl. B, Sciences and Engineering,* **51:** 4150B.

Harding, J., Martin, F.W. and Kleiman, R. (1978) *Trop. Agric.* (Trinidad), **55:** 307.

Herath, H.M.W. and Fernandez, G.C.J. (1978) *Proc. 1st Int. Symp. Winged bean,* Manil, pp.161-172.

Hernandez'O.L.F., Munoz, C.A. and Castellar, P.N. (1987) *Acta Agronomica, Universidad Nacional de Colombia,* **37:** 50-59.

Hikam, S., MacKown, C.T., Poneleit, C.G. and Hildebrand, D.F. (1991) *Ann. Bot.,* **68:** 17-22.

Hildebrand, D.F., Chaven, C. and Hymowitz, T. (1981) *Trop. Grain Legume Bull.,* **23:** 23-25.

Hildebrand, D.F., Chaven, C., Hymowitz, T., Bryan, H.H. and Duncan, A.A. (1982) *Agron. J.,* **73:** 623-625.

Hlava, B. and Michl, J. (1990) *Agricultura Tropica et Subtropica.,* **23:** 47-56.

Hung, T. Le., Hubacek, J., Lachman, J., Pivec, V., Borek, V. and Rehakova, V. (1986) *Agricultura Tropica et Subtropica,* **19:** 193-207.

Ikram, A. and Broughton, W.J. (1978) *Proc. 1st Int. Symp. Winged bean,* Manila, pp. 205-210.

Iruthayathas, E.E. and Herath, H.M.V. (1981) *Sci. Hort.,* **15:** 1-8.

Islam, M.A., Boyce, A.N., Rahman, M.M., Azirun, M.S. and Ashraf, M.A. (2016) *Braz. Arch. Biol. Technol.,* **59:** 1-9.

Jugran, H.M., Banerji, B.K. and Datta, S.K. (2001) *J. Nucl. Agric. Biol.,* **30:** 116–119.

Karikari, S.K. (1978) *Proc. 1st Inst. Symp. Winged bean,* Manila, pp. 150-160.

Kesavan, V. and Erskine, W. (1978) *Proc. Ist Inst. Symp. Winged bean,* Manila, pp. 211-214.

Kesavan, V. and Khan, T.N. (1978) *Proc. Ist Int. Symp. Winged bean,* Manila, pp. 105-109.

Khalil, R.M.A., Shafekh, S.E., Norhayati, A.H., Fatahudin, I.M., Rahimah, R., Norkamaliah, H. and Azimah, A.N. (2013) *Pak J. Nutr.,* **12:** 416–422.

Khan, T.N. (1976) *Euphytica,* **25:** 693-706.

Khan, T.N. (1982) *Food Agric. Organ. Plant Prod. Prot. Pap.,* **38:** 222.

Khan, T.N., Bohn, J.C. and Stephenson, R.A. (1977) *World Crops and Livestock,* **29:** 208.

Khurana, S.C., Pandita, M.L. and Pandey, U.C. (1990) *Res. and Dev. Reporter,* **7:** 197-198.

Klu, G.Y.P. (1996) *Efforts to accelerate domestication of winged bean* (*Psophocarpus tetragonolobus* (L.) *DC.*) *by means of induced mutations and tissue culture.* Dissertation, Wageningen University.

Klu, G.Y.P. and Kumaga, F.K. (1999) *Ghana J. Sci.*, **39**: 55–62.

Klu, G.Y.P., Jacobsen, E. and Van-Harten, A.M. (1997) *Euphytica*, **98**: 99-107.

Klu, G.Y.P., Quaynor-Addy, M., Dinku, E. and Dikumwin, E. (1989) *Mutation Breeding Newslett.*, **34**: 15-16.

Kortt, A.A. (1979) *Biochim. Biophys. Acta.*, **577**: 371–382.

Kortt, A.A. (1984) *Eur. J. Biochem.*, **138**: 519–525.

Kortt, A.A., Strike, P.M. and Jersey, J.D. (1989) *Eur. J. Biochem.*, **181**: 403–408.

Koshy, E.P., Alex, B.K. and John, P. (2013) *The Bioscan*, **8**: 763–766.

Kumari, S., Yadav, R.C., Jha, M.N. and Jha, S. (2018) *J. Pharmacog. Phytochem.*, **7**: 513-517.

Lam-Sanchez, A., Durigan, J.F., de Oliveira, E.T., Negrão Serigatto, W.J. and Faggioni, J.L. (1983) *Nutr.*, **33**: 874-883.

Lee, C.T. (1987) *Alafua Agri. Bull.*, **12**: 13-15.

Lee, C.T. (1988a) *J. Agric. Univ., Puerto Rico.*, **72**: 273-276.

Lee, C.T. (1988b) *HortScience*, **23**: 761.

Lesseps, R.J. (1988) *Productive Fmg.*, **179**: 5-9.

Lim, G. and Ng, H.L. (1977) *Plant and Soil*, **46**: 317-327.

Long, M.H., Qin, R.Y. and Liu, Z.G. (1993) *Chinese Vegetables*, **4**: 25-27.

Lubis, S.H.A. (1978) *Proc. Ist Int. Symp. Winged bean*, Manila, pp. 121-123.

Macmillan, H.T. (1949) *Tropical Planting and Gardening*, Macmillan, London.

Martin, F.W. and Delpin, H. (1978) *Vegetables for Hot Humid Tropics – The Winged bean* (*Psophocarpus tetragonolobus*), USDA Bulletin, New Orleans, p. 22.

Masefield, G.B. (1952) *Empire J. Exp. Agri.*, **20**: 175.

Masefield, G.B. (1957) *Empire J. Exp. Agri.*, **25**: 139.

Masefield, G.B. (1961) *Tropical Agric., Trinidad*, **38**: 225.

Masefield, G.B. (1973) *Field Crops Abst.*, **26**: 157-160.

Matejka, V. (1987) *Agricultura Tropica et Subtropica*, **20**: 33-45.

McKee, R. (1928) *Plant inventory no. 86. USDA, Washington.*

Michl, J. and Van Phu, N. (1985) *Agricultura Tropica, et Subtropica*, **18**: 99-100.

Mohanty, C.S., Verma, S. and Singh, V. (2013) *Am. J. Mol. Biol.*, **3**: 187–197.

Motior, M.R., Mohamad, W.O.W., Wong, K.C. and Shamsuddin, Z.H. (1998) *Experimental Agrl.* **34**: 41-53.

Motior, M.R., Mohamad, W.O.W., Wong, K.C. and Shamsuddin, Z.H. (1997) *Indian J. Plant Physiol.*, **2**: 217-220.

Motior, R.M. (1998) *Faculty of Agriculture University Putra, Malaysia.*

Naik, D.S.R., Prasad, B., Nemali, G. and Naik, A.S. (2015) *Int. J. Rec. Scient. Res.*, **6**: 3985-3987.

Nakamura, H. and Abe, J. (1989) *JARQ,*23: 71-77.

Nakanishi, K., Nawata, E. and Shigenaga, S. (1987) *Japanese J. Trop. Agric.*, **31**: 165-171.

Nangju, D. and Baudoin, J.P. (1979) *J. Hort. Sci.*, **54**: 129-136.

Nath, P., Jurgan, H.M. and Banerjee, B.K. (1986) *HCBP*, **3**: 1–2.

National Academy of Sciences (1975) *The winged bean: a highprotein crop for the tropics. Washington, USA.*

Nazri, N.A.A., Ahmat, N., Adnan, A., Mohamad, S.A.S. and Ruzaina, S.A.S. (2011) *Afr. J. Biotechnol.*, 10: 5728–5735.

Ndegwe, N.A. and Ikpe, F.N. (2000) *Afr. J. Agric. Teacher Edu.*, **9**: 83-91

Newell, C.A. and Hymowitz, T. (1979) *New Agricultural Crops, AAAS Selected Symposium,* Westview Press, Boulder, Colorado, pp. 21-40.

Okada, R., Kiyota, E., Moriyama, H., Fukuhara, T. and Valverde, R.A. (2017) Virus Genes, **53**: 141–145.

Okubo, H., Masunaga, T., Yamashita, H. and Uemoto, S. (1992) *Scientia Hortic.*, **49**: 1-8.

Pandita, M.L., Dahiya, M.S. and Vashistha, R.N. (1989) *Haryana J. Hort. Sci.*, **18**: 136-141.

Parthipan, S. and Kulasooriya, S.A. (1989) *MIRCEN Journal*, **5**: 335-341.

Parthipan, S. and Senanayake, Y.D.A. (1989b) *Legume Res.,*12: 183-185.

Peyachoknagu, S., Tantisuwichwong, N., Pongtongkam, P., Suputtitada, S. and Ohno, T. (1994) In: *Proc. Internatl. Colloq. on Impact of Plant Biotechnology on Agriculture* (Javornik, B., Bohanec, B. and Kreft, I. (Eds.), Rogla, Slovenia, December 5th-7th 1994, pp. 121-130.

Poi, S.C. and Ghosh, G. (1986) *Trop. Grain Legume Bull.*, **33**: 34-36.

Poi, S.C. and Kabi, M.C. (1982) *Trop. Grain Legume Bull.*, **24**: 24-26.

Poole, M.M. (1979) *Kew Bull.*, **34**: 211–220.

Pospisil, F., Karikari, S.K. and Boamah-Mensah, E. (1971) *World Crops*, **23**: 250.

Price, T.V. (1980) *In: The winged bean: proceedings of the 1st international symposium on developing the potential of the winged bean. Manila*, pp. 241-243

Purseglove, J.W. (1968) *Tropical Crops: Dicotyledons*, Vol. 1, John Wiley and Sons Inc., New York.

Raai, M.N., Zain, N.A.M., Osman, N., Rejab, N.A., Sahruzaini, N.A. and Cheng, A. (2020) *Ciência Rural*, **50**: 1-7.

Rathfelder, E.L. (1981) *Ann. Appl. Biol.*, **98**: 143-148.

Ravindran, G., Palmer, J.K. and Gajameragedara, S.M. (1989) *J. Agril. Food Chem.*, **37**: 327-329.

Ray, S., Tah, J. and Sinhababu, A. (2017) *Int. J. Chem. Tech. Res.*,**10**: 178-185.

Reddy, P.P. (2015) *Plant protection in tropical root and tuber crops. Springer India*, pp. 293-303.

Rosario, R.R. Del., Lozaro, Y., Noel, M.G. and Flores, D.M. (1981) *Philip. Agric.*, **64**: 143-153.

Sakata, Y., Chiba, Y., Fukushima, H., Matsubara, N., Habu, Y., Naito, S. and Ohno, T. (1997) *Plant Mol. Biol.*, **34**: 191-197.

Sakata, Y., Fukushima, H., Habu, Y., Furuya, S., Naito, S. and Ohno, T. (1994) *Biosci. Biotech. Biochem.*, **58**: 2104-2106.

Sarnaik, D.A. and Baghel, B.S. (1988) *PKV Res. J.*,**12**: 79-80.

Sastrapradza, S. and Lubis, S.H.A. (1975) In: William, I.T., Lamoureux, C.H. and Wulijarnisoetjipta, N. (Eds.), *South-East Asian Plant Genetic Resources*, IBPGR, Bogor, Thailand, pp. 147-151.

Satyanarayana, A., Rajendran, R. and Biswas, S.R. (1978) *Proc. Ist Int. Symp. Winged bean*, Manila, pp. 140-144.

She, C., Jingyu, L., Zhiyong, X. and Song Y. (2004) *Caryologia*, **57**: 387–394.

Silva, H.N. De. and Omran, A. (1987) *Field Crops Res.*, **16**: 209-216.

Singh, S.P., Shukla, S., Khanna, K.R., Dixit, B.S. and Banerji, R. (1995) *Fett WissenschaftTechnol.*,**97**: 425-427.

Smartt, J. (1980) *Euphytica*, **29**: 121–123.

Solanki, S.S. and Saxena, P.K. (1989) *Progressive Hort.*,**21**: 364-367.

Solanki, S.S., Saxena, P.K and Shah, A. (1989) *Progressive Hort.*,**21**: 263-267.

Sri Kantha, S. and Erdman, J.W. (1984) *J. Am. Oil Chem. Soc.*, **61**: 515–525.

Stephenson, R.A., Kesavan, V., Claydon, A., Bala, A.A. and Kaiulo, J.V. (1982) *Proc. Int. Symp. Root Crops*, Manila., pp. 147-152.

Suma Bai, D.I. and Sunil, K.P. (1993) *Madras Agril. J.*, **80**: 541-546.

The Wealth of India (1969) *Vol. III. Publication and Information Directorate, New Delhi.*

Thuan, N.V. (1975) *Rev. Gen. Bot.*, **82**: 157.

Tindall, H.D. (1968) *Commercial Vegetable Growing*, Oxford University Press, London, Ibadan, Nairobi.

Tixier, P. (1965) *Rev. Cytol. Biol. Veg.*, **28**: 133.

Tuquero, Joe and Takai, Glen (2018) In: Food Plant Production, UOG-CNAS, FPP-06.

Uemoto, S., Fujieda, K., Noaka, M. and Nakamoto, Y. (1982) *Bull. Inst. Trop Agri., Kyushu Univ.*, **5**: 59-70.

Valicek, P., Buresova, M. and Ta Kim Binh (1988) *Agricultura Tropica et Subtropica.***21**: 124-134.

Valle, R. Del Jr., Lugo-Lopez, M.A. and Scott, T.W. (1980) *J. Agric. Univ., Puerto Rico.,***64**: 211-218.

Vatanparast, M., Shetty, P., Chopra, R., Doyle, J.J., Sathyanarayana, N. and Egan, A.N. (2016) *Sci. Rep.,* **6**: 29070.

Veeraragavathatham, D., Venakatachalam, R. and Jayashankar, S. (1988) *South Indian Hort.,* **36**: 8-20.

Veeresh, L.C., Shivashankar, G. and Hittalmani, S. (1995) *Mysore J. Agri. Sci.,* **29**: 1-4.

Venketeswaran, S., Dias, M.A.D.L. and Weyers, U.V. (1990) *Acta Hortic.,* **280**: 202-205.

Verdcourt, B. and Halliday, P. (1978) *Kew Bull.,* **33**: 191–227.

Wilson, V.M., Haq, N. and Evans, P.K. (1985) *Plant Sci., Ireland,* **41**: 61-68.

Wong, K.C. (1983) *Effects of daylength, temperature, light intensity and applied growth substances on the growth, flowering and tuberization of winged bean (Psophocarpus tetragonolobus (L.) DC.). University Putra, Malaysia.*

Wong, K.C. and Schwabe, W.W. (1979) *Effects of daylength and day/night temperature on the growth, flowering and tuber formation of winged bean (Psophocarpus tetragonolobus (L.) DC.).*

Wong, Q.N., Tanzi, A.S. and Ho, W.K. (2017) *Genes,* **8**: 100.

Woomer, P.,Guevarra, A. and Stockinger, K. (1978) *Proc. Ist Int. Symp. Winged bean,* pp. 197-204.

Yang, S., Grall, A. and Chapman, M.A. (2018) *Afr. J. Tradid. Complement Altern. Med.,* **4**: 59-63.

Yoneyama, T., Fujita, K., Yoshida, T., Matsumoto, T., Kambayashi, I. and Yazaki, J. (1986) *Plant Cell Physiol.,* **27**: 791–799.

Zakri, A.H. (1983) *15th Int. Cong. Genet., New Delhi* (Abst.), p. 434.

Zakri, A.H. (1984) In: *Efficiency in plant breeding, Proceedings of the 10th Congress of the European Association for Research on Plant Breeding* (Lange, W., Zeven, A.C. and Hogenboom, N.G. eds.), *EUCARPIA,* Wageningen, the Netherlands, 19-24 June 1983, 1984, p. 363.

Zeven, A.C. and De Wet, J.M.J. (1982) *Centre for Agricultural Publishing and documentary, Wageningen,* p. 227.

Zusevics, J.A. (1981) *J. Pl. Nutr.,* **3**: 789-802.

JACK BEAN

A. Chattopadhyay, V.A. Parthasarathy,
B.P. Medhi and P.K. Maurya

1.0 Introduction

Jack bean (*Canavalia ensiformis*) and sword bean (*C. gladiata*) are commonly grown for the young pods and immature seeds which are used as food for human and animals. It is also cultivated for forage or for green manure. It is widely distributed in India, Argentina, China and United States and is mainly used for animal fodder and for human nutrition in a limited scale. On the south-west coast of India, *C. cathartica* is widely distributed in coastal sand dunes (CSDs) (Arun *et al.*, 1999, 2003; Seena *et al.*, 2005) and mangroves. In CSDs, *C. cathartica* establishes as a mat-forming creeper, with patchy distribution (frequency of occurrence: 22.2 per cent) (Arun *et al.*, 1999), whereas in mangroves it is a tree climber. It is also cultivated in North Eastern region of India (CSIR, 1950). In Philippines, the matured seeds are primarily used for 'ginisa' (salted) dishes while immature pods and immature leaves are used as vegetable in dishes like salad 'pinakbet', 'lumpia', 'ginataan', and as vegetable component of meat dishes such as 'sinigang' and 'nilaga' (Mendoza, 1990). In Japan, young pods of sword bean are sliced and pickled in soy sauce. It is a minor legume vegetable and work on this crop is very scanty. In Nigeria *C. ensiformis* is currently grown as ornamental plant, planted near houses and allowed to trail on walls and trees, believed to repel snakes (Abitogun and Olasehinde, 2012).

2.0 Composition and Uses

The nutritional composition of fresh sword bean fruits per 100 g edible portion is: water 83.6 g, energy 247 kJ (59 kcal), protein 4.6 g, fat 0.4 g, carbohydrate 10.7 g, fibre 2.6 g, Ca 33 mg, P 66 mg, Fe 1.2 mg, vitamin A 40 IU, thiamin 0.2 mg, riboflavin 0.1 mg, niacin 2 mg, ascorbic acid 32 mg. Dry seeds contain per 100 g: water 10.7 g, energy 1453 kJ (347 kcal), protein 24.5 g, fat 2.6 g, carbohydrate 59 g, fibre 7.4 g, Ca 158 mg, P 298 mg, Fe 7.0 mg, thiamin 0.8 mg, riboflavin 1.8 mg, ascorbic acid 1 mg (Rubatzky and Yamaguchi, 1997). The seed protein is poor in methionine, but rich in lysine. The immature pods and seeds contain about 75.2 per cent water, 6.9 per cent protein, 0.5 per cent fat, 13.3 per cent carbohydrate, 3.3 per cent fibre and 0.8 per cent ash (Indira and Peter, 1988). Mendoza *et al.* (1986) extensively investigated composition of jack bean and reported that green pods are rich sources of protein (28.31 per cent) and seeds contain carbohydrate (44.56 per cent) with a fairly digestible protein. Xia *et al.* (2017) found seeds of early cultivar showed 80.81 per cent of dry matter and 24.45 per cent of total protein contents. The pods of Jack bean can be a potential source of edible stuff as well as a source of protein, vitamins, minerals, carbohydrate and energy supplement in livestock feeds (Patel *et al.*, 2016). The seeds of jack bean are consumed in different parts of India. The mature seeds are consumed by the Indian tribal sects, Kurumba, Malayali, Erula and other Dravidian groups, after cooking (Mitre, 1991). This legume is used as a cover crop and the roasted seeds are ground to prepare drink like coffee in western countries (Bressani *et al.*, 1987). Many researchers have been investigating on how to make jack beans edible to human beings by using various treatments like heating, fermentation and extrusion but without obtaining favourable results (Justo *et al.*, 1994). This is because there are a number of antinutritional factors present in jack beans, which restricts its utilization as human food. These factors include; thermo-stable factors (canavanine, concanavalin, canavalin, canatoxin) and thermo-labile factors protease inhibitors, lectins and phytic acid (Carlini and Udedibie, 1997; Udedibie and Carlini, 1998). The concentration of these growth inhibiting substances increases age and maturity of plant tissue. Hence, only tender foliage and pods are edible. Detoxified jack bean seed has been used successfully as a high protein fish meal substitute in tilapia aquaculture (Martinez-Palacios *et al.*, 1987). Jack bean seed flour is good functional foods for nutrition, food formulation and utilization (Marimuthu and Gurumoorthi, 2013; Karoli *et al.*, 2017).

A few traditional uses of *C. cathartica* have been documented in interviews with coastal dwellers of Goa, Karnataka and Kerala of the west coast of India. In agricultural fields in and around estuaries and mangroves, *C. cathartica* are allowed to grow deliberately as a cover crop after harvest of paddy and sugarcane, with the intention of improving the soil nitrogen budget. Beans of immature or ripened pods are consumed in some parts of west coast of India during scarcity of food. Boiled or roasted tender pods or ripened beans are eaten along the south-west coast of India. Soaking tender or ripened beans in water for a long period (*e.g.* overnight), followed by cooking, may eliminate toxic factors. Such practices are likely to support microbial fermentation, which assists in detoxification to a safe level. The vines of *C. cathartica* grown on CSDs and estuarine and mangrove habitats serve as fodder for livestock, particularly rabbits, hares and cattle.

In Korea it is used in the treatment of vomiting, abdominal dropsy, kidney-related lumbago, asthma, obesity, stomach-ache, dysentery, coughs, headache, intercostal neuralgia, epilepsy, schizophrenia, inflammatory diseases and swellings (Zhu and Wang, 2002; Duranti, 2006; Li *et al.*, 2007). In Japan it is effective in treating ozena, haemorrhoids, pyorrhea, otitis media, boils and cancers, all kinds of inflammatory diseases and atopic dermatitis. In Korea soap is marketed based on extracts of sword bean; it is used for the treatment of athlete's foot and acne. Sword bean seeds are also used in traditional Chinese medicine (Chen *et al.*, 2001).

3.0 Origin, Classification and Morphology

3.1.0 Origin and Classification

The jack bean (*Canavalia ensiformis* L.) is reported to be the native of West Indies (Anon., 1950). Then it was distributed to Central and South America. It is also cultivated to a limited scale in India. Sword bean (*Canavalia gladiata* (Jacq.) DC.), sometimes also called beach bean, originates from India and China (Black *et al.*, 2006; Zhang and Yang 2007; Yang and Ma, 2010; Luo *et al.*, 2015) and has spread throughout the tropics (Moteetee, 2016). *Canavalia* is a small genus consisting of 48 species of which four species are reported from India, *viz.*, *C. ensiformis*, *C. gladiata*, *C. maritima*, and *C. virosa*. Of these four species *C. ensiformis* (jack bean) and *C. gladiata* (sword bean) occur in NE region of India and are being cultivated for the edible pods. Work done on the genetics and taxonomy of this genus is very little, hence considerable doubts exist on the species status of this genus. Parthasarathy and Singh (1991) reported that the somatic chromosomes of *C. ensiformis* and *C. gladiata*

Tea from jack bean

Type of Seed, Flowering, Cultivation and Harvesting of Jack Bean

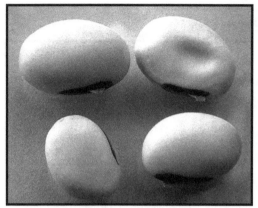

Jack bean seeds

Jack bean flower

Jack bean cultivation

Jack or sword bean cultivation

Jack bean harvested pods

are 22. While variation on flower colour (white and pink), pod characters and seed size occur in *C. gladiata*, *C. ensiformis* does not exhibit much variation. Interestingly, they observed natural introgression between the two species. Subsequent controlled crosses between these two species confirmed the crossability of *C. ensiformis* with *C. gladiata* and *vice versa*. Meiotic studies revealed no irregularities confirming the fertility of the hybrids. The studies indicate that the separation of the two species based on morphological characters warrants a change and these two actually belong to the same species.

3.2.0 Morphology

Jack bean, an annual tropical and sub-tropical plant, belongs to the family Leguminosae. It is highly drought resistant and its roots penetrate deep into the soil. The root is a well branched tap root, penetrating deep into the soil down up to 150 cm. The stem is grassy, cylindrical, green and branched (5 to 7 lateral sprouts). The cultivated plants may reach from 60 to 120 cm in height. Moreover, its height may vary to some extent from region to region depending upon the topography and the climatic condition of the region. The climbing form is called sword bean. The leaves (the first two true leaves are simple) are cordate, green and large. The true leaves are ternate, alternate, light green, spherical-elongated slightly downy. The central sprout has from 9 to 15 and each lateral one from 3 to 5 ternate leaves. The flowers lie in axillary racemes (about 10 racemes on one plant). Each raceme contains 30 to 40 flowers which are bisexual and either white or pink-purple in colour. The pods are large from 20 to 32 cm in length, flat broad, sword shaped, hooked, there are 8 to 18 seeds in each pod. The pods are indehiscent. They are yellowish in colour. The seeds are large, kidney shaped or oval elongated, 1 to 2 cm in diameter and either white or pink in colour (Xia *et al.*, 2017). The scar may be up to 1 cm long, 1000 seeds weigh from 1000-2000 g (Ustimenko-Bakumovsky, 1982).

4.0 Soil and Climate

It is a deep rooted plant with surface feeding nature. The crop is successfully cultivated on various tropical soils. It prefers a pH range of 5.0-7.0. It grows well even on nutrient depleted soils and on acid soils, even with a pH as low as 4.5. Humus is necessary for vigorous plant growth and better seed quality. The jack bean requires adequate soil moisture during early vegetative stage and flowering. Jack bean can be grown in soils with high lead concentration and has potential to be used for restoration of lead-contaminated soils (Faria Pereira *et al.*, 2010). It can grow in poor droughty soils, and does not grow well in excessively wet soil. It will drop its leaves under extremely high temperatures, and may tolerate light frosts (Florentin *et al.*, 2004).

The crop requires temperatures of 20–30°C and is cultivated from sea-level up to 1000 m altitude. It is tolerant of drought once established and also tolerant of waterlogging, shade and salinity, making it one of the hardiest tropical legumes. It prefers an evenly distributed annual rainfall of 900–1500 mm. The crop is typically tropical in adaptation and grows well in subtropics also. It is a typical short day plant. It grows well with a 10 to 12 hour day length. The seeds germinate at 25 to 27°C and jack bean seeds consume much water up to 200 per cent (Ustimenko-

Bakumovely, 1982). Sprouts appear at 20°C on the 8th-12th day, at 22 to 24°C on the 6th-8th day and at 27°C on the 4th-5th day. A distinguishing characteristic of this plant is its ability to continuously grow under severe environmental conditions (Udedibie and Carlini, 1998), even in nutrient-depleted, highly leached, acidic soils (NAS/NRC, 1979). Jack bean is drought-resistant and immune to pests (FAO, 2012; Bunch *et al.*, 1985).

5.0 Cultivation

Jack bean should be planted in full sun to light shade in seed beds roughly 1.5–3 ft apart, in late spring/early summer, at 50–60 lb/ac. Seedlings can also be transplanted after raising in polythene bags. For weed suppression, researchers have planted 4–5 seeds per square meter (80 lbs/acre) (Bunch *et al.*, 1985). This is a warm season crop which can be grown in all type of soils. The seeds are sown directly in the pit before onset of monsoon. In mid-hill of Meghalaya seeds are sown during April to July. A spacing of 1 m ´ 75 cm is usually given for planting in terrace cultivation. Singh and Yadav (1991) reported that maximum plant height was recorded in 45 cm ´ 30 cm and 45 cm ´ 40 cm but no significant variation was recorded in number and weight of pod/plant. Generally, 2 seeds are sown in a pit of 15 cm ´ 15 cm ´ 15 cm size which is filled with well rotten farmyard manure. If inter-seeding jack bean with corn or sorghum, seed should be drilled 15–30 days after sowing the main crop, and after being soaked in water for 24 hours (Bunch *et al.*, 1985). Sowing the intercrop at least 15 days after the main cash crop will limit the potential negative effects of plant competition (Caamal-Maldonado *et al.*, 2001). It does not require staking.

A basal dressing of 10 t FYM/ha is applied with 100 to 150 kg superphosphate. Two to three weedings at 3 weeks' interval with first weeding after one month of sowing are done. Farmer Participatory Research (FPR) in Uganda showed that jack bean intercropped with banana, coffee, sweet potatoes, or cassava is a preferred green manure among farmers (Fischler and Wortmann, 1999).

When used as a cover crop or green manure the plant should be terminated mechanically with a roller-crimper or chemically when it first begins to flower. Because jack bean is an annual or very short-lived perennial, it may have to be re-established each year if used as a weed smothering plant.

6.0 Harvesting and Yield

As in other beans, the tender pods are harvested for consumption and marketing. Young bean fruits can be harvested frequently from 3–4 months after sowing when they are 10–15 cm long, before they swell and become fibrous and tough. The marketable pods are available from 68 to 74 days in case of early type, whereas 110 to 120 days for pole types. Mature seed can be harvested after 5–10 months (FAO, 2012). As the fruits shatter their seeds when ripe, harvesting should be done timely.

Yields of green pod can be up to 40 q/ha. Forage yields of up to 600 q/ha have been reported. Seed yields of up to 54 q/ha are possible, but a seed yield of 15 q/ha is more common. Xia *et al.* (2017) obtained 20.6 q/ha seed yield of the sword bean in Poland. It produces about 2.67 to 3.14 kg green pod/plant in case of early flowering type and about 2.92 to 3.50 kg green pod/plant in case of pole type, whereas for production of seed, harvesting is done only when pods are fully mature. At full maturity, the pods become yellow in colour. Pods are harvested and seed is extracted manually by breaking the pod.

7.0 Diseases and Pests

Sword bean is fairly resistant to diseases and pests. The most serious fungal disease is root rot caused by *Colletotrichum lindemuthianum*. Sword bean is a host of tomato spotted wilt virus (TSWV). *Canavalia* is known to reduce nematode populations. Incorporating jack bean dry matter into the soil has reduced root galling in tomatoes and increased tomato plant height and weight (Morris and Walker, 2002). Jack bean has been successfully used to increase yield and to suppress nematode (*Pratylenchus zeae*) populations and root necrosis when intercropped with maize (Arim *et al.*, 2006). However, McIntyre *et al.* (2001) has shown that intercropping banana with jack bean had no beneficial effect on nematode populations. However, it is susceptible to the soybean cyst nematode (*Heterodera glycines*) that has not yet been recorded in Africa. Major pests are fall army worm (*Spodoptera frugiperda*), pod borer and beetle grubs that bore into the stems (Grubben, 1977). However, hairy caterpillar (*Anisecta* sp.) create problems during early stage of the plant, which can be controlled by spraying chlorpyriphos @ 2.5 ml/l. Sword bean seeds are fairly resistant to storage pests.

8.0 Crop Improvement

Worldwide collection of *Canavalia* germplasm is urgently needed. Small germplasm collections of *Canavalia* are maintained at the Australian Tropical Crops and Forages Genetic Resources Centre, Biloela, Queensland. A few accessions are available in Ethiopia (ILRI), Nigeria (IITA), South Africa, Brazil, China, Colombia and India.

Sword bean is not known from the wild and must have undergone selection during centuries. Selection has favoured increased pod and seed size but has not resulted in a reduction in biochemical toxins. This would be consistent with selection for use as fodder or as a green fruit vegetable rather than as a pulse crop. Breeding is difficult as the flowers are very sensitive to damage during emasculation and emasculated flowers usually abscise; therefore bud pollination is recommended. In South-East Asia sword bean cultivars have been developed with reduced toxicity. Hybrids of *Canavalia gladiata* with both *Canavalia africana* and *Canavalia ensiformis* have occurred from natural crosses. Breeding programmes should use this wide base of germplasm.

An attempt has been made by screening 15 genotypes of jack bean collected from NBPGR, India to assess genetic variability of 15 yield component traits and to identify important selection indices (Lenkala *et al.*, 2015). They recorded high genetic variability, high heritability in conjunction with higher genetic advance for the characters plant height at first harvest, number of primaries per plant at first harvest, plant height at last harvest, number of primaries per plant at last harvest, pod weight, number of pods per plant, number of seeds per pod, 100 seed weight and pod yield per plant indicating the predominance of additive gene action on the expression of these traits and hence direct selection will be rewarding for improvement of these traits in Jack bean. This complex set of association between various yield components clearly indicated that selection of early genotypes would be reliable to increase the pod length, pod weight, number of pods and ultimately pod yield per plant.

To increase the use of *Canavalia* green fruits and young seeds as a vegetable in tropical world, improved cultivars should be made available either by introducing Asian cultivars or by breeding. There has been limited cultivar development of this species. Most plant development has been performed by agricultural experiment stations and has focused on selecting cultivars with low toxicity (NAS/NRC, 1979). Both viny and bushy varieties exist. Major limitations for increased use of the dry seeds of *Canavalia* in human nutrition are the poor taste, the unappealing texture and antinutritional factors that make laborious preparation necessary. In these respects *Canavalia* faces the same acceptability problems in Africa as soya bean. However, as *Canavalia* is a very tough and resilient crop it could play a larger role if seeds are produced in quantity and processed on an industrial scale. Breeding and selection could play a role in developing cultivars with reduced toxicity.

9.0 References

Abitogun, A.S. and Olasehinde, E.F. (2012) *IOSR J. Appl. Chem.*, **6**: 36-40.

Anonymous (1950) *The Wealth of India*, Vol. II, Council of Scientific and Industrial Research, New Delhi p.56.

Arim, O.J., Waceke, J.W., Waudo, S.W., and Kimenju, J.W. (2006) *Plant Soil*, **284**: 243–251.

Arun, A.B., Beena, K.R., Raviraja, N.S. and Sridhar, K.R. (1999) *Curr. Sci.*, **77**: 19–21.

Arun, A.B., Sridhar, K.R., Raviraja, N.S., Schmidt, E. and Jung, K. (2003) *Plant Foods for Human Nutrition*, **58**: 1–13.

Black, M., Bewley, J.D. and Halmer, P. (2006) *The Encyclopedia of Seeds. Science, Technology and Uses. CAB International, London, UK*, p. 696.

Bressani, R., Brenes, R.S., Garcia, A. and Elias, L.G. (1987) *J. Sci. Food Agric.*, **40**: 17-23

Bunch, R.and staff, E.C.H.O. (1985) http://people.umass.edu/~psoil370/ Syllabusfiles/Green_Manure_Crops.pdf.

Caamal-Maldonado, J.A., Jimenez-Osornio, J.J., TorresBarragan, A. and Anaya, A.L. (2001) *Agron. J.*, **93**: 27–36.

Carlini, C.R. and Udedibie, A.B. (1997) *J. Agric. Food Chem.*, **45**: 4 372-4377.

Chen, H., Feng, Z.G. and Masayuki, Y. (2001). *Zhejiang Science and Technology Publishing House, Zhejiang Province, China*, **40**: 147-159.

CSIR. (1950) *The wealth of India, CSIR, New Delhi*, Vol. II, p. 56.

Duranti, M. (2006) *Fitoterapia*, **77**: 67-82.

FAO. (2012) *http://www.fao.org/ag/AGP/AGPC/doc/Gbase/DAT A/PF000012.HTM*.

Fischler, M. and Wortmann, C.S. (1999) *Agrofor. Sys.*, **47**: 123–138.

Florentin, M.A., Penalva, M., Calegari, A. and Derpsch, R. (2004) *http://www.fidafrique. net/IMG/pdf/NoTill_SmProp_Chptr_1AF912.pdf*.

Grubben, G.J.H. (1977) *Tropical vegetables and their genetic resources. IBPGR, Rome, Italy*, p. 191.

Indira, R. and Peter, K.V. (1988) *Jack bean, under exploited tropical vegetable*, KAU, Kerala, pp. 46-48.

Justo, A.T., Rutilo, C.M. and Lario, A.S. (1994). *J. Sci. Food Agric.*, **66**: 373-379.

Lenkala P., Rani K.R., Sivaraj. N., Reddy. K.R. and Prada, M.J. (2015) *Electron. J. Plant Breed.*, **6**: 625-629.

Li, N., Li, X., Feng, Z.G. and Masayuki, Y. (2007) *J. Shenyang Pharma. Uni.*, **24**: 676-678.

Luo, S.F., Hu, L.H., Chen, Y.Y. and Li, P.X. (2015) *Food Sci.*, **17**: 1-8.

Marimuthu, M. and Gurumoorthi, P. (2013) *J. Chem. Pharma. Res.*, **5**: 221-225.

Martinez-Palacios, C.A., Cruz, R.G., Olvera Novoa, M.A. and Chavez-Martinez, C. (1987) *Aquaculture*, **68**: 165–175.

McIntyre, B.D., Gold, C.S., Kashaija, I.N., Ssali, H., Night, G. and Bwamiki, D.P. (2001) *Fertil. Soils*, **34**: 342–348.

Mendoza EMI, Barrago, C.F., Laurena, A.C., Rodriguez, F.M., Raville Za, M.J.R., Sembrano; A.G.I. and Villena, E.T. (1986) *Sam-Philippine Crop Science*, **5**: 57.

Mendoza, EMI (1990) *Survey of use of seven Philippine indigenous legumes: a preliminary report*, Inst. Plant Breeding, Philippine Univ., Laguna (Philippines).

Mitre, V. (1991) In: Jain, S.K. (Ed.). *Contribution to ethnobotany of India. Scientific publishers, Jodhpur, India*. pp. 37-58.

Morris, J.B. and Walker, J.T. (2002) *J. Nemat.*, **34**: 358–361.

Moteetee, A.N. (2016) *South Afr. J. Bot.*, **103**: 6-16.

Nakaaya, K., Jakaya, O.S. and Hasheem, M. (2017) *Int. J. Agric. Food Sec.*, **3**: 039-049.

NAS/NRC. (1979) *National Academy of Sciences and the National Research Council, Washington, DC*.

Parthasarathy, V.A. and Singh, B. (1991) *Natl. Sem. on resource management for hill agriculture* 1991 p. 23. Pub. By Secy. IAHF, Barapani.

Patel, R., Singh, R.K.R., Tyagi, V., Mallesha and Raju, P.S. (2016) *Int. J. Bot. Studies.*, **1**: 18-21

Pereira, B.F.F, de Abreu, C.A., Herpin, U., de Abreu, M.F. and Berton, R.S. (2010) *Sci. Agric. (Piracicaba, Braz.)*, **67**: 308–318.

Rubatzky, V.E. and Yamaguchi, M. (1997) *World vegetables: principles, production, and nutritive values. Chapman and Hall, New York*.

Seena, S. and Sridhar, K.R. (2004) *Int. J. Food Sci.*, **55**: 615-625.

Seena, S., Sridhar, K.R. and Jung, K. (2005) *Food Chem.*, **92**: 465–472.

Singh, S.P. and Yadav, D.S. (1991) *Natl. Sem. on resource management for hill agriculture*, IAHF, Barapani. p. 83.

Udedibie, A.B.I. and Carlini, C.R. (1998) *Anim. Feed Sci. Technol.*, **74**: 179-184.

Ustimenko-Bakumovsky, G.V. (1982) *Plant growing in the tropics and sub-tropics*, translated by M.K. Viktorova, Mir Publishers, Moscow, pp. 145-46,

Xia, X., Yin, R., He, W., Górna, B. and Hołubowicz, R. (2017) *Not. Bot. Horti. Agrobo*, **45**: 561-568.

Yang, X.H. and Ma, S.B. (2010) *Vegetables*, **4**: 19.

Zhang, H.Y. and Yang, S.S. (2007) *Crops*, **6**: 38-39.

Zhu, J.B. and Wang, H.X. (2002) *Chinese Wild Plants Res.*, **3**: 17-19.

⑧

YAM BEAN

S.K. Mukhopadhyay, H. Sen,
R. Nath and A. Chattopadhyay

1.0 Introduction

The Yam bean which subsumes various members of the genus *Pachyrhizus* is a leguminous root crop of commercial importance in tropical and subtropical countries of the world. Yam bean is grown and yields abundantly over a wide range of environmental conditions, including land of low fertility with low inputs and in near drought condition. As a legume crop it returns to the soil a substantial amount of fixed nitrogen if the vegetative above ground parts are left in the field and thus forms an integral part of sustainable land-use system. Yam bean or jicama (*Pachyrrhizus erosus*), jacatupe bean (*P. tuberosus*) and Andean yam bean (*P. ahipa*) are leguminous tuber crops of commercial importance. All the three cultivated species are often termed as yam bean. The crop is surprisingly diverse with local given names such as ahipa, ashipa, and chuin in Peru; jicama in Mexico; bunga in the Philippines; bangkoewang in Indonesia; ram-kaseru, sankalu,Yam bean is also called 'potato bean' or 'Mexican turnip' in English, 'Jicama' in Spanish, and 'doushÔ and liangshÔ' in China. In India, the most popular name in Hindi is 'Misrikand' and in Bihar, it is called 'Kesaru' while in Eastern Uttar Pradesh as 'Kesaru' or 'Rain Kesaru'. It is commonly called 'Shankalu' or 'Sankesh alu' in West Bengal, Assam and Orissa. The edible *Pachyrrhizus* shares a unique combination of the general qualities present in most cultivated legumes, which makes them more attractive to the consumers, the producers and the environment for the following reasons:

1. Good adaptability to a wide climatic and edaphic range,
2. Yield reliability of the root/tuber,
3. Well balanced and nutritious composition of their protein/starch contents,
4. Agreeable taste,
5. Good postharvest/storage characteristics,
6. Biological nitrogen fixation (sustainability), and
7. Low fuelwood demand (most cultivars produce tubers that are consumed fresh).

2.0 Composition and Uses

2.1.0 Composition

The young tubers have a crisp, juicy and refreshing flesh and can be eaten raw or cooked. They can also be sliced and made into chips. Analysis of peeled tubers gave the following values in Table 1.

Table 1: Composition of Yam Bean Tuber

Moisture	82.38%	Copper	10.43 mg
Protein	1.47 g	Fe	1.13 mg
Fat	0.09 g	Ca	16.0 mg
Starch	9.72 g	Thiamine	0.05 mg

Reducing sugar	2.17 g	Riboflavin	0.02 mg
Non-reducing sugar	3.03 g	Niacin	0.2 mg
Fibre	0.64 g	Ascorbic acid	14 mg
Ash	0.5 g		

*Sorensen (1996)

The tubers contain adenine, arginine, choline and phytin. The juicy flesh is rich in ascorbic acid and has an antiscorbutic action on rats.

Analysis of the edible portion (90 per cent) of young pods gave the following values in Table 2.

Table 2: Composition of Yam Bean Pod

Moisture	86.4%	Ca	221 mg
Protein	2.6 g	P	39 mg
Fat	0.3 g	Fe	1.3 mg
Carbohydrate	10.0 g	Vitamin A	575 IU
Fibre	2.9 g	Thiamine	0.11 mg
Ash	0.7 g	Riboflavin	0.09 mg
Niacin	0.8 mg		

*Sorensen (1996)

Tuberous roots of the Mexican yam bean contained large quantities of two acidic glycoproteins which accounted for more than 70 per cent at the total soluble proteins (Gomes *et al.*, 1997). Yam bean seeds are characterized by high oil (from about 20 to 28 per cent) and protein contents (from about 23 to 34 per cent) and the seed oil contains high concentrations of palmitic (from about 25 to 30 per cent of the total fatty acids), oleic (21 to 29 per cent) and linoleic acids (35 to 40 per cent) (Gruneberg *et al.*, 1999).

Pachyrrhizus erosus seeds have high content of protein, lipids, iron and calcium, compared with other legumes. Glutelins constituted the highest protein fraction followed by globulins. Antinutritional substances, tannins, haemagglutinius and trypsin inhibitors are in low concentrations. Seeds are also processed to obtain flour which had good *in vitro*-digestibility, significant reduction in rotenoids and is rich in essential amino acids except methionine (Santos *et al.*, 1996). Seed proteins show an excellent balance of all essential amino acids; albumins contain the highest amount of essential amino acids (Morales Arellano *et al.*, 2001).

2.2.0 Uses

The flesh of the yam bean root is white and has a crisp texture and a slightly sweetish taste. It is commonly eaten raw as fruits and is often included in salad dishes. However, it retains its watery and crispy texture when cooked or boiled. Yam bean can also be sliced and made into chips. The tubers are used in Mexico in a number of different ways: (i) as a fruit – fresh tubers are cut into sticks and sprinkled with lime juice and chilli (these are often sold by street vendors); (ii) as

Uses of Yam Bean

Juice (left), Fries (middle) and Chips (right) from yam bean

Pickles from yam bean

Frozen blanched yam bean

Different Species of Yam Bean

Pachyrhizus erosus

Pachyrhizus ahipa

Pachyrhizus tuberosus

a vegetable - fresh tuber slices are used in various salad dishes; (iii) cooked tubers are used to prepare a soup, on their own or with other vegetables; (iv) tuber slices may be stirfried, or (v) sliced or diced tubers may be preserved in vinegar with onion and chilli and used as a snack with drinks. Martínez (1936) also reported that the sliced pieces might be preserved in vinegar together with the immature legumes. The crop is also used to make flour, juice, baby food, preserved sweets, candy, fermented products and mixed with milk in porridge.

The tubers and leaves are also used as fodder and superior quality starch is obtained from mature tubers. Provided that an optimal method of preserving the crisp texture of the processed tubers can be developed, yam bean tubers may well be marketed as an attractive product to be used in various dishes, and also as a snack (Mudahar and Jen, 1991). Tender tubers are cooked as vegetable or made into pickles and chutney in Latin America. In China, mature dried roots are reported to be used as cooling agent for people suffering from high fever. In Mexico, in addition to the use of tubers and their young pods or beans for human or animal consumption, the dried hay which remains after harvest provides a source of animal fodder. Yam bean tubers stay fresh for a long period but over matured tubers are fibrous and are not suitable for consumption. Tough and fibrous yam bean stem is used for making fish nets in Fiji (Sorensen, 1996). According to Kundu (1969), *P. erosus* tubers are used in the production of high grade flour in India. The young pods can be used as vegetable but they are reported to become poisonous as they become mature, causing diarrhoea, due to irritation of the hairs. Young and immature pods are used as vegetable as a substitute of French bean in Thailand. Nutritionally, these pods may be compared with soybean (Nag *et al.*, 1936). However, mature pods are not recommended for consumption as they contain rotenone, which is toxic to humans, causing diarrhoea. The rotenone itself may also be put to good use. It can be sold as a high-value naturally derived chemical or, using simpler extraction methods, be employed locally as a plant protective agent (Halafihi, 1994). Once both rotenone and oil have been removed, the remaining seed cake has protein content comparable to that of soybean cakes (Cruz, 1950). Powdered seeds are applied against prickly heat and are used in the treatment of skin diseases. They are also used as laxative and vermicide. The tincture made from seeds is used to treat pruritis, mange, head lice and cattle louse. In Cuba, tuber flour is used for treatment of dysentery and haemorrhoids. In China, tubers are consumed to treat fever, inflammation in throat or tonsils and hangovers. The yam bean juice is an effective diuretic and is commonly used in the treatment of nephritis in Brazil. The *P. ahipa* tubers are used for treating lung infection and coughs in Bolivia (Sorensen, 1996).

Presence of an insecticidal compound called rotenone ($C_{23}H_{22}O_6$) is the common generic characteristic of *Pachyrhizus* (Norton, 1943). Both rotenone and the rotenoids may also be used as a piscicide according to Greshof (1893). These ingredients are to be found in the mature seeds but not in toxic amounts in the tuber itself or in any other part of the plant. The leaves contain pachyrhizin which is poisonous to cattle rather than horses. Leaves and stems also contain free steroids. The powdered seeds are useful as insecticide and fish poison in tropical countries. They act as contact poisons and are reported to be highly effective against leaf eating caterpillars on cowpea (Adjahossou and Sogbenon, 1994) and cabbage (Halafihi, 1994). A process has been patended which employs ground seeds (or extracts) in various formulations with pyrethrin, for use in insecticidal sprays and dusts. Yam bean tuber uptakes low amount of arsenic (0.2 µg/g) from soil and can be considered safe to consume when it is grown in arsenic affected soil.

3.0 Origin, Distribution and Taxonomy

Pachyrrhizus is a native of Mexico and Central America from pre-Columbian period and cultivated widely but not intensively in Mexico, Guatemala, El Salvadore and to a limited extent in Hodurus. It is also grown in China, Vietnam, Philippines, Indonesia, Nepal, Bhutan, Burma and the South Pacific Islands with notable success. The crop was probably introduced to south-east Asia via the Philippines. It is also introduced in Tanzania, Senegal, Sierra Leone, Cameroon, Benin and Zaire in Africa. The crop is also grown in Mauritius and French Guiana. It is perennial in nature, as far as underground tubers are concerned but an annual with respect to aerial portion.

In India it is mostly grown in north Bihar and the cultivation has been extended to parts of West Bengal, Assam and Odisha. However, information on area and production of this crop has not yet been documented. Bihar is the largest producing state of yam bean in India from where it is marketed throughout the country. It is becoming a popular crop in the Gangetic alluvial zone of West Bengal due to its higher production potentiality and better market demand.

The genus *Pachyrhizus* comprises five species (Sorensen, 1988) and is placed taxonomically in the subtribe Diocleinae, tribe Phaseolae, within the family Leguminoceae and sub family fabaceae (*Papilionaceae*). Five species of the genus are: *P. erosus* (L.) Urban (Mexican yam bean), *P. ferrugineus* (Piper) Sorensen, *P. panamensis* Clausen, *P. tuberosus* (Lam.) Spreng (Amazonian yam bean) and *P. ahipa* (Wedd.) Parodi (Andean yam bean). Three of these five species (*P. erosus, P. tuberosus* and *P. ahipa*) are cultivated for their edible tubers and *P. erosus* is cultivated most widely throughout the world. *P. ahipa* and *P. tuberosus* have a South American distribution. At present, *P. ahipa* is only recorded as being cultivated, and the crop is grown by small communities located in the subtropical east Andean valleys of Bolivia and northern Argentina. Within *P. tuberosus* there are three distinct cultivar groups: the Ashipa and the Jiquima cultivar groups with low dry matter content of the tubers (below 20 per cent) and the Chuin cultivar group with high dry matter content (above 30 per cent) of the tubers (Sorensen *et al.*, 1997). The Chuin cultivar group is used like cassava. The remaining two species (*P. ferrugineus and P. panamensis*) are only to be found in the wild.

Pachyrrhizus species are twining herbs; leaves pinnately 3-foliate, stipellate; leaflets often angular or sinuately lobed. Flowers violet-blue in clusters on large glandular swellings, papilionaceous; standard broadly obovate, keel incurved, obtuse; pods long, linear. The tuberous roots are often large and used for food and as a source of starch. A number of seed characters are also specific to these species: these include the colour which ranges from olive green to brown or reddish brown and the shape, which is flat, and square to rounded, but never reniform (Sorensen, 1988, 1990).

Cytological understanding is an important parameter to understand the genetic architecture of yam bean. The ploidy level and genome size of two cultivated species of yam bean (*Pachyrhizus erosus* and *P. tuberosus*) were estimated using flow cytometric analysis of young leaf tissue, with propidium iodide as a fluorescent dye (Pati *et al.*, 2019). They analysed six genotypes of *P. erosus* and three genotypes of *P. tuberosus*. Rice (*Oryza sativa* cv. *nipponbare*) and Mung bean (*Vigna radiata* cv. Berken) were used as internal reference standards. Variation of 2C nuclear DNA content among the six *P. erousus* lines was 4.18 per cent, ranging from 1.17 to1.22 pg, whereas only 1.8 per cent variation was observed among the three *P. tuberosus* lines, which ranged from 1.07 to 1.09 pg. Moreover, it was found that the nuclear DNA content of *P. tuberosus* was lower than that of *P. erosus*. The result of the flow cytometric analysis showed that all the species were diploid (2n = 2x). This result will be helpful for yam bean genome sequencing and crop improvement programmes. This result was supported by the chromosome number of yam bean 2n = 2x = 22 (Santayana *et al.*, 2014).

4.0 Species, Types and Cultivars

The important morphological characters of different species of *Pachyrrhizus* are summarized below:

P. erosus

A herbaceous vine with great variation in the outline of the leaflets, from dentate to palmate. The species is defined by the lack of hairs on the petals, the number of flowers (4-11) per lateral inflorescence axis, *i.e.* by complex racemes, and an inflorescence length of 8-45 cm. Furthermore, morphological characters of the legumes (pods), qualitative as well as quantitative, are used to separate the species. Size (6-13 cm × 8-17 mm), reduction of the strigose hairs at maturity and colour (from pale brown to olive-green/brown) are characters specific to the legume of *P. erosus.* A number of seed characters are also specific: these include the colour, which ranges from olive-green to brown or reddish brown, and the shape, which is flat, and square to rounded, but never reniform (Sorensen 1988, 1990).

P. ahipa

It is distinguished morphologically from the other species by being a herbaceous plant with entire leaflets (a few individual plants possessing dentate leaflets have been recorded), with short racemes (48-92 mm) and the general absence of lateral axes, *i.e.* simple racemes. The number of flowers per lateral raceme, if present, is as low as 2-6. The wing and keel petals are usually glabrous, but slightly ciliolated specimens have been seen. The wings curl outwards following anthesis, a feature seen only in *P. ahipa.* The legume is 13-17 cm long and 11-16 mm wide, and almost circular in cross-section when immature, *i.e.* only slightly dorsiventrally compressed. Seeds are black or black and white/cream mottled in colour, kidney shaped, and measure 9 × 10 mm. The 100-seed weight is 29.2 (range 17.3-41.2 g). This species is additionally unique in that both twining/trailing, semi-erect to short bushy erect growth habits are found: *i.e.* both determinate and indeterminate genotypes exist.

Erect genotypes are 15-40 cm tall, semi-erect 30-60 cm, and twining types 60-200 cm long.

P. tuberosus

This species is recognized by the following morphological characters. The largest species have stems of more than 7 m and terminal leaflets of 280 × 260 mm. The legumes are also larger than those of the other species, measuring 255 × 23 mm, and are conspicuously compressed between seeds. The seeds are black, black and white mottled or orange-red in colour, kidney-shaped and 12 × 14 mm in size.

P. ferrugineus

This wild species is the only one in the genus which is evergreen (with the exception of wild accessions of *P. tuberosus*) and where the parts of the plant above ground are perennial (all species have perennial tuberous roots). The root is less tuberous, although plants grown in greenhouses have produced fair-sized tubers (± 0.5 kg). The leaflets are somewhat leathery (subcoriaceous), and they may occasionally be relatively pubescent, with reddish-brown strigose hairs. The wing and keel petals are prominently ciliolate. The seeds are rounded (13 × 13 mm), laterally compressed, and thus are quite distinct from the seed shapes observed in the remaining species, with the exception of the wild populations of *P. tuberosus*. The species is ecologically associated with evergreen to deciduous rain forest with soils low in available phosphorus (2-4 ppm in soil), a remarkable characteristic in a tuberous plant (Sorensen, 1990). The distribution area of this species is mainly along the Atlantic coast of Central America.

P. panamensis

Morphologically, the species is recognized by the white to light brown pilose hairs covering all parts of the plant including the wing and keel petals, and the low number of flowers per lateral raceme (4-7). The seeds are the smallest in the genus, rounded to slightly reniform in shape and measuring 6 × 7 mm. In Benin and Costa Rica, *P. panamensis* has produced fair-sized tubers in the greenhouse as well as in field trials (± 200 g). The species may have originally been distributed all the way from its northernmost present distribution area in Panama to the relatively dry coastal plains of the Ecuadorian province of Guayas.

Twenty one accessions of yam bean (17 *P. erosus*, 2 *P. ahipa* and 2 *P. tuberosus*) were tested in yield trials under short days in Tonga in the South Pacific (Grum *et al.*, 1994). Fresh tuber yield ranged from 5.3 to 72.0 t/ha, dry matter content from 8.0 to 20.8 per cent and dry matter yields from 1.20 to 8.57 t/ha. Crude protein content of peeled tubers ranged from 5.1 to 9.8 per cent of dry matter, yielding 71.540 kg/ha of crude protein.

Significant differences among 6 accessions of *P. tuberosus* and 1 of *P. erosus* were found for tuber yield, roundness, dry matter and soluble solids content of tubers, top yield, dry matter of top and number of tubers developed per plant. Tubers from reproductively pruned plants had a better shape than tubers from control plants, *i.e.*, more regular shaped and smooth tubers (Nielsen *et al.*, 1998).

In Ecuador, Arevalo and Sorensen (1998) evaluated the response of nine accessions of *P. erosus* at the Jardin Tropical (Esmeraldas-Ecuador) with regard to adaptation, yield, weed management, pests and diseases. Highly significant difference among treatments for total yield of tuberous roots, with accessions EC537, EC114 and EC204 showing the highest average values (32.8, 32.3 and 19.5 t/ha, respectively). In another trial, highly significant differences were also found for tuber yield with accessions EC541, EC503 and EC201 presenting the highest averages (44.0, 43.8 and 42.4 t/ha, respectively).

In Venezuela, Espinoza *et al.* (1998) studied the yield and quality of 14 accessions of *P. erosus* and reported significant differences between results for the production of tuberous roots. Accessions EC551, EC201 and EC120 produced 25, 18 and 17 t/ha of tuberous roots, respectively, and values of crude protein varying between 8.7 per cent (EC528) and 14.9 per cent (EC523). In the second experiment, the highest yields were obtained for the accessions EC565, EC525 and EC501 with 59, 26 and 19 t/ha of tuberous roots, respectively, and contents of crude protein varied between 10.4 per cent (TC239) and 17.2 per cent (EC004).

The yield potential of 24 accessions of *P. tuberosus* grown in a humid tropical climate at the island of Vava'u, Tonga, South Pacific was recorded. Tuber yields ranged from 3.3 to 37.4 t/ha. The tubers from the Chuin accessions had the highest dry matter content. Unlike the majority of *P. tuberosus* tubers, Chuin tubers when cooked compare favourably with dasheen-taro rhizomes for flavour (Nielsen *et al.*, 1998).

In Thailand, yam bean introductions and local landraces showed the existence of considerable diversity in yield, quality of plant parts and nutritional value (Ratanadilok *et al.*, 1998).

In Sierra Leone, Bedford *et al.* (2001) observed significant variation between the varieties for tuber fresh weight, leaf and vine fresh weight, pod fresh weight, and flower fresh weight including the percentage germination at 7 and 14 days after sowing. The variances between the replications were homogenous except for tuber fresh weight which showed a significant difference (7.75 per cent). The 3 best tuber yielding varieties were EC 201, EC 550 and EC 114 with tuber yields of 22870, 18520 and 17320 kg/ha, respectively.

Today the Mexican yam bean (*P. erosus*) is known to be cultivated in large regions outside its original distribution area, *e.g.* in Southeast Asia, India and the Pacific (Sorensen, 1990). In India two types of cultivars (Mexican and local) are cultivated. Mexican types are larger in size and attain a diameter of 10-15 cm and weigh up to 1.5-2.0 kg. The Mexican types are less sweet compared to local ones and have a tendency to develop cracks on the tubers. The local types have smaller tubers (200-300 g), moderate to high sweetness, less fibre conical shape, white flesh and are soft with creamy skin. They do not develop cracks on the tubers.

Rajendra Misrikand-1 (RM-1)

An improved cultivar, released by Rajendra Agricultural University, Bihar, India is very popular in Bihar and West Bengal. Its average yield is 400-550 q/ha in 110-140 days (Anon., 1995). The individual tuber weighs 0.6-0.7 kg, is sweeter, comparatively free from cracking with smooth surface, napiform shape, cream coloured tuber skin and white fleshed.

Rajendra Misrikand-2 (RM-2)

This cultivar has also been developed by Rajendra Agricultural University, Bihar, India and is very popular in Bihar. Tuber shape is fusiform, tuber surface is rough with few tuber rings. Seed is brown coloured.

Other promising Mexican line, L-19, gives better yield in Bihar, West Bengal and Orissa (Anon., 2006). Attempts have already been made to develop hybrids of Mexican and local types and some of the lines (DPH-6, 10, 20, 63; EC 100546, L No.3, 8 × 9 were found promising (Anon., 1996, 2018).

5.0 Growth and Development

The developmental stages of the genus *Pachyrrhizus* according to Sorensen *et al.* (1993) are presented in Table 3. The duration of the various stages varies according to species and genotype, growing environment and the cultivation practices. The developmental pattern of *P. erosus* was studied by Fernandez *et al.* (1997) in Mexico. There was an increase in all characteristics of foliage (fresh and dry weight, number of leaves per plant), main stem length, nodes and internodes of the main stem and in all root characteristics (fresh and dry weight, diameter and length). Chemical analysis of roots at different plant ages showed that the range values for dry matter were 16.19-22.28 per cent, protein 1.11-1.62 per cent, fat 0.553-0.867 per cent, crude fibre 0.3048-0.3943 per cent, and ash 0.669-0.89 per cent. The chemical constituents fluctuated with age but without specific trends.

Table 3: Developmental Stages of *Pachyrrhizus*, Based on Morphology and Physiological Changes*

Stage	Description
Germination	Appearance of the primary root
Emergence	Primary leaves and epicotyl emerge
Primary leaves totally opened	Primary leaves fully unfolded
First trifoliate leaf opens	Second trifoliate leaf appears
Third trifoliate leaf opens	Buds at lower nodes produce branches
Pre-flowering	First flower bud or reproductive shoot appears
Flowering	First flower opens
Pod formation and filling	At the end of this stage the pods feel hard
Post-pod growth	Pods filled and are hard
Physiological maturity	Leaves senesce and defoliation starts

* Sorensen (1993).

Venthou-Dumaine and Vaillant (1998) studied the development of *P. tuberosus* (TC355), *P. ahipa* (AC203) and *P. erosus* (EC226) during a 120-day period after sowing (DAS) in 4-litre pots. Flowers buds, flowers and pods appeared first on AC203 (at

Promising Cultivar of Yam Bean

Rajendra Misrikhand-1

40, 60 and 75 DAS), then on EC 226 (at 75, 90 and 110 DAS) and finally on TC355 (at 120, and >120 DAS). The total plant dry matter production after 120 days was similar in EC226 and TC355 and 54 per cent higher than in AC203. Partitioning of dry matter into the various organs was quite different. Leaves represented 28, 20 and 61 per cent of the DM, respectively in EC226, AC203 and TC355. AC203 essentially developed pods (63 per cent of DM at 120 DAS). EC226 started tuber bulking as early as 45 DAS, and stopped bulking at the time of flower and pod development. TC355 mainly developed leaves and roots (78 per cent of the plant DM at 120 DAS) and was late flowering and tuber forming. The DM content of the tap roots were similar in the three species at 45 DAS (about 18 per cent), then it increased to 30 per cent and 20 per cent, respectively, in TC355 and AC203, but decreased to 11 per cent in EC226 at 105 DAS. Although EC226 tuber had the lowest DM content, it had the highest fresh weight at 120 DAS (3.5 times that of TC355).

Environment plays an important role on the growth of the crop and tuber development. Long days, high humidity and high day and night temperatures promote shoot growth at the cost of root growth. Yam beans are short-day plants, *i.e.*, flowering, and tuber production will take place only under decreasing day length. Usually flowering starts when the day length approached 12.5 hours. Tuber formation is promoted by short-day condition and cool nights, therefore, time of sowing is important for faster tuber bulking and shortening crop duration. Comparatively warm and humid climate is needed for the initial vegetative growth but bright sunny days and cool nights are favourable for tuber development. It is not much suited for cultivation in a very wet climate.

Day length sensitivity in *P. erosus* has been studied by several scientists. Paull *et al.* (1988) reporting from experiments conducted under field conditions in Hawaii using local cultivars observed a significant overlap between flowering and tuberization during short days. Also, the field examinations conducted at the INRA experimental station in Guadeloupe FWI have increased knowledge on the response to different planting dates and tuber growth (Robin *et al.*, 1990: long-day; Sorensen *et al.*, 1993; Vaillant and Desfontaines, 1995, Zinsou *et al.*, 1987a, 1987b, 1988; Zinsou and Venthou-Dumaine, 1990). The results from Guadeloupe confirm the findings by Paull *et al.* (1988) thus demonstrating the strong competition between shoot growth, flowering, pod formation and tuber growth. Conversely, during long days, tuber growth was seen to begin after 4-6 weeks, but the vigorous shoot growth had a limiting effect on tuberization. Flowering was first initiated when the day length approaches 12.5 hours. Cotter and Gomez (1979) studied the behaviour of two Mexican cultivars under both short-day (9 hours natural light) and long-day greenhouse conditions (9 hours natural light + 4 hours artificial light). They also observed increased tuberization during short days.

P. ahipa accessions have been observed to have the most rapid flower initiation, *i.e.* from 87 days after sowing (in plants with determinate growth habit) to 140 days (in indeterminate plants), regardless of the season. Furthermore, the tuber growth did not appear to be influenced by variations in day length (Sorensen, 1996); hence this species may for all intents and purposes be regarded as day length neutral, and neutral lines may be bred from interspecific hybrids involving the species.

Alvarenga and Valio (1989) evaluated the effect of different temperature and photoperiodic regimes on the initiation of flowering and tuberous root formation in genotypes belonging to the ashipa cultivar group. They observed that flowering was only induced in this species at intermediate day lengths, *i.e.* 9-16 hours. They stated that the crop might be considered a short-day plant with respect to tuberization, as this process only occurred at photoperiods of less than 16 hours. The day/night temperatures of 30/25°C delayed and reduced flowering, and completely inhibited the formation of the tubers and processes were increased at day/night temperature regimes of 25/20°C and 20/15°C. Faria *et al.* (2000) evaluated the effect of different photoperiods (9, 12, 15 and 18 h) on the morpho-physiological characteristics of *P. tuberosus*. An increase in dry matter weights of tuber roots (average of 76.97 g) from 9.00 to 15.00 h was observed until the 180th day after planting. In contrast, harvest index decreased with progressive increase in photoperiod. Plants cultivated under 9-h photoperiod also recorded the highest chlorophyll contents (chlorophyll a and b). The results indicate that shorter photoperiods result in high yields of *P. tuberosus*.

Application of GA_3 on *P. ahipa* caused an increase in shoot formation and internode length. The dry matter of stem increased to the detriment of tubers, pods and nodules (Peltier *et al.*, 1996).

6.0 Soil and Climate

Yam bean grows well on light, rich sandy or sandy loam soil with good drainage. The normal tuber development is restricted in clay or clay loam soil. The optimum pH range is between 6.0 and 7.0.

Yam bean is grown in tropical and mild subtropical climate with moderate rainfall. It can be grown right from sea level to 100 m. A frost-free condition during vegetative growth is its main climatic requirement. Very wet climate is not favourable for its cultivation. Excess rainfall resulted in low yields and tuber rot (Vaz *et al.*, 1998). Tuber development is favourably influenced when there is an even distribution of rainfall throughout the growing period. Tuberization is adversely affected in low temperature particularly during early vegetative growth and it is delayed if the vegetative growth is extended.

The species *P. erosus* is recorded from 0 to 1750 m a.s.l. (above sea level), with the majority of records from 500 to 900 m a.s.1. The rainfall range varies from 250-500 mm to over 1500 mm m.a.p.r. (mean annual precipitation rate). The optimal day/night temperature range is between 30 and 20°C. In cultivation, well-drained, sandy, alluvial soils within a pH range of 6.0-7.0 are preferred, especially when the crop is irrigated. Srivastava *et al.* (1973) observed that *P. erosus* produced economic yield in the Ecuadorian province of Esmeraldas at precipitation rates of +6000 mm m.a.p.r. At Dakar, Senegal, with 240 mm m.a.p.r., tuber yields between 15 and 34 t/ha were obtained for the species (Annerose and Diouf, 1995a, 1995b).

P. ahipa appears to be well adapted to an altitudinal range of 1800-2600 m a.s.l., though the crop has been recorded at +3000 m a.s.l. in sloping north-facing fields fully exposed to the sun. The region of cultivation is located along the border between the warm ('terra templada') and cold tropics ('terra fria'). The average temperature within the region is 16-18°C. The temperature oscillates between a minimum of 0-5°C and a maximum of 30-35°C. The climate is semi-arid. The average annual precipitation rate is between 400 and 700 mm and occurs within 4-6 months of the year. The remaining months form the dry season. The *P. ahipa* plant will tolerate long dry spells, but in order to increase tuber yield, an additional water supply is essential. Cultivation is predominantly carried out along loamy river banks, although in some cases, sloping hillsides with loamy soil may also be used. A well drained soil type with a pH value of 6-8 will meet the edaphic requirements of the crop (Orting *et al.*, 1996).

For *P. tuberosus* the optimal altitude is 1200 m a.s.l., but cultivation is possible within an altitudinal range of from 550 to 2000 m a.s.1. (Munos Otero, 1945). The precipitation range is 640-5000 mm m.a.p.r., with temperatures varying between 21.3 and 27.4°C and a soil pH range of 4.3-6.8 (Munos Otero, 1945; Duke, 1981; Sorensen, 1990; Sorensen *et al.*, 1996).

7.0 Cultivation

7.1.0 Sowing

Yam bean can be propagated by tubers or seeds but commercially grown through seeds. The seeds are campylotropous and the seedling development is hypogeal. In case of *P. erosus* the seed rate used varies according to the cultivation system and desired tuber size, and seeding rates of 60-240 kg/ha have been recorded in Mexico and Thailand (Ratanadilok and Thanisawanyangkura, 1994). Deshaprabhu (1966) stated that *P. erosus* may be propagated by tubers as well as by seed. The seed rate varies considerably depending on the spacing and a seed rate of 20-60 kg/ha is generally practiced in different parts of India. Normally, the seeds are sown in well prepared soil at a distance of 30 × 30 cm to a depth of 2 cm. In order to obtain medium-size tubers which are preferred by consumers, closer planting at a distance of 15 × 15 cm is suggested. Increase in plant population also results in higher yield. The seeds germinate within 7 days but it may be delayed or rotted due to excess moisture (Sen, 2003). A spacing of 60 cm × 45 cm is also practiced in India when thinning of seedlings is not done. Sometimes 3-4 yam bean seeds are sown on hills of 15 cm high and spaced 0.75-1.0 m apart. In Mexico, the recommended plant density/distance for irrigated fields is in double rows on ridges with 0.75 m between furrows, 0.25 m between rows and 0.15-0.20 m between plants. The seeds should be planted at a depth of ± 30 mm when dry planting and ± 60 mm when irrigated. A seed rate of approximately 35-40 kg/ha is required at this density depending on the cultivar/seed size. In case of *P. ahipa* seeding rates of 21-105 kg/ha have been recorded, but general rates are 40-65 kg/ha. Again, factors such as preferred tuber size, soil fertility and obviously seed weight play a major role in determining the rates (Orting *et al.*, 1996). In *P. tuberosus* Munos Otero (1945) reports either 15 kg/ha when planting at 0.60 m between rows and 0.30 m between plants, or 35-40 kg/ha when broadcast, depending on the cultivation system used. When planted by hand, three seeds are usually sown at each planting station (Munos Otero 1945). Germination is somewhat slow, usually occurring 10-12 days after planting (Munos Otero, 1945). The seed will remain viable for one year at the most under humid conditions and for 2.5-3 years when kept dry (Munos Otero, 1945), but significantly longer when stored under optimal conditions in seed banks.

In India, seeds are sown during post-monsoon period, particularly from end of August to whole of September. Early sowing favours excess vegetative growth with profuse flowering at the expense of root growth. Similarly, late sowing (November-December) inhibits vegetative growth due to lower temperature prevailing during this period and growth and development are adversely affected. Sen *et al.* (1996) observed that higher dry matter production of vines was obtained from September sowings due to favourable temperature and humidity conditions; from October onward dry-matter production decreased due to fall in temperature. Further, crops sown in September and October showed a considerable amount of tuber formation within 45 days, but later sowings failed to produce sufficient tubers due to poor vegetative growth. For seed purpose seeds are sown in June-July at a spacing of 30 cm × 30 cm (Mohankumar *et al.*, 2000; Nath *et al.*, 2008). In Orissa, India, seeds are sown in June-July and the plants are given one or two prunings after about two months in order to restrict vegetative growth and encourage better tuber development. The seed rate varies considerably depending on the spacing and a seed rate of 20-60 kg/ha is generally practised in different parts of India. Normally, the seeds are sown in well prepared soil at a distance of 30 × 30 cm to a depth of 2 cm. In order to obtain medium-size tubers which are preferred by consumers, closer planting at a distance of 15 × 15 cm is suggested. Increase in plant population also results in higher yield. The seeds germinate within 7 days but it may be delayed or rotted due to excess moisture.

For the cultivation of *P. erosus*, in Uttar Pradesh, India the crop should be sown at the beginning of the rainy season, *i.e.* September to October. If grown for starch production, it should be left unpruned, but if the crop is for fresh consumption, the plants should be pruned when 1.5-2 months old and then again after a month. The crop will be ready to harvest in February to March, and the recommended seeding rate is 62-74 kg/ha (Srivastava *et al.*, 1973). In Maharastra, India the plant population when grown on ridges should be 133 000 plants/ha, and the crop is best sown in June-July, with harvest following in December to January. The immature pods/ legumes to be used as a vegetable can be picked continuously once they start to form. If the crop is to be used for fodder, it should be harvested when 50 per cent of the plants are flowering, to ensure maximum yield and optimal nutrition (Bhag Mal and Kawalkar, 1982).

The plant population used in Malaysia is approximately 95 500/ha (with three seeds sown at each planting station, 20-25 kg seed/ha). The crop is trellised and is sown at the beginning of the rainy season; the average yields are 7-10 t/ha, but may be as high as 95 t/ha (Sahadevan, 1987).

Planting methods of *P. erosus* by transplanting or direct sowing influenced root growth. Establishment method did not affect root FW, but root diameter was greater in direct-sown crops at 26 and 45 weeks after sowing/planting. Root system morphology was affected by both plant age and establishment method (Fernandez *et al.*, 1997).

7.2.0 Manuring and Fertilization

Organic matter improves the fertility and physical condition of soil which favourably influences the tuber bulking and quality of tuber. Investigations on the nutrient requirement of yam bean revealed that treatment with N, P and K promote vegetative growth and early tuber development. An experiment conducted in Bihar, India showed that increasing rates of N and K_2O increased tuber yield progressively and the crop responded favourably up to 80 kg N and 120 kg K/ha (Anon., 1978). Under West Bengal situation, the maximum tuber yield was obtained with 120 kg N and 80 kg K_2O/ha along with a normal dose of 60 kg P/ha (Sen and Mukhopadhyay, 1989). The AICRP on Tuber Crops at Rajendra Agricultural University, Bihar, India has recommended the application of 15-20 tonne FYM or compost along with 80: 40: 80 kg NPK/ha (Mohankumar *et al.*, 2000). They also observed that fertilizer application had practically no prominent influence upon TSS content of the tubers. Ramaswami *et al.* (1980) suggested a fertilizer dose of 80: 60: 80 kg N, P_2O_5 and K_2O/ha in Tamil Nadu, India. Dry matter accumulation was higher with higher rates of N and K but P showed no effect. However, Roy *et al.* (1976) observed that dry matter content of tuber was not affected due to fertilizer treatment. The TSS percentage of tuber at harvest decreased with the increase in fertilizer doses. Half of N and K and whole of P should be applied at the time of final land preparation and the remaining half of both N and K are to be top dressed at 35 DAS should be increased the fertilizer efficiency under irrigated condition. However, it is advised to apply whole of N, P and K (lower doses) at the time of sowing under rainfed situation. Mishra *et al.* (1994) summarised that application of 150 or 200 kg K gave similar tuber yields which were greater than with 100 kg K. Split application of K gave higher market grade tuber yield than basal application. The highest tuber yield was obtained from applying K in 3 equal splits (sowing + 30 and 60 days after planting). Cracking in tubers was significantly reduced at higher K rates. Nath *et al.* (2007) concluded that yam bean responded well to nitrogen application and 120 kg N/ha was found to be optimum for both tuber and seed production. However, seed production of yam bean was more profitable as compared to tuber production in red and laterite ecosystem of West Bengal, India. In Muzaffarpur, Bihar, India, influence of graded levels (100, 150 and 200 kg/ha) of K and its methods of application on market grade tuber yield, cracking behaviour and chemical constituents (dry matter, crude protein, starch and sugar) of yam bean (*P. erosus*) tubers. K significantly increased the market grade tuber yield from 31.98 t/ha at 100 kg K_2O/ha to 36.04 t/ha at 200 kg K_2O/ha. The application of K in 3 split doses resulted in the highest market grade tuber yield (38.04 t/ha) compared to all-basal application (27.93 t/ha). Higher levels of K_2O significantly reduced the cracking in tubers. Neither the K levels nor the application methods could affect the quality constituents of the tubers. Nersekar *et al.* (2018) obtained the maximum protein content of tuber at a spacing of 60 × 20 cm under fertilizer dose 80: 40: 80 kg NPK/ha. Closer spacing and graded levels of the potash promote physico-chemical parameters of yam bean tubers (Mali *et al.*, 2018). Laxminarayana (2017) observed that conjunctive use of organic manure and soil test–based inorganic fertilizers not only enhanced the crop productivity and biochemical constituents of yam bean but also improved the microbial activities and other soil quality parameters. Soil and crop management systems enhanced the organic matter in the soil surface and provided a source of readily available carbon substrates that could support increased microbial activity. Dehydrogenase and acid phosphatase activities were highly influenced due to integrated application of lime, FYM, and balanced chemical fertilizers, and these enzymes had greater contribution towards the yield and proximate composition of yam bean under acidic Alfisols.

In a greenhouse experiment in Brazil, applications of K and Mg to yam bean at 120 kg K and 60 kg MgO/ha significantly increased nodule dry matter production but had no effects on plant dry matter yield and nitrogen content (Figueiredo *et al.*, 1996).

7.2.1 Nitrogen Fixation

The genus *Pachyrrhizus* has an efficient symbiosis with nitrogen-fixing *Rhizobium* and *Bradyrhizobium* bacteria. These bacteria provide the plants with a source of nitrogen and, as a result, there is no need for an additional supply of nitrogen fertilizer (Sorensen, 1996). However, Mondal (1993) stated that yam bean responded well to N-fertilization because of its poor or ineffective nodulation. In contrast with many of the grain legumes, a substantial amount of the fixed nitrogen is returned to the soil if the vegetative aboveground parts are left in the field. The crop therefore forms an integral part of a sustainable land-use system, from both an ecological and a socioeconomic standpoint. The plants can be grown as green manure with advantage because grazing animals eat it most unwillingly. Kjaer (1992), Halafihi (1994) and Halafihi *et al.* (1994) examined the efficiency of the biological nitrogen fixation of yam bean under both greenhouse and field conditions.

Castellanos *et al.* (1997) conducted the first field test quantifying the actual amount of nitrogen fixed by two accessions of *P. ahipa* (58-80 kg N/ha) and three cultivars of *P. erosus* (162-215 kg N/ha). Approximately 50 per cent of the N harvested, *i.e.* ±130 kg/ha, or close to 800 kg protein/ha was accumulated in the tuberous roots in *P. erosus*. This is a value which equals or outyields the amount of protein harvested in grain legumes. The amount of N recorded in the residue (hay) of *P. erosus* was 120-150 kg/ha, twice the amount recorded in the *P. ahipa* residue and was higher than the quantity recorded in practically all grain legumes (The plant population of both species in the trials were 110 000 plants/ha and the plants were reproductively pruned).

Vansuyt and Zinsou (1989) reported that the plant formed effective nodules with indigenous strains of *Bradyrhizobium* in neutral to slightly alkaline soils (pH 6.9 to 8.4) but not in very acid soils (pH 4.6). Lynd and Purcino (1987) found that adequate amounts of P, Ca and K increased the nodule mass, nitrogenase activity and tuber yields. In other studies conducted in *P. ahipa* under glasshouse conditions, Kjaer (1992) reported that the symbiotic effectiveness of the *P. ahipa-Bradyrhizobium* association was high, as the profuse nodulation provided the inoculated plants with adequate amounts of N. Woomer (1979) tested 23 strains of *Bradyrhizobium* and found two with high efficiency, yielding more total N than the treatment in which nitrates were supplied. Tamez (1987) and Halafihi *et al.* (1994) reported that when adequate amounts of *Bradyrhizobium* were present in the soil, seed inoculation of *P. erosus* was not necessary to produce nodules to fix N required by the crop.

The pattern of N-fixation during the growth cycle was summarized by Kjaer and Sorensen (1992): i. N-fixation remained high till anthesis, ii. Pod formation caused a decline in N-fixation, iii. N-fixation reached absolute minimum during the pod filling stage and iv. N-fixation like vegetative growth including tuber growth, increased again following pod maturity. In *P. erosus* and *P. ahipa*, growth of tubers was delayed until pod maturity had been attained (Zinsou and Venthou-Dumaine, 1990). This finding confirms the need of reproductive pruning in order to obtain increased yield as reported by Heredia (1971) and Noda and Kerr (1983).

A study of symbiosis over all five species of *Pachyrhizus* revealed that nodules were found on about two thirds of the *Pachyrhizus* accessions collected and 75 per cent of the strain belonged to the genus *Bradyrhizobium* and 25 per cent to genus *rhizobium* (Grum and Sorensen, 1998), and They observed that the best mycorrhizal strains caused more than double plant growth on nutrient poor soils. Nitrogen fixation by inoculation of *Bradyrhizobium* and urea application and the role of P was studied by Cruz *et al.* (1997). All P fertilizers increased dry matter production, number and weight of nodules, nitrogen activity and P and N uptake. Of the P fertilizers tested for smag thermophosphate (FPT) combined with *Bradyrhizobium* inoculation showed the best results, similar to single superphosphate plus 200 kg N/ha. Urea applied up to 200 kg N/ha did not appear to inhibit nodulation and nitrogenase activity on yam bean. Nitrogen application had no effect on dry matter production but inhibited nodulation in the absence of P fertilizers. It is concluded that there is no need to add mineral nitrogen to yam bean when plants are inoculated with selected *Bradyrhizobium* strains. *Bradyrhizobium* (yb-10) along with phosphate solubilizing bacteria (*Pseudomonas striata*) recorded higher nodular number, plant fresh weight, nodular dry weight, plant dry weight and tuber weight per plant of yam bean than other treatments comprising *Bradyrhizobium* (yb-10) with potash mobilizer (*Fraturia aurantea*) or *Bradyrhizobium* (yb-10) alone. N, P and K @ 20: 40: 30 kg/ha as basal dose was used in all the treatments.

A total of 25 isolates from root nodules of yam bean (*Pachyrhizus erosus* L. Urban), were characterized by Fuentes *et al.* (2002). All isolates formed effective nodules mainly on lateral roots while edible tubers were developed on the taproot. The root nodules formed were identified as the typical determinate type. By an analysis of the partial sequences of the 16S rRNA gene (approximately 300 bp) of 10 strains which were selected randomly, the isolated root nodule bacteria of yam bean were classified into two different genera, *Rhizobium* and *Bradyrhizobium*. Two strains, YB2 (*Bradyrhizobium* group) and YB4 (*Rhizobium* group) were selected and used for further analyses. The generation time of each strain was shown to be 22.5 h for strain YB2 and 0.8 h for strain YB4, respectively. Differences between strains YB2 and YB4 were also reflected in the bacteroid state in the symbiosome. Symbiosome in nodule cells for the strain YB4 contained one bacteroid cell in a peribacteroid membrane, whereas a symbiosome for strain YB2 contained several bacteroid cells.

Castellanos-Ramos *et al.* (2009) observed that *P. erosus* reached the highest number of nodules at 170 DAP, while *P. ahipa* at 123 DAP. *P. ahipa* had a higher number of nodules than *P. erosus*, but of smaller size, thus, nodule dry mass was significantly higher ($p<0.05$) in *P. erosus* in most sampling dates.

Grum *et al.* (1998) observed that the best mycorrhizal strains caused more than double plant growth on nutrient poor soils.

7.2.1.1 *Flower Pruning and Nitrogen Fixation*

In an investigation on flower pruning in *P. ahipa* and *P. erosus* on symbiotic nitrogen fixation, Castellanos *et al.* (1997) concluded that the amounts of nitrogen fixed ranged from 58 to 80 kg N/ha for *P. ahipa* and from 163 to 216 kg N/ha for *P. erosus*, and this variable was not significantly affected by flower pruning treatments, even though with flower pruning tubers were harvested earlier. With pruning of flowers the amount of N in the residue ranged from 55 to 151 kg N/ha and their N concentration ranged from 32 to 35 g/kg. They also reported that the percentage of N derived from the atmosphere (Ndfa) was estimated to range from 55 to 69 per cent for *P. ahipa* and from 68 to 77 per cent for *P. erosus*, while the amount of N fixed ranged from 74 to 95 kg N/ha and from 172 to 190 kg N/ha, respectively, for these species. Positive net N balances in the system ranged from 12 to 18 kg N/ha for *P. ahipa* and from 55 to 81 kg N/ha for *P. erosus*.

Badillo *et al.* (1998) observed that flower pruning increased the vegetative cycle of the plant without changing the quantity of N fixation. With flower pruning, N fixation varied from 55 to 70 per cent (74 to 96 kg/ha) for *P. ahipa* and from 70 to 77 per cent (163 to 203 kg/ha) for *P. erosus*. The absolute quantity of N fixation for *P. erosus* was 228 kg/ha. Vaz *et al.* (1998) reported that good yields of *P. ahipa* needed specific *Bradyrhizobium* inoculum, while application of N fertilizer interfered with biological nitrogen fixation.

7.3.0 Irrigation

Normally, yam bean is grown as a rainfed crop but application of one or two irrigations particularly in the drier months (November-December) promotes tuber development. Frequent irrigation with higher doses of fertilizers makes the tubers more succulent and reduces the keeping quality.

Bedford *et al.* (2001) observed that leaf relative water content in stressed plants during flowering (before pod formation) varied between 75 and 84 per cent in EC 117 and EC 201 and between 81-91 per cent in their controls. Leaf water potential for pruned plants were not affected by water stress. The total pod yield in stressed plants was 67 and 59 per cent lower than that produced in the controls of EC 117 and EC 201, respectively. In contrast, stress plants exhibited a higher total tuber production. Tuber yield increases were 41 and 48 per cent in EC 117 and EC 201, respectively.

Investigation on physiological parameters, such as relative water content, water potential and stomatal conductance, were measured on leaves of two genotypes of *P. erosus* and one genotype of *P. tuberosus* under water stressed condition by withholding watering after tuberous roots had developed. There were also differences in development of stomatal conductance under water stress (Adjhosson and Ade, 1998a).

Mulching is also practiced for conserving soil moisture under rainfed cultivation of yam bean. Paddy straw mulching recorded higher total dry matter production of yam bean than one, two or three irrigations (Jana, 2005). From the very initial stage, the crop growth rate (CGR) was appreciably high in mulching. Growth attributes increased when the crop received three irrigations at 30, 60 and 90 DAS. Paddy straw mulching encouraged number of branches/plant, crop growth rate, total dry matter production, tuber bulking rate and average tuber weight followed by single irrigation given at 30 DAS. Straw mulch recorded the highest tuber yield (26.3 t/ha), which was 26.4 per cent more than rainfed situation, but the tuber yield decreased by 22.2 per cent with increased number of irrigations (3) at 30, 60 and 90 DAS. Single irrigation at 30 DAS increased the tuber production by 24.2 per cent over rainfed situation. Harvest index and sugar content of tuber was found maximum in rainfed cultivation.

7.4.0 Interculture, Intercropping and Reproductive Pruning

One weeding is necessary at the initial stage to check the weeds and promote the crop growth otherwise the tuber development is markedly reduced. Crop growth and the canopy that developed after one weeding and successive top dressing of nutrients at 35 days prevent weed growth at the later stages. Mulching with paddy straw is also suggested for suppression of weeds (Jana, 2005). Earthing up is not essential like other tuber crops.

In Mexico, the traditional intercropping involving maize and bean produces yields of approximately 0.5 t/ha of bean harvested at 60 days after sowing, 1.0 t/ha of maize harvested at 110-120 days and 35- 45 t/ha of yam bean harvested at 140-150 days after planting. This cultivation method produces sufficient beans and maize for tortillas to feed a family of 6-8 persons for a year, and the sale of the yam beans constitutes a major source of income. One important aspect to be kept in mind when studying the different cultivation practices in Mexico and Central America is that the yam bean crop is never fertilized. In Mexico, the crop is grown in the same field for two consecutive seasons, producing a higher yield in the second than in the first. In the third season maize, common bean or onion (*Allium cepa* L. var. *cepa*) are grown, as, yam bean does not perform well, owing to the build-up of insect and nematode pests. The crop is next planted in the field again after a break of 3-4 years. The average plant population used in the irrigated crops in Nayarit and Guanajuato is 110 000 plants/ha (Heredia and Heredia, 1994; Castellanos *et al.*, 1997). In a field study in Costa Rica, yam beans (*P. erosus*) cv. EC041 and EC509 were grown in pure stands or intercropped with cassava (Morera *et al.*, 1998). Increases in fresh and dry weight of the tuberous roots were larger in a pure stand than when intercropped, giving a maximum difference between systems at 90-135 days after sowing. Shade from the cassava plants had a negative effect on the growth and development of yam beans, especially on the neighbouring ridges.

In China, *P. erosus* or soya bean (*Glycine max* (L.) Merr.) is usually planted on the ridges between rice paddy fields. This practice is unknown in Thailand, where these ridges are usually kept cleared as a precaution against rats. In Amazon region, yam bean is always intercropped in slash-and-burn fields, and are typically mixed with crops such as plantain (*Musa paradisiaca* L.), maniac and pineapple (*Ananas comosus* (L.) Merrill.), *etc.* Fields are rotated and left for fallowing after a few years' cultivation. It is common practice to sow a few ashipa seeds together at about 5 cm depth on hills 3 m apart. The plants are often trellised on 2-m poles or alternatively allowed to climb the intercropped manioc plants. In Equador, according to local sources, monocropping was and still is a common practice, but intercropping with chilli (*Capsicum* spp.), sesame (*Sesamum indicum* L.), ground nut (*Arachis hypogaea* L.) and tomato (*Lycopersicon esculentum* Miller) is also practised. The jiquima is cultivated in small back gardens each containing between 60 and 100 plants in plots of 250-400 m², *i.e.* a plant density of 1500-2400 plants/ha when intercropped. Approximately 70-90 per cent of the plants from these plots are harvested when the tubers have reached a marketable size.

In West Bengal, India intercropping yam bean with pigeonpea in 3:1 proportion proved to be remunerative which computed the highest net return with a benefit cost ratio of 4.62 (Panda *et al.*, 2003). In north Bihar, yam bean is intercropped with maize, where maize plants utilized as trailing. But in other parts of India, normally trailing is not adopted for growing of yam bean.

Panda *et al.* (2003) also reported that cowpea (fodder)-jute-yam bean, maize-rice-yam bean, sesame-rice-yam bean, green gram-upland *taro*-yam bean, green gram-elephant foot yam-yam bean and groundnut-rice-yam bean crop sequences were feasible and the yields of yam bean were not affected by the systems. The sequences *viz.* cowpea (fodder)-jute-yam bean and groundnut-rice-yam bean improved soil nutrient status whereas green gram-elephant foot yam-yam bean, sesame-rice-yam bean and Green gram-upland *taro*-yam bean were proved to be exhaustive with respect to residual fertility status of the soil.

Field experiments were conducted in West Bengal, India to evaluate the effects of NK application on the productivity of yam bean (*Pachyrhizus erosus*)-pigeonpea (*Cajanus cajan*) intercropping system and its residual effect on the succeeding mung (*Vigna radiata*). Marketable tuber yield of yam bean increased linearly with increasing NK levels, with the highest being recorded with NK at 80 kg/ha applied in 2 splits (22.94 t/ha) closely followed by 100 kg NK/ha applied in 2 splits (22.42 t/ha). For pigeonpea, the maximum grain (14.38 q/ha) was recorded with 80 kg NK/ha applied in 2 splits. The highest level of NK (100 kg/ha) applied in 3 splits to yam bean-pigeonpea intercropping system registered the maximum grain yield of the succeeding mungbean (9.43 q/ha), which was 33 per cent higher than the untreated control (Panda *et al.*, 2003).

Reproductive pruning is necessary to obtain maximum yields. Prasad and Prakash (1973) observed that flowering in yam bean commenced from 58-68 days after sowing and lasted between 92-103 days. If the flowers were allowed to develop and formed pods, the tuber yield was drastically reduced. Caro *et al.* (1998) reported that in the yam bean production zone in Nayarit, removal of flowers is practised to increase tuber production by diverting photosynthates from flowers to roots. Mohankumar *et al.* (2000) observed that there was significant negative correlation between tuber yield and pod formation. As has been demonstrated in numerous field trials, the effect of the operation depends on the cultivar, the season used (whether long or short day) and the climate. In some areas, not only are the reproductive shoots removed, but the top half of the vegetative part is also pruned. In most Mexican

P. erosus cultivars, two reproductive pruning will suffice to ensure a good yield. When examining genotypic and environmental responses to reproductive pruning in 32 accessions of *P. erosus*, it was revealed that reproductive pruning increased tuber yield uniformly across accessions. Furthermore, although the accessions differed in tuber shape, soluble sugar and dry matter tuber content, reproductive pruning did not have any influence on these quality traits (Grum *et al.*, 1996).

Vaz *et al.* (1998) reported that flower pruning resulted in a nearly 100 per cent increase in tuber yield as compared to the tuber yield of non pruned plants. Nielsen *et al.* (1999) concluded that reproductive pruning gave an overall increase in tuber yield of 51.5 per cent in camparison with the control (*P tuberosus*: 29.2 per cent and *P. erosus*: 102.1 per cent). Vegetative pruning had no effect on tuber yield. Significant differences among accessions were found for tuber yield, roundness, dry matter and soluble solid content of tubers, top yield, dry matter of top, and the number of tubers developed per plant. Tubers from reproductively pruned plants had more regularly shaped and smooth tubers than from control plants. Martinez (1936) stated that increase of up to seven times the yield over non-reproductively pruned plants had been obtained in Mexico. Castellanos *et al.* (1997) reported yield increased from 140 to 340 per cent when testing the three highest yielding Mexican cultivars, whereas Grum *et al.* (1996) recorded yield increase from 7 to 39 per cent when testing 32 accessions in Tonga. Significant differences among accessions were found for tuber yield, roundness, dry matter (DM), and soluble solids content of tubers, top yield, DM of top, and the number of tubers developed per plant. Tubers from reproductively pruned plants had more regularly shaped and smooth tubers than from control plants. Improvement of tuber yield of all cultivars by reproductive pruning of yam bean was also reported by Adjahosson and Ade (1998b), but the size varied with cultivar and harvesting date. Bedford *et al.* (2001) reported that reproductive pruning had a significantly higher yield than the control plants as well as for plant treated with fertilizer only, fresh tuber yield from the combined application ranged from 39012 kg/ha (EC 114) to 53580 kg/ha (EC 201). Arevalo (1998) also confirmed that reproductive pruning showed statistically superior average values for yield, biomass production and sugar content. In an investigation on pruning on yield and quality of tuber, Nielsen *et al.* (1999) concluded that reproductive pruning gave an overall increase in tuber yield of 51.5 per cent in comparison with the control (*P. tuberosus*: 29.2 per cent and *P. erosus*: 102.1 per cent). Vegetative pruning had no effect on tuber yield.

It is essential to remove the buds of yam bean before they flower. Removal of buds by hand is the usual practice. Spraying of 2,4-D (50 ppm) at flower initiation stage induces flower dehiscence and results in better tuber yield (Mohankumar *et al.*, 2000). Panda and Sen (1995) reported that manual deflowering seemed to be most efficient up to 10 days after removal, but the efficiency steeply declined thereafter due to emergence of new flushes of flower buds. Matos *et al.* (1998) found that manual flower pruning (all flowers removed weekly) greatly increased tuber yield in three landraces of *P. ahipa*. However, application of 2,4-D (50 ppm) decreased pod and seed yield, but it did not affect tuber size or production. 2,4-D increased protein and sugar content in one landrace (AC524) but was not recommended as an

Cultivation Techniques of Yam Bean

Seeds and Germination of yam bean

Seed production of yam bean

Field view of yam bean cultivation

Harvesting and Selling of Yam Bean

Harvesting of yam bean tuber

Selling of yam bean in the market

alternative to manual flower pruning. However, Mishra and Mishra (1985) found that deflowering by spraying 2, 4-D was more effective than hand removing. Panda and Sen (1995) recorded the highest mortality of flowers and tuber yield by spraying of NAA (1500 and 1000 ppm) at flower bud initiation stage. Plants of 3 landraces of *P. ahipa* were pruned manually (all flowers removed weekly), or sprayed with 50 ppm 2,4-D at the start of the flowering period or unpruned. Matos *et al.* (1998) found that manual flower pruning greatly increased tuber yield. Application of 2,4-D decreased pod and seed yield, but it did not affect tuber size or production. 2,4-D increased protein and sugar content in tubers in one landrace (AC524), but is not recommended as an alternative to manual flower removal.

8.0 Harvesting and Yield

Normally, yam bean produces a single tuber which can be lifted successfully from 100 days onwards depending on the market demand and the harvesting can be delayed for a longer period but fibre content, chances of rat damage and termite attack may be increased. Harvesting the tubers just after irrigation decreases the keeping quality. Harvest takes place once the tuberous roots have attained marketable size, *i.e.* depending on whether small, medium or large tubers are preferred by the consumers. The Mexican cultivars (*P. erosus*) are harvested from 5 to 7 months after planting, as there are both early and late-maturing cultivars available. The preferred tuber size is around 0.7 kg. In Thailand, harvesting of the same species follows at 4.5-6 months after planting, as the local consumers like small 'onion'- sized tubers. Again, both early- and late-maturing cultivars exist. *P. ahipa* tubers are generally harvested after 7-9 months in Bolivia (as confirmed in field trials in Portugal), but the species has been found to be the earliest of all genotypes tested in the field trials in Mexico, *i.e.* marketable tubers were produced after only 4 months. Lastly, *P. tuberosus* needs the longest growth period of all: 8-11 months, and as described earlier, this species may even be cultivated in a 'perennial' production system.

In India yam bean (*P. erosus*) becomes ready for harvest 150 days after sowing (Mohankumar *et al.*, 2000). Delayed harvest leads to fibrous flesh along with cracks in tubers. This causes deterioration in tuber quality in market. In India the above ground portion are trimmed before digging out the tubers. The vegetative top is incorporated in the soil in Bolivia (Orting *et al.*, 1996). The plants which are allowed to flower and fruit are normally harvested after the ripening of pods. In India, the seed pods are usually harvested 240 days after sowing *i.e.* during March-April. The seed pods can be harvested when they start drying and beans obtained by beating the pods with sticks (Mohankumar *et al.*, 2000; Nath *et al.*, 2007).

In field trials in Mexico (flood irrigated), Costa Rica (dry land) and Tonga (dryland), record yields of 100-145 t/ha (fresh tubers) have been obtained (Heredia, 1994 Morera, 1994a; Nielsen, 1995). From these experiments, the consistently high yields of the Mexican cultivars were confirmed. In India 30-40 tonne of fresh tubers can be obtained within 100-120 days. In the state of Bihar, India, the yields of the so-called 'Desi' cultivars and 32 introduced Mexican lines were tested (Singh *et al.*, 1981). The average yields of the local cultivars were reported to be between 6.7 and

9.4 t/ha, with a production period of 225-250 days. The selected Mexican lines were reported to yield on an average 10.2-12.3 t/ha, with 180-200 days to harvest. Average storage root yields of 4.9 t/ha for *P. ahipa*, 31.8 t/ha for *P. erosus*, 25.4 t/ha for *P. tuberosus*, and 15 t/ha for hybrids were reported at CIP, Peru (Heider *et al.*, 2011).

In West Bengal, India, Mondal and Sen (2006) observed that fertilizer levels had no significant influence on seed yield, however, Nath *et al.* (2007) reported that seed yield of yam bean increased with increasing levels of N up to 120 kg/ha (1.46 t/ha) and declined thereafter. Mondal and Sen (2006) also reported that time of sowing had significant effect upon seed production and they obtained seed yield to the tune of 2.38 t/ha) from June sown crop fertilized with 50: 25: 50 N, P_2O_5 and K_2O kg/ha.

8.1.0 Storage and Value Addition

To increase the keeping quality of tubers, it should be harvested under relatively dry field condition and tubers are slightly dried so that the damaged surfaces develop a protective cork layers. After this treatment the tubers are stored in dry cool places for few weeks. The chilling sensitivity of the tuberous roots of *P. erosus* has been demonstrated (Bruton, 1983; Barile and Esguerra, 1984; Paull and Jung Chen, 1988; Cantwell *et al.*, 1992). Low-temperature storage has been found to reduce storage life considerably, and the optimal storage temperatures are between 12.5 and 17.5°C. The only post-harvest treatment reported is washing, trimming (removal of the non-tuberous part of the root and the basal part of the stem) and dipping in a high-concentration chlorine solution to obtain sterilizing and bleaching effects (Cantwell *et al.*, 1992). Prolonged storage alters the starch/sugar ratio. Paull and Jung Chen (1988) found that after 3 months of storage at 12.5°C, the sucrose content tripled and only one-sixth of the starch remained. In Bolivia, as many consumers prefer a sweeter tuber, some producers leave the tubers of *P. ahipa* in a sunny place for up to 2 weeks prior to marketing (Orting *et al.*, 1996). If the entire plot containing the crop is to be harvested in one operation, the tubers of the jiquima cultivar group (*P. tuberosus* complex) may be stored for 22-30 days during the post-harvest stage, provided well-ventilated indoor conditions are available. One apparent advantage to the farmers is that the jiquima may be handled like the manioc grown in the tropical lowlands, *i.e.* it can be 'stored' in the field and harvested when needed, and it is almost unaffected by poor management. A popular post-harvest treatment is to dry the tubers in the sun for 2-3 days before consumption. This may serve to increase the sugar content, as in the similar treatment of the *P. ahipa* tubers practiced in Bolivia (Sorensen *et al.*, 1996). In India, the tubers are graded after through washing and can be stored for 3-5 days without any deterioration. They are packed in jute bags and brought to market. Harvesting the tubers just after irrigation decreases the keeping quality.

One of the main constraints in the cultivation of yam bean is the rapid decrease in germination when the seeds are stored under humid conditions (Sorensen, 1996). In an experiment in China, yam bean (*Pachyrhizus erosus*) seeds were stored in a refrigerator at 4 degrees C, in a sealed drier containing saturated solution of potassium sulfocyanate in room temperature at 15 degrees C and 45 per cent relative humidity, and in ambient temperature after the seeds were dried at 6.43

per cent seed moisture (ultra-dry treatment). The germination rate of yam bean seeds declined to 50 per cent in room temperature after 12 months of storage, and completely lost the germination capacity after 24 months. The best temperature for storage of yam bean seeds was 4 ° C under which the activities of peroxidase, catalase and superoxide dismutase in seeds were still high after storage, and the values of electroconductivity and ultraviolet absorption remained low. The production of volatile aldehydes was also low during seed germination. Storage effect of ultra-dry treatment was nearly the same as treatment at 15 °C with relative humidity of 45 per cent (Qin and Zheng, 2004).

Investigations have been carried out on the effects of temperature, wrapping and coating with wax on storage of tubers of yam beans. Cantwell *et al.* (1992) reported that roots stored at 12.5° for >4 weeks remained in excellent conditions. Increases in respiration rates during storage and after transfer and ionic conductivity measurements corroborated the visual quality and decay assessments. Upon transfer to 20°C, chill-damaged roots showed an increase in respiration that did not subsequently decline to normal low levels (5-7 ml/kg/ha CO_2 at 20°C).

The effects on *P. erosus* roots at storage temperatures of 5, 10, 15 or 20°C and by wrapping with PVC film and coating with paraffin and carnauba wax were investigated by Bergsma and Brecht (1992). Paraffin wax was the most effective treatment in reducing moisture loss but resulted in the highest losses due to decay; roots coated with carnauba wax lost nearly as much moisture as controls. Wrapping with PVC film reduced moisture loss by nearly 50 per cent compared with the control without any losses due to decay. Sensory evaluations indicated no differences between the PVC film-wrapped and paraffin-coated roots in terms of crunchiness, but the PVC film-wrapped roots were judged to be crunchier and juicier than the controls and sweeter, with a more typical flavour and aroma than paraffin-coated roots. Total soluble sugars contents (which correlated well with soluble solids contents) were about 45 per cent of the initial DW and increased during the first 7 days of storage at all temperatures, but declined at 20°C between 7 and 19 days while it continued to increase at 5, 10 or 15°C. They also found that chilling injury symptoms occurred after 7 days at 1, 5 or 10°C with an additional 2 days at 20°C to allow symptoms to develop, but were not evident up to 49 days at 15°C.

In an experiment on storage of roots of 5 cultivars, Mercado-Silva *et al.* (1998a) observed that roots of all cultivars used in the experiment were very chilling-sensitive, with symptoms of injury occurring after one week at 10° plus one week at room temperature. Roots of cultivars Vega de San Juan and San Miguelito were the most and least tolerant to 10°C storage, respectively. Roots stored at 13°C showed few internal quality changes over a 5-month period, although weight loss exceeded 35 per cent. Mercado-Silva *et al.* (1998b) also reported that roots of cv. Aqua Dulce did not show chilling injury at 13°C. Storage at 5°C resulted in external signs of decay and greying of the pulp; storage at 10°C did not generally affect external quality but severe internal discoloration occurred.

Product development efforts resulted in a number of products processed from yam bean roots: fresh storage roots as raw or cooked vegetable, starch, flour, chips and gari. Gari prepared from yam bean using a traditional cassava procedure had a much browner colour as compared to cassava gari. However, protein, calcium and zinc levels in yam bean gari were two to three times higher than that of cassava gari (Wassens, 2011). Protein on dry matter basis ranged between 3.7 per cent (EC-533) and 4.7 per cent (ECKEW). On average, 6.3 mg Fe and 1.2 mg Zn were observed in 100 g of gari made from EC-KEW while 8.8 mg Fe and 2.1 mg Zn were found in gari prepared from EC-533 (Heider *et al.*, 2011). Wassens (2011) noted that output/input ratios of yam bean gari developed from *P. erosus* were 50– 75 per cent lower than those of cassava. Food processing studies by Heider *et al.* (2011) have shown that the output/input ratios of food products were much higher for *P. tuberosus* chuin accessions compared to *P. erosus* or *P. ahipa* accession. They reported yam bean gari is a food product that could provide clearly more protein, iron, and zinc to the food supply than cassava gari.

9.0 Diseases and Pests

The majority of both the diseases and insect pests are common to all three species. The differences recorded between the species are thus due to geographic, climatic, ecological and edaphic conditions. This has been confirmed in field trials in Tonga, Costa Rica, Ecuador and Thailand. When all five species were cultivated in one location, bean common mosaic virus (BCMV) infected all three cultivated species and also infected the wild species *P. panamensis,* although with some delay, due to the prominent hairiness of all vegetative parts, which has a repelling effect on the aphid vectors. Only the other wild species, *P. ferrugineus,* appeared to possess any resistance to this virus.

Another disease spread by insect vectors is the witch's broom disease, first diagnosed by Thung and Hadiwidjaja (1957) and probably caused by mycoplasma-like organisms. Sincama mosaic virus (SMV) or BCMV may become a serious problem locally, particularly in fields bordering on wild vegetation with frequent *P. erosus* escapes. The typical symptoms are irregular chlorosis of the leaves, young shoots becoming brittle and the seed set being reduced as a result of atrophied pollen (the pollen fertility of infected plants is reduced from 95-100 per cent to less than 10 per cent). Tuber growth is also affected and yield will decrease by 20-40 per cent.

In Mexico, the bacterial disease called bean halo blight, caused by *Pseudomonas syringae* (Burkh.) Dowson pv. *Phaseolicola* has been observed on several occasions in *P. erosus* (Diaz, 1979). The same disease was reported in Hawaii by Birch *et al.* (1981). However, the disease does not appear to reduce yields significantly (Diaz, 1979).

Several fungi have been reported to cause severe damage in *P. erosus.* High mortality rate in young plants as a result of 'root attacks' by *Pythium* spp., *Corticium* spp. and *Macrophomina* spp. was reported in multilocational field trials in Senegal (Sorensen, 1996). In China, *Pythium aphanidermatum* (Edson) Fitz. is the cause of root rot in *P. erosus* (Yu *et al.,* 1945). Mohanty and Behera (1961) reported a severe leaf spot disease observed in Bhubaneswar, India and succeeded in identifying the fungus as *Cercospora canescens* Ellis *et* Martin. Angular spots were observed on leaves of yam bean in Anhui province in China by Xu *et al.* (1999) and this was considered as a new disease caused by a non-fluorescent, new pathovar of *Pseudomonas syringae,* named as *P. syringae* pv. *pachyrrhizus* nov.

Diseases and Pest of Yam Bean

Mosaic infection in yam bean

Yam bean rust urediniospores on leaf

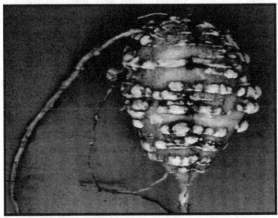

Root-knot nematode on tuber

In all species, the most serious pests endangering cultivation are various bruchids, which may destroy as much as 80 per cent of the locally stored seeds. A number of insect pests are reported to cause leaf, tuber and seed damage in *P. erosus*. Species belonging to the genus *Diabrotica* are serious pests in many humid areas, and the leaf damage is often quite extensive. In Mexico and Central America, flower buds and the young pod may be damaged by Thecla jebus. If localized, *Phyllophaga* Harris (n.v. 'gallina ciega') may cause severe tuber damage (Heredia, 1985). Termites (Termitidae) hollowed the stems of young plants, destroying 15-20 per cent of the plants. *Andrector* spp. (Gelerucidae) and *Nezara viridula* (Pentatomidae) caused severe leaf damage, and an as yet unidentified larva also attacked a considerable number of the tubers (Sorensen, 1996).

The nematode *Meloidogyne marioni* (Cornu) Chitwood et Oteifa (syn. *Eterodera marioni* (Cornu) Marcinowski) is cited by Duke (1981) as the cause of tuber damage. Singh and Yadav (2006) reported pod borer (*Maruca vitrata* G.) infestation varied widely from 0.7 to 40 per cent in 60 genotypes of yam bean in Bihar, India and widely cultivated variety RM-1 was found to be highly susceptible to the pest with more than 20 per cent pod damage. In Tonga, seed damage (*P. ahipa*) caused by the coffee bean weevil (*Araecerus fasciculatus* De Geer) has been recorded in seed belonging to all species during storage (Sorensen, 1996).

10.0 Crop Improvement

The regeneration potential of *Pachvrhizus erosus* using single node culture has been investigated on MS medium alone, and supplemented with each of Benzyladenine (BA), 2-isopentyladenine (2-ip), Kinetin (Kin), Zeatin and Adenine, at a 1 mg/1 concentration (Forbes and Duncan, 1990). Callus induction using leaf explants of different physiological ages was investigated on MS medium supplemented with 2, 4-dichlorophenoxyacetic acid (2,4-D); 3 indoleacetic acid (IAA) plus Kin; 1 naphthylacetic acid (NAA) plus Kin; and 2,4-D plus Kin at 1 mg/L concentration. The effect of light on callus initiation was noted. Callus was initiated on MS with 2,4-D plus Kin, also NAA plus Kin incubated in complete darkness. The quantity of callus varied with leaf age. Callus initiated with 2,4-D and Kinetin was placed on MS supplemented with BA using either fructose or glucose as the carbon/energy source, and incubated in light and dark environments. Chlorophyll formation was apparent in callus exposed to light, when the carbon source was fructose. Squash preparations of callus on BA supplemented medium indicated the presence of tracheids. Plantlet regeneration from callus has not yet been achieved earlier.

In vitro propagation of yam bean was reported by Munoz *et al.* (1998). The best culture method for the phase of establishment was MS + BAP 2.5 mg/l; for multiplication the best medium was MS + BAP 1.25 mg/l + IBA 0.01 mg/l. At the rooting phase, low concentrations of NAA (0.5 mg/l) effectively induced the development of roots. Cadiz *et al.* (2000) established an *in vitro* system for the induction and study of nodulation in *Pachyrhizus erosus* (jicama) via a hairy root-*Rhizobium* coculture. *In vitro*-grown *P. erosus* plantlets were infected with *Agrobacterium rhizogenes* (ATCC No. 15834) and two hairy root lines were established.

Hairy roots were grown in a split-plate system in which compartment I (CI) contained MS medium with nitrogen and different sucrose levels (0-6 per cent), while CII held MS medium without nitrogen and sucrose. Nodule-like structures developed in transformed roots grown in CI with 2-3 per cent sucrose, inoculated with *Rhizobium* sp. and transferred to CII. Nodule-like structures that developed from hairy roots lacked the rigid protective cover observed in nodules from plants grown in soil. Western blot analysis of nodules from hairy roots and untransformed roots (of greenhouse-grown jicama) showed expression of glutamine synthetase, leghemoglobin and nodulins. Leghemoglobin was expressed at low levels in hairy root nodules.

Large number of germplasm accessions are being maintained at ICAR-CTCRI (63 nos.) and AICRP Centres (205 nos) of different SAUs in India. The germplasm accessions comprise of land races and exotic collections. Most of the exotic collections are from Mexico. All the accessions from CTCRI were characterized and evaluated. Except one with white flowers, the remaining accessions produce velvet flowers. The tuber yield of the accessions ranged between 10.33-25.78 t ha^{-1}. Five accessions were identified with yield more than 25 t ha^{-1}. Analysis of the biochemical constituents of the tuber revealed, variations in dry matter (9.33-29.78 per cent), starch (3.02-7.96 per cent) and sugar (3.02-7.96 per cent) contents (Vimala, Personal communication).

Yam bean research in India has not received much attention from the national research system. Hence farmers in the country still rely on traditional land races. In India, research on yam bean is being undertaken at Central Tuber Crops Research Institute (CTCRI), Thiruvananthapuram, Kerala, India and two centers (Bihar and West Bengal) of the All India Co-ordinated Research Project on Tuber Crops (AICRPTC).

With respect to breeding, all cultivated yam beans are mainly self-pollinating, but up to 30 per cent out crossing occurs depending on the presence of pollinators (mainly bees) so that line breeding is practiced (Sorensen, 1996). There has been little attempt for genetic improvement through breeding. Breeding in yam bean is limited to selection only. Nature of compatibility was studied in a set of diallel crosses involving eight genotypes. The breeding objectives include earliness, high dry matter, improved nutrition, drought tolerance and pest and disease resistance. Attempts have already been made to develop hybrids of Mexican and local types and some of the lines were promising (Mukhopadhyay *et al.*, 2008).

The information available on the breeding of this crop is very much limited in India. Genetic variability is limited in yam bean. For inducing specific genetic changes, an exploratory gamma irradiation was carried out in seed samples of a superior collection (Sreekumari *et al.*, 1983; Nair and Abraham, 1988, 1989). Yam bean seeds treated with gamma radiation (5 - 25 kR) or ethylmethane sulphonate (EMS, 0.5 – 1.25 per cent) induced greater variability with regard to shoot length, number of branches, number of leaves and tuber yield. Treatment with gamma radiation greater than 7.5 kR significantly reduced vegetative vigour and yield. Yam bean seeds treated with gamma radiation (5 kR) stimulated vegetative vigour, induced greater shoot length, number of branches, number of leaves and tuber yield than control plants. The occurrence of multiple shoot (twins, triples and quadruplets seedlings) which accounted for nearly 2 per cent of germinated seeds has also been reported in yam bean (Sreekumari and Abraham, 1980).

Sixty accessions of yam bean (*P. erosus*) were tested in yield trials in Tonga by Nielsen *et al.* (2000). There was a highly significant correlation between the number of developed tubers per plant and the total yield. Among the 20 highest yielding accessions, yield varied between 77.0 and 125.9 t/ha. The accessions representing wild genotypes had the highest dry matter (DM) contents of tubers. A significantly negative correlation existed between tuber weight and DM content, while soluble solids content was positively correlated with the DM contents. The study of Agaba *et al.* (2016) indicated substantial genetic variation for yield and quality traits in yam bean, demonstrating potential for adaptability to growing conditions and consumer needs in East and Central Africa and for genetic improvement through selection. An unexpected finding was that it appears possible to breed for high dry matter yam beans by using low dry matter yam beans due to the observed genetic variation among low dry matter yam beans without having access to the high dry matter chuin material.

Both quantitative and qualitative characters are also useful in conducting Distinctiveness (D), Uniformity (U) and Stability (S) test, a mandatory requirement for protecting varieties under Protection of Plant Varieties and Farmers' Rights (PPA and FR) Act of India, have been developed by the task force in consultation with ICAR-CTCRI, Thiruvanathapuram and ICAR-CTCRI, Regional Centre, Bhubaneswar, India during 2018 constituted by the PPV and FR Authority for yam bean (*Pachyrhizus erosus* (L.) which is vogue in other countries. Zanklan *et al.* (2018) demonstrated that within each cultivated species a similar amount of diversity was found and that the genetic distance between species was limited. Moreover, considerable diversity existed within *P. ahipa* and *P. tuberosus* grown at both sides of the Andean mountain range. Since interspecific hybridisation is possible (Grum, 1990; Gruneberg *et al.*, 2003), all three cultivated yam bean species may constitute an important source for breeding. The close relationship among species further supports the proposition that only a few highly heritable characters are required to describe the diversity within the yam bean gene pool. Zanklan *et al.* (2018) concluded that the cultivated yam bean species represented distinct genepools and each exhibited similarly large amounts of genetic diversity. Thirty germplasm accessions of yam bean having different geographical origin were evaluated at Dholi, Bihar to identify important selection indices for tuber and seed yield, and to assess the genetic divergence based on 12 quantitative characters (Deepshikha, 2018). Path analysis revealed that tuber weight had high positive direct effect on tuber yield while pods per plant and pod weight per plant were found to have high positive direct effects on seed yield. All the thirty genotypes were grouped into 6 clusters using D^2 statistics. She suggested selection of parents from different clusters might be done for desired traits on the basis of higher cluster mean values. Tuber yield followed by tuber weight and seed yield were having maximum contribution towards total divergence.

Genetic variation and relationship in important species of yam bean clearly indicated that polymorphisms produced, proved to be useful markers for discriminating between accessions of *P. erosus* and *P. tuberosus*. Within *P. erosus*, a discrete grouping of Mexican cultivars was revealed, which is compatible with the proposal that *P. erosus* in Central Mexico and the Yucatan Peninsula is derived from a highly limited introduction of germplasm. RAPD markers were very useful for discriminating between accessions belonging to the jiquima and ashipa cultigen types of the *P. tuberosus* (Estrella *et al.*, 1998). Conservation of crop genetic resources hinges on the availability of efficient molecular tools to characterize population genetic structure and decipher the dynamics of crop genetic diversity. The case of *Pachyrhizus* illustrates the spillover benefits to be reaped from next-generation sequencing and research on model plants for the study of minor crops (Varshney *et al.*, 2010). The SSR markers developed by Deletre *et al.* (2013) showed high levels of polymorphism and enough discriminant power for distinguishing among varietal groups within species. Markers also revealed a surprisingly low level of genetic variability in the Bolivian root crop, *P. ahipa*. While the wild parent of the crop has yet to be identified, they will use the new markers to investigate the origin of *P. ahipa*. Results might shed new light on the evolutionary history of the *Pachyrhizus* genus.

Heredia (1998) obtained plants from F_1 of interspecific crossings of *P. erosus* × *P. ahipa*, *P. erosus* × *P. tuberosus*, and their respective reciprocal crosses. From these F_2 plants individual selections were made using an alternative method for improving the cultivars through clonal selection and sexual reproduction. After four and five cycles of selection 42 elite segregations in the F_5 generation, 48 outstanding F_4 segregations and 650 selection both in the F_4 and F_5 generations were obtained, ensuring that the desired characteristics have become stabilized/fixed in the more advanced generations.

Nair and Abraham (1989) reported that in the M_2 from seeds were gamma-irradiated (10, 20 and 25 kR) or treated with ethyl methanesulfonate (EMS; 0.5, 0.75, 1.0 and 1.25 per cent) dwarf, erect and spreading plant types were recorded. Mutations affecting leaves included changes in leaf number, size and shape. Plants with crinkled leaves were also obtained. Lengths of inflorescences and pods were profoundly affected with increases and decreases in size. Tuber number increased in the M_2, particularly from the gamma ray treatment. Nair and Abraham (1990) also observed a 20 cm tall mutant (vs 82 cm for the control) by treatment with 1.25 per cent EMS, with a tuber yield of 367.6 g (vs. 157.5 g) and tuber protein content of 5.31 per cent (vs 4.13 per cent) but sugar content was slightly lower than in the control. It bred true in the M_3-M_5. In another experiment, in the M_3 from seeds treated with EMS, mutants with higher tuber yields and better quality bred true in subsequent generations, were reported by Nair and Abraham (1992). One of the mutants from the 1.25 per cent EMS treatment produced tubers nearly twice the weight of those of the control (300 vs. 157 g) and had a higher starch content (10.8 vs. 8.7 per cent). Another mutant, from the 0.75 per cent EMS treatment, had a higher tuber weight (250 g) and a higher protein content (6.1 vs. 4.1 per cent) than the control.

In order to broaden the genetic base of yam bean available at CIP, Peru a cross breeding experiment was established in 2010. The experiment was designed as a completely diallelic cross between three *P. ahipa* and three *P. tuberosus* accessions and between three *P. erosus* und three *P. tuberosus* accessions (Heider *et al.*, 2011).

Interspecific hybridisation resulted in 18 F_1 interspecific and 12 F_1 intraspecific cross populations. Nine *P. ahipa* × *P. tuberosus* chuin F_1 cross populations as well as 9 *P. erosus* × *P. tuberosus* chuin F_1 cross populations were developed. Theses population were used to generate a large number of F_2 lines that serve to select genotypes with high dry matter, high starch, and adaptation to the environmental conditions of the Central African high lands or the savanna zones of West, Central, and Southern Africa. Since the beginning of the crossbreeding experiment a total of 2120 (AC × TC) and 9921 (EC × TC) pollinations were made, of which 294 and 1120, respectively were successful and seeds harvested. The hybrid populations of *P. ahipa* × *P. tuberosus* chuin and *P. erosus* × *P. tuberosus* chuin could broaden the very narrow genetic diversity of high dry matter *Pachyrhizus* spp.

A study was undertaken by Jean *et al.* (2017) to determine gene actions, combining ability and genetic parameters for earliness in F_1 (*P. ahipa* × *P. tuberosus*) hybrids under Rwandan East African highland conditions. Three *P. ahipa* accessions (early maturing, tropical high lands) as females and three accessions of *P. tuberosus* (late maturing and high yields) as males were crossed using a North Carolina II mating design. They suggested that it would be possible to recombine the earliness attribute from *P. ahipa* with storage root yield and seed yield attributes of *P. tuberosus* in new F_1 hybrid lines from which segregating populations cab be developed in future.

11.0 References

Adjahosson, D.F. and Ade, J. (1998a) *Proc. of the 2nd International Symposium on Tuberous Legumes,* Celaya, Gaunajuato, Mexico, 5-8 August, 1996, pp. 91-114.

Adjahosson, D.F. and Ade, J. (1998b) *Proc. of the 2nd International Symposium on Tuberous Legumes,* Celaya, Gaunajuato, Mexico, 5-8 August, 1996, pp. 361-375.

Adjahossou, D.F. and Sogbenon, H. (1994) *Proceedings of the First International Symposium on Tuberous Legumes* (M. Sbattu2rensen, ed.), Guadeloupe, Jordbrugsforlaget, Copenhagen, Denmark. FWI, April 21-24 1992. pp. 199-214.

Agaba, R., Tukamuhabwa, P., Rubaihayo, P., Tumwegamire, S., Ssenyonjo, S., Mwanga, R.O.M., Ndirigwe, J. and Gruneberg, W. J. (2016) *HortScience*, **51**: 1079–1086.

Alvarenga, A.A. and Valio, I.F.M. (1989) *Ann. Bot.*, **64**: 411-414.

Annerose, D.J.M. and Diouf, O. (1995a) Second Annual Progress Report, STD3 (Ed.M. Sorensen). pp. 137-151.

Annerose, D.J.M. and Diouf, O. (1995b) Fifth Bi-annual Progress Report, STD3. pp. 81-87.

Anonymous (1995) *Indian Minimum Seed Certification Standards,* Central Seed Committee, Min. Food Agric., Comm. Dev. Co-op., New Delhi, p. 102.

Anonymous (1996) *Plant Varieties J.,* **9: 4,** 25.

Anonymous (2006) *Annual Report, AICRP on Tuber Crops,* BCKV, Kalyani, India.

Anonymous (1978) *Annual Report, AICRP on Tuber Crops*, Rajendra Agril. Univ., Dholi, Bihar.

Anonymous (2018) *Genetic diversity studies in Yam Bean, M.Sc. Thesis, RAU, Pusa, Samastipur.*

Arevalo, T.A. (1998) *Proc. of the 2nd International Symposium on Tuberous Legumes,* Celaya, Gaunajuato, Mexico, 5-8 August, 1996, pp. 125-130.

Arevalo, T.A. and Sorensen, M. (1998) *Proc. of the 2nd International Symposium on Tuberous Legumes,* Celaya, Gaunajuato, Mexico, 5-8 August, 1996, pp. 115-125.

Badilllo, V., Castellanos, J.Z. and Rios-Ruiz, S.A. (1998) *Proc. of the 2nd International Symposium on Tuberous Legumes,* Celaya, Gaunajuato, Mexico, 5-8 August, 1996, pp. 389-397.

Barile, T.V. and Esguerra, E.B. (1984) Postharvest Research Notes, 1,2. Dept. Horticulture, Univ. Philippines, Los Banos. pp. 23-25.

Belford, E.J.D., Karim, A.B. and Sierra Leone (2001) *J. Appl. Bot.*, **75:** 31-38.

Bergsma, K.A. and Brecht, J.K. (1992) *Acta Hortic.*, **318**: 325-332.

Bhag, Mal and Kawalkar, T.G. (1982) *Indian Fmg*, **31**: 13-14.

Birch, R.G., Alvarez, A.M. and Patil, S.S. (1981) *Phytopathology*, **71:** 1289-1293.

Bruton, B.D. (1983) *J. Rio Grande Val. Hortic. Soc.*, **36**: 29-34.

Cadiz, N.M., Vivanco, J.M. and Flores, H. E. (2000) *Plants*, **36** (4): 238-242.

Cantwell, M., Orozco, W., Rubatzky, V. and Hernandez, L. (1992) *Acta Hort.*, **318**: 333-343.

Caro, V.F. de. J., Ca Sillas, R.D. and Rios-Ruiz, S.A. (1998) *Proc. of the 2nd International Symposium on Tuberous Legumes,* Celay, Guanajuato, Mexico, August 5-8, 1996, pp. 131-137.

Castellanos, J.Z., Zapata, F., Badillo, V., Pena-Cabriales, J.J., Jensen, E.S. and Heredia-Garcia, E. (1997) *Soil Biol. Biochem.*, **29**: 973-981.

Castellanos-Ramos, J.Z., Acosta-Gallegos, J.A., Rodriguez Orozco, N. and Munoz-Ramos, J.J. (2009) *Agricultura Tecnica en Mexico*, **35**: 277-283.

Cotter, D.J. and Gomez, R.E. (1979) *HortScience*, **14**: 733-734.

Cruz, A.O. (1950). *Philipp. J. Sci.*, **78**: 145-147.

Cruz, G.N., Stamford, N.P., Silva, J.A.A. and Chamber, Perez, M. (1997) *Tropical Grasslands*, **31**: 538-542.

Deepshikha. (2018) M.Sc thesis, RAU, Pusa, Samastipur, Bihar.

Deletre, M., Soengas, B., Utge, J. Lambourdiere, J. and Sorensen, M. (2013) *Appl. Plant Sci.*, **1** (7): 1-5.

Deshaprabhu, S.B. (1966) *Publ. Inf. Dir. Counc. Sci. Ind. Res.*, New Delhi. **7:** pp. 208-210.

Diaz A. (1979) El cultivo de la Jícama en el Estado de Guanajuato. SARH/CIAB publ. Guanajuato, Mexico. **116**: pp. 1-4.

Duke, J.A. (1981) Handbook of Legumes of World Economic Importance. Plenum Press, New York and London.

Espinoza, F., Diaz, Y., Argenti, P., Perdomo, E. and Leon, L. (1998) *Proc. of the 2nd International Symposium on Tuberous Legumes,* Celaya, Gaunajuato, Mexico, 5-8 August, 1996, pp. 139-154.

Estrella, J., Phillips, S., Abbott, R., Gillies, A. and Sorensen, M. (1998) *Proc. of the 2nd International Symposium on Tuberous Legumes, Celaya,* Gaunajuato, Mexico, 5-8 August, 1996, pp. 43-59.

Faria, L.L., Alvarenga, A.A. de, Castro, E.M. de, Salvador-Sobrinho, J.C., de Alvarenga, A.a. and de Castro, E.M. (2000) *Ciencia e Agrotecnologia,* **24:** 688-695.

Fernandez, M.V., Warid, W.a., Loaiza, J.M. and Montiel, C.A. (1997) *Plant Foods Human Nutrition,* **50:** 279-286.

Fernandez, M.V., Warid, W.A., Loaiza, J.M., Martinez, J.J. and Serrano, A. (1996) *Japanese J. Trop. Agric.,* **40:** 26-28.

Figueiredo, M.V.B., Medeiros, R., Stamford, N.P., Santos, C.E.R.S. dos, Dos Santos, C.E.R.S. (1996) *RevistaBrasileira de Ciencia do Solo,* **20:** 49-54.

Forbes, W.C. and Duncan, E.J. (1990) 26th Annual Meeting, July 29 to August 4, 1990, Mayaguez, Puerto Rico from Caribbean Food Crops Society.

Fuentes, J.B., Abe, M., Uchiumi, T., Suzuki, A. and Higashi, S. (2002) *J. Gen. Appl. Microbiol.,* **48:** 181–191.

Gomes, A.V., SirijuCharran, G. and Barnes, J.A. (1997) *Phytochemistry,* **46:** 185-193.

Gonzlez, Rivas, C. and Rios Ruiz, S.A. (1998) *Proc. of the 2nd International Symposium on Tuberous Legumes,* Celay, Guanajuato, Mexico, August 5-8, 1996, pp. 155-156.

Greshof, M. (1893) *Beschrijving der giftige en bedwelmende planten bij de vischvangst in gebruik. Meded. S Lands Plantentuin, Landsdrukkerij,* Batavia, **10:** 65.

Grum, M. (1990) *PhD Thesis,* Department of Crop Husbandry and Plant Breeding, The Royal Veterinary and Agricultural University, Copenhagen, Denmark.

Grum, M. and Rios Ruiz, S.A. (1998) *Proc. of the 2nd International Symposium on Tuberous Legumes* Celay, Guanajuato, Mexico, August 5-8, 1996, pp. 413-418.

Grum, M. and Sorensen, M. (1998) *Proceeding of 2nd International Symposium on Tuberous Legumes,* Celaya, Guanajuato, Mexico, MacKeenzie Press, Copenhagen (Denmark). 5-8 August 1996. pp. 419-429.

Grum, M., Halafihi, M., Stolen, O. and Sorensen, M. (1994) *Exper. Agric.,* **30:** 67-75.

Grum, M., Sorensen, M. and Rios Ruiz, S.A. (1998) *Proc. of the 2nd International Symposium on Tuberous Legumes,* Celay, Guanajuato, Mexico, August 5-8, 1996, pp. 523-531.

Grum, M., Stolen, O., Halafihi, M. and Sorensen, M. (1996) Urban - Experimental Agriculture, August 1995. pp. 11.

Gruneberg, W.J., Freynhagen-Leopold, P. and Delgado-Vaquez, O. (2003) *Genet. Resour. Crop Evol.,* **50:** 757-766.

Gruneberg, W.J., Goffman, F.D. and Velasco, L. (1999) *J. Amer. Oil Chemists Soc.,* **76**: 1309-1312.

Halafihi, M. (1994) *Proceedings of the First International Symposium on Tuberous Legumes* (M. Sbattu2rensen, ed.), Guadeloupe, Jordbrugsforlaget, Copenhagen, Denmark. FWI, April 21-24 1992. pp. 191-198.

Halafihi, M., Grum, M., Stolen, O. and Sorensen, M. (1994) *International Symposium on Tuberous Legumes* (M. Sorensen, ed.), Guadeloupe, Jordbrugsforlaget, Copenhagen, Denmark, FWI, April 21- 24, 1992. pp. 215-226.

Heider, B., Tumwegamire, S., Tukamuhabwa, P., Ndirigwe, J., Bouwe, G., Bararyenya, A., Hell, K., Leclercq, J., Lauti, E. and Wassens, R. (2011) *Afr. Crop Sci.,* **10**: 93–95.

Heredia Z.A. (1971) *Proceeding of the Tropical Region, XIX Annual Meeting,* 18-24 July 1971, Managua, Nicaragua, C.A., **15**: 146-150.

Heredia Z.A. (1985) Guia para cultivar jicama en el Bajio. Folleto para Productores, Secretaria de Agricultura y Recursos Hidraulicos. p. 11.

Heredia, Z.A. (1998) *Proc. of the 2nd International Symposium on Tuberous Legumes,* Celaya, Gaunajuato, Mexico, 5-8 August, 1996, pp. 343-350.

Heredia-Zepada, A. and Heredia Garcia, E. (1994) *Proceedings of the First International Symposium on Tuberous Legumes;* Guadeloupe, F.W.I., 21-24 April 1992, Copenhagen, Denmark. pp. 257-272.

Jana, S. (2005) *M.Sc. (Ag) Thesis,* BCKV, Kalyani, West Bengal, India.

Jean, N., Patrick, R., Phenihas, T., Rolland, A., Placide, R., Robert, M.O.M., Silver, T., Vestine, K., Evrard, K. and Gruneberg, W. J. (2017) *Tropical Plant Biol.,* **10**: 97–109.

Kjær, S. (1992) *Ann. Bot.,* **70**: 11-17.

Kjær, S. and Sorensen, M. (1992) *Proceedings of the First International Symposium on Tuberous Legumes* (Sorensen, M. ed.), Guadeloupe, Jordbrugsforlaget, Copenhagen. FWI, 1992, April 21-24. pp. 227-236.

Kundu, B.C. (1969) In: *Proc. of the International Symposium on Tropical Root Crops* (Eds. Tai, A., Charles, W.B., Haynes, P.H., Iton, E.F. and Leslie, K.A.), St. Augustine, Trinidad, April 208, India, pp. 124-130.

Laxminarayana, K. (2017) *Communications in Soil Science and Plant Analysis,* **48**: 186-200.

Lynd, J.Q. and A.A.C. Purcino. (1987) *J. Plant Nutr.,* **10**: 485-500.

Mali, V.V., Thorat, S.B., Pawar, A.P., Parulekar, Y.R. and Shelke, P.V. (2018) *J. Pharmacog. Phytochem.,* **7**: 1454-1456.

Martinez, M. (1936) Plantas utiles de Mexico. 2nd ed. Ediciones Botas, Mexico. pp. 244-247.

Matos, M.C., Matos, A.A., Silva, J.B.V. da and da Silva, J.B.V. (1998) *Proc. of the 2nd International Symposium on Tuberous Legumes,* Celaya, Gaunajuato, Mexico, 5-8 August, 1996, pp. 181-189.

Mercado-Silva, E., Garcia, R., Heredia-Zepeda, A. and Cantwell, M. (1998a) *Postharvest Biol. Tech.*, **13**: 37-43.

Mercado-Silva, E., Rubvatzky, V. and Cantwell, M.I. (1998b) *Acta Horticulturae*, **467**: 357-362.

Mishra, S., Singh, C.P., Singh, K.P., Singh, N.K. and Singh, U.P. (1994) *J. Potassium Research*, **10**: 271-273.

Mishra, S.S. and Mishra, S. (1985) Annual Conference of the Indian Society of Weed Science (undated): **57**.

MohanKumar, C.R., Nair, G.M., James, G., Ravindran, C.S. and Ravi, V. (2000) Production technology of tuber crops. Central Tuber Crops Research Institute, Thiruvananthapuram. pp. 174.

Mohanty, N.N. and Behera, B.C. (1961) *Sci. Nat.*, **27**: 54.

Mondal, S. and Sen, H. (2006) *J. Root Crops*, **32**: 37–40.

Morales-Arellano, G.Y., Chagolla-Lopez, A., Paredes-Lopez, O. and Barba de la Rosa, A.P. (2001) *J. Agri. Food Chem.*, **49**: 1512-1516.

Morera, J.A. (1994a) First Bi-Annual Progress Report, STD3 (Sorensen, M. ed.). pp. 9-17.

Morera, M.J.A., Mora, Q.A. and Cadima, F. (1998) *Proc. of the 2nd International Symposium on Tuberous Legumes,* Celaya, Gaunajuato, Mexico, 1996, 5-8 August, pp. 205-219.

Mudahar, G.S. and Jen, J.J. (1991) *J. Food Sci.*, **56**: 977-980.

Mukhupadhyay, S.K., Nath, R. and Sen, H. (2008). In: *Underutilized and Underexploited Horticultural Crops,* (Ed. K. V. Peter). New India Publishing Agency, New Delhi. **4:** 342-377.

Munos, Otero S. (1945) *3rd Conf. Interamer. Agric.*, Caracas, **38**: 5-34.

Munoz, L., Krogstrup, P., Estrella, J., Castillo, R., Sorensen, M., Estrella, E.J., Hamann, O.J. and Rios Ruiz, S.A. (1998) *Proceedings of the 2nd International Symposium on Tuberous Legumes* Guanajuato, Mexico, 5-8 August 1996. 1998, pp 453-467.

Nag, N.C., Banerjee, H.N. and Pain, A.K. (1936) Transactions of the Bose Research Institute, **11**: 83- 89.

Nair, S.G. and Abraham S. (1988) *J. Root Crops*, **14**: 31-36.

Nair, S.G. and Abraham, S. (1989) *J. Root Crops*, **15**: 7-13.

Nair, S.G. and Abraham, S. (1990) *Mutation Breeding Newsl.*, **36**: 5-6.

Nair, S.G. and Abraham, S. (1992) *Mutation Breeding Newsl.*, **39**: 10-11.

Nath, R., Chattopadhyay, A., Mukhopadhyay, S.K., Kundu, C.K., Gunri, S.K., Majumder, A. and Sen. H. (2008) *J. Root Crops*, **34:** 15-20.

Nath, R., Chattopadhyay, A., Mukhopadyay, S.K., Kundu, Gunri, S.K., Mazumder, A. and Sen, H. (2007) *Proceedings of the National Seminar on Ecorestoration of Soil and Water Resources Towards Efficient Crop Production*, Bidhan Chandra Krishi Viswavidyalaya, West Bengal, June 6-7, 2007. pp. 171-173.

Nersekar, P. P., Parulekar, Y.R., Pawar, A.P., Haldankar, P.M. and Mali, P.C. (2018) *Int. J. Chem. Stud.*, **6**: 3265-3268.

Nielsen, P.E. (1995) Second Annual Progress Report, STD3 (Sbattu2rensen, M., ed.). pp 153-184.

Nielsen, P.E., Halafihi, M. and Sorensen, M. (1998) *Proc. of the 2nd International Symposium on Tuberous Legumes,* Celaya, Gaunajuato, Mexico, 5-8 August, 1996, pp. 253-259.

Nielsen, P.E., Halafihi, M. and Sorensen, M. (1999) *Tropical Agriculture,* **76**: 222-227.

Nielsen, P.E., Sorensen, M. and Halafihi, M. (2000) *Tropical Agriculture,* **77**: 174-179.

Noda, H. and Kerr, W.E. (1983) *Trop. Grain Leg. Bull.,* **27**: 35-37.

Norton, L.B. (1943). *J. Am. Chem. Soc.,* **61**: 2259-2260.

Orting, B., Grüneberg, W.J. and Sbattu2rensen, M. (1996) *Genet. Resour. Crop Evol.,* **43**: 435-446.

Panda, P.K. and Sen, H. (1995) *J. Root Crops,* **21**: 97-101.

Panda, P.K., Sen, H., Mukherjee, A. and Satapathy, M.R. (2003) *Legume Res.,* **26**: 235-241.

Pati, K., Zhang, F. and Batley, J. (2019) *Plant Genet. Resour.: Characterization and Utilization,* pp. 1–4.

Paull, R.E. and Chen, N.J. (1988) *HortScience,* **23**: 194-196.

Paull, R.E., Chen, N.J. and Fukuda, S.K. (1988) *HortScience,* **23**: 326-329.

Peltier, J.B., Vaillant, V., Malsa, C. and Zinsou, C. (1996) *Proc. of the 2nd International Symposium on Tuberous Legumes,* Celaya, Gaunajuato, Mexico, 5-8 August, 1996, pp. 505-512.

Prasad, D. and Prakash, R. (1973) *Indian J. Agric. Sci.,* **43**: 531-535.

Ramaswany, N., Muthukrishnan, C.R. and Shanmugavelu, K.G. (1980) *National Seminar on Tuber Crops Production Technology, Faculty of Horticulture, Tamil Nadu Agricultural University, Coimbatore,21-22 November 1980.*

Ratanadilok, N. and S. Thanisawanyangkura. (1994) *Proceedings of the First International Symposiumon Tuberous Legumes;* Guadeloupe, FWI, 21-24 April 1992. (M. Sorensen, ed.) Jordbrugsforlaget, Kbattu2benhavn. pp. 305-314.

Ratanadilok, N., Suriyawan, K. and Thanaisawanrayangkura, S. (1998) *Proc. of the 2nd International Symposium on Tuberous Legumes,* Celaya, Gaunajuato, Mexico, 5-8 August, 1996, pp. 261-273.

Robin, C., Vaillant, V., Vansuyt, G. and Zinsou, C. (1990) *Plant Physiol. Biochem.,* **28**: 343-349.

Roy, B., Mishra, S. and Mishra, S.S. (1976) *J. Root Crops,* **2**: 29-35.

Sahadevan, N. (1987) Green fingers. Sahadevan, N. (ed). Sahadevan Publications, Malaysia. pp 208-209.

Santayana, M., Rossel, G., Nunez, J., Sbattu2rensen, M., Deletre, M., Robles, R., Fernandez, V., Gruneberg, W.J. and Heider, B. (2014) *Trop. Plant Biol.*, **7:** 121–132.

Santos, A.C.P., Cavalcanti, M.S.M. and Coehlo, L.C.B.B. (1996) *Plant Foods for Human Nutrition*, **49:** 35-41.

Sen, H. (2003) In: *Vegetable Crops* (Eds. T.K.Bose, T.K. Maity, V.A. Parthasarathy, and M.G. Som), Naya Udyog, Kolkata, pp. 461-478.

Sen, H. and Mukhopadhyay, S.K. (1989) *J. Root Crops*, **15** ; 20-23.

Sen, H., Goswami, S.B., Das, P.K. and Pillai, S.V. (1996) *Tropical Tuber Crops: Problems, Prospects and Future Strategies*, pp. 315-317.

Singh, K.P., Singh, J.R.P. and Ray, P.K. (1981) *Indian Farm*, **31:** 19-21.

Singh, P.P. and Yadav, R.P. (2006) Food security and Sustainable Environment (Naskar *et al.*, Ed.) Regional Centre, CTCRI, Bhubaneswar, India, pp. 221-224.

Sorensen, M. (1988) *Nord. J. Bot.*, **8:** 167-192.

Sorensen, M. (1990) *Wageningen Papers*, **90:** 1-38.

Sorensen, M. (1991) *Caribbean Food Crops Society, Twenty-Fifth Annual Meeting, Guadeloupe, Dosier*, **25:** 597–624.

Sorensen, M. (1996) Institute of plant genetics and crop plant research, Gatersleben/ International Plant genetic Resources Institute, Rome.

Sorensen, M., Doygaard, S., Estrella, J.E., Kvist, L.P. and Nielsen, P.E. (1997) *Biodiversity and Conservation*, **6:** 1581-1625.

Sorensen, M., Grum, M., Paull, R.E., Vaillant, V., Venthou-Dumaine, A. and Zinsou, C. (1993). Underutilized Crops: Pulses and Vegetables (J.T. Williams, ed.). Chapman and Hall, London – New York. pp. 59-102.

Sreekumari, M.T. and Abraham, K. (1980) *J. Root Crops*, **6:** 65- 67.

Sreekumari, M.T., Abraham, K. and Nayar, G.G. (1983) *J. Root Crops*, **9:** 63- 68.

Srivastava, G.S., Shukla, D.S. and Awasthi, D.N. (1973) *Indian Fmg.*, **23:** 32.

Stamford, N.P., Medeiros, R. and Mesquita, J.c.P. (1995) *RevistaBrasileira de Ciencia do Solo*, **19:** 49-54.

Tamez G., P. (1987) M.Sc. Thesis; Centro de Investigación y Estudios Avanzados del Instituton Politecnico Nacional Unidad Irapuato (Mexico). p. 108.

Thung, T.H. and Hadiwidjaja, T. (1957) De Heksenbezemziekte bij Leguminosen. T. Pl.ziekten, **63:** 58-63.

Vaillant, V. and Desfontaines, L. (1995) *Physiologia Plantarum*, **93:** 558-562.

Vansuyt, G. and Zinsou, C. (1989) *Proceedings of the 25th Caribbean Food Crops Society Meeting* (Degras, L. Ed.), Guadaloupe, Publication INRA Antilles-Suyane. 2- 8 July 1989. pp. 497-507.

Varshney, R.K., Glaszmann, J.C., Leung, H. and Ibaut, J.M.R. (2010) *Trends Biotechnol.*, **28:** 452-460.

Vaz, F., Silva, J.B.V., da, Matos, M.S., da silva, J.B.V. and Sorensen, M. (1998) *Proc. 2nd International Symposium on Tuberous Legumes*, Celaya, Mexico, 1996, pp. 285-290.

Venthou-Dumaine, A. and Vaillant, V. (1998) *Proc. of the 2nd International Symposium on Tuberous Legumes*, Celaya, Gaunajuato, Mexico, 5-8 August, 1996, pp. 291-296.

Wassens, R. (2011) *M.Sc. Thesis*, Wageningen University, The Netherlands.

Xu, Z.G., Ji, G.H. and Wei, Y.D. (1999) *Acta PhytopathologicvaSinica*, **29**: 354-359.

Zanklan, A.S., Becker, H.C., Sorensen, M., Pawelzik, E. and Gruneberg, W.J. (2018) *Genet. Resour. Crop Evol.*, **65**: 811–843.

Zinsou, C. and Venthou-Dumaine, A. (1990) *Proceeding of 24th Annual CFCS Meeting, Jamaica*, Ed. CFCS. August 15- 20. 1988. pp. 156-162.

Zinsou, C., Vansuyt, G. and Venthou-Dumaine, A. (1987a) *Agronomie*, **7**: 821-825.

Zinsou, C., Vansuyt, G. and Venthou-Dumaine, A. (1987b) *Proceedings of Joint CFCS/CAES Meeting*, Antigua, August 23-28, 1987.

Zinsou, C., Venthou-Dumaine, A. and Vansuyt, G. (1988) *Proceedings VIIth Symposium of ISTRC*, Paris. pp. 875-890.

9

FABA BEAN

T. Bhattacharjee and R. Nath

1.0 Introduction

Faba bean (*Vicia faba* L.) is a cool season grain legume crop with the potential to be grown as multi-purpose crop in areas with short growing season. It is also known as broad bean, horse bean, ticks bean (small types) and windsor bean, is a flowering species in the pea and bean family Fabaceae. Some of the other names such as Bakela (Ethiopia), Boby Kurmouvje (former USSR), Faveira (Portugal), Ful masri (Sudan), Feve (French) and Yeshil Bakla (Turkey) are used in different parts of world (Singh *et al.*, 2010). Broad or faba bean is a legume crop grown primarily for its edible seeds (beans) consumed by humans worldwide. Faba bean is a multipurpose crop used for both food and fodder (hay, silage and straw). Faba bean is a much appreciated food legume in the Middle-East, the Mediterranean region, China and Ethiopia. Various species of Fabaceae are currently growing in temperate areas, humid tropics, arid regions, high lands, savannas, and lowlands, and there are even a few aquatic legumes (Wrigley *et al.*, 2015).

2.0 Composition and Uses

2.1.0 Composition

It is rich in nutritive value which is given in Table 1.

Table 1: Composition of Faba Bean Pod (per l00g of edible portion)*

Moisture	85.4 g	Iron	2.5 mg
Protein	4.5 g	Vitamin A	14 I.U.
Fat	0.1 g	Riboflavin	0.22 mg
Fibre	3.2.g	Thiamine	0.08 mg
Carbohydrates	7.2 g	Iron	1.4 mg
Calcium	50 mg	Vitamin C	12.0 mg

*Duke (1981), Hazra and Som (2006).

2.2.0 Uses

Faba bean is a multipurpose crop used for both food and fodder (hay, silage and straw). The fresh and dry seeds of faba bean are used for human consumption; they are highly nutritious because they have a high protein content (up to 35 per cent in dry seeds), and are a good source of many nutrients, such as K, Ca, Mg, Fe, and Zn (Lizarazo *et al.*, 2015; Longobardi *et al.*, 2015; Neme *et al.*, 2015). Faba bean seeds also contain several other bioactive compounds, such as polyphenols (Turco *et al.*, 2016), carotenoids (Neme *et al.*, 2015), and carbohydrates (Landry *et al.*, 2016). Faba beans intended for human consumption are harvested when immature. The dried seeds are cooked, canned or frozen. Mature seeds are roasted and eaten as snacks in India, or ground to prepare falafel, sauces and various food ingredients such as meat extenders or skim-milk replacers (Muehlbauer and Tullu, 1997). When faba beans are intended for livestock feeding, small-seed varieties with low-tannin, low vicine-convicine and low-trypsin inhibitor contents are preferred (McVicar *et al.*, 2013). Faba beans have been suggested as an alternative protein source to soybean for livestock in Europe (Smith *et al.*, 2013; Jezierny *et al.*, 2017). Faba bean plants can be used to make good quality silage (McVicar *et al.*, 2013). Faba bean straw is valued and considered a cash crop in Egypt and Sudan (Muehlbauer *et al.*, 1997). In Sweden, it was used as a lignocellulosic biomass to produce bioethanol and biogas (Petersson *et al.*, 2007). Faba bean is grown for green manure production or as a legume ley in cereal/legume rotations (Muehlbauer and Tullu, 1997; McVicar *et al.*, 2013). Broad bean is rich in tyramine and is an amino acid that helps regulate blood pressure, thus should be avoided by those taking monoamine oxidase inhibitors. Raw broad bean also contains the alkaloids vicine and convincine which can induce haemolytic anemia in patients with the hereditary condition glucose-6-phosphate de-hydrogenase deficiency. This potentially fatal condition is called favism after fava bean. Faba bean accumulates a large amount of L -Dopa in its various parts (Etemadi *et al.*, 2015). L -Dopa, a precursor of dopamine currently used as a major ingredient in treating Parkinson's disease and hormonal imbalance (Surwase *et al.*, 2012; Inamdar *et al.*, 2013; Hu *et al.*, 2015; Etemadi *et al.*, 2018a).

Table 2: Nutrient Content of Mature Faba Bean Seed (per 100 g)*

Protein	26.2 g	Fat	1.3 g
Fiber	6.8 g	Carbohydrate	59.4 g
Thiamine	0.38 mg	Riboflavin	0.24 mg
Niacin	2.1 mg		

* Duke (1981).

3.0 Origin and Taxonomy

Knowledge of the wild progenitor and area of origin of the genus, and subsequent steps in the domestication of *V. faba* L., is scarce and disputed (Shiran *et al.*, 2014). Faba bean originated from the Middle-East in the prehistoric period. Seeds dated from 6250 BCE have been found in Jericho (McVicar *et al.*, 2013), while China seems to be a secondary center of faba bean genetic diversity (Zong *et al.*, 2010). In any case, faba bean can be considered one of the earliest domesticated crops in light of numerous archaeological findings in Eurasia and Africa which date back to the early Neolithic (Duc *et al.*, 2015a). The Chinese used them for food almost 5,000 years ago, and they were cultivated by the Egyptians 3,000 years ago, by the Hebrews in biblical times, and a little later by the Greeks and Romans (Mihailovic *et al.*, 2005; Singh and Bhatt, 2012b). Probably, it was introduced by Europeans as a garden crop into India during the Sultanic period (1206-1555), during which its cultivation has been mentioned (Naqvi, 1984; Razia Akbar, 2000). Faba bean is now widespread in Europe, North Africa, Central Asia, China, South America, the USA, Canada and Australia.

Evolution of the species was accompanied by intensified cultivation and also with selection for different desirable traits. Several wild species (*Vicia narbonensis* L. and *V. galilaea* Plitmann and Zohary) are taxonomically closely related to the

Uses of Faba Bean

Faba bean floor

Faba bean protein powder

Canned faba bean

Promising Cultivars of Faba Bean

Genotype FLIP15-196FB

Pusa Udit

cultivated crop, but they contain 2n = 14 chromosomes, whereas cultivated faba bean has 2n = 12 chromosomes (Cubero, 1974). The genotypes of *V. faba* are commonly classified into three main botanical varieties according to seed size: (a) *V. faba* var. *major* (broad beans or windsor beans) with large seeds, (b) *V. faba* var. *minor* (ticks bean or pigeon bean) with small seeds, and (c) *V. faba* var. *equina* (horse beans) with medium seeds (Cubero, 1974; Crépon *et al.*, 2010; Pietrzak *et al.*, 2016), the first two of which are relevant in European agriculture. However, faba bean germplasm is also grouped into spring and winter types, according to frost tolerance, delimiting target climatic zone, and sowing time, and according to the ability to adapt to oceanic or continental (*i.e.*, drought-prone) climates (Link *et al.*, 2010; Flores *et al.*, 2013). Cultivar groups featuring differential root system architectures exist independently of botanical variety within Europe, but that, for example, cultivars from Portugal possess greater and coarser but less frequent lateral roots at the top of the taproot in comparison with Northern European cultivars, potentially enhancing water uptake from deeper soil horizons (Zhao *et al.*, 2018). *Vicia faba* has a large genetic diversity. According to Duc *et al.* (2010, 2015b), more than 38,000 accessions of faba bean germplasm are conserved globally in numerous gene banks, as well as at the International Center for Agricultural Research in Dry Areas (ICARDA).

4.0 Cultivars

In faba bean, variety or cultivar selection requires achieving a balance between adaptability to a specific environment, disease tolerance, purpose of cultivation, and marketability. Faba bean varieties vary significantly in seed size and colour. Small-seeded types (*Vicia faba* L. subsp. *minor*) also called faba bean or tick bean are commonly used for feeding animals and cover cropping. Medium and large size seed types (*V. faba* subsp. *major* Harz) are called broad bean and commonly used as dry and green beans (Crépon *et al.*, 2010). Although modern cultivars dominate in Australia, Europe and Canada, traditional landraces are grown widely in many countries and a mix of traditional and modern cultivars are grown in other countries.

The following cultivars of vegetable faba bean have been developed by Agricultural Research Institutes, SAUs and introduced in India.

Masterpiece White Long Pod

It is a climbing perennial with fragrant, black and white flowers followed by long, leathery green pods containing edible green beans in summer. It has excellent length of pod and table quality. It has introduced from United Kingdom.

Jawahar Selection 73-31

It is an improved selection from Madhya Pradesh, India.

Snowbird

It is a medium maturing cultivar (110 to 120 days to maturity) based on early seeding. Seed size is approximately 550 grams/1000 seeds (about 2 times larger than normal field pea seed size) (Hussein and Saleh, 1985). Zero tenin faba bean cultivar is mainly used as a feed for swine and meat poultry.

A few selections made at Bihar, India are BR-1 (black seeded) and BR-2 (yellow seeded).

Pusa Sumeet

An improved cultivar of broad bean, IARI, New Delhi, India. Plants are 75 cm tall having on an average 5-7 branches/plant and bear about 100 pods. The pod length and thickness is 6.0 cm and 1.3 cm, respectively. It has attractive dark green pods and borne in cluster average yield potential is 180 q/ha.

Pusa Udit

The pods of this cultivar are extra log, flattish and light green. The fresh seeds are attractive green, good in taste. This is a dual purpose broad bean cultivar. Both tender pods and dried seeds are edible. The cultivar is suitable for packageing and transport. Average yield potential is 176.30 q/ha.

Some important introduced cultivars are Imperial White Windsor, Giant Four Seeded Green Windsor, Imperial Green Windsor (Gopalakrishnan, 2007).

In 2014, three cultivars of high-yielding faba bean were released in Sudan which can tolerate temperatures up to 35 °C including 'Hudeiba93', 'Basabeer' and 'Ed-Damer' (Etemadi *et al.*, 2019). The cultivars released in Egypt in 2011, 'Nubaria 3'are drought tolerant (ICARDA, 2018). In England, the cultivar Windsor is currently the dominant cultivar available to the growers (Etemadi *et al.*, 2017).

5.0 Soil and Climate

Faba bean can be grown as a winter or a spring crop in wetter areas. The crop is very cold hardy, but cannot withstand excessive heat at the time of flowering (Singh *et al.*, 2013). This annual legume grows best under cool and moist conditions. Hot dry weather is injurious to the crop, so early planting is an important operation for better plant establishment (Singh *et al.*, 2013). Faba bean is highly tolerant to the frost condition but faba bean is susceptible to frost during its reproductive stages (Maqbool *et al.*, 2010). Generally, seed germination of most legumes is sensitive to low soil temperature. However, faba bean is one of the few cool-season grain legumes and its germination can tolerate cold soil temperature better than most grain legumes (Etemadi *et al.*, 2019). It is a hardy plant and can withstand cold temperature as low as 4°C. The best temperature for pod setting is 15-25°C and flowers and pods may drop at very high temperatures during rainy season (Singh and Nath, 2012). Rainfall of 650 to 1000 mm per annum evenly distributed is ideal for faba bean (Abdel, 2008; Gasim and Link, 2007). In the tropics and subtropics, faba bean can be grown above 1200 m and up to an altitude of 2500 m.

Faba bean is generally considered day-neutral, while some accessions require long-day conditions in order to flower. However, thermal time is the most important contributor to flowering progress in faba bean, with approximately 83-1000 days above 0°C being required; winter faba bean genotypes require vernalization (Patrick and Stoddard, 2010).

Faba bean grows best on fine-textured soils but tolerates nearly any soil type (Jensen *et al.*, 2010). The ideal soil pH for growing faba bean is " 7 (Köpke and

Nemecek, 2010). Liming is required for faba bean cultivation when soil pH level is below 6. Faba bean is more tolerant to acid soil conditions than most legumes (Singh *et al.*, 2010). Although sandy loams are suitable for growing faba bean but more frequent irrigation will then be required. Faba bean has relatively shallow roots, thus the crop may suffer from drought stress in soils that dry quickly. Faba bean seems to be tolerant to short period of water-logging (Tekalign *et al.*, 2016).

6.0 Cultivation

6.1.0 Season and Sowing

Sowing season affects the duration of the vegetative and reproductive stages that contributes to differences in yield (Krarup, 1984). Delay in sowing from the optimum date leads to reduction in crop yield in differential degrees. Days to flowering, maturity and dry matter production of faba bean significantly decreased with delay in sowing the seeds after 30th October (Berhe, 1998). In India, the growth period of the crop varies from one agro-climatic zone to other that affects the vegetative and reproductive period leading to difference in potential yield. The optimum dates of sowing of Broad Bean in North Eastern Plane Zone (NEPZ), West Bengal are first week of October to third week of November and before fifteenth December for irrigated, timely sown crop and irrigated late sown crop respectively (Anon., 2005). In cooler agro-climatic zones, sowing is postponed until the end of winter or early spring to prevent frost damage (Sallam *et al.*, 2015).

Traditionally faba bean is seeded directly into the soil. However, to ensure early sowing in shorter-season areas, faba bean may benefit from transplanting compared with direct seeding (Etemadi *et al.*, 2015). In shorter-growing season region transplanting faba bean provides the opportunity for double cropping and avoids incident of some diseases such as chocolate spot (Etemadi *et al.*, 2018b). Transplanting seedlings also offers some other potential benefits including higher yield, better survival rate, early flowering, and earlier harvest (Lee *et al.*, 2018). Faba bean is usually sown in rows 10–30 cm apart (Bozoglu *et al.*, 2002; Yucel, 2013), using either a spacing drill (placing 2-3 seeds per hole) or seed drill. The required seed amount ranges between 70 and 200 kg/ha, dependent on seed size and planting density. The recommended sowing depth is 5-8 cm (Siddique and Loss, 1999). Germination takes place in 4-12 days, and the optimum temperature for germination is 20°C (Khamassi *et al.*, 2013).

6.2.0 Manuring and Fertilization

In general, legumes are known as self-sustaining crops regarding N. However, the efficiency and magnitude of biologically mixed N varies greatly among the legumes. Legume species that meet 50-90 per cent of their total N requirements from symbiotic N fixation are considered effective N fixer (N'Dayegamiye *et al.*, 2015). Faba bean and soybean with 75 per cent and 68 per cent, respectively are ranked as high capacity legumes for N fixation (Hoffmann *et al.*, 2007; Peoples *et al.*, 2009). Depending on the growing conditions, faba bean can fix up to 200 kg N/h (Etemadi *et al.*, 2018c).

Nitrogen fertilization is not generally required, but the application of "starter" nitrogen fertilization at a rate of 20 kg/ha seems to enhance the nodulation process in faba bean plants (Mohamed and Babiker, 2012).

Phosphorous (P) plays important roles in nodulation and biological N fixation, photosynthesis, and nutritional values of legumes including faba bean (Haling *et al.*, 2016; Makoudi *et al.*, 2018). Phosphorus application often results in increased yield and biomass of faba beans in P-limited soils (Daoui *et al.*, 2012), indicating that P fertilizer is essential for grain production in faba beans (Nebiyu *et al.*, 2016). Furthermore, legume BNF is an energy intensive process that requires large amount of phosphorus (P). Thus, P fertilization at a rate of 40 kg/ha can often enhance the nodulation process and N_1 fixation, and increase yield (Bolland *et al.*, 2000; Adak and Kibritci, 2016).

Several other studies show that faba bean crops also respond to S and K fertilization (Niewiadomska *et al.*, 2015). Nevertheless, S or K fertilizers are rarely applied, because faba bean is cultivated as a low-input crop. Faba bean often responds positively to K application, especially in dry growing seasons (Baróg *et al.*, 2018). Availability of adequate K to faba bean plants has a positive effect on N fixation (Römheld and Kirkby, 2010; Gucci *et al.*, 2019). Furthermore, micronutrient (*e.g.*, zinc, magnesium, molybdenum, sulphur and boron) deficiencies are rare and can easily be corrected through foliar sprays.

Inoculating faba bean field or seeds with *Rhizobium* is unnecessary in traditional cultivation areas. However, it is advisable to test their presence in the soil in areas where faba beans or other legumes have not been grown for several years. If absent, the crop can be inoculated with *Rhizobium leguminosarum* bv. *viciae* (Cubero, 2017). Dual inoculation with *Rhizobium* and arbuscular mycorrhizal fungi has been reported to be more efiective than inoculation with *Rhizobium* alone in promoting faba bean growth, particularly in alkaline soils; this reflects the existence of synergistic relationships between the two inoculants (Abd-Alla *et al.*, 2014).

6.3.0 Irrigation and Interculture

Faba bean usually grows without irrigation, with the exception of crops cultivated in very dry and hot climatic zones. Thus, production is highly dependent on the amount of and variation in rainfall during the growing season (Oweis *et al.*, 2005). In semiarid regions, climate changes can affect water use eficiency and growth in faba bean (Guoju *et al.*, 2016), given its sensitivity to drought (Ghassemi-Golezani *et al.*, 2009; Alghamdi *et al.*, 2015). In the Mediterranean region and similar dry and hot climatic zones, faba bean production without irrigation may be possible if cultivation takes place during the cold season. Moreover, early sowing in autumn is considered an effective strategy for avoiding water stress during the seed filling stage (Loss and Siddique, 1997). Alternatively, faba bean crops can be irrigated at the seed filling stage in order to avoid penalties in yield during drought. Additionally, Knott (1999) reports that faba bean production is usually increased by irrigating spring crops during the flowering stage and early podding. Between 231 and 297 mm of water are required to produce 3-4.4 t/ha of faba bean dry biomass (Bryla *et al.*, 2003). The development of drought-tolerant faba bean varieties is a key challenge in

Flowering and Field Growing of Faba Bean

Flowering and pod development in faba bean

Field view of faba bean cultivation

Fresh Pods of Faba Bean Selling in Local Market

(A) Fresh faba bean pods in a free market, (B) Fresh faba bean seeds,(C)Fresh faba bean pods in a bag after harvesting, and (D) Fresh faba bean long pod type after harvesting.

achieving increased and more stable production levels (Khan *et al.*, 2010; Siddique *et al.*, 2013). Several genotypes are considered tolerant to drought and can be exploited in breeding programs in order to develop drought-tolerant varieties (Ali, 2015).

Weed infestation is a major constraint in faba bean production, and can reduce yield by up to 50 per cent (Frenda *et al.*, 2013). Faba beans are poor competitors with weeds, particularly in the seedling stage (Ali *et al.*, 2000). Thus, early weed removal during the period between 25 and 75 days after sowing is necessary if a high yield is to be obtained (Tawaha and Turk, 2001). Faba bean exhibits a superior ability to compete with weeds compared with other pulse crops, such as chickpea, due to its more vigorous early growth and greater plant height (Frenda *et al.*, 2013). This makes integrated weed control essential for successful crop production. Select fields with light weed pressure. Do the primary tillage several weeks before planting and kill emerged weeds with shallow tillage just ahead of planting. Consider rotary hoeing of fields 7 to 10 days after planting and use a row cultivator if rows are 50 cm or more apart. Nevertheless, the application of herbicides is a primary method in controlling weeds in conventional faba bean production. The herbicides pendimethalin, clomazone, bentazon, and propaquizafop are registered for use on this crop. The first two are applied pre-emergence to control broadleaved and grass weeds; quizalofop-*p*-ethyl and propaquizafop are applied post-emergence to control grass weeds such as *Phalaris* spp. and *Lolium* spp., while bentazon is applied post-emergence to control broadleaved weeds. Crop rotation with spring crops can significantly reduce weed pressure, while allowing field application of herbicides that are not registered for use on faba bean (Karkanis *et al.*, 2016b).

7.0 Harvesting and Yield

Faba bean crops cultivated for fresh seed consumption may be harvested either manually or mechanically once the pods are filled, but before they start to dry. Pods are harvested by hand two to three times during the harvesting period in crops cultivated in small areas for fresh consumption. When faba bean plants are cultivated for their dry seeds, they can be harvested using a conventional cereal combine harvester. Similar to other pulses, proper selection of the harvest stage is critical if seed loss is to be minimized (Karkanis *et al.*, 2016a); seeds should be harvested when the moisture content is 14-15 per cent (Jilani *et al.*, 2012).

The dry seed yield in faba bean crops ranges between 1.6 and 5.2 tons/ha (Argaw and Mnalku, 2017; Youseif *et al.*, 2017), and the fresh pod yield ranges between 1.34 and 17.04 tons/ha (Baginsky *et al.*, 2013; Etemadi *et al.*, 2017). Faba bean yield components, such as the thousand seed weight, number of pods per plant, and number of seeds per pod, together with the duration of phenological stages and plant height, are correlated to grain yield (Sharifi, 2014; Bodner *et al.*, 2018).

8.0 Diseases and Pests

8.1.0 Diseases

Major diseases of faba bean and their control measures are described below:

Fungal Diseases

8.1.1 Chocolate Spot

Chocolate spot (*Botrytis fabae*) is one of the most important economic disease that damages the foliage, limits photosynthetic activity and faba bean production (Torres *et al.*, 2004). Long periods of high humidity and high temperature promote the switch from a non-aggressive phase to an aggressive phase of the disease (Terefe *et al.*, 2015). The aggressive phase of the disease is favoured also by low levels of K and P in the soil and high plant population density which leads to more humid conditions within the plant canopy. Symptoms for non-aggressive chocolate spot include small red brown lesions on leaves of the plant and sometimes on stems and pods (Haile *et al.*, 2016). Prevention is the most effective management strategy. Early planting is recommended to avoid high humidity and high temperatures in late spring and early summer (Kora *et al.*, 2017). Chocolate spot severity in faba bean is reduced by frequent application of mancozeb (Sahile *et al.*, 2008a). Spray the crop with 0.1 per cent carbendazim or hexaconazole or 0.2 per cent chlorothalonil at 10 days interval gives good control.

8.1.2 *Ascochyta* Blight

It is caused by *Ascochyta fabae* Speg. (teleomorph *Didymella fabae* Jellis and Punithalingam), and is one of the most serious pathogens, causing up to 30 per cent loss in yield (Omri Benyoussef *et al.*, 2012; Ahmed *et al.*, 2016). Although the application of fungicides, such as azoxystrobin and chlorothalonil, considerably reduces ascochyta blight infection, integrated management practices (*e.g.*, crop rotation, use of resistant varieties, and late sowing) are crucial to successful control (Ahmed *et al.*, 2016).

8.1.3 Powdery Mildew

This is a fungal disease caused by *Erysiphe cichoracearum*. This fungus overwinters on plant residues or alternate hosts. Disease emergence is favoured by warm, dry weather with cool nights that result in dew formation. Yellow spots on upper surface of leaves, and powdery gray-white areas coalesce to cover entire plant. If the plant is infected heavily, it may appear light blue or gray in color (Trabanco *et al.*, 2012). Use of resistant varieties and when possible use of overhead irrigation that washes the fungus from leaves and reduces its viability are preventive practices. Also, recommended is early planting to avoid high air temperature and humidity. Application of S-based fungicides may be required to control heavy infestations (Van Emden *et al.*, 1988).

8.1.4 *Fusarium* Root Rot

It is caused by *Fusarium solani*. Damage caused by the emergence of the fungal disease is worsened by warm, compacted soils, limited soil moisture, and poor soil fertility. The disease is noted by stunted plant growth and yellowing, necrotic basal leaves, brown, red, or black streaks on roots that coalesce as they mature. Lesions may spread above the soil line (Abdel-Kader *et al.*, 2011).

Diseases of Faba Bean

Ascochyta blight

Chocolate spot disease

Alternaria leaf spot

Sclerotia collar rot

Rust

Yellowing and leaf rolling of BLRV

Mosaic symptoms of BYMV

Mosaic symptoms of PSbMV

Pests and Disorders of Faba Bean

 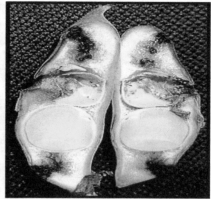

Pod borer infestation in faba bean

Black fly infestation in faba bean

Water congestion   Hail damage

Bacterial Disease

8.1.5 Leaf Blight

This bacterial disease is caused by *Xanthomonas campestris*, (syn. *Xanthomonas axonopodis*). Leaf blight is another bacterial disease that can be introduced by contaminated seeds and over-wintering of bacteria in crop residues (Belete and Bastas, 2017). This disease is favored by warm temperatures and wet, humid conditions. Water-soaked spots on leaves which enlarge and become necrotic. Spots may be surrounded by a zone of yellow discoloration, and lesions can coalesce and give the plant a burnt appearance (Buruchara *et al.*, 2010). Dead leaves remain attached to the plant and circular, sunken, red-brown lesions may be present on pods. The lesions on pod may leak during humid conditions. Use of clean seeds (Taran *et al.*, 2001; Gillard *et al.*, 2009), resistant varieties, seed treatment with an appropriate antibiotic prior to planting and spraying plants with an appropriate protective, copper-based fungicide before appearance of symptoms are used as effective methods of treatments (Omafara, 2009; Buruchara *et al.*, 2010).

8.1.6 Bacterial Brown Spot

It is caused by *Pseudomonas syringae*. The bacteria overwinter in crop residues and are more severe when the foliage is wet for a long period. Small, dark-brown necrotic spots on leaves may be surrounded by an area of yellow tissues. Water-soaked spots on pods turn brown and necrotic, and pods may twist and distort in the area of infection (Schwartz *et al.*, 2005). Use of clean seeds, proper crop rotation, and the removal of crop residues from the field after harvest are considered as preventive measures (Harveson and Schwartz, 2007).

Viral Disease

Faba bean is also susceptible to viruses, with the principal sources of infection being faba bean necrotic yellows virus (FBNYV) and bean yellow mosaic virus (BYMV) (Ortiz *et al.*, 2006; Shiying *et al.*, 2007).

8.2.0 Pests

Major insect pests of faba bean and their control measures are described below:

8.2.1 Black Bean Aphid (*Aphis fabae*)

It can cause severe damage to faba bean (Webster *et al.*, 2010). When using seedlings, check for aphids before transplanting. Reflective mulches such as silver coloured plastic can deter aphids from feeding on plants by limiting landing of the pest. Sturdy plants can be sprayed with a strong jet of water to knock aphids from leaves. Insecticides are required generally only to treat aphids if the infestation is very high since plants generally tolerate low or medium leaf infestation. Insecticidal soaps or oils such as neem or canola are usually the best insecticides for control (Birch, 1985).

Other insects that infect faba bean crops are the pea leaf weevil (*Sitona lineatus* L.) and broad bean weevil (*Bruchus rufimanus* Boh.; Evenden *et al.*, 2016; Seidenglanz and Hunady, 2016). *S. lineatus* adults feed on the foliage, while the larvae feed on faba bean and pea root nodules, affecting their ability to fix nitrogen (Cárcamo *et al.*, 2015); treating seeds with thiamethoxam could be useful in controlling this insect (Cárcamo *et al.*, 2012). Furthermore, storage pests, such as *B. rufimanus*, can cause significant yield losses in legumes; insecticides are however effective against them (Keneni *et al.*, 2011).

8.3.0 Physiological Disorder

8.3.1 Hail Damage

Symptoms occur where hail has hit the foliage white streaks and other markings appear on leaves and stems. Leaves are torn and pods smashed open. Stems may be damaged and tops of plants broken off. Exposed seeds discolour rapidly.

8.3.2 Water Congestion

Primary symptoms is black discolouration occurs at leaf tips. Water congestion causes rupturing of apical cells of developing leaflets and tips of newly formed leaves die back. Leaf tips of newest leaves are pinched and necrotic following heavy rain during the summer when plants are growing rapidly. Damage is not significant and usually affects only a single set of leaves.

9.0 Seed Production

There is no difference between the methods of raising crop for production of green pods and seeds. In case of seed crop, land in which one cultivar of faba bean was grown in the previous year should not be used for growing another cultivar in the following year to avoid contamination with the self sown plants from the previous crop. Faba bean is a self-pollinated plant with significant levels of cross-pollination (Chen, 2009). Seed production fields could be subjected to the following contaminations:

i) Genetic contamination caused by cross-pollination with other varieties of the same species growing in the field or surrounding area; and

ii) Mechanical contamination caused by mechanical mixture with seeds of other varieties of the same crop.

The contaminants may genetically derive from the crop of previous seed production field, mechanical mixture of undesirable seed in the prior production fields or in the seed lots. They may also derive from volunteer plants resulting from seed left by the prior crop and seed brought to the field by water, birds, animals, people or agricultural equipment. The effects of genetic and physical contamination can be reduced by appropriately and distantly isolating and rouging off-types and by avoiding mechanical mixtures.

The way fields should be distantly isolated varies depending on the type of crop and stage of seed production. Faba bean require isolated distance of 400 m and 200 m, for basic seed and certified seed production, respectively, from other fields of different varieties of the same crop on each side of seed production field (Keneni *et al.*, 2001). This method is effective to prevent out crossing in faba bean and mechanical mixtures. However, the distance for faba bean could be reduced if there are physical barriers between two fields cropped with different varieties of the same species that prevent easy movement of pollen from one field to another variety of the same species through pollinators. Faba bean production field can also be surrounded with species like rapeseed that does not inter-cross with faba bean but attracts the same pollen-transferring insects. The assumption is that the pollen-transferring insect first visit border plants and lose their atypical faba bean pollen to them.

However, off-types should be removed from faba bean fields before they shed pollen to prevent out-crossing. They can be identified by their deviation from the genotype *i.e.*, plant type, size, pigmentation, flower colour, *etc.* Their size and position out of rows can also help to trace and identify volunteer plants. Generally, the proportion of maximum permissible off-type plants to both crops is 0.1 per cent in basic seeds, and 0.2 per cent and 0.5 per cent in certified seeds of faba bean. Equally important is to rogue plants that look diseased and defected (Keneni *et al.*, 2001).

Late harvesting may cause shattering and, shedding and rotting of pods if untimely rain is encountered. Therefore, harvesting should be done when leaves and pods dried and the grain moisture content is significantly reduced. Under the climatic condition of Ethiopia, where the time of harvesting usually coincides with the commencement of dry season, it is possible to easily achieve low moisture contents while the crop is in a field. Faba been is not suitable for combine harvesting. However, simultaneous harvesting and threshing may cost-effectively be executed using manual labour where labour is available and cheap. Canvases and polythene sheets also can protect the crops in the field from rain after harvesting. Faba bean is indeterminate in growth habit; the lower pods mature earlier while the upper pods are immature. Therefore, the freshly cut crops should be left on ground for about three four weeks after harvesting. Then, after the crop gets dried, it should be fed to a stationary thresher to get clean seeds. The threshing floor, for both formal and informal seed production, should be clean and preferably cemented to avoid contamination by inert matter, weed seeds and other crop or variety seeds (Keneni *et al.*, 2001).

10.0 Crop Improvement

Faba bean has a long history of cultivation. A broad gene pool has therefore been developed over several centuries, including local landraces, mass selections from landraces, open-pollinated populations, inbred lines, and cultivars (Duc *et al.*, 2010). In addition, socioeconomic changes have led to decreases in cultivation and the disappearance of local genetic resources, with only small farms continuing to grow different landraces selected for their adaptation to local environmental conditions (Karaköy *et al.*, 2014).

Investment in legume breeding has been lower than for cereals (Fouad *et al.*, 2013; Duc *et al.*, 2015b) and, as a result, only a limited number of registered faba bean cultivars is available. For example, only 256 faba bean cultivars are currently registered for growing in Europe, and recorded at the EU database of registered plant varieties; for wheat (*Triticum aestivum*); however, there are 2415 registered cultivars (European Commission, 2017). "Aguadulce," "Extra," "Precose," "Tundra,"

"Fuego," "Extra Violetto," "Babylon," and "Pyramid" are some commercial varieties cultivated in Europe under a wide range of agro-climatic conditions. Registered varieties feature a range of highly differing characteristics although genetic variation is limited (Fouad *et al.*, 2013). Traits that should be targeted in selecting faba bean varieties include yield potential, quality, consistent performance, suitability for human consumption or the feed market, seed size, days to maturity, and standing ability, as well as resistance to disease and abiotic stress.

The primary focus of legume breeding, including faba bean, is yield. However, in many regions, faba bean crops are subject to different conditions of biotic and abiotic stress and, consequently, yield is ultimately dependent on cultivar resilience to multiple stress conditions. Hence, breeding new cultivars with increased resilience to abiotic stresses, such as heat and salinity, continues to challenge breeders (Siddique *et al.*, 2013; Nebiyu *et al.*, 2016). Furthermore, winter hardiness is an important trait in screening for cultivars to be cultivated during the cold season (Link *et al.*, 2010).

The genetic improvement of desired traits via breeding significantly depends on genetic variation in those traits. There is therefore an urgent need to collect and evaluate local genetic resources that can be used in well-designed breeding programs as donors of valuable features in the development of new improved varieties. In this sense, local landraces represent important sources for plant breeding, as they contain co-adapted genes that may prove valuable in future cultivation practices and in enhancing yield and quality (Harlan, 1975). The importance and wide variation of traits relating to morphology, agronomy, and quality have been previously investigated and demonstrated for local faba bean genetic resources of different origins: Turkey (Karaköy *et al.*, 2014), the Mediterranean (Terzopoulos and Bebeli, 2008), Albania (Nasto *et al.*, 2016), Palestine (Basheer Salimia *et al.*, 2014), and China (Zong *et al.*, 2009). However, only a small proportion of other faba bean genetic resources have so far been evaluated. Thus, efforts to characterize the available resources should be intensified, and the collection of new local resources is crucial, because of the genetic erosion that is currently identified. Furthermore, breeding programs need to incorporate a more complex evaluation and integrated use of traits (Duc *et al.*, 2015b). The above mentioned traits and others, such as root architecture (Zhao *et al.*, 2018), shoot architecture, parameters related to stomatal function (Khan *et al.*, 2010; Khazaei *et al.*, 2013), and multiple disease resistance (Torres *et al.*, 2006), are becoming increasingly important. It is apparent that any new variety should combine as many of the above mentioned characteristics as possible, in order to allow for a greater and more stable production of faba bean in specific agro-ecological zones.

Yield stability and quality is a major objective of faba bean breeders (Flores *et al.*, 2013), since yield instability is a common problem encountered in cultivating this species and is considered a main cause of the decline in faba bean acreage (Flores *et al.*, 2013). The stability of faba bean genotypes in different environmental conditions also needs to be examined (Temesgen *et al.*, 2015). Several approaches can be applied in evaluating yield stability. Temesgen *et al.* (2015) demonstrate that different stability parameters have varying effects on yield performance, and recommend the application of several stability parameters, rather than only considering yield in different years. Faba bean genotypes exhibit a strong interaction with environmental conditions (Flores *et al.*, 2013; Maalouf *et al.*, 2015; Temesgen *et al.*, 2015). In the study, Temesgen *et al.* (2015) reported that the environment (E) accounted for 89 per cent of yield variation, while the genotype (G) and the "G × E" interaction contributed 2 and 3 per cent, respectively. The breeding of genotypes adapted to specific climatic zones is recommended in order to increase yield stability (Flores *et al.*, 2013). G × E interactions are more common in faba bean than in most other crops; Bond (1987) reports that genotype × season interactions generally make a greater contribution than genotype × location.

10.1.0 Physiological Traits

The onset of heat stress affects the physiology of faba bean plants in several ways. Structural re-organization of chloroplast thylakoid membranes in response to heat stress was shown to be a reliable indicator of heat tolerance in faba bean (Gounaris *et al.*, 1984). Heat stress results in decreased chlorophyll variable fluorescence. McDonald and Paulsen (1997) showed that the whole-chain photosynthetic activity in thylakoids of faba bean plants was sensitive to heat stress, suggesting photosystem II to be the most labile process. Hamada (2001) showed that heat stress of 42°C decreased the growth rate, membrane stability and concentrations of photosynthetic pigments (chlorophyll a, chlorophyll b and carotenoids) in faba bean. In the latter study, extreme temperatures also resulted in considerable variations in the contents of cell wall components (pectin, cellulose, hemicellulose and lignin), cell wall-associated proteins, soluble sugars, starch, total lipids, glycolipids, phospholipids and sterols. Increase in leaf temperature above 32 °C significantly reduced the net photosynthetic assimilation rate in three Mediterranean faba bean genotypes (Avola *et al.*, 2008). Enzymes associated with oxidative stress such as superoxide dismutase showed different patterns of activity in tolerant and sensitive cultivars of faba bean under heat stress (Filek *et al.*, 1997). El-Tayeb (2006) studied the germination of five faba bean cultivars exposed to polyethylene glycol-induced water stress; Giza 40 showed the highest germination capacity and Giza 667 the lowest. In addition, the effects of low soil water content on plant growth, photosynthetic pigment contents, organic solutes, relative water content, lipid peroxidation, membrane stability index and the catalase and peroxidase activities were determined in leaves of 21-day-old faba bean cultivars Giza 40 and Giza 667. Their results indicated that Giza 40 was better protected against drought-induced oxidative stress. The adaptive physiological mechanisms under water deficit in faba bean have been categorized into drought avoidance, escape and tolerance mechanisms and a combination of traits in addition to crop management strategies have been advocated to increase yield under drought-affected environments (Khan *et al.*, 2010).

Excess soil salinity affects both growth and nitrogen fixation in faba bean plants, which are considered moderately tolerant of soil salinity (Bulut *et al.*, 2011). Salinity-tolerant faba bean genotypes are also available; one example is the line "$VF_1 12$," which has been reported as salt-tolerant because salt stress had no effect on its growth or nitrogen fixation (del Pilar Cordovilla *et al.*, 1995).

10.2.0 Disease Resistance

Foliar diseases are caused mainly by ascochyta blight (*Ascochyta fabae*), chocolate spot (*B. fabae*), rust (*Uromyces viciae-fabae*) and gall disease (*Olpidium viciae* Gusano). Chocolate spot can affect up to 61 per cent the faba bean productivity (Sahile *et al.*, 2008b); rust disease up to 30 per cent and gall disease up to 100 per cent (Abebe *et al.*, 2014). Faba bean necrotic yellow virus (FBNYV) is also considered to be the most important virus disease on faba bean causing up to 90 per cent yield losses in Egypt (Kumari and Makkouk, 2007). The first effective resistant sources for *Ascochyta* blight and chocolate spot were identified at ICARDA (Robertson, 1984; Hanounik and Robertson, 1989) and used by ICARDA and National Agricultural Research Systems (NARS) to develop breeding lines with good levels of resistance and a high yield potential. The Ethiopian Institute of Agricultural Research (EIAR) released several varieties with a high level of resistance to chocolate spot. EIAR researchers released several high yielding faba bean varieties through direct selection from germplasm supplied by ICARDA or by transferring good levels of resistance from ICARDA germplasm into locally adapted varieties. Among the faba bean varieties released (Temesgen *et al.*, 2015) with partial resistance to chocolate spot are 'Moti' (ILB 4432 × Kuse 2 27 33), 'Gebelcho' (ILB 4726 × 'Tesfa'), Obsie (ILB 4427 × CS20DK) and 'Walki' (ILB 4615 × Bulga 70). Recently, another variety named 'Gora' (ILB2717 1 × R878 1) has been released in Ethiopia with higher degree of resistance to chocolate spot and larger seed size than traditional cultivars. The annual rate of genetic gain made due to the breeding efforts in these released cultivars was "0.27 per cent for chocolate spot severity and 8.07 g/1,000 seeds (Temesgen *et al.*, 2015). In a study, Rubiales *et al.* (2012) report that the faba bean accessions V-26, V-255, V-958, V-1020, V-1085, V-1117, and L-831818 showed good levels of resistance to *A. fabae*. The accessions 132-1, 135-1, 174-1, BPL 710, ILB 4726, and ILB 5284 exhibited a good level of resistance to *B. fabae* infection (Villegas-Fernández *et al.*, 2012). Thus, these genotypes constitute an interesting genetic resource for future exploitation in breeding programs for developing chocolate spot-resistant cultivars.

Recent efforts were made to identify faba bean accessions for resistance to new Gall disease in Ethiopia. Among 14 cultivars tested under Ethiopian conditions, 'Degaga' and 'Nc 58' were identified moderately resistant to Gall disease (Yitayih and Azmeraw, 2017). Although most breeding programmes focus on developing resistant genotypes for a single disease of economic importance, efforts have recently been directed to develop faba bean lines with multiple disease resistance lines (Maalouf *et al.*, 2016), which are used currently in the ICARDA breeding programme to develop multiple disease resistant cultivars for target environments.

10.3.0 Insect Resistance

A number of insect pests such as Sitona weevil (*Sitona lineatus* L.), cowpea aphid (*Aphis craccivora* Koch) and black bean aphid (*A. fabae* Scopoli) cause damage by direct feeding as well as by transmission of viruses (Mwanauta *et al.*, 2015). There are currently integrated pest management options to control these insect pests as described by Redden *et al.* (2018). In addition, borer weevil (*Lixus algirus* L.) causes serious damage in faba bean in North Africa. Recently, new sources for resistance to this insect were identified and would be utilized to develop resistant cultivars (Ait taadaouit *et al.*, 2018).

11.0 Biotechnology

11.1.0 Tissue Culture

The investigation of *in vitro* regeneration of faba bean has been extensive since 1960s. The first attempts to cultivate faba bean *in vitro* focused on the optimal growth of callus tissue or suspension cultures rather than the induction of shoot morphogenesis and plant regeneration (Venketeswaran, 1962; Grant and Fuller, 1968; Mitchell and Gildow, 1975). The influence of media composition and explant source on the initiation and maintenance of cultures were tested and the conditions optimized for maximum increase in callus fresh weight. The morphogenic potential of faba bean cells cultured *in vitro* was low (Röper, 1979). To develop a system for plant regeneration from single cells, callus and cell suspension cultures were established over a long period of time, but all attempts to initiate shoot regeneration remained unsuccessful. A serious constraint in faba bean tissue culture is the deterioration of explant material and cultivated tissue as a result of the action of phenolic compounds. The effect of various chemical and physical parameters in axillary shoot cultures was examined and low temperatures were found to limit the formation of phenolics (Bieri *et al.*, 1984; Selva *et al.*, 1989). Plantlet regeneration from explants lacking pre-existing shoot meristems was first claimed by Thynn and Werner (1987). After the protoplasts were isolated from leaves and shoots (Binding and Nehls, 1978a; Binding and Nehls, 1978b) and suspension cells (Röper, 1981), the regeneration of faba bean plant from protoplasts was reported (Tegeder *et al.*, 1995). Somatic embryogenesis in callus and suspension cultures derived from immature cotyledons of faba bean (Griga *et al.*, 1987), a protocol in which plantlets derived from somatic embryos could be grown to maturity (Pickardt *et al.*, 1989) and an improved protocol combined with the *Agrobacterium tumefaciens*-mediated gene transfer (Pickardt *et al.*, 1991) were subsequently been achieved. Several in-vitro techniques would be very useful for faba bean breeding. By means of protoplast fusion and regeneration or by embryo-rescue assisted interspecific crossing, *e.g.* resistance to black aphid, as occurring in the related species *Vicia johannis*, could probably be introduced to *V. faba*. Still, these techniques are not yet available for faba bean. The same is true for any approach to produce doubled haploid lines.

11.2.0 Molecular Markers

The large genome size of faba bean may be largely explained by a high number of retrotransposon copies, microsatellites and even genes which are the basis of the sequence variability that can be explored in genomes (Pearce *et al.*, 2000). Morpho-agronomical genetic diversity based assessment is classic and imperative for their effective exploitation in plant breeding schemes and their efficient conservation and management. Molecular marker systems have provided a new avenue for evaluating germplasm and assessing the genetic diversity in faba bean populations. They are not affected by environmental factors or by development stages and can save time and costs in selection (Pozjaʹrkovaʹ *et al.*, 2002; Zeid *et al.*, 2003; Roman *et al.*, 2004;

Terzopoulos and Bebeli 2008; Zong *et al.*, 2009; Torres *et al.*, 2010), and can contribute to a more holistic picture of genetic diversity within a collection of populations (Curley and Jung 2004). The development of a wide range of molecular tools and their utilization will have a significant impact on the development of elite faba bean lines with superior agronomic attributes, and identification of genes/genotypes for specific traits of interest. Molecular markers and isozyme polymorphisms have been reported for faba beans (Table 3), but to a limited extent compared to other major crop species.

Table 3: Molecular Markers and their Applications in Faba Bean

Application	Marker Type	Main Output	Reference
Taxonomy	RFLP, RAPD	*V. faba* is more closely aligned to species from the sections Hypechusa and Peregrinae than to those in the narbonensis complex	Ven *et al.* (1993)
Genetic diversity	RFLP	Importance of genetically diverse faba bean parents to create a linkage map based on molecular markers	Van de Ven *et al.* (1990)
		Classification of germplasm and identification of divergent heterotic groups	Link *et al.* (*1995*)
		Varietal identification and genetic purity assessment of F$_1$ hybrid seeds	Alghamdi (2003)
	AFLP	Verification of pedigree relationships	Zeid *et al.* (2003)
		Genetic diversity assessment of Chinese winter and spring faba bean	Zong *et al.* (2009, 2010)
	ISSR	Genetic diversity analysis of Mediterranean faba bean	Terzopoulos and Bebeli (2008)
		Genetic diversity assessment in faba bean	Alghamdi *et al.* (2011)
		Genetic variation among eight faba bean genotypes differing in drought tolerance	Al-Ali *et al.* (2010)
	SSAP	Distinction between geographic origins of *V. faba* genotypes	Sanz *et al.* (2007)
Mapping	RFLP,RAPD, SSR, isozyme and allozyme markers	Construction linkage map of *V. faba*	Torres *et al.* (1993, 1995, 1998), Patto *et al.* (1999)
		Development and characterization of microsatellite markers from chromosome 1-specific DNA libraries	Poz¡a´rkova´ *et al.* (2002)
		Development of a composite map in *Vicia faba*	Roman *et al.* (2004)
		Comparative genetic map	Ellwood *et al.* (2008)
Marker development for gene tagging and MAS	CAPs and SCARs	Vicine, convicine and tannins contents in faba bean	Gutierrez *et al.* (2006, 2007, 2008)
	SSR	Growth habit	Avila *et al.* (2006, 2007)
		New loci from Orobanche resistant	Zeid *et al.* (2009)
		Development and characteri- zation of 21 EST-derived microsatellite	Ma *et al.* (2011)

Application	Marker Type	Main Output	Reference
QTL mapping	RFLP, RAPD, SSR, isozymes and allozymes markers	First QTL map in faba	Ramsey *et al.* (1995)
		Mapping of quantitative trait loci controlling broomrape resistance	Roman *et al.* (2002, 2003)
		Locating genes associated with *Ascochyta fabae* resistance	Diaz *et al.* (2004, 2005)
		RAPD markers linked to the Uvf-1 gene conferring hypersensitive resistance against rust	Avila *et al.* (2003)
		Ascochyta blight resistance	Avila *et al.* (2004)
		Frost tolerance	Arbaoui *et al.* (2008)

Several RAPD markers linked to a gene determining hypersensitive resistance to race 1 of the rust (*Uromycese viciae-fabae*) have been reported by Rojo *et al.* (2007). Molecular breeding for resistance to broomrape, ascochyta blight, rust and chocolate spot have been obtained. A major aim for any crop breeding program is the development of good quality lines with an adequate resistance/tolerance to yield-reducing stresses (Gutierrez *et al.*, 2006). The use of model legumes for comparative functional genomics may bring some new perspectives and enhances faba bean breeding efforts. In this way, identification of QTLs and/or candidate genes involved in stress tolerance and/or quality may be used to produce transgenic lines and/or these traits can be applied to breeding programs (*e.g.* MAS) (Hougaard *et al.*, 2008).

The use of marker assisted selection (MAS) can complement conventional field breeding by speeding up the selection of desirable traits and increasing selection efficiency. Recently, markers linked to a gene controlling growth habit or to select against traits affecting the nutritional value of seeds (tannins, vicine and convicine content) have also been reported (Hougaard *et al.*, 2008). Little is known about the functional correspondence of model legume genes and their putative faba bean orthologues (Hougaard *et al.*, 2008). Notwithstanding, lack of information predictions can be made based on the sequence similarities between the relatively few *M. truncatula* and faba bean gene pairs that are available and the high conservation and synteny existing between legume genomes. Whereas for highly conserved genes, favourable mutations observed in model legumes are likely to correspond to favourable alleles in faba bean, for less conserved genes (that is, many transcription factors), the relation is less reliable. Possible complications include (1) differences in gene copy number, (2) differences in transcript or protein abundance and (3) differences in specific activity (Horst *et al.*, 2007). Therefore, the information obtained in model legumes can be used as a guide to narrow down candidate genes, but proof can only come from functional studies, preferably in the homologous system. The involved steps are: (1) confirmation of candidate gene functions either directly in faba bean or indirectly in any of the model legumes, (2) identification of favourable alleles for selection and (3) variety improvement by MAS or by transformation of an elite line (Singh *et al.*, 2012). Several approaches have been developed to confirm candidate gene function at the biochemical and physiological level (Horst *et al.*, 2007). Originally, functional analysis of proteins was performed through two main techniques, protein over-expression and monitoring of promoter activity. Over-expression of a candidate gene is obtained by transferring the coding region of the gene under control of a strong promoter such as the CaMV 35S into the plant and function is assigned by scoring the phenotype of the resulting transformed line (Rojo *et al.*, 2007). Albeit with low efficiency, protocols for both *A. tumefaciens* and *A. rhizogenes* transformation have been established for faba bean and can be used for gene functional analysis in this species. Alternatively, the functional analysis could be performed in the model legumes *M. truncatula*, *L. japonicus* or soybean for which the transformation protocols are more efficient and rapid. In these model legumes, gene function can also be removed by modern molecular genetics techniques including RNAi and even TILLING (Horst *et al.*, 2007). Various molecular markers like RAPD, SSR, AFLP, RFLP and biochemical markers have been used in faba bean breeding, such as genetic diversity analysis, genetic linkage map construction, QTL mapping, *etc.*

11.2.1 Analysis of Genetic Diversity

Similarly, other studies using molecular markers could discriminate that faba bean accessions originated from different geographical area. For example, amplified fragment length polymorphism (AFLP) genotype data could separate (a) the Asian accessions as distinct as a group from those of European and North African origin (Zeid *et al.*, 2003) and (b) the Chinese germplasm from the germplasm collected outside of China and the winter types from the spring types (Zong *et al.*, 2009). In addition, SNP markers are used to study genetic diversity within and between faba bean populations and could lead to differentiate Australian accessions on the basis of geographical origin (Kaur *et al.*, 2014). However, molecular marker could not distinguish groups with different seed size groups (Göl *et al.*, 2017).

11.2.2 Construction of Genetic Linkage Map

The first genetic maps in *V. faba* were largely based on random amplified polymorphic DNA (RAPD) markers (Torres *et al.*, 1993), which are relatively difficult to score and reproduce (Penner *et al.*, 1993). The main route for exploitation of these early maps was through conversion of RAPD to sequence amplified characterized region markers (Gutierrez *et al.*, 2006). A key breakthrough was the creation of a genetic map composed entirely of sequence based markers (Ellwood *et al.*, 2008), which allowed patterns of collinearity with related taxa (*e.g.*, Medicago) to be easily traced for the first time.

The association of each marker with a low copy gene encoding sequence meant that it was possible to efficiently convert and reproduce the markers on other marker platforms, for example, the conversion of cleaved amplified polymorphic sequences (CAPS) markers to "kompetitive allele specific PCR" (KASP) assays by Cottage *et al.* (2012). Following this, molecular marker development for mapping purposes went down to two main routes. First, EST SSRs began to be systematically mined and offered both high information content per assay of SSR repeats and the ability to study synteny based on the orthologies between the ESTs in which these SSRs were embedded, and orthologous genes in other taxa (Kaur *et al.*, 2012) showed a high validation rate for SSRs mined from 454 sequencing derived EST sequences and subsequently mapped 71 of these (Kaur *et al.*, 2014). Second, in parallel with efforts targeting EST SSRs, mining of SNPs from transcriptome data began in earnest in 2014. Kaur *et al.* (2014) produced an Illumina OPA-"oligonucleotide pool assay" bead array with 768 SNP markers mined from Icarus and Ascot transcriptomes, of which 465 were mapped. Webb *et al.* (2016) reported 757 new, validated KASP assays, of which 653 newly mined SNPs and 34 legacy SNPs (Cottage *et al.*, 2012) were placed in a consensus map. A major step towards alignment of genetic linkage groups was the consensus map by Satovic *et al.* (2013), who achieved a high number of mapped loci by merging genetic maps of three inter related biparental RIL populations encompassing four diverse parents. Of the 23 linkage groups found, 13 of the largest could be tentatively assigned to parts of the six chromosomes. However, aside from the fragmentation of chromosomes, the backbone of the Satovic *et al.* (2013) consensus map was still RAPD markers, limiting the degree to which the map could be synthetically aligned with other genomes and the ease of reproducing the same markers across other populations. The first *V. Faba* genetic map where all markers fell into just six linkage groups corresponding to the six haploid chromosomes was developed by Webb *et al.* (2016). The key achievement of this study was genotyping six F_2 and RIL populations with the same set of gene based SNP assays in the KASP format, such that most polymorphic markers were present in two or more component maps. The product of this study was a map consisting of just 687 markers in six linkage groups spanning 1,403.8 cM with each linkage group assigned to a physical chromosome. It made possible for all gaps in marker coverage to be quantified, and users of the consensus map could pick an individual trait linked marker and reproduce it cheaply with little effort, or take a subset of spaced markers to sample the whole genome (*e.g.*, in a genetic diversity study), or simply use the whole set. Importantly, the 687 mapped markers contained 34 converted CAPS markers from previous studies, which meant that this new consensus map was backwards compatible (Webb *et al.*, 2016).

11.2.3 Molecular Markers Linked to Biotic Resistance

The first study to map QTLs controlling crenate broomrape response in *V. faba* was performed by Roman *et al.* (2002). Three QTLs were identified suggesting that broomrape resistance in faba bean can be considered a polygenic trait with major effects from a few single genes. One of the three QTLs explained more than 35 per cent of the phenotypic variance whereas the others accounted for 11.2 per cent and 25.5 per cent of the variation. Two of the QTLs showed considerable dominant effects in the direction of resistance. Generally, MAS programs rely on QTLs with large additive effects since they can be readily applied for breeding purposes. However, commercial production of faba bean is mainly based on improved populations as well as synthetic cultivars comprised of 5 to 8 different breeding lines (cultivars). These synthetic cultivars take advantage of heterosis as well as heterogeneity, frequently found in faba bean, to produce higher and more stable yields. The strong dominant effects of the QTLs found in this study do not preclude their use in faba bean breeding programmes since the introgression of different resistance genes into breeding lines may generate even more resistant, heterozygous plants.

Molecular marker techniques have proved to be an efficient alternative to the tedious task of screening for resistance, allowing the development of saturated maps suitable for locating and characterising individual QTLs (Lander and Bostein, 1989). Roman *et al.* (2003) identified two QTLs (*Af1* and *Af2*) for partial resistance against one isolate of *A. fabae* in a F$_2$ population from a cross between *Vf6* (resistant) and *Vf136* (susceptible). These QTLs were ascribed to chromosomes 3 and 2 and explained 25.2 and 21.0 per cent of the variability, respectively. Recently, Avila *et al.* (2004) studied the resistance on leaves and stems using two pathogenically distinct Ascochyta isolates. These authors located 6 QTLs for *A. fabae* resistance in the genetic map derived from the cross 29H × Vf136, 29H being the donor of resistance. The QTLs were named *Af3* to *Af8*, and both isolate-specific and organ-specific QTLs were detected. Thus, *Af3* and *Af4* were effective against both isolates while QTLs *Af5* to *Af8* were only effective against one of the isolates studied. Besides, *Af3*, *Af4*, *Af5* and *Af7* were effective in both leaves and stems; *Af8* was effective only in stems and *Af6* only in leaves. Nevertheless it may be possible that some QTLs detected in a single organ may be also active in the other. Kohpina *et al.* (1999) reported that the stem resistance is highly affected by environmental conditions and that this fact might prevent the detection of a real QTL.

Only a single mapping study for rust resistance in faba bean has been carried out so far (Avila *et al.*, 2003). These authors used Bulk Segregant Analysis (BSA) to identify Random Amplified Polymorphic DNA (RAPD) markers linked to a gene determining hypersensitive resistance in *V. faba* line 2N52 against *U. viciae-fabae* race 1 (Emeran *et al.*, 2001). Three RAPD markers (OPD13736, OPL181032 and OPI20900) were mapped in coupling phase to the resistance gene Uvf-1. No recombinants between OPI20900 and Uvf-1 were detected. Two additional markers (OPP021172 and OPR07930) were linked to the gene in repulsion phase at a distance of 9.9 and 11.5 cM, respectively. Line 2N52 used in this work was characterized as resistant to seven more races of the pathogen (races 2, 4, 5, 8, 9, 10 and 13) (Emeran *et al.*, 2001). Whether the resistance against these races is controlled by the same or by different genes is currently under study. Apart from 2N52, a number of additional host sources of hypersensitive resistance have been described (Sillero *et al.*, 2000). A program based on diallelic crosses among these resistant lines is in progress in our group. The final objective is to determine how many genes are involved in the hypersensitive resistance in faba bean and against how many races of the pathogen they are effective. Growing different hybrids carrying the different resistance genes or pyramiding such genes through MAS would greatly accelerate the breeding progress. This strategy may extend the life cycle of each gene providing a more durable resistance (Lawson *et al.*, 1998).

11.2.4 QTL Mapping

Quantitative trait loci or QTLs are regions of chromosomes which underline phenotypic variance of quantitative characters. Marker assisted introgression of QTLs controlling important traits have used in targeted improvement of quantitative traits of many crop plants. Several QTLs controlling important traits in faba bean such as frost tolerance (Ahmed *et al.*, 2016), stomatal characteristics (Khazaei *et al.*, 2014), ascochyta blight resistance (Kaur *et al.*, 2014), Orobanche resistance (Ramon

Diaz *et al.*, 2010) on several chromosome have been identified using linkage maps. QTLs controlling flowering and reproductive traits have also been identified in faba bean (Cruz Izquierdo *et al.*, 2012). QTL mapping has shown that resistance to aschochyta blight in faba bean is controlled by several chromosomal regions on chromosome I, II, and VI; thus improving the genetic understanding of disease resistance. Identification and validation of closely linked molecular markers to these QTLs would enable their marker assisted introgression on adapted cultivars for rapid crop improvement. QTL mapping have also facilitated identification of candidate genes controlling important traits in faba bean. These discoveries are expected to rapidly accelerate faba bean breeding.

11.2.5 Isozyme Analysis

Forms of the enzyme which differ in their primary amino acid sequences but catalyze the same chemical reaction are known as isozymes or isoenzymes. Isozymes have been used in population genetic studies of many higher plants since 1960s. In faba bean several isozyme loci have been identified and used to decipher taxonomic relationships with other *Vicia* sp., identification of cultivars including inbred lines and their F_1 hybrids. Isozymes that include Sod, 6- Pgd, Me, Est, Skdh and Gdh have been used to study the genetic diversity of faba bean, which revealed considerable phenotypes per locus. Isozyme markers have also been used to estimate level of inbreeding and outcrossing in faba bean populations (Ouji *et al.*, 2011). Isozymes along with RFLP and RAPD have also been used in construction preliminary linkage maps in faba bean (Torres, 1993). Isozyme markers combined with DNA markers can provide a powerful tool to study genetic diversity in faba bean and identification of isozyme markers associated with important traits such as resistance to Fusarium wilt and seed quality parameters can help selection of elite genotypes during breeding.

11.3.0 Genetic Transformation

Several investigators worked extensively on faba bean transformation and regeneration of transgenic plants (Schiemann and Eisenreich, 1989; Ramsay and Kumar, 1990; Saalbach *et al.*, 1994; Siefkes-Boer *et al.*, 1995; Jelenic *et al.*, 2000). The first attempts to transfer foreign genes into faba bean were attempted using *Agrobacterium rhizogenes* containing the binary vector pGSGluc1 carrying nptII and uidA genes under the control of the bidirectional TR1/2 promoter (Schiemann and Eisenreich, 1989). Beta-glucuronidase (GUS)-positive roots and subsequently transgenic calli lines were obtained. A similar study using an A. rhizogenes strain containing pBin19 for inoculation of faba bean cotyledons and stem tissue led to successful transfer of the nptII marker gene (Ramsay and Kumar, 1990). However, no transgenic faba bean plants were regenerated in any of these studies. The lack of an efficient protocol for the regeneration of transgenic plants was the main obstacle to faba bean transformation. Bacteria carrying shooty types Ti-plasmids, pGV2215 and pGV2235, were tried but neither of the two mutants appeared to improve regeneration (Jelenic *et al.*, 2000). Finally, the first transgenic faba bean plants were recovered from transformed tissues through de novo regeneration using thidiazuron (TDZ) (Böttinger *et al.*, 2001). The second successful protocol was based on direct shoot organogenesis from meristematic cells of mature or immature embryo axes (Hanafy *et al.*, 2005). The two successful studies demonstrated that transgenic approach could be used to improve protein quality in faba bean, and the bar gene (PPT acetyltransferase) could be effectively used as a selective agent of transgenic individuals, instead of the unpopular antibiotic resistance. Though the major hindrances in generating transgenic faba bean plants have been overcome, the number of transgenes that have been introduced into the faba bean genome is still limited. There are numerous traits that would be suitable for improving faba bean yield and nutritional quality. However, to complete the whole procedure of generating a commercial faba bean line, starting with the newly developed transgenic plant, then out crossing the new trait into economically valuable genotypes, and eventually getting the approval for registration of the transgenic line, could take many years to achieve (Cubero and Nadal, 2005). Furthermore, the acceptance of transgenic faba beans in the international marketplace, particularly Egypt, is unclear and therefore risky to pursue for large exporters such as Australia or the European Union.

12.0 References

Abd-Alla, M.H., El-Enany, A.E., Nafady, N.A., Khalaf, D.M., and Morsy, F.M. (2014) *Microbiol. Res.*, **169:** 49-58.

Abdel, L.Y.I. (2008) *Res. J Agric. Biolog. Sci.*, **4:** 146-148.

Abdel-Kader, M., El-Mougy, N. and Lashin, S. (2011) *J. Plant Protect. Res.*, **51:** 306-313.

Abebe, T., Birhane, T., Nega, Y. and Workineh, A. (2014) *Discourse J. Agric. Food Sci.*, **2:** 33-38.

Adak, M.S. and Kibritci, M. (2016) *Legume Res.*, **39:** 991-994.

Ahmed S., Mustapha A., Mohamed E., Nathan A. and Regina M. (2016) *Front Plant Sci.*, **7:** 1098.

Ahmed, S., Abang, M.M., and Maalouf, F. (2016) *Crop Prot.*, **81:** 65-69.

Ait taadaouit, N., El Fakhouri, K., Sabraoui, A., Rohi, L., Maalouf, F. and El Bouhssini, M. (2018) *Seventh International Food Legumes Research Conference (IFLRCVII)*, Marrakesh, Morocco. Workshop 3. p. 101.

Al-Ali, A.M., Alghamdi, S. and Alfifi, S. (2010) IFLRC V, April 26-30, Antalya, Turkey.

Alghamdi, S. (2003) College of Agriculture, Agriculture Research Center, Research Bulletin, Saudi Arabia, **95:** 5-22.

Alghamdi, S.S., Al-Faifi, S.A., Migdadi, H.M., Ammar, M.H. and Siddique, K. (2011) *Crop Pasture Sci.*, **62:** 755-760.

Alghamdi, S.S., Al-Shameri, A.M., Migdadi, H.M., Ammar, M.H., El-Harty, E.H. and Khan, M.A. (2015) *J. Agron. Crop Sci.*, **201:** 401-409.

Ali, M.B. (2015) *J. Crop Improv.*, **29:** 319-332.

Ali, M., Joshi. P.K., Pandey, S., Asokan, M., Virmani, S.M., Kumar, R. and Kandpal, B.K. (2000) ICRISAT, Patancheru-502 324, A P. India and Ithaca, New York, USA: Cornell University. pp. 35-70.

Anonymous (2005) *Pulses in India- Retrospect and Prospects,* Directorate of Pulse Development, Vindhyachal Bhavan, Bhopal p.178.

Arbaoui, M. and Link, W. (2008) *Euphytica,* 162: 211-219.

Argaw, A. and Mnalku, A. (2017) *Arch. Agron. Soil Sci.,* 63: 1390-1403.

Avila, C.M., Atienza, S., Moreno, M. and Torres, A.M. (2007) *Theor. Appl. Genet.,* 115: 1075-1082.

Avila, C.M., Nadal, S., Moreno, M.T. and Torres, A.M. (2006) *Mol. Breed.,* 17: 185-190.

Avila, C.M., Satovic, Z., Sillero, J.C., Rubiales, D., Moreno, M.T. and Torres, A.M. (2004) *Theor. Appl. Genet.,* 108: 1071-1078.

Avila, C.M., Sillero, J.C., Rubiales, D., Moreno, M.T. and Torres, A.M. (2003) *Theor. Appl. Genet.,* 107: 353-358.

Avola, G., Cavallaro, V., Patane,'C. and Riggi, E. (2008) *J. Plant Physiol.,* 165: 796-804.

Baginsky, C., Silva, P., Auza, J., and Acevedo, E. (2013) *Chil. J. Agr. Res.,* 73: 225-232.

Bar³og, P., Grzebisz, W. and £ukowiak, R. (2018) *Field Crops Res.,* 219: 87-97.

Basheer-Salimia, R., Camilli, B., Scacchi, S., Noli, E., and Awad, M. (2014) *Euro. J. Hort. Sci.,* 79: 300-305.

Belete, T. and Bastas, K.K. (2017) *J. Plant Pathol. Microbiol.,* 8: 2.

Berhe, A. (1998) *Fabis Newslett.,* 41: 13-17.

Bieri, V., Schmid, J. and Keller, E.R. (1984) *Proceedings of the 10th Congress of the European Association for Research on Plant Breeding,* Wageningen, Netherlands, 19-24 June 1983, p. 295.

Binding, H. and Nehls, R.(1987a) *Z. Pflanzenphysiol.,* 88: 327-332.

Binding, H. and Nehls, R. (1978b) *Mol. Genm. Genet.,* 164: 137-143.

Birch, N. (1985) *Ann. Appl. Biol.,* 106: 561-569.

Bodner, G., Kronberga, A., Lepse, L., Olle, M., Vagen, I.M. and Rabante, L. (2018) *Eur. J. Agron.,* 96: 1-12.

Bolland, M.D.A., Siddique, K.H.M. and Brennan, R.F. (2000) *Austral. J. Exp. Agric.,* 40: 849-885.

Bond, D. A. (1987) *Plant Breed.,* 99: 1-26.

Böttinger, P., Steinmetz, A., Schieder, O. and Pickardt, T. (2001) *Mol. Breed.,* 8: 243-254.

Bozoçglu, H., Pek˛sen, A., Pek˛sen, E. and Gülümser, A. (2002) *Acta Hort.,* 579: 347-350.

Bryla, D.R., Bañuelos, G.S. and Mitchell, J.P. (2003) *Irrig. Sci.,* 22: 31-37.

Bulut, F., Akinci, S. and Eroglu, A. (2011) *Commun. Soil Sci. Plant Anal.,* 42: 945-961.

Buruchara, R, Mukankusi, C. and Ampofo, K. (2010) Handbooks For Small-Scale Seed Producers International Center For Tropical Agriculture. (CIAT), Pan-Africa Bean Research Alliance (PABRA), Uganda, Kampala.

Cárcamo, H., Herle, C. and Hervet, V. (2012) *J. Insect Sci.*, **12**: 151.

Cárcamo, H.A., Herle, C.E., and Lupwayi, N.Z. (2015) *J. Insect Sci.*, **15**: 74.

Chen, W. (2009) *J. New Seeds.*, **10**: 14-30.

Cottage, A., Gostkiewicz, K., Thomas, J.E., Borrows, R., Torres, A.M. and O'Sullivan, D.M. (2012) *Mole. Breed.*, **30**: 1799-1809.

Crépon, K., Marget, P., Peyronnet, C., Carrouee, B., Arese, P. and Duc, G. (2010) *Field Crops Res.*, **115**: 329-339.

Cruz-Izquierdo, S., Avila, C.M., Satovic, Z., Palomino, C., Gutierrez, N., Ellwood, S.R., Phan, H.T.T., Cubero, J.I. and Torres, A.M. (2012) *Theor. Appl. Genet.*, **125**: 1767-1782.

Cubero, J.I. (1974) *Theor. Appl. Genet.*, **45**: 47-51.

Cubero, J.I. (2017) In Cultivos Hortícolas al Aire Libre, eds J.V. Maroto and C. Baxauli (Almería: Cajamar Caja Rural), pp. 703-741.

Cubero, J.I. and Nadal, S. (2005) In *Genetic Resources, Chromosome Engineering, and Crop Improvement for Grain legumes;* Singh, R.J., Jauhar, P., Eds.; CRC Press, Taylor and Francis Group: Boca Raton, FL, USA, p. 163.

Curley, J. and Jung, G. (2004) *Crop Sci.*, **44**: 1299-1306.

Daoui, K., Fatemi, Z., Bendidi, A.R., Rezouk, R., Cherguaoui, A. and Ramdani, A. (2012) In: Book of abstracts first European Scientific Conference on Agriforestry in Brussels.

del Pilar Cordovilla, M., Ligero, F. and Lluch, C. (1995) *Plant Soil*, **172**: 289-297.

Diaz, R.S. ¡ atovic,´ Z., Roman, B., Avila, C.M., Alfaro, C., Rubiales, D., Cubero, J.I. and Torres, A.M. (2005) In: Colic, J.F. and Ugarkovic, D. (eds) Second congress of Croatian geneticists. Supetar, Island of Brac, p. 8.

Diaz. R., Roman, B., S ¡ atovic,´ Z., Rubiales, D., Cubero, J.I. and Torres, A.M. (2004) In: AEP (Eds.) Conference of the 5th european conference on grain legumes with the ICLGG, Dijon, France, p. 122.

Duc, G., Agrama, H., Bao, S., Berger, J., Bourion, V. and De Ron, A.M. (2015b) *Crit. Rev. Plant Sci.*, **34**: 381- 411.

Duc, G., Aleksi´ c, J.M., Marget, P., Miki´ c, A., Paull, J. and Redden, R.J. (2015a) In: Grain Legumes, ed. A.M. De Ron (New York, NY: Springer), pp. 141-178.

Duc, G., Bao, S., Baum, M., Redden, B., Sadiki, M. and Suso, M.J. (2010) *Field Crops Res.*, **115**: 270-278.

Duke, J.A. (1981) Plenum Press, New York.

Ellwood, S.R., Phan, H.T., Jordan, M., Hane, J., Torres, A.M., Avila, C.M. and Oliver, R.P. (2008) BMC *Genom.*, **9**: 380.

El-Tayeb, M.A. (2006) *Acta Agrono. Hungar.*, **54**: 25-37.

Emeran, A.A., Sillero, J.C. and Rubiales, D. (2001) *4th European Conference on Grain Legumes.* Towards the Sustainable Production of Healthy Food, Feed and Novel Products, 8-12 July 2001, Cracow, Poland, pp. 263.

European Commission (2017) Plant Variety Database, Available at: http://ec.europa. eu/food/plant/plant_propagation_material/plant_ variety_ catalogues_ databases/search/public/index.cfm.

Etemadi, F., Hashemi, M., Mangan F. and Weis, S. (2015) Growers guide in New England.

Etemadi, F., Hashemi, M., Shureshjani, R.A. and Autio, W.R. (2017) *Agron. J.*, **109**: 1225–1231.

Etemadi, F., Hashemi, M., Autio, W., Mangan, F. and Zandvakili, O. (2018a) *J. Crop Sci.*, **6**: 426-434.

Etemadi, F., Hashemi, M., Zandvakili, O. and Mangan, F. (2018b) *Int. J. Plant Prod.*, **12**: 1-8.

Etemadi, F., Hashemi, M., Zandvakili, O., Dolatabadian, A. and Sadeghpour, A. (2018c) *Agron. J.*, **110**: 455-462.

Etemadi, F., Hashemi,M., Barker, A,V., Zandvakili, O.R. and Liu, X. (2019) *Hortic. Plant J.*, **5**: 170-182.

Evenden, M., Whitehouse, C.M., Onge, A.S., Vanderark, L., Lafontaine, J.P. and Meers, S. (2016) *Can. Entomol.*, **148**: 595-602.

Filek, M., Baczek, R., Niewiadomska, E., Pilipowicz, M. and Kos´cielniak, J. (1997) *Acta Biochi Pol.*, **44**: 315–322

Flores, F., Hybl, M., Knudsen, J. C., Marget, P., Muel, F. and Nadal, S. (2013) *Field Crops Res.*, **145**: 1-9.

Fouad, M., Mohammed, N., Aladdin, H., Ahmed, A., Xuxiao, Z. and Shiying, B. (2013) In; *Genetic and Genomic Resources of Grain Legume Improvement*, (Eds M. Singh, H.D. Upadhyaya, and I. S. Bisht), London: Elsevier, pp.113-136.

Frenda, A.S., Ruisi, P., Saia, S., Frangipane, B., Di Miceli, G. and Amato G. (2013) *Weed Sci.*, **61:** 452-459.

Gasim, S. and Link, W. (2007) *J. Central Eur. Agric.* **8:** 121-127.

Ghassemi-Golezani, K., Ghanehpoor, S., and Mohammadi-Nasab, D. (2009) *J. Food Agric. Environ.*, **7**: 442- 447.

Gillard, C.L., Conner, R.L., Howard, R.J., Pauls, K.P., Shaw, L. and Taran, B. (2009) *Can. J. Plant Sci.*, **89**: 405-416.

Göl, Þ., Doðanlar, S. and Frary, A. (2017) *Mol. Genet. Genom.*, **292:** 991-999.

Gopalakrishnan, T.R. (2007) New India Publishing Company, New Delhi, p.193.

Gounaris, K., Brain, A.R.R., Quinn, P.J. and Williams, W.P. (1984) *Biochim. Biophys. Acta.*, **766**: 198-208.

Grant, M. and Fuller, K.W. (1968) *J. Exp. Bot.*, **19**: 667-680.

Griga, M., Kubalakova, M. and Tejklova, E. (1987) *Plant Cell Tiss. Org. Cult.*, **9**: 167-171.

Gucci, G., Lacolla, G., Summo, C. and Pasqualone, A. (2019) *Sci. Hortic.*, **243**: 338-343.

Guoju, X., Fengju, Z., Juying, H., Chengke, L., Jing, W. and Fei, M. (2016) *Agric. Water Manage.*, **173**: 84-90.

Gutierrez, N., Avila, C., Moreno, M. and Torres, A. (2008) *Aust. J. Agric. Res.*, **59**: 62-68.

Gutierrez, N., Avila, C., Rodriguez-Suarez, C., Moreno, M. and Torres, A. (2007) *Mol. Breed.*, **19**: 305-314.

Gutierrez, N., Avila, C.M., Duc, G., Marget, P., Suso, M.J., Moreno, M.T. and Torres, A.M. (2006) *Theor. Appl. Genet.*, **114**: 59-66.

Haile, M., Adugna, G. and Lemessa, F. (2016) *Afr. J. Agric. Res.*, **11**: 837-848.

Haling, R.E., Yang, Z., Shadwell, N., Culvenor, R.A., Stefanski, A., Ryan, M.H. and Simpson, R.J. (2016) *Plant Soil*, **407**: 67-79.

Hamada, A.M. (2001) *Acta Physiol. Plant.*, **23**: 193-200.

Hanafy, M., Pickardt, T., Kiesecker, H. and Jacobsen, H.J. (2005) *Euphytica*, **142**: 227-236.

Hanounik, S.B. and Robertson, L.D. (1989) *Plant Dis.*, **73**: 202-205.

Harlan, J.R. (1975) *Science*, **188**: 618-621.

Harveson, R.M. and Schwartz, H.F. (2007) *Plant Health Prog.*

Hazra, P. and Som, M. G. (2006) In: *Vegetable Science*, Kalyani Publishers, New Delhi, p. 667.

Hoffmann, D., Jiang, Q., Men, A., Kinkema, M. and Gresshoff, P.M. (2007) *J. Plant Physiol.*, **164**: 460- 469.

Horst, I., Welham, T., Kelly, S., Kaneko, T., Sato, S., Tabata, S., Parniske, M. and Wang, T.L. (2007) *Plant Physiol.*, **144**: 806-820.

Hougaard, B.K., Madsen, L.H., Sandal, N., de Carvalho Moretzsohn, M., Fredslund, J., Schauser, L., Nielsen, A.M., Rohde, T., Sato, S., Tabata. S., Bertioli, D.J. and Stougaard, J. (2008) *Genetics*, **179**: 2299-2312.

Hu, J., Kwon, S.J., Park, J.J., Landry, E., Mattinson, D.S. and Gang, D.R. (2015) *Funct. Foods Health Dis.*, **5**: 243-250.

Hussein, L.A. and Saleh, M. (1985) In: *Proceedings of the International Workshop on faba bean*, Saxena, M.C. and Verma, S. (Eds.), Kabuli Chickpeas and Lentils in the 1980s. ICARDA, 16-20 May, 1983. Aleppo, Syria, pp. 257-269.

ICARDA. (2008) https://www.icarda.org/crop/fabaBeans. July/10/2018.

Inamdar, S.A., Surwase, S.N., Jadhav, S.B., Bapat, V.A. and Jadhav, J.P. (2013) *Springer Plus*, **2**: 570.

Jelenic, S., Mitrikeski, P.T., Papes, D. and Jelaska, S. (2000) *Food Technol. Biotech.*, **38**: 167-172.

Jensen, E., Steen, P., Mark, B. and Hauggaard-Nielsen, H. (2010) Faba bean in cropping systems, pp. 203-216.

Jezierny, D., Mosenthin, R., Sauer, N., Schwadorf, K., and Rosenfelder-Kuon, P. (2017) *J. Anim. Sci. Biotechnol.*, **8**: 4.

Jilani, M., Daneshian, J. and Rabiee, M. (2012) *Adv. Environ. Biol.*, **6**: 2502-2504.

Karaköy, T., Baloch, F. S., Toklu, F. and Özkan, H. (2014) *Plant Genet. Res.*, **12**: 5-13.

Karkanis, A., Ntatsi, G., Kontopoulou, C.K., Pristeri, A., Bilalis, D. and Savvas, D. (2016a) *Not. Bot. Horti Agrobot. ClujNapoca.*, **44**: 325-336.

Karkanis, A., Ntatsi, G., Lepse, L., Juan, A., Fernández, I., Vågen, M., Rewald, B., Alsin, I., Kronberga, A., Balliu, A., Olle, M., Bodner, G., Dubova, L., Rosa, E. and Savvas, D. (2018) *Front. Plant Sci.*, **9**: 1115.

Karkanis, A., Travlos, I.S., Bilalis, D. and Tabaxi, E.I. (2016b). In: Travlos, I.S., Bilalis, D.J. and Chachalis, D. (Eds.), *Integrated Weed Molecular Biology, Practices and Environmental Impact*, New York, NY: Nova Science Publishers, Inc., pp. 1-15.

Kaur, S., Rohan, B.E., KimbereNoel, O.I., Cogana, M., John, M., Forsterad. W and Paull J.G. (2014) *Front Plant Sci.*, **217**: 47-55.

Kaur, S., Pembleton, L.W., Cogan, N. O., Savin, K.W., Leonforte, T., Paull, J. and Forster, J.W. (2012) *BMC Genom.*, **13**: 104.

Keneni, G., Asmamaw, B., Gonu, D. and Fessehaie, R. (2001) *Ethiopian Agricultural Research Organization*, Technical Manual No -14. p. 34.

Keneni, G., Bekele, E., Getu, E., Imtiaz, M., Damte, T. and Mulatu, B. (2011) *Sustain*, **3**: 1399-1415.

Khamassi, K., Harbaoui, K., Teixeira Da Silva, J.A. and Jeddi, F.B. (2013a) *Agric. Conspec. Sci.*, **78**: 131-136.

Khan, H.R., Paull, J.G., Siddique, K.H.M. and Stoddard, F.L. (2010) *Field Crops Res.*, **115**: 279-286.

Khazaei, H., Street, K., Santanen, A., Bari, A. and Stoddard, F.L. (2013) *Genet. Resour. Crop. Evol.*, **60**: 2343-2357.

Khazaei, H., O'Sullivan, D.M., Sillanpää, M.J. and Stoddard, F.L. (2014) *Theor. Appl. Genet.*, **127**: 2371-2385.

Knott, C.M. (1999) *J. Agric. Sci.* **132**: 407-415.

Kohpina, S., Knight, R. and Stoddard, F.L. (1999) *Aust. J. Agric. Res.*, **50**: 1475-1481.

Kora, D., Hussein, T. and Ahmed, S. (2017) *J. Plant Sci.*, **5**: 120-129.

Köpke, U. and Nemecek, T. (2010) *Field Crops Res.*, **115**: 217-233.

Krarup, H.A. (1984) *Lens.*, **11**: 18-20.

Kumari, S.G. and Makkouk, K.H.M. (2007) *Plant Viruses*, **1**: 93-105.

Lander, E.S. and Bostein, D. (1989) *Genetics*, **121**: 185-199.

Landry, E.J., Fuchs, S.J. and Hu, J. (2016) *J. Food Compos. Anal.*, **50**: 55-60.

Lawson, W.R., Goulter, K.C., Henry, R.J., Kong, G.A. and Kochman, J.K. (1998) *Mol. Breed.*, **2**: 227-234.

Lee, M.K., Kim, D.G., Kim, J.M., Ryu, J., Eom, S.H., Hong, M.J. and Kwon, S.J. (2018) *Plant Breed. Biotechnol.*, **6**: 57-64.

Link, W., Dixkens, C., Singh, M., Schwall, M. and Melchinger, A. (1995) *Theor. Appl. Genet.*, **90**: 27-32.

Link, W., Balko, C. and Stoddard, F.L. (2010) *Field Crops Res.*, **115**: 287-296.

Lizarazo, C.I., Lampi, A.M., Sontag-Strohm, T., Liu, J., Piironen, V. and Stoddard, F.L. (2015) *J. Sci. Food Agric.*, **95**: 2053-2064.

Longobardi, F., Sacco, D., Casiello, G., Ventrella, A. and Sacco, A. (2015) *J. Food Qual.*, **38**: 273-284.

Loss, S.P. and Siddique, K.H.M. (1997) *Field Crops Res.*, **52**: 17-28.

Ma, Y., Yang, T., Guan, J., Wang, S., Wang, H., Sun, X. and Zong, X. (2011) *Am. J. Bot.*, **98**: 22-24.

Maalouf, F., Nachit, M., Ghanem, M. E. and Singh, M. (2015) *Crop Pasture Sci.*, **66**: 1012-1023.

Maalouf, F., Ahmed, S., Shaaban, K., Bassam, B., Nawar, F., Singh, M. and Amri, A. (2016) *Euphytica*, **211**: 157-167.

Makoudi, B., Kabbadj, A., Mouradi, M., Amenc, L., Domergue, O., Blair, M. and Ghoulam, C. (2018) *Acta Physiol. Plant*, **40**: 63.

Maqbool, A.,Shafiq, S., and Lake, L. (2010) *Euphytica*, **172**: 1-12.

McDonald, G.K. and Paulsen, G.M. (1997) *Plant Soil*, **196**: 47-58.

McVicar, R., Panchuk, D., Brenzil C., Hartley S., Pearse, P. and Vandenberg, A. (2013) Gov. Saskatchewan, Agriculture, Crops.

Mihailovic, V., Mikic, A., Cupina, B. and Eric, P. (2005) *Grain Legumes*, **44**: 25-26.

Mitchell, J.P. and Gildow, F.E. (1975) *Physiol. Plant.*, **34**: 250-253.

Mohamed, S. S.E. and Babiker, H.M. (2012) *Adv. Environ. Biol.*, **6**: 824-830.

Muehlbauer, F. and Tullu, A. (1997) Purdue Univ., Cent. New Crops Plants Prod., New Crop Fact sheet.

Mwanauta, R.W. Mtei, K.M. and Ndakidemi, P.A. (2015) *Agric. Sci.*, **6**: 489- 497.

N'Dayegamiye, A.,Whalen, J.K., Tremblay, G., Nyiraneza, J., Grenier, M., Drapeau, A. and Bipfubusa, M. (2015) *Agron. J.*, **107**: 1653-1665.

Naqvi, H.K. (1984) *Indian J. History. Sci.*, **19**: 329-340.

Nasto, T., Sallaku, G. and Balliu, A. (2016) *Acta Hortic.*, **1142**: 233-238.

Nebiyu, A., Diels, J. and Boeckx, P. (2016) *J. Plant Nutr. Soil Sci.*, **179**: 347-354.

Neme, K., Bultosa, G. and Bussa, N. (2015) *Int. J. Food Sci. Technol.*, **50**: 2375-2382.

Niewiadomska, A.,Bar³óg, P., Borowiak, K. and Wolna-Maruwka, A. (2015) *Fresenius Environ. Bull.*, **24**: 723-732.

Omafra (2009) *Agronomy Guide for Field Crops Publication*, p. 811.

Omri Benyoussef, N., Le May, C., Mlayeh, O. and Kharrat, M. (2012) *Phytopathol. Mediter.*, **51**: 369-373.

Ortiz, V., Navarro, E., Castro, S., Carazo, G. and Romero, J. (2006) *J. Agric. Res.*, **4**: 255-260.

Ouji, A., Suso, M.J., Mustapha, R., Abdellaoui, R. and Gazzah, M.E. (2011). *Genes Genom.*, **33**: 31-38.

Oweis, T., Hachum, A., and Pala, M. (2005) *Agric. Water Manag.*, **73**: 57-72.

Patrick, J.W. and Stodddard, F.L. (2010) *Field Crops Res.*, **115**: 234-242.

Patto, M.C.V., Torres, A., Koblizkova, A., Macas, J. and Cubero, J. (1999) *Theor. Appl. Genet.*, **98**: 736-743.

Pearce, S., Knox, M., Ellis, T., Flavell, A. and Kumar, A. (2000) *Mol. Gen. Genet.*, **263**: 898-907.

Penner, G.A., Bush, A., Wise, R., Kim, W., Domier, L., Kasha, K., Laroche, Al., Scoles, S. J., Molnar, J. and Fedak, G. (1993) *Genom. Res.*, **2**: 341-345.

Peoples, M.B., Brockwell, J., Herridge, D.F., Rochester, I.J., Alves, B.J.R., Urquiaga, S., Boddey, R.M., Dakora, F.D., Bhattarai, S., Maskey, S.L., Sampet, C., Rerkasem, B., Khans, D.F., Hauggaard-Nielsen, H. and Jensen, E.S. (2009) *Symbiosis*, **48**: 1-17.

Petersson, A., Thomsen, M.H., Nielsen, H.H. and Thomsen, A.B. (2007) *Biomass Bioenerg.*, **31**: 812-819.

Pickardt, T., Huancaruna Perales, E. and Schieder, O. (1989) *Protoplasma*, **149**: 5-10.

Pickardt, T., Meixner, M., Schade, V. and Schieder, O. (1991) *Plant Cell Rep.*, **9**: 535-538.

Pietrzak, W., Kawa-Rygielska, J., Król, B., Lennartsson, P.R. and Taherzadeh, M.J. (2016) *Bioresour. Technol.*, **216**: 69-76.

Pozarkova, D., Koblizkova, A., Rom´an, B., Torres, A.M., Lucretti, S., Lysak, M., Dolezel, J. and Macas, J. (2002). *Biol. Plantarum.*, **45**: 337-345.

Ramón, D.R., Torres, A.M., Satovic, Z., Gutierrez, M.V., Cubero, ·J.I. and· Román, B. (2010) *Theor. Appl. Genet.*, **120**: 909-919.

Ramsay, G. and Kumar, A. (1990) *J. Exp. Bot.*, **41**: 841-847.

Ramsey, G., Van de Ven, W., Waugh, R., Griffiths, D. and Powell, W. (1995) In: AEP (Eds.), *Proceedings of the 2nd European conference on grain legumes,* Copenhagen, Denmark, pp. 444-445.

Razia, Akbar. Tr. (2000) In: *17th Century by the Mughal Prince Dara Shikoh* (Agri-History Bulletin No. 3) Munshiram Manoharlal Publishers Pvt Ltd. p. 98.

Redden, R., Zong, X., Norton, R.M., Stoddard, F.L., Maalouf, F., Ahmed, S. and Rong, L. (2018) In: *Achieving sustainable cultivation of grain legumes Volume 2: Improving cultivation of particular grain legumes* (Eds.S. Sivasankar, D. Bergvinson, P. M. Gaur, S. Kumar, S. Beebe, and M. Tamo), Cambridge, UK: Burleigh Dodds Science Publishing. pp. 269-297.

Robertson, L.D. (1984) In: *Systems for cytogenetic analysis in Vicia faba* (Eds. L G. P. Chapman, and S. A. Tarawali), Dordrecht, the Netherlands: Springer Netherlands, p.79.

Rojo, F.G., Reynoso, M.M., Ferez, M., Chulze, S.N. and Torres, A.M. (2007) *Crop Protect.*, **26**: 549-555.

Roman, B., Satovic, Z., Avila, C., Rubiales, D., Moreno, M. and Torres, A. (2003) *Aust. J. Agric. Res.*, **54**: 85-90.

Roman, B., Satovic, Z., Pozarkova, D., Macas, J., Dolezel, J., Cubero, J.I. and Torres, A.M. (2004) *Theor. Appl. Genet.*, **108**: 1079-1088.

Roman, B., Torres, A.M., Rubiales, D., Cubero, J.I. and Satovic, Z. (2002) *Genome*, **45**: 1057-1063.

Romheld, V. and Kirkby, E.A. (2010) *Plant Soil*, **335**: 155-180.

Röper, W. (1979) *Z. Pflanzenphysiol.*, **93**: 245-257.

Röper, W. (1981) *Z. Pflanzenphysiol.*, **101**: 75-78.

Rubiales, D., Rojas-Molina, M.M. and Sillero, J.C. (2016) *Front. Plant Sci.*, **7**: 1747.

Rubiales, D., Avila, C.M., Sillero, J.C., Hybl, M., Narits, L. and Sass, O. (2012) *Field Crops Res.*, **126**: 165-170.

Saalbach, I., Pickardt, T., Machemehl, F., Saalbach, G., Schieder, O. and Müntz, K. A. (1994) *Mol. Gen. Genet.*, **242**: 226-236.

Sahile, S., Ahmed, S., Fininsa, C., Abang, M.M. and Sakhuja, P.K. (2008b) *Crop Prot.*, **27**: 1457-1463.

Sahile, S., Fininsa, C., Sakhuja, P. K. and Ahmed, S. (2008a) *Crop Prot.*, **27**: 275-282.

Sallam, A., Martsch, R. and Moursi, Y.S. (2015). *Euphytica*, **205**: 395-408.

Sanz, A.M., Gonzalez, S.G., Syed, N.H., Suso, M.J., Saldan~a, C.C. and Flavell, A.J. (2007) *Mol. Genet. Genom.*, **278**: 433-441.

Satovic, Z., Avila, C.M., Cruz-Izquierdo, S., Díaz-Ruíz, R., García-Ruíz, G.M., Palomino, C. and Cubero, J. I. (2013) *BMC Genom.*, **14**: 932.

Schiemann, J. and Eisenreich, G. (1989) *Biochem. Physiol. Pflanzen.*, **185**: 135-140.

Schwartz, H.F., Steadman, J.R., Hall, R. and Forster, R.L. (2005) APS Press, CO.

Seidenglanz, M. and Hun.ady, I. (2016) *Czech J. Genet. Plant Breed.*, **52**: 22-29.

Selva, E., Stouffs, M. and Briquet, M. (1989) *Plant Cell Tiss. Org. Cult.*, **18**: 167-179.

Sharifi, P. (2014). *Genet. Belgrad.*, **469**: 905-914.

Shiran, B., Kiani, S., Sehgal, D., Hafizi, A., Chaudhary, M. and Raina, S.N. (2014) *Genet. Resour. Crop Evol.*, **61**: 909-925.

Shiying, B., Xiaoming, W., Zhendong, Z., Xuxiao, Z., Kumari, S. and Freeman, A. (2007) *Austr. Plant Pathol.*, **36**: 347-353.

Siddique, K.H.M. and Loss, S.P. (1999) *J. Agron. Crop Sci.*, **182**: 105-112.

Siddique, K.H.M., Erskine, W., Hobson, K., Knights, E.J., Leonforte, A. and Khan, T.N. (2013) *Crop Pasture Sci.*, **64**: 347-360.

Siefkes-Boer, H.J., Noonan, M.J., Bullock, D.W. and Conner, A.J. (1995) *Israel J. Plant Sci.*, **43**: 1-5.

Sillero, J.C., Moreno, M.T. and Rubiales, D. (2000) *Plant Pathol.*, **49**: 389-395.

Singh, A.K. and Bhatt, B.P. (2012b) *Hort. Flora Res. Spectrum*, 1: 267-269.

Singh, A.K., Bhat, B.P., Sundaram, P.K., Chndra, N., Bharati, R.C. and Patel, S.K. (2012) *Int. J. Agric. Stat.*, **8**: 97-109.

Singh, A.K., Bhat, B.P., Sundaram, P.K., Gupta, A.K. and Singh, D. (2013) *J. Environ. Biol.*, **34**: 117-122.

Singh, A.K., Chandra, N., Bharati, R.C. and Dimree, S.K. (2010) *Environ. Ecol.*, **28**: 1522-1527.

Singh, D.N. and Nath, V. (2012) Satish Serial Publishing House, Delhi, pp. 658-662.

Smith, L.A., Houdijk, J. G. M., Homer, D. and Kyriazakis, I. (2013) *J. Anim. Sci.*, **91**: 3733-3741.

Surwase, S.N., Patil, S.A., Jadhav, S.B. and Jadhav, J.P. (2012) *Microbial. Biotechnol.*, **5**: 731-737.

Tar'an, B., Michaels, T.E. and Pauls, K.P. (2001) *Genome*, **44**: 1046-1056.

Tawaha, A.M. and Turk, M.A. (2001) *Acta Agron. Hung.*, **49**: 299-303.

Tegeder, M., Gebhardt, D., Schieder, O. and Pickardt, T. (1995) *Plant Cell Rep.*, **15**: 164-169.

Tekalign, A., Derera, J., Sibiya, J. and Fikre, A. (2016) *Indian J. Agric. Res.*, **50**: 295-302.

Temesgen, T.T., Keneni, G. and Mohammad, H. (2015) *Aust. J. Crop Sci.*, **9**: 41.

Terefe, H., Fininsa, C., Sahile, S., Dejene, M. and Tesfaye, K. (2015) *Global J. Pests. Dis. Crop Protect.*, **3**: 113-123.

Terzopoulos, P. and Bebeli, P. (2008) *Field Crops Res.*, **108**: 39-44.

Thynn, M. and Werner, D. (1987) *Angewandte Botanik*, **61**: 483-492.

Torres, A.M., Avila, C., Gutierrez, N., Palomino, C., Moreno, M.T. and Cubero, J. (2010) *Field Crops Res.*, **115**: 243-252.

Torres, A.M., Patto, M.C.V., Satovic, Z. and Cubero, J.I. (1998) *J. Hered.*, **89**: 271-275.

Torres, A.M., Román, B., Avila, C.M., Satovic, Z., Rubiales, D. and Sillero, J.C. (2006) *Euphytica*, **147**: 67-80.

Torres, A.M., Roman, B., Avila, C.M., Satovic, Z., Rubiales, D., Sillero, J.C. Cubero, J.I. and Moreno, M.T. (2004) *Euphytica*, **147**: 67-80.

Torres, A.M., Satovic, Z., Canovas, J., Cobos, S. and Cubero, J.I. (1995) *Theor. Appl. Genet.*, **91**: 783-789.

Torres, A.M., Weeden, N. and Martin, A. (1993) *Theor. Appl. Genet.*, **85**: 937-945.

Trabanco, N.O.E.M.I., Perez-Vega, E.L.E.N.A., Campa, A., Rubiales, D. and Ferreira, J.J. (2012) *Euphytica*, **186**: 875-882.

Turco, I., Ferretti, G. and Bacchetti, T. (2016) *J. Food Nutr. Res.* **55**: 283-293.

Van de Ven, M., Powell, W., Ramsay, G. and Waugh, R. (1990) *Heredity*, **65**: 329-342.

Van Emden, H.F., Ball, S.L. and Rao, M.R. (1988) Springer, Dordrecht, pp. 519-534.

Ven, W., Duncan, N., Ramsay, G., Phillips, M., Powell, W. and Waugh, R. (1993) *Theor. Appl. Genet.*, **86**: 71-80.

Venketeswaran, S. (1962) *Phytomorphol.*, **12:** 300-306.

Villegas-Fernández, A.M., Sillero, J.C. and Rubiales, D. (2012) *Eur. J. Plant Pathol.*, **132**: 443-453.

Webb, A., Cottage, A., Wood, T., Khamassi, K., Hobbs, D., Gostkiewicz, K. and O'Sullivan, D.M. (2016) *Plant Biotechnol. J.*, **14**: 177-185.

Webster, B., Bruce, T., Pickett, J. and Hardie, J. (2010) *Anim. Behav.*, **79**: 451-457.

Wrigley, C.W., Corke, H., Seetharaman, K. and Faubion, J. (2015) *Academic Press, Oxford*.

Yitayih, G. and Azmeraw, Y. (2017) *Crop J.*, **5**: 560-566.

Youseif, S.H., Fayrouz, H.A.E.M. and Saleh, S.A. (2017) *Agron.*, **7**: 2.

Yucel, D.O. (2013) *Pak. J. Bot.*, **45**: 1933-1938.

Zeid, M., Mitchell, S., Link, W., Carter, M., Nawar, A., Fulton, T. and Kresovich, S. (2009) *Plant Breed.*, **128**: 149-155.

Zeid, M., Scho"n, C.C. and Link, W. (2003) *Theor. Appl. Genet.*, **107**: 1304-1314.

Zhao, J., Sykacek, P., Bodner, G. and Rewald, B. (2018) *Plant Cell Environ.*, **41**: 1984-1996.

Zong, X., Liu, X., Guan, J., Wang, S., Liu, Q., Paull, J.G. and Redden, R. (2009) *Theor. Appl. Genet.*, **118**: 971-978.

Zong, X., Ren, J., Guan, J., Wang, S., Liu, Q., Paull, J.G. and Redden, R. (2010) *Plant Breed.*, **129**: 508-513.

Root Crops

10

CARROT

A. K. Sureja, M. K. Sadhu and K. Sarkar

1.0 Introduction

Carrot (*Daucus carota* L.; 2n = 2x = 18), a member of Apiaceae family, is a cool season root vegetable grown for its edible storage tap roots the worldover. It is grown all over the world in spring, summer and autumn in temperate countries and during winter in tropical and subtropical climate. The uniqueness of its flavour and health benefits are mainly responsible for its worldwide acceptance. Carrot has two distinct groups, *i.e.*, tropical and temperate. The tropical types are annual, whereas, the temperate types are biennial. Undoubtedly carrot is one of the most ancient vegetables. Its history has been confused with that of parsnip, because the Romans ate it as pastinaca, a name later transferred to the parsnip when carrot became *Carota* (Burkill, 1935).

During 2018, world production of carrots and turnips was 39.99 million tonnes from 1.1. million ha area. China is the leading producer of carrot and turnips (18.01 million tonnes), followed by Uzbekistan (2.18 million tonnes), USA (1.49 million tonnes), Russian Federation (1.40 million tonnes), Ukraine (0.84 million tonnes) and UK (0.82 million tonnes) (FAOSTAT, http://www.fao.org/faostat). The major carrot producing regions in China are Inner Mongolia, Shandong, Hebei, and Fujian. The four major producing regions having alternating carrot harvest periods to enable carrot production throughout the year. The Fujian region harvests its carrots between January and April, which is followed by Shandong region whose carrots enter the market between May and June. Carrots from the Hebei region enter the market from August to October, and then Inner Mongolia's carrots are in the market between October and December. The major export markets for China's carrots are Saudi Arabia, Canada, Thailand and Malaysia (http://www.carrotmuseum.co.uk).

During 2017-18, carrot production in India was 1.64 million tonnes from an area of 96.51 thousand hectare. The major carrot producing states were Haryana (0.44 million tonnes), followed by Punjab (0.19 million tonnes), Uttar Pradesh (0.16 million tonnes) and Bihar (0.14 million tonnes) (Anon., 2018).

Carrot is valued for its taste, good digestibility, and high contents of provitamin A, other carotenoids, and fibers. It is widely grown both for fresh market and processing and provides an excellent source of vitamin A and fiber in the diet. The popularity of carrot has also been influenced by the introduction of the convenient prepackaged "cut and peel" or "baby carrots," making it a leading vegetable snack item. In the USA in 2019, out of the total sales of carrot, baby carrots had a carrot sales share of 54 per cent, followed by bunch carrots (27.8 per cent), value added carrots (13.4 per cent), loose carrots (2.8 per cent), French carrots (1.5 per cent) and other carrots (0.4 per cent) (http://www.carrotmuseum.co.uk).

2.0 Composition and Uses

2.1.0 Composition

Carrot is valued as a nutritive food mainly because it is a rich source of α- and β-carotene. Analysis of the edible portion of the carrot root is given in Table 1.

Table 1: Composition of Carrot (per 100 g of edible portion)*

Composition	Carrot Root (Orange)	Carrot Root (Red)	Composition	Carrot Root (Orange)	Carrot Root (Red)
Moisture (g)	87.69	96.07	Zeaxanthin (µg)	13.93	15.49
Protein (g)	0.95	1.04	Lycopene (µg)	157.0	871.0
Ash (g)	1.16	1.22	α-carotene (µg)	2654.0	1128.0
Total fat (g)	0.47	0.47	β-carotene (µg)	5423.0	2706.0
Total dietary fibre (g)	4.18	4.49	Total carotenoids (µg)	9377.0	7570.0
Carbohydrate (g)	5.55	6.71	Calcium (mg)	35.09	41.06
Energy (KJ)	139.0	160.0	Iron (mg)	0.60	0.71
Thiamine (B$_1$) (mg)	0.04	0.04	Magnesium (mg)	16.73	18.83
Riboflavin (B$_2$) (mg)	0.03	0.03	Phosphorus (mg)	43.06	25.81
Niacin (B$_3$) (mg)	0.22	0.25	Potassium (mg)	273.0	267.0
Pantothenic acid (B$_5$) (mg)	0.30	0.27	Selenium (µg)	0.22	0.29
Total B$_6$ (mg)	0.11	0.07	Zinc (mg)	0.25	0.34
Biotin (B$_7$) (µg)	1.50	1.30	Total available carbohydrate (g)	4.48	5.35
Total folates (B$_9$) (µg)	24.04	23.67	Total starch (g)	1.24	1.39
Total ascorbic acid (mg)	6.22	6.76	Fructose (g)	0.11	1.08
Alpha tocopherol (mg)	0.19	0.19	Glucose (g)	1.15	1.13
Gamma tocopherol (mg)	0.18	0.20	Sucrose (g)	1.98	1.75
α-tocopherol Equivalent (mg)	0.21	0.22	Total free sugars (g)	3.23	3.96
Phylloquinones (K$_1$) (µg)	18.35	18.75	Total oxalate (mg)	17.45	16.41
Lutein (µg)	257.0	224.0	Total polyphenols (mg)	49.44	50.69

* Longvah *et al.* (2017).

In carrot roots, sucrose is the most abundant endogenous sugar, 10 times higher than glucose or fructose. The last two are generally present in 1: 1 ratio, although this ratio may vary considerably between cultivars (Lester and Baker, 1978).

The nutritive quality of carrot roots has been found to be influenced by N. The best quality roots are produced when sown in rows on the flat, and N at the rate of 60 kg/ha were applied (Kumar *et al.*, 1974). Draglano (1978) showed that roots of different sizes (50-250 g) did not differ in taste or in the contents of sucrose, glucose, carotene, nitrate or Mg. However, the largest roots contain more N, P and K and the percentage of dry matter tended to be higher. In a field trial in Buenos

Aires province, Argentina, 6 carrot cultivars were harvested 236, 297 or 334 days after sowing. P, Na and K contents differed significantly among harvesting dates and cultivars, and there was a significant interaction between harvesting date and cultivar for P and Na. Ca content differed significantly among harvesting dates. The overall trend was an increase in mineral content with delay in harvesting (Bianchini and Eyherabide, 1999). Carotenoids content increases significantly with delay in harvesting date, while reducing sugar content decreased (Madrid de Canizalez and Chacin Lugo, 1994).

Taste panel evaluation of cooked carrots showed that reducing sugar content, although present in small amount, has a marked bearing on flavour and, in general, roots with high level of reducing sugar are preferred (Gormley *et al.*, 1971). Effect of various blanching processes on the retention of different nutrients in carrot juice products was investigated by Rehman and Shah (1999). Maximum brix value of the juice was 9.3 when carrots were blanched in 0.05 N acetic acid solution containing 0.2 per cent $CaCl_2$. However, brix values of the juice obtained from unblanched and water blanched carrots were 8.1 and 8.5, respectively. The highest amount of carotene and ascorbic acid was found in the juice which was obtained from carrots after blanching in acetic acid and 0.2 per cent $CaCl_2$ solution. Other blanching processes showed adverse effects on the retention of carotene in carrot juice products compared with unblanched carrot juice.

2.1.1 Phytochemicals and Bioactive Compounds in Carrot

Carrots are rich source of antioxidants and phytochemicals, such as carotenoids, anthocyanins, and other phenolic compounds. They are good source of fiber. Soluble sugars are the main form of storage compounds in carrots and account for 34–70 per cent of the dry weight (Soujala, 2000). Fresh carrots contain 8.17 per cent free sugar, consisting of 3.39 per cent sucrose, 1.89 per cent α-glucose, 1.45 per cent β-glucose, 1.05 per cent fructose, and 0.39 per cent unknown (Alabran and Mabrouk, 1973). Potassium is the most abundant mineral in carrots, with a range from 443 to 758 mg/100 g fresh weight (FW) (Nicolle *et al.*, 2004).

Carrots are a significant or potential source of dietary nutrients in the form of plant pigments, including carotenoids, anthocyanins, and other flavonoids. In a situation when there is no accumulation of pigments, the carrots are white. The health benefits of these compounds, including protection against certain forms of cancer, reduction of the risk of cardiovascular disease, and scavenging of free radicals, have led to consumer interest in natural products rich in carotenoids and anthocyanins. High carotene carrots are used as a raw product for carotene extraction and food colouring. In 2007, carrots supplied an estimated 37 per cent of the available fresh vegetable β-carotene, the major provitamin A carotenoid in the U.S. diet (Arscott and Tanumihardjo, 2010). In addition, carrots contain several vitamins, phenolic substances, and volatile compounds that influence the carrot flavour.

2.1.1.1 Carotenoids

Orange carrot colour is primarily due to α- and β-carotene. Yellow and red carrot colours are due to carotenoids lutein and lycopene, respectively. Carrot contains highest amount of β-carotene among common vegetables. Raw fresh carrots contain α-carotene 8285 µg/100 g FW, β-carotene 3477 µg/100 g FW, and lutein/zeaxanthin 256 µg/100 g FW. β-carotene constitutes 60–80 per cent of the carotenoids in carrots followed by β-carotene (10–40 per cent), lutein (1–5 per cent), and the other minor carotenoids (0.1–1.0 per cent) (Chen *et al.*, 1995).

The carotenoids contained in the edible portion in carrot roots can range from 6000 to more than 54,000 µg/100 g (60-540 ppm) (Simon and Wolff, 1987). The α- and β-carotenes are present in carrot root at concentrations of 530 to 35,833 and 1161 to 64,350 µg/100 g root tissue, respectively.

Lutein has no provitamin A activity, but is found localized in the macular region of the eye in humans and may have importance in eye health and protection from age-related macular degeneration. Lutein is an important pigment in yellow carrots and they can contain from 1 to 5 ppm lutein (Arscott and Tanumihardjo, 2010).

Lycopene is the carotenoid primarily responsible for the colour of red carrots. Though lycopene also has no provitamin A activity, it has high antioxidant activity *in vitro* and has been implicated in protection from certain cancers. Lycopene content of red carrots ranges from 50 to 100 ppm, which is similar to or greater than typical tomato concentrations (Arscott and Tanumihardjo, 2010).

Purple carrots with a white core contain very low levels of carotene (4 to 6 ppm), whereas purple–orange carrots (38 to 130 ppm) can contain as much or more total carotene as typical orange carrots (Arscott and Tanumihardjo, 2010).

2.1.1.2 Phenolic Compounds

Both phenolic compounds and carotenoids are strong *in vitro* antioxidants. The main phenolic compounds found in carrots are chlorogenic acids which contribute to the organoleptic properties. Purple carrots contain the highest total phenolics and white the lowest. Total concentration of phenolic acids and chlorogenic acid is greatest in the purple carrots, followed by orange, white, and yellow carrots. The compound 6-methoxymellein (6-MM) found in carrot root is a phenolic oxidation product.

2.1.1.3 Anthocyanins

The anthocyanins in purple carrots contributed significantly to their *in vitro* antioxidant capacity. Total anthocyanin concentration in the roots of purple carrots can vary widely between cultivars and even within a cultivar based on the degree of root colouring. Total anthocyanin content has been reported to range from 0 mg/100 g FW in orange carrots to 350 mg/100 g FW in dark purple carrots (Simon *et al.*, 2008; Sun *et al.*, 2009). Total anthocyanin content in black carrot 'Pusa Asita' is 520 mg/100 g FW. Breeders have initiated development of purple carrot breeding lines. Dietary anthocyanins have low bioavailability, recoveries in urine ranging from 0.004 per cent to 0.1 per cent of intake (Manach *et al.*, 2005; McGhie and Walton, 2007).

2.1.1.4 Compounds Imparting Carrot Flavour

The characteristic "fresh carrot" flavour is primarily attributable to the volatile compounds mono- and sesquiterpenes, and also to sugars. Terpenes, generally, impart a harsh flavour. The volatile terpenoids content varies greatly between genotypes and ranges from less than 10 to over 500 ppm (Senalik and Simon, 1987), and above levels of approximately 50 ppm, most consumers note a harsh turpentine-like chemical flavour (Rubatzky *et al.*, 1999). Purple carrot cultivars are reported to have relatively low terpinolene content. A group of compounds called polyacetylenes is responsible for the bitter off-flavour of carrots. More recently, polyacetylenes have been implicated as bioactive compounds with potential effects on human physiology and disease. Four polyacetylenes have been identified in carrot root; the most abundant are falcarinol, falcarindiol (FaDOH), and falcarindiol 3-acetate. Fresh weight concentrations of polyacetylenes range from 20 to 100 mg/kg (Czepa and Hofmann, 2004; Zidorn *et al.*, 2005; Christensen and Kreutzmann, 2007).

2.2.0 Uses

Carrot roots are used as a vegetable for soups, stews, curries and pies; grated roots are used as a salad, tender roots as pickles. Carrot juice is a rich source of carotene and is sometimes used for colouring butter and other food articles. Tropical red carrots have use as carrot puddle, juice and *Gajar halwa* a delicious sweet dish. Small roots may be processed intact whereas large roots are diced or cut into slices, wedges, strips and other shapes. Carrots are also canned. Carrot puree is used in infant food preparations. Dehydrated products include carrot chips, flakes and powder. Carrot jam is also popular.

A new "minimally processed" value-added product known as 'baby' or 'cut and peel' carrots is very popular in USA and Europe. These are prepared by peeling and cutting roots into 3-7 cm length, followed by machine abrasive peeling that result in small slender carrot pieces with smooth surfaces and rounds-off cut ends. The processed roots resemble small carrots and are suitable for snack food and other fresh uses. Normally small and/or immature carrots are also marketed as baby carrots and are an important specialty product.

The red carrots are rich in lycopene, an important pigment for healthy eyes and general body growth, besides anticancerous properties. Black carrot, rich in anthocyanins, is used for the preparation of a sort of beverages called *Kanji*, which is a good appetizer. The orange varieties are rich in carotene, a precursor of vitamin A and contain appreciable quantity of thiamine and riboflavin. Carotene extracted from carrot is used for colouring margarine and as natural carotene source. Carrot extracts is also added to poultry feed to intensify skin and egg yolk colour.

Carrot leaves are said to be eaten in many countries; its tops can be used as a good source for extraction of leaf proteins. Moreover, carrot tops are used as fodder and also for preparation of poultry feed.

An infusion of carrot has long been used as a folk medicine for threadworms. Carrot increases the quantity of urine and helps elimination of uric acid. Addition of large amount of carrot to the diet has a favourable effect on the nitrogen balance.

Besides, its value as a vegetable, carrot is cultivated in some countries, notably in France, for its seed which is the source of an essential oil – the carrot seed oil. The fruits are collected, dried and the seeds are separated. The essential oil from

Uses of Carrot

Juice (left) and Halwa (right) from carrot

Sweet Burfi (left) and Halwa Trifle (right) from carrot

Cake pie bars (left) and fresh baby carrot (right)

Carrot seed oil (left) and dehydrated carrot (right)

carrot seeds has the odour of orris and is used for flavouring liqueurs and all kinds of food substitutes. Carrot seeds are aromatic, stimulant and carminative. The essential oil extracted from carrot roots has antibacterial property. The carrot seed oil is also useful in diseases of the kidney and in dropsy (Chopra, 1933; Kirtikar and Basu, 1935).

3.0 Origin, Distribution and Evolution

The present day's cultivated carrots have most probably arose as a natural variation of the Eurasian wild flower Queen Anne's lace in Afghanistan in the region where the Himalayan and *Hindukush* mountains are confluent. Its domestication was in Afghanistan and adjacent regions of Russia, Iran, India, Pakistan, and Anatolia (Vander Vossen and Sambas, 1993). Most wild *Daucus* forms are found in the South-Western Asia and the Mediterranean, a few in Africa, Australia, and America (Banga, 1976). All wild forms found in Asia, Asia Minor, Japan and USA have the same chromosome number as their European relatives. Neither polyploidy nor structural changes in the chromosome seem to have played any role in the differentiation of the species (Whitaker, 1949). According to Banga (1963a), mutation plus selection rather than hybridization has been responsible for the development of cultivated carrots.

According to Mackevic (1929), Afghanistan is the primary center of carrot because of the largest diversity in morphological characters of this species here, the roots vary in the degree of ramification, fleshiness, and colour ranges from white to anthocyanin to orange to red. Wild types are also found in southwest Asia and eastern Mediterranean regions, which are considered secondary centres of diversity and domestication. Anthocyanin carrots are still grown in eastern countries together with the western orange cultivars (Banga, 1976).

The cultivation of carrots may have begun 2000 to 3000 years ago as is evident from the indications emerged from seeds found by archaeologists in the Swiss Lake dwellings. Carrot cultivation has been traceable to the 10th century in Asia Minor. Purple carrots from place of origin, together with a yellow variant, spread to the Mediterranean area and Western Europe in the 11th-14th centuries. Introduction into India and China occurred during the 13th-14th century and to Japan in about 17th century from Western Asia. The orange types appeared as a chance mutation in the Netherlands during the 17th century and because of improved colour and flavour became popular in Europe. Then Settlers brought it to North America, where Indians and colonists adopted it. It is said to have been introduced into India from Persia and were probably first grown in America in the Salem gardens about 1620 AD (Shoemaker, 1947).

Long before carrot was domesticated, wild carrot has become widespread, as seeds were found in Europe dating back nearly 5000 years. Today, wild carrot is found around the world in temperate regions with adequate precipitation on disturbed sites, such as vacant lots, roadsides, and agricultural lands.

Carrot was introduced into India early in its history. Its Sanskrit equivalent called "*Girijana*" probably indicates the derivation of Hindi name "Gajar". Even now the primitive 'black' and 'red' types are popular in India. The 'orange' European types having higher carotenoid content were believed to have been introduced later during the visits of European traders. Obviously, the black and red types might have come earlier from the Central Asian region of its origin-probably Afghanistan, Persia and Southern Russia. These primitive of 'desi' types with low carotenoid content, were mostly annual, juicy types, forming roots under comparatively higher temperature of months in North India. Possibly its use in confectionery in North Western India, was responsible for the preference to sweeter, juice and brighter red coloured roots (Aligarh Red) than for orange coloured roots. Further, the seed production of these 'desi' types has been promoted in the North Indian plains during winter-spring season. Ain-i-Akbari (Anon) records carrot in Mughal cuisine, which explains the spread from Central Asia. In some parts of Kashmir, people still eat the wild carrots (Yawalkar, 1980). In north India, highly coloured red types of carrot are found which are not available in Europe. Types varying from absolute colourless to light lemon, light orange, orange and deep orange, light purple, deep purple and almost black are available.

Ma *et al.* (2016) reported that Chinese orange carrots are Eastern type and maintain the primitive traits of strong and pubescent leaves, and early flowering. Though Chinese orange carrots and Western orange varieties are morphologically similar, the structure and phylogenetic analysis based on SSR markers indicated that Western orange were clearly separated from Chinese carrots. They suggested that the first Chinese carrots seem to be yellow, and Chinese orange were derived from Chinese red according to the mixed distribution of red and orange accessions. Chinese orange carrots may have undergone a specific, independent process different from that of Western orange.

4.0 Taxonomy and Domestication

Approximately, 20 *Daucus* species are recognized today by the taxonomists, where *Daucus carota* includes the cultivated carrot and several wild subspecies (Heywood, 1983). Almost all species of carrot are diploid and the basic chromosome number varies from $x = 9$ to $x = 11$. The carrot species, *Daucus capillifolius* and a few other rare southern Mediterranean species have a basic chromosome number of $x = 9$ (Saenz Lain, 1981). The carrot has successful interspecific cross with *Daucus capillifolius* (McCollum, 1975) and *D. sahariensis*. Wild carrot, *Daucus carota* var. *carota*, also known as Queen Anne's Lace, is thought to be the ancestor of the present-day carrot. It is an annual and readily crosses with the cultivated carrot. Other wild carrot species are *D. maritimus, D. commutates, D. hispanicus, D. gummifer, D. fontanesii, D. bocconei,* and *D. major.*

Carrot roots occur in different colours, sizes, and shapes. Primitive carrots were purple due to anthocyanin. White- and orange-fleshed carrots appeared after repeated selections from the yellow types. Plant breeding during the 17th century in Europe resulted in development of cultivars known as Long Orange and Horn types which served as the basis of the germplasm of modern cultivars. Mutation and selection were more responsible for the development of the cultivated carrot than hybridization with wild germplasm.

Carrot domestication transformed the relatively small, white, forked, strong-flavoured taproot of a plant with annual or biennial flowering habit to the large, orange, smooth, good-flavoured storage root of uniformly biennial or "winter" annual crop we know today. European cultivars are firm textured, sweet, highly flavoured, yellowish orange to strongly orange in colour, slow bolting, and acclimatised to cool temperatures. Asiatic cultivars have a slightly softer texture, are juicy, low flavoured, bolt easily, are adapted to warm temperatures, and often have scarlet or reddish orange coloured roots. Generally, temperate cultivars are biennials; tropical cultivars display an annual habit and are grown during short-day conditions.

5.0 Cytogenetics

The carrot is diploid with nine pairs of chromosomes. These are short but slightly variable in length having only 1 pg DNA per 1C nucleus. Approximately, 40 per cent of this DNA is highly repeated. Of the nine pairs of chromosomes, four are metacentric, four are submetacentric, and one is satellite. According to Sharma and Bhattacharyya (1954), secondary constrictions on three chromosomes and a satellite on one have been noticed with a large gap separating the satellite and the remaining chromosomes in some stock. Generally, the rDNA of carrot has a high G+C content, but high A+T satellites have been noticed in some stock.

The chromosome number in various *Daucus* species is given below:

2n = 18 (*D. carota, D. capillifolius, D. sahariensis, D. syrticus*),

2n = 20 (*D. broteri, D. guttatus, D. littoralis, D. muricatus*),

2n = 22 (*D. aureus, D. involucratus, D. tenuisectus, D. crinitus, D. montevidensis, D. pusillus*),

2n = 44 (*D. glochidiatus*), or

2n = 66 (*D. montanus*).

It is assumed that × = 11 is the basic chromosome number in the Apiaceae, and × = 10 and 9 are its derivatives.

The DNA polymorphism of carrot observed in a limited population so far is about 25 per cent for cDNA and genomic clone and 70 per cent of the clones are polymorphic across the entire *Daucus* genus (Rubatzky *et al.*, 1999). There are 22 species of *Daucus* most of which have 11 pairs of chromosomes except for two with nine pairs. A modified squash technique for karyotypic analysis developed in carrot (Xing Jin *et al.*, 1994) was also found to be effective in five other species of Apiaceae.

6.0 Morphology

The carrot is an annual or biennial herb, with an erect to much branched stem, 30-120 cm high arising from a thick, fleshy taproot, 5-30 cm long. Usually the stem elongates and produces rough, hispid branches during the second year.

6.1.0 Leaves

Leaves are pinnately compound with long petioles sheathed at its base numbering 8-12. These grow in a rosette, glabrous, green with 2-3 pinnate leaf blades and the segments are divided into often-linear ultimate lobes.

6.2.0 Root

The edible portion of carrot root is actually an enlarged fleshy taproot. This is straight, conical to cylindrical, 5-50 cm long and 2-5 cm in diameter at top, orange, reddish, violet, purple, yellow or white. It consists primarily of phloem or cortex and core or xylem. Good quality carrots have a maximum of cortex and a minimum of core. In the so-called 'coreless' cultivars the core is small and deeply pigmented so that the cortex and core are evenly coloured. The periderm is the outermost tissue of the root followed by phloem, cambium, and xylem tissues. Oil ducts in the intercellular spaces of the pericycle contain essential oils that are responsible for the typical carrot odour and flavour. Roots are conical, cylindrical, round, or of intermediate shape.

6.3.0 Inflorescence

Carrot plants bear compound umbel. The development of the umbel begins with a broadening of the floral axis and internode elongation (Borthwick *et al.*, 1931). The umbels are encircled by long-lobed bracts, and umbellets are similarly surrounded by bracts. Each umbel comprises of 50 or more umbellet, each of which has approximately 50 flowers. It is the king umbel or umbel of the first order (or primary umbel) that flowers first. The umbels terminating the branches of the main stem are known as secondary umbels or umbels of the second order. In succession, third and fourth order umbels may develop in the same fashion. The first, second and other order umbels usually flower at an interval of 8-12 days from each other. The primary umbel bears more flowers at maturity usually over 1000, whereas secondary, tertiary, and quaternary umbels bear successively fewer flowers. Anthesis in a single umbel is completed in 7-9 days, the peripheral umbellate flowers open first followed by the inner umbellates. The flowers are perfect with small petals, usually white or yellowish. The calyx is entire. There are 5 stamens. The ovary is inferior and consists of two locules, each with a single ovule. On the upper surface of the ovary there is a swollen nectar secreting disk, which supports the style and stigma.

Braak and Kho (1958) revealed that though the primary umbel consists mainly of bisexual flowers, but male flowers can occur frequently (between the edge and centre of umbellet) in subsequent umbels. The pollen from flowers at the centre of an umbellet is larger and more frequently fertile than that from peripheral flowers (Nair and Kapoor, 1973). Carrot has protandry and the petals separate and the filaments begin unrolling to release the anthers at anthesis. After straightening of filament, the pollen is shed and the stamen is quickly abscised. Thereafter, the petals open fully and the style elongates. The carrot has a split style, which separates when the flower is receptive to pollination. The petals of male fertile plants fall soon after the split stigma is receptive. However, the petals of petaloid, but not of brown anther, male sterile plants are persistent until the seed ripens. The conspicuous umbels and flowers having nectaries encourage insect visits. The carrot flowers are epigynous with five small sepals, five petals, five stamens and two carpels.

Koul *et al.* (1989) reported that both wild and cultivated forms of carrot are andromonoecious and protandrous. Separation of male and female phases is complete at the level of the flower and the umbel, but the two phases overlap at the level of the whole plant, creating conditions for geitonogamy. At the same time, insect visits between umbels of adjacent plants lead to xenogamy. The extent of overlap between the male and female phases of various umbel orders is greater in wild than in cultivated forms.

6.4.0 Seed

The bilocular fruit is a schizocarp. Each carpel has two ovule primordia during early stage of development, but only the lower one continues to grow. In fact, two mericarps pair to form the schizocarp, the true carrot fruit, which develops from a two-loculed ovary. The mericarp or the seed is somewhat flattened on one side and the opposite side has longitudinal ribs with bristly hairs; the fruit is oblong, about 3 mm long. Borthwick (1931) found that carrot embryo sac is monosporic (developing from the chalazal macrospore) and 8-nucleate. Seeds are flat, ribbed, spiny, and size range from 500 to 1000 seed per gram.

7.0 Growth and Development

7.1.0 Shape and Size

Generally, the carrot storage root develops from secondary thickening of the hypocotyl and taproot. In most cultivars the carrot root first grows in length and then increases faster in diameter than in length, so that it becomes thicker with age. The increase in diameter is most rapid at the crown end and comparatively slows down towards the tip. This differential rate of increase in diameter at the crown end and the tip generally determines whether the root will be stumped (rounded) or tapering. Nevertheless, in all cultivars, young roots are always tapering. Many physical and biological factors, which cause defects such as crooking, forking and stubbed-off roots through the injury or interruption of natural taproot growth may act during this period of early growth.

The shape and size of carrot roots are influenced by several factors. Branching of roots is quite common in carrot and may result from certain hereditary factors, from presence of undecomposed organic matter or plant refuse in the soil, from injury to young taproot and from any impediment to its downward growth. Heavy soils develop more mis-shapen roots than lighter soils. Zotto and Muller (1990) found that compacted clay soil frequently caused root malformation (bifurcation). Temperature also affects markedly the shape of roots.

7.2.0 Pigmentation of Roots

Enormous diversity exists in pigmentation of wild and cultivated carrot roots; white, yellow, orange, red, purple and pink types are known. Actually, the visual colour of carrot roots is determined by the level of total carotenoids present, the accumulation of specific pigments and the distribution of pigments between phloem and xylem. In some yellow and purple cultivars, the root colour is due to anthochlor and anthocyanin (Banga, 1962). The red rooted carrot 'Kintoki' contains 150-270

Different Coloured Carrot

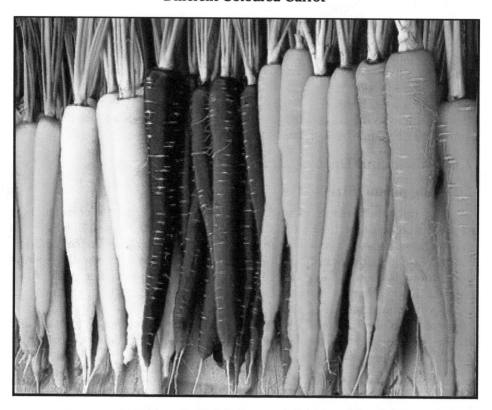

Variability of root colours in carrot

µg/g total carotenoids, of which about 3 per cent are carotenols. The major pigments in this cultivar are lycopene; β-carotene is present in small amounts. Many other pigments such as α-carotene, β-carotene, and zeta-carotene and phytofluene are also detected (Umiel and Gabelman, 1971). Genes responsible for pigment content and composition have been identified (Imam and Gabelman, 1968; Laferriere and Gabelman, 1968; Umiel and Gabelman, 1972).

Total carotenoids range from 80 to 120 ppm in cultivars. The most predominant carotenoids in orange and yellow carrots are α- and β-carotenes. β-carotene constitutes 60–80 per cent of the carotenoids in carrots followed by α-carotene (10–40 per cent), lutein (1–5 per cent), and the other minor carotenoids (0.1–1.0 per cent).

Flesh colours can be white, yellow, orange, red, or purple. Anthocyanin is responsible for the reddish purple colour. α- and β-carotene, responsible for yellow and orange colour, respectively are the major carotenoid pigments. β-carotene usually is 50 per cent or more of the total carotenoid content; the ratio of α to β is usually about 1:2. Red colouration in the flesh of some cultivars is from lycopene. The carotenoids are not uniformly distributed in the root. Carotene synthesis proceeds from the proximal to the distal tissues of the taproot. Phloem tissue usually contains about 30 per cent more pigment than the xylem. The carotene content of the most widely grown carrot cultivars ranges from 60 to more than 120 µg/g fresh weight.

Usually the young roots are yellowish white in colour, which changes to whitish yellow, light yellow, dark yellow, orange or orange-red, as a result of accumulation of carotenoids in varying amounts. Carotenes are first deposited in the oldest cells of the phloem and then in the oldest cells of the xylem. Since a carrot root grows from the cambium, the oldest cells are those immediately adjacent to the epidermis and in the centre of the core. More carotenes are deposited in the oldest cells as the root grows older, thus establishing a colour gradient, which radiates from the centre of the root outward and from the epidermis inward, leaving a light coloured zone at the cambium. It is more evident in relatively young and in fast growing roots. The vertical development of colour is from the top of the root to the tip. In general, the intensity of colour (carotene concentration) in roots increased with the age of the root until a maximum is reached, and the colour either remains almost unchanged throughout the life of carrot or may show some decline. It has been found that large roots generally contain more carotene than small roots. The top part of the root contains the highest carotene level, followed by the middle; the distal end contains the lowest level. Similarly, the outer part of the root contains more carotene than the inner part irrespective to root size. Evers (1989a) also recorded improvement in carotene content in response to PK and NPK placement as compared with broadcast fertilization or single application of irrigated or NPK fertigation. It was further observed that carrot roots fertilized with supraoptimal N showed a tendency to have higher carotene content than those fertilized with optimal N (Evers, 1989c). In general, the carotene content was more affected by the stage of maturity than by fertilization. The most effective method for the retention of total carotenoids during both processing and storage was freezing, followed by canning (Ustun *et al.*, 1999). Barba-Espin *et al.* (2017) observed that foliar-applied ethephon enhances the content of anthocyanin of black carrot roots. The roots of ethephon-treated carrot plants exhibited an increase in anthocyanin content of approximately 25 per cent, with values ranging from 2.25 to 3.10 mg g"1 fresh weight, compared with values ranging from 1.50 to 1.90 mg g"1 fresh weight in untreated roots. They noted most rapid accumulation rate for anthocyanins, phenolic compounds, soluble solids and dry matter between 10 and 13 weeks after sowing in both untreated and ethephon-treated carrots.

8.0 Cultivars

Many cultivars, some indigenous but mostly introduced from Europe and America, are cultivated in India. Important among the exotic types grown in India are Chantenay, Danvers, Nantes, Early Horn and Early Gem. Chantenay and Danvers are known for their long tapering roots and excellent quality. Early Horn and Early Gem are famous for their earliness and for their tender and mild flavoured roots. Usually the indigenous cultivars are coarser in texture, and possess less flavour than the exotic types. These types are hardier than exotic types, but colour development is poor in indigenous cultivars.

The carrot cultivars may be classified on the basis of shape of their roots or on their temperature response to flowering.

A. Classification Based on Shape of Roots

1. Long rooted: Roots may be 25 cm or more in length, generally tapering; these perform best in comparatively light soil.

2. Half-long rooted: Root length does not usually exceed 20 cm.

   (i) Roots cylindrical with straight or sloppy shoulder, *e.g.*, Nantes

   (ii) Roots tapering with blunt or semi-blunt tip, *e.g.*, Chantenay, Imperator.

3. Short-stump rooted: These cultivars are suitable for growing in heavy soils.

   (i) Heart-shaped, *e.g.*, Oxheart

   (ii) Oval, *e.g.*, Early Scarlet Horn

   (iii) Round, *e.g.*, French Forcing.

B. Classification Based on Temperature Response to Flowering

1. **Tropical, Asiatic or annual types:** These cultivars do not require any low temperature treatment for flowering; they seed freely in the plains of India. Most of these cultivars are slightly softer texture with more moisture content, are less sweet, low flavoured, bolt easily, are adapted to warm temperatures, and often have scarlet or reddish orange coloured roots.

2. **Temperate or European or biennial types:** These cultivars are biennial in characters and require low temperature (4.8-10°C) treatments for flowering to occur. These do not produce seeds in the plains of India, *e.g.*, Pusa Yamdagni, Nantes, Chantenay, Imperator, *etc.* Most of the temperate cultivars are firm textured with less moisture content, sweet, highly flavoured, yellowish orange to strongly orange in colour with slow bolting habit, and acclimatised to cool temperatures. They have higher concentration of α-carotene and are nutritionally superior.

Most of the European types have chilling requirement for bolting and they do not bolt under plains. The Red Asiatic types generally do not have any chilling requirement and seed production is taken up in the tropical plains.

8.1.0 Tropical Cultivars

Pusa Kesar

This is a selection from a cross between Local Red and Nantes Half Long, released in 1963 from IARI, New Delhi. It has short tops. Roots are red, tapering, small and contains high amount of carotene (38 mg/100 g edible portion) than Local Red (26 mg/100 g). Roots stay longer in field without bolting and set seeds freely in north Indian plains. Yield 250 q/ha in 100-120 days of crop duration.

Pusa Meghali

It is a selection from a cross of Pusa Kesar and Nantes, released in 1985 and developed by IARI, New Delhi. It has short tops. Roots are smooth with orange flesh, self-coloured core and rich in α-carotene (11571 IU/100 g edible portion). It sets seeds freely in north Indian plains. Yield 250-260 q/ha in 100-120 days of crop duration.

Pusa Vrishti

It is developed by recurrent breeding from local collections at IARI, New Delhi. It is suitable for early sowing beginning August under hot and humid conditions in north Indian plains. It takes 90-100 days to marketable maturity with average yield of 200 q/ha.

Pusa Rudhira

It is developed by recurrent breeding from local collections from U.P. at IARI, New Delhi. It was released and notified in 2008. It gives more than 50 per cent higher yield over Pusa Kesar and is self core red coloured cultivar with delayed bolting. It takes 75-90 days to marketable maturity with average yield of 300-330 q/ha.

Pusa Asita

It is developed by recurrent breeding in local material from Haryana at IARI, New Delhi. It was released and notified in 2008. It gives more than 30 per cent higher yield over Pusa Kesar and is self core black coloured late bolting cultivar. It takes 100 days to marketable maturity with average yield of 300 q/ha.

Hisar Gairic

Hisar Gairic has been developed through mass selection from diverse Indian germplasms. It is suitable for early sowing, which matures in 120 days. Compared to Pusa Kesar, its average yield of long, tapering roots is 290 q/ha, as against 240 q/ha in Pusa Kesar, with 9 per cent longer and 40 per cent thicker roots, which have less forking and fibre formation. Total carotenoid content is 96 mg/100 g fresh weight (Kalloo *et al.*, 1993).

PC 34

It is selected from plant materials collected from Ropar, Punjab and developed at P.A.U, Ludhiana. It was released in 2005. Leaves are dark green. Roots red, long with a small core, have excellent quality characteristics, juice yield, α-carotene and total sugar content.

Sel. No 29

It is a selection from local material and developed at P.A.U, Ludhiana. Roots are long, tapering, thin and light-red in colour.

Sel. No 233

It is a derivative of a cross of Nantes and Sel. No 29 and developed at P.A.U, Ludhiana. It was released in 1978. Tops are small. Leaves are 30-35 cm in size. Roots are 15 cm long, smooth, semi-cylindrical and orange. Core is light orange in colour. Root formation in 90-100 days. It has less cracking and forking.

Punjab Carrot Red

The roots of this cultivar mature in about 85-90 days after sowing. Its leaves are dark green and average plant height is 65 cm. Roots are dark red in colour, 27 cm long and 3.40 cm in diameter. This cultivar has an excellent quality characters. Roots have high juice content (515 ml/kg of roots), sweet (TSS 9.2 per cent, sugar content 8.04 per cent), rich in β-carotene (8.44 mg/100 g) and dry matter content (11 per cent). Average root yield is 230 q/ha.

Punjab Black Beauty

The roots of this cultivar attain edible maturity after 90-95 days of sowing. Leaves are dark green and petioles are purple in colour. Roots are purple-black, 26 cm long and 3.20 cm in diameter. Roots have ability to stay in the field over a fortnight after reaching edible maturity. This cultivar has high nutraceutical values and excellent quality characters. It is rich in anthocyanins (182 mg/100 g) and phenols (73 mg/100 g). It has high juice content (580 ml/kg), calcium (50 mg/100 g), iron (1.10 mg/100 g), TSS (7.5 per cent) and dry matter (11 per cent). Fresh carrots are suitable for salad, juice, pickle and kanji. Average root yield is 196 q/ha.

Kashi Krishna

It has green colour leaves, petiole colour purple to green acropetally, root shape Danvers type (triangular), bright black colour roots and self-coloured medium sized core. The roots are ready to harvest in around 95-105 days after sowing. The roots are of 22-25 cm length with root weight of 110-120 g and shoulder diameter of 3.2-3.5 cm having longer field stand for 30-35 days. The marketable root of 85-90 per cent having harvest index of 65-70 per cent with yield potential of 210-225 q/ha.

Arka Suraj

Roots of this cultivar are orange with self colour core, smooth root surface and conical shape. Root length is 15-18 cm, root diameter 3-4 cm, TSS 8-10 per cent and carotene content 11.27 mg per cent. It is tolerant to powdery mildew and nematodes.

PC 161

It is developed by PAU, Ludhiana. It is a tropical cultivar and its roots mature in 90 days after sowing. Its foliage is dark green. Roots are deep red in colour, 30 cm long, slender and 2.84 cm in diameter. The roots have more juice content (575.50 ml/kg), sweet (TSS 9.5 per cent and sugar content (8.75 per cent)) and rich in α-carotene (8.88 mg/100g). Its average root yield is 256 q/ha.

Pusa Vasuda

It is first public sector tropical carrot hybrid developed using CMS system. Roots are smooth, attractive, vigorous, self-coloured, red, juicy, rich in total carotenoids, lycopene, TSS and minerals. It is suitable for salad, juice extraction, cooking and industry for carotenoid extraction. It matures within 80-90 days. Average yield 350-400 q/ha.

8.2.0 Temperate Cultivars

Nantes

It has small and brittle top which makes pulling difficult. The roots are well shaped, small, orange, perfectly cylindrical, abruptly ending in small thin tail, sweet and self-coloured core with orange flesh. Average root yield is 120 q/ha.

Nantes Half Long

The roots are cylindrical in shape, blunt tip, orange in colour along with self-coloured core. It is suitable for sowing from mid October to early December, matures in 100-120 days and yield potential is 100-125 q/ha.

Early Nantes

The roots are almost cylindrical in shape, blunt tip, orange in colour along with self-coloured core and suited for canning. It is fit for sowing from mid October to early December, matures in 90-100 days and yield potential is 100-120 q/ha.

Chantenay

It is an excellent cultivar for canning and storage. Roots are 11.5-15 cm long and 3-5 cm in diameter, with tapering to blunt end, deep orange cortex and core.

Royal Chantenay is a widely adapted cultivar of this type. It is primarily grown for processing, but is well suited for home garden. Red Cored Chantenay is another cultivar of this type, which is quite popular and extensively grown.

Danvers

This cultivar is grown for fresh market as well as for processing. It is a mid-season cultivar with large and strong foliage. Roots are 12.5-15 cm long, 2.8-4.5 cm in diameter with tapering to short-tapering or slightly rounded end, deep orange cortex and a slightly more yellow core.

Imperator

It is said to be a cross between Nantes and Chantenay and is extensively grown for fresh market. It is a mid-season to late maturity cultivar with large and strong foliage. Roots are 15-17.5 cm long and 2.5-4.5 cm in diameter, with short tapered end, deep-orange cortex and slightly less pigmented core. Long Imperator 58 is one of the best cultivar of this type.

Pusa Yamdagni

It is derived from the cross between EC 9981 × Nantes Half Long. Roots long slightly tapering, cylindrical with small tops and orange coloured roots with self-coloured core. It is sown in April-August in hills and in October-December in plains. Maturity 90-100 days after sowing. Average root yield is 200-250 q/ha. Compared to Nantes Half Long it has 8.87-12.29 per cent more carotene and is richer in minerals including Fe (Gill *et al.*, 1987).

Chaman

It was developed at SKUAST, Srinagar. Foliage is dark green and semi erect. Roots are long, cylindrical, semi-blunt, tolerant to cracking and forcing. Flesh is orange, sweet, fine textured. Yield potential is 250 q/ha.

Ooty 1

It is a selection from half-sib progeny of local type DC 3 developed at Horticlutural Research Station, TNAU, Ooty and released in 1997. Suitable for growing in hilly areas with an altitude above 1800 m. Roots are long, slightly tapering, deep orange with self coloured core. Free from pre-mature bolting and resistant to powdery mildew, leaf spot and drought. It matures within 100-110 days with average yield of 300 q/ha.

Zeno

An introduction from Germany. Roots are 15-17 cm long, slightly tapering toward the end, deep orange with self coloured core. Root formation in 110-120 days. It is popular in Nilgiri hills.

Solan Rachna

It is a direct introduction from Netherlands. Roots are stumpy to semi-stumpy, long, smooth, orange colored with self core. It is early in maturity. Root yield is higher than Early Nantes. Average yield is 220-250 q/ha.

Pusa Nayanjyoti

It is the first orange colour hybrid developed by Public Sector in India. Roots are orange, smooth, uniform, cylindrical, stumpy with small indistinct self-coloured core. Roots possess high β-carotene content (7.552 mg/100 g fresh weight). It is sown in April-August in hills and in November-December in the Northern Indian Plains. Average yield is 300-350 q/ha.

Texas Gold Spike

Texas Gold Spike is an open pollinated carrot developed to replace Imperator 58, and is derived from the cross Pioneer × WI 33. It has short open foliage and has

Important Carrot Cultivars

Pusa Rudhira

Pusa Vasuda

Pusa Vrishti

Pusa Asita

Pusa Nayanjyoti

Pusa Meghali

Pusa Yamdagni

Nantes

similar crown diameter and root length to Imperator 58, but roots are cylindrical and blunt. Roots are smooth, uniformly orange in cross section and have higher carotenoid contents than Imperator 58 (Pike *et al.*, 1991).

Prima

Prima was selected by several cycles of recurrent selection from cv. Brasilia. It is resistant to bolting during spring and autumn cultivation and to *Alternaria dauci*. It has dark orange, cylindrical roots with 18-20 cm length and 3.0-3.5 cm diameter and vigorous dark green, upright leaves. Prima can be sown from mid-September to early April in the south, southeast and centre west region of Brazil (Cardoso *et al.*, 1992).

Beta III

This cultivar, developed from breeding population USDA B 951-1, was released in 1986 as a garden cultivar (Peterson *et al.*, 1988). Mature roots are 2-3 cm in diameter at the shoulder, 17-23 cm long and uniformly dark orange externally. The tops are vigorous, strong at the shoulder for mechanical harvesting and tolerant of *Alternaria dauci*.

Volzhskaya 30

It is a mid-season cultivar suitable for culinary use and canning. The roots are cylindrical, 15-18 cm long and 4.5-5.2 cm in diameter, and are smooth and readily extractable from the soil. The flesh is deep orange, firm and tasty, with an orange core. The roots weigh 89-217 g and contain 4.9-6.4 per cent sugar, 10.2-11.5 per cent dry matter and 9.2-11.5 mg carotenoids per 100 g fresh weight. They are suitable for mechanical harvesting and have good keeping quality. Mean yield is 450-650 q/ha (Anon., 1991).

Vytenumanto

Vytenumanto was developed by mass and recurrent selection from a local population. Roots are cylindrical, blunt at the end, smooth skinned and orange in colour. Flesh and pith are orange. Mean root yield is 76.1 t/ha with a marketable root percentage of 71.1 per cent. Root dry soluble solids content is 8.2 per cent and sugar content 6.6 per cent (Armolaitiene, 1999).

Xin Huluobu No.1

Xin Huluobu No. 1 is a new carrot cultivar selected from a natural hybrid progeny of a local cultivar. Its fleshy orange red root is cylindrical, 14-16 cm long, 4-5 cm wide and 120-140 g in weight. The flesh is uniformly orange red. It is suitable for fresh consumption and processing. The average yield is 645 q/ha (Jin *et al.*, 1999).

Mustang

It is a medium-bodied, Berlicum/Nantes type F_1 hybrid from Russia. The rosette of leaves is semi-erect. The root crop is smooth, cylindrical with a shoot to the base, obtuse, 17–19 cm long, orange in color. The core is small orange. Root crop weight is 180–210 g. The content of dry matter 10.5–15.0 per cent, sugars 8.5–11.0 per cent, carotene up to 20 mg per 100 g of raw material. Commercial yield is 600–730 q/ha (Khovrin and Kosenko, 2020).

Eight cultivars were evaluated in coastal Odisha (India) by Panda *et al.* (1994), who found the longest roots (15.88 cm) in Early Sweet Tender and shortest (12.21 cm) in Early India. Chantenay had the greatest root diameter (3.63 cm). The highest root fresh weight (45.7 g) was recorded with Early Nantes, which gave the greatest yield (6.49 t/ha). Bishnoi and Dhaliwal (2017) evaluated three carrot varieties in farmer participatory trial in Sri Muktsar Sahib district of Punjab and observed that cultivar Punjab Carrot Red recorded the highest plant height (51.6 cm), number of leaves (11.2), root length (23.6 cm), root diameter (2.90 cm) and higher yield (452.7 q/ha) compared to cultivar Desi Red of Sungrow (360.1 q/ha) and Gayatri of Global Seeds (341.3 q/ha).

Synnevag (1989) found Napoli as the best cultivar for early production under plastic, giving high yield of class I carrots. The cv. Nanthya was also reliable for early growing. Panther was a little later but was a good alternative and could be left in the ground without splitting or growing too large.

Twelve carrot cultivars (for marketing as bunched carrots) were sown on 15 June 1999 in Belgium, and harvested on 30 August. Dark and good quality foliage was observed in cv. Merida and Evora. The foliage of Jeanette was long while that of Evora was short. The best cylindrical shape was obtained from cv. Evora, Calibra, Nantino and B 1901, while cv. Evora, Aradia and SG 6386 scored best for smoothness (Callens and Callewaert, 1999).

9.0 Soil and Climate

9.1.0 Soil

Although carrot can be grown on all types of soil, it thrives best on a deep, loose, friable well-drained loamy soil. For early crop, a sandy loam soil is preferred, but for large yields silt or silt-loam soil is desirable. The long, smooth slender roots desired for fresh market can successfully be grown on deep well-drained light soils. Carrots grown on heavy soils tend to be more rough and coarse than those grown on light sandy soils. Hence, heavy, stony, compacted or poorly-drained soils are less suitable. Crusting of clay soil results in poor seedling emergence in carrot. In heavy, humus-rich soils, carrot develops excessive leaf growth, and form forked hairy roots.

Carrots do not grow well on a soil that is highly acidic. Best soil pH for carrot crop is 6.0-7.0. Gupta and Chipman (1976) obtained maximum root and top yields at pH 6.6 and 7.1, respectively; at pH 4.3 and 8.1 yields were low. At higher pH, Mn toxicity may result, causing chlorosis of leaves. Racz (1973) noticed a direct relationship between penetration resistance and root deformity.

9.2.0 Climate

Carrot is grown in winter in the plains of India. Its climatic requirement for market crop is more or less like radish. Seed germination and root growth are greatly influenced by soil temperature. A temperature range of 8 to 24°C is considered optimum for seed germination. As emphasized earlier, very high or low temperature is not conducive for maximum growth of roots and for proper development of root colour. Wendt (1979) found that the growth was poorer at 15°C than at 20-30°C; the highest root weight was recorded at the temperature range of 20-25°C. However, the tropical types can form roots even at higher temperatures of 25°C. Top growth is reduced at mean temperature of 28°C and roots become very strong flavoured. At low temperature the colour is poor and roots become enlarged and tapered below 16°C. At high temperatures the roots tend to be shorter, often with a poor flavour. At 4-10°C there is very little enlargement of roots and very little top growth. Cool night temperatures are essential for carrot production in the tropics. Day length of 9-14 hr has no effect on root colour.

Climate is considered much more often as a limiting factor than the soil. Although oriental types of carrots are well adapted to the climatic conditions of the plains of India and seed freely under tropical condition, the temperate or European types require low temperature (4.8-10°C) for 4-6 weeks at any time during the development of roots or after they mature, either in storage or under field condition. Seedstalk formation takes place only when these plants are subjected to a subsequent temperature of 12.2-21.1°C. It is for this reason that the seeds of European types of carrot are produced only in the hills of India, where the winter is severe.

Gray *et al.* (1988) found that carrot seeds grown at day/night temperature of 30/20°C had the largest embryos and the highest N, DNA and RNA contents; they germinated and emerged earlier and gave higher germination percentage. Elballa and Cantliffe (1996), on the other hand, reported that high (33/28°C) day/night temperature during pollination, fertilization or early stages of seed development can greatly reduce carrot seed yield and quality. Seeds that developed at 33/28°C day/night produced seedlings with lower vigour. The highest germination and seedling vigour were recorded in seeds developed at 20/15°C day and night.

10.0 Cultivation

10.1.0 Land Preparation

For cultivation of carrot, the field should be worked deep to a good tilth a depth of at least 30 cm and properly manured. If the soil is not thoroughly prepared and if it contains soil clods, or undecomposed organic matter, good quality well shaped roots cannot be produced. Root deformity usually occurs in fields which are underprepared. Sub-soiling helps in breaking compacted soil layers. Whitaker *et al.* (1970) indicated that although the land surface should be as smooth as possible before planting, excessive tillage should be avoided since it is costly and affects soil structure adversely. In an investigation, White (1978) reported that rolling after ploughing decreased carrot yields. Orzolek and Carrol (1978) demonstrated that no-till carrot production is feasible and may be more advantageous than conventional production methods. Yield and quality of carrots grown under no-till system appear to be promising. Pietola (1995) found that external quality of carrot root was adversely affected by soil compaction; hence use of light and loose soil for carrot has been suggested.

Millette and Vigier (1981) found that raised beds offered no yield advantage in a wet season and no yield reduction in a dry season. In a wet season, however, bed system favoured harvesting operation by reducing soil moisture and rendering field surface firmer. Evers *et al.* (1997), on the other hand, reported that the total and marketable yields were higher on flat land and narrow ridges than on broad ridges and compacted broad ridges of fine sandy soil. On ridges, the fine sand dried quickly.

10.2.0 Sowing

Carrots are grown from seeds sown directly in the field where the crop is to mature. In South and Central India they are cultivated largely on the hills and the seeds are sown during January-February, June-July and October-November. When the climatic conditions are favourable, carrots can be grown almost all the year round and successive sowing may be done every fortnight to obtain a continuous supply of roots. Seeds are sown from August to December in the northern plains of India; in the hills it is sown during March-June. In the southern Bengal, last week of October seems to be the best sowing time (Pariari and Maity, 1992).

Carrot seeds are sown by either broadcasting or drilling in lines. Seeds are mixed with fine sand before sowing to facilitate even distribution. The seed rate varies from 5 to 6 kg per hectare. Bathkal and Patil (1968) compared the flat bed planting method (25 cm between rows) with planting on one side of a ridge (50 cm between ridges) and on both sides of such ridges. The yield from planting on both sides of the ridges was significantly superior to that from flat bed planting at equivalent density. Planting on one side of the ridge encouraged root forking. Germination is slow and usually 10-12 days are required for seedlings to appear. It is desirable to give light irrigation immediately after sowing or alternatively to soak the seeds in water for 12-24 hours before sowing to hasten germination. The beneficial effect of irrigation applied to the seed bed on the uniformity of seedling emergence has been demonstrated by Finch-Savage (1990).

Carrot seeds contain a germination inhibitor, which has been named as carrotal by Aki and Watanabe (1961). Immature seeds contain maximum level of carrotal which usually does not show any change during storage of seeds (Aki and Watanabe, 1962).

Jacobsohn (1978) and Corbineau *et al.* (1995) reported that large seeds germinated better than small seeds and gave a higher yield in both spring and summer planted carrots in Israel. Hoyle (1973), on the other hand, found that small seeds gave as good root yield as large seeds. Germination and subsequent seedling growth were found to be positively correlated with the embryo size. Seeds with small embryos sometimes failed to produce seedlings (Gray *et al.*, 1991).

Many external factors affect seed germination in carrots, of which soil temperature and moisture seem to be most important. Wendt (1979) reported that at a soil temperature of 35°C or above seeds did not germinate. Corbineau *et al.* (1995) reported that germination in carrot seeds was reduced at low temperatures (5-10°C) and at temperatures above 30°C. Germination was also reduced at low oxygen tensions. They further noticed that seeds produced from umbels of first and second order germinated faster and to a higher percentage than those collected from third order umbels. Wiebe (1968) found that the optimum soil moisture content for seed germination was 40 per cent of field capacity. Seeds sown in groups were affected considerably more by soil moisture than single seed. Grouped seeds require warm temperature for comparatively longer time for germination. Shallow sowing (1 cm deep) of grouped seeds resulted in greater emergence than 2 or 3 cm deep sowing. Report from Norway revealed that irrigation regimes (watering when soil tension reached 30 or 60 mm) did not greatly affect germination (Haland, 1975).

Several pre-sowing seed treatments are found to increase seed germination. Cantliffe and Elballa (1994) suggested use of primed seeds in polythelene glycol (PEG) at –10 bar for 7 or 14 days at 15°C during the warm part of the growing season in Florida for improved carrot stand establishment.

Shishkina and Galeev (1974) noticed that when seeds of cv. Nantes were soaked in Mn (0.02 per cent) and/or B (0.01 per cent) solution for 24 hours, then dried and sown, the treated seeds absorbed 7-9 per cent more water within the first 4-hour, germinated better in the field and gave a thicker stand of larger roots than the control. The yield increases over control were 20.8 per cent, 23 per cent, 14.2 per cent for seeds treated with Mn, B and Mn + B, respectively. The treatments also increased the sugar and carotene content of roots.

Pantielev *et al.* (1976) studied the effects of the following seed treatments: (i) grading, (ii) soaking in water, (iii) soaking in macronutrient + micronutrient solution, and (iv) application of dry shredded cowdung + peat (1: 1) to germinating seeds on field germination of carrots. Soaking in water increased field germination by 13 per cent and increased yield upto 60 per cent. Grading was also found to increase production of marketable roots by 17-23 per cent.

Yield and quality of carrot roots have been shown to improve following presowing seed treatments. Drozdov (1962) reported that yield of carrot cv. Chantenay increased by about 38 per cent when seeds were treated with succinic acid at 1 g in 80 litre of hot water (45°C) as compared to dry seeds; the treatment increased the carotene content of roots by 14.6 per cent over control. Maurya (1986) obtained the highest germination (98 per cent), root yield (20.64 q/ha), root carotene (8.8 mg/100 g) and ascorbic acid (8.85 mg/100 g) contents when the seeds of cv. Pusa Kesar were soaked in IBA at 50 ppm for 12 h.

Increased yield and sugar, carotene and dry matter contents were also obtained by Alekseeva and Rasskazov (1976) by using carrot seeds soaked in solutions containing B, Mn, Co and Mo, the best results being obtained with Co.

Pill and Evans (1991) observed that under greenhouse conditions, seeds of carrot cv. Danvers 126 given a hydration treatment (soaking in water for 1 day at 20°C followed by incubation for 3 days in fluid drilling gel at 20°C) generally emerged more rapidly and gave greater seedling fresh weights than osmotically primed seeds.

Seed pelleting was found to advance root maturation and improve crop quality. Mahmood-ur-Rehman *et al.* (2020) reported that lower concentration of salicylic acid (0.1 mM) can be effectively used to boost carrot seed germination even under high temperature conditions.

The spacing for the market crop as well as for the seed crop varies widely depending on the cultivars, types of soil, fertility of soil, environmental condition, *etc.* Frohlich *et al.* (1971) reported that the total and marketable yields of late carrots increased as a result of reduction in row spacing from 41.7 to 25 cm. Increasing plant density enhanced the marketable yield more than the total yield. They noticed further that seed placement resulted in higher yield of roots than normal drilling. An investigation conducted by Buike and Lepse (2000) to ascertain the most suitable sowing system for Nantes type carrots under Latvian climatic conditions indicated that three-row line produced the highest total and marketable yield (68.3 t/ha).

In a trial with 15 carrot cultivars, Pavlek (1977) noticed a positive correlation between the number of plants and yield per hectare. Draglano (1978) showed that both total and marketable yields increased with increasing plant density from 40 to 100 plants/m². Root size decreased with increasing plant density but the number of split or branched roots was not affected significantly by the plant spacing. Kepka *et al.* (1978) also demonstrated that total as well as marketable yield of carrots rose with decreasing inter-row spacing from 45 to 10 cm and intra-row spacing from 6 to 2 cm. A population of 111 or 222 plants/m² gave the highest and most uniform yield. Crown greening was maximum when seeds were sown in bands 45 cm apart. Mack (1980) reported that total root yields as well as roots of 25 and 25-38 mm in diameter were increased in carrot cv. Red Cored Chantenay as row spacings were reduced from 60 to 15 cm. Different within-row spacing did not have a significant impact on total yield but affected yield of various size grades. In general, the root length decreased as the plant population was increased. Salter *et al.* (1979) reported that total root yield was not significantly affected by plant arrangements or plant density in most cases. However, the yield of canning size roots (20-30 cm diameter) was influenced by plant density; with the lowest density treatments the highest yield of canning roots was obtained from the earliest harvest.

Like market crop, the plant spacing varies widely in seed crop. The optimum spacing for seed crop was reported to be 20 × 90 cm in South Africa (Maree and Lotter, 1971). Reports from Chile showed that seed yield per plant were highest from the widest spacing (0.7 × 1.2 m) and per hectare from the closest spacing (0.3 × 0.8 m). The closest spacing did not adversely affect seed germination which in all cases was 84 per cent. The 0.3 or 0.5 × 0.8 cm spacing was recommended for practical seed production in Chile (Krarup and Montealegre, 1975). Saini and Rastogi (1976) and Kumar and Nandpuri (1978) also obtained the highest seed yield with closest spacings (20-30 × 30 cm). Oliva *et al.* (1988) noticed that the seed yield/plant decreased continuously as population increased, but seed yield/unit area increased to a maximum at 12 plants/m², while the seed quality remaining unaffected by plant density.

Gurr and Surina (1980) found that close spacing accommodating up to 55000 plants per hectare advanced seed ripening by 2-3 days and increased seed yield per hectare by 51 per cent compared to wider spacing. Gray (1981) and Gray *et al.* (1983) examined the responses of carrots to plant densities from 100000 to 800000 plants per hectare for the root-to-seed method and from 110000 to 2560000 plants per hectare for the seed-to-seed method. Yields of seeds from the root-to-seed method increased from 1100 to 1500 kg/ha, and in the seed-to-seed method the seed yield rose from 700 to 2400 kg/ha over the densities used. At the highest density, 60 per cent of the seed yield was produced from primary umbels compared to less than 20 per cent at the lowest density. There was, however, no significant effect of plant density on plant height, time of flowering or crop maturity.

Gill *et al.* (1981) found that plants at the widest spacing (60 × 45 cm) produced the largest number of umbels of different orders; these umbels yielded larger amount of seeds than those at a closer spacing (40 × 30 cm or 50 × 30 cm). They noted also that the secondary umbels produced most of the seeds of best quality. Gray *et al.* (1983), on the other hand, found that seeds from primary umbels and low density crop were heavier than those from secondary umbels and high density crop. Noland *et al.* (1988) in USA found no consistent relationship between plant density and seed quality parameters such as per cent germination, speed of germination and seed weight. However, manipulation of plant density may be used to maximize carrot seed yield.

10.3.0 Manuring and Fertilization

Carrot responds well to manures and fertilizers. Both organic and chemical fertilizers are applied. Undecomposed farm yard manure (FYM) or compost should be avoided, because application of such manures usually lowers the quality of roots for table purpose. Application of fresh manure to carrots may result in branching of roots, which makes the carrot unfit for market. However, according to Kumar and Mathur (1965), application of plentiful organic matter is essential for carrot cultivation. They recommended that 23 tonnes of well-rotted farm yard manure should be incorporated into the soil to a depth of about 20-25 cm much before planting. When organic manure was applied, the need for fertilizers was reduced and an application of 50 kg N, 40 kg P and 50 kg K per hectare was considered sufficient to obtain optimum yield. In the absence of organic manure, application of 80 kg N, 60 kg P and 60 kg K per hectare was desirable. Half of the amount of N should be applied along with phosphorus and potassium before sowing and the rest half with second irrigation. Kropisz and Wojciechowski (1978) reported that composts in combination with NPK had a beneficial effect, but when used alone had no effect on increasing the yield of carrots. Evers (1988) observed smaller yield for organically grown carrot. Evers (1989b) noticed further that the sugar content in organically cultivated carrots did not differ from that of conventionally grown carrots, but the taste was worse. However, an increase in soil organic matter content was accompanied by an increase in nitrate level in the carrot roots (Sady *et al.*, 1999). Kováèik *et al.* (2018) reported that the inoculation of soil containing vermicompost by earthworms *Eisenia foetida* increased the root yield and total antioxidant activity in roots. It increased the content of vitamin C and total polyphenols in leaves.

Bajpai *et al.* (1972) reported that application of 60 kg N and 60 kg P through a fertilizer mixture resulted in maximum yield of 193.33 quintal per hectare. In the sandy loam soil of the Gangetic alluvium, the optimum doses of N and P were found to be 54 and 50 kg/ha, respectively (Choudhury *et al.*, 1972). Sharangi and Paria (1995, 1996), on the other hand, found that application of 80 kg N and 50 kg

K/ha produced the largest (21.59 cm), thickest (5.82 cm) and heaviest root (120.25 g). The optimum N, P_2O_5 and K_2O rates for cv. Nantes in Assam were found to be 60, 32 and 125 kg/ha, respectively (Maurya and Goswami, 1985). Sowing in rows on the flat (22.5 cm apart) and application of 60 kg N/ha were found optimum for harvesting nutrient rich roots (Kumar *et al.*, 1974). Lenka *et al.* (1990) obtained the highest yield (8.96 t/ha) with 75 kg N + 2 t lime/ha. The full rate of lime and half of the N were applied before sowing together with P + K at 60 kg/ha. The remainder of the N was applied one month after seed germination. Hassan *et al.* (1992) reported from Egypt that the average root weight, root length, total plant fresh weight and yield in carrot cv. Red Cored Chantenay increased with increasing rates of fertilizers upto 95.2 kg N + 85.7 kg P_2O_5 + 116.7 kg K_2O/ha. Abo-Sedera and Eid (1992) also from Egypt, on the other hand, recorded best results in terms of vegetative growth, yield and quality in cv. Red Cored Chantenay on a clay loam soil with 142.9 kg N + 171.42 kg K_2O/ha and NAA at 50 ppm. Thus, fertilizer requirement for carrot seems to be much less under Indian condition as compared with other countries, especially European countries. Da Silva *et al.* (2017) reported the use of nitrogen doses up to 103 kg ha^{-1} increased the commercial production of carrot. Potassium doses influenced root length, root diameter, commercial classification, shoot dry matter and root yield in addition to commercial production up to 87.75 kg ha^{-1}. The doses of nitrocalcium positively influenced shoot height, shoot fresh matter, commercial production and non-commercial production of the carrot crop to the 311 kg ha^{-1} dose of the fertilizer.

Subhan (1988) suggested foliar application of 3.0 g Gandasil D (containing 14 per cent N, 12 per cent P_2O_5, 14 per cent K_2O and 1 per cent Mg)/l at 2-week intervals for highest yield of carrot. Silva and Silva (2017) determined leaf indices and reference values using the compositional nutrient diagnosis (CND) method, at three phenological stages of carrot crops [40 and 70 days after sowing (DAS) and at harvest]. They created a database of leaf contents of the nutrients N, P, K, Ca, Mg, S, B, Cu, Mn and Zn and root yields, in commercial carrot crops. At 70 DAS and at harvest, Mg and Mn were the most limiting nutrients by deficiency, while Zn was the most limiting by excess at 70 DAS, and P and Ca were the most limiting by excess at harvest. Their results indicated that the foliar diagnosis performed with the CND method is only effective for the correction of nutritional disturbances in subsequent crops.

Biacs *et al.* (1995) studied the effect of certain microelements like Mo, Zn, Cr and Se on the yield and carotenoid content of carrot grown in a calcarious loamy chernosem. They found that Zn application slightly increased root yield, while Mo had no effect and Se decreased root yield. The highest Mo application resulted in a significant decrease in total root carotenoid content. Amending the soil with Se and Zn stimulated an increase in alpha-carotene and lutein (xanthophyll) synthesis and a decrease in beta-carotene with a slight increase in the total carotenoid content. LeBlanc and Gupta (1994) also found that application of trace elements such as B, Mo, Cu and Zn did not affect the marketable yield of carrot cv. Spartan Fancy grown on a virgin sphagnum peat bog soil. No response to Mo was also reported by Gupta *et al.* (1990).

Fertilizer application affects the quality of roots. Kanwar and Malik (1970) reported that application of high levels of NPK fertilizers caused bolting and root splitting. They also noticed that micronutrients had no significant effect on root quality, although B increased the root yield when applied in combination with lower dose of NPK. The highest protein content in roots was recorded when N and P at 60 kg each per hectare were applied (Bajpai *et al.*, 1972). Batra and Kalloo (1990) found that percentage splitting, forking and bolting in carrot cv. Gurgaon Selection increased with increasing rates of N application from 30 kg/ha to 90 kg/ha. Nitrate content of the root is influenced by the rates of N application. For high yield, acceptable root nitrate content and good soil fertility, N application rate should be kept at 40 kg/ha (Drlik and Rogl, 1992). Habben (1973) found that high N levels promoted carotene formation, but K had little effect. The total sugar content varied little with fertilizer levels. Pankov (1976a) reported that although P deficiency reduced the yield, the dry matter, sugar and carotene contents increased in P-deficient roots. Burdine and Hall (1976) noticed that the percentage of sucrose and alcohol insoluble solids in roots increased linearly with K. The nitrate content in carrot has been found reach the highest level when ammonium nitrate is applied at 75 kg N/ha (Mazur, 1992). However, slow release fertilizers tend to lower root nitrate content.

Fertilization affects the storage quality of carrot roots. Vleck and Polach (1978) noticed that high N rates increased rotting in sand-stored carrots, but not in carrots stored in crates. Conversely, higher rates of both P and K improved storage life of carrots in crates but not in sand. The nutrients had no marked effect on dry matter, total and reducing sugar or β-carotene content.

A close correlation was found between carrot root yield and N, P and K content in leaves. Optimum leaf N content in relation to plant growth and yield was found to be 2.8 and 4 per cent of dry weight, sampled 96 and 49 days after emergence, respectively (Hipp, 1978). The leaf P and K levels for optimum yields were 0.25-0.26 per cent and less than 1.7 per cent, respectively (Pankov, 1976a, b).

In studies on seed production, Alekseeva and Kutsenko (1976) found that N, P and K at 60, 180 and 180 kg per hectare, respectively gave the highest seed yield, increased the production of standard grade mother plants by 8 per cent and reduced storage losses by 6.5 per cent compared to the non-fertilized control. Krarup and Moretti (1976) obtained the highest seed yield (192 kg/ha) from plants receiving N, P and K at 192, 150 and 100 kg per hectare, respectively. Nazeer *et al.* (1989) recorded the highest seed yield (17.19 q/ha) in cv. Early Nantes with 90 kg N/ha, applied in two doses, one at bolting and the other at the pre-flowering stage. P had a less pronounced effect as seed yield. Several other workers showed the beneficial effect of N fertilizer on seed yield in carrot (Polach, 1987). Addition of micronutrients was not found to be beneficial for seed production in carrots (Malik, 1970).

10.4.0 Irrigation

The effect of irrigation depth was studied by Saparov (1992) who found that wetting soil to a depth of 40 cm before and 60 cm after root formation produced higher yield than wetting to 30 cm before and 40 cm after root formation. Barta and

Kalloo (1991) noticed the highest water use efficiency in carrot with the lowest rate of irrigation. Karim *et al.* (1997) also reported from Bangladesh that the water use efficiency of carrot was highest when the crop was irrigated at 60 per cent depletion of available soil water. Prabhakar *et al.* (1991), on the other hand, reported that root yield, total DM content, leaf area index, N uptake and water use efficiency increased with increasing rate of irrigation. The highest yields were obtained with 100 per cent evaporation replenishment and 120 kg N/ha. Pleskachev *et al.* (2018) reported a combination of deep chisel plowing with ammonium nitrate fertigation in 1-4 applications and NS 30: 7 fertilizer in 5-8 applications resulted in the highest carrot yield in 2015–2017 and averaged 90.6 t/ha.

Carrot requires irrigation at frequent intervals for proper growth of roots. Schmidhalter and Oertli (1990) estimated that under optimum condition and favourable nutrient supply, evapotranspiration coefficients varied from 300 to 400 l/kg storage root dry weight of carrot. Core diameter, bolting and percentage of split roots increased with increasing irrigation depth (Batra and Kalloo, 1991). White (1992) recorded the greatest number of marketable roots at the medium soil water content (54 per cent). The high soil water content (60 per cent) reduced marketable root yield and root length. The average irrigation application should vary from 60 to 80 mm. In general, frequent irrigations during emergence and growth increase the growth rate and gave higher yield. Orzolek and Carrol (1978) noticed that supplemental irrigation increased yields and reduced secondary root formation in all the six cultivars tested. In another study, they found that irrigation decreased secondary root growth in conventionally tilled and rye mulched plots, but not with soybean stub-bles (Orzolek and Caroll, 1978). A study conducted in Poland reveals that irrigation of carrot significantly enhanced both total and marketable yield (Cebulak and Sady, 2000). They also noticed that the nitrate content was 104 per cent lower in irrigated than non-irrigated crops (Sady and Cebulak, 2000).

On medium or heavy loamy soils in the region between the rivers Volga and Don, the soil moisture content during germination to start of root development should be maintained at not less than 80-85 per cent and thereafter at 70 per cent. For this, 15 irrigations (4400 m^3 water/ha) should be applied in dry years and 8 applications (2500 m^3/ha) in wet years.

Tomic (1976) studied the effect of seed pelleting and irrigation on root yield of carrots (cv. Nantes). With irrigation the pelleted and unpelleted seeds gave comparable yields but without irrigation unpelleted seeds produced better results.

Marouelli *et al.* (1988) recorded best results with regard to seed yield and to labour and energy economy when the seed plants were irrigated with on the 14th, 21st and 28th day after planting. In a subsequent report, however, they recorded highest seed yield (1300 kg/ha) by irrigation until 120 days after planting of stecklings (Marouelli *et al.*, 1990a). They reported further that the highest seed yield (1052 kg/ha) could be obtained, when irrigation was applied as the soil water tension reached 75 k Pa (Marouelli *et al.*, 1990b). Steiner *et al.* (1990) reported improvement in seed quality, measured as seedling root length, with increasing water application.

10.5.0 Interculture

Seedlings are thinned to a spacing of 5 cm to maintain the plant population of about 6-8 lakh/ha. Plant population also depends on the carrot type grown. In USA, per hectare plant population suggested for processing (dicing) type, cello (fresh market type), slicer type and cut-and-peel (baby carrots) is 4.4 to 6.0, 8.5 to 11.0, 11.0 to 13.5 and 32.0 to 46.0 lakhs, respectively. Timely hoeing is done for proper soil aeration. Weeds are a serious problem in carrot fields and timely control of weeds is essential to avoid heavy loss in yield of top quality roots due to weed competition. However, Deuber *et al.* (1976) showed that any weed competition for 15-50 days from sowing caused significant yield reductions, Pitelli *et al.* (1976) reported that if weed control did not start until 30 days after germination, yields were significantly reduced.

Both mechanical weeding and chemical weed control are practised in India, while chemical weed control is the usual method employed in advanced countries. In India, the between-row soil is hoed several times to check the weeds. Hoeing not only checks weeds but also facilitates soil aeration and results in better root growth. However, very deep cultivation should be avoided. Earthing up of roots along with weeding is also done to prevent exposure of roots to atmosphere and consequent discolouration.

Spraying of pendimethalin @ 3.5 litres/ha immediately after sowing is recommended to control the weeds in carrot field. Proper soil moisture should be there at the time of Pendimethalin application. Carrots can recover from competition when weeds are removed early. However, yield can shrink 30 to 60 per cent under severe competition. Carrots can tolerate weeds only up to 5 weeks after emergence. Unchecked weeds can cause yield losses from 65 and 75 per cent. A pre-emergent spray of linuron at 0.5 to 1.0 kg/ha is most useful for control of weeds without lowering yield of carrot. Good weed control and enhanced yield can also be achieved by using oxadiazon, metribuzin, pendimethalin and oxyfluorfen. One earthing up is done during root formation to cover the exposed roots and prevent greening.

Montemurro *et al.* (2016) reported that Bismark (clomazone + pendimethalin) sprayed at 1.5 and 2 L/ha showed optimal control of *Ammi majus, Galium aparine, Polygonum persicaria, Chenopodium polyspermum* and *Datura stramonium*, while only the higher rate (2 L/ha) was effective on *Solanum nigrum* and *Polygonum persicaria*. Bismark with Song 75 WG (metribuzin), sprayed in mixture or in post-emergence, was effective on *Amaranthus retroflexus*, whereas only the mixture of Bismark + Song 75 WG was effective on *Diplotaxis erucoides*. Bell *et al.* (2000) reported about six times greater carrot yield where linuron was applied as pre-emergence and post-emergence herbicide. Geminiani *et al.* (2008) found metribuzin to be also useful as post-emergence herbicide for controlling weeds except solanaceous weeds in carrot. Prometryn is applied as both pre- and post-emergence herbicide in carrot. Trifluralin is applied to control many annual grasses and certain broadleaf weeds in carrot fields.

Singh and Jaysawal (2018) showed that the black polythene mulch was found to be the best among the various treatments and recorded maximum plant height

Different Field Growing Operations of Carrot

Seed sowing (left, middle) and seedling emergence (Right) in carrot

Thinning (left) and furrow irrigation (right) in carrot

Field view of carrot growing

Different Stages of Harvesting Carrot Roots

Harvesting of carrot roots and making bundles for marketing

Marketing and Post-harvest Management of Carrot

Packing of carrot for marketing

Carrot displayed in the market for selling

Washing, grading and sorting of carrot

(61.70 cm), leaf length (26.78 cm), leaves number (9.84/plant), fresh weight of leaves (39.38 g/plant), dry weight of leaves (5.83 g/plant), fresh weight of root (225.33 g/plant), dry weight of roots (17.88 g/plant), fresh weight of plant (264.72 g), dry weight of plant (23.71 g), total root length (23.45 cm) and total root diameter (5.54 cm) in carrot. It also recorded the maximum yield (1.43 kg/m^2 and 54.69 t/ha), which was followed by blue polythene mulch for these parameters. They did not observe any significant effect of different mulches on TSS of the freshly harvested carrot.

A novel method of physical weed control, where an intensive wave of heat is used to kill weeds without disturbing the soil or harming the crop root system, has been reported by Lacko-Bartosova and Rifai (1998) from Poland. The method has resulted in a significant reduction in the number of weeds and consequently in labour requirements for hand weeding. Yields from the flamed trial, however, showed a slight reduction.

11.0 Harvesting, Yield and Storage

11.1.0 Harvesting and Yield

Carrots should be harvested at proper stage of maturity when they reach the desired size and maturity but are still tender and succulent, otherwise it will become fluffy and unfit for consumption. Delay in harvest results in firm roots and splitting. The crop is usually ready for market within 3 to 3½ months. Tropical types attain marketable maturity stage when roots are 2.5-4 cm in diameter at the upper end. Roots are dug with a spade or *khurpi* when the soil is sufficiently moist. The roots are trimmed, washed, graded and send to market. Carrots with leaves attached are made into bunches having about 5 to 10 or more roots and are packed in crates. The freshness and quality of leaf gives an indication of the freshness of the product, especially at the retail outlet. Usually carrots are sent to markets without leaves and packed into mesh pockets. Perforated plastic bags, or cardboard boxes with an inner plastic covering are also used for packing carrot.

In India, the carrot crop was digged manually and it required high labour and time for digging. Narender and Rani (2019) evaluated the performance of a tractor operated digger for the carrot crop. They observed best performance of the digger for diging of carrot crop at a speed of 2.2 km/hr and blade angle of 23 degree with a digging efficiency of 100 per cent, cut percentage of 46, bruised percentage of 2.28 and exposed percentage of 92.12. The field capacity of the machine was 0.18 ha/h. The saving of digger as compared to manual digging was Rs. 7359 per hectare. Kumar *et al.* (2017) developed a low cost tractor drawn carrot bed loosening implement having 97.56 per cent efficiency to assist in carrot harvesting. Nath *et al.* (2019) determined the physical and mechanical properties of carrot for designing combine harvesting mechanism.

In general, the percentages of root splitting, firmness, dry matter, carotene and sucrose increase with the age of roots, whereas the contents of glucose and fructose decrease (Fritz and Habben, 1977). Salter *et al.* (1979) found that the root yield was affected by plant density and harvesting time, the highest absolute yield being obtained from the latest harvests from the medium and high plant density, while with the lowest density treatments the highest yield of canning roots was obtained from the earlier harvests. In general, the common cultivars attain the marketable stage when their diameter is 2-4 cm at the upper end. Experiments were performed to justify the use of the term 'maturity' in connection with the development of carrot roots (Rosenfeld, 1998). Carrots cv. Panther F_1 were grown in climate chambers for up to 100 days at 5 constant temperatures (9, 12, 15, 18 or 21°C), and harvested on 9 successive dates. Dry matter, and glucose, fructose, sucrose and α- and β-carotene contents were analysed together with root shape. The results were tested by means of Principal Component Analysis and Partial Least Square Regression. None of the chemical variables investigated, indicated the existence a definable stage of biological development to indicate root maturity. The computed term 'cylindricity' showed the closest connection with chemical variables and might be used as a criterion for fully developed roots, together with the root weight.

In India, harvesting is done manually, while mechanical harvesting is the rule in advanced countries. It is advised to give a light irrigation before harvesting, because that will permit easy pulling of roots from the soil without any damage.

After harvesting, the roots are washed, cleaned, graded and tied in bunches of 6 or 12 roots. Carrots are sometimes marketed with their tops attached to indicate freshness. Zink (1961) reported that application of BA extended the storage life of the tops for 2-3 days over that of the control.

Average yield of tropical types of carrot varies between 250 and 350 quintals per hectare, whereas temperate types yield 100-150 q/ha. Singh and Saimbhi (1985) concluded that sowing carrot seeds in double rows on 45 cm ridges gave the highest yields of 705.6 quintals/ha in Sel. 233 and 748.2 quintals/ha in No. 29. Sagiv *et al.* (1994) reported that highest yield of carrot (cv. Nantes) [8.3 t and 7 t total and marketable yield/dunam (0.1 ha), respectively] was obtained with combination of organic manure and N at 30 kg/dunam (0.1 ha).

11.2.0 Storage

Fresh carrots cannot be stored for more than 3-4 days under ordinary condition. However, long-term storage in cold stores is possible without any appreciable change in quality. At temperature of 0-4.4°C with 93-98 per cent relative humidity (RH), carrots can be stored in good condition for 6 months. Berg and Lentz (1973) reported that storage at 98-100 per cent RH resulted in less decay than at 90-95 per cent RH. At a higher humidity, moisture loss is considerably less, as a result the produce remains crisper, firmer and of better colour. The optimum temperature for minimizing decay has been found to be 0-1°C. Gnaegi *et al.* (1977) noticed no change in the nutritive and organoleptic properties of carrots stored in crates covered with perforated plastic films at 0°C and 93-96 per cent RH for 7-8 months. The stored carrots yielded juice of acceptable commercial quality. Cernaianu *et al.* (1975) found that carrots keep better in cold stores than in stores with forced ventilation. Losses in cold stores were 6.4 per cent lower than in ventilated stores. Moreover, such roots were of better quality and could be stored for one month longer. Packing carrots in polyethylene bags reduced storage losses by 15.5 and 12.0 per cent in cold and ventilated stores, respectively. Vakis *et al.* (1975) reported that carrots packed in cotton nets lined with perforated polyethylene showed 2-4 per cent less weight loss and were substantially fresher in appearance and contained no infected roots than those in unlined nets. Lining bulk bins with polyethylene also reduced dehydration to an acceptable extent. Quality of carrots treated in this way remained satisfactory even after 5 months of storage at 1.1°C (Derbyshire and Crisp, 1978). Lingaiah and Huddar (1991) recorded maximum shelf-life (19 days) with uncooled, unwaxed carrots stored in polyethylene bags of 100 or 200 gauge thickness with 18 vents. In another study, Lingaiah and Reddy (1996) noticed the high percentage of disease incidence in carrots packed in non-ventilated polyethylene bags of 300 gauge thickness. Non-packed carrot did not show any disease incidence during storage under ambient condition. Augustinussen (1975) studied the effect of carbon dioxide and oxygen concentration on losses of carrot in storage by storing carrots at 1-1.5°C in air-tight boxes supplied with different levels of O_2 and CO_2. It was found that at a CO_2 level exceeding 4 per cent the quality of roots was reduced and the number of marketable roots fell considerably with increasing CO_2; the sugar percentage seemed to be unaffected by $O_2 : CO_2$ ratio but the carotene content tended to be lower at 4 per cent CO_2.

Quite a few investigators studied the changes in chemical composition of stored carrot roots. Toivonen *et al.* (1993) studied the effect of low temperature (1°C) pre-conditioning prior to being placed under simulated shelf-condition (13°C, >95 per cent RH). Carrots that were pre-conditioned at 1°C for 4 or 14 days showed significantly less weight loss (30 per cent less) and maintained a brighter orange appearance than unconditioned carrots. Pre-conditioning enhanced deposition of suberin on the surface of the periderm and lignification of subsurface cell layers. These changes are suggested as possible mechanism for reduction of weight loss and discolouration of carrots held under shelf condition. Pal *et al.* (1991) suggested pre-storage dip in 100 ppm Cl (bleaching powder) for extension of shelf-life and better root quality in carrot (cv. Nantes).

Carrot cultivars Eagle and Paramount were grown in muck soil in 6-litre pots (8 carrots/pot) in a greenhouse at the University of British Columbia, Vancouver, Canada. The plants were watered to field capacity every second day for 5.5 months prior to exposure to 100 (low), 75 (medium), 50 (high) or 25 per cent (severe stress) field capacity water stress treatments for 4.5 weeks. Postharvest moisture loss of carrots stored at 15°C with 32 per cent RH was monitored every second day for 3 weeks. The percentage moisture loss was low in the least stressed carrots, and was high in the severely stressed carrots of both cultivars. Root crown diameter, weight, and water and osmotic potentials decreased whereas specific surface area and relative solute leakage increased, with increasing preharvest water stress. Root water potential, followed by relative solute leakage, were the variables which accounted for most of the variation in moisture loss. It is presumed that preharvest water stress lowers membrane integrity of carrot roots, and this may enhance moisture loss during storage (Shibairo *et al.*, 1998).

Sagar *et al.* (1997) studied on sun drying of two varieties of carrot *viz.* Nantes and Pusa Yamdagni. Pusa Yamdagni showed that dry matter yield was higher in

Nantes due to lesser moisture content of 84.93 per cent as compared to 88.25 per cent in case of Yamdagni. Among various treatments of steam, salt and water blanching, higher dry matter was obtained in both the varieties in steam blanching followed by salt and water blanching. It took 10 hrs to dry up carrot slices of 5 mm thickness at a tray temperature of 50-55°C. Reconstitution ratio of the dried material was found to be best in steam blanching followed by salt and water blanching. Similar trend was obtained in organoleptic score of the dried product. Hence, steam blanching is best to get a better dried product of carrot. Sagar *et al.* (2008) evaluated 24 carrot genotypes for their suitability for preparation of dehydrated slices based on drying and dehydration ratio, α-carotene, lycopene, anthocyanin and ascorbic acid content and sensory score in respect of colour, flavour and texture. Cultivar IPC-13 and IPC-133Y in yellow colour and IPC-33 and IPC-133R in red colour were found to be superior genotypes for dehydration.

Many factors have been shown to affect the storage life of carrots. Varietal differences in the storage quality of roots have been reported (Sokol *et al.*, 1979). The effects of crop rotation on the occurrence of storage diseases (licorice rot, *Mycocentrospora* and *Sclerotinia* rot, *S. sclerotiorum*) of carrots were investigated in Finland. Disease incidence was dependent on the frequency of carrot cultivation in the previous 4-5 years; in carrots from fields without a history of carrot cultivation in the previous 5 years, incidence was 24 per cent compared with an average of 49 per cent from fields with a history of carrot cultivation. When carrots had been cultivated for >1 year in the previous 5 years, then an increase in storage diseases, particularly licorice rot, occurred (Suojala and Tahvonen, 1999).

Fertilizers, especially nitrogenous fertilizers, affect the storage quality of roots. Vleck and Polach (1978) noticed that high N rates increased rotting in sand-stored carrots but not in carrots stored in crates. Conversely, higher rates of both P and K improved storage life of carrots in crates but not in sand.

The effect of harvest time on storage ability of carrot was investigated during 1995-96 in the Netherlands by Suojala and Pessala (1999). The crop was harvested in September/October at 2-week intervals. Results showed that the storage performance of carrot was greatly affected by the harvesting time. An optimum harvest time reduced storage losses by 20 per cent on an average; losses occurred mainly with the earlier harvested crops, which were affected by storage diseases. Further, Suojala (1999) reported that delay in harvest decreased storage losses which mainly comprised spoilage due to storage diseases. Infections by the two major pathogens, *Mycocentrospora acerina* and *Botrytis cinerea*, were reduced towards the end of the harvest period. No further improvement in storability was observed after early October, but storage losses did not increase with a later harvest in most cases. Changes in storability were generally not related to weather conditions at or prior to harvest.

Bertolini and Restaino (1978) reported that exposure to 20 krad-irradiation for 230 days successfully prevented sprouting and rooting but attack by *Alternaria*, *Sclerotinia* and other root organisms was much increased. The effect of hot water treatment (55°C for 1 minute) on the shelf-life of carrot at room temperature (27 + 3°C) and ambient RH (65 ± 5 per cent) was studied by Karunaratne (1999) who reported that decay initiation was delayed and the shelf-life of carrot stored at 100 per cent RH was reduced by about half compared with those at ambient RH due to early decay initiation despite the hot water treatment.

Laboratory experiments demonstrated that steam treatments substantially reduced decay of stored organically grown carrots. Prior to being packaged, winter carrots were exposed for 3 seconds to steam. After 60 days of storage at 0.5°C plus an additional week at shelf conditions (20°C), 2 per cent of the carrots were decayed, compared with 23 per cent in the non-treated control. When carrot was inoculated with the fungi _Alternaria alternata_, _A. radicina_ and _Sclerotinia sclerotiorum_, percentages of decay, after similar periods of storage and shelf-life, were 5 per cent for steam-treated carrot and 65 per cent for the non-treated control. In semi-commercial experiments carrots were treated with steam during the sorting process, and similar results were obtained. Higher decay was found in spring-grown carrots because of the presence of the bacterium, _Erwinia carotovora_ ssp. _carotovora_ (Afek _et al._, 1999).

Microwave, steam- or water-blanched material was frozen and then stored at 24°C. Steam blanched carrots were subjected to blast freezing or cryogenic freezing at different temperatures before frozen storage (Kidmose and Martens, 1999). The influence of these process conditions on the texture, microstructure, dry matter, sugars, carotene and drip loss was investigated. Microwave blanching differed from the other blanching methods by resulting in a heterogenic cell structure. The content of dry matter, carotene and sucrose was higher following microwave blanching. Blast freezing resulted in low maximum load which seemed to be caused by major tissue damage. Concerning cryogenic freezing, lowering the temperature from –30°C to –70°C resulted in better preservation of the native microstructure together with an increase in maximum load, which was most pronounced after one month of storage. No significant effect was observed when lowering the temperature from –30°C to –70°C for any of the other measured parameters. Washed or unwashed carrots were stored for up to 120 days at 2°C in polyethylene (30 or 60 µm thick with 0, 100, 200 or 400 pinholes). Washing reduced the percentage of decay. Packing in polyethylene (60 µm) reduced water loss and maintained firmness, regardless of the number of pinholes, but high concentrations of CO_2 were observed in treatments with no pinholes. The best treatment for maintaining firmness and market quality and reducing water loss and decay was storage in 30-µm polyethylene with no pinholes (Lim _et al._, 1998).

Carrots may enter storage with an incipient infection which may develop into crown rot in storage and spread to adjacent healthy roots by hyphal growth (Goodliffe and Heale, 1975). Dehydration and deterioration from rots may occur in cold storage. The commonly occurring rot organisms are _Botrytis cinerea_, _Microcentrospora acerina_, _Sclerotinia sclerotiorum_, _Stemphylium radicinum_, _Erwinia_ spp. and _Fusarium_ spp.; of these _Botrytis cinerea_ seemed to be the most important and wide spread. Losses from attack with _B. cinerea_ after 4-5 months cold storage may range from 5 to 80 per cent (Lockhart, 1975).

12.0 Diseases, Pests and Disorders

12.1.0 Foliar Diseases

Fungal Diseases

Diseases are not a very serious problem in carrot; however, a number of diseases have been reported to occur in different parts of the world. The common diseases of above ground portion of carrots are alternaria blight, leaf spot, powdery mildew and viruses. Besides, various decays and rots caused by different pathogens may occur while the roots are held in storage.

12.1.1 Alternaria Leaf Blight

Leaf blight is caused by *Alternaria dauci* (Kuhn.) Groves and Skolko. On the foliage, small dark brown to black spots with yellow edge appear at first mostly along the leaf margin. The number of spots gradually increases and the interveinal tissues die. Eventually the entire leaflet dies and shrivels. In moist weather, the blackening and shrivelling progress so rapidly that the entire field resembles frost injury. Sometimes elongated spots appear on the petiole and the entire leaf dies without showing any foliate spots. The outer mature leaves are more susceptible than younger inner leaves. Young carrot roots are not susceptible despite severe foliar infection (Soteros, 1979). The disease has no effect on carrot roots in the field or in storage.

A. dauci can also infect the inflorescences and developing seeds in umbels. The disease is seed-borne; the infected debris can provide a further source of infection. The spores and mycelium are spread by irrigation water, splashing rains and agricultural implements. Moist climate favours rapid spread of this disease.

The disease can be kept under check if a well-drained soil is selected and suitable crop rotation is adopted. Seed treatment with captan or thiram at the rate of 3 g per kg of seed before sowing is helpful.

Japanese varieties such as PI 261648, PI 226043 and Imperial Long Scarlet showed resistance against alternaria blight (Standberg *et al.*, 1972). Pryor *et al.* (2000) evaluated the susceptibility of 46 carrot cultivars to infection by *Alternaria radicina* (black rot disease); relatively resistant cultivars were Panther and Caro-pak, while susceptible cultivars included Royal Chantenay and Nogales. Moderate resistance is reported in Brasilia (Boiteus *et al.*, 1993) and Bolero (Lecomte *et al.*, 2012).

12.1.2 Cercospora Leaf Spot

Leaf spot initiated by *Cercospora carotae* is a wide spread and one of the important foliage diseases of carrot.

The symptoms appear first as elongated lesions along the edge of the leaf segment, resulting in a lateral curling. Non-marginal lesions appear as small, pinpointed chlorotic spots which soon develop into a necrotic centre surrounded by a diffuse chlorotic border. In dry weather, the spots are light tan in colour, whereas in humid weather the spots are darker in colour. Coalescence of spots often occurs. Dark lesions may develop on the petiole also, sometimes girdling it and killing the leaf. The optimum temperature range for infection is 20–28°C and high relative humidity.

Carrot cultivars resistant to _Cercospora_ leaf spot are not available but some are more tolerant than others. Control measures include seed treatment with fungicides captam or thiram at 2.5 g/kg seed. Spraying with mancozeb (2.5 g/liter), copper oxychloride (3 g/liter) of chlorothalonil (2 g/liter) at an interval of 7-10 days satisfactorily controls the diseases in areas where it is serious.

12.1.3 Powdery Mildew

Powdery mildew caused by _Erysiphe heraclei, Leveillula lanuginosa, L. taurica_ is sometimes becomes a major disease of carrot. It appears first on the leaves, but later may spread to flowers, stems and fruits. Two types of powdery mildew occur on carrot.

(I) _Oidium type_: It is caused by _Erysiphe heraclei_ (synonyms _E. polygoni_ and _E. umbelliferarum_) and is most common. It is severe in warm and semiarid regions. The disease severity depends on the weather conditions, crop growth stage, production practices and cultivar. The pathogen produces white mycelium and sporulation on leaves, petioles, flower stalks, bracts and umbels. Severely infected foliage becomes chlorotic and leaves senesce prematurely.

(II) _Oidiopsis type_: It is caused by _Leveillula lanuginosa_ and _L. taurica_ and is less common. The pathogen is favoured by very warm and very dry climates. The pathogen produces mycelium that is both endophytic and ectotopic. Pale yellow lesions appear on the upper leaf surface and white sporulation on the lower leaf surface. Gradually sporulation can develop on the upper leaf surface and chlorotic areas become necrotic.

Finely divided sulphur dust is the cheapest and most effective control of mildew. It may be applied even after appearance of the disease because this fungicide is not eradicative and protective. Spraying of Karathane at 2 g/liter is also effective in controlling powdery mildew. Application of Azoxystrobin is also found to be effective for powdery mildew control.

Bacterial Diseases

12.1.4 Bacterial Leaf Blight (BLB)

It is caused by the bacterial pathogen _Xanthomonas hortorum_ pv. _carotae_. Bacterial leaf blight is seedborne and the lesions on petioles, umbels, and seed stalks are sometimes accompanied by production of a gummy bacterial exudate. It can infect the foliage, stems, umbels, and seed. Partial resistance to BLB has been reported in PI 418967, PI 432905 and PI 432906 (Christianson _et al._, 2015).

12.1.5 Phytoplasma

The bacterium '_Candidatus Liberibacter solanacearum_' (CaLso) is a gram-negative, phloem-limited bacterium, mainly transmitted by psyllid _Bactericera trigonica_ Hodkinson. Symptoms include leaf curling with yellow, bronze, and purple discolouration, twisting of petioles, stunted growth of shoots and roots, and proliferation of secondary roots (Satta *et al.*, 2016).

Viral Diseases

Several viral diseases such as yellows, carrot thin leaf virus (CTLV), carrot motley dwarf virus (CMDV), carrot mosaic, *etc.* may cause extensive damage to carrot. Of these, yellows (also known as aster yellows) seem to be the most important viral disease of carrot.

The symptoms of carrot yellows first appear on leaves which become yellow, sometimes accompanied by vein clearing. Dormant buds in the crown grow out into chlorotic shoots which give a witches' broom appearance on the top. Older leaves are reddish, twisted and may eventually break off. Size and quality of roots are reduced and malformed, fibrous roots are commonly produced. The internal colour, texture and flavour show marked changes causing reduction in value of carrots for fresh market as well as for processing. Roots of infected plants have a bitter, astringent flavour.

The infected seed plants are stunted in growth with pale green or yellowish leaves. The seed stalk develops earlier than usual and the pedicels may be usually long. Sterility occurs in the flowering umbels, resulting in no or very little seed setting.

The disease is mainly transmitted by the six spotted leaf hopper (*Macrosteles divisus*) although other species may also be involved and a number of plants including some vegetables such as lettuce, tomato, potato, celery *etc.* may be infected with this disease.

The most effective method of control of this disease is to spray insecticides to control the leaf hopper. Weed control, specially of those acting as alternate host, also eliminates the disease.

12.2.0 Diseases of Root

A number of diseases causing various decays and rotting may occur while the root is held in storage. However, the following diseases are most common and economically important.

Fungal Diseases

12.2.1 Watery Soft Rot/White Mold

Watery soft rot caused by *Sclerotinia sclerotiorum* is one of the most common decay of carrot in storage. Small translucent spots appear on roots and are rapidly covered by white, flocculent mycelium, which develops into black sclerotia (reproductive bodies). The infected roots become soft and watery. The sclerotia can survive up to 10 years in soils. The optimum temperature for watery soft rot is 23°C (Walker, 1952). *Sclerotinia* sometimes infects the roots while in the field.

Practice deep summer ploughing and a three-year rotation with non-susceptible crops. Remove and destroy infected crop residues. Improve drainage and do not encourage wet conditions as it favour the disease development. No resistance sources have been identified so far in the carrot germplasm. Drench the soil near root zone with carboxin 2 g/liter or chlorneb 15 g/3 liter.

12.2.2 Gray-Mold Rot

Gray-mold is caused by a fungus, *Botrytis cinerea* and is quite widespread. The affected tissue is water-soaked and light brown in colour which later becomes spongy. The fungus appears on the decay as a white mold upon which the graymold appears in moist atmosphere. The disease occurs over a wide range of temperatures from 0 to 25°C.

12.2.3 Black Rot

Black rot, caused by *Alternaria radicina*, is a widespread seedborne disease affecting the foliage as well as the fleshy roots. The symptoms of the foliage are just like those caused by *Alternaria* blight. It is a seed transmitted postharvest disease causing seed rot, poor seedling establishment and/or damping-off. It also causes a black decay on the foliage, petioles, and umbels and is characterized by black, sunken necrotic lesions on the taproots and crowns. On the roots, black, sunken areas, irregular to circular in outline, may develop. The decayed tissue is greenish black to jet black due to presence of masses of black spores. This disease affects the roots in the field as well as in storage. Sanitation and rotation may keep the disease under control. Storage at 0-2°C is necessary to keep storage decay to a minimum. The pathogen can persist in the soil for long periods (as long as 8 years).

12.2.4 Cavity Spot

Cavity spot is caused by several species of *Pythium*, the most common and the most virulent on carrot being *P. violae* and *P. sulcatum*. Other species associated with cavity spot are *P. ultimum*, *P. intermedium*, *P. irregulare* and *P. sylvaticum*. Cavity spot rarely causes a reduction in root yield but make shallow lesions on the root surface thereby affecting market value. A cavity appears in the cortex; in most cases the subtending epidermis collapses to form a pitted lesion. The pathogen infects carrot roots within the first four to six weeks after seeding and continues to develop on roots in storage. Root lesions can be infected by secondary microorganisms, including bacteria. The severity of cavity spot disease is influenced by soil temperature, soil moisture, soil microflora, the species of *Pythium*, age of carrot roots, *etc*. Higher soil moisture, particularly flooding, cool soil temp. (15°C) is optimal for development of cavity spot. Severity of cavity spot increases with the length of time roots remain in the soil. Screening for resistance to cavity spot is difficult because of uneven distribution of inoculum in fields and very sporadic nature of disease within and among fields.

12.2.5 *Rhizoctonia* spp.

Three soilborne species of *Rhizoctonia* cause diseases of carrot *viz*. *R. solani*, *R. crocorum* and *R. carotae*.

R. solani cause damping-off of carrot seedlings and crown rot of mature carrots. Damping-off results in poor seed germination (seed rot) and death of seedlings.

In crown rot, the petioles and crowns rot, turn dark brown and sunken lesions develop near the crown. Affected roots become unmarketable. Secondary invasion of lesions by bacteria can initiate soft rot. The lesions continue to expand when roots are kept in storage.

R. crocorum causes violet root rot of carrot in which dark purple-brown, firm lesions develop on roots, on the surface of which a violet to dark brown and leathery dense mat of mycelium of the fungus forms. The pathogen can infect carrot roots at soil temperatures ranging from 5 to 30°C. No resistance source has been reported so far.

R. carotae causes crater rot of carrot, a postharvest storage disease. Dry, sunken craters or pits appear on the root surface under very humid and cool conditions in storage, with white mycelium and dark brown sclerotia appearing on the root surface. Crater rot is a dry rot, but infected roots can become colonized by bacteria, leading to soft rot. The fungus can develop on roots at temperatures as low as 2–3°C and high relative humidity. No resistance source has been reported so far.

12.2.6 Phytophthora Root Rot or Rubbery Brown Rot (*Phytophthora* spp.)

Phytophthora root rot is a minor disease of carrot but can cause significant losses in waterlogged soils. The causal organism includes several species of *Phytophthora*, including *P. cactorum*, *P. cryptogea*, *P. megasperma*, and *P. porri*. The affected roots become dark brown to black and rubbery. Firm, water-soaked lesions develop in the middle and crown areas of roots. Secondary invasion of the lesions by bacteria and fungi can lead to a soft rot. Cool to moderate conditions favour inoculum production and spread of the pathogen. The pathogen spreads readily in storage. No source for resistance to this disease has been reported in carrot.

12.2.7 Carrot Sour Rot

In 1996, carrot sour rot was first found in the traditional markets in Taipei, Taiwan (Wu *et al.*, 1999). The causal organism was identified as *Geotrichum candidum* and the disease occurred mostly at the tip of the tap root. The early symptom was a water-soaked appearance on the infected tissue followed by a thin, cheese-like layer of white mycelium on the infected areas during the late stages of disease development. These diseased tissues were crumbly and rotted rapidly, exuding a sour smell. The optimal temperature for disease development was 28°C, and at <20°C the disease severity was reduced. Carrot sour rot did not occur below 4°C. Temperature was shown to be the critical environmental factor affecting the development of carrot sour rot. Growth of *G. candidum* was significantly inhibited by 10 ppm fenpropimorph and flutriafol. Antagonistic microorganism isolates <hash>52 and <hash>224 were isolated from the surfaces of carrots and were able to cause the hyphae of *G. candidum* to become deformed. The amount of sporulation was also reduced significantly. The 2 antagonistic isolates (<hash>52 and <hash>224) were identified as *Pseudomonas fluorescens* and *Bacillus amyloliquefaciens*, respectively. They had the capability to reduce the disease severity in this test.

The vapours of the essential oils of caraway (*Carum carvi*), spearmint (*Mentha spicata*), thyme (*Thymus vulgaris*), basil (*Ocimum basilicum*) and garlic were tested *in vitro* for antifungal properties against *Mycocentrospora acerina*, *Fibularhizoctonia carotae* [*Rhizoctonia carotae*] and *Sclerotinia sclerotiorum*, 3 important postharvest pathogens of carrots (Horberg, 1998). The mycelial growth of all 3 carrot pathogens was completely inhibited by the vapours from garlic soil, added in amounts from 10-80 ppm, volume per volume empty containers. Various effects were obtained with basil soil, which was tolerated by *M. acerina* but not to the vapours from thyme oil, *R. carotae* being completely inhibited at all doses and exposure times, *S. sclerotiorum* at doses of 80 ppm or higher and *M. acerina* at 160 ppm and a long exposure time. Spearmint and caraway had a similar influence on all 3 pathogens. The results also indicated that high dosage levels were more important than exposure time for the fungicidal activity of the plant extracts.

Bacterial Diseases

12.2.8 Bacterial Soft Rots

Soft rots of carrot is caused by several bacteria *viz. Pectobacterium carotovorum* subsp. *carotovorum* (formerly *Erwinia carotovora* subsp. *carotovora*), *Dickeya dadantii* (formerly *D. chrysanthemi* = *Pectobacterium chrysanthemi* = *Erwinia chrysanthemi*) and *Pectobacterium atrosepticum* (formerly *P. carotovorum* subsp. *atrosepticum* = *E. carotovora* subsp. *atroseptica*). It is a destructive disease of carrots in storage and transit. These bacteria infect carrot roots through wounds or natural openings, causing small, water-soaked lesions that enlarge rapidly. Root degradation is most rapid under warm conditions (20–25°C for *P. carotovorum* subsp. *carotovorum* and 30–35°C for *D. dadantii*). The infected tissue softens, becomes mushy, watery or slimy. The interior tissue may ooze through cracks on the root surface. A foul odour from the decayed roots distinguishes it from the fungal soft rot. The bacteria causing soft rot live in the soil and in decaying refuse. Wounds on the root surface caused during harvesting or by insects, help in disease incidence. Abundant moisture on the root surface favours disease incidence. The genetic basis of resistance of carrot lines to bacterial soft rot pathogens has not been determined so far.

Control measures include careful handling of roots during harvesting, grading or transit so that all bruises on root surface can be avoided. The root surface should be kept dry and stored at 0°C and 90 per cent relative humidity.

12.3.0 Pests

Insects are not a major problem in carrot production and only a few species of insects may damage the crop to any serious extent.

12.3.1 Carrot Rust Fly

Psila rosae, the adult fly is about 42 mm long, dark green to black in colour. The maggot (larva) is slender, yellowish white in colour and about 6.3 mm in length. The larva burrows into the roots, often causing it to become misshapen and subject to decay and render them unfit for marketing. If the roots are injured severely, the leaves become rusty or even dried. Apply chlorpyrihos 20 EC at 2.5 liter/ha with irrigation water.

Diseases of Carrot

Alternaria leaf blight (left) and *Cercospora* leaf spot (right) infection in carrot

 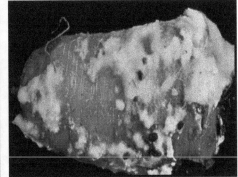

Sclerotinia infection in carrot plant and root

Black rot infection

Pests of Carrot

Carrot rust fly

Root knot infestation

Kettunen *et al.* (1988) found that ash sprinkled on carrot rows reduced egg numbers of carrot rust fly to about one-third of the control. They also noticed that cv. Sytan received fewer eggs than others, while Chantenay showed significantly less root damage than Nantes 20 Notabene.

12.3.2 Lygus Bug

It is a serious pest of the seed crop. It is thought that low viability or low germination percentage of seeds of Umbelliferae is associated with injury caused by lygus bug. The pest damages the embryo of the seed; sometimes the seed is without any embryo. Seed treatment with insecticides may control this pest.

12.3.3 Other Insect Pests

Other insect pests of carrots of less importance are turnip moth (*Agrotis segetum*), carrot weevil (*Desiantha maculata*), carrot spotted spider mite (*Tetranychus truncatus*) *etc.* Turnip moth could be controlled by spraying with orthene (acephate) at 1.5 kg/ ha or with chlorpyrifos at 4 kg/ha (Ramert, 1976).

12.4.0 Nematode

Besides the above insect pests, nematodes may also cause serious damage to the crop, making the root completely unfit for market. The important nematode damaging carrots are *Heterodera carotae* and *Meloidogyne* sp., *Meloidogyne hapla* Chitwood, *M. javanica* (Treub) Chitwood and *M. incognita* (Kofoid and White) Chitwood may cause losses up to 100 per cent along with with yield reduction and shape deformation (forking and galling of the taproot). *M. hapla* is the predominant species in cooler production areas whereas *M. javanica* and *M. incognita* are major pests in warmer areas (Parsons *et al.*, 2015). Nematode number can be reduced to as low as 97 per cent by growing *Tagetes* (Weischer, 1961). A non-chemical management of *Meloidogyne hapla* on carrot was investigated by Svikumar and Mehta (1998). They found that application of *Paecilomyces lilacinus* at 10×106 spores/m^2 resulted in a gall index of 1.6 as against 5.0 in the untreated control. The treatment reduced the crop damage by 92.4 per cent and increased yield by 147.6 per cent. Nagachandrabose (2018) reported that *P. fluorescens*-ST at 100 mL/kg of seed and *P. lilacinum*-SD at 5 L/ha are highly effective for the management of *M. hapla* in carrot. Maistrello *et al.* (2018) used aqueous tannin solutions to control carrot cyst nematode *Heterodera carotae*.

Hussain *et al.* (2020) applied ten isolates from seven different fungi, *Arthrobotrys oligospora, Dactylella oviparasitica, Clonostachys rosea, Stropharia rugosoannulata, Lecanicillium muscarium, Trichoderma harzianum* and *Pleurotus ostreatus*, along with two chemicals, Vydate and Basamid (G), were evaluated against northern root-knot nematodes, *Meloidogyne hapla*, on carrots in a greenhouse. All fungi and chemicals efficiently reduced the infestation level of *Meloidogyne hapla* and provided better growth of carrots compared to their controls. Maximum reductions in nematode population were observed in the plants treated with *Lecanicillium muscarium* and both chemicals.

Khan *et al.* (2019) reported that the leaves of *Phyllanthus amarus* plant can be used for the management of *M. incognita* and could be a possible replacement for synthetic nematicides.

12.5.0 Physiological Disorders

12.5.1 Carrot Splitting or Cracking

Besides genetic factor, the tendency of root splitting in carrot is controlled by a number of factors. Splitting of roots occur when there is a sudden change in the soil moisture and also in case of heavy application of nitrogen. The splitting is reduced by low N. Ammonium form of N causes more splitting than other forms of N. Large roots are more likely to split than small ones. Wider spacing results in greater amount of splitting.

12.5.2 Carrot Root Forking and Stubbing

When the main root cannot elongate further, it results in production of very short roots called 'stubbing'. Forking is branching of roots. Both forking and stubbing may result from soil compaction, cold, excess water, nematode or fungal attack.

13.0 Seed Production

From seed production point of view carrot has two races: one is biennial or temperate or European type and another is annual or oriental or tropical type. In the plains of Northern India, the oriental types produce seeds freely, whereas the temperate cultivars being biennial in character remain vegetative in the first year and become reproductive in the following growing season. For biennial cultivars, the exposure of the plant to temperatures below a critical level (vernalization) is essential for the induction of the reproductive stage, *i.e.*, the formation of the seedstalk and flower development (bolting). However, the extent of bolting depends upon cultivar, prevailing low temperature and maturity. Late planting and higher densities both resulted in smaller roots and therefore, less cold responsive plants are produced (Dowker and Jackson, 1975).

The rate of seedstalk elongation and its final height are greatly influenced by the post-vernalization temperatures, provided the vernalization process is completed. Storage of cv. Chantenay stecklings for 10 weeks at 5°C is suggested. However, for cv. Nantes a shorter cold exposure may suffice (Hiller and Kelly, 1979). Post-vernalization temperatures exceeding 15°/20°C (night/day) cause reduction in seedstalk growth (Elisa and Wallace, 1969). Hiller and Kelly (1979) demonstrated a reduction of 80-90 per cent in seedstalk height with a post-vernalization temperature regime of 27°/32°C.

Several workers have demonstrated that the low temperature requirement for bolting can be replaced by a foliar application of GA₃ (Dickson and Peterson, 1960) or by dipping the roots in GA₃ solution (Globerson, 1972). Spraying the plants with GA₃ at 250 ppm at the 8- to 10-leaf stage hastened bolting, flowering and maturity; seed yield was also increased due to increase in seed number (Galmarini *et al.*, 1995a, b). In contrast, suppression and delay of the bolting process in carrot has been

Disorders in Carrot

Forked (left) and cracked (right) carrot root

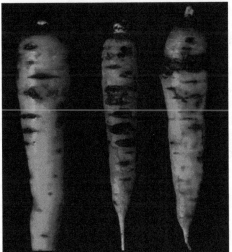

Cavity spot in carrot roots

Nitrogen (left), phosphorus (middle) and potassium (right) deficiency in carrot

achieved by using the plant growth retardants like Daminozide and Chlormequat (Jacobsohn *et al.*, 1980).

13.1.0 Time of Sowing

The oriental types are sown in the plains of India during August-September for seed production. Temperate cultivars are sown from mid-July to mid-August in the hills. About 2.5-3 kg seed of temperate types and 4-5 kg seed of oriental types are sufficient for one hectare of land, which, in turn, will produce well developed roots for planting 4-6 hectares.

13.2.0 Method of Seed Production

Both seed-to-seed and root-to-seed methods are followed for carrot seed production; the latter should, however, be preferred for production of high quality seeds.

Brar and Nandpuri (1972) have worked out the seed production techniques of Asiatic types. Seed production technique of temperate type has been reported by Singh *et al.* (1960). During October-November when the roots are mature, the crop is uprooted. After selection of true to type roots, their tops and tips are cut and replanted in well prepared soil in rows 75 cm apart. Nath and Kalvi (1969a, 1969b) suggested that the stecklings should be given one third top (shoot) cut and one fourth to one half root cut to obtain higher yield of better quality seeds in carrot. Digole and Shinde (1990) obtained the highest seed yield (22 g/plant or 15 q/ha) in cv. Pusa Kesar when shoots and roots of 75-80-day-old stecklings were cut back by a quarter and half, respectively. The distance between plants may vary from 10 to 30 cm. Lal and Pandey (1986) obtained the highest average seed yield (1.104 t/ha) from plants spaced at 30 × 30 cm using intact roots and with foliage cut to two-thirds of the full size. Verma *et al.* (1993), on the other hand, obtained highest seed yield (3796.7 kg/ha) in cv. Pusa yamdagni from the largest stecklings (125 g). Madan and Saimbhi (1986) reported the carrot seed yield increased from 5-5.6 q/ha up to 90 kg/ha of nitrogen. No such responses were observed in case of P and K. Report from Egypt shows that cutting plant foliage to 2/3 of its length before planting, using transplants with large roots (15-17 cm long) and spraying plants 4 times during the growing season with 200 ppm GA_3 resulted in increased plant survival, seed yield components like number of flowering shoots and umbels/plant and total seed yield and quality (Eid and Abo-Sedera, 1992). Farghali and Hussein (1994) recorded improvement in seed yield when roots of cv. Chantenay Red Core were stored at 5°C for 30 days and then soaked in IBA at 40 ppm. Shaimanov *et al.* (1996) found that seed quality in carrot depends on age of stecklings at transplanting and plant density. They have recommended that stecklings weighing 20 g be planted at a density of 8-10 plants/m² for better seed production. During planting, care should be taken so that the crown remains exposed. Noor *et al.* (2019) reported application of potash significantly affected yield as well as seed quality attributes of carrot. Application of potash at 80 to 100 kg ha⁻¹ produced highest seed yield, 1000-seed weight of primary and secondary umbels. Potash application at 80 to 100 kg ha⁻¹ also enhanced the seed quality traits, including final germination percentage (17 per cent and 15 per cent) and vigour index of seeds (49 per cent and 42 per cent), respectively, from primary and secondary umbels. It also enhanced the yield and quality, in terms of 1000-seed weight, germination (per cent) and vigour of tertiary umbel seeds up to 60 per cent, 45 per cent and 80 per cent, respectively, compared to control.

Kumar *et al.* (2017) recorded highest 100-seed weight of primary umbel, secondary umbel and tertiary umbel; seed germination percentage of primary umbel, secondary umbel and tertiary umbel; seedling vigour index I and II of primary umbel, secondary umbel and tertiary umbel at 60 × 75 cm spacing and with large steckling size (125-250 g).

13.3.0 Selection and Roguing

The selection of roots is made on the basis of character of tops (small and heavy), colour of skin, shape and size of root, colour of flesh, colour and size of core, *etc.* Thorough examination and selection of true to type plants are absolutely necessary for raising quality seeds. Both early and late bolters should be removed. Similarly, all diseased, forked or hairy roots should be removed.

A minimum of four inspections shall be made as follows:

1. The first inspection shall be made before flowering in order to determine isolation, volunteer plants, outcrosses, planting ratio, errors in planting and other relevant factors.

2. The second and third inspections shall be made during flowering to check isolation, off-types, the number of pollen shedding umbels and other relevant factors.

3. The fourth inspection shall be made at maturity to verify the true nature of umbels.

Specific Requirements for Carrot Seed Ptroduction

Factor	Maximum Permitted	
	Foundation	Certified
Roots not conforming the varietal characteristics including forked roots	0.10 per cent (by number)	0.050 per cent (by number)
Off types (plants) in seed parent at and after flowering	0.010 per cent	0.050 per cent
Off types (plants) in pollinator at and after flowering	0.010 per cent	0.050 per cent
Plants of pollen shedding umbels in seed parent at flowering	0.050 per cent	0.10 per cent

13.4.0 Storage of Stecklings

The ideal storage condition for carrot stecklings is a temperature of 0°C and a relative humidity of 90-95 per cent.

13.5.0 Flowering and Seed Setting

Carrot plants bear compound umbels. It is the king umbel (primary umbel) or the umbel of the first order that flowers first. The secondary, tertiary and other orders of umbel flower at an interval of 8-12 days from each other. Anthesis in a single umbel is completed in 7-9 days. The peripheral umbellets flower first, followed by the inner umbellets.

The temperate cultivars start bolting by third week of April in the hills of India and flowering starts by the end of May. The crop is harvested during July-August. Bolting and flowering in tropical types occur during early spring and the crop is harvested by the end of May.

Studies on seed setting in cv. Imperator revealed that umbels of the first two orders set seeds in 100 per cent umbels and the third, fourth and onward orders set in 82.7, 6.29 and 2.78 per cent umbels, respectively. The seed setting in late formed umbels was extremely poor (Singh *et al.,* 1960). Singh *et al.* (1994), on the other hand, found that the average contribution to the total seed yield by the main, first and second order umbels were 21, 67 and 12 per cent, respectively. Seed weight of umbels of all orders was significantly improved by 80 kg N/ha. The quality of seed of main and first order umbels was better than that of second order umbels. Higher root yield and better root quality were noticed in plants raised from primary umbels (Krarup and Villanueva, 1977). Cardoso (2000) also found that primary umbels gave highest seed yield (7.5 g/umbel), followed by secondary umbels (3.0 g/umbel) and tertiary umbels were the least productive (0.5 g/umbel). Primary umbel seeds were heavier (2.1 g/1000 seeds) than secondary (1.6 g/1000 seeds) and tertiary ones (1.4 g/1000 seeds). Seeds from primary and secondary umbels were of better quality than tertiary umbel seeds. By contrast, Elballa and Cantliffe (1997) found best seed quality from tertiary umbels. Delaying carrot seed harvest until tertiary umbel seeds reached maturity was recommended to increase seed yield and gain improved seed quality. Zhang *et al.* (1998) also recorded that the 1000-seed weight was greater in the main flower head than in the subsidiary heads. The more compound the umbel, the less productive are the plants raised from its seeds (Shevtsova and Korol, 1976).

The carrot flower is protandrous, hence requires cross-pollination. Insects are the pollinating agents in carrot. Bohart and Nye (1960) noticed occurrence and pollinating activities of 334 spp. from 71 families of insects in carrot fields. Of these, honeybees and houseflies seem to be the most important. Sharma and Sharma (1968) found that houseflies were more prominant and active than honeybees in the pollination of carrot in Kalpa valley (Himachal Pradesh). Sinha and Chakrabarti (1992) recorded greatest pollinator activity at 10.00 h at Karnal, which was closely related to temperature and humidity. Bohart and Nye (1960) recommended a combination of honeybee colonies adjacent to carrot seed fields and elimination of competing flowers as the most practical methods of pollination of carrot. Abrol (1997) found that insect-pollinated plots produced significantly more and heavier seeds than plots enclosed to exclude insects.

Carrot cultivars should be isolated to a distance of 1000 m to produce foundation seed (Anon., 1971).

13.6.0 Harvesting

Although shattering of seeds occurs to some extent, it is not a serious problem in carrot. Usually the crop is harvested when the secondary umbels are fully ripe and the third order umbels have started to turn brown. The crop may be harvested either mechanically or manually. The harvested plants are then kept in wide rows in small piles of 3-5 plants for curing. Curing may continue from 5 to 14 days or even more depending on weather condition. Rain during curing may cause deterioration of seed quality. The seed is then threshed out, cleaned and graded either mechanically or manually. The spines on the seeds are removed by breaking off the spines through some sort of rubbing action in mills. Since there is significant correlation between seed weight and germination (Nascimento and Andreoli, 1990), seeds should be graded according to seed weight.

13.7.0 Seed Yield and Seed Standards

A yield of 330-550 kg seed per hectare was obtained from the European cv. Nantes in Kulu valley (Singh *et al.*, 1960), whereas Nath and Kalvi (1969b) reported an average yield of 2054 kg seed/ha in the Asian cultivars when proper root and shoot cuts were given.

Seed Standards for Carrot

Factors	Standard for each Class	
	Foundation	Certified
Pure seed (minimum)	95.0 per cent	95.0 per cent
Inert matter (maximum)	5.0 per cent	5.0 per cent
Other crop seeds (maximum)	5/kg	10/kg
Weed seeds (maximum)	5/kg	10/kg
Other distinguishable varieties (maximum)	5/kg	10/kg
Germination (minimum)	60 per cent	60 per cent
Moisture (maximum)	8.0 per cent	8.0 per cent
For vapour-proof containers (maximum)	7.0 per cent	7.0 per cent

13.8.0 Seed Storage and Seed Treatment

Carrot (cv. Brasilia) seeds were held in aerated solutions of polyethylene glycol (PEG 6000, 200 g/l), potassium nitrate (0.3 M), potassium monophosphate (0.3 M) or distilled water for 24, 48 or 72 h at 25°C. Osmotic priming increased the speed and uniformity of germination (at both 15 and 35°C), with PEG 6000 giving the best results. The viability of primed seeds decreased by 11-13 per cent after 150 days' storage in paper bags under ambient laboratory conditions, whereas the viability of control non-treated seeds remained constant. From the above results Sampaio and Sampaio (1998) suggested that primed seeds should be stored in hermetically sealed packs at low temperatures after treatment.

13.9.0 Hybrid Seed Production

Hybrid cultivars have the advantage of relatively uniform roots and have been produced by the use of two systems as described by Riggs (1987). The first system is with cytoplasmic male sterility (CMS) in which the pollen does not develop beyond the microspore stage, sometimes referred to as 'brown another form.' The other type is petaloid form in which the five anthers are transformed into petaloid structures during their early development and do not produce early pollen. The final morphology of the petaloid anthers varies from petal-like to filamentous (Eisa and Wallace, 1969). According to Riggs, most commercial F_1 carrot hybrids are produced from the petaloid CMS. Production of parental lines and the hybrid seed by either system should be done in accordance with the instructions of the maintenance breeder. Less insect activity because of smaller petals on the male sterile flowers of the seed producing lines results in to low seed yield. Seed producers frequently use colonies of bees to supplement the natural level of pollinating insect activity when producing hybrid carrot seed.

Seed production of F_1 hybrid cultivars is based on cytoplasmic male sterility (CMS) of one of the parent-inbred lines. F_1 hybrids in carrot may be three-way hybrids, where the female parent is an F_1 hybrid itself, or two-way hybrids where the female is a CMS "A" inbred line. The number and arrangement of pollen parent plant rows relative to seed parent rows in fields varies depending mainly on inbred characteristics and grower practices. Generally the ratio of female to pollinator rows as reported by Takahashi (1987), is from 2: 1 to 4: 1. But a common male: female ratio is 4: 1 which is often grown is an 8: 2 arrangement with four two-row beds of female alternating with a single two-row bed of the pollen parent in an isolated field for production of F_1 hybrid seeds. The seeds are harvested from male sterile line and the pollinator plants are removed before collecting the seeds from the male sterile female parent.

14.0 Crop Improvement

14.1.0 Genetic Resources

Present day orange carrots have a narrow genetic base considering that they all have been derived from a few 18th century Dutch cultivars. The exploitation of the genetic variation existing in wild *Daucus* germplasm in the Mediterranean and South-Western Asian region has recently begun. The germplasm collections of carrot and other species are relatively small. About 5600 accessions are held worldwide with over 1000 of these at the Vavilov Institute in Russia (VIR) (Frison and Serwinski, 1995). The genetic resources unit of Horticultural Research International, Wellesbourne, Warwick (UK) and NBPGR, New Delhi is also maintaining small working collection. The most accessions of which are *Daucus carota*, but samples of *D. broteri*, *D. glochidiatus*, *D. gracilis*, *D. hispidifolius*, *D. involucratus*, *D. littoralis*, *D. montevidensis*, and *D. muricatus* are also held. Approximately, 800 accessions are

Different Stages of Carrot Seed Production

Carrot roots selected for preparation of steckling

Flowering in carrot

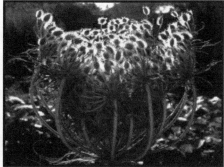

Carrot flowering under nylon net cage for seed production

Schizocarps of carrot

maintained in USDA collection at the North Central Regional Plant Introduction Station in Ames, Iowa. In these collections, 95 per cent are *D. carota* and the other species are *D. aureus, D. broterii, D. capillifolius, D. crinitus, D. durieua, D. glochidiatus, D. guttatus, D. littoralis, D. muricatus,* and *D. pusillus.* Besides, France, Czech Republic, Germany, Poland, Ukraine, The Netherlands and Japan also hold small working collections.

Most of the germplasm collections of carrot have focused on open-pollinated local cultivars and little germplasm of wild carrot has been collected. Most of the *Daucus* species other than carrot are primarily located around the Mediterranean Sea, except for as Australian species (*Daucus glochidiatus*), a wide-spread American species (*Daucus pusillus*), and a more isolated American species (*Daucus montanus*) (Rubatzky *et al.*, 1999).

14.2.0 Breeding Objectives

1. Early and high root yield.
2. Combining desirable characters of both European and Asiatic groups, especially high carotene content and ability to set seeds in plains.
3. Developing F_1 hybrids using cytoplasmic male sterility.
4. Desirable uniform root size, shape (cylindrical), dark orange external and internal colour, *i.e.*, uniform in xylem and phloem, top-root ratio, *i.e.*, small tops and smooth heavy tender roots. Yielding capacity increases with root length but root length should not usually exceed 25-30 cm because roots longer than this limit are liable to damage during harvesting and handling.
5. Improvement in quality, especially carotene, lycopene and anthocyanin content, flavour, texture, high sugar and dry matter in roots. High dry matter content in roots is preferred because this favours good storage and transportation without much loss.
6. Resistance to cracking and breaking of the root during harvesting and post harvest handling.
7. Broad shouldered, uniform tapering or stump rooted thin and self-coloured slow bolting carrots.
8. Tolerance to environmental stresses and wider adaptability.
9. Resistance/tolerance to diseases, insect pests and nematodes, especially leaf blight (*Alternaria* and *Cercospora*), black rot, powdery mildew, bacterial soft rot, carrot yellows, caterpillars, carrot fly.
10. Resistance/tolerance to defects such as excessive secondary root development, splitting, secondary growth and cavity spot.
11. Resistance to bolting or premature flowering.

14.3.0 Breeding Methods

Initial selection is based upon individual root performance in the field as observed for horticultural characteristics deemed important by the breeder. Field performance and quality attributes such as flavour and carotene content can be evaluated before seed production begins. Only those plants making satisfactory each of field and quality performance should be selected for seed production.

Line selection involving only self-pollination is most effective in fixing traits. However, because of inbreeding depression a broader genetic base for carrot inbred development is required to achieve uniformity and maintain vigour. In carrots, starting at the F_3 or even the F_2 generation, several individual roots from a given family are self-pollinated, whereas others from the same family are sib-mated in two to five plant masses. Seed produced from self-pollination and sib-mating are grown in adjacent rows the next year, and if vigour is adequate, the best performing self-pollinating plants are selected to continue the line. If severe inbreeding depression is observed in the form of small plants, then plants from the sib-mated sub-population are selected. This process is repeated two to five generations beyond the F_2 generation until plant characteristics are satisfactory and uniform. This constitutes a carrot 'inbred' which is used as a pollinator in a crossing cage to create hybrid combinations for evaluating combining ability.

The cultivated carrot, *Daucus carota*, intercrosses freely with almost all of its wild forms. It is an out-breeder. No incompatibility has been observed but selfing of individual flower and also of individual umbel is practically prevented by protandry (Banga, 1976). Systematic carrot breeding work with a view to developing high yielding cultivars with good adaptability has started very recently. The method previously employed was mass selection or combined mass-pedigree selection. In the early 17th century 'Long Orange' cultivar was developed in the Netherlands. From this cultivar, three orange coloured cultivars differing in earliness and size were developed *viz.*, Late Half-Long (the biggest), Early Half-Long and Early Scarlet-Horn (the smallest). All present day cultivars of the western carotene carrots have been developed from these four closely related cultivars, either by simple selection or by intercrossing different types (Banga, 1963b).

14.4.0 Genetics of Traits

Inheritance of colour, sugar and dry matter content was studied by several workers (Carlton and Peterson, 1963; Imam and Gabelman, 1968; Leferriere and Gabelman, 1968). Timin and Vasilevsky (1997) investigated inheritance of a number of traits in carrot. They found that the petaloid form of cytoplasmic male sterility (CMS) is determined by a sterility factor in the cytoplasm (Sp) and 3 dominant nuclear genes (*Ms 3, Ms 4* and *Ms 5*). Inheritance of root colour is determined by the interaction of 5 genes, while flower colour depends on the fertility/sterility of plants and is dependent on two complementary genes. Inheritance of glabrousness of the seedstalk and purple colouring of the surface roots and the leaf stalk are dependent on single dominant genes (*Gls, Pr1* and *Pr*, respectively). The identification and characterization of simply inherited visually assessable quality and quantitative traits in carrot has resulted several loci (Table 2).

Table 2: Genes for various traits reported in carrot*

Gene Symbol	Character Description	Gene Source
A	α-carotene synthesis (may be identical to *Io* or *O*)	'Kintoki"
Io	Intense orange xylem	Miscellaneous
L	Lycopene synthesis	'Kintoki"
Ms-1, Ms-2, Ms-3	Maintenance of male sterility	'Tendersweet'
Ms-4, Ms-5	Maintenance of male sterility	'Tendersweet','Imperator 58', PI 169486
O	Orange xylem	Miscellaneous
P1	Purple root	PI 173687
P2	Purple node	PI 175719
P3, P4	Purple root	Miscellaneous
rp	Reduced carotenoid pigmentation	W 266 Wisconsin inbred
rs	Reducing sugar in root	Miscellaneous
y	Yellow xylem	Miscellaneous
Y1, Y2	Differential xylem/phloem carotene levels	Miscellaneous
Ce	*Cercospora* leaf spot resistance	WCR-1 Wisconsin inbred
Eh	Downy mildew (*Erysiphe heraclei*) resistance	*D. carota* ssp. *dentatus*
mh-1, mh-2	*Meloidogyne hapla* resistance	'Rotin', Wisconsin inbreds
Mj-1	*Meloidogyne javanica*, resistance	'Brasilia'
Cr	*Cracking* roots (dominant to non-cracking)	'Touchon'
g	g = *Green* petiole, G = Purple petiole	'Tendersweet'
gls	*Glabrous* seedstalk	W-93 Wisconsin inbred
sp-1, sp-2	Seed *spine* formation	'Amkaza'
Ms-1, Ms-2, Ms-3	Maintenance of *male sterility*	'Tendersweet'
Ms-4, Ms-5		'Tendersweet', 'Imperator 58', PI 169486

* Simon (2000).

Kust (1970) described three dominant alleles–*Y, Y1, Y2*–which prevented the formation of orange colour in root xylem tissue. Buishand and Gabelman (1979) characterized the effects of the series of *Y* alleles on carotenoid content in phloem and xylem. The *Y* allele, which conditions lack of pigmentation (or white roots), was dominant to orange which was considered *yy*. The presence of *Y* and *Y2* always resulted in white roots, again demonstrating the dominance of lack of pigmentation. Orange carrots are *yyy2y2*, while white carrots are *YYY2Y2*, with yellow and pale orange colour in other genotypes.

The genetic control of β-carotene synthesis in carrot root tissue has indicated that the presence of pigment (*i.e.*, orange roots) is recessive to white or non-pigmented roots. However, Goldman and Breitbach (1996) identified and characterized a recessive gene that causes a 93 per cent reduction in carotenoid content, suggesting a new interpretation of carotenoid biosynthetic pathway, and may provide clues as to the details of this important process in carrot. The first several leaves of *rprp* plants are white and speckled, suggesting an effect of the allele on chlorophyll through the reduction of carotenoid pigmentation. Leaves appear identical to wild type appearance by the sixth leaf, suggesting development effects of the *rp* allele.

Through analysis of this mutant, Koch and Goldman (2005) determined that carrots produce tocopherols, particularly α-tocopherol or provitamin E. In experiments designed to assess the impact of the *rp* allele (Koch and Goldman, 2005), the reduced pigment (*rp*) mutation of carrot exhibited a 96 per cent reduction in levels of α- and β- carotene and a 25-43 per cent reduction in α- tocopherol when compared to a near-isogenic line. In plants homozygous for *rp*, a substantial increase was observed in phytoene, a precursor to carotenoids, suggesting the location of the *rp* lesion in the carotenoid synthesis pathway. In xylem tissue, α- tocopherol was significantly (pd"0.001) positively correlated with α-carotene (r=0.65) and with β-carotene (r=0.52). This positive correlation indicates possibility of selection for both increased á-tocopherol as well as carotenoids in carrot.

Purple carrots bring a different category of nutrients to consumers, the anthocyanins. A single dominant gene, *P1*, conditions purple root colour, while a second gene, *P2*, is hypostatic to *P1* and conditions purple pigmentations in aerial parts (Simon, 1996). The wide range of genetic variation in purple variation for purple pigment type in carrot, its amount, and distribution awaits future genetic analysis and application in breeding programmes.

A single gene controls sugar type (sucrose vs. reducing sugar) in carrots (Freeman and Simon, 1983). When the dominant allele (*Rs*) is present, the carrot root will accumulate the reducing sugars *fructose* and *glucose*. When both recessive alleles (*rsrs*) are present, the carrot root will accumulate primarily *sucrose*.

Broad-sense heritability estimates have been made for total dissolved solids (40-45 per cent, Stommel and Simon, 1989), and alternaria leaf blight resistance (45-82 per cent, Vieira *et al.*, 1991). Narrow sense heritability of 16-48 per cent is reported for Javanese nematode resistance (Huang *et al.*, 1986), and 40 per cent for alternaria resistance (Boiteux *et al.*, 1993).

14.5.0 Combining Ability, Gene Action and Heterosis

The F_1 hybrids show a great promise in getting uniform and high yields in crops. The first report of heterosis in carrot upon crossing of two inbred lines was made by Poole (1937). Inbreds producing roots averaging 12.8 and 24.9, respectively, produced hybrids with roots weighing 80.5 g upon hybridization. The degree of sterility in carrot depends more on cultivar than meteorological conditions and male sterility was mostly confined to primary umbels. The hybrids obtained from male sterile forms yielded 34.4-44.9 percent more and contained higher amount of dry matter and sugars compared to intervarietal pollination. Katsumata and Yasui (1965) reported that F_1 hybrids among the cultivars Yokono, male sterile Sansun and Kuroda possessed good root form, high carotene content and were high in quality. Most of the F_1 hybrids exhibited heterosis for carotene content. The male sterile forms of the cultivars Nantes 14 and Chantenay 2461 were successful in the production of heterotic hybrids (Litvinova, 1979) and heterosis was 20-22 percent higher than in hybrids produced without using CMS (cytoplasmic male sterility). These F_1 hybrids showed better uniformity of morphological characters. On the contrary, Bonnet (1978) reported from France that root characters such as shape and colour were controlled by the genotype of the line and were not affected by the type of cytoplasmic male sterility. F_1 hybrids out-yielded the cv. Touchon by 21 per cent and had smallest percentage of rejects.

Carrot is an out-breeding species due to being protandrous. It has no self-incompatibility system, but inbreeding is severe. The distinctive umbels and floral nectaries attract insects, which are most responsible for performing pollination. The heterosis breeding in carrot has been facilitated by the cytoplasmic male sterility (CMS) which is of two types, _viz._ (i) brown anther type, in which the anthers degenerate and shrivels before anthesis, based on S-cytoplasm and at least two recessive genes with complementary action and (ii) petaloid, in which five additional petals replace anthers, based on S-cytoplasm and at least two dominant genes with complementary action. The brown anther type was first discovered by Welch and Grimball (1947) and this report was followed in 1953 by the discovery of a sterile wild carrot (petaloid type) by Munger of Cornell University in USA. In hybrid development petaloid steriles are employed more widely than the brown anther type. If genetic and environmentally stable brown anther steriles were available, then they would be preferred over petaloids because of their higher seed yielding potential. Basically, there are three lines in heterosis breeding, namely the male sterile, male fertile sister line and the pollinator line which is male fertile and has a good combining ability with the male sterile line. The male sterile and the pollen parent lines are inbred for several generations for attaining uniformity. The loss in vigour in these can be restored by hybridization. The hybrids in carrot are normally three way crosses, (A × B) × C, because the hybrid vigour in a single cross F_1 female seed parent normally results in much greater seed production than that of inbred male sterile parent. Single cross hybrids, A × B, are on an average more uniform than three way crosses. Moreover, they do not require an extra year to produce F_1 seed parent stock. So the single crosses can be used, if their productivity is adequate. Reduced uniformity of three way crosses can be overcome if backcrosses are utilized as seed parents in hybrids as a result the final product attains the form [(A × B) × B] × C. Although, compared to three way crosses, it consumes an additional year of seed parent production, but it permits utilization of less similar seed parent inbreds, A × B, than what is required.

Kalia (2004) studied the performance of 100 selections of tropical carrot for quality and root characters and found lines IPC 40, IPC 99, IPC 35, IPC 42 and IPC 36 have long root length, IPC 16, IPC 100, IPC 40, IPC 39 and IPC 60 have high root weight and IPC 14, IPC 11, IPC 13, IPC 10 and IPC 31 have high β-carotene content. Kalia and Longvah (2008) studied the biochemical genetic diversity in tropical carrot. They recorded highest total carotenoids and lycopene in IPC 13 (4887 μg/100g, 3030 μg/100g), β-carotene in IPC 122 (994 μg/100g), lutein in IPC 25 (378 μg/100g) and niacin in IPC 31 (0.80 mg/100g).

Imam (1966) found that inbreeding sharply reduced vigour and viability in carrots and inbred lines could not be maintained after five generations of inbreeding. Bhagchandani and Choudhury (1971) studied the heterosis and inbreeding depression in Asiatic carrot. Saini *et al.* (1981) studied the variability and genotypic environment interactions in Carrot. Low genetic co-efficient of variability moderate and high heritability and low genetic advance as percentage of mean was observed for root diameter, whereas these estimates were comparatively high for all other attributes. Singh *et al.* (1987) worked out the genetic variability, heritability and genetic advance in 40 exotic and indigenous genotypes. It is reported that selection based on phenotypic values may prove quite effective because of high heritability estimates. Chandel and Rattan (1988) conducted studies to determine the role of epistasis and evaluate the importance of additive versus dominant effects for traits not influenced by epistasis. They observed that both additive and dominance variance were significant for other characteristics.

Bujdoso and Hrasko (1990) also showed that hybrids were superior in yield and quality to the open-pollinated varieties. F_1 hybrids showed strong hybrid vigour in quantitative characters such as leaf and root weight. The increase of root weight in F_1 hybrids was more influenced by increase in root diameter than that of root length (Su *et al.*, 1999). A late maturing hybrid Spartan North has also been developed from single cross between parental lines MSU 5931A and MSU 5986B (Baker, 1978c). This hybrid cultivar has long, slim roots with good internal and external characters.

14.6.0 Cytoplasmic Male Sterility

The 'brown anther' (Sa) type of male sterility is characterized by presence of rudimentary brown anthers. The 'brown anther' CMS type was first discovered in the cultivar 'Tendersweet' (Welch and Grimball, 1947). Later on 'brown anther' sterility was selected in several other cultivars.

The 'petaloid' (Sp) male sterility was first discovered in 1953 by Dr. Henry M. Munger in a North American wild carrot (*D. carota* subsp. *carota*) near Orleans, Massachusetts (USA) and was later termed 'Cornell-CMS' (Thompson, 1961). Petaloidy is characterized by a replacement of the stamen by petals or petal-like structures. It was used to produce the majority of hybrid carrots in the United States (Goldman, 1996). Thompson (1961) developed a complex model for inheritance of pollen sterility from the study of wild carrot (petaloid type) along with the brown anther material from Welch and three additional brown anther sources from Gabelman. He concluded that there are three duplicated dominant genes *Ms1, Ms2, Ms3* necessary to maintain male sterility and an epistatic locus to restore fertility in both CMS systems. The useful maintainer line would, therefore, have to be free from the restorer and homozygous dominant at one of the Ms loci. The studies of Morelock (1974) supported the hypothesis that petaloid sterility in wild carrot cytoplasm is controlled by two dominant nuclear genes (15 fertile: 1 sterile in F_2). The brown anther type is controlled by two recessive genes as evident from 15 sterile: 1 fertile in domestic cytoplasm. Alessandro *et al.* (2013) identified and mapped a single dominant nuclear gene determining restoration of petaloid cytoplasmic male sterility (*Rf1*) to chromosome 9.

McCollum (1966) detected petaloid structures, staminodes, and sterile stamens in a wild carrot population obtained from Sweden. Petaloidy has been also reported in other North American–'Wisconsin-CMS' (Morelock *et al.*, 1996) and Canadian wild carrots–'Guelph-CMS' (Wolyn and Chahal, 1998). Three new CMS sources were reported in the wild relatives *D. carota* subsp. *gummifer*, *D. carota* subsp. *maritimus*, and *D. carota* subsp. *gadecaei*. Flowers of the CMS-GUM type is characterized by a nearly complete loss of petals and stamen in an early stage of organ development. The CMS-MAR type is comparable to the common petaloid CMS flower types. The CMS-GAD type has only short filament-like stamen rudiments in the flowers (Linke *et al.*, 1999; Nothnagel *et al.*, 1997, 2000)

Dame *et al.* (1988) suggested that male sterile forms of the petaloid type should be more widely used in the production of hybrid varieties than brown anther types. CMS lines, maintainer lines and pollinator lines for use in breeding hybrid varieties were produced from the varieties such as Natskaya 4 and Moskovskaya Zimnyaya A 515 through various breeding methods by Timin and Vasilevskii (1995). The hybrids produced by them out-yielded standard varieties by 30-50 per cent and produced uniform roots with a high content of carotene.

14.7.0 Breeding for High Nutritional Quality

Carrots are a natural target for carotenoid biofortification due to their high levels of consumption, affordability to consumers, widespread cultivation worldwide, and potential for increased carotene biosynthesis. Because carotenes in carrots are not necessary for growth, they have greater genetic variation for provitamin A content than for storage carbohydrate or protein contents. Early varieties of orange carrots during 1960s in the USA had carotene contents in the range of 70 ppm, but breeding efforts, through phenotypic recurrent selection of roots with a darker orange colour, increased that value to 90 ppm by the 1970s. Modern spectrophotometric detection of high levels of root carotenes has allowed breeders to elevate the concentration above 130 ppm in typical orange and dark orange carrots, that is, high-carotene mass, can reach 500 ppm (Simon *et al.*, 1989).

High-carotene transgressive segregants were observed in crosses among European cultivars and also in European by Asian crosses. The elevated carotene content in one European by Asian cross was especially interesting, since the high incidence of transgressive high-carotene segregants resulted in F_2 mean carotene content greater than the higher parental stock (Simon *et al.*, 1989). Derivatives of this population have been used as a source of high carotene content in the development of new breeding stocks and hybrids. Although the heritability of solids and carotenoid levels is not particularly high, it is desirable to eliminate undesirable segregants from the population before pollination begins.

Carotene content exhibited the highest value of genotypic and phenotypic coefficient of variation, heritability (broad sense) and genetic advance as percentage of mean, indicating that this character can be effectively improved through selection (Kaur *et al.*, 2005). Hussain *et al.* (2005) observed medium heritability for total soluble solids content in carrot. Singh *et al.* (2005) reported negative correlation of root weight with TSS and carotene content at the phenotypic and genotypic levels. Gill and Kataria (1974) studied the biochemical aspects of European and Asiatic varieties of carrot. A marked variation in dry matter, total soluble solids and carotene content amongst the varieties was found. European types of carrot were superior to Asiatic types in respect of all the attributes included in the study. The bigger size of core and high percentage of moisture was mainly responsible for poor nutritive value of Asiatic type of carrot.

Additive × dominance and dominance × dominance gene interactions were significant for ascorbic acid, dry matter, β-carotene and total sugar contents. Additive × additive gene interactions were also significant for ascorbic acid and total soluble solids contents. Overdominance was displayed in the case of total sugars and ascorbic acid content, while a partial degree of dominance was observed for dry matter and β-carotene contents (Chandel *et al.*, 1997). Chandel *et al.* (1994) reported both additive and nonadditive genetic variances were significant for total soluble solids and β-carotene content. The additive genetic components of variation were more pronounced for ascorbic acid content. Dominance in the range of overdominance was recorded for total soluble solids, while it was partial in the expression of other traits. Rattan and Jamwal (1991) reported both additive and dominance effects for β-carotene content. Direct selection for traits showing dominance of additive gene effects may lead to improvement of quality.

Heritability and number of genetic factors have also been estimated for total carotenes, β-carotene, α -carotene, lycopene, and phytoene (Santos and Simon, 2006). Heritabilities for each trait were around 90 per cent. The estimated number of factors was 4 for 1–2 for lycopene and total carotenes, and 1 each for β -carotene, and phytoene. Interestingly, while several factors account for variation of individual carotenes, most of them were clustered, so that discrete inheritance of as few as 2 major loci separate white from orange carrot.

Nitrates have anti-nutritional value, especially for carrots used to make baby food. The inheritance of nitrate content in carrot is complex within complete dominance so that low-nitrate parents are necessary to obtain low-nitrate hybrids. Progress in genetic selection of low-nitrate carrots has been successful in Poland.

14.8.0 Disease Resistance

Breeding efforts for disease and pest resistance in carrot has gained importance in recent years. In North America, primary focus is selection for resistance to leaf diseases that include Alternaria blight, Cercospora blight, aster yellows and motley dwarf, whereas in Europe selection for cavity spot, soft rot, carrot fly, nematode, powdery mildew and *Pythium* resistance is more emphasized.

Alternaria leaf blight caused by *Alternaria dauci* is one of the most important foliar diseases, considerably reducing the carrot yield around the world. Immunity to leaf blight has not been reported, but several sources of resistance have been incorporated into cultivars. The genetics of resistance is conditioned by several genes with varying levels of dominance (Strandberg *et al.*, 1972; Boiteux *et al.*, 1993; Simon and Strandberg, 1998). Asian and Brazilian cultivars have been used as the source of resistance genes for alternaria leaf blight (Vieira *et al.*, 1983). Asian germplasm is an important source of alternaria leaf blight resistance for Western carrots (Strandberg *et al.*, 1972). Efforts are done for incorporating resistance into breeding stocks (Vieira *et al.*, 1991; Boiteux *et al.*, 1993; Simon and Strandberg, 1998). *D. carota* subsp. *capillifolius* and *D. carota* subsp. *maximus* and *D. crinitus* may provide new sources of resistance to Alternaria leaf blight (Arbizu *et al.*, 2017).

Spartan Bonus is a three way hybrid (Parent MSU 1558, MSU 5831B and MSU 9541B) with medium long tops that are resistant to *Cercospora carotae* and *Alternaria dauci*. In comparison with Danvers, Spartan Bonus has given higher yields and has less oily flavour, and superior external and internal root colour (Baker, 1978b).

Development of high quality hybrids, high yielding hybrids and disease resistant hybrids has also been reported by Baker (1978a) from the U.S.A. Report from Michigan State University indicated release of 3 three-way carrot hybrids: Spartan Classic from MSU 5931A, MSU 5931B and MSU 5988C; Spartan Premium from MSU 5931A, MSU 5986B and MSU 5988C and Spartan Winter from MSU 5931A, MSU 6000B and MSU 5988C. All the 3 hybrids are early maturing, smooth rooted, moderately resistant or tolerant to *Cercospora carotae* and resistant to rusty root (*Pythium* sp.), and are suitable for fresh market and processing. In comparison with open pollinated cv. Nantes, Spartan Winter gives high yield (Baker and Valk, 1978).

Cavity caused by *Pythium* sp. is among the most destructive post-harvest storage diseases of carrots. Information on the inheritance pattern and selection efficiency for soft rot (Michalik *et al.*, 1992) is reported. Resistance to cavity spot in carrot is generally is quantitative based on relatively minor differences in severity among cultivars. Partial resistance to cavity spot has been reported in Six Pak, SR-481, Eagle, Red Core, Panther, Caropride, Fannia, Navajo, CS 732 and CS 736 (McDonald, 1994; Benard and Punja, 1995; McDonald *et al.*, 2017).

Carrot cultivars 'Panther' and 'Caropak' are found to be relatively resistant to black rot (Pryor *et al.*, 2000). Grzebelus *et al.* (2013) isolated somaclonal variants within protoplast cultures that were challenged by fungal culture filtrates and obtained regenerated plants with greater tolerance.

Breeding for resistance to cercospora leaf spot (Angell and Gabelman, 1968), caused by *Cercospora carotae*, has also been conducted. Monogenic resistance to Cercospora blight and powdery mildew has been reported, and field evaluation has been expanded for both diseases (Lebeda and Coufal, 1987; Lebeda *et al.*, 1988).

Bonnet (1983) identified that the resistance to powdery mildew in *D. carota* subsp. *dentatus* controlled by a single dominant gene, Eh. He reported that the cultivar Bauers Kieler Rote is a potential source of resistance to powdery mildew. Lebeda and Coufal (1987) reported cultivar 'Gavrilovskaya' to be resistant. Report from France indicates that male sterile lines show tolerance to *Erysiphe herachi*. Wild carrot species such as *D. siculus*, *D. manimus* and *D. dentatus* are also reported to be resistant. In crosses between Touchon (susceptible) and *D. dentatus*, the resistance was found monogenic and dominant (Anon., 1977-78).

Aster yellows (AY), an aster leafhopper transmitted disease caused by a mycoplasma-like organisms (MLO), is a destructive carrot disease in the U.S.A, Canada and other parts of the temperate world. Germplasm derived from University of Wisconsin inbreds and four open-pollinated cultivars provides significant protection to aster yellows (Gabelman *et al.*, 1994). European open-pollinated carrot cultivars have served as the germplasm source of aster yellows resistance (Gabelman *et al.*, 1994).

14.9.0 Pest Resistance

Carrot fly (*Psila rosae*) is the most important insect pest of carrot in Europe and Canada. *Daucus capillifolius* (Gilli), a wild relative of carrot found in Northern Africa, is resistant to carrot fly. Ellis *et al.* (1991) at Horticulture Research International (HRI) in the UK have extensively examined carrot fly resistance in *D. capillifolius* and derivatives of its crosses with cultivated carrot. Introgression of resistance trait has led to the release of the carrot fly resistant cultivar "Flyway" (Ellis and Hardman, 1992). Michalik and Wiech (2000) assessed the resistance of 9 breeding lines, 7 cultivars and 13 other accessions of carrots to carrot fly (*Psila rosae*). Among them, only five breeding lines (HRI 93171, HRI 96732, HRI 93104, HRI 96382 and AR 694) and cv. Puma were found tolerant, while others were susceptible.

Several species of nematodes attack carrots. Northern root-knot nematode, *Meloidogyne hapla*, is prevalent in cool temperate areas of northern Europe and North and South America, while the tropical and Southern root-knot nematodes *M. javanica* and *M. incognita* are widespread in warmer carrot production areas of southern Europe, southern U.S.A, northern South America, and Asia. A two-gene model for *M. hapla* resistance has been reported by Wang and Goldman (1996) whereas a single gene confers resistance to *M. javanica* and *M. incognita* (Simon *et al.*, 2000). Several sources of resistance to nematodes *M. hapla* and M. *javanica* have been identified and used in development of new carrot inbreds (Frese, 1983; Huang *et al.*, 1986; Wang and Goldman, 1996; Simon *et al.*, 2000). Cultivar 'Brasilia' resistant to nematode species *M. incognita* and *M. javanica* have been developed (Huang *et al.*, 1986). Asian and Brazilian cultivars have been used as the source of resistance genes for *M. javanica* (Vieira *et al.*, 1983). European open-pollinated carrot cultivars have been used as a source of Northern root knot nematode *M. hapla* resistance (Gabelman *et al.*, 1994). Charchar and Vieira (1994) tested some 384 lines of carrot for resistance to *Meloidogyne* spp. in a naturally infested field. After 4 selection cycles, six lines were selected having infestation rates ranging from 3.7 to 25.2 per cent. In a second set of experiment, they selected 27 lines with infestation rates of 5.8-23.7 per cent. Wild carrot subspecies *hispanicus* has also been used as a source of genes for resistance to *M. hapla* (Frese, 1983). Yunhee *et al.* (2014) screened Korean carrot lines and reported 61 lines resistant to *M. incognita* race 1. 'DR-333', a cultivar in north India was tolerant to *M. incognita* (Siddiqui *et al.*, 2011). Huang *et al.* (1986) reported low narrow-sense heritabilities of resistance to *M. javanica*, from 0.16 to 0.48 for root galling and from 0.31 to 0.35 for egg mass production, depending on the cultivar.

Broad-sense heritability for resistance to *M. incognita* was 0.33 and 0.25 in two carrot populations derived from a cross among three sources of resistance from Syria, South America and Europe (Parsons *et al.*, 2015). Wang and Goldman (1996) identified two homozygous recessive genes with epistatic control of *M. hapla* resistance, suggesting that this resistance may be relatively simply introgressed into inbred lines via backcrossing. Yunhee *et al.* (2014) reported that resistance to *M. incognita* may be governed by one or a few genes.

14.10.0 Hybridization of Tropical and Temperate Types

Efforts have been made to combine the desirable characters of European and Asiatic types in suitable crosses and a new carrot cv. Pusa Kesar has been developed, which has good qualities of both European and Asiatic types with an ability to produce seeds in the plains of India (Singh, 1963). A high yielding early cultivar PusaYamdagni has been developed from a cross between EC 9981 and Nantes Half Long (Gill *et al.*, 1987).

15.0 Biotechnology

Significant progress has been made in biotechnological work in carrot, which provides ample scope for the development of the crop with different desirable attributes and is summarized below.

15.1.0 Haploid and DH Plant Production

Hybrid carrot seeds are produced using inbred populations obtained after several generations of self- or sib-pollination. Carrot has a strong inbreeding depression. The use of completely homozygous doubled haploid (DH) lines would considerably shorten breeding processes. Haploids and DH lines can be obtained in various ways, including anther, microspore, and ovule culture. DH lines have been used successfully in breeding programs of many crops. The species of Apiaceae family are considered to be recalcitrant to DH technologies, although carrot is a model plant for tissue culture and regeneration protocols are available for several species.

15.1.1 Anther or Microspore Culture

Practical utilization of the DH technique is hindered by the low efficiency of embryogenesis in carrot anther cultures. Only 0.8 per cent of 20,400 carrot anthers generated callus (Andersen *et al.*, 1990), the highest efficiency accession was HCM A.C. about 5.6 embryos per 100 anthers among five cultivars (Go´recka *et al.*, 2005), and only six accessions demonstrated successful microspore embryogenesis among 39 different genotype donors (Zhuang *et al.*, 2010). Moreover, carrot anthers are very small hence are difficult to manipulate, making the avoidance of interfering sporophytic anther wall tissues difficult when establishing calli or embryos. Compared with anther culture, the procedure of isolated microspore cultures (IMC) is simpler and more efficient. The growth course of microspores can be observed directly and interference with somatic tissue is avoided.

Studies on anther culture in carrots showed that most of the regenerated plants were of somatic origin, caused by frequent indirect embryogenesis and callus formation from anthers. Culture of isolated microspores in liquid media has also been reported, but the frequency of carrot haploid development via direct embryogenesis from microspores was low. Secondary embryogenesis occurring from both primary embryos and calli, increased the final number of regenerants and resulted in the regeneration of plants of various ploidies, including haploids,

Anther Culture in Carrot

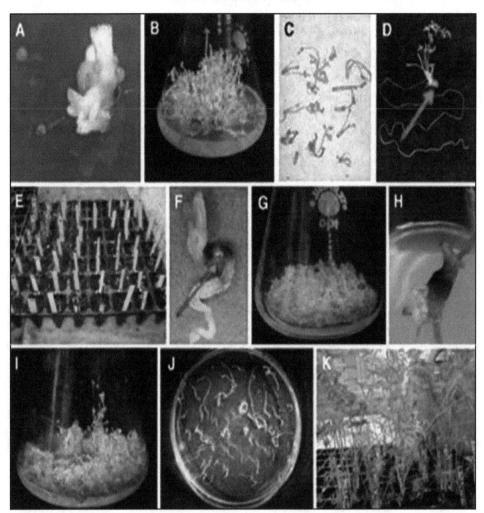

Carrot Plant Regeneration from Embryos Obtained in Anther Cultures.

A: Embryos on the anther of carrot; B and C: Shoots obtained during the two-stage regeneration on the MS-2 medium; D: Plant with roots obtained on a rooting medium containing auxins during two-stage regeneration; E: Adaptation for *ex vitro* conditions of carrot plants obtained from anther cultures and regenerated in two stages; F: Secondary embryogenesis on a carrot embryo obtained from anther cultures; G: Numerous secondary embroys; H: Secondary embryo on carrot plant obtained through conversion of embryo; I: Conversion of carrot embryo; J: Carrot plantlets obtained by the conversion secondary embroys; K: Adaptation for *ex vitro* condition of carrot (Gorecka *et al.*, 2009).

diploids, and triploids. The response from isolated microspore cultures is highly dependent on the donor plant genotype.

Li *et al.* (2013) developed protocols for the generation of haploid and doubled haploid plants from isolated microspores of carrot. They screened 47 carrot accessions, including six inbred lines, 11 cultivars, 20 F_1s, two BC_1F_1s, four F_2s, one F_3, and three F_4s, to evaluate the genotype influence on isolated microspore embryogenesis over 4 years. Twenty-eight accessions responded by producing embryos and/or calli. A cytological analysis showed that two modes of carrot microspore embryogenesis exist: an indirect route via calli (C mode), and a direct route via embryos (E mode). Eleven accessions were in the C mode, and 17 were in both modes. The highest production rates were in 10Y25 (a European Nantes cultivar) with 27 calli and 307 embryos, and 100Q6 (a semi-Nantes F_1 hybrid) with 176 calli and 114 embryos. The time period to produce embryos or calli differed significantly between 2 and 6 months. Cold and heat pretreatment generally had a negative impact on the induction of microspore embryogenesis, but a short pretreatment showed a positive influence on some accessions. Twenty eight lines regenerated plants from the primary individual embryos or calli of three accessions were established to analyze the ploidy level. The percentage of spontaneous diploidization showed very wide differences among the accessions and lines. Differences in leaf color intensity, leaf size, and leaf dissection were found among haploid, doubled haploid, and triploid plants.

Hu *et al.* (1993) cultured immature anthers of cv. Senkou 5 Sun on MS solid medium containing various combinations of growth regulators. Calli were formed only on media containing 2,4-D and kinetin. The highest rate of embryoid formation (15 per cent) was obtained on the medium containing 1.0 mg/12,4-D without kinetin. Embryoids transferred to solid MS medium produced plantlets. Among the 18 plants obtained, 16 were haploid (n=9) and the other two were aneuploid (n=10 and 11).

Matsubara *et al.* (1995) induced calli and adventitious embryoids development from carrot microspores after one month of anther culture. Callus regenerated from anthers of different stages on the media tested and many adventitious buds and roots were produced.

15.1.2 Ovule or Ovary Cultures

Ovule or ovary cultures are an alternative method for haploid production. Haploid production from the female gametophyte occurs via either gynogenesis or parthenogenesis induced by wide pollination. In gynogenesis, haploids arise from an unfertilized egg cell that is stimulated to divide by donor plant pretreatment and *in vitro* culture conditions. In parthenogenesis, haploids develop as a result of stimulation by application of irradiated pollen or pollen of other species or genera. Parthenogenesis has also been described for carrots (Kielkowska and Adamus, 2010) after stimulation of ovule development by pollination by other Apiaceae species like parsley or celery. Haploid and diploid plants were obtained from unfertilized ovules cultured in vitro. The efficiency of this laborious and time-consuming process was low: only 1.85 per cent of cultured ovules responded and more than half of them showed callus development rather than embryo development. Thus, the usefulness of the method was limited.

Kielkowska *et al.* (2014) developed an improved protocol for induced parthenogenesis and ovule culture of carrot. They studied the effects of pollination with parsley pollen and/or 2,4-dichlorophenoxyacetic acid (2,4-D) treatment on the stimulation of parthenogenesis using heterozygous donor plants of 30 varieties and breeding populations of carrots. The application of 2,4-D on pollinated flowers stimulated callus development but did not increase the frequency of embryo development from ovules and, thus, was not useful for increasing the frequency of haploid plant recovery. The efficiency of embryo development was accession dependent and varied from 0 to 24.29 per cent. In optimized conditions, most accessions responded by embryo development exclusively. The highest frequency of embryo development was observed from ovules excised from ovaries 20–22 days after pollination with parsley pollen. Among several media used for ovule culture, 1/2-strength Murashige and Skoog medium with 0.06 μM indole-3-acetic acid (IAA) was the best. It allowed the production of embryos at a similar frequency as on the media supplemented with kinetin, gibberellic acid, putrescine, or thidiazuron, but restricted callus development. Most plants obtained were haploids and diploids derived from parthenogenesis, as evidenced by homozygosity at three independent loci based on isozyme and PCR analyses. In total, considering haploids and embryo-derived homozygous diploids together, 72.6 per cent of regenerated plants were of gametic origin.

15.2.0 Protoplast Culture and Fusion

Protoplast culture has been reported by many workers and protoplasts have been successfully isolated from roots, hypocotyl and callus tissues (Wan *et al.*, 1987; Shea *et al.*, 1989). The capacity to produce embryoids varies between genotypes. Badawi (1978) initiated cell cultures of 5 carrot cultivars from actively growing callus, and from cell cultures. Protoplasts were isolated enzymatically and 15 different somatic combinations were obtained. Fused and unfused protoplasts were then left to divide and undergo embryogenesis. The best results were obtained with cv. Duwicki and its combinations where the plantlets with a mean height of 3.8 cm were obtained in 120 days from fusion, while unfused Duwicki cells gave plants of 5.2 cm in height in that period.

Carrot is considered as a model system for studying protoplast fusion. Intergeneric fusions involving protoplasts of carrot and other plants such as rice (Sala *et al.*, 1985; Kisaka *et al.*, 1994, 1996), maize (Ma *et al.*, 1981), barley (Kisaka *et al.*, 1997), tobacco (Smith *et al.*, 1989; Guo *et al.*, 1995), spinach (Hodgson and Rose, 1984), celery (Wang *et al.*, 1989) have been reported. Somatic hybrids of *Oryza sativa* (rice) and carrot has been reported by Kisaka *et al.* (1994), when protoplasts isolated from a cytoplasmic male sterile carrot were fused with idoacetamide-treated protoplasts isolated from a 5 methyltryprophen (5 MT) resistant rice suspension culture by electrofusion. Somatic hybrids of barley (*Hordeum vulgare*) and carrot were also reported by Kisaka *et al.* (1997).

Cells from glyphosate resistant cell line PR were fused with lines resistant to DL-5-methyltryptophan (5MT) and azetidine-2-carboxylate (A2C), somatic hybrids expressed resistance to 5MT in a semidominant fashion while A2C and glyphosate resistance was expressed as a dominant or semi-dominant trait depending on which parental line was expressed as a dominant or semi-dominant trait depending on which parental line was used (Hauptmann *et al.*, 1988).

Transfer of cytoplasmic male sterility by means of protoplast fusion has been tried by Ichikawa *et al.* (1988), Suenaga *et al.* (1991) and Bach *et al.* (1997). Three potential novel CMS sources with different phenotypes have been identified in carrot based on the cytoplasms of *D. carota* subsp. *gummifera, maritimus* and *gadacei*. These new CMS sources are associated with brown anther and petaloid types of sterility (Nothnagel *et al.*, 1997), which could be used for production of parents for hybrid seed production.

15.3.0 Somatic Embryogenesis

Plant regeneration by somatic embryogenesis was first reported in carrot by Steward *et al.* (1958). Somatic embryo production in a bioreactor was found comparable to that obtained in shake flasks (Ducos *et al.*, 1993). A production of about 50×103 embryos/litre/day could be achieved with an inoculum density of 0.1 per cent volume of tissue/volume of medium. Regular changing of the medium resulted in increased embryo viability. Teng *et al.* (1994) found that bioreactor cultured embryos germinated with relatively short cotyledons and long roots, whereas flask embryos germinated with relatively long cotyledons and short roots. An effective technique of cell tracking system to determine the capability of individual single suspension cells of carrot to develop into somatic embryo was reported (Toonen *et al.*, 1994). Suspension cells were immobilized into phytagel and then somatic embryos were recovered from single cells.

15.4.0 Molecular Markers for Diversity Analysis and Domestication Studies

The RAPD markers for genotype identification of carrot lines and F_1 hybrids have been reported to be useful although it appeared to be limited for hybrid purity testing due to the high level of heterogeneity within the analyzed hybrids. Grzebelus *et al.* (1997) screened nine breeding lines and 3 F_1 hybrids using 33 decamer primers. Of these, primer, 15 produced in total 47 stable polymorphic bands which were used to assess genetic distance between the tested plant material. Nakajima *et al.*, taxonomically identified mitochondrial and nuclear genomic diversities of 8 carrot varieties including 6 pure lines and 2 cytoplasmic male-sterile (CMS) lines using PCR with 19 RAPD primers.

Different molecular markers are being used to detect polymorphism and linkage. A 109-point linkage map consisting of three phenotypic loci (P1, Y2 and Rs), six RFLPs, two RAPDs, 96 AFLPs and two selective amplifications of microsatellite polymorphic loci (SAMPL) was constructed for carrot (Vivek and Simon, 1999). The incidence of polymorphism was 36 per cent for RFLP probes, 20 per cent for RAPD primers and 42 per cent for AFLP primers.

In carrot, traits important for primary domestication included the ability to form fleshy roots, minimal lateral root branching and biennial growth habit. Generally, Western carrots appear as a more advanced group, better adapted for commercial production and processing and usually develop roots of cylindrical or tapered cylindrical shape which is favoured by the food processing industry, have less pubescent leaves and show little tendency for bolting. They have high content of pro-vitamin A carotenoids, mostly α-carotene, are sweeter, having on average 18 per cent higher sugar content than Eastern carrots. Eastern carrots commonly grown in Asia produce thicker, shorter roots or narrow conical roots with a tendency for branching in some varieties. They often have pubescent leaves and tend to flower early, and hence exhibit more primitive traits. Their roots are poor in provitamin A carotenoids and have yellow (lutein), purple (anthocyanin) or red (lycopene) color. In contrast to Western carrots, anthocyanin rich purple rooted carrots are richer in phenolic compounds, resulting in higher radical scavenging activity. However, despite the apparently high selective pressure imposed on the domesticated population, Iorizzo *et al.* (2013) observed no reduction in genetic diversity resulting from a domestication-related bottleneck in Eastern and Western gene pools of cultivated carrot. Iorizzo *et al.* (2013) provided clear evidence for diversification between wild and cultivated accessions, supporting previous reports based on amplified fragment length polymorphism markers.

Analysis of SSR markers by Baranski *et al.* (2012) showed evidence for the separation of the cultivated germplasm into two distinct groups, the Eastern (Asian) and Western (European and American) gene pools. Genetic diversity of the Asian gene pool was higher than that of the Western gene pool. A recent study by Iorizzo *et al.* (2013) based on *D. carota* plants of different origin genotyped with more than 3,300 SNP markers suggested that Central Asia is the center of origin of cultivated carrot, and that orange-rooted carrots of the Western type were selected from the yellow domesticated carrots.

Grzebelus *et al.* (2014) developed a Diversity Arrays Technology (DArT) platform for wild and cultivated carrot and used it to investigate genetic diversity and to develop a saturated genetic linkage map of carrot. They analyzed a set of 900 DArT markers in a collection of plant materials comprising 94 cultivated and 65 wild carrot accessions. The accessions were attributed to three separate groups: wild, Eastern cultivated and Western cultivated. They identified 27 markers showing signatures for selection which showed a directional shift in frequency from the wild to the cultivated, likely reflecting diversifying selection imposed in the course of domestication. Two of the 27 DArT markers with signatures for selection segregated in the F_2 mapping population and were localized on chromosomes 2 and 6. Chromosome 2 was previously shown to carry the Vrn1 gene governing the biennial growth habit essential for cultivated carrot.

Rong *et al.* (2014) sequenced the root transcriptomes of widely differing cultivated and wild carrots and identified 11,369 SNPs, of which 622 were validated and used to genotype a large set of cultivated carrot, wild carrot and other wild *Daucus carota* subspecies, primarily of European origin. Phylogenetic analysis indicated that eastern carrot may originate from Western Asia and western carrot may be selected from eastern carrot. Different wild *D. carota* subspecies may have contributed to the domestication of cultivated carrot. Genetic diversity was significantly reduced in western cultivars, probably through bottlenecks and selection. However, a high proportion of genetic diversity (more than 85 per cent of the genetic diversity in wild populations) is currently retained in western cultivars. Model simulation indicated high and asymmetric gene flow from wild to cultivated carrots, spontaneously and/or by introgression breeding. The up-regulation of water-channel-protein gene expression in cultivars might be involved in changing water content and transport in roots. The activated expression of carotenoid-binding-protein genes in cultivars could be related to the high carotenoid accumulation in roots. The silencing of allergen-protein-like genes in cultivated carrot roots suggested strong human selection to reduce allergy. These results suggest that regulatory changes of gene expressions may have played a predominant role in domestication.

15.5.0 Molecular Mapping and Marker Assisted Breeding

15.5.1 Mapping of Root Carotenoids Content

In addition to single genes conditioning important traits, several QTL have been identified in carrot through segregation analysis. To date, QTL conditioning synthesis in carrot roots of provitamin A á-and α-carotenes, the carotene lycopene, and precursors in the carotene pathway have been mapped. Most modern carrot breeding effort has exclusively involved intercrosses among orange carrots and the numerous QTL involved in that color class.

QTL have been mapped for carrot total carotenoids and five component carotenoids; phytoene, α-carotene, zeta-carotene, and lycopene (Santos and Simon, 2002) and the majority of the structural genes of the carotenoid pathway is now placed into this map (Just *et al.*, 2007). Bradeen and Simon (1998) studied 103 F_2 individuals of the cross B9304 × YC7262, which segregated for core color. Using bulked segregant analysis combined with F_2 mapping, they identified six AFLP markers linked to and flanking the *Y2* locus. Markers were located between 3.8 and 15.8 cM from the gene. Using the same F_2 mapping population, Vivek and Simon (1999) subsequently identified a single AFLP marker located 2.2 cM from the *Y2* locus, assigning the locus to one end of linkage group B.

Just *et al.* (2009) reported that two major interacting loci, *Y* and *Y2* on linkage groups 2 and 5, respectively, control much variation for carotenoid accumulation in carrot roots. These two QTLs are associated with carotenoid biosynthetic genes zeaxanthin epoxidase and carotene hydroxylase and carotenoid dioxygenase gene family members as positional candidate genes. Dominant *Y* allele inhibits carotenoid accumulation. When *Y* is homozygous recessive, carotenoids that accumulate are either only xanthophylls in *Y2* plants, or both carotenes and xanthophylls, in *y2y2* plants. These two genes played a major role in carrot domestication and account for the significant role that modern carrot plays in vitamin A nutrition.

Few molecular markers in or linked to carrot major genes or QTL have been developed. Examples have been reported for carotene QTL (Santos and Simon, 2002) and the *Y2* gene (Bradeen and Simon, 1998) and the *Rs* sugar type gene (Yau and

Simon, 2003; Yau *et al.*, 2005), with marker-assisted selection exercised successfully in the latter case. Yau *et al.* (2005) suggest that these identifications can be done on leaf tissue from early growth, thereby removing the need for growing mature roots and analyzing sugar content. The identification of markers for soluble solids, carotenoids, and tocopherol, should be feasible and useful as the genetics of these important traits are studied. As codominant markers are more widely developed and maps are joined, the application of these genomic tools can have immediate application in marker-based breeding.

15.5.2 Mapping of Root and Leaf Anthocyanin Pigmentation

Anthocyanins, a subclass of water-soluble colored flavonoids, provide red, blue and purple pigmentation to different organs of a wide range of higher plants and confer a range of health-related benefits, including protection against oxidative stress, coronary heart disease, inflammation, some types of cancer and other age-related diseases. Besides, they play important roles, such as attraction of pollinators and seed dispersers and, given their antioxidant properties, protection against ultraviolet (UV) and high intensity light, drought, wounding, cold temperatures and phytopathogen attack. However, the health effects of dietary anthocyanins depend on amounts consumed and on their bioavailability. The bioavailability and excretion of anthocyanins and other polyphenols is highly influenced by their chemical structure, including the nature of the sugar conjugate, the phenolic aglycone and acylation. Carrot almost exclusively accumulates cyanidin derivatives. Two classes of genes are involved in anthocyanin biosynthesis: structural genes encoding the enzymes that directly participate in the formation of anthocyanin pigments, and regulatory genes that control the transcription of structural genes. Many of the genes involved in anthocyanins biosynthesis, both structural and regulatory, have been identified and characterized for several model species, such as petunia, snapdragon and maize, and both regulatory and structural genes vary widely across species. Consequently, information regarding the genetic control of anthocyanin biosynthesis may not be reliably extrapolated across species.

Anthocyanins may undergo a series of chemical modifications including glycosylation, acylation and methylation which are usually performed by glycosyl-, acyl and methyl-transferase enzymes, respectively. In purple carrots, acylation reduces anthocyanin bioavailability. But acylated anthocyanins are more stable than non-acylated anthocyanins, providing the former with an advantage for their use as colorants in the food industry.

Carrot can accumulate up to 17–18 mg/100 g fresh weight quantities of anthocyanins in its storage roots. Purple or "black" carrots accumulate almost exclusively cyanidin glycosides, both acylated and non-acylated, with five cyanidin pigments reported. Tissue distribution of root purple pigmentation varies greatly across carrot genotypes, ranging from a few pigmented cell layers in the periderm to a completely- and intensively colored root. Current breeding programs in purple carrot aim at increasing total anthocyanin content as well as achieving favorable ratios of acylated versus nonacylated anthocyanins depending on the end-market purpose, with a preference for high content of acylated anthocyanins for their use as food colorants, but conversely, high level of non-acylated forms for increasing bioavailability and nutraceutical value.

To date, a single dominant gene controlling anthocyanin accumulation in carrot roots, P1, has been described. Yildiz *et al.* (2013) genetically mapped P1 and several anthocyanin biosynthetic genes [five structural genes *i.e.* anthocyanin biosynthesis genes {FLS1 (flavonol synthase), F_3H, LDOX2, PAL3, and UFGT} and three regulatory genes *i.e.* anthocyanin transcription factors (DcEFR1, DcMYB3 and DcMYB5)], but no tight linkage was found between P1 and any of the candidate structural genes evaluated. P1 was mapped to chromosome 3 and of the eight anthocyanin biosynthesis genes, only F_3H and FLS1 were linked to P1. However, the five structural genes (CHS1, DFR1, F_3H, LDOX2, PAL3) were expressed differentially and in decreasing order among those listed when comparing solid-purple (purple in phloem and xylem), purple orange (purple phloem and orange xylem) and orange carrots, with transcript accumulation coinciding with anthocyanins accumulation. These studies suggested a coordinated regulatory control of anthocyanin biosynthesis in the carrot root, but the molecular and biochemical basis of genetic factors controlling the presence vs. absence of carrot root anthocyanin pigmentation remain unknown. Furthermore, very little is known about the genetics of anthocyanin accumulation in carrot organs and tissues other than the tap root. Variable expression of purple pigmentation in leaves, nodes and flowers was reported but not genetically characterized, and a simply inherited dominant locus controlling purple versus green pigmentation in petioles, P2, was also described by Simon (1996) and linkage between P1 and P2 was suggested but not mapped.

Since mapping of P1 was performed in a single genetic background, comparative analysis of the loci controlling anthocyanin pigmentation across diverse carrot backgrounds would be of interest, considering the broad variation for phenotype, genotype and geographical origin observed among purple carrot genetic stocks.

High density linkage maps constructed with informative sequence-based markers, such as SNPs and SSRs, are essential for fine mapping of QTL, comparative analysis of synteny, searching for candidate genes, facilitating genome sequence assembly, and for marker assisted breeding. The majority of the carrot maps constructed to date were unsaturated and used anonymous dominant markers, such as amplified fragment length polymorphisms and randomly amplified polymorphic DNAs, although in some cases a few codominant SSR or restriction fragment length polymorphisms markers were included. A DArT map with the highest map resolution (1.1 cM) achieved to date in carrot was recently reported by Grzebelus *et al.* (2014). However, the anonymous and dominant nature of DArT markers does not allow straight forward comparative map analysis or identification of candidate genes associated with QTL.

Cavagnaro *et al.* (2014) developed a high resolution gene-derived SNP-based linkage map of carrot with 894 markers covering 635.1 cM with a 1.3 cM map resolution, using a carrot population segregating for anthocyanin pigmentation in root and leaves. A total of 15 significant QTL for all anthocyanin pigments and for RTPE were mapped to six chromosomes. Eight QTL with the largest phenotypic effects mapped to two regions of chromosome 3 with co-localized QTL for several anthocyanin glycosides and for Root total pigment estimate (RTPE). They identified and mapped A single dominant gene conditioning anthocyanin acylation. Comparative mapping with two other carrot populations segregating for purple color indicated that carrot anthocyanin pigmentation is controlled by at least three genes, in contrast to monogenic control reported previously. Purple petiole pigmentation was conditioned by a single dominant gene that co-segregates with one of the genes conditioning root pigmentation. They found a similar genetic control for purple pigmentation in an unrelated (Chinese) genetic background. They mapped the genetic loci controlling both root and petiole purple pigmentation in three unrelated carrot backgrounds and discovered that root and petiole anthocyanin pigmentation in diverse genetic backgrounds mapped to not only P1 but also other loci. Bannoud *et al.* (2019) identified a MYB transcription factor, *DcMYB7*, and two cytochrome CYP450 genes with putative flavone synthase activity as candidates regulating both the presence/absence of pigmentation and the concentration of anthocyanins in the root phloem. They found concomitant expression patterns of *DcMYB7* and eight anthocyanin structural genes, suggesting that *DcMYB7* regulates transcription levels in the latter. Sharma *et al.* (2020) found that the simultaneous expression of *AmRosea1* and *AmDelila* transcription factors from snapdragon can activate the anthocyanin pathway in orange carrots, leading to the synthesis and accumulation of anthocyanins in the taproots.

Xu *et al.* (2014) studied anthocyanin biosynthesis at the molecular level in carrots and isolated six novel structural genes involved in anthocyanin biosynthesis among which, PAL3/PAL4, CA4H1, 4CL1, CHS1, CHI1, F_3H1, $F_3'H1$, DFR1, and LDOX1/LDOX2 may participate in anthocyanin biosynthesis in the taproots of purple carrot cultivars. CHS1, CHI1, F_3H1, $F_3'H1$, DFR1, and LDOX1/LDOX2 may lead to loss of light-independent anthocyanin production in orange and yellow carrots. Numerous structural genes were involved in anthocyanin production in the taproots of purple carrot cultivars and in the loss of anthocyanin production in non-purple carrots. Unexpressed or scarcely expressed genes in the taproots of non-purple carrot cultivars may be caused by the inactivation of regulator genes.

15.5.3 Mapping of Root Sugar Content

The *rs* allele has recently been found to be a naturally occurring knockout mutant of a carrot invertase isozyme which produces no functional enzyme (Yau and Simon, 2003). Vivek and Simon (1999) mapped *Rs* to one end of linkage group C, 8.1 cM away from an AFLP marker. Mapping results are consistent with inheritance data indicating that *Rs* is genetically unlinked to *Y2* and *P1* (Simon, 1996). Marker-assisted selection has been reported for *Rs* gene (Yau *et al.*, 2005).

Two sucrose synthase genes, *Susy1*Dc1* and *Susy1*Dc2* have been isolated from carrot. These two sucrose synthase genes differ markedly in their expression patterns. Northern analyses revealed that *Susy1*Dc1* is expressed in leaves, roots, flowers, and developing seeds, but *Susy1*Dc2* is expressed exclusively in carrot flowers. Several experiments have concluded that sucrose synthase appears to play an important role in metabolic activities associated with plant growth, however it is not significantly involved in sucrose partitioning in carrot.

A direct role of soluble acid invertase in sucrose partitioning and flavor development in the carrot root was established by Yau and Simon (2003) using a candidate gene approach. The dominant *Rs* allele conditions accumulation in carrot roots of the simple hexose sugars, fructose and glucose. Homozygous *rs/rs* individuals accumulate sucrose in the roots. The *Rs* locus is a key determinant of carrot root sugars and flavor. Towards characterization of the molecular basis of the *Rs* locus, Yau and Simon (2003) generated near isogenic *Rs/Rs* and *rs/rs* lines from the carrot inbred B4367. They used RT-PCR to detect transcripts of key enzymes in the sucrose/simple sugar pathway including sucrose synthase, extracellular acid invertase, and two isoforms (designated isoform I and II) of soluble acid invertase. They observed a perfect linkage between the genetically defined *Rs* locus and the observed mutant soluble acid invertase isoform II allele, providing strong evidence that the *Rs* locus encodes for soluble acid invertase isoform II.

15.5.4 Mapping of Vernalization Requirement and Fertility Restoration Genes

Temperate carrot is normally classified as a biennial species, requiring vernalization to induce flowering. Tropical cultivars adapted to warmer climates require less vernalization and can be classified as annual. Carrot roots quickly become very lignified after vernalization, even before the floral stalk elongates, and for that reason the initiation of flowering results in a complete loss of commercial value (Rubatzky *et al.*, 1999).

Most modern carrot cultivars are hybrids which rely upon cytoplasmic male-sterility for commercial production. One major gene controlling floral initiation and several genes restoring male fertility have been reported but none have been mapped. Alessandro *et al.* (2013) developed the first linkage map of carrot locating the genomic regions that control vernalization response and fertility restoration. Using an F_2 progeny, derived from the intercross between the annual cultivar 'Criolla INTA' and a petaloid male sterile biennial carrot evaluated over 2 years, both early flowering habit (named as *Vrn1*), and restoration of petaloid CMS (named as *Rf1*), were found to be dominant traits conditioned by single genes. On a map of 355 markers covering all 9 chromosomes with a total map length of 669 cM and an average marker-to-marker distance of 1.88 cM, *Vrn1* mapped to chromosome 2 with flanking markers at 0.70 and 0.46 cM, and Rf1 mapped to chromosome 9 with flanking markers at 4.38 and 1.12 cM. *Vrn1* and *Rf1* were the first two reproductive traits mapped in the carrot genome, and their map location and flanking markers provide valuable tools for studying traits important for carrot domestication and reproductive biology, as well as facilitating carrot breeding. Wohlfeiler *et al.* (2019) proposed a model of two genes (*Vrn-A* and *Vrn-B*) with three alleles controlling the vernalization requirement in carrot. Dominance of annuality is clear for both genes, with *A1* allele having an epistatic effect over *Vrn-B*. *Vrn-A* and *Vrn-B* interact generating different vernalization requirement levels.

15.5.5 Resistance Breeding

15.5.5.1 Biotic Stress

Le Clerc *et al.* (2009) reported the resistance to Alternaria leaf blight is polygenic and 3 QTL regions in a population of F_2: 3 progeny with the phenotypic variation explained by each QTL ranged from 10 to 23 per cent. Le Clerc *et al.* (2015) detected 11 QTLs for resistance to leaf blight (*Alternaria dauci*) in carrot through combining biparental and multiparental connected population analyses. Four QTLs were consistent across years, seven were detected within a single year. The heritabilities for both populations PC2 and PC3 were high (75 and 78 per cent, respectively), suggesting that the resistance of carrot to *A. dauci* was little affected by these environmental conditions, but the instability of QTL over years may be due to changing environmental conditions.

Takaichi and Oeda (2000) developed transgenic versions of the carrot cultivars Kurodagosun and Nantes Scarlet using *A. tumefaciens* to transfer a plasmid containing the human lysozyme under control of the constitutive CaMV 35S promoter. Two of the transgenic plants of 'Nantes Scarlet' displayed partial resistance to powdery mildew, and one was also partially resistant to Alternaria leaf blight. The increase in resistance in these lines was correlated with an increase in the production of the human lysozyme.

In warmer areas of world, *Meloidogyne incognita* is a major pest of carrot causing direct economic loss of the crop due to galling and forking of the carrot roots, rendering an infected carrot unmarketable. Considering the application of soil-applied nematicides as expensive and harmful to the environment, the genetic resistance provides an ideal solution to RKN control. The Mj-1 locus of carrot was discovered in a 'Brasilia' cultivar, line 'Br1252,' and it imparts resistance to *M. javanica* as a monogenic dominant trait (Simon *et al.*, 2000). The resistance in the cultivar Brasilia was associated with retarded nematode penetration, development, and egg production, and fast plant growth that culminated in a low nematode population density (Huang, 1986). Simon *et al.* (2000) also observed partial resistance to *M. incognita* in Br1252 and derivatives, in which determination of resistance could not be explained by Mj-1 alone. Recently, a second *M. javanica* resistance locus, Mj-2, was mapped in PI 652188 from China and found to be distantly linked to Mj-1, on chromosome 8 (Ali *et al.*, 2014).

Parsons *et al.* (2015) identified three diverse sources of resistance to *Meloidogyne incognita* in carrot, from Syria (HM), Europe (SFF) and South America (Br1091) and mapped the genetic resistance to *Meloidogyne incognita*. They developed two F_2 mapping populations using these parents, (Br1091 × HM1) and (SFF × HM2), as well as a segregating population derived from the self-pollination of a HM plant (HM3). Analysis revealed four QTLs conditioning resistance in Br1091 × HM1, three in SFF × HM2, and three in HM3. A consensus genetic map of the three populations revealed five non-overlapping QTLs for *M. Incognita* resistance, one each on carrot chromosomes 1, 2, 4, 8, and 9. One QTL was present in all three populations, in the same region of chromosome 8 as Mj-1 which imparts resistance to *M. javanica*.

The application of RNA interference (RNAi) to confer resistance to host plants engineered to express dsRNA and small interfering RNAs (siRNA) to target and silence specific nematode genes can be a promising strategy (Roderick *et al.*, 2018).

15.5.5.2 Abiotic Stress Tolerance

Osmotin and osmotin-like proteins belong to the PR-5 pathogenesis-related group of proteins and are induced in response to various types of biotic and abiotic stresses in several plant species. Ali *et al.* (2014) transformed carrot with a tobacco osmotin gene that encodes a protein lacking the vacuolar sorting motif that is composed of a 20-amino-acid sequence at the C-terminal end, under the control of the cauliflower mosaic virus 35S promoter, using Agrobacterium-mediated transformation. Under drought stress conditions, all transformants exhibited slower rates of wilting compared with the wild-type plants and recovered faster when the drought stress was alleviated. Transformants showed lower levels of hydrogen peroxide accumulation, reduced lipid peroxidation and electrolyte leakage, and higher leaf water content under drought stress.

Environmental stresses significantly influence the growth and development of plants. To overcome these stresses, higher plants have evolved a variety of complicated molecular mechanisms. Basic helix-loop-helix (bHLH) transcription factors (TFs) play important roles in plant growth and development in response to environmental stresses. Chen *et al.* (2015) identified 146 DcbHLH TFs from carrot, based on a genomic and transcriptomic database. Based on the previous classification system of *Arabidopsis thaliana*, the DcbHLH TFs were divided into 17 subfamilies. Multiple sequence alignment of bHLH conserved domains indicated that 109 DcbHLH proteins were bound to DNA (83 proteins were E-box binders and 52 DcbHLHs were G-box binders). From evolutionary analysis, bHLH TFs selected from plants, metazoans, and fungi demonstrated that the number of bHLH TFs increased during the evolution of these species. The expression profiles of eight DcbHLH genes from subfamily 15 showed differences in three tissues and four abiotic stresses in two carrot cultivars, Junchuanhong and Kurodagosun.

AP2/ERF is a large transcription factor family that regulates plant physiological processes, such as plant development and stress response. Li *et al.* (2015) identified 267 putative AP2/ERF factors from the whole-genome sequence of carrot. These AP2/ERF proteins were phylogenetically clustered into five subfamilies based on their similarity to the amino acid sequences from *Arabidopsis*. The distribution and comparative genome analysis of the AP2/ERF factors among plants showed the AP2/ERF factors had expansion during the evolutionary process, and the AP2 domain was highly conserved during evolution. They identified 60 orthologous and 145 coorthologous AP2/ERF gene pairs between carrot and *Arabidopsis* constructed the interaction network of orthologous genes. The expression patterns of eight AP2/ERF family genes from each subfamily (DREB, ERF, AP2, and RAV) were related to abiotic stresses. Yeast one-hybrid and α-galactosidase activity assays confirmed the DRE and GCC box-binding activities of DREB subfamily genes.

15.6.0 Transcriptome Analysis

The carrot root undergoes a notable size change during its growth process, but only limited information regarding carrot root development is available. Transcriptome sequencing has been a useful method of identifying novel transcripts and splice isoforms as well as performing expression analysis. Numerous sequences and plant molecular information have been obtained using transcriptome sequencing in several crops.

The molecular mechanisms underlying hormone-mediated root growth of carrot have not been illustrated. Wang *et al.* (2015) conducted transcriptome analysis of the carrot root, and extensively investigated the genes involved in hormone biosynthesis and signaling pathways to help elucidate the hormonal control of root growth based on digital gene expression. They collected carrot root samples from four developmental stages (Stage 1, 22 days old; Stage 2, 40 days old; Stage 3, 56 days old; and Stage 4, 95 days old), and performed transcriptome sequencing to understand the molecular functions of plant hormones in carrot root growth. From the transcriptome a total of 160,227 transcripts were generated, which were assembled into 32,716 unigenes with an average length of 1,453 bp. A total of 4,818 unigenes were found to be differentially expressed between the four developmental stages. They identified 87 hormone-related differentially expressed genes were identified. They suggested that plant hormones may regulate carrot root growth in a phase dependent manner.

In plants, the homeobox gene family is represented by transcription factors, which has been implicated in secondary growth, early embryo patterning, and hormone response pathways. Que *et al.* (2018) identified a total of 130 homeobox family genes, classified into 14 subgroups, in the carrot genome. Whole genome and proximal duplication participated in the homeobox gene family expansion in carrot. Purifying selection also contributed to the evolution of carrot homeobox genes. The members of the WOX and KNOX subfamilies are likely implicated in carrot root development.

15.7.0 Genetic Transformation

Genetic transformation of carrot has been carried out employing various methods such as *Agrobacterium* mediated transformation with *Agrobacterium tumefaciens* and *A. rhizogenes*, using electrical field, electroporation, microinjection and protoplast transformation.

Hardegger and Sturm (1998) described an efficient protocol for the transformation of carrot by *A. tumefaciens*. The binary vector contained the GUS marker gene (*uidA*), driven by the CaMV 35S promoter and the *nptII* marker gene. The highest T-DNA transfer rates were obtained by cocultivating bacteria with hypocotyl segments of dark-grown seedlings on solidified B5 medium containing NAA and BA.

Malik *et al.* (1999) has demonstrated the role of one small heat shock protein encoding *Hsp 17.7* in the ability of carrot cells and plants to survive thermal stress. By manipulating this gene, resistance to temperature stress can be induced. Direct gene transfer into carrot may help in developing herbicide resistance (Droge *et al.*, 1992), insect resistance (Murray *et al.*, 1991; Adang *et al.*, 1993) and pathogen resistance (Punja and Raharjo, 1996; Dolgov *et al.*, 1999).

Adang *et al.* (1993) designed a *B. thuringiensis cryIIIA* delta-endotoxin gene for optimal expression in plants. The modified *cry* gene has the codon usage pattern of an average dicot gene and does not contain AT-rich nucleotide sequences typical of native *Bt cry* gene. Modified *cryIIIA* gene expression was compared with native gene expression by electroporation of carrot protoplasts. CryIIIA-specific RNA

Genetic Transformation in Carrot

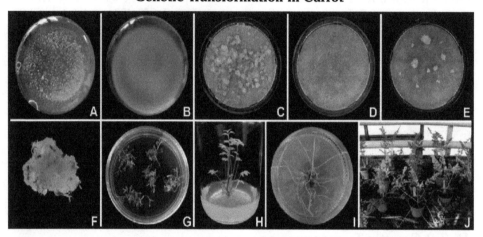

***Agrbacterium tumefaciens*-Mediated Transformation of Carrot Cell Suspension and the Development of Transgenic Carrot Plants.**

One-week-old initial cell suspension derived from callus (A), established 12 week-old cell suspension (B), growth of the non-transformed callus on a solid medium (C), died nontransformed cells on the kanamycin selection medium (D), selection of the transgenic calli on the kanamycin enriched medium (E), shoot regeneration from the kanamycin resistant callus (F), micropropagation (G), shoot development (H), rooting (I) and transgenic plants in the glasshouse (J) (Baranski, 2008).

and protein were detected in carrot protoplasts only after electroporation with the rebuilt gene. There was good correlation between insect control and the levels of delta-endotoxin RNA and protein.

Transgenic carrot plants expressing a thaumatin-like protein from rice showed significantly enhanced tolerance to *S. sclerotiorum* when detached petioles and leaflets were inoculated under controlled environmental conditions (Punja and Chen, 2004). Punja and Raharjo (1996) transformed carrot cultivars, Nanco and Golden State, with 2 chitinase genes from tobacco and found that the rate and extent of lesion develop-ment due to inoculation with *Botrytis cinerea*, *Rhizoctonia solani* and *Sclerotium rolfsii* were significantly lower in the transgenic carrot than in non-transgenic carrot.

Meyer *et al.* (1999) isolated a gene encoding an antifreeze protein from carrot using sequence information derived from the purified protein. The carrot antifreeze protein AFP is highly similar to the polygalacturonase inhibitor protein (PGIP) family of apoplastic plant leucine-rich repeat (LRR) proteins. Low temperatures rapidly induce expression of the *afp* gene. Furthermore, expression of the *afp* gene in transonic *Arabidopsis thaliana* plants leads to an accumulation of antifreeze activity. It is concluded that this new type of plant antifreeze protein has been recently evolved from PGIPs.

The tissues transformed with wild-type or cyt-Ti-plasmids not only synthesized larger amounts of IAA but also converted a large amount of free IAA to conjugated IAA than non-transformed tissues. Transformation by wild-type *Agrobacterium tumefaciens* results in the insertion of IAA synthesis genes in the genome, one enco-ding tryptophan monooxidase and another encoding amidohydrolase. This allows transformed cells to synthesize IAA along the prokaryotic pathway *i.e.* first tryptophan monooxidase converts tryptophan into naphthylacetamide (NAAm) and then amidohydrolase catalyzes the NAAm conversion into IAA (Gamburg *et al.*, 1999).

One of the drawbacks in inducing the synthesis of ketocarotenoids in transgenic plants is the accompanying decrease in α-carotene level, especially in transgenic plants overexpressing only a α-carotene ketolase. In an attempt to overcome this drawback, Ahn *et al.* (2012) isolated a α-carotene ketolase (*HpBkt*) cDNA from *Haematococcus pluvialis* and generated transgenic carrot plants overexpressing *HpBkt* cDNA under the control of the *ibAGP1* promoter and its transit peptide sequence. Overexpression of *HpBkt* caused an increase in the transcript levels of the endogenous carotenogenic genes, including phytoene synthase 1 (*PSY1*), phytoene synthase 2 (*PSY2*), lycopene α-cyclase 1 (*LCYB1*), and α-carotene hydroxylase 1 (*CHXB1*), which resulted in elevated α-carotene levels in the *HpBkt*-transgenic plants (range 1.3- to 2.5-fold) compared to the wild-type plants. Thus, *HpBkt*- overexpressing carrot plants under the control of the *ibAGP1* promoter and its transit peptide are capable of both newly synthesizing ketocarotenoids and enhancing their α-carotene level.

15.8.0 *In vitro* Conservation

Many workers have reported various techniques of developing artificial seeds (encapsulated) in carrots (Shigeta, 1995; Wake *et al.*, 1992, 1995). Liu *et al.* (1992)

described a method of producing dry type of artificial seeds in carrot. In this method, somatic embryos from cell suspension cultures were singly encapsulated with alginic acid. The capsules were then dehydrated in an air stream to reduce the moisture content by more than 80 per cent. When rehydrated, 97 per cent of the capsules dried to 12 per cent moisture content germinated. Treating the embryos with abscisic acid (ABA) for 10 days before encapsulation enhanced germination of capsules. Tsuji *et al.* (1993) obtained hardness of capsules by suspending the somatic embryos in 1 per cent sodium alginate mixture and soaking in 1 per cent $CaCl_2$ for 5 minutes, which was associated with high percentage of germination, optimum plant growth and improved ease of hardening of plantlets. Dupuis *et al.* (1994) developed pharmaceutical type capsules, composed of polyvinyl chloride or polyvinyl acetate, to allow nutrient supply and subsequent development of somatic embryos.

15.9.0 Biopharmaceuticals Production

Since carrot produces edible tissues or cells, it is an attractive source of biomass for oral vaccine formulation requiring minimum processing, *e.g.*, freeze drying and packing in gelatin capsules. In 2012, the introduction into the market of the first plant-made pharmaceutical product approved for human use, consisting of the enzyme glucocerebrosidase for Gaucher's disease treatment produced in carrot (*Daucus carota* L.) (Protalix. Retrieved from http://www.protalix.com/index.asp.).

Gaucher's disease (GD) is a lysosomal storage genetic disorder caused by the lack of the glucocerebrosidase (GCD) enzyme. Enzyme replacement therapy (ERT) is the current treatment for this disease, which results in very high costs since the conventional recombinant enzyme, named CerezymeR, is obtained in Chinese-hamster ovary (CHO) cells and subsequently modified *in vitro* to expose the mannose residues that are required for recognition by macrophage mannose receptors once administered to patients (Grabowski *et al.*, 1995).

Three α-glucocerebrosidase products have been developed using distinct expression platforms: (i) Imiglucerase (Genzyme Corporation, a Sanofi company), produced in CHO cells and modified in vitro with exoglycosidases to expose mannose residues; (ii) Velaglucerase alfa (Shire Pharmaceuticals Inc.), produced in a human fibroblast carcinoma cell line in the presence of a mannosidase I inhibitor (kifunensine) to obtain proper glycosylation profiles; (iii) Taliglucerase alfa (Protalix Biotherapeutics), produced in carrot cells (Pr°CellExR platform) in which terminal paucimannosidic type N-glycans are attained by targeting to the plant storage vacuoles where the terminal residues are removed.

Taliglucerase alfa carrot-based production platform was developed using the human gene modified to replace the native signal peptide by that of the *Arabidopsis thaliana* basic endochitinase gene to facilitate translocation into the endoplasmic reticulum (ER).

15.10.0 Carrot Genome Sequencing

Daucus carota is a typical biannual diploid (2n = 2x = 18) outcrossing species with a relatively small genome estimated as 473Mb. Iorizzo *et al.* (2016) sequenced the genome of a double-haploid Nantes-type orange carrot using Illumina and BAC end sequences, reporting a genome assembly representing 89.1 per cent of the 473-Mb estimated genome size. The assembly statistics suggest that this is a high-quality genome, comparable to other recently reported plant genomes of similar size. The carrot genome assembly consists of 46 per cent repetitive sequences and 32,113 gene models. Iorizzo *et al.* (2016) also resequenced 35 accessions representing the existing variability in carrot, including wild and cultivated accessions of both eastern and western origin. The observed phylogeny of these accessions is in agreement with placing the primary center of carrot domestication in Central Asia. Iorizzo *et al.* (2016) described several chromosomal regions showing differentiation between cultivated eastern varieties and wild accessions, suggesting that these regions were strongly selected for during the process of domestication. Interestingly, two of these regions overlap quantitative trait loci (QTLs) for carotenoid accumulation, a major domestication trait in carrot.

Wang *et al.* (2018) published the genome sequence of DC-27, an inbred line derived from 'Kurodagosun', a major carrot variety in Japan and China. They identified a total of 31,891 predicted genes. These assembled sequences provide candidate genes involved in biological processes including stress response and carotenoid biosynthesis. Genomic sequences corresponding to 371.6 Mb was less than 473 Mb, the estimated genome size. The 'DC-27' genome sequence data will advance our knowledge on the biological research and breeding of carrot as well as other Apiaceae and also provide a new resource to explore the evolution of other higher plants.

15.11.0 Carrot Genome Editing

Clustered Regularly Interspaced Short Palindromic Repeats (CRISPR)/CRISPR-associated (Cas9) is a powerful genome editing tool used for precise genome editing in many plant species. Klimek-Chodacka *et al.* (2018) reported application of the CRISPR/Cas9 system for efficient targeted mutagenesis of the carrot genome. They tested multiplexing CRISPR/Cas9 vectors expressing two single-guide RNA (gRNAs) targeting the carrot flavanone-3-hydroxylase (F_3H) gene for blockage of the anthocyanin biosynthesis in a model purple-colored callus using Agrobacterium-mediated genetic transformation. This approach allowed fast and visual comparison of three codon-optimized Cas9 genes and revealed that the most efficient one in generating F_3H mutants was the Arabidopsis codon-optimized AteCas9 gene with up to 90 per cent efficiency. Knockout of F_3H gene resulted in the discoloration of calli, validating the functional role of this gene in the anthocyanin biosynthesis in carrot as well as providing a visual marker for screening successfully edited events. Most resulting mutations were small Indels, but long chromosome fragment deletions of 116–119 nt were also generated with simultaneous cleavage mediated by two gRNAs. Their results demonstrate successful site-directed mutagenesis in carrot with CRISPR/Cas9 and the usefulness of a model callus culture to validate genome editing systems.

Four sgRNA expression cassettes, individually driven by four different promoters and assembled in a single CRISPR/Cas9 vector, were transformed into carrots using Agrobacterium-mediated genetic transformation by Xu *et al.* (2019). The have chosen four sites of *DcPDS* and *DcMYB113*-like genes as targets. Knockout of *DcPDS* in orange carrot 'Kurodagosun' resulted in the generation of albino carrot plantlets, with about 35.3 per cent editing efficiency. They also successfully edited *DcMYB113*-like in purple carrot 'Deep purple', resulting in purple depigmented carrot plants, with about 36.4 per cent rate of mutation. Sequencing analyses showed that insertion, deletion, and substitution occurred in the target sites, generating heterozygous, biallelic, and chimeric mutations. The highest efficiency of mutagenesis was observed in the sites targeted by AtU6-29-driven sgRNAs in both *DcPDS*- and *DcMYB113-like*-knockout T0 plants, which always induced double-strand breaks in the target sites. Their study proved that CRISPR/Cas9 system could be used for generating stable gene-editing carrot plants.

16.0 References

Abo-Sedera, F.A. and Eid, S.M.M. (1992) *Asiut J. Agric. Sci.*, **23**: 209-225.

Abrol, D.P. (1997) *Insect Environ.*, **3**: 61.

Adang, M.J., Brody, M.S., Cardineau, G., Eagan, N., Roush, R.T., Shewmaker, C.K., Jones, A., Oakes, J.V. and McBride, K.E. (1993) *Plant Mol. Biol.*, **21**: 1131-1145.

Afek, U., Orenstein, J. and Nuriel, E. (1999) *Crop Protect.*, **18**: 639-642.

Ahn, M.J., Seol Ah Noh, Sun-Hwa Ha, Kyoungwhan Back, Shin Woo Lee and Jung Myung Bae (2012) *Plant Biotechnol. Rep.*, **6:** 133–140.

Aki, S. and Watanabe, S. (1961) *J. Jap. Soc. Hort. Sci.*, **30**: 311-317.

Aki, S. and Watanabe, S. (1962) *Tech. Bull. Fac. Agric.*, Kagawa, pp. 14-19.

Alabran, D.M. and Mabrouk, A.F. (1973) *J. Agric. Food Chem.*, 21: 205–208.

Alekseeva, A.M. and Kutsenko, E.M. (1976) *Nauchnye Trudy Voronezh. S-kh. Institut*, **83**: 109-116.

Alekseeva, A.M. and Rasskozov, M.A. (1976) *Nauchnye Turdy Voronezh, S-kh. Institut*, **85**: 5-13.

Alessandro, M.S., Galmarini, C.R., Iorizzo, M. and Simon, P.W. (2013) *Theor. Appl. Genet.*, **126:** 415–423.

Ali, A., Mathews, W.C., Cavagnaro, P.F., Iorizzo, M., Roberts, P.A. and Simon, P.W. (2014) *J. Hered.*, **105**: 288–291.

Ali, A., Rathore, K. and Crosby, K. (2014) *In Vitro Cell. Dev. Biol. Plant*, **50**: 299–306.

Andersen, S.B., Christiansen, I. and Farestveit, B. (1990) In: *Biotechnology in Agriculture and Forestry* I. *Haploids in crop improvement* (ed. Y.P.S. Bajaj), vol 12. Springer, Berlin, pp. 393–402.

Angell, F.F. and Gabelman, W.H. (1968) *J. Amer. Soc. Hort. Sci.*, **93**: 434-437.

Anonymous (1971) *Indian Minimum Seed Certification Standards*, Central Seed Committee, Min. Food Agric. Comm. Dev. Co-op., New Delhi.

Anonymous (1977-78) *Annual Report*, Vegetable Crops Breeding Station, Avignon-Montfavet, France.

Anonymous (1991) *Kartofel' iOvoshchi*, No. 5, p. 13.

Anonymous (2018) Indian Horticulture Database, National Horticulture Board, Gurugram, India.

Arbizu, C.I., Tas, P.M., Simon, P.W. and Spooner, D.M. (2017) *Crop Sci.*, **57**: 2645–2653.

Armolaitiene, J. (1999) *SodininkysteirDarzininkyste*, **18**: 57-61.

Arscott, S.A. and Tanumihardjo, S.A. (2010) *Comprehensive Reviews in Food Science and Food Safety*, **9**: 223-239.

Augustinussen, E. (1975) *Tidsskrift for Planteavl.*, **79**: 326-336.

Bach, I.C., Madsen, O.M. and Olesen, A. (1997) *J. Appl. Genet.*, **38A**: 178-185.

Badawi, M.A. (1978) *Res. Bull. Ain Shams University, Faculty of Agriculture.*, **903**: 15.

Bajpai, M.R., Dhakar, L.L. and Jhorar, L.R. (1972) *Proc. 3rd Int. Symp. Subtrop. And Trop. Hort.*, Bangalore, pp. 141-142.

Baker, L.R. (1978a) *BiuletynWarzywniczy*, **22**: 127-136.

Baker, L.R. (1978b) *Res. Rept. Michigan State Univ. Agric. Exp. Sta.*, East Lansing, No. 359, pp. 2-3.

Baker, L.R. (1978c) *Res. rept. Michigan State. Univ. Agric. Exp. Sta.*, East Lansing, No. 359, p. 6.

Baker, L.R. and Valk, M. (1978) *Res. Rept. Michigan State. Univ. Agric. Exp. Sta.*, East Lansing, No. 354, pp. 3-4.

Banga, O. (1962) *Main Types of the Western Carotene Carrot and their Origin*, W.E. TjeenkWillink, Zwolle, Netherlands.

Banga, O. (1963a) *Genet. Agron.*, **17**: 357-370.

Banga, O. (1963b) *Main Types of Western Carotene Carrot and their Origin*, W.E. Tjeenkwillink Zwolle, Netherlands.

Banga, O. (1976) *Evolution of Crop Plants* (Ed. N.W. simmonds), Longman, New York.

Bannoud, F., Ellison, S., Paolinelli, M., Horejsi, T., Senalik, D., Fanzone, M., Iorizzo, M., Simon, P.W. and Cavagnaro, P.F. (2019) *Theor. Appl. Genet.*, **132**: 2485-2507.

Baranski, R. (2008) *Transg. Plant J.*, 2: 18-38.

Baranski, R., Maksylewicz-Kaul, A., Nothnagel, T., Cavagnaro, P.F., Simon, P.W. and Grzebelus, D. (2012) *Genet. Resour. Crop Evol.*, **59**: 163–170.

Barba-Espin, G., Glied, S., Crocoll, C., Dzhanfezova, T., Joernsgaard, B., Okkels, F., Lutken, H. and Muller, R. (2017) *BMC Plant Biol.*, **17**: 70.

Bathkal, B.G. and Patil, C.B. (1968) *Poona Agric. Coll. Mag.*, **58**: 13-17.

Batra, B.R. and Kalloo, G. (1990) *Veg. Sci.*, **17**: 127-139.

Batra, B.R. and Kalloo, G. (1991) *Veg. Sci.*, **18**: 1-10.

Bell, C.E., Boutwell, B.E., Ogbuchiekwe, E.J. and McGiffen, M.E. Jr (2000) *HortScience,* **35:** 1089–1091.

Benard, D. and Punja, Z.K. (1995) *Can. J. Plant Pathol.,* **17:** 31–45.

Berg, L. Van den and Lentz, C.P. (1973) *J. Amer. Soc. Hort. Sci.,* **98:** 129-132.

Bertolina, P. and Restaino, F. (1978) *Frutticoltura,* **40:** 40-51.

Bhagchandani, P.M. and Choudhury, B. (1971) *Prog. Hort.,* 2: 65-75.

Biacs, P.A., Daood, H.G. and Kadar, I. (1995) *J. Agric. Food Chem.,* **43:** 589-591.

Bianchini, M.R. and Eyherabide, G.A. (1999) *Revista-de-la Facultad-de-Agronomia-Universidad-de-Buenos-Aires,* **19:** 69-74.

Bishnoi, C. and Dhaliwal, N.S. (2017) *Environ. Ecol.,* **35:** 3556-3558.

Bohart, D.K., Bohart, G.E. and Nye, W.P. (1960) *Bull. Utah Agric. Exp. Sta.,* **419:** 16.

Boiteux, L.S., Della Vecchia, P.T. and Reifschneider, F.J.B. (1993) *Plant Breed.,* **110:** 165–167.

Bonnet, A. (1978) *BiuletynWarzywniczy,* **22:** 147-150.

Bonnet, A. (1983) *Agronomie,* **3:** 33-37.

Borthwick, H.A. (1931) *Bot. Gaz.* (Chicago), **92:** 23-44.

Braak, J.P. and Kho, Y.O. (1958) *Euphytica,* **7:** 131-139.

Bradeen, J.M. and Simon, P.W. (1998) *Theor. Appl. Genet.,* **97:** 960-967.

Brar, J.S. and Nandpuri, K.S. (1972) Cultivation of root crops. Punjab Agricultural University, Ludhiana.

Buike, F. and Lepse, L. (2000) *SodininkysteirDarzininkyste,* **19:** 103-112.

Buishand, J.G. and Gabelman, W.H. (1979) *Euphytica,* **28:** 611-632.

Bujdoso, G. and Hrasko, I. (1990) *ZoldsegtermesztesiKutatoIntezetBulletinje,* **23:** 109-116.

Burdine, H.W. and Hall, C.B. (1976) *Proc. State Hort. Soc.,* **89:** 120-125.

Burkill, I.H. (1935) *A Dictionary of the Economic Plants of Malay Peninsula,* Grown Agents, London.

Callens, D. and Callewaert, D. (1999) *Proeftuinnieuws,* **9:** 17-18.

Cantliffe, D.J. and Elballa, M. (1994) *Proc. Florida State Hart. Soc.,* **107:** 121-128.

Cardoso, A.I.I. (2000) *Bragantia,* **59:** 77-81.

Cardoso, A.I.I., Della-Vecchia, P.T., Takazaki, P.E. and Terenciano, A. (1992) *Hortic. Bras.,* **10:** 44-45.

Carlton, B.C. and Peterson, C.E. (1963) *Proc. Amer. Soc. Hort. Sci.,* **82:** 332-340.

Cavagnaro, P.F., Iorizzo, M., Yildiz, M., Senalik, D., Parsons, J., Ellison, S. and Simon, P.W. (2014) *BMC Genomics,* **15:** 1118.

Cebulak, T. and Sady, W. (2000) *Folia Hort.,* **12:** 77-84.

Cernaianu, A., Amarintei, A., Wekerle, I. and Martin, I. (1975) *Stiintifice Institutal de Cercetaripentru Valorificarea Legumelorsi Fructelor*, **6**: 99-107.

Chandel, K. S., Singh, A. K. and Rattan, R. S. (1994) *Indian J. Genet.*, **54**: 389-394.

Chandel, K.S. and Rattan, R.S. (1988) *Veg. Sci.*, 15: 31-37.

Chandel, K.S., Singh, A.K. and Rattan, R.S. (1997) *South Indian Hort.*, **45**: 22-25.

Charchar, J.M. and Vieira, J.V. (1994) *Hortic. Bras.*, **12**: 144-148.

Chen, B.H., Peng, H.Y. and Chen, H.E. (1995) *J. Agric. Food Chem.* **43**: 1912–1918.

Chen, Yi-Yun, Meng-Yao Li, Xue-Jun Wu, Ying Huang, Jing Ma and Ai-Sheng Xiong (2015) *Mol. Breed.*, **35**: 125.

Chopra, R. N. (1933) *Indigenous Drugs of India*. The Art Press, Calcutta.

Choudhuri, B., Som, M. G. and Das, A. K. (1972) *Proc. 3rd Int. Symp. Subtrop and Trop. Hort.*, Bangalore, p. 150.

Christensen, L.P. and Kreutzmann, S. (2007) *J. Sep. Sci.*, **30**: 483–490.

Christianson, C.E., Jones, S.S. and du Toit, L.J. (2015) *HortScience*, **50**: 341–350.

Corbineau, F., Picard, M.A, Bonnet, A. and Come, D. (1995) *Seed Sci. Res.*, **5**: 19-135.

Czepa, A. and Hofmann, T. (2004) *J. Agric. Food Chem.*, **52**: 4508–4514.

da Silva, L.M., Basilio, S. deA., Silva Jr., R.L., Benett, K.S.S. and Benett, C.G.S. (2017) *Revista de Agricultura Neotropical, Cassilandia-MS*, **4**: 69-76.

Dame, A., Bielau, M., Stein, M. and Weit, E. (1988) *Arch. Gartenbau. Berlin*, **36**: 345–352.

Derbyshire, D.M. and Crisp, A.F. (1978) *Exp. Hort.*, **30**: 23-28.

Deuber, R., Forster, R. and Singnori, L. H. (1976) *Resmos XI SeminarioBrasileiro de Herbicidas e ErvasDaninhas Londrina*, pp. 21-22.

Dickson, M.H. and Peterson, C.E. (1960) *Canad. J. Plant Sci.*, **40**: 468-473.

Dickson, M.H., Rieger, B. and Peterson, C.C. (1961) *Proc. Amer. Soc. Hort. Sci.*, **77**: 401-405.

Digole, P.T. and Shinde, N.N. (1990) *Veg. Sci.*, **17**: 20-24.

Dolgov, S.V., Lebedev, V.G., Firsov, A.P., Taran, S.A. and Tjukavin, G.B. (1999) In: *Genetics and Breeding for Crop Quality and Resistance, Proc. XV Eucarpia Cong.*, Viterbo, Italy, September 20-25, 1998, pp. 111-118 and 165-172.

Dowker, B. D. and Jackson, J. C. (1975) *Ann. Appl. Biol.*, **79**: 361-65.

Draglano, S. (1978) *Forskningog. ForsokiLandbruket*, **29**: 161-74.

Drlik, J. and Rogl, J. (1992) *Zahradnictvi*, **19**: 39-46.

Droge, W., Broer, I. and Puhler, A. (1992) *Planta*, **187**: 142-151.

Drozdov, N. A. (1962) *Zemledelie,Mosk*, **25**: 56-57.

Ducos, J. P., Bollon, H. and Pettard, V. (1993) *Applied Microbiol. Biotech.*, **39**: 465-470.

Dupuis, J. M., Roffat, C., Derose, R. T. and Molle, F. (1994) *Biotechnology*, **12**: 385-389.

Eid, S.M.M. and Abo-Sedera, F.A. (1992) *Assiut J. Agric. Sci.*, **23**: 227-242.

Eisa, H.M. and Wallace, D.H. (1969) *J. Amer. Soc. Hort. Sci.*, **94**: 545-548.

Elballa, M. M. A. and Cantliffe, D. J. (1996) *J. Amer. Soc. Hort. Sci.*, **121**: 1076-1081.

Elballa, M.M.A. and Cantliffe, D.J. (1997) *J. Veg. Crop Production*, **3**: 29-41.

Elisa, H. M. and Wallace, D. H. (1969) *J. Amer. Soc. Hort. Sci.*, **94**: 657-649.

Ellis, P.R, Saw, P.L. and Crowther, T.C. (1991) *Ann. Appl. Biol.*, **119**: 349-357.

Ellis, P.R. and Hardman, J.A. (1992) In: Vegetable Crop Pests (ed. R.G. Mckinlay), CRC Press, Boca Raton, Florida, pp. 327-378.

Evers, A.M. (1988) *J. Agric. Sci. Finland*, **60**: 135-152.

Evers, A.M. (1989a) *J. Agric. Sci. Finland*, **61**: 7-14.

Evers, A.M. (1989b) *J. Agric. Sci. Finland*, **61**: 113-122.

Evers, A.M. (1989c) *J. Agric. Sci. Finland*, **61**: 123-134.

Evers, A.M., Tuuri, H., Hagg, M., Plaami, S. Hakkinen, U. and Talvitie, H. (1997) *Plant Foods for Human Nutrition*, **51**: 283-294.

Farghali, M.A. and Hussein, H.A. (1994) *Assiut J. Agric. Sci.*, **25**: 89-97.

Finch-Savage, W. E. (1990) *Acta Hort.*, **267**: 209-216.

Freeman, R.E. and Simon, P.W. (1983) *J. Amer. Soc. Hort. Sci.*, **108**: 50-54.

Frese, L. (1983) *J. Plant Dis. Protect.*, **81**: 396-403.

Frison, E.A. and Serwinski, J. (1995) Directory of European Institutions holding Crop Genetic Resources Collections. IBPGRI, Rome.

Fritz, D. and Habben, J. (1977) *Gartenbauwissenschaft*, **42**: 185-190.

Frohlich, H., Schroder, E. and Stopperka, F. (1971) *DtscheGartenb.*, **18**: 69-72.

Gabelman, W.H., Goldman, I.L., and Breitbach, D.W. (1994) *J. Amer. Soc. Hort. Sci.*, **119**: 1293-1297.

Galmarini, C., Borgo, R., Gaviola, J.C. and Tizio, R. (1995b) *Horticultura Argentina*, **14**: 74-86.

Galmarini, C., Gaviola, J.C. and Tizio, R. (1995a) *Horticultura Argentina*, **14**: 87-89.

Gamburg, K.Z., Gamanetz, L.V., Markova, T.A. and Barykova, O.M. (1999) *Russian J. Plant Physiol.*, **46**: 745-748.

Geminiani, E., Bucchi, R. and Rapparini, G. (2008) *Giornate Fitopatologiche*, **1**: 455–462.

Gill, H. S., Lakhanpal, K. D. and Sharma, S. R. (1987) *Indian Hort.*,**31**: 25-26.

Gill, H.S. and Kataria, A.S. (1974). *Curr. Sci.*, **43**: 184-185.

Gill. S. S., Singh, H. and Singh, J. (1981) *Veg. Sci.*, **8**: 6-11.

Globerson, D. (1972) *J. Hort. Sci.*, **47**: 69-72.

Gnaegi, F., Perraudin, G. and Schopfer, J. F. (1977) *Revne Suisse de Viticulture d' Arboriculture et d'Horticulture*, **9**: 173-177.

Go´recka, K., Krzyzanowska, D. and Go´recki, R. (2005) *J. Appl. Genet.*, **46**: 265–269.

Goldman, H. (1996) *HortScience*, **31**: 882-883.

Goldman, I.L. and Breitbach, D.N. (1996) *J. Hered.*, **87**: 380-382.

Goodliffe, J. P. and Heale, J. B. (1975) *Ann. Appl. Biol.*, **80**: 243-246.

Gorecka, K., Krzyzanowska, D., Kiszczak, W. and Kowalska U. (2009) *Acta Physiol. Plant*, **31**: 1139–1145.

Gormley, T. R., O'Riordain, F. and Prendiville, M. D. (1971) *J. Food Technol.*,**6**: 393-402.

Grabowski, G.A., Barton, N.W., Pastores, G., Dambrosia, J.M., Banerjee, T.K., McKee, M. A. (1995) *Annals of Internal Medicine*, **122**: 33–39.

Gray, D. (1981) *Acta Hort.*, **11**: 159-165.

Gray, D. Steckel, J. R. A., Drew, R. L. K. and Keefe, P. D. (1991) *Seed Sci. Tech.*, **19**: 655-664.

Gray, D., Steckel, J. R. A. and Ward, J. A. (1983) *J. Hort. Sci.*, **58**: 83-90.

Gray, D., Steckel, R. A., Dearman, J. and Broacklehurst, P. A. (1988) *Ann. Appl. Biol.*, **122**: 367-376.

Grzebelus, D., Iorizzo, M., Senalik, D., Ellison, S., Cavagnaro, P, Macko-Podgorni, A., Heller-Uszynska, K., Kilian, A., Nothnagel, T., Allender, C., Simon, P.W. and Baranski, R. (2014) *Mol. Breed.*, **33**: 625–637.

Grzebelus, D., Szklarczyk, M. and Michalik, B. (1997) *J. Appl. Genet.*, **38A**: 33-41.

Grzebelus, E., Kruk, M., Macko-Podgorni, A. and Grzebelus, D. (2013) *Plant Cell Tissue Organ Cult.*, **115**: 209–222.

Guo, X.C., Wu, B.J., Jiang, H. and Zheng, G.C. (1995) *Acta Bot. Sin.*, **37** ; 339-345.

Gupta, U. C. and Chipman, E. W. (1976) *Plant Soil*, **44**: 559-566.

Gupta, U.C., Leblane, P.V. and Chipman, E.W. (1990) *Canadian J. Soil Sci.*, **4**: 717-721.

Gurr, R. E. and Surina, A. F. (1980) *SbornikNauchStateiKaragandGos. S. Kh. Opyt. St.*, **6**: 101-105.

Habben, J. (1973) *Landwirtschaftlicheforschung*, **26**: 156-172.

Haland, A. (1975) *Gastnearyrket*, **65**: 280, 282, 284.

Hardegger, M. and Sturm, A. (1998) *Mol. Breed.*, **4**: 119-129.

Hassan, M. N. M., Adbel-Ati, Y. Y. and Farrag., M. M. (1992) *Assint J. Agric. Sci.*, **23**: 121-135.

Hauptmann, R.M., Smith, A.G., Kishore, G.M., Widholm, J.M. and Cioppa, G. della (1988) *Mol. Gen. Genet.*, **211**: 357-363.

Heywood, V.H. (1983) *Israel J. Bot.*, 32: 51-65.

Hiller, L. K. and Kelly, W. C. (1979) *J. Amer. Soc. Hort. Sci.*, **104**: 253-257.

Hipp, B. W. (1978) *HortScience*, **13**: 43-44.

Hodgson, R.A.J. and Rose, R.J. (1984) *J. Plant Physiol.*, **115**: 69-78.

Horberg, H. (1998) *Vaxtskvddsnotiser,* **62**: 87-89

Hoyle, B. J. (1973) *Calif. Agric.,* **27**: 4-5.

Hu, K.L., Matsubara, S. and Murakami, K. (1993) *J. Jap. Soc. Hort. Sci.,* **62**: 561-565.

Huang, S.P. (1986) *J. Nematol.,* **18**: 408–412.

Huang, S.P., Vecchia, P.T. and Ferreira, P.E. (1986) *J. Nematol.,* **18**: 496–501.

Hussain, K., Singh, D. K., Ahmed, N., Nazir, G., and Rasool, R. (2005) *Environ. Ecol.,* **23**: 644-647.

Hussain, M., Manasova, M., Zouhar, M. and Rysanek, P. (2020) *Pakistan J. Zool.,* **52**: 199-206.

Ichikawa, H., Suenaga, L.T. and Imamura, J. (1988) *Plant Cell Tissue Organ Cult.,* **12**: 201-204.

Ichikawa, H., Tanno-Suenaga, L. and Imamura, J. (1989) *Theor. Appl. Genet.,* **77**: 39-43.

Imam, M. K. (1966) *Diss. Abstr.,* **26**: No. 6330-6331.

Imam, M. K. and Gabelman, W. H. (1968) *Proc. Amer. Soc. Hort. Sci.,* **93**: 419-428.

Iorizzo, M., Ellison, S., Senalik, D., Zeng, P., Satapoomin, P., Huang, J., Bowman, M., Iovene, M., Sanseverino, W., Cavagnaro, P., Yildiz, M., Macko-Podgórni, A., Moranska, E., Grzebelus, E., Grzebelus, D., Ashrafi, H., Zheng, Z., Cheng, S., Spooner, D., Van Deynze, A. and Simon, P. (2016) *Nat. Genet.,* **48**: 657–666.

Iorizzo, M., Senalik, D., Ellison, S., Grzebelus, D., Cavagnaro, P.F., Allender, C., Brunet, J., Spooner, D., Van Deynze, A. and Simon, P.W. (2013) *Amer. J. Bot.,* **100**: 930–938.

Jacobsohn, R. (1978) *Special Publication,* Volcani Centre, pp. 93-99.

Jacobsohn, R., Sachs, M. and Kelmen, Y. (1980) *J. Amer. Soc. Hort. Sci.,* **105**: 801-805.

Jhang, T., Kaur, M., Kalia, P. and Sharma, T.R. (2010) *J. Agric. Sci.,* **148**: 171–181.

Jin, W.L., Lu, X.D. and Zhang, R. (1999) *China Veg.,* **1**: 31-32.

Just, B.J., Santos, C.A.F., Fonseca, M.E.N., Boiteux, L.S., Oloizia, B.B. and Simon, P.W. (2007) *Theor. Appl. Genet.,* **114**: 693–704.

Just, B.J., Santos, C.A.F., Yandell, B.S. and Simon, P.W. (2009) *Theor. Appl. Genet.,* **119**: 1155–1169.

Kalia, P. (2004) *Umbelliferae Improvement Newsletter,* **14**: 1-3.

Kalia, P. and Longvah, T. (2008) In: Abstracts – 3rd Indian Horticulture Congress, New R and D Initiatives in Horticulture for Accelerated Growth and Prosperity, 6-9 November, 2008, Bhubaneswar, Abstract No 4A.85, pp. 152.

Kalloo, G., Amer Singh, Balyan, D.S., Baswana, K.S., Pratap, P.S. and Singh, A. (1993) *Indian Hort.,* **37**: 4, 21.

Kanwar, J. S. and Malik, B. S. (1970) *Indian J. Hort.,* **27**: 48-53.

Karim, A.J.M.S., Egashira, K., Quadir, M.A., Choudhury, S.A. and Majumdar, K.M. (1997) *Ann. Bangladesh Agric.,* **6**: 117-123.

Karunaratne, A.M. (1999) *Ceylon J. Sci. Bio. Sci.*, **27**: 67-72.

Katsumata, H. and Yasui, H. (1965) *Engei Shikonjo hokoku/Bull. Hort. Res. Sta. Ser. D,Kurume*, **3**: 79-112.

Kaur, P., Cheema, D.S. and Chawla, N. (2005) *J. Appl. Hortic.*, **7**: 130-132.

Kepka, A., Umiccka, L. and Fajkowska, H. (1978) *Acta Hort.*, **72**: 21724.

Kettunen, S., Havukkala, I., Holopainen, J.K. and Knuuttila, T. (1988) *Ann. Agri. Fenniae*, **27**: 99-105.

Khan, F., Asif, M., Khan, A., Tariq, M., Ansari, T., Shariq, M. and Siddiqui, M.A. (2019) *Current Plant Biol.*, **20**: 100115.

Khovrin, A.N. and Kosenko, M.A. (2020) *Potato and Vegetables*, **7**: 24-27.

Kidmose, U. and Martens, H.J. (1999) *J. Sci. Food Agric.*, **79**: 1747-1753.

Kielkowska, A. and Adamus, A. (2010) *Plant Cell Tissue Org. Cult.*, **102**: 309–319.

Kielkowska, A., Adamus, A. and Baranski, R. (2014) *In Vitro Cell. Dev. Biol. Plant*, **50**: 376–383.

Kirtikar, K.R. and Basu, B.D. (1935) In: *Indian Medicinal Plants* (Ed. Lolit Mohan Basu), Allahabad, India.

Kisaka, H., Kisaka, M., Kanno, A. and Kameya, T. (1997) *Theor. Appl. Genet.*, **94**: 221-226.

Kisaka, H., Kosaka, M. and Kameya, T. (1996) *Breed. Sci.*, **46**: 221-226.

Kisaka, H., Lee, H., Kisaka, M., Kanno, A., Kang, K. and Kameya, T. (1994) *Theor. Appl. Genet.*, **89**: 365-371.

Klimek Chodacka, M., Oleszkiewicz, T., Lowder, L.G., Qi, Y. and Baranski, R. (2018) *Plant Cell Rep.*, **37**: 575–586.

Koch, T. and Goldman, I.L. (2005) *J. Agric. Food Chem.*, **53**: 325-331.

Koul, P., Koul, A.K. and Hamal, I.A. (1989) *New Phytologist*, **112**: 437-443.

Kováèik, P., Šalamún, P., Smoleñ, S., Škarpa, P., Šimanský, V. and Moravèík, ¼. (2018) *Potravinarstvo Slovak J. Food Sci.*, **12**: 520-526.

Krarup, A. and Montealegre, J. (1975) *Agro. Sur.*, **3**: 50-53.

Krarup, A. and Moretti, J. (1976) *Agro. Sur.*, **4**: 21-24.

Krarup, A. and Villanueva, G. (1977) *Agro. Sur.*, **5**: 42-44.

Kropisz, A. and Wojciechowski, J. (1978) *BiuletynWarzywniczy*, **21**: 127-142.

Kumar, A., Afroza, B., Faheema, S., Nabi, A. and Mufti, S. (2017) *Annals Biol.*, **33**: 117-119.

Kumar, J.C. and Nandpuri, K.S. (1978) *J. Res. India*, **15**: 38-42.

Kumar, J.C., Sharma, B.N., Sharma, P.B. and Paul, Y. (1974) *Indian J. Hort.*, **31**: 262-267.

Kumar, V. and Mathur, M.K. (1965) *Fertl. News*, **10**: 1-4.

Kumar, V., Kumar, A., Naresh, Rani, V., Mukesh, S. and Poonia, R. (2017) *Annals Biol.*, **33**: 135-138.

Kust, A.F. (1970) Ph.D thesis University of Wisconsin, Madison, Wisconsin.

Lacko-Bartosova, M. and Rifai, M.N. (1998) *Annals Warsaw Agric. Univ.*, **32**: 43-48.

Laferriere, L. and Gabelman, W.H. (1968) *Proc. Amer. Soc. Hort. Sci.*, **93**: 408-418.

Lal, S. and Pandey, U.C. (1986) *Seed Res.*, **14**: 140-143.

Le Clerc, V., Marques, S., Suel, A., Huet, S., Hamama, L., Voisine, L., Auperpin, E., Jourdan, M., Barrot, L., Prieur, R. and Briard, M. (2015) *Theor. Appl. Genet.*, **128**: 2177-2187.

Le Clerc, V., Pawelec, A., Birolleau-Touchard, C., Suel, A. and Briard, M. (2009) *Theor. Appl. Genet.*, **118**: 1251–1259.

Lebeda, A. and Coufal, J. (1987) *Arch Züchtungsforsch* Berlin, **17**: 73–76.

Lebeda, A., Coufal, J. and Kvasnicka, P. (1988) *Euphytica*, **39**: 285-288.

LeBlanc, P.V. and Gupta, U.C. (1994) *J. Pl. Nutrition*, **17**: 199-207.

Lecomte, M., Berruyer, R., Hamama, L., Boedo, C., Hudhomme, P., Bersihand, S., Arul, J., N'Guyen, G., Gatto, J., Guilet, D., Richomme, P., Simoneau, P., Briard, M., Le Clerc, V. and Poupard, P. (2012) *Physiol. Mol. Plant Pathol.*, **80**: 58–67.

Lenka, P.C., Dash, P.K., Das, J.N. and Das, D.K. (1990) *Orissa J. Hort.*, **18**: 57-61.

Lester, G.E. and Baker, I.R. (1978) *HortScience*, **13**: 3, 11.

Li, Jin-Rong, Fei-Yun Zhuang, Cheng-Gang Ou, Hong Hu, Zhi-Wei Zhao and Ji-Hua Mao (2013) *Plant Cell Tiss Organ Cult.*, **112**: 275–287.

Li, Meng Yao, Zhi Sheng Xu, Ying Huang, Chang Tian, Feng Wang and Ai Sheng Xiong (2015) *Mol. Genet. Genomics*, **290**: 2049-2061.

Lim, B., Lee, C., Choi, S., Kim, Y., Lin, B.S., Lee, C.S., Choi, S.T. and Kim, Y.B. (1998) *RAD J. Hort. Sci.*, **40**: 83-88.

Lingaiah, H.B. and Huddar, A.G. (1991) *Mysore J. Agric. Sci.*, **25**: 231-235.

Lingaiah, H.B. and Reddy, T.V. (1996) *Hort. J.*, **9**: 33-40.

Linke, B., Nothnagel, T. and Börner, T. (1999) *Plant Breed.*, **118**: 543–548.

Litvinova, M.K. (1979) *Tsitoplazmatich, muzhsk. Steril'nost, I selektsiyarastkiev. Ukranian SSR*, pp. 200-203.

Liu, J.R., Leon, J.H., Yang, S.G., Lee, H.S., Song, N.H. and Jeong, W.J. (1992) *Sci. Hort.*, **5**: 1-11.

Lockhart, C.L. (1975) *Grower*, **84**: 401.

Longvah, T., Ananthan, R., Bhaskarachary, K. and Venkaiah, K. (2017) Indian Food Composition Tables 2017. National Institute of Nutrition, Hyderabad, India, p.535.

Luz, J.M.Q., Pasqual, M., Souza, R.V. de and De Souza, R.V. (1993) *Hort. Bras.*, **11**: 129-130.

Ma, C., Lin, Z.P., Zhao, Y.J. and Liu, H.J. (1981) *Acta Bot. Sinica*, **23**: 27-30.

Ma, Z., Kong, X., Liu, L., Ou, C., Sun, T., Zhao, Z., Miao, Z., Rong, J. and Zhuang, F. (2016) *Euphytica*, **212**: 37-40.

Mack, H.T. (1980) *HortScience*, **15**: 144-145.

Mackevic, V.I. (1929) *Bull. Appl. Bot. Genet. Pl. Breed.*, **20**: 517-557.

Madan, S.P. and Saimbhi, M.S. (1986) *Haryana J. Hort. Sci.*, **15**: 147-148.

Madrid-de-Canizalez, C. and Chacin-Lugo, F. (1994) *Revista de la Facultad de Agronomia, Univ. Central de Venezuela*, **20**: 47-72.

Mahmood-ur-Rehman, M., Amjad, M., Ziaf, K. and Ahmad, R. (2020) *Pak. J. Agri. Sci.*, **57**: 351-359.

Maistrello, L., Sasanelli, N., Vaccari, G., Toderas, I. and Iurcu-Straistaru, E. (2018) *Scientific Papers. Series A. Agronomy*, Vol. LXI, No. 1, pp. 316-321.

Malik, B.S. (1970) *Indian J. Agron.*, **15**: 402-405.

Malik, M.K., Slovin, J.P., Hwang, C.H. and Zimmerman, J.L. (1999) *Plant J.*, **20**: 89-99.

Manach, C., Williamson, G., Morand, C., Scalbert, A. and Remesy, C. (2005) *Amer. J. Clin. Nutr.*, **81**: 230S–242S.

Maree, P.C. and Lotter, B.F. (1971) *Farming in South Africa*, **47**: 35, 37, 39.

Marouelli, W.A., Carrijo, O.A. and Oliveira, C.A.S. (1990a) *Pesquisa Agropecuaria Brasileira*, **25**: 299-303.

Marouelli, W.A., Clieira, C.A.S. and Silva, W.L.C. (1990b) *Pesquisa Agropecnaria Brasileira*, **25**: 339-343.

Marouelli, W.A., Oliveira, C.A.S. and Silva, W.L.C. (1988) *Hort. Bras.*, **6**: 13-16.

Matsubara, D., Dohya, N. and Murakami, K. (1995) *Acta Hort.*, **392**: 129-137.

Maurya, K.R. (1986) *Indian J. Hort.*, **43**: 118-120.

Maurya, K.R. and Goswami, R.K. (1985) *Prog. Hort.*, **17**: 212-217.

Mazur, Z. (1992) *BiuletynWarzywniczy*, **38**: 123-139.

McCollum, G.D. (1966) *Swed. Econ. Bot.*, **2**: 361–367.

McCollum, G.D. (1975) *Botanical Gazette*, **136**: 201-206.

McDonald, M.R. (1994) *Ph.D. thesis*, University of Guelph, Ontario, p. 314.

McDonald, M.R., van der Kooi, K. and Simon, P. (2017) In: *Muck Crops Research Station Annual Report*, University of Guelph, Ontario, p.6. https://www.uoguelph.ca/muckcrop/pdfs/Muck.

McGhie, T.K. and Walton, M.C. (2007) *Mol. Nutr. Food Res.*, **51**: 702–713.

Meyer, K., Keill, M. and Naldrett, M.J. (1999) *FEBS-Letters*, **447**: 171-178.

Michalik, B. and Wiech, K. (2000) *Folia Horticulturae*, **12**: 43-51.

Michalik, B., Simon, P.W. and Gabelman, W.H. (1992) *HortScience*, **27**: 1020-1022.

Millette, J.A. and Vigier, B. (1981) *J. Amer. Soc. Hort. Sci.*, **106**: 491-493.

Montemurro, P., Capella, A., Cazzato, E., Crivelli, L., D'Ascenzo, D., Fusiello, R., Guastamacchia, F. and Olivieri, D. (2016) *ATTI Giornate Fitopatologiche*, **1**: 503-510.

Morelock, T.E. (1974) *Ph.D. Thesis*, University of Wisconsin, Madison, USA.

Morelock, T.E., Simon, P.W. and Peterson, C.E. (1996) *HortScience*, **31**: 887–888.

Murray, E.E., Rocheleau, T., Eberle, M., Stock, C., Sekar, V. and Adang, M. (1991) *Plant Mol. Biol.*, **16**: 1035-1050.

Nagachandrabose, S. (2018) *Crop Protet.*, **114**: 155-161.

Nair, P.K.K. and Kapoor, S.K. (1973) *J. Palynol.*, **9**: 152-159.

Nakajima, Y., Yamamoto, T., Muranaka, T. and Oeda, K. (1999) *Theor. Appl. Genet.*, **99**: 837-843.

Narender and Rani, V. (2019) *Internat. J. Agric. Engg.*, **12**: 217-222.

Nascimento, W.M. and Audreoli, C. (1990) *RevistaBrasileira de Sements*, **12**: 28-36.

Nath, Prem and Kalvi, T.S. (1969a) *Farm J.*, **11**: 9-12.

Nath, Prem and Kalvi, T.S. (1969b) *Punjab Hort. J.*, **9**: 81-89.

Nath, S., Kumar, A., Mani, I., Singh, J.K. and Sureja, A.K. (2019) *Indian J. Agric. Sci.*, **89**: 1011-1016.

Nazeer, Ahmed, Tanki, M.I. and Ahmed, N. (1989) *Veg. Sci.*, **16**: 107-112.

Nicolle, C., Simon, G., Rock, E., Amouroux, P. and R´em´esy, C. (2004) *J. Amer. Soc. Hort. Sci.*, **129**: 523–529.

Noland, T.L., Maguire, J.D., Oliva, R.N., Bradford, K.J., Nelson, J.L., Grabe, D. and Currans, S. (1988) *J. Appl. Seed Produc.*, **6**: 36-43.

Noor, A., Ziaf, K., Amjad, M. and Ahmad, R. (2019) *Pak. J. Agri. Sci.*, **56**: 83-94.

Nothnagel, T., Straka, P. and Budahn, H. (1997) *J. Appl. Genet.*, **38A**: 172-177.

Nothnagel, T., Straka, P. and Linke, B. (2000) *Plant Breed.*, **119**: 145–152.

Oliva, R.N., Tissoni, T. and Bradford, K.J. (1988) *J. Amer. Soc. Hort. Sci.*, **113**: 532-537.

Orzolek, M.D. and Carrol, R.B. (1978) *J. Amer. Soc. Hort. Sci.*, **103**: 236-269.

Pal, R.K., Roy, K.S. and Wasker, D.P. (1991) *Maharastra J. Hort.*, **5**: 98-105.

Panda, J.M., Sahoo, A., Sethi, P.K. and Dora, D.K. (1994) *Orissa J. Hort.*, **22**: 84-86.

Pankov, V.V. (1976a) *Agrokhimiya*, **6**: 26-30.

Pankov, V.V. (1976b) *Turdy Gor'kov S-kh. Instituta*, **94**: 3-12.

Pantielev, Ya. KH, Solov'eva, V.K., Kamyanina, T.N. and Smirnov, I.M. (1976) *Intensifik Zemledeliya v Tseutr. Re-ne Nechernozem Zony. Moscow*, USSR, pp. 117-123.

Pariari, A. and Maity, T.K. (1992) *Crop. Res.*, **5**: 158-162.

Parsons, J., Matthews, W., Iorizzo, M., Roberts, P. and Simon, P. (2015) *Mol. Breed.*, **35**: 114.

Pavlek, P. (1977) *PolijoprivrednaZnanstvenaSmotra*, **42**: 67-73.

Pawalicki, N., Sangwan, R.S. and Sangwan Norell, B.S. (1993) *Acta Botanica Gallica*, **140**: 17-22.

Peterson, E.C., Simon, P.W., Rubatzky, V.E. and Strandberg, J.O. (1988) *HortScience*, **23**: 917.

Pietola, L. (1995) *Agric. Sci. Finland*, **4**: 144-237.

Pike, L.M., Maxwell, R.V., Horn, R.S., Rogers, B.A. and Miller, M.E. (1991) *HortScience*, **26**: 1230-1231.

Pill, W.G. and Evans, T.A. (1991) *J. Hort. Sci.*, **66**: 67-74.

Pitelli, R.A., ChuratanMasca, M.G.C. and Oliveira, A.F. (1976) *Resumos XI Seminario Brasileiro de Herbicidas e Ervas Daninhas*, London, p. 22.

Pleskachev, Y.N., Chamurliev, O.G. and Gubina, L.V. (2018) *RUDN J. Agron. Animal Industries*, **13**: 360-365.

Polach, J. (1987) *Bull Vyzkumny a SlechtitelskyUstavZelinarsky Olomouc*, **31**: 125-135.

Poole, C.F. (1937) Improving the root vegetables. U.S. Dept. Agr. Yearbook, pp. 300-325.

Prabhakar, M., Srinivas, K. and Hegde, D.M. (1991) *Gartenbauwissenschaft*, **56**: 206-209.

Pryor, B., Davis, R.M. and Gilbertson, R.L. (2000) *HortScience*, **35**: 1099–1102.

Punja, Z.K. and Chen, W.P. (2004) *Acta Hort.*, **637**: 295–302.

Punja, Z.K. and Raharjo, S.H.T. (1996) *Plant Dis.*, **88**: 999-1005.

Que, F., Wang, G., Li, T., Wang, Y., Xu, Z. and Xiong, A. (2018) *Funct. Integr. Genomic.*, **18**: 685–700.

Racz, Z. (1973) *PolyoprivrednaZnanstvenaSmotra*, **30**: 263-276.

Ramert, B. (1976) *Vaxtskyddosnotiser*, **40**: 96-98.

Rattan, S.P. and Jamwal, R. S. (1991) *Veg. Sci.*, **18**: 184-191.

Rehman, Z. and Shah, W.H. (1999) *Pakistan J. Sci. Indust. Res.*, **42**: 133-136.

Riggs, T.J. (1987) In: Hybrid production of selected cereal oil and vegetable crops (Eds. W.P. Feistritzer and A.F.Kelly) FAO, Rome, pp. 149-173.

Roderick, H., Urwin, P.E., Atkinson, H.J. (2018) *Plant Biotechnol J.*, **16**: 520–529.

Rong, J., Y. Lammers, J.L. Strasburg, N.S. Schidlo, Y. Ariyurek, T.J. de Jong, P.G.L. Klinkhamer, M.J.M. Smulders and K. Vrieling (2014) *BMC Genomics*, **15**: 895.

Rosenfeld, H.J. (1998) *Gartenbauwissenschaft*, **63**: 87-94.

Rubatzky, V.E., Quiros, C.F. and Simon, P.W. (1999) Carrots and related vegetable umbelliferae. CAB International, Wallingford, UK.

Sady, W. and Cebulak, T. (2000) *Folia Horticulturae*, **12**: 35-41.

Sady, W., Crys, R. and Rozek, S. (1999) *Folia Horticulturae*, **11**: 105-115.

Saenz Lain, C. (1981) *Anal Jardin Botanico Madrid*, 37: 481-533.

Sagar, V.R., Kalia, P. and Rajesh Kumar (2008) *Veg. Sci.*, 35(2): 156-159.

Sagar, V.R., Maini, S.B., Rajesh Kumar and Netra Pal (1997) *Veg. Sci.*, 24(1): 64-66.

Sagiv, B., Hadas, A. and Bar-Yoset, B. (1994) *Hassadeh*, **74**: 631-634.

Saini, S.S. and Rastogi, K.B. (1976) *Veg. Sci.*, **3**: 17-20.

Saini, S.S., Korla, B.N. and Rastogi, K.B. (1981) *Veg. Sci.*, **8**: 93-99.

Sala, C., Biasini, M.G., Morandi, C., Nielsen, E., Parisi, B. and Sala, F. (1985) *J. Plant Physiol.*, **118**: 409-419.

Salter, P.J., Currah, I.E. and Fellows, J. Rs. (1979) *J. Agric. Sci.*, UK, **93**: 431-440.

Sampaio, N.V. and Sampaio, T.G. (1998) *Revista Cientifica Rural*, **3**: 38-45.

Santos, C.A.F. and Simon, P.W. (2002) *Mol. Genet. Genomics*, **268**: 122–129.

Santos, C.A.F. and Simon, P.W. (2006) *Euphytica*, **151**: 79–86.

Saparov, U.B. (1992) *Problems of Desert Development*, **4**: 67-70.

Satta, E., Ramirez, A.S., Paltrinieri, S., Contaldo, N., Benito, P., Poveda, J.B. and Bertaccini, A. (2016) *Phytopathologia Mediterranea*, **55**: 401"409.

Schmidhalter, U. and Oertli, J.J. (1990) *Acta Hort.*, **278**: 203-211.

Senalik, D. and Simon, P.W. (1987) *Phytochem.*, **26**: 1975-1979.

Shaimanov, A.A., Leunov, V.I. and Shaimanova, L.A. (1996) *Kartofel' i Ovoshchi*, 3: 46-47.

Sharangi, A.B. and Paria, N.C. (1995) *Hort. J.*, **8**: 161-164.

Sharangi, A.B. and Paria, N.C. (1996) *Environ. Ecol.*, **14**: 408-410.

Sharma, A.K. and Bhattacharyya, N.K. (1954) *Genetica*, **30**: 1-68.

Sharma, P.L. and Sharma, B.R. (1968) *Indian J. Hort.*, **25**: 216.

Sharma, S., Holme, I.B., Dionisio, G., Kodama, M., Dzhanfezova, T., Joernsgaard, B. and Brinch-Pedersen, H. (2020) *Plant Mol. Biol.*, **103**: 443-456.

Shaw, M.W., Allen, R.M. and Inkson, R.H.E. (1961) *Plant Pathol.*, **10**: 110-115.

Shea, E.M., Gibeant, D.M. and Carpita, N.C. (1989) *Planta*, **179**: 293-308.

Shevtsova, A.M. and Koral, O.N. (1976) *VoprosyEstestvoznaniya Minsk, Belorussain SSR*, pp. 134-138.

Shibairo, S.I., Upadhyay, M.K. and Toivonen, P.M.A. (1998) *J. Hort. Sc. Biotech.*, **73**: 347-352.

Shigeta, J. (1995) *Biotechnal. Tech.*, **9**: 771-776.

Shishkina, L.A. and Galeev, N.A. (1974) *BiologiyaiAgrotekhnikaSel-skokhozyaistvennykhKul'tur. Ufa*, USSR, pp. 157-160.

Shoemaker, J.S. (1947) *Vegetable Growing*. John Wiley and Sons, New York.

Siddiqui, Z.A., Nesha, R. and Varshney, A. (2011) *J. Plant Pathol.*, **93**: 503–506.

Silva, F.D.S. and Silva, N.O. (2017) *Pesq. Agropec. Trop., Goiânia*, **47**: 399-407.

Simon, P.W. (1996) *J. Hered.*, **87**: 63-66.

Simon, P.W. (2000) *Plant Breed. Rev.*, **19**: 157-190.

Simon, P.W. and Strandberg, J.O. (1998) *J. Amer. Soc. Hort. Sci.*, **123**: 412-415.

Simon, P.W. and Wolff, X.Y. (1987) *J. Agric. Food Chem.*, **35**: 1017-1022.

Simon, P.W., Matthews, W.C. and Robers, P.A. (2000) *Theor. Appl. Genet.*, **100**: 735-742.

Simon, P.W., Tanumihardjo, S.A., Clevidence, B.A. and Novotny, J.A. (2008) In: Culver CA, Wrolstad RE, editors. Color quality of fresh and processed foods. ACS Symposium Series 983. Washington, D.C.: ACS Books.

Simon, P.W., Wolff, X.Y., Peterson, C.E., Kammerlohr, D.S., Rubatzky, V.E., Strandberg, J.O., Bassett, M.J. and White, J.M. (1989) *HortScience, 24*: 174.

Singh, B and Saimbhi, M.S. (1985) *J. Res. PAU*, Ludhiana, **21**: 629-641.

Singh, B., Pandey, S., Pal, A.K., Singh, J. and Rai, M. (2005) *Veg. Sci., 32*: 136-139.

Singh, G. and Jaysawal, N. (2018) *Annals Biol.*, **34**: 181-186.

Singh, H. (1963) *Indian Hort.*, **7**: 20-21.

Singh, H.B., Thakur, M.R. and Bhagchandani, P.M. (1960) *Indian J. Hort.*, **17**: 38-47.

Singh, R., Sukhjia, B.S. and Hundal, J.S. (1987) *Veg. Sci.*, **14**: 33-36.

Singh, S., Nehra, B.K. and Malik, Y.S. (1994) *Crop Res.* Hissar, **8**: 543-548.

Sinha, S.N. and Chakrabarti, A.K. (1992) *Seed Res.*, **20**: 37-40.

Smith, M.A., Pay, A. and Duditz, D. (1989) *Theor. Appl. Genet.*, **77**: 641-644.

Sokol, P.F., Naslov, A.P. and Panomareva, G.N. (1979) *Doklady Vsesoynznoi Ordena Lenina Akademii Sel'skokhozyaistvennykh NaukImeni* V. I. Lenina, **8**: 18-20.

Soteros, J.J. (1979) *New Zealand J. Agric. Res.*, **22**: 191-196.

Soujala, T. (2000) Academic dissertation, University of Helsinki, p. 47.

Standberg, J.O., Basset, M.J., Peterson, C.E. and Berger, R.D. (1972) *HortScience,* **7**: 345.

Steiner, J.J., Hutmacher, R.B., Mantel, A.B., Ayars, J.E. and Vail, S.S. (1990) *J. Amer. Soc. Hort. Sci.*, **115**: 722-727.

Steward, F.C., Mapes, M.O. and Mears, K. (1958) *Amer. J. Bot.*, **45**: 705-708.

Stommel, J.R. and Simon, P.W. (1989) *J. Amer. Soc. Hort. Sci.,* **114**: 695-699.

Strandberg, J.O., Bassett, M.J., Peterson, C.E., and Berger, R.D. (1972) *HortScience*, **7**: 345.

Su, Y.K., Youn, G.H., Cho, Y.H. and Paek, K.Y. (1999) *J. Korean Soc. Hort. Sci.*, **40**: 697-701.

Subhan (1988) *BuletinPenelitianHortikultura*, **16**: 76-82.

Suenaga, L.T., Nagao, E. and Imamura, J. (1991) *Jap. J. Breed.*, **41**: 25-33.

Sun, T., Simon, P.W. and Tanumihardjo, S.A. (2009) *J. Agric. Food Chem.*, **57**: 4142–4147.

Suojala, T. (1999) *J. Hort. Sci. Biotech.*, **74**: 484-492.

Suojala, T. and Pessala, R. (1999) In: *Agri. Food Quality* II, (eds. M. Hagg, R.Ahvenainen, A.M. Evers, and K. Tiilikkala), The Royal Society of Chemistry, Cambridge, UK, pp. 227-231.

Suojala, T. and Tahvonen, R. (1999) In: In: *Agri. Food Quality* II, (eds. M. Hagg, R.Ahvenainen, A.M. Evers, and K. Tiilikkala), The Royal Society of Chemistry, Cambridge, UK, pp. 76-78.

Synnevag, G. (1989) *Gartneryrket*, **79**: 10-11.

Takahashi, O. (1987) Utilization and seed production of hybrid vegetable varieties in Japan. In: Hybrid production of selected cereal oil and vegetable crops (Eds. W.P. Feistritzer and A.F. Kelly). FAO, Rome, pp. 313-328.

Takaichi, M. and Oeda, K. (2000) *Plant Sci.*, **153**: 135–144.

Teng, W.L., Liu, Y.J., Tsai, Y.C. and Soong, T.S. (1994) *HortScience*, **29**: 1349-1352.

Thompson, D.S. (1961) *Proc. Amer. Soc. Hortic. Sci.*, **78**: 332-338.

Timin, N.I. and Vasilevskii, V.A. (1995) *KartofeliOvoshchi*, **3**: 27-28.

Timin, N.I. and Vasilevsky, V.A. (1997) *J. Appl. Genet.*, **38A**: 232-236.

Toivonen, P.M.A., Upadhyaya, M.K. and Gaye, M.M. (1993) *Acta Hort.*, **343**: 339-340.

Tomic, F. (1976) *PoljoprivrednaZnanstvenaSmotra*, **35**: 197-200.

Toonen, N.A.J., Hendriks, T., Schmidt, E.D.L., Verhoeven, H.A., Kammen, A. van and Vries, S.C. de (1994) *Planta*, **194**: 565-572.

Tsuji, K., Nagaoka, M. and Oda, M. (1993) *JARQ, Japan Agric. Res. Quarterly*, **27**: 116-121.

Tyukavin, G.B., Shmykova, N.A. and Monakhova, M.A. (1999) *Russian J. Plant Physiol.*, **46**: 767-773.

Umiel, N. and Gabelman, W.H. (1971) *J. Amer. Soc. Hort. Sci.*, **96**: 702-704.

Umiel, N. and Gabelman, W.H. (1972) *J. Amer. Soc. Hort. Sci.*, **97**: 453-460.

Ustun, N.S., Tosun, J. and Ozyavuz, B. (1999) *Ondokuzmayis Universitesi, Ziraat Fakultesi Dergisi*, **14**: 25-32.

Vakis, N., Marriott, J. and Proctor, F.J. (1975) *J. Sci. Food Agric.*, **26**: 609-615.

Van der Vossen, H.A.M. and Sambas, E.N. (1993) In: *Plant resources of south-east Asia* (Eds. J.S. Siemonsma and Kasem Piluek), No. 8, Vegetables, Pudoc Scientific Publishers, Wageningen.

Verma, T.S., Ramesh-Chand, Sharma, S.C., Lakhanpal, K.D., Joshi, S. and Chand, R. (1993) *Indian J. Agric. Sci.*, **63**: 574-577.

Vieira, J.V., Casali, V.W.D., Milagres, J.C., Cardoso, A.A. and Regazzi, A.J. (1991) *Rev. Brasil. Genet.*, **14**: 501-508.

Vieira, J.V., Della Vecchia, P.T., and Ikuta, H. (1983) *Hort. Bras.*, **1**: 42.

Vivek, B.S. and Simon, P.W. (1999) *Theor. Appl. Genet.*, **99**: 58-64.

Vizzotto, V.J. and Muller, J.J.W. (1990) *AgropecuariaCatarinense*, **3**: 39-40.

Vleek, F. and Polach, J. (1978) *KerteszetiEgyetemKozlemenyei*, **42**: 45-54.

Wake, H. Umetsu, H. Matsunaga (1995) *Somatic Enbryogenesis and Synthetic Seeds*, **31**: 170-182.

Wake, H., Arasaka, A., Umetsu, H., Ozeki, Y., Shimomura, K. and Matsunaga, T. (1992) *Appl. Microbial. Bechnol.*, **36**: 684-688.

Walker, J.C. (1952) *Diseases of Vegetable Crops*, McGraw-Hill Book Co., Inc., New York.

Wan, X.S., Wang, F.D., Song, Y.G. and Xia, Z.A. (1987) *Plant Physiol. Commun.*, **1**: 30-33.

Wang, F., Wang, G., Hou, X., Li, M., Xu, Z. and Xion, A. (2018) *Mol. Genet. Genomics*, **293**: 861–871.

Wang, F.D., Wang, X.S., Ye, S.f. and Xia, Z.A. (1989) *Plant Physiol. Commun.*, **6**: 26-29.

Wang, G.L., Xiao Ling Jia, Zhi Sheng Xu, Feng Wang and Ai Sheng Xiong (2015) *Mol. Genet. Genomics*, **290**: 1379–1391.

Wang, M. and Goldman, I.L. (1996) *J. Hered.*, **87**: 119–123.

Weischer, B. (1961) *Nachr. Bo. Dtsch. PflschDienst.*, Braunschweig, **13**: 134-140.

Welch, J.E. and Grimball, E.L. (1947) *Science*, **106**: 594.

Wendt, T. (1979) *Gemuse*, **15**: 289-296.

Whitaker, T.W. (1949) *Proc. Amer. Soc. Hort. Sci.*, **53**: 305-308.

Whitaker, T.W., Shert, A.F., Lange, W.H., Nicklow, C.W. and Redeward, J.D. (1970) *USDA Hanbbook*, p.375.

White, J.M. (1978) *J. Amer. Soc. Hort. Sci.*, **103**: 433-435.

White, J.M. (1992) *HortScience*, **27**: 105-106.

Wiebe, H.J. (1968) *Gemuse*, **4**: 267-269.

Wohlfeiler, J., Alessandro, M.S., Cavagnaro, P.F. and Galmarini, C.R. (2019) *Euphytica*, **215**: 37.

Wolyn, D.J. and Chahal, A. (1998) *J. Amer. Soc. Hort. Sci.*, **123**: 849–853.

Wu, H.C., Chen, T.W. and Wu, W.S. (1999) *Plant Patho. Bull.*, **8**: 1-8.

Xing Jin, H., Tang-Lesheng, He, X.J. and Tang, Z.S. (1994) *J.S.W. Agric.Univ.*, **16**: 488-491.

Xu, Z., Feng, K. and Xiong, A. (2019) *Mol. Biotech.*, **61**: 191–199.

Xu, Zhi-Sheng, Ying Huang, Feng Wang, Xiong Song, Guang-Long Wang and Ai-Sheng Xiong (2014) *BMC Plant Biol.*, **14**: 262.

Yau, Y.Y. and Simon, P.W. (2003) *Plant Mol. Biol.*, **53**: 151–162.

Yau, Y.Y., Santos, K. and Simon, P. (2005) *Mol. Breed.*, **16**: 1–10.

Yawalkar, K.S. (1980) Vegetable Crops of India. Agri-Horticultural Publishing House, Nagpur.

Yildiz, M., Willis, D.K., Cavagnaro, P.F., Iorizzo, M., Abak, K. and Simon, P.W. (2013) *Theor. Appl. Genet.*, **126**: 1689–1702.

Yunhee, S., Park, J., Kim, Y.S., Park, Y. and Kim, Y.H. (2014) *Plant Pathol. J.*, **30**: 75–81.

Zhang, H.M., Wang, L.J. and Ren, N.H. (1998) *J. Henan Agric. Sci.*, **3**: 28.

Zhuang, F.Y., Pei, H.X., Ou, C.G., Hu, H., Zhao, Z.W. and Li, J.R. (2010) *Acta Hortic. Sinica*, **37**: 1613-1620.

Zidorn, C., Johrer, K., Ganzera, M., Schubert, B., Sigmund, E.M., Mader, J., Greil, R., Ellmerer, E.P. and Stuppner, H. (2005) *J. Agric. Food Chem.* 53: 2518–2523.

Zink, F.W. (1961) *J. Agric. Food Chem.*, **9**: 304-307.

11

RADISH

A. K. Sureja, M. K. Sadhu and K. Sarkar

1.0 Introduction

Radish is a popular vegetable in both tropical and temperate region of the world and is best recognized for the crisp texture and pungent flavour of its enlarged storage roots and hypocotyls. It is primarily cultivated as a winter season vegetable. However, now-a-days it is being grown round the year. Radish is an important vegetable crop in China, Japan, India, Korea, Egypt and many other countries of the world. In India, the major radish producing states are West Bengal, Bihar, Uttar Pradesh, Punjab, Assam, Haryana, Gujarat and Himachal Pradesh. Being a short duration and quick growing crop, it is easily grown as a companion crop of intercrop between the rows of other vegetables. Radish is also grown as a cover crop to avoid soil erosion and to suppress weeds.

Radish is one of the most ancient vegetables. Inscriptions on the inner walls of pyramids show that radish was an important vegetable in Egypt about 2000 B.C. Certain remarks of Herodotus reveal that it was cultivated about 27 B.C. (Becker, 1962). It has spread to China in about 500 B.C. and to Japan in 700 A.D. (Sirks, 1957).

2.0 Composition and Uses

2.1.0 Composition

Composition of radishes is given in Table 1. Radish leaves are more nutritious compared to the roots, particularly in total ascorbic acid, carotenoids, lutein, β-carotene and mineral nutrients content. Radish is a good source of vitamin C (ascorbic acid), containing 14-65 mg per 100 g of edible portion, and supplies a variety of minerals. Trace elements in radish include aluminium, barium, lithium, manganese, silicon, titanium, fluorine and iodine. Vitamin C content of radish roots is greatly influenced by light condition. For example, Sid'ko et al. (1975) found that vitamin C content of roots was higher in plants grown under blue light, while Lichtenthaler (1975) noted enhanced synthesis of β-carotene under red light. Vitamin C content of radish roots was also influenced by fertilizer application, P significantly increasing the ascorbic acid content (Roy and Seth, 1971; Joshi and Patil, 1988). Patil and Patil (1986) recorded an inverse relationship between N application and vitamin C contents in both roots and tops. Gaweda et al. (1991) reported reasonably higher vitamin C content in early cultivars.

Radish root contains glucose as the major sugar and smaller quantity of fructose and sucrose. Pectin and pentosans are also reported to be present. The characteristic pungent flavour of radish is due to the presence for volatile isothiocyanates (4-methyl thio-3-butenyl isothiocyanate), and the colour of the pink cultivars is due to the presence of anthocyanin pigments. In trials conducted in 1994, 1995 and 1996, a total of 27 red radish cultivars were grown at 2 locations (Corvallis and Hermiston, Oregon, USA) and harvested at 2 maturity stages. Pigment content was dependent on cultivar, root weight and location, higher amounts being obtained at Hermiston. Spring cultivars had pigmentation in the skin, ranging from 39.3 to 185 mg ACN/100 g skin. Red-fleshed winter cultivars had pigment content ranging from 12.2 to 53 mg ACN/100 g root. ACN profiles were similar for different cultivars, the major pigments being pelargonidin-3-sophoroside-5-glucoside, mono- or di-acylated with cinnamic and malonic acids; individual proportions varied among cultivars. Among the spring cultivars, cv. Fuego had the highest pigment content and among the winter cultivars the Chinese Red Meat cultivar had the highest pigment content. Estimated pigment yields ranged from 1.3 to 14 kg/ha, suggesting that production of pigments on a commercial scale could be viable (Giusti *et al.*, 1998).

Table 1: Composition of Radish (per 100 g of edible portion)*

Composition	Radish Leaves	Radish Root			
		Elongate, Red Skin	Elongate, White Skin	Round, Red Skin	Round, White Skin
Moisture (g)	91.19	89.32	89.05	89.68	89.76
Protein (g)	2.22	0.67	0.77	0.89	0.80
Ash (g)	1.50	0.73	0.82	0.91	0.80
Total fat (g)	0.51	0.13	0.15	0.16	0.14
Total dietary fibre (g)	1.82	2.46	2.65	2.29	2.37
Carbohydrate (g)	2.77	6.71	6.56	6.07	6.13
Energy (KJ)	109.0	134.0	135.0	130.0	129.0
Thiamine (B^1) (mg)	0.06	0.03	0.02	0.03	0.03
Riboflavin (B_2) (mg)	0.13	0.02	0.02	0.02	0.02
Niacin (B_3) (mg)	0.47	0.31	0.30	0.30	0.24
Pantothenic acid (B_5) (mg)	0.19	0.13	0.15	0.18	0.15
Total B_6 (mg)	0.16	0.07	0.07	0.07	0.07
Biotin (B_7) (µg)	4.39	2.65	2.48	2.92	2.59
Total folates (B_9) (µg)	53.14	24.65	29.75	24.59	22.60
Total ascorbic acid (mg)	65.76	17.63	19.91	15.69	14.00
Phylloquinones (K_1) (µg)	185.0	2.10	2.50	2.60	1.90
Lutein (µg)	1741.0	8.68	5.34	7.80	6.36
α-carotene (µg)	2591.0	1.62	-	1.20	-
Total carotenoids (µg)	9339.0	17.61	10.60	13.07	23.69
Calcium (mg)	234.0	28.44	30.20	35.76	34.23
Iron (mg)	3.82	0.37	0.36	0.42	0.41
Magnesium (mg)	57.96	13.34	16.07	22.25	15.46
Phosphorus (mg)	50.08	27.51	30.10	28.27	29.47
Potassium (mg)	304.0	255.0	288.0	308.0	287.0
Selenium (µg)	33.05	0.13	0.10	0.22	0.13
Zinc (mg)	0.49	0.16	0.22	0.18	0.17
Total starch (g)	1.80	0.41	0.59	0.64	0.30
Total free sugars (g)	0.29	1.15	0.95	0.86	1.16
Total oxalate (mg)	53.83	12.73	12.72	12.27	15.49
Total polyphenols (mg)	27.57	49.75	46.74	44.93	46.39

* Longvah *et al.* (2017).

Ishii and Saijo (1987) noticed that isothiocyanate content of the roots decreased with time after sowing, being higher in roots harvested in early summer than in those harvested in autumn. Isothiocyanate content was higher in roots grown in an alluvial soil than in roots grown in an Ando soil. Okano *et al.* (1990) measured the contents of 4-methylthio-3-butenyl isothiocyanate (MTB-ITC) in 38 cultivars in Japan and found that in half of the cultivars, MTB-ITC concentration was low, ranging from 200 to 300 μmol/100 ml juice. In other cultivars it was higher, reaching 1735 μmol/100 ml in Karami, followed by the Shinshu-jidaikon group with 400-700 μmol/100 ml. The Chinese cultivars had the lowest concentration (100-200 μmol/100 ml). Lee *et al.* (1996) also found the thiocyanate contents varied with cultivar and growth site. Long and slender Japanese cultivars had higher contents of thiocyanate than short and solid Korean cultivars. Portion adjacent to the root tip had higher thiocyanate than portions near to the root top or shoulder. Peel contained higher concentrations (10-50 per cent) than the inner root flesh. It was further noticed that thiocyanate contents were not significantly reduced in roots stored in a cool and moist cellar for up to 2 months. The small amount of thiocyanate in storage roots of early and Japanese cultivars is reflected in their milder taste when compared with winter radish. In general, thiocya-nate content is high in leaves of all cultivars (Capecka *et al.*, 1998).

The radish seeds are a potential source of a non-drying fatty oil suitable for soap making, illuminating and edible purposes.

The pods of rat-tail radish contain 92.3 per cent moisture, 1.3 per cent protein, 0.3 per cent fat, 1.1 per cent fibre, 4.3 per cent carbohydrates and 0.7 per cent minerals. They contain 7.8 mg calcium and 2.4 mg phosphorus per 100 g of edible portion.

2.2.0 Uses

Radish is grown for its young tender tuberous roots which are cooked as a vegetable or preserved by salting, pickling, canning and drying, and is also eaten fresh as grated radish, garnish and salad. Radish has a unique pungent flavour. It is relished for its pungent flavour and is considered as an appetizer. It is also used in parathas which are taken with curd for breakfast in North India. The young soft leaves are also cooked as leafy vegetable. The leaves are more nutritious than its roots. Traditionally in the East Asian countries, and increasingly in the West, the sprouted seeds, young leaves and immature seed pods of radishes are eaten. The 'Japanese Bisai' type is particularly suited for sprouts and leaves (Crisp, 1995). Sprouted seedlings at the cotyledon stage (microgreens) are also used as garnish in many countries. In radish, microgreens are harvested 7-9 days after germination. They are cut along with the stem and attached cotyledons/seed leaves with the help of scissor. If left for longer, they rapidly elongate and loose colour and flavour. "Kosena-daikon" is a Japanese radish landrace which has a small root and is particularly grown as a leafy vegetable. In Japan, the young seedling sprouts known as 'kaiware-daikon' is very popular. Some types of mini-radish such as 'Hatsuka-Daikon' which requires 20 days for harvest in Japan are also cultured for salad in Europe and other countries (Kaneko and Matsuzawa, 1993).

Uses of Radish

Radish juice

Radish leaf soup

Radish Salad

Radish as microgreens

Radish Pickle

Radish seed oil

Fried radish

Radish Poriyal

The landraces of daikon, Japanese radish, *Raphanus sativus* L. var. *hortensis* Becker, are consumed for several Japanese dishes such as Kayu (rice gruel with vegetables or red beans), Nitsuke (boiled foods with soy source), Sunomono (vinegared foods), Oroshi (grated radish), Takuan (pickled radish), and Sengiri or Kiriboshi (radish strips dried for winter storage) (Yamaguchi and Okamoto, 1997). In Korea, Chinese cabbage and radish are the most widely used vegetables in making kimchi, a lactic acid fermentation product (Lee, 2003).

Radish has refreshing and diuretic properties. In Indian system of medicine, radish preparations are useful in liver and gall-bladder troubles. In homoeopathy, they are used for neuralgic headaches, sleeplessness and chronic diarrhoea. Roots, leaves, flowers and pods are quite effective against gram-positive bacteria. The roots are said to be useful in urinary complaints, piles and in gastrodynia. A salt, extracted from roots, dried and burnt to white ash, is said to be used in stomach troubles. The juice of fresh leaves is used as diuretic and laxative. The seeds are said to be peptic, expectorant, diuretic and carminative (Kirtikar and Basu, 1935).

Radish contains two important medicinal compounds – peroxidase and isothiocyanates. Isothiocyanates, a dietary glucosinolate (GL) have anticarcinogenic properties (Barillari *et al.*, 2002). 4-(Methylthio)-3-butenyl isothiocyanate is a principal antimutagen in daikon (Japanese white radish) (Nakamura *et al.*, 2001). Hyperlipidaemia, a condition associated with excess fat in the blood, is the main cause of coronary heart diseases. Radish roots are a rich source of peroxidase, an oxidoreductase, which can scavange harmful free radicals (Curtis, 2003). Black radish (*Raphanus sativus* var. *nigra*) has antioxidant properties and the juice has beneficial health effect in the hyperlipidaemia.

In India and Egypt, *R. sativus* var. *oleifer* is cultivated for vegetable oil and green manure (Iwasa, 1980). Radish has also been grown successfully as a fodder crop in countries like U.K. and South Africa. In South Africa, the giant radish of Japan is grown as a fodder crop, yielding more than 60 tonnes of roots/ha and 12-25 tonnes of leaves/ha (Kolbe and Voss, 1952). Fodder-radish bears little or no fleshy root and is grown for its foliage, used as fodder or green manure. It is gaining popularity in Western Europe due to its rapid growth.

The Indian 'rat-tail' or 'mougri' radish and German 'München Beir' types, which are very similar to common radish, but does not produce the characteristic fleshy root, are grown for their long (about 30 cm) slender pods which are eaten fresh, cooked or pickled, and as snack foods.

3.0 History, Origin, Evolution, Taxonomy and Cytology

3.1.0 History, Origin and Evolution

Radish is one of the most ancient vegetables. Pliny records that radish was extensively cultivated in Egypt at the time of pharaohs and was also known to and highly prized by Greeks. The Greeks served them on dishes of gold in sacrificial offers to Apollo (Gill, 1993). Radish (probably niger type) was depicted on the inner walls of the pyramids in Egypt about 2000 B.C. (Banga, 1976). Herodotus (c. 484-424 BC) suggested that it was already an important food crop in Egypt nearly

5000 years ago *i.e.*, about 2700 B.C. (Becker,1962). It has spread to china in about 500 B.C. and to Japan about 700 A.D. (Sirks, 1957). Radish was known in the eastern Mediterranean at least two millennia before it was known in China (Banga, 1976). In China and Korea there exist the oldest records on domestication, dated 400 and 100 BC, respectively (Kaneko and Matsuzawa, 1993). Radish was a well-established crop in eastern China over 2000 years ago, prior to the establishment of the 'Silk Road' to central Asia (Li, 1989) down which several occidental crops were introduced to China (Crisp, 1995). This crop was cultured in Europe in the fifteenth or sixteenth century and was introduced to England, France and America in the beginning of 16th century. In 1806, eleven sorts were known in America (Gill, 1993).

Early work suggesting that variability of the crop is greatest in Europe and least in Japan (Sinskaia, 1928; Sirks, 1957) now seems incorrect. There may be a much greater diversity of physiological and culinary types in Asia, particularly in China, than in Europe (Zhao, 1989), thus indicating east of Mediterranean, most probably China as the centre of origin of radishes. Some taxonomists consider that probably radishes have originated as crop plants in both Asia and Europe, with subsequent gene flow between the two major gene centres commencing with the establishment of the Silk Road (Crisp, 1995). However, some taxonomists do not consider China to be the centre of origin of radish.

Radish (*Raphanus sativus* Linn.) is not known in its wild state. It was previously thought to have been evolved from *R. raphanistrum* Linn., a widely distributed weed in Europe, but in view of considerable ecological and morphological differences existing among the cultivated radishes in the different regions of the world, they are now presumed to have been evolved from more than one source. The most important wild species of radish are *R. raphanistrum* Linn., *R. maritimus* Smith, *R. landra* Morett. and *R. rostratus* DC., all are found in the Mediterranean region. It is thought that the European types have originated from the above-mentioned wild species. The area from the eastern Mediterranean to the Caspian sea is likely origin of European types, but Werth (1937) have argued for a most easternly source. The Japanese types are thought to have been derived from *R. sativus* f. *raphanistroides* Makino (syn. *R. raphanistroides* Sinsk.), which occurs wild in the coastal regions of Japan. Indian group of radishes including rat-tail type is believed to have originated in the Near Eastern and Indo-Burma centre of origin though their ancestors have been lost (Bailey and Bailey, 1956; Mansfield, 1959). The rat-tail radish *R. caudatus* L. is considered to be closely related to the Indian group of radishes but more tropical in distribution (Mansfeld, 1959).

The cultivated radish is thought to originate from a single wild species [*R. raphanistrum* (De Candolle, 1883, Kobabe, 1959) or *R. maritimus* (Henslow, 1898, Schulz, 1919, Thellung, 1927)], or through interspecific hybridization of two wild species [*R. landra* × *R. maritimus* (Schulz, 1919, Kitamura, 1958, Clapham *et al.*, 1962)]. The Japanese types are thought to have been derived from East Asian wild radish *R. sativus* L. var. *hortensis* f. *raphanistroides* Makino (syn. 'Hama-Daikon', *R. raphanistroides* Sinsk.) (Kumazawa, 1956; Yamagishi and Terachi, 1996). Another idea is that *R. sativus* descended from a common ancestral species with *R. raphanistrum*, but that the ancestral species is now extinct. However, none of these ideas provide any decisive evidence, because they were mostly based on morphological observations and have no convincing genetic, cytological, or molecular data. Furthermore, the fact that all of the taxa of wild and cultivated species might be included in *R. sativus* (Pistrick, 1987) makes the study more complicated. Recently, Yamagishi and Terachi (2003) suggested that cultivated radish was of multiple origin as all of the mtDNA types present in the wild species (*R. landra*, *R. maritimus*, and *R. raphanistrum*) were also present in the modern cultivars of radish. AFLP markers variation showed a very close genetic relationship between *R. raphanistrum* and *R. sativus*, confirming the assumption that *R. raphanistrum* might be involved in the origin of *R. sativus* (Huh and Ohnishi, 2002). However, Yamane *et al.* (2005) reported that *R. raphanistrum* have a unique chloroplast haplotype not present in cultivated radish, hence was not the maternal ancestor of cultivated radish. Cultivated Chinese radish (var. *hortensis*) has originated from East Asian wild radish (Sinskaya, 1931, Yamane *et al.*, 2005) through a long history of gene flow between the two (Yamane *et al.*, 2005). Japanese cultivated radish varieties have a long history of more than a thousand years, and a large number of local varieties with various morphological characteristics have been developed, thus making the phylogenetic relationships among them so complicated that the evolutionary process has been difficult to analyze (Yamagishi and Sasaki, 2004).

The wild taxa of *Raphanus* occur through much of Europe and Asia, and are introduced weeds in America (Crisp, 1995). The greatest diversity of the important wild species of radish *R. raphanistrum* Linn., *R. maritimus* Smith, *R. landra* Morett. and *R. rostratus* DC., is found in the area between the Mediterranean and Caspian Seas, but 'wild radish' also occurs in China (Zhao, 1989) and Japan (Makino, in Pistrick, 1987). European wild radish, *R. raphanistrum* is widely distributed in Eurasia, Africa and North America, but is absent from China and India (De Candolle, 1883, Trouard-Riolle, 1914, Hegi, 1935). However, *R. raphanistrum* L. has been recently reported as a weed in wheat fields in Himachal Pradesh, India and is probably a recent introduction to the Western Himalayas from Europe (Sharma and Dhaliwal, 1997). *Raphanus raphanistrum* L. naturally occurs from Europe eastwards to the Volga and Mediterranean eastwards to the Caspian; *R. maritimus* Smith (sea radish) and *R. landra* Morett. (an inland form of *maritimus*) are distributed from Mediterranean eastwards to the Caspian and, in Europe, along the coasts of France, Belgium, Holland and Great Britain; and *R. rostratus* DC from Greece eastwards to the Caspian (Banga, 1976).

East Asian wild radish, *Raphanus sativus* L. var. *hortensis* f. *raphanistroides* Makino (=*R. raphanistroides* (Makino) Sinsk.) is a self-incompatible outcrossing species (Hinata and Nishio, 1980) and is called 'Hamadaikon' in Japanese. It grows primarily on dunes, sandy beaches, sandy cliffs or barren fields near the sea in East Asia, mainly China, Korea and Japan (Kitamura, 1958, Aoba, 1967, 1989, Lee *et al.*, 1992, Huh, 1995, Yamane *et al.*, 2005). Natural populations of East Asian wild radish have played important roles in the establishment of Japanese landraces of Chinese radish (Kumazawa, 1961, Aoba, 1993), hence, is very important study material in investigations of the origin of cultivated radish varieties and the genetic relationship between them (Kaneko and Matsuzawa, 1993) and from the viewpoint of crop evolution.

According to Banga (1976), the European or table types (small types) are thought to have originated much later than the large types. A long white form of it first appeared in Europe at the end of the 16th century, most probably from the large types (Helm, 1957). White globular forms were developed in 18th century and later on the red ones were also developed. In the earlier phases of the evolution of radish there was great variation in root shape (long, half-long, globular, pear-shaped, even flattened) and colour (white, red, yellow and black) (Banga, 1976). Garden radish and three groups of north Chinese small radish, occidental small radish and black radish were native to the western region of the Central Asia or southwest China. The wild form grown in coastal areas of Japan and south China played a part in the origination of this group (Kaneko and Matsuzawa, 1993). North Chinese radish was differentiated in the secondary centre in north China, where the wild and cultivated forms from north China, Central Asia and the plains of southwest China have contributed in its origin (Kumazawa, 1965).

Yang *et al.* (2002) sequenced the chloroplast noncoding region between trnT and trnF and ascertained that *Raphanus* is closely related to the rapa lineage. They hypothesized that *Raphanus* was derived from hybridization between the nigra lineage and rapa lineage, with the latter as the female parent. Lysak *et al.* (2005) estimated the time of the split between nigra and rapa lineages to be 7.9–14.6 million years ago. Mitsui *et al.* (2015) estimated the divergence between *R. sativus* and *B. rapa* as 16.7 million years ago.

3.2.0 Taxonomy

The genus *Raphanus* is classified into two sections: Raphanis DC. and Hesperidopsis Boiss. The section Raphanis DC. consists of six species, *R. sativus*, *R. raphanistrum*, *R. rostratus*, *R. landra*, *R. maritimus* and *R. microcarpus* (Kitamura, 1958), which can cross reciprocally with each other (George, 1986). However, the recent nomenclature categorize the wild species as *R. raphanistrum* and place three subspecies within it, namely ssp. *raphanistrum*, *landra*, and *maritimus* (Sahli *et al.*, 2008; Warwick and Francis, 2005). The section Hesperidopsis Boiss contains only one species, *R. aucheri* (Kitamura, 1958). Of the seven *Raphanus* species, only *R. sativus* is cultivated.

According to Banga (1976), there are four types of radish commonly cultivated in various regions of the world: (i) small, cool season radish, (ii) large radish with wide range of temperature adaptation, (iii) rat-tail or mougri radish forming no fleshy root but forming long slender (20-60 cm) pods, and (iv) fodder radish also producing no fleshy root. All four types of radish belong to the species *Raphanus sativus* L. with 2n=18, and botanically they are known as varieties *radicula, niger, mougri* (or *caudatus*) and *oleifera*, respectively. All the four types intercross freely with each other and also with related wild species.

Radish cultivars can be ecologically classified into 5 main varieties: *R. sativus* var. *niger* (Mill.) Pers.; var. *radicula* DC.; var. *raphanistroides* Makino; var. *caudatus* L. and var. *oleifer* Netz (Hida, 1990; Kaneko and Matsuzawa, 1993). According to Rubatzky and Yamaguchi (1997) there are seven botanical varieties of *R. sativus*:

var. *radicula* (garden or European radish); var. *longipinnatus* (Daikon, Chinese winter radish, varied types); var. *raphanistroides* (Japanese winter radish); var. *niger* (Black or Spanish radish); var. *caudatus* (Rat-tail radish, grown for its long edible tender seed pods); var. *moughi* (fodder radish, no enlarged storage root) and var. *oleifera* (oilseed radish). However, a division of *sativus* into small-rooted (sometimes referred to as var. *radicula*) versus large-rooted types (including names such as var. *nigra*, *niger*, *sinensis*, *acanthiformis* or *longipinnatus*) is too artificial to be useful as taxonomic character, although such a distinction may have some phylogenetic relevance, and reflects commercial development (Crisp, 1995).

Pistrick (1987) divided the cultivated forms into three groups: convar. *Oleifera* – the oilseed and fodder radishes; convar. *Caudatus* – the rat-tail radish (Syn var. mougri); and convar. *Sativus* – all forms with edible roots, which he divided into four geographic types - European, Chinese, Indian and Japanese. Recent study has shown that all of the basic variation extant in the Japanese types is derived from the Chinese gene base (Crisp, 1995).

Some authors divide *R. sativus* into more than one species because of the existence of well marked groups of radishes differing in morphological and ecological characters, the European types and Japanese types forming the two extremes, while the Indian types occupying an intermediate position. Thus, the wild radish found growing in the coastal areas of Japan and neighbouring regions, as well as the turnip-shaped giant radish cultivated in the Sakurajima Island of Japan, are together included under *R. raphanistroides* and the Indian types are grouped under *R. indicus* Sinsk., while the European types are put under *R. sativus*. *Raphanus sativus* var. *caudatus* (syn. *R. caudatus*), the rat-tail radish is considered to be closely related to the Indian group of radishes but more tropical in distribution (Mansfeld, 1959).

3.3.0 Cytology

Raphanus sativus and its wild relatives *Raphanus raphanistrum, R. landra, R. maritimus* and *R. rostratus* have 2n=2x=18, and are cross-fertile (Lewis-Jones *et al.*, 1982).

Cytological studies of interspecies hybrids between *R. sativus* and *B. rapa* and between *R. sativus* and *B. oleracea* has demonstrated close chromosome homology as indicated by the presence of up to six bivalents in the amphihaploid produced from each cross (Richharia, 1973). Recent work has shown close affinities between the *R. sativus* and *B. oleracea* genomes.

4.0 Morphology

Radish is an annual or biennial herb depending on types and belongs to the family Brassicaceae.

4.1.0 Leaves

The rosette leaves are alternate, lyrate-pinnatifid, and may vary from 10 to 15 cm in some small-rooted cultivars to as much as 45 cm long in some large-rooted cultivars; leaves are usually covered with stiff bristles.

4.2.0 Root

The edible portion of radish root develops from both primary root and the hypocotyl. Radish roots vary greatly in size, shape and other external characters, as well as the length of time they remain edible. The size may vary from 2.5 cm to 90 cm in length depending on cultivars. European small radish has a small round root of about 2–3 cm in diameter, whereas Japanese cultivar "Sakurajima-daikon" has a large (more than 20 kg) round root of more than 30 cm. Most widely grown radish cultivars in Asia have cylindrical shaped roots of around 10 cm in diameter and 40 cm in length. However, Japanese cultivar "Moriguchi-daikon" has a thin and long cylindrical root of more than 2 m with a diameter of about 3 cm. A radish landrace called "Newar" or "Jaunpur Giant", grown in Jaunpur district of Uttar Pradesh, India grows up to 75 -90 cm in length and 50 – 60 cm in girth. The shape varies from oblate to long tapering.

The radish roots have various colours. The majority of Asian radish has a white thick root and is called "East Asian big long radish," "white radish," or "daikon" whereas European small radish has mostly red root surface. The exterior colour may also be purple, green, black or various shades of scarlet. The flesh colour of most cultivars is white. However, some cultivars in China have red or green colour inside the roots.

Weaver and Bruner (1927) found that the distribution of the network of absorbing roots differs with the cultivars. The taproot of a seed plant may penetrate to a depth of 2 m and the lateral spread may be as much as 1 m. However, the bulk of the network of smaller roots of radish is confined mostly to the uppermost 30 cm of the soil.

The pithiness determines the edible quality of radish roots. Park and Fritz (1990) studied the effect of growing season, harvest date, fertilizer rate and soil moisture content on pithiness of radish. It was found that the degree of pithiness was greater in summer than in spring or autumn, and increased with delay in harvesting. Pithiness was closely correlated with average temperature, average sunshine duration/day and solar radiation levels during the 3 weeks before harvest. Pithiness increased as the amount of NPK fertilizer applied was increased and the soil moisture content decreased from 90 per cent to 50 per cent.

4.3.0 Inflorescence

The inflorescence of radish is a typical terminal raceme of Brassicaceae. The flowers are small, white, rose or lilac in colour with purple veins in bractless racemes; sepals erect; petals clawed.

4.4.0 Fruit

Fruit is indehiscent corky siliqua, rounded at base, with a long conical beak at apex, and smooth or rarely slightly constricted between seeds. Silique size differs among cultivar types, but is usually 3-7 cm long and up to 1.5 cm in diameter with 6-12 seeds. The pods of *R. caudatus* (rat-tail radish) may be 20-100 cm in length but usually 20-30 cm.

Siliques of the family Brassicaceae have two parts, a beak formed from the sub-stylar region and a valvar portion having valves. Most species of Brassicaceae have seeds in the valvar portion and not in the beak. However, radish silique has no valvar portion and seeds are present in the beak. In radish, seeds are larger (about 5 times in weight that of turnip) and arranged in a line in the silique whereas in most other Brassicaceae species, the seeds are arranged in two lines in a silique.

4.5.0 Seed

Radish seeds are globose or ovoid, about 3 mm in diameter. Radish seed when mature is at first yellowish but turning reddish brown with age. There will be about 120 seeds per g. The seeds take 7-14 days to germinate under optimum environmental conditions and remain viable for 5 years.

The principle difference used by taxonomists to separate the two species *R. sativus* and *R. raphanistrum* is pod morphology (de Candolle, 1886). In *R. sativus* the silique is continuous and forms a single cavity. *Raphanus raphanistrum*, in contrast, has siliques that are articulated with a single seed in each division. The seeds are dispersed when the divisions break apart. *Raphanus sativus* var. *hortensis* f. *raphanistroides* is herbaceous, diploid biennial, with 2n=18 (Kitamura and Murata, 1957). Three leaf types were found, of which the pinnatifid (PF) form is predominant, the lyrate (LY) form is less common and the pinnatisect form rare (Yamaguchi, 1987). The plant height is 30-60 cm with purple or whitish-pink flowers. Flowers are visited by insects and are cross-pollinated. Each silique contains 1-10 seeds and the matured siliques do not burst (Huh and Ohnishi, 2001).

4.6.0 Cultivar Groups

There are four broad cultivar groups of culinary radishes, small-rooted European radish, and large-rooted Asiatic radish or Chinese radish (daikon), leaf radish and rat-tailed radish.

(a) **Small-rooted or European radish:** The small, short season European types are exclusively grown in the field as garden salad crop throughout temperate regions, and can be produced under plastic or glass in Arctic regions. They are used as relishes, appetizers and to add variety and colour to green leaf salads. Intensive and early production under glass is done by specialist growers. The storage root sizes range from 2 to 5 cm in diameter and length less than 15 cm. They differ in root shape - from highly elongated to flattened spheres, the most common being globe shape. Skin colours vary from red to white, or a combination of both. They are harvested within a short period that may range between 20 and 40 days. Most radish cultivars have a relatively short post-harvest life of 7-10 days.

(b) **Large-rooted or Asiatic radish:** The large-rooted radishes are mostly Asian and is an important vegetable in Japan, Korea, China, India and other eastern countries where it is eaten raw or cooked or preserved by storage, pickling, canning or drying (Banga, 1976). They are far more important and diverse than the small European types and include types suited to temperate or tropical conditions. Root shapes vary from highly elongated to flattened spheres. Skin colours include white, red, green, yellow, purple, and black. Flesh can be white, red, purple or green. In Tibet and north-east China they are stored as intact roots as winter food and in Korea they are pickled with other vegetables in large underground vats. A few distinct winter storage types occur in Europe, but most have now virtually disappeared (Crisp, 1995). The large-rooted radish has a wider range of temperature adaptation, storage ability and reaction to various pests and diseases, which has not been fully exploited yet.

Chinese winter radish (daikon) is a major vegetable in Japan, China, Korea and many other Asian countries and is highly appreciated in Japanese diets. Daikon is eaten fresh, dried, pickled and can be sliced, diced and cooked. They require a longer growth period to produce the large storage root and foliage biomass. Leaf blades are often strongly notched and may be erect, spreading, or prostate. Daikon root size range from 10 to more than 50 cm in length, 4 to 10 cm in diameter, and 1 and 4 kg in weight. Roots, as heavy as 20 kg (like 'Jaunpuri mooli' or 'Newari' of U.P, India), are also obtained. However, edible quality diminishes with excessive size. Faster growth rate can result in early pithiness. Under refrigerated storage conditions, many daikon cultivars can be stored for 3-4 weeks.

Cultivars or varieties of Chinese radish are classified into 4 ecological types, *i.e.* south Chinese, middle Chinese, north Chinese and western plateau Chinese. The south Chinese radish has low starch content and watery roots, hence not suitable for storage. However, they are suitable for salted vegetables and cooking. The north Chinese radish has a relatively smaller root with high starch content and a greenish or reddish skin colour, and can be consumed after storage (Kaneko and Matsuzawa, 1993).

The Japanese radishes are divided into 15 groups (Kumazawa, 1965), the major ones being 'Minowase', 'Nerima', 'Miyashige', 'Shogoin', 'Hatsuka' and 'Ninengo' (Kaneko and Matsuzawa, 1993). Japanese radish is derived from south Chinese radish and introduced to Japan from China. The north Chinese radish and Hatsuka radish were grown for salad at the beginning of The Meiji period (1868) (Aoba, 1979). The derivatives of north Chinese radish were domesticated predominantly in the restricted mountainous regions of central Japan such as Shinshu, Sanin, Hokuriku and Tohoku provinces, and now form some local varieties called 'Jidaikon' groups (Kaneko and Matsuzawa, 1993).

(c) **Leaf radish or fodder radish:** Fodder radish which bears little or no fleshy root and may be a derivative of an ancient oilseed form (*R. sativus* var. *oleiferous*), is grown in Southeast Asia, Africa and Europe for leafy fodder, and as green manure (Banga, 1976, Crisp, 1995). Stewart and Moorhead (2004) reported a unique smooth-leaved, low-crowned, late-flowering fodder radish, 'Ceres Graza', originated from a complex series of crosses and selections, carried out over 17 years, involving vegetable garden radish (*Raphanus sativus*), cabbage (*Brassica oleracea*) and perennial seaside radish (*Raphanus maritimus*).

(d) **Rat-tail radish or Mougri radish:** Rat-tail radish or Mougri-radish is grown in Southeast Asia specifically for its leaves and its young long immature seed pods, which is usually 30 cm but may grow to 80-100 cm long.

Besides above, other cultivar groups include Spanish radish which is an annual winter radish type and produces round, usually black-skinned roots of 10 to 15 cm in diameter, maturing in 50-90 days after planting. Spanish radishes can be stored for 2-3 months under properly refrigerated condition.

5.0 Cultivars

A large number of cultivars, indigenous as well as introduced, are cultivated in the different regions of India. The indigenous cultivars are usually white with conical shape, attaining 25-40 cm in length and are said to be more pungent than the introduced European types. European types are not very common and popular in India. A type cultivated in and around Jaunpur in Uttar Pradesh, known as 'Newari' or 'Jaunpuri Giant' reaches an enormous size with a length upto 75-90 cm and a girth of 50-60 cm and may weigh up to 5-15 kg or even more. 'Baramasi' is another indigenous type that can be grown throughout the year. Pusa Desi is an improved indigenous type. Important characteristics of some of the recommended cultivars are given below:

5.1.0 Tropical Cultivars

Pusa Desi

It is a tropical or subtropical cultivar, derived by mass selection from heterozygous materials collected from old Sabzi Mandi, Delhi, suitable for sowing from middle of August to October in the northern plains of India. Roots are pure white, 30-35 cm long, tapering with green stem end, pungent and heavy yielder. It matures 50-55 days after sowing (Sirohi *et al.*, 1992). Average yield is 300 q/ha.

Pusa Chetki

It is a selection made from a material collected from Denmark (Choudhury and Sirohi, 1975). Since it can tolerate high temperature it has been found suitable for sowing from the middle of March to the middle of August. This cultivar sets seeds profusely in the plains, because it bolts very early in the month of late October or early November. Roots are medium large and stumpy, pure white, tender and smooth and mildly pungent. It becomes ready for harvesting in 40-45 days after sowing. It may be grown as a summer and monsoon season crop from April to September. During summer-sowing, though the yield will be comparatively low, it is compensated by the high price it fetches at that time. With the introduction of the Pusa Himani and and Pusa Chetki, it is now possible to grow radishes all the year round under North Indian condition. Its root is medium (20-25 cm), stumpy, pure white, tender, smooth and mildly pungent. It matures in 40-45 days and has yield potential of 200-250 q/ha.

Pusa Reshmi

It is suitable for main season (October to November). Leaves are small to medium size and green. The roots are mildly pungent, 30-35 cm long, tapering, and white with green tinge on root shoulder. Good size roots become ready for harvesting in 55-60 days. It has yield potential of 350 q/ha.

Japanese White

It was originally named as Shiroaguri-Kyo and released by IARI under the name Japanese White. The roots are cylindrical, 22-25 cm in length and 5 cm in diameter. The skin is snow-white, flesh crisp, solid and mildly flavoured. It matures in 45 days. It is suitable for sowing during October-December in the plains and July-September in the hills. Yield potential is 350-400 q/ha.

Pusa Mridula

Its leaves are dark green and tender. The roots are globular with root size of 2.9 × 4.0 cm. The roots are red skinned and white fleshed with soft texture and mild pungent taste. It is quick growing cultivar and reached marketable maturity in 20-25 days after sowing. It is suitable for sowing from September to February. Average yield is 135 quintals per hectare.

Pusa Shweta

It is characterized by medium long, white, cylindrical roots suitable for sowing from September to November in North Indian Plains. It takes 50-55 days to marketable maturity. The average root weight is 200 g with average root yield about 400 q/ha.

Pusa Jamuni

It is first purple fleshed unique trait nutritionally rich radish cultivar. It has higher anthocyanins and ascorbic acid. It matures within 55-60 days. Average yield 300 q/ha.

Pusa Gulabi

It is first entire pink fleshed unique trait nutritional rich radish cultivar. It has medium root size with cylindrical. It matures within 55-60 days. Average yield 300-350 q/ha.

Punjab Safed

Roots are white, tapering, smooth, mild in taste, free of forking, 30-40 cm in length and 3-5 cm thick. It is a quick growing type with roots remaining edible for 10 days after attaining full size.

Punjab Pasand

Developed by repeated crossing of selected plants from a base population of cv. Pusa Chetki. It is a main season, quick growing cultivar, attaining marketable maturity within 40 days after sowing. Roots are long, pure white, semi-stumped and free from hair. Mean root yield is 534.8 q/ha. It has 5.75 per cent dry matter, 3.4 per cent total sugars, 0.67 per cent crude fibre and 16.48 mg/100 g ascorbic acid (Singh *et al.*, 1998).

Punjab Ageti

It is also developed at PAU, Ludhiana. Roots are bicolour *i.e.* red skinned at the top and white at the lower half, tapering, long (24-28 cm), medium thick (2.5-3.0 cm), less pungent and smooth with few hairs. It is most suitable for sowing during April to August.

Kashi Sweta

It is developed through selection from Chetki population at Indian Institute of Vegetable Research (IIVR), Varanasi. It is recommended for cultivation in the states of U.P., Punjab, Bihar and Jharkhand. The leaves are green colour and sinuate type. Roots are 25-30 cm long, 3.5-4.0 cm in diameter, almost straight, bulging above ground, tapering with pointed tip. Roots become ready for harvesting in 35-40 days after sowing. Average root weight is 150-200 g. It gives yield of 450-470 q/ha.

Kashi Hans

This cultivar is also developed through selection at IIVR, Varanasi. It is recommended for cultivation in the states of U.P., Punjab, Bihar and Jharkhand. It is suitable for sowing form September to February, and first harvesting can be done within 40-50 days after sowing. It can stand in the field without pithiness up to 10-15 days after edible maturity, hence favours delayed harvesting without impairing root quality. The leaves are soft, smooth and lyrate type. Roots are mild pungent, straight, tapering, 35-40 cm long and 3.5-4.2 cm in diameter. Average root weight is 140-200 g. The yield potential is 430-450 q/ha.

Kashi Mooli-40 (VRRAD-203)

Tolerant to high temperature (35-42°C), suitable for summer cultivation, delayed bolting less pithiness, iciclical root shape, attractive white colour, having root yield of 30-35 t/ha in normal season and 20-23 t/ha during summer season. It is recommended for cultivation in Uttar Pradesh.

Kashi Lohit (VRRAD-131-2)

Attractive red colour roots, iciclical root shape, suitable for salad dressing, excellent source of anti-oxidants (80-100 per cent higher than white radish), good source of vitamin C and phytochemicals *i.e.* Anthocynin and phenolics content. Tolerant to pithiness. Root yield is 40-45.0 t/ha. It is recommended for cultivation in Uttar Pradesh.

Kalianpur No. 1

Recommended for Punjab, Uttar Pradesh, Bihar, Jharkhand. Roots are 22-23 cm long, smooth, white, crispy, thick, tapering with green shoulder. Top is heavy, leaves and stem green, long, broad and less lobed. It is free from mustard sawfly, aphids and white rust.

Arka Nishant

Recommended for Karnataka, Tamil Nadu, Kerala. Roots are long, marble-white, crisp in texture with mild pungency. It matures within 45-55 days. Roots are medium sized (25 cm × 3-4 cm), pleasant aroma. It sets seeds freely in plains. Resistant to pithiness, root branching and forking. Yield is 300 q/ha.

IHR-1-1

Roots are thin, long (30 cm), free from pre-mature bolting, pithiness, splitting and forking and mildly pungent, each root weighing about 300 g. Root surface is smooth and shining white. It takes about 45 days to reach harvest maturity. It yields is 200-250 q/ha.

CO 1

Recommended for Karnataka, Tamil Nadu, Kerala. Plants are vigourous with green leaves and wavy margins. Roots are milky white, 23 cm long, cylindrical and less pungent. It sets seeds freely in plains. Yield is 20 t/ha in a crop duration of 45 days. Susceptible to rust and aphids.

Hisar Sel.-1

It is developed by CCSHAU, Hisar. Roots are mild pungent, medium long, smooth, soft, white and straight. It is grown in September-October and requires 50-55 days to mature. Its root to shoot ratio is about 1.70. Average yield is 300-325 q/ha.

Mino Early

It is a quick growing cultivar and an introduction from Japan. The roots attain harvesting maturity 40 days after sowing. Roots are pure white, 30-40 cm long, slightly tapering and mildly pungent.

Nadauni

It is a popular cultivar of Himachal Pradesh. Its root is long, tapering, light pink in colour, top dark green with cut leaves.

Kalyani White

The roots are 25-30 cm long along with blunt tip and uniform in size. The skin is pure white and flesh mildly pungent. Plants are having light green top. It can be grown throughout the year except during summer. The edible maturity appropriates in 45-50 days. It was developed at BCKV, West Bengal, India.

Chinese Pink

Recommended for Jammu and Kashmir, Himachal Pradesh, Uttarakhand, Punjab, Uttar Pradesh, Bihar and Jharkhand. The roots are 12-15 cm long, semi-stumpy to stumpy, pink with white colour towards the tip. The roots are smooth, crisp, solid, and mildly pungent along with shining red exterior root and white interior root. It matures in 50-55 days. This is most suited for hills but grows well in plains with mild climate.

Palam Hriday

Recommended for Jammu and Kashmir, Himachal Pradesh and Uttarakhand. Strap-like entire margin leaves. Roots are oblong having green shoulders and cream globe with 6-7 cm diameter and small top. Flesh is attractive pink, crisp, juicy. Late bolting and free from pithiness. Roots reach edible maturity in 50-60 days. Suitable for sowing from October to November in north Indian plains and September in mid hills, March-May and June-July in dry temperate areas. Average yield is 300 q/ha.

5.2.0 Temperate/European Cultivars

Rapid Red White Tipped

The roots are small, smooth, round or globular, upper portion of the roots are red and lower portion is white. Flesh is pure white, crisp, snappy and sweet in taste. It is suitable for sowing in kitchen garden or container garden from October to January. It is extra early, matures in 25-30 days. Yield is 50 q/ha.

White Icicle

Roots small (8-10 cm long), thin, pure white, icicle-shaped, tapering, crisp, juicy and sweet flavoured. This cultivar is suitable for sowing in kitchen garden or container garden from October to January. It matures in short duration *i.e.* 30 days and have yield potential of 50q/ha. The main disadvantage of this cultivar is that the roots become pithy within a week after maturity.

Pusa Himani

It is an attractive cultivar for market and home use developed at IARI Vegetable Research Station, Katrain (Kulu valley) by hybridization between a temperate type (Black) and a popular Asiatic type (Japanese White). This cultivar is suitable for growing all the year round in the hilly regions of North India. It can also be grown successfully in spring in the plains with mild climate. The roots of Pusa Himani are 30-35 cm in length and 10-12 cm in girth, white with green stem end. They are semi-stump to tapering with short tops. The skin is pure white; the flesh is crisp and sweet-flavoured with mild pungency. The top is short, leaves green having less hair than Japanese White. It takes 60-65 days from sowing to maturity. It has wide adaptability and is grown almost throughout year in the North Indian hills and December to February in North Indian plains. Yield is 300-325 q/ha.

Scarlet Globe

It is an introduce material. Recommended for Jammu and Kashmir, Himachal Pradesh and Uttarakhand. Its roots are round, small, 2.0-2.5 cm in diameter and bright red in colour. The foliage length is 10-15 cm. The root is crisp, soft and red root exterior along with white root interior. It takes 25-30 days from sowing to complete root formation. The delayed harvesting results in pithiness.

Scarlet Long

It is an introduce material. Recommended for Jammu and Kashmir, Himachal Pradesh and Uttarakhand. It is an early cultivar. The leaves are 15-20 cm long and light green in colour. Roots long, tapering to a point with red skin. Flesh is white and crisp. It is ready for harvesting in 30-40 days after sowing.

Kvarta

It is a short duration cultivar suitable for protected cultivation. The roots are red, weighing 8-15 g, and the flesh is white, tender and juicy. Content of dry matter is 5.2 per cent, sugars 0.88 per cent and vitamin C 32.3 mg/100 g. Harvest maturity is attained in 24-30 days after emergence and yield is 1.7 kg/m² (Antonova, 1991).

The main-season radish crop does not require any special technique for its cultivation, but for spring and summer season crops the growers should keep in mind the following points:

i) The right cultivar should be selected and sown at the right time.

ii) There should be adequate moisture supply.

iii) Spring and early summer crops are attacked by certain insects, especially aphids, hence adoption of proper plant protection measures is essential.

iv) Closer spacing should be given as compared with the main-season crop.

v) Pre-harvesting light irrigation is very useful for the summer crop, because it keeps the roots fresh and makes it less pungent.

Radish is very specific in its climatic requirement, especially temperature. In general, the temperate cultivars start bolting without forming marketable roots, if sown in the later part of January, because of increasing day length. There are specific cultivars for different periods of the year and no particular cultivar of a group (tropical, subtropical or temperate) is suitable for growing all the year round. Therefore, it is very important to select the right cultivar for sowing in a particular season. Choudhury and Sirohi (1972) reported a time schedule for growing radish throughout the year in the northern plains, recommending suitable cultivars for every season.

In a trial on the performance of radish cultivars in Maharashtra, Japanese White and Pusa Himani gave considerably higher yield and White Icicle had the best flavour (Gupta *et al.*, 1974). Under Bihar condition, of the 12 radish cultivars tried, Pusa Himani performed best (Pujari *et al.*, 1977). In Varanasi (U.P.), both Pusa Reshmi and Jaunpuri were found to perform well (Maurya *et al.*, 1972, 1977). Nautiyal *et al.* (1974) obtained highest yield from Newari (88.8 q/ha), followed by Kalianpur Type I (85.9 q/ha). Under Nagpur (Maharashtra) condition, Pusa Reshmi proved as the best cultivar (Deotale *et al.*, 1994). For the Karnataka region, Arka Nishant was found as the most promising of the cultivars tested (Anjanappa *et al.*, 1998).

Some 22 radish (*Raphanus sativus*) cultivars were evaluated at 6 locations in Germany during 1996-99. Content, Lucia and Gina were best for early and Donar, Estared, Super Red and Tarzan for very early harvests. Estared, Lucia, Metro, Rudi, Superred and Vitared had good root quality (Heine and Dreyer, 2000).

Nguyen *et al.* (1999) studied the effect of planting time on the growth and quality of 7 Japanese white radish cultivars in Australia. Results showed that the processing cultivars always achieved a standard root size of 1000 g about 5-10 days later than fresh market cultivars. The occurrence of pithiness and pungency varied among cultivars. Generally, processing cultivars showed a significantly higher level of pungency than fresh market cultivars.

Important Radish Cultivars

Kashi Sweta

Kashi Hans

Kashi Mooli

Japanese White

Chinese Pink

Rapid Red White Tipped

Important Radish Cultivars

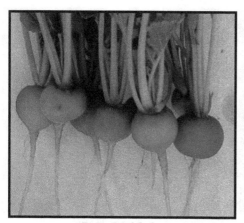

Pusa Mridula

Pusa Sweta

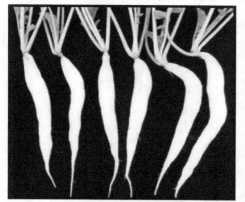

White Icicle

Pusa Chetki

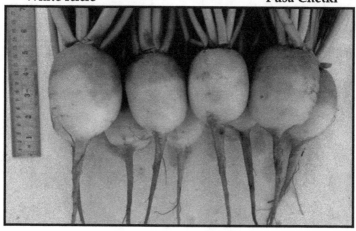

Palam Hriday

Sharma *et al.* (2002) evaluated 28 cultivars of radish for root yield and component traits in Palampur, Himachal Pradesh, India. Based on root weight, length, root diameter, girth and yield per plot, the superior Asian cultivars Mino Early White, Sutton's Long White and Nadauni, and European cultivar Palam Hirday were identified. Langthasa and Barah (2000) analysed the performance of nine radish cultivars (Long White Icicle, Japanese White, Pusa Chetki, Pusa Rahsmi, Pusa Himani, Bombay Long White, Crimson French Breakfast, Punjab Selection and Bombay Red) and recommended cultivars Pusa Chetki and Punjab Selection as most suitable for the hill zone of Assam. Among radish cultivars Pusa Himani, Pusa Chetki, Japanese White, Pusa Reshmi, Bombay Red, Jaunpuri and Punjab Safed, the cultivar Bombay Red has almost consistent performance in both dry and rainy seasons in South Andaman (Dhar, 1998). The F_1 hybrid KSSC-03 has crimson-white highly pungent roots (Badiger *et al.*, 2001).

6.0 Soil and Climate

6.1.0 Soil

Radishes can be grown on nearly all types of soils, but the best results are obtained on light, friable loam soil that contains ample humus. It is moderately tolerant to soil acidity (pH 6.8-5.5), and the root and shoot growth is adversely affected below pH 5.5. Acidic condition (pH <5.5) adversely affects the plants *viz.* damaging the leaves, leaching the essential plant nutrients from soils, releasing aluminium ions attached to insoluble soil particles, releasing ions of metals such as cadmium, lead and mercury that are toxic to plants, thus making the plants more susceptible to biotic and abiotic stresses. Since it is a short duration crop, it can be grown on soils that are not considered satisfactory for other root crops. For early crop, sandy or sandy loam soils are preferred; however, for summer crop a cool and moist soil gives best result. Usually, heavy soils ad soils having unrotten humus with hard and stony subsoil produce rough, misshapen/deformed roots with a number of small fibrous laterals, and such soils should be avoided. High levels of soil salinity (>4 dS m⁻¹) inhibit seed germination, plant growth, root development and yield of radish due to the effects of high osmotic potential and specific ion toxicity by disturbing the nutrient uptake and other metabolic activities of the plant.

6.2.0 Climate

Radish is best adapted to a cool or moderate climate but tropical varieties can tolerate higher temperature and have greater temperature adaptation. The tropical Indian varieties, particularly the long rooted, can better withstand heat and rain than the temperate types. The seeds germinate in 3-4 days with soil temperatures of 18-30°C and adequate soil moisture. However, the minimum and maximum temperatures for germination varied from 5-35°C. The temperature of 10-15°C is ideal for proper growth and root development, and its texture and flavour. Maximum root growth occurs initially at 20-30°C but later at temperature of 10-14°C, particulary in the temperate varieties than the tropical ones. Higher temperature above 25°C favours more foliage growth. Wendt (1977) reported that increase in soil temperature from 5 to 20°C favoured leaf development. At 20-25°C uptake of nutrients was greater and dry matter content was higher but at a still higher temperature (25-30°C) the water requirement of the plant increased considerably and root deformation occurred. Nieuwhof (1976, 1978) obtained higher leaf and dry matter production at 20-23°C, maximum root growth initially at 20-23°C but later at lower temperatures (10-14°C). Radish yield, shape, smoothness, quality, flavor and taste decline in hot weather. During hot weather the roots become tough and pungent before reaching the edible size. When temperatures are too hot, the root growth and development is suppressed. High temperature also causes pithiness, hot flavour (pungent), bitterness and hollow root before they reach the harvestable/ edible maturity.

Root shape is also influenced by photoperiod and light intensity. In general, roots grown under high light intensity are thicker and shorter than under low light intensity (Hall, 1991). He also found that roots grown at 27°C were longer than those grown at 21°C. Photoperiod also influenced root development in radish. Sarkar *et al.* (1978) obtained the largest roots under 9 hour photoperiod. It appears that all cultivars do not respond equally to photoperiod.

Knecht (1975) observed increased growth of both plant tops and roots of radish cv. Cherry Belle due to elevated CO_2 concentration (1200 ppm) in environmentally controlled plastic growth chambers. The root-shoot ratio was favourably influenced by CO_2 enrichment, compared with controls at 400 ppm CO_2. Elevated CO_2 increased the accumulation of dry matter, but had no effect on the rate of photosynthesis. Sink activity of the storage root was also increased by elevated CO_2. Up-regulation of sucrose synthase activity in the hypocotyl might be responsible for the increased sink activity (Usuda *et al.*, 1998).

Flowering of radish, especially of biennial types, is greatly influenced by temperature. Low temperature is a critical factor causing flowering, which is accelerated by long photoperiod (Park *et al.*, 1976). Long day photoperiod along with high temperature triggers early bolting even prior to proper development of roots (premature bolting), which is an undesirable trait. Normally, the plants bolt when days are 8-10 h long. Radish crop can easily withstand frost.

Eguchi *et al.* (1963) reported that flower bud differentiation, flower stalk development and flowering in Japanese radish were hastened by low temperature treatment. The older the seedling, the greater was the sensitivity to low temperature. An increase in the period of exposure from 3 to 30 days resulted in earlier and more uniform flowering, fewer abnormal flowers, a higher percentage of fruit set, larger fruits, heavier seeds and a higher percentage of large seeds. Yoo and Uemoto (1976) reported that flowering in Japanese radish (cv. Wase-Shijunichi) was hastened by vernalization of seeds at 5°C for 10 days before sowing. Yoo and Uemoto (1975) found that immediately preceding flower-bud formation, the level of endogenous growth substances in radish leaves declined. The level of endogenous growth substances was lower in plants raised from seeds which had been treated at 5°C for 20 days than in plants raised from untreated seeds.

Park *et al.* (1976) reported that bolting and flowering of oriental radishes were more sensitive to temperature than day length, whereas European radishes responded equally to both factors. Koyama *et al.* (1968) reported that growth, floral development and seed formation of vernalized plants were accelerated by high temperature, whereas insufficiently vernalized plants were sensitive to low temperature.

The effect of 3 levels of shading (0, 30 and 50 per cent) on the development and tuberous root yield of radish (*Raphanus sativus* L.) cv. Vermelho Redondo was studied by Souza *et al.* (1999) under field conditions at Sao Manuel Country, Sao Paulo, Brazil. The plants were evaluated at 7, 14, 21 and 28 days after emergence. The 50 per cent level of shading increased the life cycle and foliar area, and reduced the leaf chlorophyll content and the tuberous root yield. The 30 per cent level of shading did not reduce the size or weight of the roots.

7.0 Cultivation

7.1.0 Sowing and Seed Treatment

The soil for radish should be thoroughly prepared so that there is no clod to interfere with the root development. The soil should not contain any undecomposed organic matter, because that may result in forking of roots or mis-shapen roots.

In the northern hills, the seeds are usually sown in the first fortnight of March and continued till late October or the beginning of November depending upon the temperature. In low temperature the germination is poor. The main season of sowing of indigenous types in the northern plains is from August to January and for temperate types from September to March. April–June and October–December are best sowing time for Maharashtra, while March-August for Karnataka, Tamil Nadu and Andhra Pradesh. In areas with mild climate radish can be sown all the year round. In Bangalore, it is grown almost throughout the year, but the best quality roots are produced from November to December sowing. For regular supply of fresh and tender roots, the sowing should be done at 15 or 20-day interval. Bold seeds germinate better, producing vigorous and healthy plants.

Radish seeds count about 100-125 seeds per gram, and about 9-12 kg of seed is sufficient to sow one hectare depending upon the type and spacing. Seeds are treated with a fungicide (captan or thiram @ 3g/kg seeds) to reduce the incidence of seed borne or soil borne pathogens causing seedling diseases.

Seeds are treated first with *Trichoderma* @ 4g/kg of seeds the night before sowing and dried in shade. Next morning the treated seeds are coated with bio-fertilizer slurry (Azotobacter 200 g + 200 g Phosphobacteria + 200 g K mobiliser) before sowing.

The spacing varies widely depending on the type of radish. For the Indian/tropical cultivars, relatively wider spacing is used; the seeds are usually sown on ridges 45 cm apart and the plants are later thinned out to keep them at a distance of 6-8 cm in a row. Depth of sowing is 0.5 to 1.0 cm. Temperate varieties which are ready to harvest in 25-30 days are given closer spacing. Sounda *et al.* (1989) obtained the highest yield with a spacing of 30 × 10 cm.

It is advisable to sow the seeds either on ridges or raised beds to improve root growth, shape and appearance. Pandey and Joushua (1987) found that ridge sowing was better than sowing on flat ground.

Radish seeds under good storage condition may retain viability for several years, however, under tropical and humid condition the viability is lost within a year. The seeds usually give a good and rapid germination. Singh *et al.* (1990) found that seed treatment with GA$_3$ at 30 ppm increased growth and yield attributes in cv. Japanese White. Sundaravelu and Muthukrishnan (1993) recorded longer and thicker roots and a greater root yield when the seeds of cv. Japanese White were treated with *Azospirillum* sp. at 75 g/kg seed and GA$_3$ at 10 ppm.

7.2.0 Technology for Year Round Cultivation of Radish

Radish is one of the important salad vegetable crops required throughout the year. Being a long day crop, its one variety cannot be employed for growing roots throughout the year. Division of Vegetable Science, ICAR-IARI, New Delhi has developed different cultivars with specific growing requirements for cultivation in different seasons. There are five such cultivars below which can be grown under North Indian plains for sowing right from August and continued for the entire year so as to make fresh radishes available to the consumers (Table 2).

Table 2: Improved Radish Cultivars for Growing in different Seasons

Cultivars	Period of Sowing	Period of Harvesting
Pusa Desi	Middle of August to Middle of October	Mid-September to Mid-December
Pusa Mridula	1st fortnight of September to Mid-November	2nd fortnight of October to 1st fortnight of January
Japanese White	Mid-October to 2nd fortnight of December	Mid-December to 1st fortnight of March
Pusa Himani	2nd fortnight of December to end of February	Mid-February to 3rd week of April
Pusa Chetki	1st week of April to mid-August	1st fortnight of May to 2nd fortnight of September

7.3.0 Manuring and Fertilization

Radish is a short-duration and fast growing crop, hence judicious and proper use of manures and fertilizers is essential to get good yield and excellent root quality. It has already been stated that application of fresh undecomposed manures should be avoided to prevent forking or branching of edible roots. Judicious application of NPK will favour early vegetative growth and rapid root thickening leading to high and quality root yield in radish. About 15-20 tonnes of well rotten FYM should be mixed with soil at last tilling or at least 15 days before sowing. Depending upon the soil fertility status and variety, 60-80 kg N, 50 kg P and 50 kg of K should be applied per hectare. The complete doses of P and K, and half of N should be added to the soil before sowing as basal dressing. The remaining half of N is top dressed in 2 split dose *viz.* during early plant growth and at root formation. Moderate amounts of boron, molybdenum, zinc, copper and manganese should also be applied. The boron deficiency, hollow root, could be managed by soil application of borax at 15-20 kg/ha before sowing. Molybdenum deficient plants develop narrow and leathery leaves and growth is checked. Application of sodium molybdate or ammonium molybdate at 1.0-1.2 kg/ha of controls this disorder.

Haag and Minami (1987) estimated that a population of 12,50,000 plants/ha (cv. Early Scarlet Globe) removed a total of 276 kg N, 38.7 kg P, 574.6 kg K, 105.1 kg Ca, 45.8 kg Mg, 23.4 g Cu, 319.5 g Fe, 366.8 g Mn and 341.9g Zn, 48 days after germination.

Misra (1987) reported increased seed yield in cv. Pusa Reshmi with increasing rates of N and P up to 80 kg/ha in calcareous soils of North Bihar. Rastogi *et al.* (1987) obtained the highest seed yield (1.28 t/ha) in cv. Japanese White with 100 kg N/ha applied 20 cm below soil surface. Sharma (1987) also recorded the highest seed yield by application of 100 kg N/ha at 20 cm below the soil surface. In another study, Sharma (2000) found that the highest N rate (200 kg/ha) delayed flowering and decreased pod and branch numbers/plant compared with intermediate N rates (100 or 150 kg/ha). Gill *et al.* (1995) reported that seed yield increased up to 75 kg/ha and as P rate increased up to 20 kg/ha. A study conducted in Korea revealed that 40 kg N/ha resulted in the highest rate of pod set and seed production with line 25 of hybrid cultivar Pyungji Yeoreum, while line 27 exhibited the similar trend for pod set but number of flower stalk increased with N rate up to 160 kg/ha, resulting in the highest seed yield (Yoon *et al.*, 1996). Earlier, Kumar *et al.* (1994) recorded the highest seed yield in cv. Pusa Chetki when the plants received N and K at 150 and 40 kg/ha, respectively.

Sharma and Raina (1993) reported from Himachal Pradesh that seed yield in cv. Japanese White increased with increasing N application up to 75 kg/ha. N recovery and N use efficiency were optimal at 75 kg N/ha and irrigation to 90 per cent cumulative pan evaporation (CPE). With irrigation of 30 per cent CPE, N use efficiency fell with increasing rate of N used. Lovato *et al.* (1994) obtained optimum seed yield in annual radish with an application of 150 kg N/ha, while Sharma and Kanaujia (2000) recorded highest seed yield in cv. Japanese White with 200 kg N/ha. Rastogi *et al.* (1987), on the other hand, recorded the highest seed yield (11.36 q/ha) in cv. Japanese White spaced at 60 × 45 cm and receiving 60 kg N/ha.

Sharma (2000) found that 200 kg N rate delayed flowering and decreased pod and branch numbers/plant compared with intermediate N rates (100 or 150 kg N/ha). Jadhav *et al.* (1998) noticed that the seed yield in cv. Pusa Chetki was lowest (5.08 q/ha) in plants supplied with 50 kg N and 50 kg P_2O_5/ha. Kanaujia and Sharma (1998) obtained the highest seed yield (12.36 q/ha) and net returns of Rs. 13413/ha in cv. Japanese White with the combination of transplanting on 25 November and application of 200 kg N/ha in Himachal Pradesh. Bhople *et al.* (1998) recorded the highest seed yield, 1000-seed weight and germination percentage in cv. Pusa Chetki by spraying plants with 100 ppm NAA as bolting began. Rankov *et al.* (1988) reported that fertilizer (NPK) rates and proportions had little effect on germination rate, germination energy or 1000-seed weight.

At Rajouri (Jammu and Kashmir), Bhat (1996) found no significant benefit from the application of P, but the root yield increased with increasing N rates, the highest yield (38.4 t/ha) being recorded with the highest N rate (70 kg/ha) (Table-3). Bhople *et al.* (1998) were of the view that application of 100 kg N and 50 kg P$_2$O$_5$/ha in the form of urea and single superphosphate was superior with respect to number of leaves per plant, leaf length, root length, root diameter, root weight and root yield of radish cv. Pusa Chetki at Akola, Maharashtra (India). While Ndang and Sema (1999) opined that N at 25 kg/ha and K at 60 kg/ha were the best treatment for promoting growth and yield. Nitrogen significantly influenced number of leaves and leaf area index but not dry matter content of leaves. Increasing rates of K significantly increased yield, whereas all N treatments increased yield although no trend was observed. Sanchez *et al.* (1991) found that radish in Florida did not respond to K application because the soil histosols contain sufficient amount of K. Results of leaf tissue analysis reveal that the critical P concentration in radish leaves is 0.45 per cent. According to Jadhav *et al.* (1998), application of 100 kg N and 50 kg P$_2$O$_5$/ha resulted in the best growth and highest seed yield (7.67 q/ha) of radish cv. Pusa Chetki at Akola, Maharashtra (India). Kovacik and Lozek (1999) noticed that excessive N fertilization (160 kg/ha) resulted in the increased production of above ground foliar biomass with reduced root biomass production. The beneficial effect of N application has been shown by several other workers in different cultivars (Ijaz *et al.*, 1997; Djurovka *et al.*, 1997; Verma *et al.*, 1997; Lu and Shen, 1996).

Table 3: Effect of Nitrogen and Phosphorus on Yield of different Cultivars of Radish*

Treatment	No. of Leaves/ Plant	Root Diameter (cm)	Root Length (cm)	Shoot Weight (g)	Root Weight (g)	Shoot: Root Ratio	Yield (t/ha)
Cultivars							
V1	12.9	9.1	15.4	40.3	34.8	1.14	31.3
V2	11.6	7.8	17.8	52.8	50.5	1.03	33.9
V3	13.8	10.2	20.3	62.8	66.5	0.93	37.3
C.D. (P = 0.05)	2.11	1.03	2.01	3.35	4.28	0.10	2.42
Nitrogen levels (kg/ha)							
N1 30	12.7	9.9	15.9	45.0	48.9	1.08	34.0
N2-50	13.8	8.9	18.6	56.6	55.9	1.01	35.6
N3-70	14.9	11.3	22.3	68.7	63.5	1.17	38.4
C.D. (P = 0.05)	1.01	0.3	2.24	3.84	3.57	1.07	1.23
Phosphorus levels (kg/ha)							
P1-15	11.8	7.9	16.5	58.7	56.3	1.09	34.3
P2-30	12.9	11.1	18.7	56.2	54.3	1.10	34.6
C.D. (P = 0.05)	N.S.	N.S.	N.S.	2.35	N.S.	N.S.	N.S.

V1 = Japanese White, V2 = Chinese Pink, V3 = PusaHimani. N.S. = Not significant.

* Bhat (1996).

Chang and Chang (2000) found that K increased the content of protein by 7.14-12.38 per cent, amino acid by 17.7-91 mg/kg, vitamin C content by 144-240 mg/kg, while the nitrate content decreased by 6-116 mg/kg as compared with the control.

Inoculation of radishes with *Pseudomonas fluorescens* 20 and *Glomus mosseae* was investigated with regard to applications of mineral fertilizers. Activity of micro-organisms (evaluated in terms of CO_2 gas exchange) depended on the form of mineral N and quantity of NPK fertilizers applied. Inoculation of radishes with *P. fluorescens* 20 (against a background of nitrate fertilizers) resulted in higher yields than inoculation with *G. mosseae*. Mycorrhization of radishes was more effective in conjunction with ammonium fertilizers than with nitrate fertilizers. Inoculation with microorganisms in conjunction with increasing applications of NPK fertilizers did not increase yields. Further inoculation with microorganisms allowed applications of NPK fertilizers to be reduced by 1.5 times without causing reduction in yield (Shabaev *et al.*, 1998).

A 2-year (1995-96) small-plot greenhouse experiment was carried out at Nitra, Czech Republic to study the effect of N rate (0, 120 or 160 kg/ha) and foliar application of sucrose solution (500 l/ha) on the yield of above- and below-ground radish biomass. The soil substrate was prepared from loamy orthic luvisol and Terravita garden soil. Sucrose solution was applied to radish leaves 12 days before harvest. Increasing the N rate improved the formation of total radish biomass. Excessive N fertilization resulted mainly in the production of above-ground foliar biomass, with reduced root biomass production. Sucrose application had a positive influence on root growth. The best economic yield was observed when radish was fertilized at the recommended rate for greenhouse radishes, with sucrose application (Kovacik and Lozek, 1999).

As regards uptake of nutrients, Singh *et al.* (1995) noticed that higher rates of N and P increased the uptake of these elements, whereas K uptake was not influenced by any fertilizer treatment, but higher K rates decreased N uptake. Nieuwhof and Jansen (1993) found that higher rates of N application increased the N content, nitrate content and N index (percentage N present as nitrate) in leaves and roots, which may cause health hazards to humans. Parthasarathi *et al.* (1999) noted that cv. Pusa Chetki had the highest uptake of N, P and K at the highest fertility level (125: 100: 75 kg/ha).

As regards the effects of micronutrients on yield of radish, Prasad *et al.* (1987) recorded the highest yield in cv. Japanese White with B at 2.5 kg/ha applied to the soil. Improvement in yield and root quality in terms of sugar and vitamin C content, acidity and organoleptic rating as a result of foliar application of 0.1 per cent borax 3 weeks after sowing was noticed by Kumar *et al.* (1996).

The foliar spray of 0.1 per cent boric acid and a soil application of 10 kg $ZnSO_4$ per hectare were found most effective for increasing the number of pods per plant, diameter of the main shoot, seed yield, 1000-seed weight and germination percentage (Sharma *et al.*, 1999).

Mishra and Yadav (1989) recorded the highest seed yield and seed germination (83-88 per cent) in cv. Pusa Reshmi by spraying the plants with 5 per cent $MnSO_4$

20 and 40 days after transplanting the stecklings. A basal application of 5.0 kg $MnSO_4$/ha significantly was found to increase the yield of Japanese White on an acidic soil (pH 4.5) of Nagaland (Prasad and Gupta, 1989). The foliar spray of 0.1 per cent boric acid and a soil application of 10 kg $ZnSO_4$ per ha were found most effective for increasing number of pods per plant, seed yield, 1000-seed weight and germination percentage (Sharma *et al.*, 1999).

7.4.0 Irrigation

It is difficult to make definite recommendation regarding the number of irrigations to be given, which depends upon season, type of soil and amount of organic matter present. For rapid germination of seeds and production of tender and attractive roots, the soil should contain plenty of moisture. If the soil is not moist enough at the time of sowing, the first irrigation is given immediately after sowing. During summer months, frequent irrigation is necessary, otherwise the growth will be checked and the roots will be pungent and tough making them unfit for the market. Care should be taken that the field should not become dry and compact which can check root development. Subsurface drip irrigation gives better results than the surface, especially in the sandy loam soils. The irrigation requirement for radish was found to be about 210-250 mm on sandy loam soils of Delhi (Davenport, 1962); irrigation should be given at 5-7-day intervals. However, Verma *et al.* (1997) found that irrigation every 3 days resulted in better shoot growth, root size and total yield compared with irrigation at 6- or 9-day intervals. Hedge (1987) reported that frequent irrigations at 0.2 and 0.4 bar soil water potential resulted in higher dry matter and root yield, lower nitrate-N content of the root and higher N uptake and water use efficiency than with irrigation at 0.6 bar. Ghanti *et al.* (1989) also found that irrigation at 0.50 atm tension gave a yield of 372.74 q/ha which was significantly higher than yields with other irrigation treatments (0.25 or 75 atm tension). Method of irrigation (basin or furrow) had no effect.

Mangal *et al.* (1988) showed the beneficial effect of irrigation on seed yield in cv. HR-1, the highest seed yield (21.5-22.1 q/ha) being recorded under 0.4 PEC (pan evaporation coefficient).

The beneficial effect of mulching with dried weeds on radish was demonstrated by Joung Keun and Semisi (1995). Application of mulch increased soil moisture status and improved growth and yield over no mulch treatment. A combination of mulching and irrigation showed a remarkable improvement over the mulch alone treatment. Mulching and irrigation reduced percentage of boron deficiency and root cracking.

7.5.0 Interculture

Thinning at 15 days after sowing is must to make proper plant population and better root development. Regular weeding is necessary to check the growth of weeds. During rainy season, two weedings will be required to keep the growth of weeds under check. Fluchloralin (1.35 kg/ha) and pendimethalin (1.2 kg/ha) have been reported to be quite effective in reducing weed density in radish cv. Pusa Himani (Sharma, 2000). One earthing up and one weeding during the early stages of growth

Different Field Operations of Radish

Radish seeds (left), seedling emergence (middle) and thinning (right) of seedlings

Fertilizer application and interculture in radish

Inspection of farmers in radish field

Radish crop getting ready for harvesting

Harvesting of Radish Roots

Washing of radish roots

Harvested roots getting ready
for marketing

are necessary for proper development of roots. Radish has a tendency to bulge out of the soil as it grows in size; therefore, thorough covering by means of earthing up is recommended to produce quality roots. For the seed crop, another earthing up during flowering and fruiting is advocated to prevent lodging of the plants.

7.6.0 Cropping System

Winter radish is a quick growing crop, hence can be intercropped with slower growing crops such as potato, cabbage, *etc.* However, summer radish can be intercropped with amaranth. Maize is grown as intercrop or border crop for summer crop of radish in the northern plains of India, to protect it from heat wave. The cropping sequence of radish-tomato-bitter gourd and radish-onion-okra is practiced in north-western India.

8.0 Harvesting, Yield and Storage

8.1.0 Harvesting

Depending on the cultivars, the roots become ready for harvesting in about 25-55 days after sowing. Early, rapid maturing temperate cultivars reach harvest maturity in 25-30 days after sowing. They become bitter and pithy if the harvesting is delayed. The tropical types take 35-50 days.

Generally roots are harvested when they are tender, sizable, relatively young and about 2.0-2.5 cm in diameter. In India, harvesting is done manually. A light irrigation may be given before harvesting to facilitate lifting of roots. Gently hold the tops, twist and pull out vertically. The harvested roots along with tops are properly washed, allowed for some time to remove water, graded and tied in bundles. Generally, 3-6 roots are tied in a bunch depending upon the size of roots. These bundles are loose packed in baskets and transported to the local markets. In advanced countries, commercial radish growers use a single row harvester that pulls the plants from the soil, cuts the roots from the tops, then places them in burlap bags for transportation to a packing shed. Here the roots are washed, graded and bagged in plastic bags for market.

8.2.0 Yield

The average root yield of Indian cultivars varies from 150 to 200 quintals per hectare, whereas the temperate cultivars produce 50-70 quintals per hectare. Singh *et al.* (1971) reported that the average yield of Pusa Himani may be as high as 390-471 quintals per hectare both in spring-summer and winter sowing.

8.3.0 Storage

Harvested radish roots can be stored for 3-4 days at room temperature without impairing its quality. However, it can be stored up to 2 months in cold storage at 0 °C with 90-95 per cent relative humidity. Radish roots are pre-cooled rapidly to 5 °C temperature to maintain their crispy texture. Water is used to pre-cool the roots quickly. Pre-packaging and Zero Energy Cool Chamber also increases the shelf-life of radish.

In general, late cultivars produce roots of best storage potential with greater dry matter and contents of sugars, crude fibre, P, Ca and Mg (Gaweda *et al.*, 1991).

Results of packaging trials showed that foodtainers preserved radish quality better than untreated or coated with a protective film. Radishes in foodtainers had a shelf-life of 5 days (Schreiner *et al.*, 2000).

Wang *et al.* (2000) suggested a possible correlation between the activity of peroxidase and level of reducing sugar during storage of radish.

9.0 Diseases and Pests

9.1.0 Diseases

Radish develops so quickly and the entire period of culture, even of a seed crop, is so much short that the crop matures before any slow developing disease can show its serious effect. Some of the important diseases of radish are black root (*Aphanomyces raphani*), alternaria blight (*Alternaria raphani*), crown gall disease (*Agrobacterium tumefaciens*), white rust (*Albugo candida*), radish mosaic virus, *etc.* Damping off caused by *Rhizoctonia solani* also occurs sometimes. Henis *et al.* (1978) from Israel suggested inoculating soil with *Trichoderma harzianum* at 0.04-0.15 g/kg of soil to protect radish seedlings from damping off. *T. harzianum* caused reduction of *R. solani* population and adding PCNB (quintozene) at the rate of 4 μg/g soil to the *T. harzianum* inoculum had an additive effect.

Wilson (1962) reported that radish yellows, caused by *Fusarium oxysporum* f. *raphani*, was most active in high soil and air temperatures of mid- and late-summer. Successive sowing generally resulted in greater loss from this disease; hence radishes should be grown once in every 3 years in plots showing this infection.

Segall and Smoot (1962) reported incidence of black spot disease, caused by a variety of *Xanthomonas vesicatoria* on both red and white cultivars of radish packaged in polythene bags. Injuries during mechanical harvesting, washing and grading predisposed radishes to infection. The infection can be reduced by adding 40-60 ppm chlorine to the washing water and by holding packaged radishes at 5-10°C during transit and storage.

In India, the most common diseases are alternaria blight, leaf spot, white rust, downy mildew and radish mosaic virus.

Fungal Diseases

9.1.1 *Alternaria* Blight

It is caused by *Alternaria brassicae, A. brassicicola, and A. raphani.* Symptoms usually first appear on the leaves of seed stem in the form of small, yellowish, slightly raised lesions. These enlarge many times as they become older. Lesions appear later on the stems and seed pods. Infection spreads rapidly during rainy weather, and the entire pod may be so infected that the styler end becomes black and shrivelled. The fungus penetrates in pod tissues, ultimately infecting the seeds. The infected seeds may lose their viability. It can be managed by adopting crop rotation; use of disease free seeds; hot water seed treatment (50 °C for 20 to 30 min); and spray

Diseases of Radish

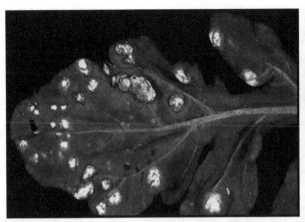 

White rust infection in radish

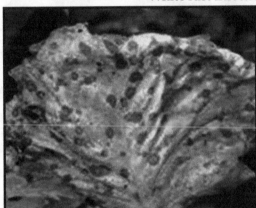 

Alternaria leaf spot (left) and downy mildew (right) in radish

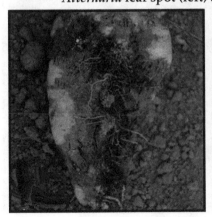 

Root rot (left) and Rhizoctonia leaf blight in radish

Pests of Radish

Flea beetle in radish

Root maggots in radish

A) Aphids B) Diamondback moth caterpillar C) Diamondback moth adult in radish

of the crop with metriam (0.15 per cent), ridomil (0.15 per cent) or blitox (0.20 per cent) at 7-10 days interval. Vannacci (1991) found that combined application of an antagonist like *Trichoderma harzianum* and sodium propionate at 0.1 per cent gave significantly higher percentage of healthy seedlings.

9.1.2 White Rust

It is caused by *Albugo candida*. In some areas this disease assumes a serious problem of radish. Etebarian (1993) found that oospores of *A. candida* were abundant in infected leaves, stems and sheaths but not in seeds of radish. Varietal difference in the resistance of this disease has been reported by Cruz *et al.* (1973) from Brazil. IHR-1-1, an Asiatic cultivar, released from Bangalore, is reported to be highly resistant to white rust. Infection is favoured by cool, moist weather and fungicides should be applied when these conditions are forecast. Sprinkler irrigation should be avoided and sanitation measures followed to reduce inoculum (Johnson and Gabrielson, 1990). Disease symptoms appear on the leaves and flowering shoots which become deformed and bear only malformed flowers. It produces a white powdery substance in patches on the under-surface of the leaves. Destruction of diseased crop debris helps in reducing the inoculum. Weed control and other sanitary precautions are also essential. Use of resistant cultivars and regular spraying of 0.8 per cent Bordaux mixture or with Dithane Z-78 at a concentration of 0.2-0.3 per cent have been suggested as its control measures.

9.1.3 Leaf Spot

This is caused by *Cercospora carotae*. The elongated lesions along the edges of the leaf segment appear which results in a lateral curling. In severe cases, the leaves turn black. Its incidence could be minimized through adoption of crop rotation, use of disease free seeds, seed treatment in hot water at 50 °C for 25-30 min, and spraying of bavistin (0.1 per cent) or dithane M-45 (0.25 per cent) at 7-10 days interval.

9.1.4 Downy Mildew

This disease is caused by *Peronospora parasitica*. In addition to radish, the fungus also attacks cauliflower, cabbage, turnip and other brassicas. The disease is characterized by the appearance of the purplish brown spots on the under surface of the leaves. These spots may remain small or enlarge considerably. The upper surface of the leaf above the lesion is tan to yellow. Downy growth usually appears on the under surface of these lesions. Crop rotation, clean seed beds, destruction of weed and other sanitary measures are important to check the spread of the disease. Intensity of the disease can be reduced by spraying 0.25 per cent metalaxyl-mancozeb or cymoxanil-mancozeb 2-3 times at 10 days interval after the disease has been noticed in the field.

Viral Diseases

9.1.5 Radish Mosaic Virus (RMV)

The symptoms first appear as small, circular to irregular, chlorotic lesion in between and adjacent to the veins. Little or no leaf distortion is noticed and stunting or abnormal formation rarely occurs. Severe yield loss in susceptible cultivars of radish caused by RMV has been reported from Japan (Sakai and Kono, 1975). It is transmitted through aphids. The disease can effectively be checked by controlling aphids with insecticides and eliminating weed hosts.

9.1.6 Radish Phyllody

It is another disease of the seed crop which appears at flowering. This was reported by Misra and Gupta (1977). The sepals, petals and carpels of affected flowers showed phyllody and the stamens became sepaloid. The degree of phyllody increased in the direction of carpels.

9.2.0 Pests

Insects often do more harm to radish than diseases. The important insect pests of radish are described below:

9.2.1 Aphids (*Myzus persicae, Brevicoryne brassicae, Lipaphis erysimi*)

Aphids are the most serious pest of radish. It attacks both seedlings and mature crops. Cloudy and humid conditions favour the spread of their infestation. They are often found in large colonies on the under-surface of leaves. A colony consists of winged and wingless adults and various sizes of nymphs. Aphids may be black, yellow or pink, but mostly are various shades of green. Aphids feed by sucking plant sap. In case of heavy infestation the plants are completely devitalized; leaves and shoots curl up, become yellowish and finally die. The plants may be covered by a sticky substance and honey dew, which is excreted by the aphids. The aphids are vectors of several viruses. The aphid population can be managed by growing trap crops like nasturtium, mint and mustard; mulching with coloured and reflective plastics; parasitic wasps, *Diaeretiella rapae*; and seed treatment with thiram @ 3 g/ kg of seeds. Additionally, spraying of acetamiprid @ 15 g a.i., imidacloprid @ 20 g a.i., azadirachtin (5 per cent) @ 200 ml, dimethoate @ 700 g a.i. and fipronil @ 40 g a.i. per ha in 800-1000 lit of water is recommended for effective control of aphids.

9.2.2 Mustard Sawfly (*Athalia lugens proxima*)

It sometimes causes severe damage to radish and turnip. It attacks the crop in both vegetative and flowering stage. The grub feeds on the leaves and fruits. Hand picking of larvae when the area involved is small is recommended. Spray the crop with Malathion @ 2 ml/l of water.

Patel *et al.* (2000) found that the larval population of *A. lugens proxima* started to increase during the fourth week of July and first week of November in the seedbed plot and transplanted seed crops, respectively. The population was negatively correlated with bright sunshine hours and positively correlated with rainfall, temperature (minimum and mean), vapour pressure and relative humidity.

9.3.0 Physiological Disorders

9.3.1 Akashin/Brown Heart

It is a disorder in radish due to B deficiency and also known as hollow root. Initially, dark spots appear on the thickest part of the root. The plant growth is

Disorders of Radish

Cracking of radish

Forking of radish

Nitrogen deficiency in radish

Potassium deficiency in radish

Phosphorus deficiency in radish

Magnesium deficiency in radish

Seed Production of Radish

Flowering in radish

Pod formation (left) and dried pods (right) ready for seed extraction

checked and it remains stunted. The leaves are smaller than the normal, lesser in number and later on show variegated appearance with yellow and purplish red blotches. The leaf stalks show longitudinal splitting. The root remains small, showing distorted and grayish appearance. The B deficiency can be managed by soil application of borax 15-20 kg/ha or foliar application of borax @ 0.1 per cent.

9.3.2 Pore Extent or Pithiness

Pithiness is formed by the collapse of paranchymatous tissue/cells in root tissue caused by excessive root growth in comparison with the corresponding assimilation ability of leaf tissue. Pithiness of root is more common in summer crop than spring or autumn. It is correlated with high temperature and maturity. This disorder is more pronounced when harvesting is delayed. It is also triggered by excess application of NPK and soil moisture stress. As a result, the quality and commercial value of radish is reduced. Over-maturity of roots also causes pithiness. Hence, harvest the crop at appropriate maturity and select proper variety to grow during off-seasons.

9.3.3 Elongated Root or Forking

Forking is an undesirable character as it deteriorates the quality and market value. There is secondary elongating growth in the roots that gives a look of fork like structure to the root. Elongated root relates closely to soil adaptation. Varieties with vigorously elongated root, short root or round root have been selected in areas of shallow arable soil and varieties with poorly elongated root, long root and huge root have been cultivated in areas with deep soil conditions. The disorder is due to the excess moisture during the root development of radish. It also occurs in heavy and gravelled soil due to the soil compactness and injury. Un-decomposed organic manure also favours forking of roots.

9.3.4 Splitting

It is caused due to heavy irrigation after long dry spell and wide difference in the temperature regimes during later stage of root development *i.e.* wide fluctuation in soil moisture and temperature. The frequency of splitting is more in large rooted varieties and sparsely sown crops.

9.3.5 Greening

Exposure of crown/shoulder to sunlight leads to greening. It tends to be a problem in light soils prone to wind erosion and when sown on raised beds. An earthing-up will manage the problem effectively. It is also a genetic trait in few varieties.

9.3.6 Bolting

Development of seed stalk without proper development of roots is termed as bolting. Once the plant produces a flower stalk, radish flavour and quality deteriorate. The degree of bolting ability has been studied because of its disadvantage in cultivation, and breeding has been directed towards late bolting.

10.0 Seed Production

Radish cultivars available in India can be classified into three groups so far as seed production is concerned: (i) winter radish or Japanese radish (biennial) which produces seeds only in the temperate hills of India; these cultivars require low temperature for flowering and are generally sown in autumn, *i.e.*, during the second fortnight of September, (ii) the second group includes summer radishes of temperate regions (*e.g.*, Rapid Red White Tipped, Woods Long Frame, *etc.*). These cultivars, though very quick in root development, behave just like winter radishes for seed production. In the hills, the seeds of these cultivars can be produced both from autumn and spring sown crops. The autumn sown crop gives higher seed yield which matures earlier than the spring sown crop, and (iii) the third group includes cultivars which produce seeds freely in the plains but can produce good seeds in the hills also. Generally, seeds of cultivars belonging to the first two groups are produced in the hills.

10.1.0 Sowing Time

Temperate varieties of radish are biennial, which require low temperature (0-4 °C) for a period of 40-60 days to induce bolting and flowering. Sowing of seed of these varieties is done in September and their seed production is restricted to temperate hilly regions. Temperate summer varieties are grown during summer in the hills whose seed production can be done both in summer and autumn seasons in the hills. Tropical varieties are sown in September or October and they produce seed both in tropical and temperate region of India.

The autumn sowings are done from last week of September to middle of October. Ghormade *et al.* (1989) reported that seed yield in cv. Pusa Chetki, Pusa Reshmi, H.R.1 and Japanese White tended to decline with later planting after 24 October. In the lower Gangetic plains, however, Chatterjee (1989) obtained the highest seed yield (12.83 q/ha) in cv. Improved Chinese Pink, when the crop was sown in the 2nd week of October. Seeds produced from October sown crops showed the highest germination percentage. For *in situ* crop, however, late sowing is beneficial. The spring sowing is done in the first fortnight of March.

Trials conducted at 4 sites in Nepal revealed that the best sowing date for radish seed production is 22 September. Earlier or later sowing results in lower seed yield (Jaiswal *et al.*, 1997).

The cultural requirements of radish as a seed crop are similar to those of the market crop.

10.2.0 Methods of Seed Production

Both seed-to-seed and root-to-seed methods are employed for seed production in radish. The latter method is preferred for raising nucleus seeds (Karivaratharaju *et al.*, 1988). It has been found that seeds weighing less than 5 mg should not be used for steckling production. If the stock seed is of high quality, commercial seeds can be produced by seed-to-seed method. Root-to-seed method cannot be employed in case of European cultivars (group 2) because they do not stand transplantation well.

In root-to-seed method, when the roots are fully mature, the crop is harvested, true-to-type roots are selected and after giving proper root and shoot cut, they are transplanted in a well-prepared field. The selection and roguing are done on the basis of foliage characters, root characters (colour, shape, size, flesh colour, core size, pungency, *etc.*) and bolting time. Small, mis-shapen, diseased or any other undesirable roots are discarded. Hairy and forked roots are also eliminated. Early or late bolters are also removed. That planting date, spacing and size of steckling have significant effect on plant characteristics and seed yield has been reported by many workers in a number of radish cultivars. Sharma and Kanaujia (1992) reported that transplanting the stecklings of cv. Japanese White on 25 November, combined with application of 200 kg N/ha produced the highest values for the days to 50 per cent bolting and flowering, plant height, number of branches/plant, diameter of the main shoot and seed yield. Srivastava *et al.* (1992) recorded significantly higher number of primary branches/plant in cv. Pusa Chetki with widest spacing (60 cm × 60 cm). However, the best treatment for seed yield seemed to be 60 cm × 45 cm spacing with 80 kg N/ha. Jamwal and Thakur (1993) also recorded that planting 3/4 size stecklings (15 cm long) of cv. Pusa Himani at a spacing of 60 cm × 45 cm gave the highest seed yield (1.44 t/ha). Maurya *et al.* (1990), on the other hand, found that seed yield/ha in cv. Pusa Chetki was not influenced by spacing, but seed yield from the late planting of stecklings was poor regardless of spacing. Sharma and Lal (1990) reported that the 1000-seed weight in cv. Pusa Reshmi increased with increasing plant spacing and the highest values were obtained at the widest spacing (60 cm × 60 cm), while steckling size had no effect on 1000-seed weight. The highest 1000-seed weight was obtained with 150 kg N/ha and a plant spacing of 60 cm × 60 cm. For a high yield of good quality seeds, Sharma and Lal (1987) suggested use of 15 cm long stecklings planted at 60 cm × 30 cm and fertilized with 100-150 kg N/ha. In a subsequent study, Sharma and Lal (1994) found that application of 150 kg N/ha with a plant spacing of 60 cm × 60 cm gave the highest 1000-seed weight, germination percentage, seedling length, vigour index and seed yield. Steckling size had no effect on these parameters. For high yield of good quality seeds of cv. Pusa Reshmi, Sharma and Lal (1987, 1991) recommended planting of 15 cm long stecklings at a spacing of 60 cm × 60 cm and fertilized with 100-150 kg N/ha. In another study, Saharan and Baswana (1991) found that the widest spacing (60 cm × 75 cm) and earlier planting of steckling (15 November) resulted in the highest number of branches and pods per plant and consequently seed yield. While Chatterjee and Som (1991) recorded the highest seed yield in cv. Improved Chinese Pink at 30 kg N and 80 kg K_2O/ha and a spacing of 40 cm × 10 cm, Singh *et al.* (1990) reported that N at 80 kg/ha and a plant spacing of 60 cm × 45 cm resulted in the highest seed yield in cv. Japanese White.

After elimination of off-types, the roots are transplanted at a distance of 20 cm between plants in the rows spaced 90 cm apart. Before transplanting, proper root and shoot cuts are given. The root stecklings are prepared by cutting lower 2/3rd portion of root and trimming of the tops in the same proportion. Jauhari and Purandare (1959) reported that one-fourth root cut gave the highest yield of seeds (136.2 g/10 plants) and vegetative dry matter, followed by intact roots; crown of the root gave the lowest yield. Kalvi and Nath (1970a) reported that plants with one-third top cut and one-half root cut or one-fourth root cut proved to be superior to control plants with intact top and root in respect of new vegetative growth and seeds talk emergence. In a subsequent report, Kalvi and Nath (1970b) mentioned that plants with one-third top cut and one-half root cut produced the highest seed yield of 7.4 q/ha (as against an yield of 3.4 q/ha in control plants). However, the highest germination was recorded in plants receiving one-third top cut and one-fourth root cut. Sarnaik *et al.* (1987) obtained the highest seed yield in cv. Pusa Reshmi from half-size stecklings. Jandial *et al.* (1997) suggested that to obtain high seed yield in cv. Japanese White, 1/2 roots of stecklings should be trimmed off before planting at 30 cm × 30 cm. Gill and Gill (1995), on the other hand, reported that whole root stecklings (cv. Punjab Safed) planted early (15 November or 1 December) produced plants with luxuriant vegetative growth and a high number of branches, the major seed yield component.

10.3.0 Flowering and Fruit Setting

Honeybees are the chief pollinating agents. Kremer (1945) demonstrated that seed yield in radish was greatly influenced by the number of honeybees visiting the flowers. Nectar secretion, pollen formation and bee activity are influenced by environment. It has been suggested that radishes grown for seed production should not be located too close to fields of major honey producing plants such as clover, because the bees normally tend to visit these plants in preference to radish.

Kremer (1945) further reported that temperature of 32°C or higher caused the stigma to become dry and the pollens failed to germinate. A period of dry weather might cause formation of undeveloped pods. Kulwal *et al.* (1975) reported that the combination of cold storage and GA_3-spray (100 ppm) after replanting led to the highest percentage of flowering in cvs. White Icicle, Pusa Himani and Japanese White. Pyo *et al.* (1976) found that GA_3 promoted bolting and flowering and was most effective when applied at 100 ppm to the seed or at the end of the seedling stage. IAA delayed bolting by 4 days, but the date of flowering was not affected. Treatment with 200 ppm Ethephon or 1500 ppm SADH did not affect bolting or flowering, but 0.25 per cent MH delayed it.

Nishijima (2000) also reported that undesirable bolting in Japanese radish could be controlled by reducing the concentration, of endogenous gibberellins during the long day induction. Following detailed study with cv. Harumaki-Minowase, Matsubara *et al.* (1990) recommended vernalization of radish plants at 3°C for 2 or 3 weeks for seed production. Hawlader *et al.* (1997) found that both genotype and vernalization period had significant effect on flower induction and seed setting in radish. Vernalization of radish seedlings at 5°C for 4 weeks induced early flowering and maturity and more pods per plant, more seeds per pod and higher seed yield per plant.

10.4.0 Care after Bolting

All weeds, especially wild radish, wild turnip, wild mustard should be removed from the radish field to avoid possible cross-pollination. One or two irrigation may be given after flowering which results in better seed yield. Sometimes staking is done to provide support to the seed stalks.

10.5.0 Isolation

Radish cultivars should be isolated at a distance of about 1600 m and 1000 m for nucleus or foundation and certified seeds, respectively.

10.6.0 Harvesting and Threshing

There is no problem of shattering of seeds in radish, because the pods do not dehisce. Since there is no natural dehiscence, the pods are allowed to mature and ripe fully before they are harvested. The radish seed crop is harvested from the end of June to 15 July in the Kullu valley (Singh *et al.*, 1960).

When most of the pods are brown, the seed stalks are cut and kept in small piles to dry. Seed extraction is difficult in radish, and thorough drying of seed pods facilitates easy extraction of seeds. The seed immediately after extraction is thoroughly dried in the sun; otherwise it will lose its viability. Thorough drying immediately after threshing is essential. The seeds after threshing and drying are so cleaned as to make them free from small seeds, other crop or weed seeds or any other materials.

Bichta and Wegrzyn (1994) suggested that harvesting and threshing of radish seed plants should be done when the moisture content of the seeds, siliquas and stems is 11-18, 14-24 and 25-30 per cent, respectively. Due to high capacity of moisture absorption by the siliquas, harvesting and threshing should be carried out on sunny warm days.

Seed Standards for Radish

Factors	Standards for each Class	
	Foundation	Certified
Pure seed (minimum)	98.0 per cent	98.0 per cent
Inert matter (maximum)	2.0 per cent	2.0 per cent
Other crop seeds (maximum)	None	None
Weed seeds (maximum)	None	None
Germination (minimum)	70 per cent	70 per cent
Moisture (maximum)	6.0 per cent	6.0 per cent
For vapour-proof containers (maximum)	5.0 per cent	5.0 per cent

10.7.0 Yield

Average seed yield of temperate varieties is 500-550 kg/ha, while tropical varieties give seed yield of 700-800 kg/ha. The seed is cleaned, dried and graded properly. The seed is dried to 6 per cent moisture level. Seed yield of radish was found to be positively correlated with plant height, number of primary branches/plant, diameter of the main shoot, pod length, number of pods/plant, number of seeds/pod and number of secondary and tertiary branches/plant (Sharma and Kanaujia, 1995; Sharma and Lal, 1989).

Ghormade *et al.* (1989) reported that Pusa Chetki is the heaviest seed yielding cultivar, followed by Pusa Reshmi, HR-1 and Japanese White.

11.0 Hybrid Seed Production

It involves two steps, (1) Development of F_1 hybrids, and (2) Commercial production of F_1 hybrid seed.

11.1.0 Development of F_1 Hybrids

Parental materials are carefully selected keeping in view the objectives of breeding programme. Nearly homozygous inbred lines are developed by bud pollination or sibbing selected plants for 6-8 generations. Combining ability of the lines is tested to select the best combiners. Based on the estimates of combining ability, promising hybrid combinations are evaluated in a replicated yield trial along with check(s), followed by multilocation testing. Seeds of promising hybrids are produced commercially.

11.2.0 Commercial Production of F_1 Hybrid Seed

Commercial seed production of F_1 hybrid is commonly done using Ogura CMS system. Single cross hybrid, three-way cross hybrid or top cross hybrid can be made in radish. An isolation distance of 3000 m for breeder seed, 1600 m for foundation seed and 1200 m for certified seed is recommended. Generally 2: 1, or 3: 1 or 4: 1 row arrangement for planting of CMS lines and male parent, respectively is followed. Agronomic practices as described above for seed producion is followed. Four rougings are done, at vegetative stage, at uprooting and replanting stage, at bolting stage and finally at flowering stage to remove the offtype plants. Plants are staked to prevent the inflorescence from lodging. About 15 beehive boxes are kept in one hectare area for increasing pollination, seed set, seed yield and seed vigour. Pods mature about 60-80 days after fertilization. Average seed yield is 600-900 kg/ha.

12.0 Crop Improvement

Radish breeding has been in progress for some centuries by means of mass selection or combined mass pedigree selection (Banga, 1976). At the Indian Agricultural Research Institute, New Delhi, high yielding hybrids of radish were obtained which yielded 30-60 per cent more than the better parent (Pal and Sikka, 1956). Kumar *et al.* (2002) made 56 crosses in radish, and identified the best F_1 hybrids *viz.* Mino Early White × Pusa Himani, Japanese White × DPR-1 (Palam Hirday), and Chinese Pink × Nadauni, manifesting maximum heterosis for root yield and its component traits. Kutty and Sirohi (2003) identified a superior hybrid Acc. No. 30205 × Acc. No. 192 for commercial exploitation. Lal and Srivastava (1975) crossed Kalianpur No. 2 with long root and small fruit and rat-tail radish (*R. sativus* var. *caudatus*) with small root and found that all characters except root thickness gave high estimates of heritability. High values for genetic advance were recorded for cortex thickness and fruit length. Improvement in root or fruit length can be achieved through selection breeding. Murali *et al.* (1998) reported that root yield had a significant positive correlation with root girth and root dry weight at the phenotypic and genotypic levels. A significantly correlation of root weight was noted both at genotypic and phenotypic levels, with the root length, number of leaves per plant, average leaf size, leaf weight per plant and plant weight, indicating selection on the basis of these characters will be effective (Danu and Lal, 1998). Kumar and Chandel (2003) observed higher SCA estimates of than GCA for root yield, dry matter, ascorbic acid and nitrate content, revealing the preponderance of non-additive gene action. The crosses Mino Early White × Pusa Himani and Japanese White × DPR-1 exhibited high heterosis over Japanese White and Chinese Pink with significant SCA effects. Sirohi and Kutty (2000) reported overdominance for characters like root weight, shoot weight, leaf length, leaf width and yield per plant, whereas dominance was observed for petiole length and root diameter.

Radish breeding in Europe has concentrated mainly on the red globe, the oval red and white types, which have been bred by means of mass selection or combined mass and pedigree selection methods. Japanese radish now comprises many varieties bred by means of pedigree selection and natural or artificial hybridization made on the original base genotypes. Most radish cultivars in Japanese and Korean markets today are F_1 hybrids. The first F_1 radish cultivar, 'Spring Cross,' was released in 1961 by Takii Seed Company in Japan. Subsequently, different F_1 varietal groups *viz.* 'Miyashige', 'Tokinashi', 'Minowase', 'Nerima' *etc.* have been developed in Japan. Today, many cultivars have been developed throughout the world, keeping specific objectives in view. The major breeding objectives are outlined below.

a) To develop high-yielding varieties capable of an increased marketable yield of white, long/stumpy, nonpithy roots (longer standing period in field without becoming pithy) with thin tap root and non-branching habit.

b) To breed early-rooting short duration varieties fitting in sequence cropping system.

c) To breed varieties with characteristic cylindrical root of appropriate length for mechanical harvesting.

d) To develop varieties and F_1 hybrids having wider adaptability to different agro-climatic conditions.

e) To develop F_1 hybrids utilizing self-incompatible lines.

f) To evolve varieties tolerant to abiotic stresses, particularly tolerance to heat, low temperature, drought, excessive rain and saline, alkaline and acid soils.

g) To develop varieties having superior edible quality such as desired pungency, low nitrate content, less pore formation, *etc.* For varieties eaten fresh as salad increased sweetness and the root colour are important quality parameters.

h) To breed varieties and F_1 hybrids with multiple resistance to diseases and insect pests such as alternaria blight, white rust, yellows, clubroot, soft rot, radish mosaic virus and aphids.

i) To develop varieties with slow bolting and late-flowering habit.

12.1.0 Self-Incompatibility

One of the breeding objectives is the exploitation of hybrid vigour. Incompatibility and male sterility phenomena are used to develop hybrid seeds.

Radishes are normally self-incompatible and insect pollinated. The incompatibility system is of sporophytic type with multiple S-alleles located at a single locus in which the papillae on the stigma surface form the barrier to pollen tube penetration. By disrupting the stigma surface or by applying pollen immediately to the conducting tissue of the style or bud pollination it is possible to get normal seed set by selfing (Roggen and Van Dikj, 1973; Dickinson and Lewis, 1973). Tatebe (1974) reported from Japan the behaviour of compatible pollen grains in fertilization of Japanese radish. Following cross-pollination of radish some compatible pollen grains on the stigma began to germinate 40 minutes later, but the majority did so after 50-60 minutes at 22°C. Eight hours after pollination some tubes penetrated the style and the largest of them reached the ovary. Fertilization was complete within 24 hours. In another report, Tatebe (1977) stated that high temperature (25 or 30°C), compared to room temperature (15-18°C), slightly reduced self-incompatibility. Treatment of the stigma with a small amount of either ether or KOH (10 per cent) solution before pollination has been found to reduce self-incompatibility in Japanese radish (Tatebe, 1968). Siddique (1983) recorded best fruit-set when pollination was done on the day of flower opening.

In radish, self-incompatibility (SI) is governed by the S-locus, which consists of a series of multiple alleles. Self-incompatibility is being employed by radish breeders to produce F_1 hybrids. In this context, the identification of S-genotypes of the plants before flowering assumes significance. Recently the PCR-RFLP method has proven useful for the identification of S alleles in inbred lines and for listing S haplotypes in *R. sativus* (Niikura and Matsuura, 2001, Lim *et al.*, 2002). Hawlader and Mian (1997) investigated the self-incompatibility mechanism in 10 local radish cultivars in Bangladesh through seed set analysis and pollen tube growth behaviour. The objective was to indentify self-incompatible radish lines for hybrid seed production. Based on a low seed set ratio as an estimate for self-incompatibility, cultivars Tangail Local, Tasaki, Kuni, Aushi and Indian Aushi were identified as self-compatible. An intermediate seed set ratio was observed in Red Mollika, Pinky and Red Bongi. Moderately strong self-incompatibility was observed in Red Kalpin and Kuni White. Pollen tube growth behaviour revealed that self-incompatible lines had a relatively lower number of germinated pollen grains per stigma and pollen tubes per style than self-compatible lines. The two self-incompatibility test methods gave comparable results. Nikornpun *et al.* (2004) bred F_1 hybrids exhibiting good horticultural characteristics (such as uniformity, dark green leaf without hair, cylindrical shape, smooth skin, firm root texture and 360 to 400 g root weight), utilizing self-incompatible inbred lines of Chinese radish (*R. sativus* var. *longipinnatus*).

Monakhos and Barasheva (1999) crossed 8 self-incompatible lines in a full diallel and parents and evaluated 56 F_1 hybrids for root mass. The results indicated that the trait was controlled by several dominant genes, with lines differing greatly in dominance and additive effects as well as cytoplasmic effects. Heterosis due to overdominance was observed in the F_1 hybrids. Antonova (1991) reported development a new radish variety called Kvarta for protected cultivation. Selected in a hybrid population derived from Ilka and Teplichnyi, this early, uniformly maturing variety can form roots under low light and temperature regimes. It has a slow transition to be reproductive stage.

12.2.0 Male Sterility

Male sterility has been reported in Japanese radish. Cytoplasmic male sterility (CMS) is a maternally-inherited trait that is characterized by the inability of a plant to produce functional pollen. Male-sterile cytoplasms have long been of interest because of their use in hybrid seed production and also as an invaluable to study mitochondrial function and interaction between mitochondrial and nuclear genes. The CMS trait is mitochondrially encoded and always found associated with mitochondrial DNA (mtDNA) rearrangements. Male sterility in radish was first reported by Tokumasu (1951), who identified male sterile plants in the 'Tokinashi-Daikon' variety. Ogura CMS is the most widely used CMS in Brassicaceae. Ogura (1968) described a type of male sterility in radish, controlled by the interaction of a recessive gene, ms and S-cytoplasm, which has been confirmed by Bonnet (1970). In Ogura radish, the abortion of male gametogenesis occurs at the early microspore stage due to degeneration of the tapetal tissues (Ogura, 1968). The mitochondrial genome of Ogura radish is highly rearranged relative to its normal (fertile) counterpart, and contains sequences not present in the normal radish genome (Makaroff and Palmer, 1988). An open reading frame, designated orf138, which is specific to Ogura cytoplasm (Bonhomme *et al.*, 1992; Krishnasamy and Makaroff, 1993), is the actual determinant of cytoplasmic male sterility (CMS) for Ogura male sterility (Grelon *et al.*, 1994; Krishnasamy and Makaroff, 1994). The orf138 gene is one of the most intensively studied "CMS genes" (Makaroff, 1995) and is found both in wild and cultivated radishes (Yamagishi and Terachi 1994a, b, 1996). The Kosena CMS (kos CMS) isolated from a population of the Japanese radish cultivar Kosena (Ikegaya, 1986) is related to Ogura CMS (Iwabuchi *et al.*, 1999). Nahm *et al.* (2005) detected NWB CMS having non-viable pollen in South Korean line of radish.

Several studies indicate the origin of Ogura-type male-sterile cytoplasm in Japanese wild radishes (Yamagishi and Terachi, 1994a,b, 1996, 2001). In contrast, Ogura-type cytoplasm is found in only a few cultivars of Japanese cultivated radishes and have orf138 and its variant, orf125 (Yamagishi and Terachi, 2001).

Many mitochondrial genes that determine CMS can be suppressed or counteracted by the products of one or more nuclear genes known as fertility restorer genes (Rf). CMS-Rf systems are used for commercial-scale F_1 hybrid seed production in many crops. Besides, the study of CMS and Rf genes helps in increasing our understanding of nuclear-cytoplasmic interactions. Nieuwhof (1990) suggested that male sterility is probably determined by one dominant and 2 recessive independently acting genes, but minor genes may also be involved. Hybrid cultivars can now be produced using either incompatibility or male sterility. Yamagishi and Terachi (1994b) reported two dominant restorer genes controls the Ogura CMS in radish. Koizuka *et al.* (2000) demonstrated that two dominant restorer genes (Rf1 and

Rf2) were necessary for fertility restoration in kos CMS radish. The kosena CMS-Rf system is genetically the same as that of the Ogura CMS-Rf system (Koizuka *et al.*, 2000). Japanese breeders have been producing hybrid seeds in radish for the last 40 years. Frost (1923) was the first to develop hybrid seeds in radish, the crosses between selfed lines were very vigorous and usually exceeded the better parent in root size and all plant characters. Keppler (1941) also reported manifestation of hybrid vigour, especially in crosses within superior families. At the Indian Agricultural Research Institute, New Delhi, high yielding hybrids of radish were obtained which yielded 30-60 per cent more than the better parent (Pal and Sikka, 1956).

Zhang *et al.* (1999) showed that the male sterility in spring-summer radish (*R. sativus* var. *radiculus*) is cytoplasmic-genetic male sterility, which is determined by cytoplasm and two genes in nucleus. The genotype of ms line is S (ms 1 ms 1 ms 2 ms 2); the maintainer's is N (ms1 m1 ms 2 ms 2). Hawlader *et al.* (1997) crossed Ogura radish, a cytoplasmically genetic male sterile line, with four local and three Japanese cultivars to identify maintainer lines. Out of seven F_1 families, one cross involving a local cultivar, Aushi, produced 100 per cent male sterile (MS) progeny. The crosses involving the other two local cultivars, Tangail Local and Kuni, produced about 90 per cent MS progeny, indicating the presence of maintainer gene(s) for male sterility. All three exotic cultivars (Japanese) appeared to possess the chromosomal gene(s) for controlling the male sterility. The successful transfer of Ogura CMS to three cultivars of radish (Japanese White, Pusa Chetki and Pusa Desi) has been reported by Sodhi *et al.* (1993). These three cultivars producing flowers with pollenless anthers are now available as male sterile lines for hybrid seed production in radish. Yamagishi (1998) found that the restorer gene is widely distributed in the Japanese wild radish (*R. raphanistrum*) irrespective of cytoplasm type. Among the cultivated radishes, the European and Chinese varieties have the restorer gene, while most of the Japanese varieties do not have restorer gene.

Li *et al.* (1997) reported from China development and selection of a new F_1 hybrid cultivar named Lu Luobu 8 by crossing male steile line 111A with inbred line 110. It is characterized by late bolting, early maturity, high yields, good quality and disease resistance. Its fleshy root is 50 cm long, 500 g in weight with white skin and flesh. It can be harvested 50-55 days after sowing. Yield is 80 t/ha.

Sazonova (1973) reported from the USSR that although F_1 hybrids of radish were superior to parental cultivars in earliness and productivity, they were generally inferior in quality in terms of dry matter, sugar and ascorbic acid contents. Ling *et al.* (1986) reported from China that 11 of the 12 hybrids showed heterosis for root diameter, but only one was better than the better parent for root length. Although the hybrids hold much promise, such seeds are still expensive to produce.

12.3.0 Breeding for Greater Adaptability

The other breeding objectives are to produce cultivars with greater adaptability to growing seasons of different lengths, with improved resistance to pests and diseases and with improved marketing qualities. Selection of cultivars suitable for early production has led to small size, high growth rate and annual life cycle. Banga and Van Bennekom (1962) selected strains suitable for winter production under glass. In India, Pusa Himani, a cultivar with wider adaptability, has been developed from a cross between Radish Black (a late bolter) and Japanese White (Singh *et al.*, 1971).

Cultivars suitable for earlier production of radish in greenhouse were selected in the USSR (Kvasriikov and Nikonova, 1976). Similarly, tetraploid radish yielding 1.5-2 times as much as the diploids was developed (Savos'kin, 1966; Savos'kin and Ceredeeva, 1969).

12.4.0 Breeding for Quality

In radish, quality is related to pungency, sugar content, water content and pore extent, degree of bolting and also external appearance like root length, shape, colour, gloss, smoothness of skin, *etc.* Lal and Srivastava (1975) crossed Kalianpur No. 2 with long root and small fruit and rat-tail radish (*R. sativus* var. *caudatus*) with small root and found that all characters except root thickness gave high estimates of heritability. High values for genetic advance were recorded for cortex thickness and fruit length. Improvement in root or fruit length can be achieved through selection breeding.

Raising the growth rate to promote earliness may increase sensitivity to pithiness (Hagiya, 1952-59). Many pithiness-resistant cultivars have been developed, the first being A.R. Zawan's Cherry Belle, which has spread all over the world.

In China, genetic resources of superior radish types for fruit are 'Beijing Heart Beauty' and 'Tianjin Green'; for cooking purposes 'Beijing Red Robe' and 'Late Long White of Taihu'; for processing 'Yan-zhong', 'Dutch Egg Head' and 'One Cut Variety'; for early spring production 'March Radish'; for early harvest varieties with prolonged sowing date 'Short Leaf No. 13' and for bigger root 'Lasa Winter' (Depel, 1989). A high quality, non-pungent variety 'Yidaozhong Luobo' with a low water content, tolerance to cold and high fertilizer doses, and disease resistance has been identified in China. The preserved radishes are bright yellow and crisp (Zhou, 1991). The salad types of Chinese radish (Xin Li Mei, with red flesh; and Tianjin, with green flesh), possess the visual appeal and eating qualities. Within Europe, old types such as the Italian Treviso Red, long red and white skinned cultivars (*e.g.* cv. French Breakfast), and German Rettish types are marketed as high-value exotics (Crisp, 1995).

The red, pink, violet and purple colour of radish is due to the presence of anthocyanins, responsible for the red, purple and blue colours of many vegetable, fruits and cereals. They have nutraceutical, colourant and antioxidant properties. They act as free radical scavengers, help to lower the LDL cholesterol and the risk of cardiovascular disease, prevent obesity, inhibits the formation and progression of atherosclerosis and oxidative stresses. They also protect the cells against environmental stresses such as ultraviolet and high intensity light, wounding, cold temperature and water stress. Among the various anthocyanins, pelargonidine (acylated pelargonidin-3-sophoroside-5-glucoside) and cyanidine (acylated cyaniding-3-sophoroside-5-glucoside) are responsible for red/pink and purple/violet colour of the roots in radish, respectively (Giusti and Wrolstad, 1996; Tatsuzawa *et al.*, 2010). Absence of anthocyanins resulted in a white colour.

In recent years, the use of coloured radish in the salads or as garnish is gaining popularity in India.

Generally, the pigmented radish contains higher amounts of anthocyanins, phenolic compounds and antioxidant activity than non-pigmented radish. Singh *et al.* (2017) studied 24 lines of coloured radish and reported that they differed by 4.98 fold for total phenolics, 36.16 fold for anthocyanins content, 4.96 fold for FRAP activity and 4.03 fold for CUPRAC activity. They estimated 20-250 per cent higher antioxidants in coloured-rooted radishes as compared to white-rooted commercial/national check cultivars, namely Japanese White and Kashi Shweta. Improved lines of coloured radishes possessed 80-250 per cent higher ascorbic acid (18.5–26.5.0 mg/100g FW), total phenolics (32.5–65.0 mg/100 g FW), anthocyanin content (90–175 µg/g FW), antioxidant-FRAP value (3.15–5.90 µmol/g FW) and antioxidant-CUPRAC value (5.25–11.50 µmol/g FW as compared to white-coloured commercial cultivars (Singh *et al.*, 2016).

12.5.0 Breeding for Stress Resistance

Differences in resistance to pests and diseases have been reported in the USA, Europe and Japan. Lines resistant to *Fusarium oxysporum* f. *raphani* race 2 and *Albugo candida* have been bred by William and Pound (1963) in the USA and lines resistant to virus diseases by Shimizu *et al.* (1963) in Japan and Sheen and Wang (1980) in Taiwan. Ashizawa *et al.* (1979) reported screening of radish cultivars for *Fusarium* resistance. The cultivars most resistant to *F. oxysporum* f. *raphani* were Red Prince and White Spike in the European group, and Kotabe in the Japanese group. Three new Japanese hybrid cultivars have also been bred for *Fusarium* resistance; these are Natsutomi (Summer Fortune), Natsuminowase-3 and Kazusanatsuminowase-3 (Ashizawa *et al.*, 1979).

The promising germplasm identified for various diseases in Japanese radish, Chinese radish and Indian radish are presented in Tables 4, 5 and 6, respectively.

Table 4: Promising Germplasm for Various Traits in Japanese Radish*

Trait	Genotype
Resistance/tolerance to biotic stress/Disease	
Mosaic virus	Minowase group: Bansei-Mino, Motohashi-Mino (resistant) Nerima group: Aki-wase, Takakura, Tosai, Nishimachi-riso (resistant)Awa-Bansei group: Choko-hatsusyu (resistant)
Yellows (*Fusarium oxysporum* f. sp. *conglutinans*)	Hatsuka group: Red Prince (resistant) Minowase group: Otabe (resistant) Motohashi-Taibyo-Minowase, Natsutomi (Moderately resistant) Nerima group: Tosai (moderately resistant) Miyashige group: Utsugi-Gensuke (moderately resistant)
Soft rot (*Erwinia carotovora* subsp. *carotovora*)	Shiroagari group: Yokokadokei (highly resistant); Nerima group: Takakura, Miura, Tosai (resistant) ; Miyashige group: Asahi-shirokubi, Taibyo-miyashige (moderately resistant) Ninengo group: Ninengo (moderately resistant)
Plasmodiophora brassicae (race 2) and *Erwinia carotovora*	Waincha, Kuro Daikon A, Yokomonkei (moderately resistant)

* Kaneko and Matsuzawa (1993); Bunin *et al.* (1991).

The disease resistance traits from *Raphanus* have also been utilized by the *Brassica* breeders, *viz.* resistance to turnip mosaic virus (Fujisawa, 1990), resistance to clubroot (Hagimori *et al.*, 1992) and resistance to beet cyst nematode (Lelivelt *et al.*, 1993).

Table 5: Promising Germplasm for Various Traits in Chinese Radish*

Trait	Genotype
Resistance/tolerance to biotic stress/Disease	
Radish mosaic virus	Rose China Winter, Zheltyi (Resistant)
Yellows (*Fusarium oxysporum* f. sp. *conglutinans*)	China Rose, Kinmon-Aka (Resistant)
Soft rot (*Erwinia carotovora* subsp. *carotovora*)	North Chinese radish group: Pekin Shinribi (Highly resistant) Kiriba-Matsumoto (Moderately resistant)

* Kaneko and Matsuzawa (1993); Mavlyanova (1986).

Table 6: Promising Germplasm for Various Traits in Indian Radish*

Trait	Genotype
Resistance/tolerance to Disease	
Alternaria blight (*Alternaria brassicae*)	Accessions 7401, 6802, 8801, 8803, 8805, Pusa Desi and Jaunpuri (Resistant), Accessions 7212, 7110, 7108, 7208, 7218, 8804, S-271 and IHR-1 (Moderately resistant)
White rust (*Albugo candida*)	IIHR-1-1 (Resistant)

* Kumar and Singh (2003).

12.6.0 Interspecific and Intergeneric Hybridization

Although distantly related, intergeneric hybrids can be obtained between radish and *Brassica* species. Radish is a useful genetic resource for breeding of *Brassica* crops for disease resistance and stress tolerance. Karpechenko (1924) crossed *R. sativus* (R) was with *B. oleracea* (C) to produce amphidiploid Raphanobrassica (RRCC). Since then interspecific or intergeneric crosses involving *Raphanus* have been used for transfer of beet cyst nematode resistance (Lelivelt *et al.*, 1993), transfer of clubroot resistance (Hagimori *et al.*, 1992), transfer of restorer gene (Heyn, 1976), transfer of shattering resistance to rapeseed (Agnihotri *et al.*, 1990) and production of monosomic chromosome addition lines (Kaneko *et al.*, 2001). Now, forage varieties of Raphanobrassica (RRCC) have also been developed for commercial production. Choudhary *et al.* (2000) synthesized Amphidiploid Raphanofortii by colchicinization of the F_1 hybrid *Brassica tournefortii* (TT, 2n=20) × *Raphanus caudatus* (RR, 2n=18) which has great potential as a new commercial crop, as well as a bridge species for the transfer of economically important attributes of both the species to other *Brassica* species.

13.0 Biotechnology

The importance and application of biotechnological work in crop improvement is well known. Sen and Deka (2001) have summarised the results of investigations on different aspects of biotechnology in radish. Hanning (1904) was the pioneer in successfully growing excised embryo of *Raphanus* to maturity on mineral salts and sugar solution under aseptic condition.

13.1.0 Tissue Culture

13.1.1 Protoplast Culture and Fusion

Plant regeneration from leaf protoplasts of radish has been demonstrated by several workers. Lee *et al.* (1996) showed that protoplasts isolated from 3- to 4-week-old leaves of radish formed callus in liquid medium. The callus developed shoot in agar medium supplemented with 0.1 mg NAA, 1 mg benzyladenine and 3 mg $AgNO_3$/l. Addition of $AgNO_3$ was found essential for shoot regeneration, while roots regene-rated on MS medium without growth regulators. Hagimori and Nagaoka (1992) found that nurse culture technique using cells of *Nicotiana, Brassica, Daucus, Lactuca* and *Asparagas* resulted in the improvement of cell division and colony formation of mesophyll protoplasts of Japanese radish.

Tonosaki *et al.* (2013) studied the efficiency of hybrid production without embryo culture between *B. rapa* and *R. sativus* and identified three QTLs controlling hybrid formation ability.

The important role of cytoplasmic male sterility (CMS) has been studied in radish. Effective transfer of CMS from radish to rape was reported by Paulmann and Robbelen (1988). The incorporation of radish CMS in vegetable species by protoplast fusion was also reported by Kameya *et al.* (1989). Two of the regenerated plants were found to be somatic hybrids with the nucleus of cabbage and the cytoplasm of radish. All the regenerated plants flowered and all were sterile.

Hagimori *et al.* (1992) reported production of somatic hybrids between the Japanese radish and cauliflower. Somatic hybrids were produced by electrofusion of protoplast to transfer clubroot disease resistance of radish into cauliflower. Among 40 regenerated plants, 37 were morphologically hybrid-type and the hybrids were further confirmed by isozyme and RFLP analysis. All of the hybrid-type plants showed resistance to clubroot disease and the level of resistance was comparable to radish.

Sakai *et al.* (1996) reported introduction of a gene from fertility-restored radish into *Brassica napus* by fusion of X-irradiated protoplasts from a radish restorer line. X-irradiated protoplasts isolated from shoots of radish were fused with iodoacetamide treated protoplast of *B. napus* CMS cybrid. Among 300 regenerated plants, six were male-fertile of which three were aneuploids.

When a trait of interest is not available in the primary or secondary pool, the breeders look to tertiary gene pools to transfer valuable genes controlling traits from different wild species into related crop species. Interspecific crosses or somatic cell fusions can be used to create alloplasmic plants, in which the cytoplasm of one plant species is present in combination with the nuclear genome of a different species. These cytoplasmic substitutions often result in the expression of mitochondrial genes determining CMS (Chase, 2006). CMS cybrids derived from protoplast fusion are useful tools for the identification of CMS-related regions because of frequent recombinations among the parental mtDNAs cause rearrangement of the mt genomes in the fusion products. CMS was transferred from *Raphanus sativus* cv. Kosena (CMS line) to *B. napus* rape cv. Westar through cytoplast-protoplast fusion (Sakai and Imamura, 1990). Motegi *et al.* (2003) obtained Ogura CMS plants by asymmetrical protoplast fusion between red cabbage (fertile) and normal radish (fertile).

13.2.0 Somatic Embryogenesis

Many workers have been able to grow stomatic embryos from explants like hypocotyl segments (Matsubara and Hegazi, 1990; Jeong *et al.*, 1995). Hypocotyl segments of radish, when cultured on MS medium supplemented with 1 mg/l 2,4-D, produced yellow compact calli. On transfer to a medium containing 6-BA and NAA 0.1 mg/l, somatic embryos were produced from these calli. The embryo reached the cotyledonary stage after transferring to a medium containing 0.1 mg/l 2,4-D and 1 mg/l ABA. Subsequently, 90 per cent of the embryos developed into plantlets in a half-strength MS basal medium. These plantlets could be successfully transplanted into soil, which grow to maturity in a phytotron. Pua *et al.* (1996) investigated the synergistic effect of ethylene inhibitors and putrescine on shoot regeneration from hypocotyl explants of radish. Although the explants were recalcitrant in culture, exogenous application of ethylene inhibitor [20-30 µM aminoethoxyvinylglycine (AVC) or $AgNO_3$] enhanced shoot regeneration from the explant grown on media supplemented with 2 mg/l BA and 1 mg/l NAA and it was further promoted by putrescine.

13.3.0 Genomics and Molecular Mapping

The rapid developments in DNA technologies have opened the door for marker-assisted breeding of complex traits in several crops, including radish. In radish, the RAPD, AFLP, ISSR and RFLP molecular markers have been used for cultivar identification (Pradhan *et al.*, 2004a), phylogenetic analysis (Kong *et al.*, 2005), genetic diversity analysis (Muminovic *et al.*, 2005) and construction of genetic maps (Bett and Lydiate, 2003; Tsuro *et al.*, 2005). Pradhan *et al.* (2004b) have identified several RAPD markers associated with high or low seed weight, germination proportion, seedling length and fresh weight in radish, which can be further used for marker-assisted breeding and selection of improved radish cultivars.

Genetic diversity of 30 radish (*Raphanus sativus*) accessions was investigated at the phenotypic level with morphological characters and at the DNA level using RAPD technique (Rabbani *et al.*, 1998). Thirty-six morphophysiological traits were recorded from seedling stage to harvest. The 31 primers used generated 22 RAPD bands, of which 158 (78.2 per cent) were polymorphic. Dendrograms were generated for the Euclidean distance of RAPD markers. Phenotypically, the accessions were classified into four major groups corresponding to the different forms of cultivated radish. The morphological diversity existing within each of these groups suggested that they should be discriminated into the three botanical convarieties, *sativus* (large-rooted), *caudatus* (pod-type) and *oleifer* (oil seed-type). The phylogenetic relationships among Japanese wild radishes, cultivated radishes and wild *R. raphanistrum* species were examined using RAPD analysis. PCR products of 61 plants belonging to fourteen strains or varieties within the three categories were compared by using seven 10-mer random primers analysis (Yamagishi *et al.*, 1998).

Based on RFLP analysis, two evolutionary lineages for *Brassica* diploid species have been proposed by Yang *et al.* (1998). These are: (i) the *nigra* lineage, and (ii) the *rapa oleracea* lineage. The phylogenetic relationship of *Raphanus* species to these two lineages is still unclear because chloroplast and mitochondrial DNA genomic restriction site variation.

Bett and Lydiate (2003) have developed the first genetic map of the *Raphanus* genome, based on meiosis in a hybrid between *R. sativus* (cultivated radish) and *R. raphanistrum* (wild radish). The genetic markers and the reference map of *Raphanus* is of considerable value for mapping quantitative trait loci (QTL) controlling important traits in radish such as root shape and size, skin colour, nutritional value, quantity of glucosinolate, development time, self- incompatibility, restoration of CMS and resistance to various diseases.

Though a monogenomic species (n = 9), most genes of radish have three similar copies in a genome as in the *Brassica* monogenomic species, suggesting that genome triplication has occurred after divergence from an ancestor of *Arabidopsis thaliana*. Genome size of radish has been estimated to be from 526 (Arumuganathan and Earle, 1991) to 573 Mb (Johnston *et al.*, 2005). The sequences of the whole radish genome were first reported by Kitashiba *et al.* (2014). Using short-read genomic sequences of 191.1 Gb, they predicted 61,572 genes and assigned 1345 scaffolds to a high-density linkage map of 2553 DNA markers. Shirasawa *et al.* (2020) sequenced the genome of Japanese radish caultivar 'Sakurajima Daikon' possessing giant root and identified a total of 89,915 genes, of which 30,033 were newly found. The genome information will contribute to the establishment of a new resource for the radish genomics and also provide insights into the molecular mechanisms underlying formation of the giant root.

By comparative mapping, radish breeders can easily and cheaply access genetic information from related species. The comparative maps of the *Raphanus* and Brassica genomes can be efficiently utilized for efficient interspecies transfer of *Raphanus* genes to *Brassica* crops. In future, comparative mapping with *Arabidopsis thaliana* will greatly facilitate map-based gene cloning, candidate gene identification and help in understanding the evolutionary relationships between *Raphanus* and its genetically well-characterized relatives.

Kaneko *et al.* (2007) identified a QTL-associated resistance to fusarium wilt that was linked to the RAPD markers, OPJ14 and CC18 locus on linkage group 1 of radish. Xu *et al.* (2014) performed QTL mapping using F_2 populations derived from a downy mildew resistant line, NAU-dhp08, and a susceptible line, NAU-qtbjq-06 and determined that resistance to downy mildew DM at the seedling stage in radish was controlled by a single dominant allele *RsDmR*. Duan (2014) reported 14 QTLs associated with resistance to the black rot pathogen Xcc8004 in radish.

Isothiocyanates are hydrolyzed products of glucosinolates which are responsible for pungency of radish. The major glucosinolate in radish roots is glucoraphasatin. QTLs controlling glucosinolate content in radish root have been reported (Zou *et al.*, 2013), and candidate genes have been inferred. A mutant having a high amount of glucoerucin without glucoraphasatin has been selected (Ishida *et al.*, 2015). Later, the gene responsible for this mutation has been identified (Kakizaki *et al.*, 2017).

13.4.0 Male Sterility

The distribution of Ogura male-sterile cytoplasm among Japanese wild radish populations and Asian cultivated radish was studied by means of PCR-aided assays using mitochondrial atp6 and orf138 loci as molecular markers (Yamagishi and Terachi, 1996). Among 217 wild radish plants, 93 had both Ogura-type atp6 and orf138; whereas 124 had normal-type atp6. A complete linkage between Ogura-type atp6 and orf138 loci was also reported in Japanese wild radish. Among the 44 Asian cultivars analyzed, 40 were determined to have normal cytoplasm. The mutiple and independent introduction of Ogura-type cytoplasm from the wild radish into these cultivars was also suggested.

Nahm *et al.* (2005) have characterized a new CMS, NWB CMS, obtained from a novel male sterile radish line, collected from South Korea. NWB CMS is unique and has more profound degree of sterility than any of the other CMS types, including Ogura CMS. It induced complete sterility in progenies, when crossed with any of the tested breeding lines. The CMS phenomenon of the NWB CMS line may be induced by cytoplasmic factor(s). NWB CMS differs from other CMS types with regard to its degree of sterility and mitochondrial genome arrangement. The Ogura CMS specific factor, orf138 is absent from NWB CMS line. A specific 2 kbp DNA fragment confers male sterility in the NWB CMS line. The "WonBaek" CMS line from China is different from Ogura CMS. The orf138 gene is not present in the WonBaek CMS. The crucial factors that determine male sterility for the Ogura and Kosena CMS, NWB CMS, and WonBaek CMS lines are different (Nahm *et al.*, 2005).

The fertility restorer genes for Ogura CMS are widely distributed in the *Raphanus raphanistrum* and Japanese wild radish (Hamadaikon). Among the cultivated radishes, the restorer gene is frequently found in the European and Chinese varieties, but not in Japanese varieties (Yamagishi, 1998). It has been suggested that the Ogura CMS restorer gene(s) alters the expression of orf138 at a translational or post-translational level and in an organ-specific manner (Krishnasamy and Makaroff, 1994; Bellaoui *et al.*, 1999). In Kosena CMS radish, the fertility restorer gene orf687 has recently been cloned by positional cloning (Koizuka *et al.*, 2003). Orf687 functions either directly or indirectly to lower the levels of Kosena CMS-associated gene, orf125, thereby restoring the fertility (Koizuka *et al.*, 2003). The genomic structures of Kosena CMS-associated gene orf125 and Ogura CMS-associated gene orf138 are almost identical (Bonhomme *et al.*, 1992; Iwabuchi *et al.*, 1999).

Various molecular markers linked to the Rf gene have been explored. AFLP markers linked to the Rf gene have been identified in radish (Murayama *et al.*, 2003), which can be further utilized for identification of QTLs using molecular maps. However, the molecular mechanism underlying the restoration of fertility still remains unclear. Currently there is great interest in the map-based cloning of the Rf gene.

The complete nucleotide sequences of the mitochondrial genomes of eight radish cultivars have been studied (Tanaka *et al.*, 2012; Chang *et al.*, 2013; Park *et al.*, 2013; Jeong *et al.*, 2016). Recent genomics studies indicate that the structural variations among radish mitochondrial genomes are not only due to inversions and translocations, but are associated with the acquisition of a new sequence of unknown origin.

13.5.0 Genetic Transformation

Radish is recalcitrant in cell and tissue cultures. Krsnik-Rasol *et al.* (1992) reported infection of leaf, excised from 3-6 cm plantlets, regenerated from lateral bud of a root explant, with *A. tumefaciens* or *A. rhizogenes*. Tumours and hairy roots emerged within two weeks. Total peroxidase activity was reported to be higher in all transformed tissues than in normal ones.

Genetic transformation of radish by using foreign DNA has been studied by several workers. Kang *et al.* (1990) reported genetic stability of characters transformed by the microinjection of foreign DNA in radish plants. A detailed procedure for *Agrobacterium* injection, tissue preparation and GUS histochemistry was given by Castle and Morris (1990). Shishkova *et al.* (1994) also discussed *in vitro* transformation of radish by *A. rhizogenes*. However, till now genetic transformation of radish cannot be utilized commercially.

Radish is one of the most recalcitrant crop plants in culture and has a very low shoot regeneration frequency (Curtis, 2003). Since an efficient tissue culture system does not exist for the production of transgenic radish in culture (Curtis, 2003), the first transgenic radish (*Raphanus sativus* var. *longipinnatus*) was produced in Korean cultivar 'Jin Ju Dae Pyong', using an alternative method known as 'floral-dip' (Curtis and Nam, 2001). Floral-dip method is currently the only procedure of gene transfer available for producing transgenic radish (Curtis, 2003).

Korean ecotypes of radish are cold sensitive, hence readily bolt at 5-6°C during autumn, making production of high quality roots not possible (Curtis, 2003). Hence, transfer of late-flowering genes into radish is important. GIGANTEA (GI), a late-flowering gene in Arabidopsis, has been cloned recently. The 'floral-dipping' technique has been used successfully to delay both bolting and flowering in radish by co-suppression of the photoperiodic gene, GIGANTEA and a late-flowering transgenic radish produced (Curtis *et al.*, 2002). The production of late-flowering germplasms of radish will allow cultivation of cold-sensitive long-day radish over an extended period. The development of an efficient gene transfer system in radish would enable the radish breeders to transfer agronomically important traits, and manipulate the pharmaceutically valued endogenous chemicals for the production of novel germplasms for the benefit of mankind (Curtis, 2003).

Genetic Transformation in Radish

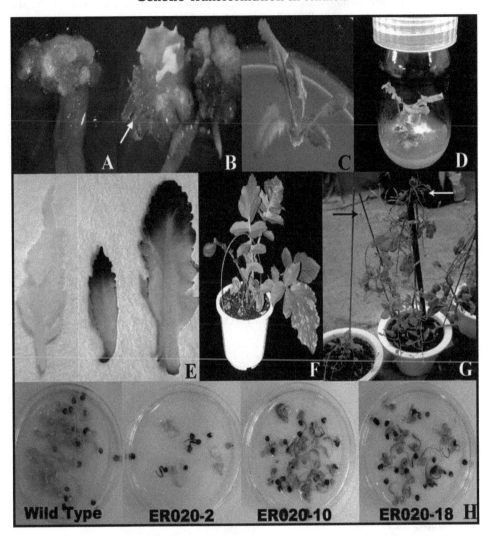

Plant Regeneration from Hypocotyl Explants of Radish Transformed with Gus Gene.

A: Yellowish-green callus formation on selection medium with 10 mg l_1 hygromycin. B: Hygromycin-resistant shoot primordia. C, D: Shoot elongated from the hygromycinresistance primordia. E: GUS negative response in leaf of nontransgenic plant (left), GUS positive response in the leaf of transgenic plant (right). F: Transgenic radish grown in soil. G: Flowering plants. H: GUS assay for transgenic radish events (T1) (Cho *et al.*, 2008).

14.0 References

Agnihotri, A., Shivanna, K.R., Raina, S.N., Lakshmikumaran, M., Prakash, S. and Jagannathan, V. (1990) *Plant Breed.*, **105**: 292-299.

Anjanappa, M., Reddy, N.S., Murali, K. and Krishnappa, K.S. (1998) *Karnataka J. Agric. Sci.*, **11**: 862-864.

Antonova, N.G. (1991) *Kartofel' iOvoshchi*, No. 1, p. 18.

Aoba, T. (1967) *J. Yamagata Agr. Forest Soc.*, **24**: 7-12 (in Japanese).

Aoba, T. (1979) *The iden.*, **33**: 55.

Aoba, T. (1989) *Noko Gijutu (Arts in Agriculture)*, **12**: 92-114 (in Japanese).

Aoba, T. (1993) Vegetables in Japan. Yasaka Shobo, Tokyo, 311 p. (in Japanese).

Arumuganathan, K. and Earle, E.D. (1991) *Plant Mol. Biol. Rep.*, **9**: 208-218

Ashizawa, M., Hida, U. and Voshikawa, H. (1979) *Bull. Veg. Ornam. Crops Res. Sta. (Japan)*, No. 6, pp. 39-70.

Badiger, S., Sivakumar, K.C., Reddy, M.A.N. and Mathew, D. (2001) *Current Res. Univ. Agric. Sci. –Bangalore*, **30**: 201-202.

Bailey, L.H. and Bailey, E.Z. (1956) Hortus second – A concise Dictionary of Gardening and General Horticulture. Macmillan Co., New York.

Banga, O. (1976) In: *Evolution of Crop Plants* (Ed. N.W. Simmonds), Longman, London and New York, p. 60-62.

Banga, O. and Van Bennekom, J.L. (1962) *Euphytica*, **11**: 311-326.

Barillari, J., Rollin, P. and Lori, R. (2002) *NutraCos*, **1**: 6-9.

Becker, C. (1962) *Handbuch der Pflanzenzuchtung*, **6**: 23-78.

Bellaoui, M., Grelon, M., Pelletier, G. and Budar, F. (1999) *Plant Mol. Biol.*, **40**: 893–902.

Bett, K. E. and Lydiate, D. J. (2003) *Genome*, **46**: 423-430.

Bhat, K.L. (1996) *Crop Res.*, **11**: 204-206.

Bhople, S.R., Bharad, S.G., Dod, V.N. and Gholap, S.V. (1998) *Orissa J. Hort.*, **26**: 34-36.

Bhople, S.R., Dod, V.N., Bharad, S.G., Gholap, S.V. and Jadhao, B.J. (1998) *J. Soils Crops*, **8**: 214-215.

Bichta, H. and Wegrzyn, A. (1994) *Ann. Univ. Marial Curie Sklodowska, Sectio EEE, Horticultura*, **2**: 67-71.

Bonhomme, S., Budar, F., Lancelin, D., Small, I., Defrance, M.C. and Pelletier, G. (1992) *Mol. Gen. Genet.*, **235**: 340-348.

Bonnet, A. (1970) *Eucarpia CRNA*, P. Versailles, pp. 83-88.

Bunin, M.S., Esikava, Kh. and Yoshikawa, H. (1991) *Doklady Vsesoyuznoi Ordena Lenina-i-Ordena Trudovogo Krasnogo Znameni Akademii Sel'skokhozyaistvennykh Nauk-im. V.I. Lenina*, **11**: 32-38.

Capecka, E., Libik, A., Thomas, G. and Monteiro, A.A. (1998) *Acta Hort.*, **459**: 89-95.

Castle, L.A. and Morris, R.O. (1990) *Pl. Mol. Biol. Rep.*, **8**: 28-39.

Chang, L.X. and Chang, J.X. (2000) *J. Hebei Agric. Univ.*, **23**: 20-24.

Chang, S., Chen, J., Wang, Y., Gu, B., He, J., Chu, P. and Guan, R. (2013) *J. Genet. Genomics*, **40**: 117–126.

Chase, C.D. (2006) *Trends Genet.*, **23**: 81-90.

Chatterjee, R. (1989) *Hort. J.*, **2**: 55-58.

Chatterjee, R. and Som, M.G. (1991) *Indian J. Hort.*, **48**: 145-147.

Cho, M.A., Min, S.R., Ko, S.M., Liu, J.R. and Choi, P.S. (2008) *Plant Biotechnol.*, **25**: 205–208.

Choudhary, B.R., Joshi, P. and Singh, K. (2000) *Theor. Appl. Genet.*, **101**: 990-999. Clapham, A. R., Tutin, T. G., Warburg, E. F. (1962) Flora of the British Isles. Cambridge University Press, Cambridge.

Choudhury, B. and Sirohi, P.S. (1972) *Indian Hort.*, **16**: 17-18.

Choudhury, B. and Sirohi, P.S. (1975) *Indian Hort.*, **20**: 15-16.

Crisp, P. (1995) Radish. In: Evolution of Crop Plants, 2nd edn, eds. J. Smartt and N.W. Simmonds. Longman Scientific and Technical, Harlow, Essex, England. p. 86-89.

Cruz, B.P.B., Silveira, A.P.Da, Demate, M.E.S.P. and silveria, S.G.Da (1973) *Biologico*, **39**: 203-205.

Curtis, I.S. (2003) *Trends Plant Sci.*, **8**: 305-307.

Curtis, I.S. and Nam, H.G. (2001) *Transgenic Res.*, **10**: 363-371.

Curtis, I.S., Nam, H.G., Yun, J.Y. and Seo, K.H. (2002) *Transgenic Res.*, **11**: 249-256.

Danu, N.S. and Lal, S.D. (1998) *Progressive Hort.*, **30**: 135-138.

Davenport, D.C. (1962) *M.Sc. Thesis*, Indian Agricultural Research Institute, New Delhi.

De Candolle, A. (1883) Origin of Cultivated Plants. D. Appleton and Co., New York.

Deotale, A.B., Belorkar, P.V., Badvaik, N.G., Patil, S.R. and Rathod, J.R. (1994) *J. Soils Crops*, **4**: 120-121.

Depel, Z. (1989) The genetic resources of radish in China. Intl. Symp. Hortic. Germ. Cult. and Wild. Part II. Vegetables.

Dhar, S. (1998) *Prog. Hort.*, **30**: 217-220.

Dickinson, H.G. and Lewis, D. (1973) *Proc. Roy Soc.*, London, B, **183**: 21-38.

Djurovka, M., Markovic, V., Ilin, Z., Jevtic, S. and Lazic, B. (1997) *Acta Hort.*, **462**: 139-144.

Duan, Y. (2014) *Ph.D. Dissertation*, Chinese Academy of Agricultural Sciences, China.

Eguchi, T., Matsumura, T. and Koyama, T. (1963) *Proc. Amer. Soc. Hort. Sci.*, **82**: 322-331.

Etebarian, H.R. (1993) *Iranian J. Pl. Pathol.,* **29**: 51-52.

Frost, H.B. (1923) *Genetics,* **8**: 116-153.

Fujisawa, I. (1990) *Japan. Agric. Res. Quarterly,* **23**: 289-293.

Gaweda, M., Kopecka, Z. and Capecka, E. (1991) *Folia Horticulturae,* **3**: 47-59.

George, R.A.T. (1986) In: *Vegetable Seed Production.* Longman, London.

Ghanti, P., Sounda, G. and Ghatak, S. (1989) *Environ. Ecol.,* **7**: 957-959.

Ghormade, B.G., Kale, P.B., Kulwal, L.V. and Deshmukh, P.P. (1989) *PKV Res. J.,* **13**: 34-38.

Gill, H.S. (1993) In: *Advances in Horticulture, Vegetable Crops* (Eds. K.L. Chadha, and G. Kalloo), Malhotra Publishing House, New Delhi, India, pp. 201-215.

Gill, S.S. and Gill, B.S. (1995) *Seed Res.,* **23**: 28-31.

Gill, S.S., Gill, B.S., Brar, S.P.S. and Singh, B. (1995) *Seed Res.,* **23**: 47-49.

Giusti, M.M. and Wrolstad, R.E. (1996) *J. Food Sci.,* **61**: 322-6.

Giusti, M.M., Rodriguez-Saona, L.E., Baggett, J.R., Reed, G.L., Durst, R.W. and Wrolstad, R.E. (1998) *J. Food Sci.,* **63**: 219-224.

Grelon, M., Budar, F., Bonhomme, S. and Pelletier, G. (1994) *Mol. Gen. Genet.,* **243**: 540-547.

Gupta, P.K., Mahakal, K.G. and Sadwarte, K.T. (1974) *Punjabrao Krishi Vidyapith Res. J.,* **5**: 24.

Haag, H.P. and Minami, K. (1987) *Anais da Escola Superior de Agricultura 'Luiz de Querioz',* Universidade de Sao Paulo, **44**: 409-418.

Hagimori, M. and Nagaoka, M. (1992) *Plant Sci.,* **86**: 105-113.

Hagimori, M., Nagaoka, M., Kato, N. and Yoshikawa, H. (1992) *Theor. Appl. Genet.,* **82**: 819-824.

Hagiya, K. (1952-59) *J. Hort. Ass. Japan,* **21**: 81-86, 165-173; **26**: 111-120; **27**: 68-77; **28**: 109-114.

Hall, C.B. (1991) *Proc. Fla Sta. Hort. Soc.,* **103**: 110-101.

Hanning, E. (1904) *Bot. Ztg.,* **62**: 34-80.

Hawlader, M.S.H., Mian, M.A.K. and Ali, M. (1997) *Euphytica,* **96**: 297-300.

Hawlader, M.S.H., Mian, M.A.K. and Quadir, M.A. (1997) *Bull. Inst. Trop. Agric., Kyushu Univ.,* **20**: 59-64.

Hedge, D.M. (1987) *Indian J. Agron.,* **32**: 24-29.

Hegi, G. (1935) Illustrierte Flora von Mittel-Europa. Band IV, Haffte IJR Lehmann's Velarg, Munchen.

Heine, H. and Dreyer, S. (2000) *Gemuse,* **36**: 7-9.

Helm, J. (1957) *Kulturpflanze,* **5**: 41-54.

Henis, Y., Ghaffar, A. and Baker, R. (1978) *Phytopath.,* **68**: 900-907.

Henslow, G. (1898) *Gardeners Chronicke*, **23**: 389.

Heyn, F. W. (1976) *Cruciferae Newsl.*, **1**: 15–16.

Hida, K. (1990) In: *Collection of Plant Genetic Resources* (Ed. T. Matsuo). p. 823.

Hinata, K. and Nishio, T. (1980) In: *Brassica Crops and Wild Allies*, (Eds. S. Tsunoda, K. Hinata and C. Gomez-Campo), Jap. Sci. Soc. Press, Tokyo, p. 223-234.

Huh, M.K. (1995) *Ph.D. Thesis*, Pusan National University, Pusan, p.145.

Huh, M.K. and Ohnishi, O. (2001) *Genes Genet. Syst.*, **76**: 15-23.

Huh, M.K. and Ohnishi, O. (2002) *Breed. Sci.*, **52**: 79-88.

Ijaz, H., Ihsanul, H., Mohammad, S. and AsifurRahaman (1997) *Sarhad J. Agric.*, **13**: 39-43.

Ikegaya, Y. (1986) *Jpn. J. Breed.*, **36**: 106-107.

Ishida, M., Kakizaki, T., Morimitsu, Y., Ohara, T., Hatakeyama, K., Yoshiaki, H., Kohori, J. and Nishio, T. (2015) *Theor. Appl. Genet.*, **128**: 2037–2046

Ishii, G. and Saijo, R. (1987) *J. Japanese Soc. Hort. Sci.*, **56**: 313-320.

Iwabuchi, M., Koizuka, N., Fujimoto, H., Sakai, T. and Imamura, J. (1999) *Plant Mol. Biol.*, **39**: 183-188.

Iwasa, S. (1980) In: *Vegetables in Tropics*, Youkendo, Tokyo, p. 293.

Jadav, B.J., Damke, M.M., Wash, A.P., Joshi, P.S. and Kulkarni, P.M. (1998) *PKV Res. J.*, **22**: 178-179.

Jaiswal, J.P., Subedi, P.P. and Bhattarai, S.P. (1997) *Working Paper, Lumle Agric. Res. Cent.*, No. 97, p. 14.

Jamwal, R.S. and Thakur, D.R. (1993) *Indian J. Agric. Sci.*, **63**: 170-172.

Jandial, K.C., Samnotra, R.K., Sudan, S.K. and Gupta, A.K. (1997) *Environ. Ecol.*, **15**: 46-48.

Jauhari, O.S. and Purandare, O.S. (1959) *Sci. Cult.*, **25**: 256-2588.

Jeong, W.J., Min, S.R. and Lin, J.R. (1995) *Plant Cell Rep.*, **14**: 648-651.

Jeong, Y.M., Chung, W.H., Choi, A.Y., Mun, J.H., Kim, N. and Yu, H.J. (2016) *Mit. DNA Part A*, **27**: 941–942.

Johnson, D.A. and Gabrielson, R.L. (1990) *Extn. Bulletin Co-operative Extn. Coll. Agric. and Home Eco., Washington State Univ.*, EB1570, p. 1.

Johnston, J.S., Pepper, A.E., Hall, A.E., Chen, Z.J., Hodnett, G., Drabek, J., Lopez, R. and Price, H.J. (2005) *Ann. Bot.*, **95**: 229–235

Joshi, P.C. and Patil, N.S. (1988) *South Indian Hort.*, **36**: 331-332.

Joungkeun, K. and Semisi, S.T. (1995) *J. South Pacific Agric.*, **2**: 23-26.

Kakizaki, T., Kitashiba, H., Zou, Z., Li, F., Fukino, N., Ohara, T., Nishio, T. and Ishida, M. (2017) *Plant Physiol.*, **173**: 1583–1593.

Kalvi, T. S. and Nath, P. (1970a) *Farm J.*, pp. 17-22.

Kalvi, T. S. and Nath, P. (1970b) *Punjab Hort. J.*, **10**: 137-148.

Kameya, T., Kanzaki, H., Toki, S. and Abe, T. (1989) *Jap. J. Genet.*, **64**: 27-34.

Kanaujia, S.P. and Sharma, S.K. (1998) *Hort. J.*, **11**: 59-62.

Kaneko, Y. and Matsuzawa, Y. (1993) In: *Genetic Improvement of Vegetable Crops*, (Eds. G. Kalloo and B.O. Bergh), Pergamon Press, New York. pp. 487-510.

Kaneko, Y., Kimizuka-Takagi, C., Bang, S.W. and Matsuzawa, Y. (2007) In: *Genome mapping and molecular breeding in plants* (Ed. C. Kole), vol 5. Springer, Heidelberg, pp. 141–160

Kaneko, Y., Yano, H., Bang, S.W. and Matsuzawa, Y. (2001) *Plant Breed.*, **120**: 163-168.

Kang, N.J., Cho, J.L. and Kang, S.M. (1990) *J. Korean Soc. Hort. Sci.*, **31**: 185-192.

Karivaratharaju, T.V., Palanisamy, V. and Vanangamudi, K. (1988) *South Indian Hort.*, **36**: 81-82.

Karpechenko, G. D. (1924) *J. Genet.*, **14**: 375-396.

Keppler, E. (1941) Z. *Pflanzenz.*, **23**: 611-684.

Kirtikar, K. R. and Basu, B. D. (1935) In: *Indian Medicinal Plants*, Published by Lalit Mohan Basu, Allahabad.

Kitamura, S. (1958) In: *Radish in Japan* (Ed. I. Nishiyama), Japan Scientific Society Press, Tokyo, pp. 1-19 (in Japanese).

Kitamura, S. and Murata, G. (1957) Colored Illustrations of Herbaceous Plants of Japan (Choripetalae). Hoikusha Publ. Co., Ltd., Osaka, Japan.

Kitashiba, H., Li, F., Hirakawa, H., Kawanabe, T., Zou, Z., Hasegawa, Y., Tonosaki, K., Shirasawa, S., Fukushima, A., Yokoi, S., Takahata, Y., Kakizaki, T., Ishida, M., Okamoto, S., Sakamoto, K., Shirasawa, K., Tabata, S. and Nishio, T. (2014) *DNA Res.*, **21**: 481–490.

Knecht, G. N. (1975) *HortScience*, **10**: 274-275.

Kobabe, G. (1959) *Zeitschrift fur Pflanzenzuchtung*, **41**: 1-10 (in German).

Koizuka, N., Imai, R., Fujimoto, F., Hayakawa, T., Kimura, Y., Kahno- Murase, J.K., Sakai, T., Kawasaki, S. and Imamura, J. (2003) *Plant J.*, **34**: 407–415.

Koizuka, N., Imai, R., Iwabuchi, M., Sakai, T. and Imamura, J. (2000) *Theor. Appl. Genet.*, **100**: 949–955.

Kolbe and Voss (1952) *Fmg. S. Afr.*, **27**: 235.

Kong, Q.S., XiXiang, L., Xiang, C.P., Yang, Q. and Shen, D. (2005) *Sci. Agric. Sinica*, **38**: 1017-1023.

Kovacik, P. and Lozek, O. (1999) *Acta Horticulturae et Regiotechturae*, **2**: 32-35.

Koyama, T., Nakamura, H. and Yamada, E. (1968) *Bull. Hort. Res. Sta.*, Hiratsuka, No. 7, pp. 237-283.

Kremer, J. C. (1945) *Mich. Agric. Exp. Sta. Quart. Bull.*, **27**: 413-420.

Krishnasamy, S. and Makaroff, C.A. (1993) *Curr. Genet.*, **24**: 156-163.

Krishnasamy, S. and Makaroff, C.A. (1994) *Plant Mol. Biol.*, **26**: 935-946.

Krsnik-Rasol, M., Jelaska, S., Serkos-Sojak, V. and Delic, V. (1992) *PeriodicumBiologogorum*, **94**: 105-109.

Kulwal, L. V., Tayde, G. S. and Joshi, A. T. (1975) *Punjabrao Krishi Vidyapith Res. J.*, **3**: 140-141.

Kumar, P. and Singh, D.V. (2003) *Farm Sci. J.*, **12**: 61-62.

Kumar, P.B., Acharya, P.K., Dora, D.K. and Behera, T.K. (1994) *Orissa J. Hort.*, **22**: 36-40.

Kumar, R., Singh, A., Singh, M. and Singh, A. R. (1996) *Recent Hort.*, **3**: 93-97.

Kumar, V. and Chandel, K.S. (2003) *Crop Improv.*, **30**: 33-38.

Kumar, V., Chandel, K.S. and Pathania, N.K. (2002) *Veg. Sci.*, **29**: 34-36.

Kumazawa, S. (1956) Sosaiengei Kakuron. Yokendo, Tokyo, 637 pp. (in Japanese).

Kumazawa, S. (1961) Horticulture, Detailed Discussion of Crops. Yokendo, Tokyo, p.637 (in Japanese).

Kumazawa, S. (1965) Radish. In: Vegetable Gardening. Yokendo, Tokyo. p. 295.

Kutty, C.N. and Sirohi, P.S. (2003) *Indian J. Hort.*, **60**: 64-68.

Kvasnikov, B. V. and Nikonova, N. A. (1976) *Nauch. Tr. NII Ovosch. Kh-va*, **6**: 222-226.

Lal, S. and Srivastava, J. P. (1975) *Indian J. Hort.*, **32**: 91-93.

Langthasa, S. and Barah, P. (2000) *Indian J. Hill Frmg.*, **13**: 85-86.

Lee, J.M., Yoo, I.O. and Min, B.H. (1996) *J. Korean Soc. Hort. Sci.*, **37**: 349-356.

Lee, S.K. (2003) *Acta Hort.*, **620**: 127-133.

Lee, W.S., Kim, S.B. and Yamaguchi, H. (1992) *Korean J. Hort. Sci. Tech.*, **10**: 62-63 (in Korean).

Lelivelt, C.L.C., Lange, W. and Dolstra, O. (1993) *Euphytica*, **68**: 111-120.

Lewis-Jones, L.J., Thorpe J.P., and Wallis, G.P. (1982) *Biol. J. Linnean Soc.*, **18**: 35–48.

Li Shuxuan (1989) In: *The origin and resources of vegetable crops in China*. International Symposium on Horticultural Germplasm, Cultivated and Wild, Beijing, China, Sept. 1988. Chinese Society for Horticultural Science. International Academic Publishers, Beijing, pp. 197-202.

Li, M., Gai, S.Y. and Gai, Y.J. (1997) *China Vegetables*, No. 4, pp. 26-28.

Lichtenthaller, H. K. (1975) *Plant Physiol.*, **34**: 357-358.

Lim, S.H., Cho, H.J., Lee, S.J., Cho, Y.H. and Kim, B.D. (2002) *Theor. Appl. Genet.*, **104**: 1253-1262.

Ling, X. L., Li, Y. A. and Wang, M. J. (1986) *Sci. Agric. Senica* No. 2, pp. 54-60.

Longvah, T., Ananthan, R., Bhaskarachary, K. and Venkaiah, K. (2017) Indian Food Composition Tables 2017. National Institute of Nutrition, Hyderabad. 535 p.

Lovato, A., Montanari, M., Miggiano, A., Quagliotti, L. and Belletti, P. (1994) *Acta Hort.*, **362**: 117-124.

Lu, Y.H. and Shen, Y.M. (1996) *J. Shandong Agric. Univ.*, **27**: 135-140.

Lysak, M.A., Koch, M.A., Pecinka, A. and Schubert, I. (2005) *Genome Res.*, **15**: 516-525.

Makaroff, C.A. (1995) In: *The Molecular Biology of Plant Mitochondria* (Eds. C. S. Levings, and I. K. Vasil), Kluwer Academic Publ, Dordrecht, The Netherlands, pp. 515-555.

Makaroff, C.A. and Palmer, J.D. (1988) *Mol. Cell Biol.*, **8**: 1473-1480.

Makino, T. (1909) *Bot. Mag.*, **23**: 59-75.

Mangal, J.L., Batra, B.R. and Singh, G.R. (1998) *Haryana J. Hort. Sci.*, **17**: 97-101.

Mansfield, R. (1959) Die Kulturpflanzen baikeft. Akademic Velag, Berlin, pp. 1-659.

Matsubara, S. and Hegazi, H. H. (1990) *HortScience*, **25**: 1286-1288.

Matsubara, S., Miki, N., Murakami, K. and Uchida, K. (1990) *J. Japanese Soc. Hort. Sci.*, **59**: 137-142.

Maurya, A.N., Pathak, M.P. and Singh, K.P. (1990) *Acta Hort.*, **267**: 169-173.

Mavlyanova, R.F. (1986) *Nauchno Tekhnicheskii Byulleten' Vsesoyuznogo Ordena Lenina-i-Ordena Druzhby Narodov Nauchno Issledovatel'skogo Instituta-Rastenievodstva-imeni N. I. Vavilova*, **161**: 61-63.

Misra, H. P. (1987) *Indian J. Hort.*, **44**: 69-73.

Misra, H.P. and Yadav, K. (1989) *Seeds and Farms*, **15**: 14-17.

Misra, S. and Gupta, M. G. (1977) *Proc. Indian Acad. Sci.*, **85**: 319-326.

Mitsui, Y., Shimomura, M., Komatsu, K., Namiki, N., Shibata-Hatta, M., Imai, M., Katayose, Y., Mukai, Y., Kanamori, H., Kurita, K., Kagami, T., Wakatsuki, A., Ohyanagi, H., Ikawa, H., Minaka, N., Nakagawa, K., Shiwa, Y. and Sasaki, T. (2015) *Sci. Rep.*, **5**: 10835.

Monakhos, G.F. and Barasheva, G.M. (1999) *Izvestiya Timiryazevskoi Sel'skokhozyaistvennoi Akademii*, No. 1, pp. 92-100.

Motegi, T., Nou, I., Zhou, J., Kanno, A., Kameya,T., Hirata, Y., Nou, I.S. and Zhou, J.M. (2003) *Euphytica*, **129**: 319-323.

Muminovic, J., Merz, A., Melchinger, A.E. and Lubberstedt, T. (2005) *J. Amer. Soc. Hort. Sci.*, **130**: 79-87.

Murali, K., Reddy, N.S., Anjanappa, M. and Krishnappa, K.S. (1998) *Karnataka J. Agric. Sci.*, **11**: 1140-1141.

Murayama, S., Habuchi, T., Yamagishi, H. and Terachi, T. (2003) *Euphytica*, **129**: 61-68.

Nahm, S.H., Lee, H.J., Lee, S.W., Joo, G.Y., Harn, C.H., Yang, S.G. and Min, B.W. (2005) *Theor. Appl. Genet.*, **111**: 1191-1200.

Nakamura, Y., Iwahashi, T., Tanaka, A., Koutani, J., Matsuo, T., Okamoto, S., Sato, K. and Ohtsuki, K. (2001) *J. Agric. Food Chem.*, **49**: 5755-5760.

Nautiyal, M. C., Prasad, A. and Lal, H. (1974) *Plant Sci.*, **6**: 106-107.

Ndang, Z. and Sema, Akali (1999) *Indian J. Hill Fmg.*, **12**: 84-87.

Nguyen, V.Q., Coogan, R.C. and Wills, R.B.H. (1999) *Acta Hort.*, **483**: 83-88.

Nieuwhof, M. (1976) *Sci. Hort.*, **5**: 111-18.

Nieuwhof, M. (1978) *Netherlands J. Agric. Sci.*, **26**: 68-75.

Niewwhof, M. and Jansen, R. C. (1993) *Gartenbauwissenschaft.*, **58**: 130-134.

Niikura, S. and Matsuura, S. (2001) *Acta Hort.*, **546**: 359-365.

Nikornpun, M., Putivoranat, M. and Zhong, L. (2004) *Southwest-China J. Agril. Sci.*, **17**: 78-83.

Nishijima, T. (2000) *Bull. National Res. Inst. Veg. Ornamental Plants Tea.*, No. 15, pp. 135-208.

Ogura, H. (1968) *Mem. Agric. Kagoshima Univ.*, **6**: 39-78.

Pal, B. P. and Sikka, S. M. (1956) *Indian J. Genet.*, **16**: 98-104.

Pandey, U. C. and Joushua, S. D. (1987) *Haryana J. Hort. Sci.*, **16**: 125-129.

Park, J.Y., Lee, Y.P., Lee, J., Choi, B.S., Kim, S. and Yang, T.J. (2013) *Theor. Appl. Genet.*, **126**: 1763–1774

Park, K.W. and Fritz, D. (1990) *J. Korean Soc. Hort. Sci.*, **31**: 1-6.

Park, Y., Pyo, H. K. and Lee, B. Y. (1976) *J. Korean Soc. Hort. Sci.*, **17**: 113-118.

Parthasarathi, V.A., Krishnappa, K.S., Gowda, M.C., Reddy, N.S. and Anjanappa, M. (1999) *Karnataka J. Agric. Sci.*, **12**: 230-233.

Patel, G.P., Jhala, R.C. and Shekh, A.M. (2000) *J. Agrometeorology*, **2**: 93-95.

Patil, H. B. and Patil, A. A. (1986) *South Indian Hort.*, **34**: 266-270.

Pistrick, K. (1987) *Kulturpflanze*, **35**: 225-321.

Pradhan, A., Yan, G. and Plummer, J.A. (2004a) *Australian J. Exptl. Agric.*, **44**: 95-102.

Pradhan, A., Yan, G. and Plummer, J.A. (2004b) *Australian J. Exptl. Agric.*, **44**: 813-819.

Prasad, K. and Gupta, R.K. (1989) *Indian J. Agric. Res.*, **23**: 143-148.

Prasad, K., Gupta, R.K. and Datta, M. (1987) *Haryana J. Hort. Sci.*, **16**: 108-114.

Pua, E.C., Sim, G.E., Chi, G.L. and Kong, L.F. (1996) *J. Korean Soc. Hort. Sci.*, **36**: 805-811.

Pujari, M.M., Jain, B.P. and Mishra, G.M. (1977) *Haryana J. Hort. Sci.*, **6**: 85-89.

Pyo, H.K., Lee, B.Y. and Park, Y. (1976) *J. Korean Soc. Hort. Sci.*, **17**: 47-54.

Rabbani, M.A., Murakami, Y., Kuginuki, Y. and Takayanagi, K. (1998) *Genetic Res. Crop Evol.*, **45**: 307-316.

Rastogi, K.B., Sharma, P.P. and Korla, B.N. (1987) *Veg. Sci.*, **14**: 105-109.

Rastogi, K.B., Sharma, S.K. and Sharma, P.P. (1987) *Indian J. Agri. Sci.*, **57**: 472-474.

Richharia, R.H. (1973) *J. Genet.*, **34**: 19-44.

Roggen, H.P.J.R. and Van Dijk, A.J. (1973) *Euphytica*, **22**: 260-263.

Rubatzky, V.E. and Yamaguchi, Y. (1997) World Vegetables. Chapman and Hall, New York, USA, p. 843.

Saharan, B.S. and Baswana, K.S. (1991) *Haryana J. Agron.*, **7**: 123-128.

Sahli, H.F., Conner, J.K., Shaw, F.H., Howe, S. and Lale, A. (2008) *Genetics*, **180**: 945–955.

Sakai, T. and Imamura, J. (1990) *Theor. Appl. Genet.*, **80**: 421-427.

Sakai, T., Liu, H.J., Iwabuchi, M., Kohno-Murase, J. and Imamura, J. (1996) *Theor. Appl. Genet.*, **93**: 373-379.

Sakai, Y. and Kono, T. (1975) *Bull. Hiroshima Prefectual Agric. Exp. Sta.*, No. 369, pp. 73-76.

Sanchez, C.A., Lockhart, M. and Porter, P.S. (1991) *HortScience*, **26**: 30-32.

Sarnaik, D.A., Baghel, B.S. and Singh, K. (1987) *J. Farming Systems*, **3**: 25-27.

Savos'kin, I.P. (1966) *Sel'hoz. Biol.*, **1**: 102-106.

Savos'kin, I.P. and Ceredeeva, V.S. (1969) *Genetika, Moskva*, **5**: 22-31.

Sazonova, L.V. (1973) *Trudy po PrikladnojBotanikeiSelekeii*, **49**: 234-242.

Schreiner, M., Huyskens, K.S. and Peters, P. (2000) *Gemuse-Munchen*, **36**: 18-21.

Schulz, O.E. (1919) In: *Cruciferae-Brassiceae* Part I: *Brassicinae and Raphaninae* (Ed. A. Engler).

Segall, R.H. and Smoot, J.J. (1962) *Phytopathol.*, **52**: 97-103.

Sen, P. and Deka, P. C. (2001) In: *Biotechnology of Horticultural Crops*, Vol. 2, (Eds. V. A. Parthasarathy, T. K. Bose and P. C. Deka), Naya Prakash, Calcutta, India, pp. 665-694.

Shabaev, V.P., Safrina, O.S. and Mudrik, V.A. (1998) *Agrokhimiya*, No. 6, pp. 34-41.

Sharma, A.K. (2000) *Agric. Sci. Digest*, **20**: 46-49.

Sharma, J.J. (2000) *Indian J. Weed Sci.*, **32**: 1-2.

Sharma, M. and Dhaliwal, D.S. (1997) *Indian Forest.*, **123**: 870-872.

Sharma, S.K. (1987) *J. Plant Nutri.*, **10**: 1581.

Sharma, S.K. and Kanaujia, S.P. (1992) *Seed Res.*, **20**: 92-95.

Sharma, S.K. and Kanaujia, S.P. (1995) *Ann. Agric. Res.*, **16**: 487-489.

Sharma, S.K. and Kanaujia, S.P. (2000) *Ann. Agric. Res.*, **21**: 465-468.

Sharma, S.K. and Lal, G. (1987) *Indian J. Agric. Sci.*, **57**: 672-674.

Sharma, S.K. and Lal, G. (1989) *South Indian Hort.*, **37**: 215-216.

Sharma, S.K. and Lal, G. (1990) *Seed Res.*, **18**: 154-156.

Sharma, S.K. and Lal, G. (1991) *Veg. Sci.*, **18**: 82-87.

Sharma, S.K. and Lal, G. (1994) *Hort. J.*, **7**: 121-124.

Sharma, S.K., Singh, H. and Kohli, U.K. (1999) *Seed Res.*, **27**: 154-158.

Sharma, V.K., Chandel, K.S., Kalia, P. and Pathania, N.K. (2002) *Himachal J. Agric. Res.*, **28**: 26-29.

Sheen, T.F. and Wang, H.L. (1980) *Scientific Research Abstracts in Rep.*, China, Taiwan, pp. 523-524.

Shimizu, S., Kanazawa, K., Kono, H. and Yokota, Y. (1963) *Bull. Hort. res. Sta., Ser. A*, No. 1, pp. 83-106.

Shirasawa, K., Hirakawa, H., Fukino, N., Kitashiba, H. and Isobe, S. (2020) *DNA Res.*, pp. 1-6.

Shishkova, S.O., Buzovkina, I.S. and Lutova, L.A. (1994) *Cruciferae Newsl.*, **16**: 67-86.

Sid'ko, F. Ya., Tikhomirov, A.A., Zolotukhin, I.G. and Polonskii, V.I. (1975) *Fizziologiya I BiokhimiyaKul'turnykhRastenii*, **7**: 181-184.

Siddique, B.A. (1983) *Acta Botanica Indica*, **11**: 150-154.

Singh, A.P., Singh, M.K., Singh, J.P. and Singh, S.B. (1990) *Veg. Sci.*, **17**: 210-212.

Singh, B.K., Koley, T.K., Karmakar, P., Tripathi, A., Singh, B. and Singh, M. (2017) *Indian J. Agric. Sci.*, **87**: 1600-1606.

Singh, B.K., Koley, T.K., Singh, B. and Singh, M.P. (2016) *ICAR News*, **22**: 12.

Singh, G., Kumar, J.C. and Dhaliwal, M.S. (1998) *J. Res.*, PAU, **35**: 124.

Singh, J.P., Swarup, V. and Arora, R.S. (1971) *Indian Hort.*, **15**; 13-14.

Singh, Mahavir, Singh, B.R. and Chattarjee, P. (1990) *Sci. Cult.*, **56**: 127-129.

Singh, V.B., Kar, P.L. and Tatung, T. (1995) *Adv. Hort. Forestry*, **4**: 127-132.

Sinskaia, E.N. (1928) *Bull. Appl. Bot. Genet. Pl. Br.*, **19**: 1-648.

Sinskaia, E.N. (1931) *Turdy Prikladnoi Bot., Genet. Seletskii (Leningrad)*, **26**: 1-58 (in Russian).

Sirks, M.J. (1957) *Genen en Phaenen.*, **2**: 2-10.

Sirohi, P.S. and Kutty, C.N. (2000) *Veg. Sci.*, **27**: 142-144.

Sirohi, P.S., Sharma, S.C. and Choudhury, B. (1992) *Indian Hort.*, **36**: 34-35.

Sodhi, Y.S., Pradhan, A.K., Mukhopadhyay, A. and Pental, D. (1993) *Indian J. Agric. Sci.*, **63**: 421-422.

Sounda, G., Ghanti, P. and Ghatak, S. (1989) *Environ. Ecol.*, **7**: 178-180.

Souza, J.R.P., Mehl, H.O., Rodriques, J.D. and Pedras, J. (1999) *Sci. Agricola*, **56**: 987-992.

Srivastava, B.K., Singh, M.P. and Jain, S.K. (1992) *Seed Res.*, **20**: 85-87.

Stewart, A.V. and Moorhead, A.J. (2004) *Agron. New-Zealand*, **34**: 1-7.

Sundaravelu, S. and Muthukrishnan, T. (1993) *South Indian Hort.*, **41**: 212-213.

Takahata, Y., Komatsu, H. and Kaizuma, N. (1996) *Plant Cell Rep.*, **16**: 163-166.

Tanaka, Y., Tsuda, M., Yasumoto, K., Yamagishi, H. and Terachi, T. (2012) *BMC Genom.*, **13**: 352.

Tatebe, T. (1968) *J. Japanese Soc. Hort. Sci.*, **37**: 227-230.

Tatebe, T. (1974) *J. Japanese Soc. Hort. Sci.*, **43**: 255-259.

Tatebe, T. (1977) *J. Japanese Soc. Hort. Sci.*, **46**: 48-51.

Tatsuzawa, F., Saito, N., Toki, K., Shinoda, K., Shigihara, A. and Honda, T. (2010) *J. Jap. Soc. Hortic. Sci.*, **79**: 103–107.

Thellung, M.A. (1927) L' origine de la carotte et du radish cultives Revue de Botanique appliqué et d' Agriculture tropicale **7**: 666-671 (in French).

Tokumasu, S. (1951) *Sci. Bull. Fac. Agric. Kyushu Univ.*, **13**: 83–89.

Tonosaki, K., Michiba, K., Bang, S.W., Kitashiba, H., Kaneko, Y. and Nishio, T. (2013) *Theor. Appl. Genet.*, **126**: 837–846.

Trouard-Riolle, Y. (1914) *Ann. Sci. Agron. Fr. Etrang*, **31**: 295-322, 346-550.

Tsuro, M., Suwabe, K., Kubo, N., Matsumoto, S. and Hirai, M. (2005) *Breed. Sci.*, **55**: 107-111.

Usuda, H., Shimogawara, K. and Garab, G. (1998) *Proc. XIth Int. Cong. On Photosynthesis*, Budapest, Hungary, 17-22 August, 1998, pp. 4031-4034.

Vannacci, G. (1991) *Agriculture Mediterranea*, **121**: 186-190.

Verma, I.M., Purohit, S.P. and Gogoi, S. (1997) *Ann. Biol.*, Ludhiana, **13**: 263-265.

Wang, L.S., Zhang, W.L., Wei L.Q. and Xu, C.S. (2000) *J. Sci. Food Agric.*, **80**: 1767-1771.

Warwick, S.I. and Francis, A. (2005) *Can. J. Plant Sci.*, **85**: 709–733.

Weaver, J.E. and Bruner, W.E. (1927) In: *Root Development of Vegetable Crops*, McGraw-Hill Book Co., New Delhi.

Wendt, T. (1977) *Gemuse*, **13**: 268-270.

Werth, E. (1937) *Angew Bot.*, **19**: 194-205.

Williams, P.H. and Pound, G.S. (1963) *Phytopath.*, **53**: 1150-1154.

Wilson, J.D. (1962) *Bull. Kanagawa Hort. Exp. Sta.*, **13**: 59-70.

Wolfe KH, Li WH, Wood, Y.H., Park, D.Y. and Lee, J.M. (1991) *Research Collection of the Institute of Food Development, KymgHee University*, Korea Rupublic, **12**: 1-11.

Xu, L., Jiang, Q.W., Wu, J., Wang, Y., Gong, Y.Q., Wang, X.L., Limera, C. and Liu, L.W. (2014) *J. Integr. Agr.*, **13**: 2362–2369.

Yamagishi, H. (1998) *Genes Genet. Syst.*, **73**: 79-83.

Yamagishi, H. and Sasaki, J. (2004) *Breed. Sci.*, **54**: 189-195.

Yamagishi, H. and Terachi, T. (1994a) *Theor. Appl. Genet.*, **87**: 996-1000.

Yamagishi, H. and Terachi, T. (1994b) *Euphytica*, **80**: 201-206.

Yamagishi, H. and Terachi, T. (1996) *Theor. Appl. Genet.*, **93**: 325-332.

Yamagishi, H. and Terachi, T. (2001) *Theor. Appl. Genet.*, **103**: 725-732.

Yamagishi, H. and Terachi, T. (2003) *Genome*, **46**: 89–94.

Yamagishi, H., Tateishi, M., Terachi, T. and Murayama, S. (1998) *J. Jap. Soc. Hort. Sci.*, **67**: 526-531.

Yamaguchi, H. (1987) *Jap. J. Breed.*, **37**: 54-65.

Yamaguchi, H. and Okamoto, M. (1997) *Euphytica*, **95**: 141-147.

Yamane, K., Lu, N. and Ohnishi, O. (2005) *Plant Sci.*, **168**: 627-634.

Yang, Y.W., Tai, P.Y., Chen, Y. and Li, W.H. (2002) *Mol. Phylogenet. Evol.*, **13**: 455–462.

Yang, Y.W., Tseng, P.F., Tai, P.Y. and Chang, C.J. (1998) *Bot. Acad. Sinica*, **39**: 153-160.

Yoo, K.C. and Uemoto, S. (1975) *Sci. Bull. Fac. Agric.*, Kyushu Univ., **29**: 147-150.

Yoo, K.C. and Uemoto, S. (1976) *Plant Cell Physiol*, **17**: 863-865.

Zhang, L., Shen, X.Q. and Shao, G.Y. (1999) *Acta Hort. Sinica*, **26**: 238-243.

Zhao, D. (1989) In: *International Symposium on Horticultural Germplasm, Cultivated and Wild*, Beijing, China, Sept. 1988. Chinese Society for Horticultural Science. International Academic Publishers, Beijing, pp. 388-393.

Zhou, M.X. (1991) *Crop Genet. Resour.*, **3**: 41-42.

Zou, Z., Ishida, M., Li, F., Kakizaki, T., Suzuki, S., Kitashiba, H. and Nishio, T. (2013) *PLoS One*, **8**: e53541.

12

BEETROOT

A. K. Sureja, M. K. Sadhu and K. Sarkar

1.0 Introduction

The garden beet or table beet or red beet or beetroot is a popular root vegetable grown in home gardens as well as in market gardens mainly for its fleshy enlarged roots which store large amounts of reserve food. In fact, beet is a useful vegetable in a number of ways. The swollen roots are eaten boiled or as a salad. They are also pickled. The tender leaves and the young beet plants are used as greens (pot-herbs). The large-sized beets are used for canning.

In Germany, it was first described in 1557 when it was referred to as Roman beet. The crop was introduced in the USA in 1800 and became known as garden beet. According to Burkill (1935), beet was introduced into India in remote times, but whilst the plains of the Ganges suited it, it did not spread to further east. It was taken by sea by the Arabs to China around 850 AD. The major beetroot growing countries are Russia, France, USA, Ukraine, China, Germany, Turkey, Poland and Italy. In India, beetroot is grown on a small scale mainly in kitchen gardens and market gardens.

2.0 Composition and Uses

2.1.0 Composition

Beetroot is rich in protein, carbohydrate, calcium, phosphorus and vitamin C. The food value of beetroot has been shown in Table 1, which reveals beetroot contains a relatively high level of the B vitamin folic acid.

Table 1: Composition of Roots of Red Beetroot (per 100 g of edible portion)*

Moisture (g)	86.95	Lutein (µg)	28.6
Protein (g)	1.95	β-carotene (µg)	10.14
Ash (g)	1.46	Total carotenoids (µg)	12.88
Total fat (g)	0.14	Calcium (mg)	17.28
Total dietary fibre (g)	3.31	Iron (mg)	0.76
Carbohydrate (g)	6.18	Magnesium (mg)	33.21
Energy (KJ)	149.0	Phosphorus (mg)	36.33
Thiamine (B_1) (mg)	0.01	Potassium (mg)	306.00
Riboflavin (B_2) (mg)	0.01	Selenium (µg)	0.25
Niacin (B_3) (mg)	0.21	Zinc (mg)	0.30
Pantothenic acid (B_5) (mg)	0.26	Total starch (g)	1.69
Total B_6 (mg)	0.07	Fructose (g)	1.67
Biotin (B_7) (µg)	2.56	Glucose (g)	1.46
Total folates (B_9) (µg)	97.37	Sucrose (g)	1.21
Total ascorbic acid (mg)	5.26	Total free sugars (g)	4.35
Phylloquinones (K_1) (µg)	2.98	Total oxalate (mg)	71.37
Total polyphenols (mg)	57.56		

* Longvah *et al.* (2017).

Beetroot contains a relatively high level (3.3-15.2 µg/g on dry mass basis) of folic acid (Wang and Goldman, 1996, 1997). According to Makarenko *et al.* (1999), palmitic acid was the predominant saturated acid, comprising 20.0-24.0 per cent of total fatty acid in the vacuolar lipids of beetroot. The red pigment in beetroot is betanin, the contents of which may range from 426-691 mg/kg fresh weight (Mariassyova *et al.*, 1999).

Kovaroviè *et al.* (2017) analysed the composition of four beetroot varieties (Cylindra, Kahira, Chioggia, Crosby Egyptian) and reported the range of total polyphenols content from 218.00 mg kg^{-1} to 887.75 mg kg^{-1}, total anthocyanins content from 14.48±0.40 mg kg^{-1} to 84.50±4.71 mg kg^{-1} and values of antioxidant activity from 8.37±0.29 per cent to 21.83±0.35 per cent.

2.2.0 Uses

The use of beet probably dates back to the prehistoric times when the leaves were used as pot-herbs. According to Anderson (1952), beet was certainly domesticated first as a leafy vegetable, then as a root crop and finally as a source of sugar. Aristotle mentioned red chards and Theophrastus mentioned light green and dark green chards used in the 4th century B.C. The Romans used beet as feeds for animals and man. Romans called it *Beta* and took it from Italy to Northern Europe. By the 16th century it was widely used for feeding animals, particularly during winter. Red beet featured in Roman recipes of the 14th century. The roots and leaves of beet have been used by the ancient Romans in folk medicines to treat fever, constipation, liver disorders and in reducing blood cholesterol. Beetroot juice was considered as an aphrodisiac. Betalain pigments present in beetroot is used to protect against oxidative stress. However, the presence of oxalic acid in root may lead to formation of kidney stones if consumed in excess. Freshly extracted beetroot juice is a rich source of nutrients and bioactive compounds and contain 62.20 per cent DPPH scavenging antioxidant activity, 990.7 mg/100 mL total phenols, 790 mg/l anthocyanins and 520.3 mg/l betanins, suggesting its use for value addition in food formulations (Arora *et al.*, 2019).

Red beet microgreens are quite popular owing to its nutrient-dense properties, such as betalains and polyphenols and also intense aromatic flavour. The microgreens are harvested 9-11 days after germination. They are cut along with the stem and attached cotyledons/seed leaves with the help of scissor. If left for longer, they rapidly elongate and loose colour and flavour.

The betalain pigments are responsible for colouring in the leaves and roots in beetroot. Beet root slices are used as salad, pickled, roasted, boiled or used to make soup. Pigments extracted from roots are used as a source of natural food colourant. It is also used in cosmetics, candy, ice cream, meat products, yogurt, and powdered drink mixes. The typical earthy flavour of beet is due to the presence of compounds called geosmins. The bioactive compounds of beetroot and its utilization in food processing, functional food and health industry has been reviewed by several authors (Chhikara *et al.*, 2019; Gengatharan *et al.*, 2015; Moreno *et al.*, 2008).

Uses of Beet Roots

Juice (left) and powder (right) from beet

Salads (left) and ice-cream (right) from beet

Pie (left) and Curry (right) from beet

3.0 Origin and Taxonomy

Garden beet (*Beta vulgaris* subsp. *vulgaris* L.) has probably originated from *Beta vulgaris* L. ssp. *maritima*, sea beet, a variable species of the Mediterranean, possibly by means of hybridization with *B. patula*, a closely related species of Portugal and the Canary Island. *Beta vulgaris* ssp. *maritima* has been found to grow wild on seashores in Britain and through Europe and Asia to the East Indies. Genus *Beta* also includes Swiss chard, mangel and sugar beet. All these crops are derived from the same species, *vulgaris*. Beetroot and Swiss chard are primarily used as vegetables whereas mangel and its derivatives as animal feed, and sugarbeet as a source of sucrose. Beetroot was first mentioned in the literature in Mesopotamia in the 9th century B.C. Initially, beets were grown mainly for their leaves. The progenitor of the beetroot was originally selected for its use as a leaf vegetable in the Mediterranean region and then later for use as a fresh or stored root vegetable (Campbell, 1976). The leaf beet is a vegetable form of beet grouped in the *vulgaris* subspecies and beet has fleshy petioles and it does not possess a swollen root. In Northern Europe, leaf beet may have resulted in the transition toward a biennial life cycle by creating selection pressure towards a swollen hypocotyl/root as an over-wintering propagule (Ford-Lloyd, 1995). Alternatively, selection for a swollen-rooted form may have taken place in the Mediterranean region prior to its movement into northern Europe. Towards the end of the Middle Ages, the garden beet had become an important vegetable in Central Europe. In Germany, it was first described in 1557 as Roman beet. In the 18th century, the use of beetroot was expanded to be used for fodder in Europe. Fodder beet is also known as forage beet, mangels, mangolds, and mangel-wurzels. All of these possess very large swollen roots of various shapes and colours and were developed as forage crop for animal feed (Ford-Lloyd, 1995). Beetroot was introduced in the United States of America in 1800.

The cultivated beets are derived from section Beta, one of the four sections in genus *Beta* (Ford-Lloyd, 1995). Section Beta includes six subspecies within *B. vulgaris* viz. *vulgaris* (sugar beet, table beet, mangel or fodder beet), *cicla* (Swiss chard, leaf beet, spinach beet), *maritima* (wild sea beet), *adanensis*, *trojana*, and *macrocarpa*. The section Beta also includes the species *B. patula* and *B. atriplicifolia* (Ford-Lloyd, 1995). Some authors classified four major groups of cultivars which differ from one another in their external features and common usage: the leaf beet group [*Beta vulgaris* L. ssp. *vulgaris* convar. *cicla* (L.) Alef.], garden beet group [*B. vulgaris* L. ssp. *vulgaris* var. *vulgaris*], fodder beet group [*B. vulgaris* L. ssp. *vulgaris* var. *rapacea* Koch], and sugar beet group [*B. vulgaris* L. ssp. *vulgaris* var. *altissima* Doll] (Lange *et al.*, 1999; Hammer, 2001). Swiss chard is a leafy vegetable. The sugar beet is grown for its high sugar content whereas the fodder beet, also known as mangold, mangel, or mangel-wurzel, is grown for its nutritious tops and roots that are used as animal fodder. Beet is a diploid with 2n=2x=18. In beet, *B* allele converts biennialism to annualism. Hence, an understanding of this allele would benefit to our understanding of domestication of beetroot.

The evolution of cultivated table beets as postulated by Campbell (1976) is as follows:

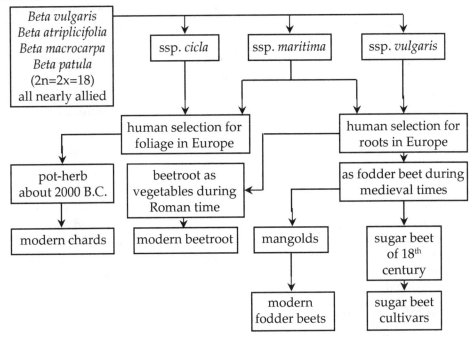

4.0 Morphology

4.1.0 Roots

The beetroot is thick and fleshy storing large amount of reserve food. Morphologically, the upper portion of the root develops from the hypocotyl and the lower portion, from which the secondary roots arise, develops from the taproot. The size, shape and colour of the root depend on several factors such as temperature, type of soil, population density, nutrition, *etc.*, apart from genetical factors. In general, greater harvest weights, greater length, middle width and length × width values are found at low plant density (Goldman, 1995). The zones or concentric rings which are visible in a median cross-section of any mature beetroot are the result of alternate formation of vascular tissues and storage parenchyma tissue during the development of the fleshy root.

The primary root colours in table beet are red and yellow. The red colour present in most table beet roots is actually a representation of two different pigments, betacyanin (reddish-purple or red-violet colour) and betaxanthin (yellow colour). When both are present, in approximate 3: 1 ratios, the red colour appears. When betacyanin is absent, yellow root coloration is found. The root interior colour is influenced by cultivar and such factors as temperature, time of the year grown, type of soil, nutrition, *etc.* Water stress condition results in poor development of root colour (Sistrunk and Bradley, 1970). The colour of beetroot is due to presence of red-violet pigment, β-cyanins and yellow pigment, β-xanthins. Varietal differences in the content of β-cyanins and β-xanthins have been reported by several workers (Nilsson, 1970; NG and Lee, 1978; Sapers and Hornstein, 1979). Pigment accumulation is affected by genotype, sowing date and growing temperature (Takacs-Hajos, 1993).

The beet has an extensive root system and the taproot may penetrate the soil to a depth of 3.0 m (Weaver and Bruner, 1927). A number of strong lateral roots grow diagonally for a distance and then downward to depths nearly as that of the taproot. The extensive root system penetrating to a great depth indicates drought resistant property of beet. The shape of the root may be globe, round, cylindrical, fusiform or napiform. Generally, round shape is more preferred.

4.2.0 Leaves

Leaves are often ovate and cordate, dark green or reddish. The rosette leaves develop in a close spiral order with the oldest ones on the outside. According to Magruder *et al.* (1940), the size, shape and colour of foliage are influenced by temperature, time of growing, spacing and soil moisture. Erwin and Haber (1934) found that at relatively low temperature (10 to 15.6°C) the leaves are broad, thick and heavily savoyed with short, thick, straight and erect petioles, whereas at a temperature range of 21.1 to 26.7°C narrow, thin, smooth surfaced leaves with long, slender and dropping petioles are produced.

4.3.0 Inflorescence

The inflorescence which normally develops in the second year is botanically a large spike. The flowers are almost sessile. These arise in clusters of 3-4 in the axils of bracts of the inflorescence axis and its secondary branches. The flowers are small, inconspicuous without corolla, but with green calyx, which become thicker towards base as the fruit ripens. Stamens 5, ovary sunk in disc, stigma 3. Although the ovary is three carpellate, there is only one ovule per ovary.

4.4.0 Fruit and Seed

Fruits are mostly aggregate formed by the cohesion of 2 or more fruits and held together by swollen perianth (calyx) base and thus forming an irregular dry cork-like body, known as seedball or the so called seed. However, if there is a single flower, a single germ seed will develop. The true seed is small, about 3 mm long, kidney-shaped and shiny to reddish brown in colour. A gram of beet seedball counts about 50 seeds, which may retain viability for 5-6 years under ordinary storage condition. Monogerm cultivars produce single rather than multiple flowers in bract axils; therefore, the fruit will contain only a single seed.

5.0 Cultivars

In sugarbeet, tetraploid and triploid cultivars have been developed whereas table beet cultivars are diploid. The primary root shapes in beet are: (1) round and globe shaped roots, which are most common type; (2) flattened globe or Egyptian types; and (3) cylindrical types, which are of specific value in the processing industry. Beet cultivars are usually classified on the basis of shape of roots:

i) Flat: Flat Egyptian.

ii) Short-top shaped: Flattened at top and bottom with rounded sides and conical or tapered base, *e.g.*, Crosby Egyptian, Early Wonder, Asgrow Wonder.

iii) Round or globular: Roots are round or globular in shape, *e.g.*, Detroit Dark Red, Crimson Globe.

iv) Half-long: Length is shorter than long types, *e.g.*, Half-Long Blood, Winter Keeper.

v) Long: Roots are long, may grow as much as 40 cm, quite popular in Europe, *e.g.*, Long Dark Blood.

Goldman and Navazio (2003) reported three clearly identifiable groups of table beet cultivars in USA. These are

i) "Egyptian" group: It comprises of Flat Egyptian, Crosby Egyptian, Light Red Crosby, and Early Wonder;

ii) "Detroit" group: It comprises of Detroit Dark Red and Ohio Canner;

iii) "Long" group: It comprises of Long Dark Red and Cylindra.

The cultivars recommended for cultivation in India are Detroit Dark Red and Crimson Globe.

Detroit Dark Red

This population was originally selected from a population known as 'Early Blood Turnip' by a man named Reeves of Port Hope, Ontario, Canada. In 1892, this cultivar was listed as 'Detroit Dark Red Turnip' beet by D.M. Ferry and Co. This population is perhaps the most important and widely-adapted variety. It has been used for fresh market, processing, and market garden production. The roots are round with smooth, uniform, deep red skin. Flesh dark blood-red with light red zoning, tender and fine grained, tops small, leaves dark green tinged with maroon. It is a heavy yielding cultivar, maturing in 80-100 days.

Crimson Globe

Roots globular to flattened, medium red with small shoulders, flesh medium dark crimson-red with indistinct zoning, top medium to tall, large, bright green leaves with maroon shades, heavy cropper.

Crosby Egyptian

This was developed by Josiah Crosby of Arlington, Massachusetts, USA, from a population of 'Flat Egyptian'. This population was first listed in seed catalogues in 1885. Its roots are flat globe with a small taproot and a smooth exterior. The internal colour is dark purplish red with some indistinct zoning. The top is medium tall, green with red veins. The cultivar reaches maturity in 55-60 days after sowing and shows pronounced white zoning when grown in warm weather.

Important Cultivars of Beet

Detroit Dark Red

Crimson Globe

Crossby Egyptian

Early Wonder

This population was first listed in 1911 by the F.H. Woodruff and Sons and S.D. Woodruff and Sons catalogues. The roots are flattened globe with rounded shoulders with a smooth, dark red skin. The interior is dark red with some lighter red zoning. The top is heavy, green with red veins. This cultivar also takes 55-60 days after sowing to reach harvest maturity.

Ooty 1

It is a selection from the local type and is suitable for growing in hills of south India. The skin is thin and the flesh colour is blood red. It has a yield potential of 310–450 q/ha. The crop duration from sowing to harvest is 120–130 days.

Pant S 10

This cultivar is recommended for cultivation in Uttar Pradesh, Rajasthan, Punjab and West Bengal. It is tolerant to *Cercospora* leaf spot and *Sclerotium* root rot Average yield is 300-350 q/ha. It is rich in sugar (sucrose about 14.5–15 per cent).

Radost

It is a cultivar developed in Bulgaria. Radost is a multigerm diploid population, obtained through crossing and following individual and individual-family selection of table beet gene pool. The form of the root is cylindrical elongated with red colouring of the outer part of the root. The inner part is red coloured, with 6–8 concentric circles with a crisp consistence. Radost has relatively good resistance to powdery mildew and mean resistance to *Cercospora* (Uchkunov, 2012).

Little research work has been done to assess the performance of different beet cultivars under different agro-climatic conditions of India. Paria (1973) reported good yield performance of Crimson Globe under West Bengal condition. Of the 5 beet cultivars compared in a 3-year trial, Pillai *et al.* (1977) obtained the highest yield from Red Ball and Crimson Globe, followed by Detroit Dark Red.

Esyunina *et al.* (1972) studied the biochemical composition in cultivars of different groups of beet cultivated in the USSR. Of the 12 red beet cultivars studied, Long Season Harris showed the highest dry matter content (16.9 per cent), Perfected Detroit the highest betanin (163 mg/100 g) and Long Season had the largest roots (381 g).

6.0 Soil and Climate

6.1.0 Soil

Good beets are produced on a wide cultivar of soils, but deep, well-drained loams or sandy loams are considered best. Heavy soils are not satisfactory for beets since the roots are likely to be unsymmetrical in shape when grown in such soils. Moreover, since such soils become hard and form a crust after a rain or irrigation, the seedlings may not come out on germination, resulting in poor stand of crops.

Beet is sensitive to soil acidity; its yield decreases considerably as the soil pH goes below 5.8. A soil pH of 5.8-6.2 is considered best for beet. Although beet can be grown successfully on alkaline soils, scab is often worst on neutral or alkaline soil. Boron deficiency is most prevalent in alkaline soils. Beet is one of the few vegetables that can successfully be grown on saline soils. If grown on alkaline soils, manganese sulphate should be added and on soils with a pH as low as 4.6, copper sulphate should be included in the fertilizer schedule. Rajkumar *et al.* (2016) obtained significantly higher beetroot yield in soil salinity block of < 4 dSm^{-1} (19.07 t/ha), followed by 4-8 dSm^{-1} (16.61 t/ha).

6.2.0 Climate

Beets are rather hardy and can tolerate some freezing. It grows best in winter in the plains of India. It attains best colour, texture and quality in a cool weather condition. Excessively hot weather, however, causes zoning –the appearance of alternating light and dark red concentric circles in the root. Lorenz (1947) found that roots developed at a relatively high temperature had poor colour while those planted in cooler months had excellent colour and quality.

The optimum temperature for proper growth, development and good quality root production is 18°C–21°C. Beet plants are very sensitive to low temperature. If they are exposed to relatively low temperature of 4.5°C-10°C for 15 days or more, bolting is likely to occur before the roots reach marketable size. At 30°C or above, the sugar accumulation in the roots is ceased and there is distinct zoning inside flesh.

Beet requires abundant sunshine for development of storage roots. Lebedeva *et al.* (1978) compared red beet plants grown under short day (12 hour) with plants grown under continuous illumination, the plant's photosynthetic productivity increased from 23 to 27 g/m^2/day to 37-40 g/m^2/day and the root yield rose from 17 to 34 kg/m^2 over a growing period of 119 days.

7.0 Cultivation

7.1.0 Land Preparation

The soil for beet should be thoroughly prepared by ploughing 15-20 cm deep followed by sufficient disking and hoeing to pulverize the clods. The soil surface should be smooth and loose, and free from all clods and trashes. Well-rotted farmyard manure or compost is also added at the time of land preparation.

7.2.0 Sowing

Beet seeds are usually planted directly in the field by sowing 'seedballs' that contain one or more seeds. The seedballs are planted at the rate of 7-9 kg per hectare in rows 45-60 cm apart and are thinned later to an in-row spacing of 8-10 cm. A population of 3,00,000 plants per hectare can be obtained by keeping 50–60 cm spacing between rows and 8–10 cm between plants. The seeds are sown at 1.5–2.5 cm depth to ensure good germination. Cultivars with small tops like Flat Egyptian are given closer spacing than those with heavy tops like Detroit Dark Red. Plant population density has significant effect on shape and size of beetroots. Goldman (1995) found greater harvest weight, a higher percentage of harvestable beets and greater length, middle width and length × width values at low plant density in all four genotypes studied. Gaharwar and Ughade (2017) observed significantly higher marketable root yield in beetroot at closer planting distance 30 cm × 10 cm but recorded significantly greater fresh weight of beet root under wider plant spacing.

In India, the usual sowing time of beet in the hills is March-July. In the northern plains it is sown during September-November, while in the southern plains its sowing is extended from July to November. Seeds sown early, when the temperature is high, may produce beetroots with coarse and woody flesh and dull colour. Sowing at an interval of one or two weeks during the sowing season ensures a steady supply of tender beets throughout the season.

To facilitate uniform and early germination, the seed bed should not be too wet or too dry; the seedballs should be placed in close contact with soil by compacting. Seed treatment with thiram or captan at 2.5 g/kg seed gives better seedling emergence, and controls pre-emergence and post-emergence damping off. Soaking the seeds in water for about 12 hours before sowing facilitates better germination in the field.

There are three major seed factors that negatively influence beet germination, *viz.* (i) a mucilaginous layer that can surround the seedball; (ii) the ovary cap tenacity; and (iii) the presence of phenolic chemical inhibitors (Taylor *et al.*, 2003). The protective corky coat of the seedball restricts water and gaseous movement during germination besides containing germination inhibitors (Junttila, 1976). Hence, removal of corky layer or pre-sowing soaking in water generally stimulates germination.

7.3.0 Thinning

Thinning is an essential operation in beet cultivation, because the seedball is actually a fruit consisting of 2-6 seeds, each of which may germinate and produce a plant. Generally, the plants emerge in groups unless segmented seed or monogerm seed is used. Hand thinning is usually done. Sometimes thinning is delayed until some of the beets are large enough to use. When the large ones are removed, the small ones are left to develop. Thinning is seldom practised with the crop grown for processing because of high cost of labour involved and small beets are preferred by processors (Thompson and Kelly, 1957).

7.4.0 Manuring and Fertilization

Beet plants must make uninterrupted rapid growth to produce best quality roots and it responds well to fertilizers. The kind and amount vary with soil type, inherent fertility and previous fertilization programme. Haag and Minami (1987) estimated that a population of 3,30,000 plants/ha removed a total of 77.8 kg N, 13.7 kg P, 202.8 kg K, 19.5 kg Ca, 29.1 kg Mg, 22.9 g Cu, 735.9 g Fe, 583.6 g Mn and 387.6 g Zn.

Beets have a fairly high nitrogen requirement. Nitrogen uptake by beet plants may be as high as 78 per cent from the soil and 22 per cent from the added fertilizer (Ovoshchevodstvo and Tenat, 1978). Lee (1973) noticed beneficial effect of applied N up to 100 kg per hectare on contents of carotene, thiamine and riboflavin. Increase in total amino acids in beet to the extent of 60 per cent was also noticed. Bagchi (1982) recorded linear increase in fresh weight and protein content of beet with increase in nitrogen application.

There is wide difference of opinion about the optimum dose of N needed for beetroot. While Paria (1973) recorded the optimum dose of N and P as 30 and 60 kg per hectare, respectively for cv. Crimson Globe under West Bengal condition, Mack (1979) reported the optimum dose of N in USSR being in the range of 130 and 160 kg per hectare for cv. Detroit Dark Red. Hipp (1977) noticed no yield response to added N when the harvesting was done quite early (less than 86 days after sowing), but application of 112 kg N per hectare increased yield by 55.8 per cent when the growing period was 140 days. N did not increase the size when the growing season was short, but increased beet size when the growing season was long. Michalik *et al.* (1995), on the other hand, found that the root size increased with higher N doses and the increase was accompanied by a fall in the contents of dry matter, total sugar and red pigment. Vukasinovic *et al.* (1991) reported from Sarajevo (Yugoslavia) that the yield of red beetroot increased with increasing N-rates upto 183 kg/ha. They noticed further that nitrate content also increased with increasing N rates. In cv. Bicor.

N-rates of 107 and 193 kg/ha produced 1144.9 and 2138.5 mg nitrate/kg, respectively and in cv. Egyptian, 1321.8 and 2559.0 mg/kg, respectively. Grzebelus (1995a,b) also recorded higher nitrate content of beetroots with increasing N application. He noticed further that the amount of nitrate in the beetroots depended predominantly on the genetic characteristics of the cultivar, some cultivars always accumulating more nitrate than others. Jarvan (1995) found that the permissible nitrate level in beetroot was exceeded by applying 130 kg N/ha prior to sowing of the crop. Least nitrate accumulation was recorded when the crop was grown on substrates rich in organic matter. Sizov and Lunev (1991) also noticed that nitrate content of beet root exceeded permitted limits (3000 mg/kg) when more than 140 kg N/ha was applied. The rate of N application seems to be the major factor determining nitrate accumulation in the roots.

Michalik *et al.* (1995) found that in 11 cultivars grown at 2 sites under 3 N fertilizer regimes, root size tended to increase with higher N doses, the increase being accompanied by a fall in the content of dry matter from 15.5 to 15.1 per cent, total sugar from 9.8 to 8.7 per cent and red pigment from 78.8 mg/100 g fresh weight under lowest N dose to 73.4 mg/100 g under the highest N dose. Yellow pigment content was not affected by N dose. Takacs-Hajos *et al.* (1997) found that the highest N rate (140 kg/ha) resulted in the lowest root betanin content, low water soluble solids content and an accumulation of nitrate.

In Slovenia, beetroot cv. Bicor was either direct sown or transplanted on raised beds and fertilized with N at rates ranging from 0 to 350 kg/ha (Cerne *et al.*, 2000). The highest yields of beetroots were obtained for the transplanted crops fertiized with 150 or 200 kg N/ha. Half of the N should be broadcast prior to sowing and the remainder should be applied in 2 equal doses 4 and 6 weeks after sowing.

In a study with beetroot cv. Bordo, Zinkevich and Novikov (1977) found the petiole of a mature leaf as the best indicator of plant nutritional status. Optimum plant development and yields were obtained with plants containing 2.28 per cent N, 0.78 per cent P and 5.58 per cent K at the 10-leaf stage. Rao and Subramanian (1991) noticed that the K content of beetroot plants at 60 days was significantly correlated with their yield at harvest, indicating that this was the best sampling date to assess K requirement of beetroot. The critical plant content of K was found to be 3.4 per cent. Subbarao *et al.* (1999), however, showed that in some beetroot cultivars, 95 per cent of the normal tissue K can be replaced by Na without any reduction in growth.

Smagina (1977) studied the effectiveness of single or combined application of micronutrients such as Cu, Zn, Mn, B, Mo and KI for seed production in red beet. The highest seed yields were obtained from plants raised from seeds treated with Zn, Mn + Cu or Mn + Mo. Boron deficiency causes internal breakdown as black spot or dry rot in beetroot. In boron-deficient soils, borax 20–25 kg/ha is applied at the time of last ploughing.

Biochar contains elevated levels of sodium salt (antagonists of potassium). Maroušek *et al.* (2017) reported that biochar in combination with sodium leads to the transformation of mineral nitrogen into soil organic matter, thereby reducing the nitrate content in the beetroot. Besides, it increased beetroot yields (by 2.4 t ha⁻1) and they concluded that biochar has the potential to increase the water retention capacity of the topsoil.

Fertilization with earthworm humus in greenhouse beet cultivation increased the efficiency of water use and can be used instead of chemical fertilization to meet the nutritional needs of the crop (da Silva *et al.*, 2018). They reported that the nitrogen dose of 2.8 g per pot, corresponding to 6.22 g of urea per plant, provided the highest water content in beet cultivation in greenhouse.

7.5.0 Irrigation

A constant water supply, either in the form of rainfall or irrigation, is essential for seed germination and high yields of good quality beetroots. A deficiency of water may lead to reduction in size. Irrigation application at 0.2 bar soil moisture tension at 15-22 cm soil depth was found optimum for beet growth in sandy loam soils in Delhi (Davenport, 1962). The water requirement of beet has been found to be about 300 mm applied in 5-6 irrigations. However, when there is winter rains, only about 3 irrigations are sufficient for this crop. The roots crack or become hard when the soil lacks sufficient moisture. Rajkumar *et al.* (2016) recorded significantly higher beetroot yield with drip irrigation at 1.2 ET (18.08 t/ha) which was at par with drip irrigation at 1.4 ET (19.69 t/ha).

7.6.0 Interculture

Weeds drastically decrease yield and should be controlled. Hand weeding is usually practised in India, whereas in advanced countries mechanical cultivation is commonly employed to control weeds between rows, and herbicides are used within the rows. Clean and shallow hoeing should be given to check weed; deep cultivation may damage the crop. One or two earthing up is also given to prevent the exposure of roots to sunlight. Two light weeding and hoeing during the early stages of crop growth are required to check the weeds growth.

Yordanova and Gerasimova (2016) reported that beetroot yield increased by 7.8–9.3 per cent and 22.8 per cent by the application of barley straw mulch (BSM) and mulch from spent mushroom compost (SMCM) during the 2 years and only for the first year, respectively.

Good rotational crops for beetroot include potatoes, grain crops, onions and legumes. A 4-5 year rotation is recommended.

8.0 Harvesting and Yield

8.1.0 Harvesting and Post-harvest Management

Beets are harvested as they attain a diameter of 3-5 cm. They are usually pulled by hand, the tops are removed and after washing the roots are graded according to size. In developed countries, the mature roots are mechanically harvested, detopped, washed, graded and finally packed in polythene bags. Removal of tops and packing in polythene bags lengthen the self-life of beets by reducing water loss during transit and storage. In beetroots, storage for seven months reduced the contents of arabinose, galactose and flucose in the insoluble fibre and of galactose in the soluble fibre (Elkner *et al.*, 1998). Small size bunched beets are also in great demand in Europe and USA. Hence, after harvesting by hand the dead and injured leaves are removed and then tied in bunches of 4-6 beets with their tops on. Oversized beets are not in demand, because they are tough and woody and cracks appear on the surface.

Beetroots are evenly waxed to avoid wilting in a dry atmosphere and to retain good quality and appearance. Uniform size roots having 4 cm or less diameter are canned without cutting. Usually, 8–10 roots are kept in a can. The roots can also be peeled and processed in light syrup. For slicing, size of beetroot should not exceed 6 cm in diameter as large roots are often coarse and poorly coloured. The roots are cut into cubes of about 1.25 cm on a side by a slicing machine.

The roots can be stored up to 2–3 days at room temperature and several weeks at 0°C temperature and 95 per cent relative humidity. Freezing injury occurs from -0.5°C. The cracked and bruised roots have higher respiration rate than undamaged beets, and hence, they lose more sucrose. Avoid large piles of roots to minimize shrivelling and decaying.

High betanin concentration and high betanin to vulgaxanthin ratio increase the commercial value of the colourant product in beet root. Lowering soluble solids content facilitates higher concentration of beetroot colour during processing. Barba-Espin *et al.* (2018) reported foliar spray of ethephon during the growth of beetroot, resulted in increased betanin (22.5 per cent) and decrease in soluble solids contents (9.4 per cent), without detrimental effects on beetroot yield. They observed most rapid accumulation rate for betanin and soluble solids between 3 and 6 weeks after sowing in both untreated and ethephon-treated beetroots. The expression of the betalain biosynthetic genes (*CYP76AD1*, *CYP76AD5*, *CYP76AD6* and *DODA1*), determining the formation of both betanin and vulgaxanthin, increased in response to ethephon treatment, and also the expression of the betalain pathway activator *BvMYB1*. Postharvest use of short-term UV-B radiation (1.23 kJm" 2) followed by storages for 3 and 7 days at 15 °C resulted in increased betanin to vulgaxanthin ratio (51 per cent) and phenolic content (15 per cent). They also showed that enhanced

Different Field Growing Operations in Beet

Harvesting of Beet Roots

betanin content in ethephon-treated beetroots is linked to increased expression of betalain biosynthetic genes.

8.2.0 Yield

The average yield of beetroot varies from 250 to 300 quintals per hectare.

9.0 Diseases, Pests and Physiological Disorders

9.1.0 Diseases

The garden beet is affected by comparatively few diseases and pests. The most serious diseases are leaf spot, downy mildew and viruses.

Fungal Diseases

9.1.1 Leaf Spot (*Cercospora beticola*)

It is a very widespread disease of garden beets. Small, circular spots of about 2 mm in diameter with purplish brown borders appear in great numbers. These spots have definite margins, usually darker in colour than the rest of the lesion. A black dot is present in the center of the lesion. The spot may drop out and the leaf may assume a shot-hole appearance. Petioles are also affected with spots tending to elongate. When numerous spots develop the whole leaf becomes senescent, dies prematurely and drops. Older leaves are mostly affected. Defoliation occurs throughout the growing season resulting in reduction in root size and yield. On seed plants all above ground parts are affected including the seed clusters. High humidity usually favours the spread of this disease.

The pathogen survives on plant refuse in the soil and serves as a source of primary inoculum. Hence, rotation with other crops and cleaning up the refuse after harvesting are thought beneficial in controlling the disease. The spores which are borne on the lesion surface are air-borne and provide the secondary inoculum. It is, therefore, important to select fields 30 m or more away from those of the previous year.

Chemical control measures include spraying of 0.3 per cent blitox (copper oxychloride) three times at an interval of 15 days. Treat the seeds with captan and bavistin at 2-3 g/kg of seed. Follow crop rotation of at least 3 years with non-host crops. Practice deep ploughing of soil during summer season. Avoid high density planting.

9.1.2 Downy Mildew (*Peronospora schachtii*)

Downy mildew sometimes becomes a serious disease of garden beets, which may occur at any stage of growth. The disease can be seed borne and is mostly prevalent during the cooler months.

All above-ground parts may be affected. On the leaves, spots of various sizes up to 4 cm in diameter appear. The affected portions become lighter green on the upper surface, while on the under surface the mildew (fungal growth) is noticed. The lesions on the older leaves are usually isolated and irregular. In dry weather, the margins of leaves may have a pale red pigmentation. The infected leaves may become small and thicker than normal, and are often curled downward at the edges. Excessive production of new leaves follows and invasion by other organisms may result in root decay.

Flower shoots on infected plants become stunted and distorted; sepals and petals become swollen and mildew appears on all parts in moist weather. The entire inflorescence has a compact appearance and excessive leaf development may give a witches' broom appearance. The seedballs vary from normal appearance to various degrees of shrivelling with sterile flower remnants.

The fungus survives on the crop residues in the soil and is also carried by the seed.

Control measures include field sanitation, crop rotation, use of resistant cultivars and seed treatment with fungicides. Spraying 0.3 per cent dithane Z-78 or 0.25 per cent mancozeb thrice at an interval of 15 days is also suggested as an effective control measure. Avoid excess irrigation and follow crop rotation with non-host crops.

9.1.3 Rhizoctonia Root Rot (*Rhizoctonia solani*)

This fungus causes damping-off and wire stem of seedlings and bottom rot, root rot and heart rot of plants in the field and storage. The petiole base of the leaf turns black, leaves collapse and die. The flesh becomes brown and the skin cracks. Affected roots start decaying and leaves turn yellow and wilt, causing loss in yield and quality. Follow crop rotation with non-host crops, treat the seed with captan or thiram @ 2.5 g/kg, apply neem cake @ 20 q/ha and drench the soil with copper oxychloride 0.03 per cent and carbendazim 0.1 per cent.

9.1.4 Powdery Mildew (*Erysiphe polygoni*)

Yellowish spots of 2.5 cm diameter appear on the upper surface of older leaves and white powdery growth on lower surface. The affected leaves become dry. In case of severe infection, both the surfaces of leaves are covered by whitish powdery growth, affected parts get shriveled and distorted, causing premature defoliation. In case of early infection, the roots remain small and produce little sugar. Spray the crop with Karathane 0.1 per cent, carbendazim 0.1 per cent, or sulphur 0.2 per cent.

9.1.5 Phoma Blight and Heart Rot (*Pleospora betae*)

It is seed borne. Water-soaked areas appear on roots, which quickly turn brown and finally become black. On seed stalks, brown to black necrotic streaks with a greyish center appears. Follow long crop rotation with non-host crops and treat the seed with captan or thiram 2.5 g/kg.

9.1.6 Sclerotium Root Rot (*Sclerotium rolfsii*)

The affected leaves turn yellow and wilt, roots start decaying and numerous small black sclerotia are embedded in the infected plant parts. Adopt phytosanitary measures to reduce the load of inoculum, treat the seed with Bavistin, captan, or thiram @ 2–3 g/kg of seed and drench the soil with 0.03 per cent solution of copper oxychloride. Apply nitrogen in the form of calcium ammonium nitrate or ammonium sulphate instead of urea.

Diseases of Beet Roots

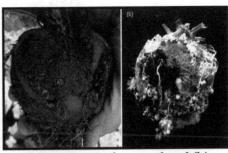

Bacterial leaf spot (left) and *Phoma* leaf spot (right) in beet

a) Common scab caused and (b) Southern Sclerotium root rot

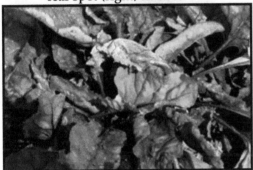

Beet curly top virus

Cercospora leaf spot in beet

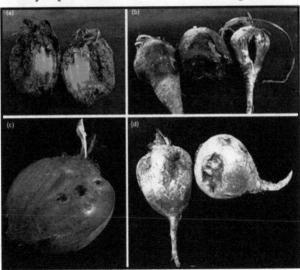

(a) **Papery mummified decay** caused by *Pythium* spp. (b) **Black rot** caused by *Rhizoctonia solani* (c) **Dry rot** caused by *Fusarium* spp. and (d) **Outer skin scarring** caused by *Aphanomyces cochlioides*.

Pests and Disorders in Beet

Leaf miner in beet

Beet leaf hopper adults

Boron deficiency in beet

9.1.7 Common Scab (*Actinomyces scabies*)

Scab is a soil-borne organism that starts attacking when the root starts forming. As the beet root grows the scab enlarges and the root surface becomes rough and irregular. Scab can live on decomposing material and does not need a host crop to survive, therefore a 4-5 year crop rotation should be followed. Scab can also be managed using irrigation. Irrigation keeps the soil moist when the roots begin to form. The disease is also prevalent in soils containing high lime.

Viral Diseases

Beets are affected by a number of viral diseases like mosaic (beet mosaic virus, BtMV), curly-top (Beet Curly Top Virus, BCTV) and yellows (Beet Western Yellows Virus, BWYV; Rhizomania - Beet Necrotic Yellow Vein Virus, BNYVV).

9.1.8 Beet Mosaic Virus (BtMV)

This potyvirus is seedborne. Appearance of conspicuous mottling with chlorotic, zonate ring spots is common. These may become necrotic with age. When these ring spots develop, their centres are usually green. Virus infected plants remain stunted and may lose some leaves. The disease is normally transmitted and spread by aphids.

9.1.9 Beet Curly Top Virus (BCTV)

The affected leaves are undersized and curled and the veins are transparent. This geminivirus is transmitted by beet leaf hoppers (*Circulifer tenellus*).

9.1.10 Beet Western Yellows Virus (BWYV)

It is a polerovirus. The older leaves of infected plants become chlorotic, noticeably thickened, leathery and brittle. The foliage becomes abnormally red or yellow and often dies. It is transmitted mainly through aphids.

9.1.11 Rhizomania - Beet Necrotic Yellow Vein Virus (BNYVV)

It is a benyvirus. Above ground symptoms appear as yellow leaves, with a characteristic erectness. Gradually, the leaf veins will become more yellow, and the leaves become necrotic. Infected roots are stunted, with masses of lateral roots, giving the root a bottled-brush appearance. BNYVV exists in the parasitic soil fungus *Polymyxa betae* which survives in soil and in plant debris. It is favoured by high soil moisture and warm soil conditions.

Control of virus diseases includes isolation from related plants which may act as secondary hosts. Keep the fields and field edges weed free. Control of insect pests and destruction of infected plants also prevent spread of these diseases. Spray the crop after germination with imidacloprid 0.03 per cent to reduce the population of vectors and disease spread.

9.2.0 Pests

Beets are also attacked by certain insect pests, of which beet leaf miner, webworms and semiloopers are most important.

9.2.1 Beet Leaf Miner (*Pegomyia hyocyami*)

The larva is a white maggot, about 8 mm in length, which feeds on the tissues between the upper and lower layers of the leaf, thus causing serious injury to the leaf making it unfit for manufacture of food, consequently the plant growth is checked. Infested leaves present a blistered appearance. The fly lays her eggs on the underside of the leaf and the maggot hatches and enters the leaf.

Control measures of beet leaf miner include destruction of all fallen leaves and other plant refuse after harvesting of roots and spraying the underside of the leaf with either malathion 0.1 per cent. or chlorpyrifos 0.25 per cent. Apply carbofuran 0.5 kg a.i./ha at sowing and at 40 days after germination.

9.2.2 Webworms (*Hymenia fascialis* Cramer or *Loxostege* sp.)

Both the above species may attack beet. The adults lay eggs on the leaves and the larvae attack the foliage, either by spinning small webs among the tender leaves or feeding on the underside, protected by small webs. This pest can be controlled by spraying the crop with Rogor at the rate of 1 ml/l water. Practice deep ploughing of soil during summer season to kill the pupae. Set up light traps and pheromone traps @ 12/ha to attract and kill the adult moths. Spray *Bacillus thuringiensis* at 2 g/litre.

9.2.3 Semiloopers (*Plusia nigrisigna*)

The green caterpillars feed on the foliage damaging green foliage badly. Dusting the crop with 5 per cent malathion powder at 25 kg/ha may control the semiloopers. Apply 5 per cent neem seed kernel extract. Systemic insecticides are also effective against this pest.

9.2.4 Root Knot Nematodes (*Heterodera schachtii, Globodera rostochinensis, Meloidogyne incognita*)

The root knot nematode problem is most serious in light sandy soils. In infected plants, the plant growth is stunted, and the leaves show chlorosis and wilting symptoms. Bahadur (2020) reported infection with root-knot nematode (*Meloidogyne incognita*) in beet-root crop in Tripura. Control measures include adoption of crop rotation with non-host crops like wheat, maize, sorghum, pearl millet, groundnut, cotton and pigeon pea up to 4–5 years. Practice deep ploughing during summer season. Grow marigold as trap crop or as intercrop to reduce the nematode population. Apply neem cake @ 20 q/ha and fumigate the soil with nematicides like Nemagon @ 25 litre/ha with irrigation or carbofuran 3 G @ 25 kg/ha at the time of planting in the furrows.

9.3.0 Physiological Disorder

9.3.1 Internal Black Spot or Brown Heart or Heart Rot or Crown Heart

Boron deficiency may cause a physiological disorder in garden beets, which is known as internal black spot or brown heart or heart rot (Walker, 1939). Boron deficient plants usually remain dwarf or stunted; the leaves are smaller than normal. The young unfolding leaves fail to develop normally and eventually turn brown or black and die. The leaves may assume a variegated appearance due to development of mixture of yellow and purplish red blotches over parts or whole, while the stalk of such leaves shows longitudinal splitting. Frequently the affected plant has twisted leaves, and exhibits a slight shortening and distorting of its leaf-stalk in the centre of its crown. The growing point may die and decay. The roots do not grow to full size and under conditions of severe boron deficiency they remain very small and distorted, and have a rough, unhealthy, greyish appearance instead of being clean and smooth (Gupta and Cutcliffe, 1985). Their surfaces often are wrinkled and cracked. Within the fleshy roots, hard or corky spots are found scattered throughout the roots, but more numerous on the light coloured zones or cambium layers.

The effects of boron deficiency are evident in the cells much before the macroscopic symptoms appear. Discolouration of cell walls and granulation of protoplasts of meristematic cells appear first. In the cambium, cell division continues with little differentiation into xylem and phloem and many cells are abnormally large. This leads to disturbed translocation and eventually distorted growth follows. In the regions of abnormal cell activity death of cells comes next, leading to development of necrotic black spots (Walker, 1952).

Many soils are naturally deficient in boron, in others boron is present in a form unavailable to the plant. Boron deficiency is common in alkaline soils or on soils relatively high in calcium. It has been found that as the calcium availability to the plant in the nutrient medium increases, the requirement for boron also increases. This fact explains in part the greater severity of boron deficiency in soils rich in calcium. Over-liming often brings about boron deficiency (Walker, 1952).

Seasonal conditions may influence this disorder, as a result the disorder may be very severe in one season, but the crop may grow well in the following season in the same field. In general, boron deficiency appears more often in a dry season, particularly when a long dry spell is followed by a wet period favouring rapid growth, most probably because the root system in the upper soil layer ceases to function during the dry period and thus intake of boron is reduced. Moreover, boron may be fixed in the dry soil.

Cultivars may differ in their susceptibility to boron deficiency in the soil (Walker *et al.*, 1945; Sleeth, 1961), Flat Egyptian being found most susceptible and Long Dark Blood most resistant, while Detroit Dark Red showed moderate susceptibility. Sleeth (1961) noticed 0.1 per cent or less incidence of black spot in Seneca Detroit, Ruby Queens, and three strains each of Detroit Red and Perfected Detroit. Tehrani-Moayed (1968) reported that boron-susceptibility is controlled by a single dominant gene.

This disorder can effectively be controlled by application of borax to the soil. The quantity of borax needed for satisfactory control of boron deficiency varies with the nature of the soil, the soil reaction and the soil moisture, and application of 5-50 lb of borax per acre is recommended. Mack (1989) found that application of 11.2 kg B and/or 224 kg N/ha reduced the number of roots with B deficiency symptoms in cv. Detroit Dark Red, compared with low or no N and B application.

9.3.2 Distinct Zoning

During hot weather, the alternating circular bands of conducting and storage tissues in beetroot becomes more distinct because the betalain production is suppressed at high temperatures. Nieuwhof and Garretsen (1974) studied the inheritance of an orange-yellow colour deviation in red table beets and reported that this disorder to be governed by a number of genes, presumably partly acting in an additive and partly in a non-additive way which can largely be solved by simple mass selection.

9.3.3 Speckled Yellows

Beetroot plants affected by manganese deficiency have yellowish-green leaves with chlorotic mottled areas, which become necrotic, resulting in breaking of the leaf lamina. The leaf margins roll upward, remains upright and turn into an arrow-shaped outline. Manganese is observed in very sandy and very alkaline soils. Apply manganese sulphate 5–10 kg/ha in soil or do two or three sprays of 0.25 per cent manganese sulphate.

10.0 Seed Production

Unlike other root crops, garden beet does not have any tropical type. All cultivars of beet are biennial temperate type and hence their seeds can only be produced in the hills of India. Beetroot is an obligate long day plant. Like temperate carrot, beet requires exposure to low temperature (vernalization) of 4.4-7.7°C for 6-8 weeks for initiation of flower stalks. Such temperatures are considered available in the hills in winter.

In Kullu valley, the sowing of beet is done from mid-July to the end of July. Late cultivars, however, can be sown from the end of June to mid-July. About 6.25-7.5 kg seed is sufficient to raise seedlings in one hectare of land which are considered sufficient to plant 10-12.5 hectares of seed crop.

Seed to seed method is not advised for beetroot seed production. In root-to-seed method, beet roots are uprooted and true to type roots are selected, then they are replanted within hours. In high snowfall areas, the selected roots are stored in a cool root cellar, or in a pit in the ground and are replanted in the following spring. This method allows selection of true to type roots, enables growers to produce beet seed in climates that are too cold for the seed-to-seed method to be used and it also avoids transmission of diseases that are more common in seed-to-seed method. Singh *et al.* (1960) reported that the usual method of seed production in Kullu valley is root-to-seed method. In this method, during November-December, the well-developed roots are dug out. After selection, the taproot and tops of the roots are trimmed taking care not to injure the crown and planted in a well-prepared field. Immediately after replanting the crop is irrigated. In case of beet, planting of whole root is recommended. Depending upon the cultivar, a spacing of 60 × 60 cm or 60 × 45 cm is followed for raising a commercial seed crop. The available nitrogen in the soil should not be too high to avoid excessive vegetative growth before bolting, as this promotes lodging of the plants during seed set.

10.1.0 Selection and Roguing

Selection of true-to-type cultivars and discarding all off-type plants are important operation in the production of quality seeds. Selection and roguing are done in several stages. All plants showing differences in leaf colour are discarded. When the roots reach maturity they are lifted from the field and examined for external root characters like size, shape and colour of roots. After ascertaining the leaf size, shape and colour, petiole colour and strength and root colour (external), shape and size, the roots are sectioned to judge their interior quality. Usually roots having uniform and rich flesh colour with indistinct rings are selected. All malformed, forked, diseased and off-type roots are discarded. Roots having a rough appearance of the crown are rejected.

True to type whole roots are planted in well prepared field. For long term storage of roots, remove most of the petioles (leaf stems) and remove portion of the tap root that is longer than 7 cm. Temperature of the storage space should be 1-2.5°C and relative humidity 95 per cent.

10.2.0 Flowering and Seed Setting

Seedstalk elongation in beet starts in early part of April in Kullu valley, and the crop is in full bloom from mid-May to mid-June. As stated already, the inflorescence is a large panicle and seed maturity begins at the base of the panicle. Beetroot flower is protrandous and is wind pollinated. Multigerm seed ball is capable of producing 2, 3, 4, and even 5 seedlings.

10.3.0 Isolation

Beet is a wind-pollinated crop. The recommended isolation distances for certified and nucleus seed production are 1000 m and 1600 m, respectively. Beetroot should be isolated from other beet fields, from Swiss chard, fodder beet (mangel) or sugar beet fields.

10.4.0 Harvesting and Seed Yield

Beetroot crop takes approximately 140–160 days for the seeds to mature from the replanting date, depending on variety, climate, and planting date. The crop ripens in July in Kullu valley. However, when the summer is hot and dry, harvesting may start as early as the last week of June, but usually it is done from the first week of July to the end of July. Rains during ripening of seeds may affect the seed maturity and quality (viability) of the seed crop. The maturity and ripening of beet seeds start from the base of the inflorescence. Generally, when 70-80 per cent of the seedballs on a plant get hardened and those at the base of the inflorescence turn brown, the crop is harvested; otherwise there is possibility for shattering of seeds during harvesting. The crop is then stacked for curing and then dried under sun. Threshing, cleaning and grading of beet seed are done as in case of other root vegetables. Seed is dried to 9 per cent moisture level or less, cleaned and stored.

The average seed yield is 10-12 quintals per hectare. Harvested, cleaned, dried and graded seeds are stored airtight in a cool (< 15 °C) and dry place.

11.0 Crop Improvement

Most of the garden beet cultivars grown today are results of mass selection rather than evolved by means of control breeding programme involving inbreeding and cross-pollination. The garden beet is a cross-pollinated crop and it is mostly self-incompatible. Self-sterility in beet can be overcome by sib-mating (Magruder, 1932). By using sib-mating for several generations, Ohio Canner, a cultivar with a uniformly dark red flesh colour, has been developed.

There are three primary gene pools of beetroot, *viz.* wild species of Beta; crop relatives such as sugarbeet and Swiss chard, and table beet populations and cultivars. These gene pools are used by breeders for the improvement of beetroot. The wild species, *Beta procumbens* (source of root-knot nematode resistance) and *Beta webbiana* have been used for sugarbeet improvement. W. H. Gabelman utilized the genes found in sugar beet to improve table beet germplasm. He transferred from sugarbeet to table beet the traits such as self-fertility, annual growth habit, cytoplasmic male sterility, and the monogerm character. Development of inbreds in beetroot will be difficult if self-incompatibility is present. Therefore, Gabelman transferred a dominant gene for self-fertility, *SF* allele (Savitsky, 1954) from sugar beet to table beet. Incorporation of *SF* allele allowed inbreeding for development of uniform inbred lines in table beet which was otherwise not possible due to self-incompatibility.

F.V. Owen (1945) described cytoplasmic male sterility (CMS) and the fertility restoration in sugar beet wherein a plant is male sterile if alleles at both the × and *z* loci are homozygous recessive. Dominant alleles at either or both loci results in male fertility. The maintainer lines has normal fertile cytoplasm and homozygous recessive alleles at both × and *z* loci and is referred to as "B" lines by vegetable breeders and "O-types" by sugar beet breeders. Gabelman introduced the × and *z* alleles in homozygous recessive condition from sugar beet into table beet germplasm. These alleles in combination with the sterile cytoplasm and the *SF* allele resulted in the development of male sterile breeding lines and their maintainer lines in table beet. Inbred A line refers to the sterile phenotype having genotype *Sxxzz*, and the B line is maintainer of sterility (fertile) phenotype with genotype *Nxxzz*.

Cheng *et al.* (2011) examined the cytoplasmic diversity in leaf and garden beet accessions using polymorphisms in four mitochondrial minisatellite loci. They identified eleven multi-locus haplotypes, of which one (mitochondrial minisatellite haplotype 4 *i.e.* min04) was associated with male-sterile Owen cytoplasm and two others (min09 and min18), with a normal fertile cytoplasm. European leaf beet germplasm exhibited the greatest haplotype diversity, with min09 and min18 predominating. They inferred that North African leaf beet accessions were descended from European genotypes. They showed that mitochondrial genome diversity was low in garden beet germplasm, with min18 being highly predominant, which might be due to the geographically restricted origin of as well as relatively short cultivation histories of garden beet.

Application of the *B* allele conditioning annual flowering habit to table beet breeding was discussed by Goldman and Navazio (2003). Biennial breeding lines carry the *b* allele in the homozygous recessive condition.

The garden beet cultivars are biennial, and if young plants are exposed to low temperature they initiate flowers prematurely. Roots of early bolters are usually smaller in size and are more lignified than roots of normal plants. Avon Early, a selection made from a population of 70-80 Detroit cultivars, has been developed in the U.K. as a bolting resistant cultivar (Holland and Dowker, 1969). This cultivar has been reported to be resistant to *Peronospora farinosa* f. sp. *betae* (Channon, 1969). Esfahani *et al.* (2020) found the sugar beet genotypes 16 (SB32-HSF-5) and 5 (SB31-HSF-2) to be resistant to cyst nematodes, *Heterodera schachtii*. They employed SSR markers which efficiently characterized the genetic diversity between resistant and susceptible sugar beet genotypes. However, the genetic diversity was not always directly related to the resistance and or susceptibility of genotypes.

Another achievement in beet breeding is the development of monogerm (unilocular seedball) seeds. Multigerm seed character will make precision seeding more difficult in beetroot. The monogerm character, conditioned by recessive alleles at the m locus, was first identified as a mutant in a commercial sugar beet field (Savitsky, 1950). W. H. Gabelman incorporated the monogerm character into table beet germplasm, resulting in the development of first table beet inbred lines carrying this trait (Goldman, 1996). Khavskaya, a monogerm cultivar, was bred at the Voronesh Experimental Vegetable Station (USSR), using elite seed plants of the cultivar Bordeaux with monogerm and digerm fruits. The breeding programme included several selection cycles and inbreeding. Khavskaya is a mid-season cultivar with spherical or near spherical root, dark red flesh and a seed germination percentage of 85 (Drobysheva and Sycheva, 1979). The variety Joniai was obtained by crossing the Russian variety Pushkinskaja Ploskaja with the Dutch variety Early Red Chief. Reciprocal crossing, and mass and recurrent selection were utilized in the breeding process along with digerm plants. Joniai is a universal mid-early Bordo type variety. Roots are of good marketable appearance and biochemical composition, of good flavour with good storability. Seeds of Joniai are suitable for mechanical sowing since 75-80 per cent of seeds are digerm (Petroniene, 2000).

In beet, the alleles at two linked loci (R and Y) condition betalain pigment production (Keller, 1936). Wolyn and Gabelman (1990) reported that three alleles at the R locus determine the ratio of betacyanin to betaxanthin in the beet root and shoot. They observed incomplete dominance for pigment ratio in *Rt* and *R* genotypes. Colour patterning in the beet plant is affected by the alleles at both *R* locus and *Y* locus. Red roots are observed only when alleles at the *R* and *Y* loci are present in dominant state, while roots are white when alleles at both loci are recessive. A *yy* condition coupled with *rr*, which is characteristic of most sugar beet cultivars, produces no betacyanin and produces betaxanthin only in the hypocotyls. Linde-Lauren (1972) suggested a third locus, the *P* allele, which is indispensable for colour formation and reported its close linkage with the *R* and *Y* loci. White-rooted beet plants carry the *p* allele in a homozygous recessive condition along with dominant alleles at the *R* and *Y* loci. At University of Wisconsin, half-sib family recurrent selection has been successfully used to increase the pigment levels by 450 per cent at the end of 15 selection cycles in table beet.

Grzebelus *et al.* (1997) showed that Okragly Ciemmoczerwony contained low levels of nitrates in the roots which were stable in different environments. It may be used as a good source of initial material for breeding cultivars with low nitrate content.

12.0 References

Anderson, F. (1952) *Plants, Man and Life, Little Borwn*, Boston.

Arora, S., Siddiqui, S. and Gehlot, R. (2019) *Asian J. Dairy Food Res.*, **38**: 252-256.

Bagchi, D.K. (1982) *Ph.D. Thesis*, Calcutta University.

Bahadur, A. (2020) *Indian Phytopath.*, https://doi.org/10.1007/s42360-020-00233-y.

Barba-Espin, G., Glied-Olsen, S., Dzhanfezova, T., Joernsgaard, B., Lutken, H. and Muller, R. (2018) *BMC Plant Biol.*, **18**: 316.

Burkill, I.H. (1935) *A Dictionary of the Economic Products of the Malay Peninsula*, Crown Agents, London

Campbell, G.K.G. (1976) In: N.W. Simmonds (ed.), Evolution of Crop Plants, Longman, London. pp. 25.

Cerne, M., Ugrinovic, K., Briski, L. and Kmecl, V. (2000) *Zbornik Biotehniske Fakultete Univerze V Lijubljani. Kmetijstvo*, **75**: 115-127.

Channon, A.G. (1969) *Plant Path.*, **18**: 89-93.

Cheng, D., Yoshida, Y., Kitazaki, K., Negoro, S., Takahashi, H., Xu, D., Mikami, T. and Kubo, T. (2011) *Genet. Resour. Crop Evol.*, **58**: 553–560.

Chhikaraa, N., Kushwaha, K., Sharma, P., Gat, Y. and Panghal, A. (2019) *Food Chem.*, **272**: 192-200.

da Silva, P.F., de Matos, R.M., Borges, V.E., Sobrinho, T.G., Neto, J.D., de Lima, V.L.A., Ramos, J.G. and Pereira, M. de O. (2018) *Australian J. Crop Sci.*, **2**: 1335-1341.

Davenport, D.C. (1962) *M.Sc.Thesis*, Indian Agricultural Research Institute, New Delhi.

Drobysheva, N.A. and Sycheva, L.V. (1979) *Referativny I Zhurnal*, 6.65.244.

Elkner, K., Horbowicz, M. and Kosson, R. (1998) *Veg. Crops Res. Bull.*, **49**: 107-120.

Erwin, A.T. and Haber, E.S. (1934) *Iowa Agric. Exp. Sta. Bull.*, p. 308.

Esfahani, M.N., Naderi, R. and Khaniki, G. B. (2020) *Physiol. Mol. Plant Path.*, **112**: 101518.

Esyunina, A.I., Lukovnikova, G.A. and Aizina, M.I. (1972) *Trudy po Prikladnoi Botanike, Genetike I selektsii*, **48**: 89-96.

Ford-Lloyd, B.V. (1995) In: Evolution of crop plants (Eds. J. Smartt, and N.W. Simmonds), 2nd edn, Longman Scientific and Technical, Essex. pp. 35-40.

Gaharwar, A.M. and Ughade, J.D. (2017) *Int. Res. J. Agril. Econom. Stat.*, **8**: 51-55.

Gengatharan, A., Dykes, G.A. and Choo, W.S. (2015) *LWT-Food Sci. Tech.*, **64**: 645-649.

Goldman, I.L. (1995) *J. Amer. Soc. Hort. Sci.,* **120**: 906-908.

Goldman, I.L. (1996) *HortScience,* **31**: 880-881.

Goldman, I.L., and Navazio, J.P. (2003) *Plant Breed. Rev.,* **22**: 357-388.

Grzebelus, D. (1995a) *Folia Horticulturae,* **7**: 43-49.

Grzebelus, D. (1995b) *Folia Horticulturae,* **7**: 35-41.

Grzebelus, D., Krajewski, P. and Kaczmarek, Z. (1997) *Proc. 10 Meeting EUCA RPIA Section Biometries in Plant Breeding,* Poland, pp. 139-142.

Haag, H.P. and Minami, K. (1987) *Anais da Escola Superior de Agricultura 'Luiz de Queiroz',* Universidade de Sao Paulo, **44**: 401-408.

Hammer, K. (2001) Chenopodiaceae. In: Hanelt P, Institute of Plant Genetics, Crop Plant Research (eds) Mansfield's encyclopedia of agricultural and horticultural crops, vol 1., Springer, Berlin. pp 235–264.

Hipp, B.W. (1977) *J. Amer. Soc. Hort. Sci.,* **102**: 598-601.

Holland, H.S. and Dowker, B.D. (1969) *J. Hort. Sci.,* **44:** 257-264.

Jarvan, M. (1995) *Agraarteadus,* **6**: 257-277.

Junttila, O. (1976) *J. Exp. Bot.,* **27**: 827-836.

Keller, W. (1936) *J. Agr. Res.,* **52**: 27-38.

Kovaroviè, J., Bystrická, J., Tomáš, J. and Lenková, M. (2017) *Potravinarstvo Slovak J. Food Sci.,* **11**: 106-112.

Lange, W., Brandenburg, W. and de Bock, T.S.M. (1999) *Bot. J. Linn. Soc.,* **130**: 81–96.

Lebedeva, E.V., Williams, M.V. and Tsvetkova, I.V. (1978) *Fizziologiya Rastenii,* **25**: 191-195.

Lee, C.W. (1973) *J. Sci. Fd. Agric.,* **24**: 843.

Linde-Laursen, I. (1972) *Hereditas,* **70**: 105-112.

Longvah, T., Ananthan, R., Bhaskarachary, K. and Venkaiah, K. (2017) Indian Food Composition Tables 2017. National Institute of Nutrition, Hyderabad. p.535.

Lorenz, O.A. (1947) *Proc. Amer. Soc. Hort. Sci.,* **49**: 270-274.

Mack, H.J. (1979) *J. Amer. Soc. Hort. Sci.,* **104**: 717.

Mack, H.J. (1989) *Communications in Soil Sci. and Pl. Analysis,* **20**: 291-303.

Magruder, R., Bosweu, V.R., Jones, H.A., Miller, J.C., Wood, J.F., Hawthorn, L.R., Parker, M.M. and Zimmenrly, H.H. (1940) U.S. Dept. Agri. Misc., Pub., P. 374.

Magruder, R. (1932) *Bimonthly Bull. Ohio Agric. Exp. Sta.,* **154**: 18-25.

Makarenko, S.P., Konenkina, T.A. and Salyaev, R.K. (1999) *Russian J. Plant Physiology,* **46**: 561-565.

Mariassyova, M., Silhar, S., Baxa, S., Kovac, M. and Tiilikkala, K. (1999) *Agri-Food Quality. II. Quality Management of Fruits and Vegetables - From Field to Tables,* pp. 314-315.

Maroušek, J., Kolábattu2, L., Vochozka, M., Stehel, V. and Maroušková, A. (2017) **168**: 60-62.

Michalik, B., Sleczek, S., Zukowska, E. and Szklarczyk, M. (1995) *Folia Horticulturae,* **7**: 137-144.

Moreno, D.A., Garcia-Viguera, C., Gil, J.I. and Gil-Izquierdo, A. (2008) *Phytochem. Rev.,* **7**: 261-280.

NG, J.J. and Lee, Y.N. (1978) *HortScience,* **13**: 581-582.

Nieuwhof, M. and Garretsen, F. (1974) *Euphytica,* **23**: 365-367.

Nilsson, T. (1970) *Lanbrttogsk ann.,* **36**: 179-219.

Ovoshchevodstvo, M. and Tenat, S. (1978) *Referativny i Zhurnal,* **49**: 349.

Owen, F.V. (1945) *J. Ag. Res.,* **71**: 423-440.

Paria, N.C. (1973) *Indian Agric.,* **17**: 289.

Petroniene, D. (2000) *Sodininkyste ir Darzininkyste,* **19**: 81-86.

Pillai, O.A.A., Irulappan, I., Doraipandian, a., Jayapal, R. and Balasubramanian, S. (1977) *South Indian Hort.,* **25** ; 162-163.

Rajkumar, R.H., Anand, S.R., Vishwanatha, J., Karegoudar, A.V. and Subhas, B. (2016) *Env. Ecol.,* **34**: 673-677.

Rao, M.H. and Subramanian, T.R. (1991) *J. Potassium Res.,* **7**: 190-197.

Sapers, G.M. and Hornstein J.S. (1979) *J. Fd. Sci.,* **44**: 1245-1248.

Savitsky, H. (1954) *Proc. Amer. Soc. Beet Technol.,* 8.

Savitsky, V.F. (1950) *Proc. Am. Soc. Sugar Beet Tech.,* **6**: 156-159.

Singh, H.B., Thakur, M.R. and Bhagchandani, P.M. (1960) *Indian J. Hort.,* **17**: 38-47.

Sistrunk, W.A. and Bradley, G.A. (1970) *Arkans. Fm. Res.,* **19**: 9.

Sizov, A.P. and Lunev, M.I. (1991) *Agrokhimiya,* No. 4, pp. 40-44.

Sleeth, B. (1961) *J. Rio Grande Valley Hort. Soc.,* **15**: 99-101.

Smagina, V.N. (1977) *Referativny i Zhurnal,* 8.55.592.

Subbarao, G.V., Wheeler, R.M., Stutte, G.W. and Levine, L.H. (1999) *J. Plant Nutri.,* **22**: 1745-1761.

Takacs-Hajos, M. (1993) *Zoldsegtermoztesi Kutalo Intezet Bulletinje,* **25**: 81-95.

Takacs-Hajos, M., Simandi, P. and Posza, I. (1997) *Horticultural Sci.,* **29**: 61-65.

Taylor, A.G., Goffinet, M.C., Pikuz, S.A., Shelkovenko, T.A., Mitchell, M.D., Chandler, K.M. and Hammer, D.A. (2003) In: The Biology of Seeds: Recent Research Advances (Eds. G. Nicolás, K.J. Bradford, D. Côme and H.W. Pritchard). CAB Internationa, Wallingford, UK. pp. 433-440.

Tehrani-Moayed, G. (1968) *Diss. Abst.,* No. 28.

Thompson, H.C. and Kelly, W.C. (1957) *Vegetable crops,* McGraw-Hill Book Co., Inc., New York.

Uchkunov (2012) *Macedonian J. Animal Sci.*, **2**: 183–185.

Vukasinovic, S., Jerkic, I., Babic, S. and Cota, J. (1991) *Radovi Poljoprivrednog Fakulteta Univerzitela u sarajevu*, **39**: 23-27.

Walker, J.C. Jolivette, J.P. and Hare, W.W. (1945) *Soil Sci.*, **59** ; 461-464.

Walker, J.C. (1939) *Phytopathology*, **29**: 120-128.

Walker, J.C. (1952) *Diseases of Vegetable Crops*, McGraw-Hill Book Co., New York.

Wang, M. and Goldman, I.L. (1996) *J. Amer. Soc. Hort. Sci.*, **121**: 1040-1042.

Wang, M. and Goldman, I.L. (1997) *Plant Foods for Human Nutrition,* **50**: 1-8.

Weaver, J.E. and Bruner, W.E. (1927) *Root Development of Vegetable Crops*, McGraw-Hill Book co., Inc., New York.

Wolyn, D.J., and Gabelman, W.H. (1990) *J. Hered.*, **80**: 33-38.

Yordanova, M. and Gerasimova, N. (2016) *Org. Agr.*, **6**: 133–138.

Zinkevich, A.S. and Novikov, V.S. (1977) *Referativny i Zhurnal*, 1.55.637.

13

TURNIP

A. K. Sureja, M. K. Sadhu and K. Sarkar

1.0 Introduction

Turnip is an important root vegetable grown as a summer crop in temperate climate and as a winter vegetable in places where the winter is not very severe. While it is not suitable for growing in low lands of wet tropics, it can successfully be grown at an elevation of 1500 m or above. Sturtevant (1919) suggested that the cultivation of turnip was probably first attempted by the Celts and Germans when they were compelled to make use of its nutritious roots.

2.0 Composition and Uses

2.1.0 Composition

The turnip root contains 91.6 g moisture, 5.2 g carbohydrates, 0.5 g protein, 0.3 g protein, 43 mg ascorbic acid, 0.04 mg thiamine, 0.04 mg riboflavin, 30 mg calcium, 40 mg phosphorus and 0.4 mg iron per 100 g edible portion. Turnip leaves are good source of calcium, iron and vitamins A, B and C.

Klopsch *et al.* (2017) have done metabolic profiling of glucosinolates and their hydrolysis products in leaves and tubers of 16 germplasm collection of *Brassica rapa* turnips and identified 13 glucosinolates (8 aliphatic, 4 indolic and one aromatic). Bonnema *et al.* (2019) observed substantial differences in glucosinolate profiles between aboveground tissues and turnip tuber, reflecting the differences in physiological role.

2.2.0 Uses

Turnip is grown for its enlarged fleshy root as well as for its foliage. Extra seedlings from thinning are often used as greens. The fresh roots are consumed in salads or cooked as a vegetable or used in pickles. The young leaves, which contain high amount of ascorbic acid and iron, and rank second in vitamin A content, are eaten cooked as greens. Sprouted seedlings at the cotyledon stage (microgreens) are also used as garnish. In turnip, microgreens are harvested 8-9 days after germination. They are cut along with the stem and attached cotyledons/seed leaves with the help of scissor. If left for longer, they rapidly elongate and loose colour and flavour.

3.0 History, Origin, Taxonomy and Cytology

In Indian Sanskrit records, oil seed form of turnip is mentioned as early as 2000-1500 BC. Turnip is under cultivation in Europe for the last 5000 years or so. The European type of turnip was grown in France as early as the first century AD. From France, Romans brought it to England where it was cultivated as a field crop during early 18th century. In Europe and Japan, a well-defined, polymorphic group of vegetable turnips were created, independently, by the 18th century. It reached Mexico in 1586, Virginia in 1610, and New England in 1628. Turnip was introduced into Canada by Jacques Cartier in 1541 and into the USA in 1609. Cultivation of turnip in India was adopted from the colonists.

Uses of Turnip

Fries (left) and Curry (right) from turnip

Roasted turnip (left) and soup (right) from turnip greens

Turnip Juice along with root slices

Turnip (*Brassica rapa* L. subsp. *rapa* L.) belongs to the family Cruciferae. Two main centres of origin of turnip are indicated. The Mediterranean area is thought to be the primary centre of European types, while eastern Afghanistan with adjoining area of Pakistan is considered to be primary centre of origin of tropical types. Asia Minor, Transcaucasus and Iran are considered as secondary centres (Sinskaia, 1928). The parents of cultivated turnip are found wild in Russia, Siberia and Scandinavia. Cartier introduced turnip in Canada in 1540 during his voyage of exploration and it was introduced in Virginia in 1609 (Shoemaker, 1953). Turnips are now cultivated throughout the world. In India, it is mainly cultivated in Jammu and Kashmir, Punjab, Himachal Pradesh and western U.P.

Takahashi *et al.* (2016) elucidated the phylogenetic relationships of 87 accessions of Eurasian turnips by examining their morphology and analyzing 6 cpSSR and 18 nuSSR loci. Examination of seed coat mucilage and leaf hairs revealed existence of geographic distinctions. Accessions from continental Asia showed various haplogroups and clusters and higher levels of genetic diversity than those from other regions. They suggested that central Asia is the sole geographic origin of turnips that Asian turnips did not originate as descendants of European turnips, and that almost all Japanese turnips were derived from central Asia.

Linnaeus (1753) published two species names *Brassica rapa* L. and *Brassica campestris* L., which were combined by Metzger (1833) and the combined name *Brassica rapa* was chosen. Hence, the classification and nomenclature of *B. campestris* is changed to *B. rapa*. *B. rapa* consist of several phenotypically diverse distinct cultivated morphotypes (leafy, root or fodder and oilseed types) which are described below.

1. *Brassica rapa* L. subsp. *rapa* L.

This subspecies is called turnip and has a wide genetic diversity with for root shape, size, and colour. It is sometimes used as fodder crop in Europe.

2. *Brassica rapa* L. subsp. *chinensis* (L.) Hanelt.

It is known as Pak-choi or nonheading Chinese cabbage. Wide morphological variation with respect to petiole and leaf characters is found in this subspecies in China, Korea, and Japan.

3. *Brassica rapa* L. subsp. *chinensis* (L.) Hanelt var. *parachinensis* (L.H. Bailey) Hanelt.

This variety is commonly known as Caixin or flowering Chinese cabbage. It is considered as a derivative of subsp. *chinensis* because of similarities in petiole morphology. Flower buds with growing stems and leaves are cooked as vegetables in southern and central China, Malaysia, Indonesia, Vietnam and Thailand.

4. *Brassica rapa* L. subsp. *chinensis* (L.) Hanelt var. *purpuraria* (L.H. Bailey) Kitam.

It is commonly called Zicaitai. It is considered as a derivative of subsp. *chinensis* because of similarities in petiole morphology. The purple branches of stems flower buds and leaves are cooked as vegetables in central China.

5. *Brassica rapa* L. subsp. *pekinensis* (Lour.) Hanelt.

This subspecies comprises mainly heading types of Chinese cabbage. It is a native of China where it probably evolved from the natural crossing of Pak Choi and turnip. According to head compactness, they are further divided into loose, semi-heading and completely heading types. Head shape is variable *e.g.* long, short, tapered, round or flat top, wrapped-over or jointed-tip leaves, *etc.*

6. *Brassica rapa* L. subsp. *narinosa* (L.H. Bailey) Hanelt.

It is commonly known as Taitsai or Wutacai or Chinese flat cabbage and is cold tolerant. The plant is compact, low and produce clusters of thick, often wrinkled leaves having white broad petioles. The crisp leaves and thick petioles are boiled as a vegetable.

7. *Brassica rapa* L. subsp. *nipposinica* (L. H. Bailey) Hanelt.

This subspecies is characterized by excess basal branching and leaves and is an unique vegetable of Japan. Two types are recognized in Japan: 'Mizuna', having deeply dissected bipinnate leaves and 'Mibuaa', with slender entire leaves.

8. *Brassica rapa* L. subsp. *sylvestris* L. Janch. var. *esculenta* Hort.

This variety is known as Brocoletto or friariello. Its origin is in south Italy. The buds somewhat resemble broccoli. The leaves, buds, and stems are edible.

9. *Brassica rapa* L. subsp. *oleifera* (DC.) Metzg.

The plants of this subspecies have many branches with well-developed siliques and seed for yielding oil. Based on growth habit they are grouped in to spring and winter types.

Turnip is a diploid having 2n = 20 and AA genome. It crosses readily with swede (*Brassica napus* L. var. *napobrassica* Peterm) and produces sterile triploid hybrids (3n = 29). Turnip shares the A parental genome and is one of the diploid progenies of the allotetraploid oilseed crops, *B. juncea* (2n = 36, AABB) and *B. napus* (2n = 38, AACC), hence can be used to broaden the genetic base of these species.

4.0 Botany

4.1.0 Root

The fleshy thickened underground portion of turnip is actually the hypocotyl, the colour and shape of which vary depending mainly on the cultivar. Its shape may vary from flat through globular to top-shaped and long. The below ground colour may be white or yellow, while the skin colour of above ground portion may be red purple, white, yellow or green. A distinct taproot and secondary roots arise from the lower part of the swollen hypocotyl. A report from Japan shows that the elongation of turnip hypocotyl stops by the 6th leaf stage in all cultivars tested, whereas the beginning of thickening depends on the cultivar. In some cultivars the thickening begins at the 4th leaf stage, while in others it is not evident till the 6th leaf stage. Thickening begins in the central part of the hypocotyl, followed by the upper and lower parts (Iwasaki and Takeda, 1977). Generally, the edible maturity of roots reaches in 40-80 days, depending on cultivar and cultural condition.

4.2.0 Leaf

Leaves and petioles of turnip are hairy and coarse to the touch and yellowish green in colour. Leaves are oblong to oval and may be entire, serrate or even pinnate depending on cultivar and time of development; leaves formed early are less likely to be pinnate. The leaves developing on the inflorescence are alternate, glaucous, oblong or lanceolate and entire or dentate.

4.3.0 Inflorescence

The inflorescence of turnip is a terminal raceme on the main stem. The flowers in structure and arrangement are practically identical with other cruciferous crops. Pedicel is 1-3 cm long; sepals yellow green; petals yellow, clawed, 6-11 mm long; stamens 6, tetradynamous; carpels 2 with superior ovary.

4.4.0 Fruit

Fruit is a linear, 4-10 cm × 0.2-0.4 cm siliquae with 0.5-3 cm long beak. Seeds globose, 20-30 per silique, 1-1.5 mm in diameter and dark brown with a distinct reticulum.

5.0 Cultivars

According to leaf traits and geographical distribution, Sinskaia (1928) classified turnips into seven geographic groups;

1. Teltow turnips,
2. West European turnips with dissected leaves,
3. Asia Minor and Palestine turnips,
4. Russian turnips of the Petrovsky type,
5. Asiatic Afghanistan turnips with glabrous leaves,
6. Japanese turnips with entire glabrous leaves, and
7. European entire-leaved turnips with pubescent leaves.

Many cultivars differing in shape and colour are now available for cultivation. Shoemaker (1953) classified turnip cultivars into several groups on the basis of morphological characteristics of the root and top.

1. White fleshed:
    (a) Purple-topped
        (i) Flat type: Purple Top Milan
        (ii) Globe type: Purple Top White Globe
    (b) Green-topped
        (i) Globe type: Green Top White, Green Globe
        (ii) Long type: Cowhorn

(c) White-topped

   (i) Flat type: White Milan, White Flat Dutch, Norfolk

   (ii) Globe type: White Stone, Quick Silver

   (iii) Half-long type: White Egg, White Gem

   (iv) Long type: Lily White

2. Yellow fleshed:

   (a) Purple-topped, globe type: Aberdeen Purple Top

   (b) Bronze or green-topped, globe type: Aberdeen Green Top, Amber Globe

   (c) Yellow-topped, globe type: Yellow Globe, Golden Ball, All Seasons

3. Foliage cultivars: Flat Japan, Shogoin, Seven Top.

From seed production point of view turnip cultivars can be classified into two groups: annual or tropical type and biennial or temperate type. These are also referred to as Asiatic or Oriental and European types, respectively. The former type can be grown for seed both in the plains and in the hills, whereas the latter produces seeds only under temperate climate. For seed production of European cultivars, these are grown in the hills of India.

Shibutani and Kinoshita (1969) compared the nutritive constituents of some Asiatic cultivars with those of the European types. The percentages of dry matter and total carbohydrates were found to be higher in the European types but the crude protein content was lower. Similarly, the carotene contents were generally higher in the European types, especially the yellow rooted cultivars, but no difference in vitamin C content was noticed between the two types.

Aissiou *et al.* (2018) described novel wild and cultivated *B. rapa* forms collected from Algeria which can be used for *B. rapa* and *B. napus* breeding. Teltow turnip and Bavarian turnip are two traditional turnip cultivars from Germany. Both these cultivars have higher dry matter content than in the modern varieties. 'Bavarian Turnip' has the black colour cortex, a distinct and unique character. The length of the 'Bavarian Turnip' is 12–15 cm and the width is 3–5 cm in the more cylindrical diameter. The 'Teltower Rubchen' turnip has a more conical shape (Reiner *et al.*, 2005).

A number of good cultivars of both tropical and temperate types are now available in India. A brief description of some of the cultivars is given below:

5.1.0 Tropical Types

Pusa Kanchan

This cultivar has been selected from crosses made between tropical type (Local Red Round) and the European type (Golden Ball). It has all the good qualities of both tropical type and European type. The skin is red, but the flesh is creamy yellow; it has excellent flavour and taste. The leaf top is shorter than the Local Red Round. It becomes ready for harvesting about 10 days later than Local Red Round, but the roots can be kept longer in the field without being spongy. It produces seeds satisfactorily in the plains, though the seed crop is harvested about a fortnight later than the Local Red Round. Unlike European types, it produces seed stalks quickly without enlargement of roots if sown later in the season, *i.e.*, December onwards. Late sowing, therefore, is not recommended for seed production. Average yield is 200-300 q/ha.

Pusa Sweti

A tropical cultivar suited for sowing from August to October. It has attractive white roots. It produces seeds in the plains of India. It matures in 45-50 days after sowing and gives an average yield of 200-300 q/ha.

L-1

This cultivar has been developed at PAU, Ludhiana from the cross PTWG × 4-White. Plant top is medium; leaves are complete with serrated margins and dark green in colour. Roots are round, pure white, smooth, rat-tailed and crisp with mild flavour. It takes 45-60 days to reach marketable maturity. Average yield is 250 quintals per hectare.

Early Milan Red Top

It is an extra early and high yielding cultivar reaching maturity in 45 days. The roots are deep flat with purplish red tops and white underneath. The flesh is pure white, well grained, crisp and mildly pungent. The tops are very small with 4-6 sessile leaves. It produces seeds freely in sub-tropical conditions.

5.2.0 Temperate Types

Pusa Chandrima

This cultivar has been developed by hybridization between tropical (Japanese White) and European type (Snowball). It is an early maturing and high yielding cultivar which possesses all the good qualities of the European type. Roots are medium to large in size, 8-9 cm in length and 9-10 cm in diameter. The tops are medium but not so deeply cut. The skin is smooth, pure white, fine grained with sweet and tender flesh. The leaves of Pusa Chandrima are superior to Snowball in ascorbic acid content in raw as well as in cooked condition. The root reaches harvest maturity in 50-55 days after sowing. It is a temperate cultivar and does not produce seeds in the plains of India. Average yield is 300 quintals per hectare.

Pusa Swarnima

This cultivar has been developed by hybridization and selection between Asiatic type (Japanese White) and European type (Golden Ball). It matures a fortnight earlier than Golden Ball and gives about 40 per cent higher yield. The roots are flattish round, 6-7 cm long and 7-8 cm in diameter with creamy yellow skin, flesh amber coloured, fine textured and mild flavoured. The tops are medium and not very deeply cut. This cultivar is superior to Golden Ball in seed yield. It matures in 65-70 days. Average yield is 280 quintals per hectare.

Important Cultivars of Turnip

Pusa Swarnima

Pusa Chandrima   Snowball

Purple Top White Globe

Purple Top White Globe

It is a heavy yielding large rooted cultivar. The roots are nearly round, upper part purple, lower portion creamy; flesh is white, firm, crisp and mildly sweet-flavoured. The top is small, erect with cut leaves. It is suitable for cooler months. Roots attain marketable size 60 days after sowing. Average yield is 200 quintals per hectare.

Golden Ball

The roots are perfectly globe shaped, medium sized and smooth. It has bright creamy yellow skin and pale, amber coloured flesh of fine texture and flavour. The top is small, erect and with cut leaves. It reaches marketable maturity 70 days after sowing. It is very shy seeders and highly susceptible to turnip malformation. Average yield is 175 quintals per hectare.

Snow Ball

It is a cultivar of 60-65 days duration. Leaves are small and light green. Roots are globe shaped, medium sized with pure white skin. Flesh is crisp, pure white and mildly sweet flavoured. This is also an introduction.

In India, little work has been done to assess the performance of turnip cultivars. Of the six different cultivars, Sel. 66 was found to produce maximum root yield under both saline and sodic soils of North India. Snow Ball and Purple Top White Globe ranked second and third among the cultivars screened for these soil, while Pusa Kanchan was found most susceptible to salinity (Mishra *et al.*, 1973). Pandey and Arora (1969) noticed no consistent yield difference between Purple Top White Globe and Pusa Kanchan under Delhi condition.

6.0 Soil and Climate

As with other root vegetables, a moderately deep, friable fertile, well drained soil is ideal for turnip. Extremely clay soils or very sandy soils should be avoided and except these extremes a satisfactory crop can be grown on moist soils. The optimum soil pH is 5.5-6.8.

Turnip is best adapted to a cool or moderate climate. It is a hardy crop and it can tolerate frost and mildly freezing temperature. High temperature adversely affects quality; the roots become woody, tough and bitter in taste in hot weather. On the other hand, temperatures below 10°C are likely to cause flowering. The most favourable temperatures for the development of the root and the ratio of root/greens are 10-13°C air temperature and 18-23°C root temperature (Hori *et al.*, 1968).

7.0 Cultivation

The method of preparing the soil for turnip is same as for radish. Since turnip does not thrive in hot weather, the seed is sown after the dry hot spell of summer is over. In the plains of northern India, the sowing is done from September to December. Pandey and Arora (1969) reported that seeds sown in the first week of October gave the highest yield under Delhi condition, and line sowing gave greater yields than broadcasting. In the lower hills, seeds are usually sown from July to

                                      Vegetable Crops, Volume 1

October, while in the higher hills the sowing time is from July to September. In the higher hills, the seeds must be sown at least 2 months before hard freezes occur.

Turnip seeds are found to retain viability for 5-6 years when stored at 18-22°C and 55-80 per cent RH; however, the highest productivity has been recorded from 3- to 5-year-old seeds (Lazukov, 1969).

Three to four kg of seeds are sufficient to sow an area of one hectare; higher seed rates affect colour development in roots (del Valle and Harmon, 1970). The seeds are sown in rows 45 cm deep. del Valle (1970) reported reduction in yield as the spacing was reduced from 15 to 7.5 cm within the row. Sowing is done either by hand or by seed drills. The seeds are sometimes mixed with fine sand or ash to facilitate uniform sowing. Sowing in rainy season or in low lying areas should be done on ridges.

In a field experiment by Bracy *et al.* (2000), open-pollinated and hybrid turnip cultivars were planted with a belt-type precision seeder at 56, 112 and 168 mm in single- or double-plant lines per row and with a bulk seeder set at the smallest opening (average seed spacing of 51 mm) utilizing 100 per cent viable seed or 50 per cent viable seed/50 per cent killed seed (by weight) mix. Overall, total yield was not affected by seeding rate or seeder; however, plant population produced a shift in yield among grades. Generally, yield of extra-large roots was greatest on plots with lower plant populations. Medium and cull yields were greater on plots with the higher plant populations (bulk seeder with 100 per cent viable seed and precision seeder with 56 mm spacing or 2-row configuration) (Table 1).

Table 1: Effect of Seeding Rate on Root Production of 'Purple Top White Globe' Turnip in Spring 1997*

Seeder	Seeding Rate	Yield (MT/ha)			Plants/ha		
		Total	Extra-large	Large	Medium	Culls	(000's)
Planet Jr.y	100 per cent viable seed	43.7	13.3 ax	25.6 b	4.8 a	2.1 bc	131.3 b
Planet Jr.y	50 per cent viable seed	40.7	14.4 a	25.3 b	1.1 c	0.8 d	57.5 d
Stanhay	56 mm spacing, 1-line	46.9	0.9 c	28.5 a	6.3 a	3.1 a	158.5 a
Stanhay	112 mm spacing, 1-line	45.7	4.8 bc	38.3 a	2.6 b	1.0 d	85.2 c
Stanhay	112 mm spacing, 2-line	46.1	2.1 bc	37.9 a	6.1 a	2.4 ab	150.0 ab
Stanhay	168 mm spacing, 1-line	44.8	8.9 ab	33.9 a	2.0 bc	0.8 d	70.4 cd
Stanhay	168 mm spacing, 2-line	46.2	5.3 bc	37.7 a	3.2 b	1.4 cd	91.3 c
Significance		N.S.	**	**	**	**	**

z: Graded by size: extra-large > 114 mm, large 63-113 mm, medium 44-62 mm, culls were <44 mm or misshaped. Total marketable yield included extra-large, large and medium grades.

y: Planter Jr. set at smallest opening for an average seed spacing of 51 mm. Killed seed was mixed with viable seed.

x: Mean separation within columns by Duncan's multiple range test, $P = 0.05$.

* Bracy *et al.* (2000).

7.1.0 Manuring and Fertilization

Under Indian conditions, the fertilizer doses recommended for turnip are 23 tonnes farm yard manures per hectare supplemented with 60-70 kg nitrogen and 40-50 kg each of phosphorus and potash (Choudhury, 1967; Yawalkar, 1980). Nath (1976) recommended 20-25 tonnes of FYM, 60-98 kg nitrogen and 44 kg each of phosphorus and potash per hectare. Based on the research work done in India and other countries, Kumar and Mathur (1965) made the following recommendations:

i) Well-rotted farm yard manure at the rate of 18-23 tonnes per hectare should be given before sowing with 28 kg of P as superphosphate. A dose of 22 kg nitrogen in the form of ammonium sulphate or calcium ammonium nitrate at the time of root formation should also be given.

ii) The nitrogenous fertilizers should be given in two split doses – half before sowing and the rest, at the time of root formation with the second or third irrigation. Phosphatic and potassic fertilizers should be placed 7-8 cm below the seed before sowing.

Dhesi and Nandpuri (1968) observed that 56 kg each of N and P with 28 kg K per hectare was the optimum dose of fertilizer for turnips growing on sandy loam soil of Punjab. Significant increase in yield and uptake had been noted with application of nitrogen up to 100 kg per hectare by Lal and Dey (1972) working under Delhi conditions.

For seed crops, Sandhu *et al.* (1965, 1966) reported that nitrogen at the rate of 56 kg per hectare was the optimum dose for length and number of seeds per pod, but 84 kg nitrogen per hectare gave the highest seed yield. The application of nitrogen and phosphorus helped in reducing the percentage of underdeveloped seeds. The most economical fertilizer combination was found to be 84 kg nitrogen, 56 kg P and 56 kg K per hectare, which gave an average yield of 1252 kg seeds per hectare.

del Valle and Harmon (1970) noticed that the nitrate source of N, alone or in combination with the ammonium source, was equally effective in increasing the yield of turnip roots and N percentage in tops. The best yield was obtained with a split application of 160 lb N per acre and the best colour development with seed rates of 2.5 or 5 lb per acre.

Saleem *et al.* (2019) reported that leaf length, number of leaf/plant, fresh leaf weight/plant, leaf yield (q/ha), root diameter, average root, root depth and root yield (q/ha) significantly increased with increasing nitrogen levels up to 75 kg/ha in turnip. A spacing of 30 × 15 cm was best for the growth and yield characters. However, increasing the row to row spacing to 45 cm increased the leaf length, number of leaf and fresh leaf weight per plant, with increase in weight of individual root but yield/unit area was significantly higher in 30 × 15 cm spacing. Yanthan *et al.* (2012) recommended application of 50 per cent pig manure and 50 per cent NPK for turnip cultivation in foot hill condition of Nagaland.

7.2.0 Irrigation

Irrigation requirement of turnip is similar to radish. Pre-sowing irrigation ensures good germination. In general, the crop is irrigated every 10-15 days.

7.3.0 Interculture

Spraying Stomp (pendimethalin) @ 1.2 litres per hectare is effective for controlling weeds in turnip. Thinning is an important operation in turnip cultivation. Germination of seeds is usually good and when the plants are well established, they are thinned out keeping a distance of 10-15 cm from plant to plant. Besides thinning, weeding should also be done once or twice during the growing season to keep the growth of turnips unchecked. One earthing up 25-30 days after sowing is also recommended. Earthing up is usually done after the application of second dose of nitrogenous fertilizers.

8.0 Harvesting, Yield and Storage

8.1.0 Harvesting and Yield

Turnips should be harvested soon after reaching suitable size, since the quality rapidly deteriorates. A desirable size is 5-7.5 cm in diameter; larger roots are often coarse in texture and bitter is taste. Early maturing varieties reach marketable maturity 35-45 days after sowing and late maturing varieties 60-70 days after sowing. A light irrigation may be given before harvesting to facilitate lifting. The roots are cleaned, tops are cut and the roots are graded according to size.

The average yield of turnip varies from 200 to 300 quintals per hectare.

8.2.0 Storage

Turnip does not store well. Turnip can be stored for 3-4 weeks in cold stores at 0°C and 90-95 per cent relative humidity. Haller (1947) reported that washing of turnips improved the appearance and reduced decay. Application of wax emulsion to the roots neither improved the appearance nor retarded weight loss, but dipping in hot paraffin reduced shrinkage and improved the appearance of roots.

9.0 Diseases and Pests

9.1.0 Diseases

Most of the diseases and pests that attack radish also attack turnip.

Alternaria Leaf Spot (*Alternaria raphani*)

Alternaria leaf spot disease is quite wide spread and when the foliage of turnip is infected, the roots may also become infected and show symptoms. The spots are nearly circular, often zonate and show various shades of brown or sometimes black. White mould or dark spores may cover the spot. The infection spreads rapidly in humid conditions.

Practice long term crop rotation, use disease free seed, treat the seed with captan or thiram @ 2.5 g per kg seed.

Different Cultural Operations in Turnip

Seed sowing and seedling emergence in turnip

Field growing and harvesting of turnip roots

Diseases of Turnip

 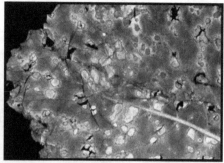

Alternaria leaf spot (left) and Anthracnose (right) in turnip

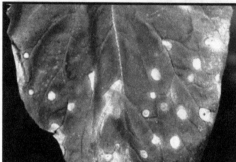 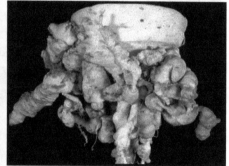

Cercospora leaf spot (left) and Club root (right) in turnip

Sclerotinia rot (left) and Bacterial soft rot (right) in turnip

Mosaic in turnip

Phyllody

Phyllody is also common and causes severe damage of seed crops in the hills. It shows malformation of flowering shoots and the infected plants usually do not produce any normal fruit. The cultivar Golden Ball is extremely susceptible to phyllody.

Turnip Yellow Mosaic Virus

Turnip yellow mosaic virus often occurs in some places. It is very infectious and is transmitted through flea beetle. It can easily be controlled by using insecticides to check the vectors. Cabbage may serve as an alternate host of this virus. Remove and destroy the diseased plants and alternative host plants from vicinity of turnip fields.

9.2.0 Pests

The common pests of turnip are aphids and mustard saw fly. The nature of damage caused by these insects and their control measures are same as in case of radish.

10.0 Seed Production

From seed production point of view turnip cultivars can be grouped into two types: (i) biennial or temperate or European type, and (ii) annual, tropical or Asiatic type. The former produces seeds only in temperate climate which is prevalent in the hills of India, but the latter can be grown for seed production in the hills as well as in the plains. However, seed production of the annual turnips in the hills is not economical and hence its seeds are produced only in the plains. Singh *et al.* (1960) gave a detailed account of seed production of biennial turnips in the Kullu valley (Himachal Pradesh).

The annual cultivars are usually sown from July to September in the plains and the biennial cultivars from last week of August to first week of September in the hills. However, the sowing may be delayed up to the end of September in case of extra early types like Early Milan Red Top and Japanese White Milan. If the seeds are to be raised *in situ*, sowing may be as late as 10th October. Sowing on ridges 45 cm apart is preferred to flat sowing, because ridges facilitate drainage and better root development. Singh *et al.* (1960) recommended application of 672.5 kg FYM along with 224 kg superphosphate per hectare during autumn transplanting of developed roots. Besides one application of ammonium sulphate at the rate of 112 kg/ha should be given before bolting and again during spring before the beginning of flowering. Sandhu *et al.* (1965, 1966) recommended application of 84 kg nitrogen, 55 kg phosphorus and 55 kg potash for seed crops.

Sowing is done in lines 45 cm apart and the plants are thinned to 7.5-10 cm apart in the lines. One weeding and one earthing up during the early stage of growth is necessary to promote root development. Another hoeing and earthing up are given before bolting so as to provide support to the seed plants, otherwise they may lodge during flowering and fruiting. Irrigation may be given according to the need of the crop and the field should not be allowed to dry out during root development, and flowering and fruiting. Diseases and insect pests should be kept under control by regular use of fungicides and insecticides.

10.1.0 Methods of Seed Production

Both seed-to-seed or *in situ* and root-to-root or transplanted method are employed for seed production in turnips. In general, *in situ* method gives better yield than transplanted one. However, this method can be practised only if quality nucleus seed is used to raise the seed crop. Root-to-seed or transplanted method is desirable because it provides an opportunity to select desirable type of roots, thus enabling maintenance of high standard of seeds.

During November-December when the roots are fully mature, the plants are uprooted and true-to-type roots are selected. All underdeveloped, deformed, diseased and off-type roots are discarded. After pruning the taproot and clipping the tops, leaving the crown intact, the selected roots are transplanted in a freshly prepared soil at a distance of 60 × 60 cm or 60 × 40 cm. Nath and Sachan (1969) obtained seed yields of 273.4 kg and 245.4 kg per hectare in treatments, such as two-third top left and half the root cut and one-third top left and half the root cut, respectively.

The treated stecklings are either transplanted immediately in the field or stored at a temperature of 0°C and a relative humidity of 90-95 per cent. In Kalpa valley of Himachal Pradesh, turnip roots are stored satisfactorily in trenches covered with a wooden plank and a thin layer of soil is spread over the plank. Sometimes the roots are furrowed down at the original place of sowing and removed during spring, selected and replanted. Tropical turnips do not need any storage.

10.2.0 Flowering and Fruiting

Singh *et al.* (1960) reported that bolting in European cultivars of turnip starts in the month of January and continues till March and the crop is in full bloom in the month of March-April depending on cultivar.

The inflorescence of turnip is a typical terminal raceme. Honeybees are the chief pollinating agents. The peak anthesis period is between 8 to 9 a.m. Anther dehiscence begins with flower opening and is maximum between 10 a.m. and 12 noon. Maximum pollen fertility is observed at and a day before anthesis. Stigma becomes receptive two days before and remains receptive for two days after anthesis. To ensure good seed set, pollination of all the flowers on inflorescence is necessary and to ensure it a beehive adjacent to the seed field should be provided. Application of growth regulators (GA_3, NAA, MH and ethrel) significantly affected bolting and flowering, plant height, number of primary, secondary and tertiary branches, number of siliqua, stalk yield and total seed yield as compared to control. However, siliqua length, number of seed per siliqua, test weight and germination percentage were less influenced by these chemicals. Among different growth regulators, MH, in general, delayed bolting as well as flowering by two weeks, decreased plant height but induced more number of branches, increased number of siliqua and the seed yield followed by ethrel (400-600 ppm) and NAA (200 ppm). The most optimum concentration of MH was 100 ppm followed by 500 and 200 ppm. Test weight and germination percentage, in general, were improved by the application of GA_3 (100-400 ppm) followed by NAA (100 ppm) in comparison to MH (Hussain *et al.*, 1998). Singh *et al.* (1989) observed that plant height, number of branches, number of siliquae, number of seeds/siliquae and total yield were favourably influenced with larger root size stecklings.

Turnips should be well isolated from fields of other cultivars as well as from Chinese cabbage, rutabaga, rape, and mustard. An isolation distance of 1600 metre and 1000 metre should be provided for production of stock (nucleus) seeds and certified seeds, respectively. At least three inspections at different plant growth stages *viz.* vegetative growth, root development and flowering are required to rogue out off-type and diseased plants.

10.3.0 Harvesting and Seed Yield

Turnip fruits dehisce easily when fully dried. Hence shattering is a serious problem in fully dried seed crops of turnip. It is, therefore, advisable to cut the whole crop when 60-70 per cent pods turn yellow brown in colour or at a stage when most of the pods turn yellow but not fully dried.

The crop can be piled in a heap, covered with a tarpaulin and allowed to cure for 4-5 days. The pile is turned upside down and allowed to cure for another 4-5 days. The period of curing, of course, depends on weather condition. The crop is then threshed by beating with a stick. The seeds immediately after threshing should be thoroughly dried in sun, otherwise it will quickly lose viability. The seeds are then cleaned so as to make them free from small seeds and other inert materials. The seed is dried to 6 per cent moisture level.

The seed yield varies with cultivar and the growing region. Average seed yield is 3-4 quintals/ha.

10.4.0 Hybrid Seed Production

For hybrid seed production, cytoplasmic male sterility system is more preferred than self-incompatibility. Isolation distance of 1000 m is maintained to avoid outcrossing in the F_1 hybrid seed production field. Takahashi (1987) suggested 1:1 ratio of female to male parent while utilizing cytoplasmic male sterility and also for seed parents planting in case of self-incompatibility. Self-incompatible lines are maintained by bud pollination.

11.0 Crop Improvement

No intensive breeding work has been done in turnip. Improvement in uniformity is an important breeding objective in turnip. However, in turnip, being a self-incompatible crop, successful inbreeding is difficult to achieve, although not impossible. This difficulty can be overcome by bud pollination. Generally much inbreeding depression is not observed until third or fourth inbred generations.

11.1.0 Breeding Objectives

The major breeding objectives of vegetable turnip are:

1. Earliness to attain marketable root size.
2. High yield of both root and greens (leaves).
3. Root colour (white, purple, golden) as per consumers' preference.
4. Stump rooted varieties with thin taproot and non-branching habit.
5. Slow bolting and no pithiness.
6. Dry matter (8-9 per cent) in roots.
7. Resistance/tolerance to club root, powdery mildew, white rust, phyllody, turnip mosaic virus, cabbage root fly, and turnip root fly.
8. Resistance/tolerance to hot and humid climate, particularly in tropical types.

11.2.0 Genetics of Traits

The genetics of flesh and skin color in turnip was studied by Davey (1931) who reported that white flesh is dominant to yellow and is determined by a single locus. Skin colour in turnip is determined by two independent loci conditioning the presence or absence of green or red pigmentation, respectively. Brar *et al.* (1969) studied the inheritance of root skin colour, root shape, and leaf shape in turnips. Root colour in the crosses cv. Purple White Top Globe (purple rooted) × Desi Red (red rooted) and cv. Purple White Top Globe (purple rooted) × cv. Desi White (white rooted) was conditioned by two genes, designated R (red colour) and P (purple colour). Root shape in the cross cv. Purple White Top Globe (round) × cv. Desi White (flat) was conditioned by a single incompletely dominant gene, designated F. Leaf shape in the cross cv. Purple White Top Globe (compound leaf) × cv. Desi White (simple leaf) was conditioned by a single dominant gene, designated C, for compound leaf. Genes for root colour and leaf shape were linked (recombination frequency: 46.3 per cent). Ahmed and Tanki (1994) studied the inheritance of root characteristics in the F_1, F_2 and backcrossed (BC) generations of crosses between 5 turnip varieties. The pink and partially purple root colours were dominant over white and they were under the control of single dominant genes P and C, respectively. The purple root colour genotype was designated as a double dominant (PPCC). Pink and partially purple genotypes were dominant over white and thought to be due to recessive alleles at 1 locus, PPcc and ppCC, respectively. White roots were produced in double recessive homozygotes. The genotypes were PPCC (purple), PPcc (pink), ppCC (partially purple), and ppcc (white). Round shape was dominant over flat-round and was monogenic (R).

Several plant characters in turnip are governed by single recessive gene, *viz.* corolla colour (cream, light yellow, dark yellow), petal (cupped petal, apetalous, striped petal, red petal margin), puckered leaf and anthocyanin-less hydrathode, anther-tip and style-tip and yellow-green plant. The characters controlled by single dominant gene are rolled petal margin, polypetalous, flesh colour and resistance to turnip mosaic virus. In purple root, the skin colour is controlled by two independent dominant genes. Clubroot resistance is governed by three independent dominant genes whereas resistance to powdery mildew is polygenic with partially recessive genes (Pink, 1993).

11.3.0 Breeding Methods and Achievements

Mass selection, hybridization followed by progeny selection has been commonly adopted for turnip improvement. Mass selection has been successfully used to overcome varietal degeneration in Purple Top White Globe (Kumar *et al.*, 2008). Recurrent selection in progeny of crosses involving cultivars such as Purple Top White Globe, Raab Salad and Shogoin turnip has resulted in considerable progress towards developing cold tolerant and aphid resistant turnips (Barnes and Cuthbert, 1975; Robbins, 1977). Clubroot resistant turnip cv. Manga has also been reported from New Zealand (Lammerink *et al.*, 1978); however, this cultivar is slightly low yielding and has a higher tendency to bolting when grown in spring. Nevertheless, this can be utilized in producing disease resistant cultivars.

At Indian Agricultural Research Institute, sub-tropical variety 'Pusa Kanchan' has been developed by hybridization between the tropical and temperate types (Singh, 1963). Pusa Kanchan has been developed from crosses between an Asiatic type, Red Round and European type, Golden Ball. It has all the good qualities of both Asiatic and European types. Another tropical variety 'Pusa Sweti' was also developed at IARI (Choudhury *et al.*, 1976). 'Punjab Safed-4' is a tropical variety from PAU, Ludhiana. Temperate group varieties, 'Pusa Chandrima' (Singh *et al.*, 1971) and 'Pusa Swarnima' (Gill, 1979) have been developed by hybridization.

Heterosis for dry matter yield has been demonstrated by Wit (1966). Synthetic cultivars, based on inbred lines of good combining ability, have been shown to out-yield the best commercial cultivars in Holland.

Gill *et al.* (1974) reported varietal variation for ascorbic acid, TSS and dry matter in turnip greens. Saimbhi and Singh (1987) evaluated turnip varieties for dehydration and found Purple Top White Globe was best, followed by Red-4 and White-4.

Brar and Gupta (1969) reported heterosis for yield and root length in turnip. Heterosis for root yield, root length, root diameter and number of leaves per plant was observed by Pathania *et al.* (1987). Ahmed and Tanki (1988) reported that seed yield in turnip is positively correlated with growth and yield contributing attributes.

Ahmed and Tanki (1999) recorded the presence of substantial amounts of heterosis in the F_1 (69.3 per cent), F_2 (31.9 per cent) and F_3 (24.7 per cent) generations. They suggested the utilization of crosses with significant residual heterosis (Pusa Chandrima × White Flat Round, Pusa Snow Ball × White Flat Round, Purple Top White Globe × Pusa Snow Ball and Pusa Chandrima × Pusa Sweti) either in the synthesis of composites or directly as advance generations of single crosses. They also described a method consisting of random selection of roots in the F_2 generation followed by random mating for restoring hybrid vigour in later generations.

11.4.0 Self-Incompatibility and Male Sterility

Self-incompatibility (SI) in turnip is controlled by multiple alleles of the S locus. More than 30 S alleles have been identified in *Brassica rapa* (Nou *et al.*, 1993). Improved 'Ogura' cytoplasm has been transferred to vegetable *B. rapa* (Heath *et al.*, 1994). Several improved F_1 hybrid cultivars of turnip have been developed in

Japan. However, in India attempts have not been made to develop F_1 hybrid for commercial cultivation.

Lin *et al.* (2019) reported that Ogura-CMS turnip has a reduction in the size of the fleshy root, and had distinct defects in microspore development and tapetum degeneration during the transition from microspore mother cells to tetrads. Due to defective microspore production and premature tapetum degeneration during microgametogenesis it has short filaments and withered white anthers, leading to complete male sterility of the Ogura-CMS line. They identified 5,117 differentially expressed genes (DEGs), including 1,339 up- and 3,778 down-regulated genes in the Ogura-CMS line compared to the male fertile (MF) line. They identified the genes regulating tapetum programmed cell death (PCD) and associated with pollen wall formation and proposed 185 novel genes that function in male organ development, of which 26 DEGs were genotype specifically expressed. In their recent study, Lin *et al.* (2020) identified 28 miRNAs to be involved in the reproductive development for the Ogura-CMS and MF lines of turnip.

11.5.0 Stress Resistance

European turnip cultivars *e.g.* Siloga, Gelria, Milan White, and Debra have been used for breeding different clubroot resistant cultivars (Crute *et al.*, 1980). Gao *et al.* (2020) reported clubroot resistance in Tibetan turnip accession 156-13 and chilling resistance in accessions 156-8 and 156-10. Cho *et al.* (2016) reported resistance to *Plasmodiophora brassicae* race 4 in European turnip (*B. rapa* ssp. *rapifera*) inbred line 'IT033820' and the resistance was controlled by two dominant genes.

Zhang *et al.* (2018) reported that the candidate gene BrRLP48 was involved in disease resistant response and the disease-inducible expression of BrRLP48 contributed to the downy mildew resistance in *Brassica rapa*.

Transparent Testa 8 (TT8) is an important transcription factor in regulatory networks controlling development, metabolism and stress responses in plants (Zhang *et al.*, 2020). Zhang *et al.* (2020) isolated BrTT8 (HQ337791) from 'Tsuda' turnip. PCR analysis revealed BrTT8 expression exhibited tissue specificity and the highest expression of BrTT8 was in the red root epidermis of turnip. They concluded that BrTT8 might play a crucial role in UV-A-induced anthocyanin biosynthesis and the BrTT8 gene is regulated by abiotic stresses, such as high- and low-temperature, hormone, osmotic and salt stresses.

11.6.0 Mutation Breeding

Basak and Prasad (2004) treated cv. Rose Red seeds with the mutagen EMS and obtained a desirable root mutant in the M_2 generation which showed 3-fold larger root compared to the parent.

11.7.0 Polyploidy Breeding

Prasad and Kumari (1996) induced autotetraploidy in turnip by treating fully grown seedlings with 0.2 per cent colchicine solution intermittently for several days using the modified cotton swab method. 28 out of 261 plants were autotetraploids. They suggested that frequency of chromocentres can be utilized as a reliable cytological criterion for the screening and confirmation of polyploidy in turnip. Prasad and Jain (1998) crossed highly fertile tetraploid line T76 with a diploid to yield a triploid form which had the smallest seeds, but satisfactory seed germination. They observed an increase in the numbers of chloroplasts per guard cell, numbers of leaves and branches and plant height with the ploidy level.

The diploid turnips outyielded their respective tetraploids for dry matter production (Gowers, 1977). Hua-bing *et al.* (2011) observed better adaptation to a high concentration salt medium (200 mmol L^{-1}) in tetraploid turnip compared to its diploid progenitor.

11.8.0 Haploid Production

Park *et al.* (2019) produced a doubled haploid (DH) line of Ganghwa turnip, an heirloom specialty crop in Korea. The whole genome of DH line G14 was sequenced on an Illumina HiSeq 4000 platform, and the reads mapped onto the *B. rapa* reference genome identified 1,163,399 SNPs and 779,700 indels. Despite high similarity in overall genome sequence, turnips and Chinese cabbage have different compositions of transposable elements (TEs). The long terminal repeat (LTR) retrotransposons are more enriched in turnips than in Chinese cabbage genomes, in which the *gypsy* elements are classified as major LTR sequences in the turnip genome. Their findings suggested that subspecies-specific TE divergence is in part responsible for huge phenotypic variations observed within the same species.

11.9.0 Distant Hybridization

In Brassicaceae family, distant hybridization plays an important role in creating new germlines, utilization of heterosis, and crop improvement. Besides, it has significance in the study of the origin, evolution, and genetic variation of species. Lou *et al.* (2017) produced 24 intergeneric hybrids from crosses between twelve turnip and eight radish cultivars. All of the hybrids were amphihaploids, with 19chromosomes in mitosis. They were pollen-sterile, had white petals, and were intermediate in most of their morphological traits with respect to their parents. Jin *et al.* (2020) performed intergeneric hybridization between radish and turnip by hand emasculation followed by tissue culture of putative hybrid seeds. They obtained total of three and one hybrids from the cross of 'Long yellow turnip' and 'Golden turnip' with radish, respectively, but could not obtain any hybrids in their reverse cross, indicating that the same cross combination has a large difference in the reciprocal crosses, and that the hybrid has certain unidirectionality. The majority of the hybrids of radish and turnip showed male sterility, white petals, and intermediate performance in most morphological traits when compared to their parents. The improvement of these amphihaploids obtained from above studies may be useful to enrich the gene pool for radish and turnip breeding.

Lu *et al.* (2008) constructed a genetic map using an F_2 population developed from a cross between Chinese cabbage (*Brassica rapa* ssp. *chinensis*) and turnip (*B. rapa* ssp. *rapifera*). The detected 18 QTLs for the 3 root traits, including 7 QTLs for taproot thickness, 5 QTLs for taproot length, and 6 QTLs for taproot weight. Individually, the QTLs accounted for 8.4–27.4 per cent of the phenotypic variation. The 2 major

QTLs, *qTRT4b* for taproot thickness and *qTRW4* for taproot weight, explained 27.4 per cent and 24.8 per cent of the total phenotypic variance, respectively. These QTLs for root traits, may provide a basis for marker-assisted selection to improve productivity in turnip improvement.

12.0 References

Ahmed, N. and Tanki, M.I. (1988) *Veg. Sci.*, **15:** 1-5

Ahmed, N. and Tanki, M.I. (1994) *Indian J. Genet.*, **54:** 247-252.

Ahmed, N. and Tanki, M.I. (1999) *Appl. Biol. Res.*, **1:** 71-74

Aissiou, F., Laperche, A., Falentin, C., Lode´, M., Deniot, G., Boutet, G., Re´gnier, F., Trotoux, G., Huteau, V., Coriton, O., Rousseau-Gueutin, M., Abrous, O., Che'vre, A.M. and Hadj-Arab, H. (2018) *Euphytica*, **214:** 241.

Barnes, W.C. and Cuthbert, F.P.Jr. (1975) *HortScience*, **10:** 59-60.

Basak, S. and Prasad, C. (2004) *Eucarpia Cruciferae Newsletter*, **25:** 9-10.

Bonnema, G., Lee, J.G., Shuhang, W., Lagarrigue, D., Bucher, J., Wehrens, R., de Vos, R. and Beekwilder, J. (2019) *PLoS ONE*, **14:** e0217862.

Bracy, R.P., Parish, R.L. and McCoy, J.E. (2000) *J. Veg. Crop Prod.*, **6:** 43-50.

Brar, J.S. and Gupta, V.P. (1969) *Plant Sci.*, **1:** 24-29.

Brar, J.S., Gill, H.S. and Nandpuri, K.S. (1969) *Punjab Agric. Univ. J. Res.*, **6:** 907-911.

Cho, K.H., Kim, K.T., Park, S., Kim, S., Do, K.R., Woo, J.G. and Lee, H.J. (2016) *Korean J. Hortic. Sci. Technol.*, **34:** 433-441.

Choudhury, B. (1967) In: *Vegetables*, National Book Trust India, New Delhi.

Choudhury, B., Sirohi, P.S. and Sharma, J.C. (1976) *Indian Hort.*, **21:** 7.

Crute, I.R., Gray, A.R., Crisp, P. and Buczacki, S.T. (1980) *Plant Breed. Abstr.*, **50:** 91–104.

Davey, V.McM. (1931) *Scott. J. Agric.*, **14:** 303.

del Valle, C.G. (1970) *HortScience*, **5** ; 228-230.

del Valle, C.G. and Harmon, S.A. (1970) *J. Amer. Soc. Hort. Sci.*, **95:** 62-64.

Dhesi, N.S. and Nandpuri, K.S. (1968) *Punjab Agric. Univ. Farm Bull.*, **2:** 128.

Gao, Y., Gong, W., Li, R., Zhang, L., Zhang, Y., Gao, Y., Lang, J., Zhao, K., Liu, K. and Yu, X. (2020) *Genet. Resour. Crop Evol.*, **67:** 209-223.

Gill, H.S. (1979) Pusa Swarnima, a turnip variety of promise. *Indian Hort.*, **24:** 7-8.

Gill, H.S., Tewari, R.N. and Singh, R. (1974) *Sci. Cult.*, **40:** 491-492.

Gowers, S. (1977) *Euphytica*, **26:** 203-206.

Haller, M.H. (1947) *Proc. Amer. Soc. Hort. Sci.*, **50:** 325-329.

Heath, D.W., Earle, E.D. and Dickson, M.H. (1994) *HortScience*, **29:** 202–203.

Hori, S., Arai, K., Hosoya, T. and Oyamada, M. (1968) *Bull. Hort. Res. Sta.*, Hiratsuka, 7, pp. 187-214.

Hua-bing, M., Si-si, J., Shui-jin, H., Xian-yong, L., Yuan-long, L.I., Wan-li, G. and Li-xi, J. (2011) *Agricultural Sci. China*, **10**: 363-375.

Hussain, M., Ahmad, N., Mir, A.H. and Mir, M.A. (1998) *Haryana J. Hort. Sci.*, **27**: 200-204.

Iwasaki, F. and Takeda, Y. (1977) *J. Japanese Soc. Hort. Sci.*, **46**: 193-200.

Jin, P., Zhu, Z., Guo, X., Chen, F., Wu, Y., Chen, J., Wu, J. and Zhu, Z. (2020) *Euphytica*, **216**: 90.

Klopscha, R., Witzela, K., Börnerb, A., Schreinera, M. and Hanschena, F.S. (2017) *Food Res. Intl.*, **100**: 392-403.

Kumar, R., Kanwar, M.S. and Korla, B.N. (2008) *Haryana J. Hortic. Sci.*, **37**: 312-313.

Kumar, V. and Mathur, M.K. (1965) *Fert. News*, **10**: 2-5.

Lal, B. and Dey, R. (1972) *Indian J. Agric. Sci.*, **42**: 156-160.

Lammerink, J., Butel, J.C. and Hart, R. (1978) *New Zealand J. Agric.*, **136**: 29.

Lazukov, M.I. (1969) *Dokl. Mosk. Sel'-hoz. Akad. K.A. Timirjazeva*, **153**: 119-124.

Lin, S., Miao, Y., Su, S., Xu, J., Jin, L., Sun, D., Peng, R., Huang, L. and Cao, J. (2019) *PLoS ONE*, **14**: e0218029.

Lin, S., Su, S., Jin, L., Peng, R., Sun, D., Ji, H., Yu, Y. and Xu, J. (2020) *PLoS ONE*, **15**: e0236829.

Linnaeus, C. (1753) Species Plantarum, 1st Edn. Stockholm.

Lou, L., Lou, Q., Li, Z., Xu, Y., Liu, Z. and Su, X. (2017) *Sci. Hort.*, **220**: 57-65.

Lu, G., Cao, J., Yu, X., Xiang, X. and Chen, H. (2008) *J. Appl. Genet.*, **49**: 23-31.

Metzger, J. (1833) Systematische Beschreibung der kultivirten Kohlarten. Heidelberg.

Mishra, B., Joshi, Y.C. and Sarin, M.N. (1973) *Indian Hort.*, **18**: 13-14.

Nath, P. (1976) *Vegetables for Tropical Region*, ICAR, New Delhi.

Nath, P. and Sachan, S.P. (1969) *Allahabad fmr.*, **43**: 267-270.

Nou, I.S., Watanabe, M., Isuzugawa, K., Isogai, A. and Hinata, K. (1993) *Sex. Plant Reprod.*, **6**: 71–78.

Pandey, S.L. and Arora, P.N. (1969) *Indian J. Agron.*, **14**: 172-174.

Park, H.R., Kang, T., Yi, G., Yu, S.H., Shin, H., Kim, G.W., Park, J.E., Kim, Y.S. and Huh, J.H. (2019) *Plant Biotech. Rep.*, **13**: 677-687.

Pathania, N.K., Rattan, R.S. and Thakur, M.C. (1987) *Veg. Sci.*, **14**: 161-168.

Pink, D.A.C. (1993) In: *Genetic Improvement of Vegetable* Crops (Eds. G. Kalloo and B.O. Bergh), Pergamon Press, New York., pp. 511-519.

Prasad, C. and Jain, S.C. (1998) *Eucarpia Cruciferae Newsletter*, **20**: 25-26.

Prasad, C. and Kumari, N. (1996) *Eucarpia Cruciferae Newsletter*, **18**: 20-21.

Reiner, L., Gladis, T., Amon, H. and Emmerling-Skala, A. (2005) *Genet. Resour. Crop Evol.*, **52**: 111-113.

Robbins, M.L. (1977) *Cruciferae Newsletter*, No. 2, p. 31.

Saimbi, M.S. and Singh, B. (1987). *Veg. Sci.*, **14**: 203-205.

Saleem, T., Ganai, N.A., Sultan, T., Tramboo, M.S. and Shah, M.A. (2019) *Agric. Biol. Res.*, **35**: 1-6.

Sandhu, K.S., Dhesi, N.S. and Kang, U.S. (1965) *Indian J. Agron.*, **10**: 279-282.

Sandhu, K.S., Dhesi, N.S. and Kang, U.S. (1966) *Indian J. Agron.*, **11**: 45-50.

Shibutani, S. and Kinoshita, K. (1969) *Sci. Rept. Fac. Agric. Okayama*, No. 33, pp. 9-20.

Shoemaker, J.S. (1953) *Vegetable Growing* (2nd *edn.*), John Wiley and Sons., Inc. New York.

Singh, C.G., Pandita, M.L. and Khurana, S.C. (1989) *Veg Sci.*, **16**: 119-124.

Singh, H.B. (1963) *Indian Hort.*, **7**: 21.

Singh, H.B., Thakur, M.R. and Bhagchandani, P.M. (1960) *Indian J. Hort.*, **17**: 38-47.

Singh, J.P., Gill, H.S. and Bhullar, B.S. (1971) *Indian Hort.*, **16**: 19-20, 23.

Sinskaia, E.N. (1928) *Bull. Appl. Bot. Genet. Pl. Breed.*, **19**: 555-607.

Sturtevant, E.L. (1919) *Notes on edible Plants* (Ed. U.P. Hendrick), New York Agric. Exp. Sta.

Takahashi, O. (1987). In: *Hybrid production of selected cereal oil and vegetable crops* (Eds. W.P. Feistritzer and A.F. Kelly). FAO, Rome, pp. 313-328.

Takahashi, Y., Yokoi, S. and Takahata, Y. (2016) *Genet. Resour. Crop Evol.*, **63**: 869–879.

Wit, F. (1966) *Qualitas Plant Mater. Veg.*, **13**: 305-310.

Yanthan, T.S., Singh, V.B., Kanaujia, S.P. and Singh, A.K. (2012) *J. Soils Crops*, **22**: 1-9.

Yawalkar, K.S. (1980) In: *Vegetable Crops of India*, Agri-Horticultural Publishing House, Nagpur, India.

Zhang, B., Li, P., Su, T., Li, P., Xin, X., Wang, W., Zhao, X., Yu, Y., Zhang, D., Yu, S. and Zhang, F. (2018) *Front. Plant Sci.*, **9**: 1708.

Zhang, Y., Wang, G., Li, L., Li, Y., Zhou, B. and Yan, H. (2020) *Sci. Hort.*, **268**: 109332.

Index

CPSIA information can be obtained
at www.ICGtesting.com
Printed in the USA
LVHW081005161122
733047LV00042B/4

9 789354 616822